mHealth
Multidisciplinary
Verticals

mHealth
Multidisciplinary
Verticals

Edited by
Sasan Adibi

CRC Press
Taylor & Francis Group
Boca Raton London New York

CRC Press is an imprint of the
Taylor & Francis Group, an **informa** business

CRC Press
Taylor & Francis Group
6000 Broken Sound Parkway NW, Suite 300
Boca Raton, FL 33487-2742

First issued in paperback 2017

© 2015 by Taylor & Francis Group, LLC
CRC Press is an imprint of Taylor & Francis Group, an Informa business

No claim to original U.S. Government works

ISBN-13: 978-1-4822-1480-2 (hbk)
ISBN-13: 978-1-138-74862-0 (pbk)

Library of Congress Cataloging-in-Publication Data

mHealth multidisciplinary verticals / editor, Sasan Adibi.
  p. ; cm.
  Includes bibliographical references and index.
  ISBN 978-1-4822-1480-2 (hardcover : alk. paper)
  I. Adibi, Sasan, 1970- , editor.
  [DNLM: 1. Telemedicine. 2. Attitude to Computers. 3. Delivery of Health Care--methods. 4. Medical Informatics Applications. 5. Telemetry. W 83.1]

  R858
  610.285--dc23   2014039177

Visit the Taylor & Francis Web site at
http://www.taylorandfrancis.com

and the CRC Press Web site at
http://www.crcpress.com

This book is dedicated to all the engineers who have been poking their noses

into the field of medical sciences and made this world a better place.

Contents

Section VII mHealth Regional, Geographical, and Public Health Perspectives

Section VIII mHealth Technology Implications

Section IX   mHealth Cloud Applications

Preface

This book provides an excellent coverage of a number of multidisciplinary verticals within the field of mobile health (mHealth), covering the following nine main domains in the mHealth space: (1) preventive and curative medicine; (2) consumer and patient-centric approaches; (3) psychological, behavioral, and mental verticals; (4) social perspectives; (5) education, adoption, and acceptance; (6) aged care and the aging population; (7) regional, geographical, and public-health perspectives; (8) technology implications; and (9) cloud applications. This book is comprised of 36 diverse chapters, which are the results of extensive research and collaborative efforts of professional authors from more than 20 countries, and have been carefully reviewed and selected by the team of reviewers to promote the main aim of the book—to provide an alternative insight to mHealth using a multidisciplinary approach.

Sasan Adibi
School of Information Technology
Faculty of Science & Build Environment
Deakin University
Burwood, Victoria, Australia

Editor

Sasan Adibi (BS'95, MS'99, MS'05, PhD'10, SMIEEE'11) holds a PhD degree in communication and information systems from the University of Waterloo, Canada, and is the recipient of the best PhD thesis award from the IEEE Society. He is currently involved in the research, design, implementation, and application of electronic health (eHealth) and mobile health (mHealth). Dr. Adibi's research publication background is mostly in the areas of quality of service (QoS), security, and eHealth/mHealth.

He is first author for more than 50 journal, conference paper, book chapter, patent, and white-paper publications and is a coeditor of four books, two in the areas of mHealth and two in the areas of 4th generation mobile networks and QoS.

Dr. Adibi is an inventor/coinventor of four patents in the area of health informatics. He has more than nine years of industry experience, having worked in a number of high-tech companies, including Nortel Networks, Siemens Canada, BlackBerry Corp., WiMAX Forum, and Huawei Technologies. He is the founder and managing director of Cyber Journals and serves as the associate editor of its Journal of Selected Areas in Health Informatics (JSHI). He served as a vice chancellor research fellow at the Royal Melbourne Institute of Technology (RMIT), Department of Business IT and Logistics for two years. He is currently a lecturer in computer networks in the school of Information Technology at Deakin University.

Contributors

Arwa Fahad Ababtain
College of Public Health and Health
    Informatics
King Saud bin Abdul-Aziz University
    for Health Sciences
Riyadh, Kingdom of Saudi Arabia

and

Information Services Department
Imam Abdulrahman Bin Faisal Hospital
Dammam, Kingdom of Saudi Arabia

Kenza Abdelaziz
Computer Laboratory
University of Pau and Pays de l'Adour
Pau, France

Hossein Adibi
School of Psychology and Social Science
Edith Cowan University
Perth, Western Australia, Australia

Alireza Ahmadvand
Non-Communicable Diseases Research
    Center
Endocrinology and Metabolism Population
    Science Institute
Tehran University of Medical Sciences
Tehran, Iran

Deana Ahmad AlMulhim
College of Public Health and Health
    Informatics
King Saud bin Abdul-Aziz University
    for Health Sciences
Riyadh, Kingdom of Saudi Arabia

and

Information Services Department
Imam Abdulrahman Bin Faisal Hospital
Dammam, Kingdom of Saudi Arabia

Mohammad Nabil Almunawar
Faculty of Business
Economics and Policy Studies
Universiti Brunei Darussalam
Brunei, Darussalam

Hassan Sadeghi Amini
School of Computer Sciences
Universiti Sains Malaysia
Penang, Malaysia

Muhammad Anshari
Continuing Education Centre
Universiti Brunei Darussalam
Brunei, Darussalam

Denise L. Anthony
Department of Sociology
Institute for Security Technology
    and Society
Dartmouth College
Hanover, New Hampshire

Mahmoud Numan Bakkar
School of Business Information Technology
    and Logistics
RMIT University
Melbourne, Victoria, Australia

Abbas Bhuiya
Centre for Equity and Health Systems
International Centre for Diarrhoeal
    Disease Research
Dhaka, Bangladesh

Dirk Boecker
MassineBoecker GmbH
Berlin, Germany

and

MassineBoecker Inc.
Palo Alto, California

Nandini Bondale
School of Technology and Computer Science
Tata Institute of Fundamental Research
Mumbai, India

Angela Brunstein
School of Medicine
Royal College of Surgeons in Ireland
Manama, Bahrain

Joerg Brunstein
Science and Technology Research Center
American University in Cairo
Cairo, Egypt

Christie Butrico
UH Operating Rooms
University of Michigan Health System
Ann Arbor, Michigan

Joshua Caballero
College of Pharmacy
Nova Southeastern University
Fort Lauderdale, Florida

Manuel Casillas
Department of Electronics and
    Telecommunications
Center for Scientific Research and Higher
    Education of Ensenada
Baja California, Mexico

Rahul Chakrabarti
Ophthalmology Registrar
The Royal Victorian Eye and Ear Hospital
The University of Melbourne
Parkville, Victoria, Australia

Andrew Chitty
Digital Life Sciences Ltd
London, United Kingdom

Kevin A. Clauson
College of Pharmacy
Lipscomb University
Nashville, Tennessee

Grant P. Cumming
Department of Obstetrics and Gynaecology
Dr Grays Hospital
NHS Grampian
and
Moray College
University of the Highlands and Islands
Elgin, Scotland

and

University of Aberdeen
Aberdeen, Scotland

Rozita Dara
School of Computer Science
University of Guelph
Guelph, Ontario, Canada

John Dinsmore
Centre for Practice & Healthcare
    Innovation
School of Nursing and Midwifery
Trinity College
University of Dublin
Dublin, Ireland

Banafsheh Farahani
Department of Hospitality and Tourism
    Management
Azad University of Noor
Mazandaran, Iran

Alejandro Galaviz-Mosqueda
Department of Electronics and
    Telecommunications
Center for Scientific Research and Higher
    Education of Ensenada
Baja California, Mexico

Ana Lilia González
Department of Electronics and
    Telecommunications
Center for Scientific Research and Higher
    Education of Ensenada
Baja California, Mexico

Nedelko Grbic
ExAudio Limited
Malmö, Sweden

Zaher Hajar
College of Pharmacy
Nova Southeastern University
Fort Lauderdale, Florida

Lucy Hederman
School of Computer Science
   and Statistics
Trinity College
University of Dublin
Dublin, Ireland

Anita E. Heywood
School of Public Health and Community
   Medicine
The University of New South Wales
Kensington, New South Wales, Australia

Mowafa Said Househ
College of Public Health and Health
   Informatics
Kingdom of Saudi Arabia
King Saud bin Abdul-Aziz University for
   Health Sciences
Riyadh, Saudi Arabia

Anpeng Huang
mHealth Lab
Peking University
Haidian, Beijing, People's Republic
   of China

Thomas D. Hull
Teachers College
Columbia University
Upper Manhattan, New York

Pantea Keikhosrokiani
School of Computer Sciences
Universiti Sains Malaysia
Penang, Malaysia

Fatema Khatun
School of Public Health and Community
   Medicine
The University of New South Wales
Kensington, New South Wales, Australia

and

Centre for Equity and Health Systems
International Centre for Diarrhoeal
   Disease Research
Dhaka, Bangladesh

Sanjay Kimbahune
TCS Innovation Lab
Tata Consultancy Services
Mumbai, India

Saso Koceski
Faculty of Computer Science
Goce Delčev University of Štip
Stip, Republic of Macedonia

David Kotz
Institute for Security, Technology
   and Society
Dartmouth College
Hanover, New Hampshire

Igor Kulev
Faculty of Computer Science and
   Engineering
Ss. Cyril and Methodius University
   of Skopje
Skopje, Republic of Macedonia

Sébastien Laborie
Computer Laboratory
University of Pau and Pays de l'Adour
Pau, France

Tayeb Lemlouma
Institute for Research in IT and Random
   Systems
University of Rennes I
Lannion, France

Dave Lewis
School of Computer Science
    and Statistics
Trinity College
University of Dublin
Dublin, Ireland

Wanqing Li
Information and Communication
    Technology Research Institute
University of Wollongong
Wollongong, New South Wales, Australia

Siaw-Teng Liaw
School of Public Health and Community
    Medicine
The University of New South Wales
Kensington, New South Wales, Australia

Megan S.C. Lim
Centre for Population Health
Burnet Institute
and
Department of Epidemiology
    and Preventive Medicine
School of Public Health and Preventive
    Medicine
Monash University
Melbourne, Victoria, Australia

Joanne S. Luciano
Web Science Research Center
Tetherless World Constellation
and
Predictive Medicine, Inc.
Rensselaer Polytechnic Institute
GE Global Research Center
University of California, Irvine
Irvine, California

Sandra MacRury
Department of Diabetes
    and Endocrinology
Raigmore Hospital
and
Division of Health Research
University of the Highlands and Islands
Inverness, Scotland
and
University of Aberdeen
Aberdeen, Scotland

Roberto Magana
Department of Electronics
    and Telecommunications
Center for Scientific Research and Higher
    Education of Ensenada
Baja California, Mexico

Marilia Manfrinato
Department of Medicine
Federal University of Paraná
Curitiba, Brazil

Jennifer Martin
School of Global, Urban and
    Social Studies
RMIT University
Melbourne, Victoria, Australia

Michael K. Martin
Department of Psychology
Carnegie Mellon University
Pittsburgh, Pennsylvania

Thomas Martin
School of Continuing Studies—Technology
    Management
Georgetown University
Washington, DC

Eilish McAuliffe
Centre for Global Health
Trinity College
University of Dublin
Dublin, Ireland

Ronan McDonnell
Department of General Practice
HRB Centre for Primary Care Research
Royal College of Surgeons in Ireland
Dublin, Ireland

Elspeth McKay
School of Business Information Technology
    and Logistics
RMIT University
Melbourne, Victoria, Australia

Douglas McKendrick
Department of Anaesthetics
Dr Grays Hospital
NHS Grampian
Elgin, Scotland

and

University of Aberdeen
Aberdeen, Scotland

Bernhard Mikoleit
MassineBoecker GmbH
Berlin, Germany

Sayedmehran Mirsafaie Rizi
School of Electrical Engineering
Universiti Sains Malaysia
Penang, Malaysia

Abbas Mohammed
ExAudio Limited
Malmö, Sweden

Saradhi Motamarri
Asia Pacific ubiquitous Healthcare Centre
The University of New South Wales
Sydney, New South Wales, Australia

Norlia Mustaffa
School of Computer Sciences
Universiti Sains Malaysia
Penang, Malaysia

Michael Mutingi
Faculty of Engineering and the Built
    Environment
University of Johannesburg
Johannesburg, South Africa

Sandy Ng
School of Economics, Finance
    and Marketing
RMIT University
Melbourne, Victoria, Australia

Duc Thanh Nguyen
Information and Communication
    Technology Research Institute
University of Wollongong
Wollongong, New South Wales, Australia

Wendy Nilsen
Office of Behavioral and Social Sciences
    Research
National Institutes of Health
and
Division of Information and Intelligent
    Systems (CISE/IIS)
Smart and Connected Health Program
National Science Foundation
Washington, DC

Phillip Olla
Center for Research
Madonna University
Livonia, Michigan

Raymond L. Ownby
Department of Psychiatry and Behavioral
    Medicine
Nova Southeastern University
Fort Lauderdale, Florida

Elin Palm
Centre for Applied Ethics
Linköping University
Linköping, Sweden

Arun K. Pande
IT Innovation for Masses
Maharashtra, India

Misha Pavel
College of Computer & Information
    Science
and
Bouve College of Health Sciences
Northeastern University
Boston, Massachusetts

Chandrashan Perera
Journal of Mobile Technology in Medicine
and
Department of Ophthalmology
Fremantle Hospital
Fremantle, Western Australia, Australia

Chaz Pichette
School of Nursing
Madonna University
Livonia, Michigan

Aarathi Prasad
Institute for Security, Technology
    and Society
Dartmouth College
Hanover, New Hampshire

Yasmine Probst
Faculty of Science Medicine and Health
Smart Foods Centre
University of Wollongong
Wollongong, New South Wales, Australia

Gilnaz Purhossein
School of Computer Sciences
Universiti Sains Malaysia
Penang, Malaysia

Abderrezak Rachedi
LIGM Lab
University of Paris-Est
Marne la Vallée, France

Pradeep K. Ray
Asia Pacific Ubiquitous Healthcare Centre
Australian School of Business
University of New South Wales
Kensington, New South Wales, Australia

Brenda M.B. Reginatto
Insight Centre for Data Analytics
University College Dublin
Dublin, Ireland

Raúl Rivera
Telematics Division
Center for Scientific Research and Higher
    Education of Ensenada
Baja California, Mexico

Megan Rollo
Faculty of Health and Medicine
Priority Research Centre in Physical
    Activity and Nutrition
University of Newcastle
Newcastle, New South Wales, Australia

Philippe Roose
Computer Laboratory
University of Pau and Pays de l'Adour
Pau, France

David Scheffer
MassineScheffer GmbH
and
Nordakademie
University of Applied Science
Elmshorn, Germany

Shakira Shaheen
Department of Electrical Engineering
Blekinge Institute of Technology
Blekinge, Sweden

Kami Shalfrooshan
Department of medicine
University College London
London, United Kingdom

Paul Solano
School of Public Policy and Administration
University of Delaware
Newark, Delaware

Jacob Sorber
School of Computing
Clemson University
Clemson, South Carolina

Donna Spruijt-Metz
mHealth Collaboratory
Center for Economic and Social Research
and
Department of Preventive Medicine
and
Responsible Conduct in Research
Keck School of Medicine
University of Southern California
Los Angeles, California

Timothy Stablein
Department of Sociology
Union College
Schenectady, New York

Kate Stephen
Moray College
University of the Highlands and Islands
Elgin, Scotland

Vladimir Trajkovik
Faculty of Computer Science
and Engineering
Ss. Cyril and Methodius University
of Skopje
Skopje, Republic of Macedonia

Chung-Li Tseng
Business School
The University of New South Wales
Sydney, New South Wales, Australia

Frédérique Vallières
Centre for Global Health
Trinity College
University of Dublin
Dublin, Ireland

Ibrahim Venkat
School of Computer Sciences
Universiti Sains Malaysia
Penang, Malaysia

Salvador Villarreal-Reyes
Department of Electronics
and Telecommunications
Center for Scientific Research and Higher
Education of Ensenada
Baja California, Mexico

Luis Villasenor
Plamex Design Center
Plamex S.A. de C.V.Tijuana, Baja California

Elena Vlahu-Gjorgievska
Faculty of Administration and Information
Systems Management
St. Kliment Ohridski University
Bitola, Republic of Macedonia

Lucia Vuong
Outware Mobile
Melbourne, Victoria, Australia

P.J. Wall
School of Computer Science
and Statistics
Trinity College
University of Dublin
Dublin, Ireland

Robyn Whittaker
National Institute for Health Innovation
University of Auckland
and
Waitemata District Health Board
Auckland, New Zealand

Faisel Yunus
College of Public Health and Health
Informatics
King Saud bin Abdul-Aziz University
for Health Sciences
Riyadh, Kingdom of Saudi Arabia

Nasriah Zakaria
School of Computer Sciences
Universiti Sains Malaysia
Penang, Malaysia

and

Medical Education Department
College of Medicine
King Saud University
Riyadh, Kingdom of Saudi Arabia

1

Introduction

CONTENTS

Mobile health (mHealth) refers to the overlapping fields between mobile technology and health care. It has evolved from electronic health (eHealth) centered on the application of smartphones. Besides the technologies revolving around mHealth, there are a number of verticals associated with mHealth and this book takes a multidisciplinary approach covering a number of these verticals. The goal of this book is to capture emerging and diverse mHealth applications under a single umbrella. These applications and technologies are linked to key market indicators, vendor players, and interdependencies and synergies between relevant stakeholders, which derive the research forces behind mHealth.

This book is composed of 9 sections and 36 chapters, covering the following multidisciplinary fields interrelated with mHealth: preventive and curative medicine; consumer-centric approaches; psychological, behavioral, and lifestyle changes; social perspectives; education, adaptation, and technology acceptance; aging population; technology implications; and cloud applications.

In this introductory chapter, we aim to define concepts and definitions discussed in this book.

1.1 Definitions

To start, the following fundamental concepts are defined:

Health informatics: The body of knowledge covering the acquisition, usage, storage, and retrieval of information concerning human health and the design and management of related information resources, devices, and methods to advance the understanding and practice of health care. Health informatics covers areas such as health-care systems and infrastructures, health-care standards, and eHealth/mHealth.

Electronic Health (eHealth): The term eHealth emerged in the late 1990s, which was referred to the health-care practices supported by telecommunication systems and electronic processes. There are multiple forms of eHealth in terms of systems and services, for example, electronic health record (EHR), personal health record (PHR), telemedicine, mHealth, health-care information systems (HIS), and consumer health informatics.

Mobile Health (mHealth): mHealth is the practice of eHealth supported by mobile devices and smartphones, which are used to capture, analyze, store, and transmit health-related information from various sources including personal inputs, sensors, and other biomedical acquisition systems.

Health-care systems: This covers the combination of stakeholders, resources, institutions, and infrastructures working together to provide health-related services to groups and individuals.

1.2 Section I: mHealth Preventive and Curative Medicine

In the first section of this book, Chapters 2 through 4 are dedicated to the application of mHealth in preventive and curative medicine. In these applications, mHealth can be used to monitor the state of health of individuals and groups and to inform the caregiver(s) of possible deteriorating health conditions when the illness is imminent.

Prevention is an essential part of the continuum of care in medicine and health. The focus of preventive services is to provide means to prevent individuals as well as communities from illness and health issues.

Curative medicine is committed to provide treatment services and the role of mHealth systems and medical applications running on them are to assist medical practitioners for both preventive and curative phases.

The aim of the chapters in this section is to describe the current state-of-the-art mHealth applications in preventive and curative medicine and bring forth the challenges of the deployment of mHealth devices and gadgets (e.g., wearable sensors), which are caused by a number of factors, such as the increasing trends of available sensor data and currently missing guidance for health behavior changes. We will focus on diabetes as an example of a chronic disease and discuss how mHealth can help manage diabetes in the affected patients.

1.3 Section II: Consumer- and Patient-Centric Approaches in mHealth

This section is composed of Chapters 5 through 7 about consumer- and patient-centric approaches in mHealth. The next-generation health-care systems is expected to be more patient-centric since health-care providers will be focused on managing the health of individuals through emerging personalized approaches, including EHR and PHR.

Modern medical approaches are influenced by patient-centered or personalized medicine. Patient-centric environments enable health-care personnel to be able to timely access, review, update, and send patient information from wherever they are, whenever they want.

In that way, pervasive health care takes steps to design, develop, and evaluate computer technologies that help citizens participate more closely in their own health care, on the one hand, and on the other to provide flexibility in the life of patients who lead an active everyday life with work, family, and friends.

Chapter 5 is dedicated to the application of mHealth for monitoring diet, which is also a subset of preventive medicine. The risk factors for many chronic diseases such as cardiovascular disease, diabetes, and some cancers are associated with poor diet and/or obesity. Preventing diseases through improving nutrition is now a global health priority. However, before people can improve the quality of their diet, they need to have greater awareness of what constitutes healthy eating. Many are unaware of the nutritional content of their diet. Further education is needed to address this issue and induce healthy dietary and lifestyle changes. Crucial to lifestyle change is the accurate and timely assessment of food intake so that individuals can be better informed on what constitutes a healthy diet consistent with dietary guidelines.

Mobile devices have increasingly been used to guide individuals in the selection of foods, to self-monitor dietary intake in relation to the management of diseases, in monitoring of intake to inform assessment, and to assist in the provision of nutrition intervention by health professionals.

Mobile devices, such as smartphones and tablets, can be used to capture food intake and deliver dietary feedback using text messages. This use of mobile devices has great potential to improve accuracy and reduce participant burden when assessing nutritional intake. Presently, some of the fastest-growing mobile application markets focus on providing weight loss tips, nutritional information, and calorie-tracking capability. Recording dietary information and tracking daily behaviors via technology usage and data analysis also forms a major component of movements such as the quantified self (QS).

An overview of the technology required as well as current developments and future challenges and opportunities for the use of mobile devices for informing, guiding, monitoring, and improving nutritional intake of individuals, groups, and populations is presented in Chapter 5.

Such an increasing trend in consumer-based health applications running on their smartphones is another indication of the shift toward health-care personalization. In Chapter 6, which focuses on wellness applications, a discussion is made on how these apps help consumers achieve their health behavior goal and improve their general well-being and lifestyle. Such apps are cost effective for the health-care industry and can effectively be used to disseminate current public health information and services on health prevention and awareness messages and provide an avenue for health service providers to form closer relationships with their consumers.

Chapter 7 is focused on a personalized health-care model, which creates an opportunity for individuals to have a 24 h medical monitoring at the comfort of their homes resulting with reduction of overall costs for consumers and health-care institutions. This model is based on novel technologies utilizing mobile, web, and broadband technologies, enabling users to have ubiquity service support everywhere.

1.4 Section III: Psychological, Behavioral, and Mental Verticals of mHealth

mHealth has also found intriguing applications within the psychology space through the power of motivation, suggestion, and mood identification, provided through applications aimed to engage with the users on various levels. These applications can assess the mood and behavioral aspects of the user using challenge-and-response and reaction-based interactions and provide necessary mental stimulations. Furthermore, measures of the implicit personality systems of users provide a powerful basis for customized communication to and motivation of specific personality types. Users can be grouped into specifically designed peer groups where the members are homogenous in certain traits like extraversion and heterogeneous in other traits (e.g., action/state orientation). Interventions based on personality differences have been proven to provide significantly increased effectiveness.

In recent decades, the fields of health care and patient treatment have seen enormous technological progress and success. Conversely, advances in physician/patient communication and support of patient self-management are minor and rare. Health plans and disease management organizations are still contacting their large member populations with boilerplate information or via costly call centers. All in all today's direct member communication is limited and inefficient.

Most major constituents wish to enhance compliance and behavior of their members and customers with an easy and cost-effective solution.

We report on the development of a new and more effective way for communicating with large populations. We use a novel combination of automated yet individualized outreach. mHealth technologies and communication channels are used to reach large populations and our unique process of individualizing interventions offers effective impact despite the substantially lowered cost compared to conventional methods.

We configure our interventions to the specific profile of each participant. We start by determining the psychographic and personal characteristics of individual patients or representative population segments; then, we create a tailored communication profile. We stimulate the patient's internal motivation ensuring better opportunity for long-lasting engagement, compliance, and sustained self-management.

Our special method has long proved itself in neuromarketing, that is, the use of recent findings from psychology and neurosciences in marketing and sales, and now enables health-care professionals and organizations to communicate with patients in a personalized and personality type–specific manner. Individualized adaptation of all standard interventions offers a renewed while cost-effective approach in population management.

The system provides a direct method for reducing health-care cost through improving self-care and management, compliance, and overall engagement. It will help individuals to slow down disease progression and improve their overall condition and happiness and most importantly reduce utilization and hospitalization. We describe our developed solutions and services for payers, health plans, care management organizations, and companies active in the pharma and diagnostic industry.

1.5 Section IV: mHealth Social Perspectives

In this section, Chapters 13 through 15, the focus is on the social perspectives of health with targeted social patterns on health.

Although commercial interest is the main driver of technological inventions, yet unintended social outcomes may overtake the commercial interest in long term. Traditional design analysis typically focuses on the system evaluation from the technological point of view, whereas social analysis looks at the way technology is incorporated into the users' activities. Based on social viewpoints, social and technical issues are interdependent and therefore cannot be separated.

Social factors, such as aging, gender, and culture, can affect user's behavior toward the use of mHealth; on the other hand, characteristics of individual personal attitude and personal philosophy may override these factors. Recognizing effective social factors in mHealth allows mobile health-care providers to understand users' requirements and improve the design of the new mobile health-care systems. There are many social-related applications of mHealth, including gender, social class, and aged care, which the chapters in this section will explore.

1.6 Section V: mHealth Education, Adoption, and Acceptance

This section, which is composed of Chapters 16 through 19, covers the education, adoption, and acceptance of mHealth technologies among various individuals and users. This becomes more critical among elderly population as they may face more challenges in utilizing technology for health monitoring. Therefore, investigating the factors associated with user acceptance of the new technology has been the center of attention in the recent years. Several models are developed to investigate the factors affecting the acceptance of new technologies in organizations, which is one of the main focuses of this section.

From the mHealth education perspective, health-care personnel may use mHealth services for the purpose of medical education/training activities to access the latest information, retrieve concepts and phrases quickly, share information, identify how to use different resources, and provide real-time patient advice. Medical practitioners are using smartphones on daily basis to install instructional applications to enhance the visualization of their work. Other medical-related associates also use smartphones to install various medical-related applications to enhance their learning skills, assist them with diagnosis, and collaborate with other medical staff.

Smartphones and tablets enable nurses to provide direct patient care based on the most current, evidence-based research available using the most advanced technology our society is embracing. The innovative mobile-based medical devices and applications, which exist today, and those planned for the near future, aim to provide immediate care, even remotely, in unimaginable ways. Nurses now have the ability to chart patient information, retrieve lab results, access medical reference materials, and educate their patients from mobile devices, all of which contribute in revolutionizing the health-care industry. Nurses in particular have the ability to use smartphones during their practices in a variety of ways. Smartphones can become diagnostic devices, store data, and maintain nursing applications.

1.7 Section VI: mHealth Aged Care and Aging Population

This section is composed of Chapters 20 through 22, which explore the potentials of mHealth solutions to positively disrupt health-care provision for the elderly. These solutions not only have the potential to enable better management of chronic conditions but also support older adults to live longer and healthier, in the comfort of their place of choice. Additionally, such mHealth solutions have the potential to support family caregivers, professional caregivers, and other health-care professionals to better care for the elderly.

Information and communications technology (ICT) provides powerful tools for assisting the older generation with changing biological, psychological, and social needs, while ultimately remaining connected to their families and communities. To maintain this connectivity, there are a limited number of free mobile device applications that fall into the following categories: social networking, medical, reference, utilities, aged care standards and accreditation, health and fitness, and lifestyle. Therefore, one of the main aims of this section is to identify the needs of the elderly and explore how mHealth can impact upon the ICT tools that are adopted in the general community and more specifically targeted at the elderly.

1.8 Section VII: mHealth Regional, Geographical, and Public Health Perspectives

Health-care practices have been framed and shaped based on the regional, geographical, and geopolitical influences, resulting in the creation of regional rules and policies. The application of smartphones in health-care delivery in this situation follows the same influential factors and is considered in this section, which is composed of Chapters 23 through 27.

Public health systems, including both rural and urban populations, can benefit greatly from mHealth approaches by targeting specific groups within the public society that match certain selection criteria. Smartphones can be efficiently used in a scalable way to effectively filter out such selection criteria.

In many emerging economies, digital divisions exist between rural and urban populations. The relevance of remote monitoring aspect of mHealth, especially for rural population and the significance of mHealth for rural public health are also covered in this section.

There is a growing awareness that Internet-based health interventions work; however, there is a debate about the conditions and audience of its operation.

To be effective in improving access to health-care and medical information, mHealth tools must meet the community's socioeconomic needs especially for lower socioeconomic and disadvantaged groups. While designing such tools, developers take social conditions, level of education, and other cultural aspects of the targeted groups into account. One of these tools is provided by human–computer interface (HCI) technologies, which offer special significance for rural population to access quality health care. For mHealth-based rural public health to be sustainable, one would require innovations in business models as well.

The potential of mHealth in taking services to the masses especially to the disadvantaged people is promising. Further research however is needed to understand what is possible with the technology and what infrastructure and regulatory frameworks are needed to optimize the potential of mHealth to be effective.

1.9 Section VIII: mHealth Technology Implications

A number of issues, challenges, and implications associated to mHealth are considered in this section, Chapters 23 through 27, which include decision enhancements, security, and quality of service (QoS) relevant to health-care practices, optimized by smartphone deployments. Besides, mHealth services are expected to improve personalization in services, clients' empowerment, social support, and sharing through social networks.

One of the major barriers to advancing mHealth adoption is presenting information to policy makers or decision makers in an adequate fashion. The easiest and most impactful method to reveal consumer or constituent preference is to assess benefits and present information from an economic perspective. The challenge presented by mHealth is that many associated benefits with the use of technology are less than tangible to many analysts. However, many methods currently used in health-care program assessment are applicable, with some minor modifications, for use in assessing mHealth.

Most frameworks and infrastructures supporting mHealth applications have undergone major technology shifts in the recent years, shifting from a stand-alone bedside technology to end-to-end cloud-ready systems. These technology shift aims to support security, open platforms, regulatory services, quality of care (QoC), bring your own device (BYOD), mHealth ecosystem and health service redesign for technology adoption, and data and knowledge management. In addition, mHealth shifts a role of clients from recipient to partner of care. It can offer clients to have a greater role in the decision-making process related to their health as they can be empowered with the ability to access and control information that fits with their personalized needs. The proposed mHealth's model extends the role of the clients as an individual health actor, a social health agent, and a medical care partner. The usage of mHealth enables a shift in care locus. Health-care delivery systems are transforming away from their dependence on traditional care units toward *anywhere, anytime* support. In many ways, this shift enables and facilitates care recipients to self-manage their conditions or disabilities with greater effectiveness. Furthermore, the term *mHealth* additionally implies that care recipients can be monitored continuously. A constant measuring and controlling of health-related data may influence care recipients' privacy, self-understanding, and relationships with health-care providers. Thus, careful analysis of mHealth implications and balancing of benefits and drawbacks of mHealth are needed.

1.10 Section IX: mHealth Cloud Applications

In this section, Chapters 33 through 36, the concept of cloud applications interrelated with mHealth is introduced. Technology-assisted methods for medical diagnosis and biomedical health monitoring are rapidly shifting from classical invasive methods to mobile-based noninvasive approaches. Managing health-related data flow and characterization of medical traffic have become huge challenges in the past few years as the number of health-based cloud applications has increased exponentially. Medical imagining is an example of another prominent medical application, which contributes hugely to such increasing data flow trends. In this section, Chapter 35 considers the medical imaging schemes for mHealth, including technologies, physical properties, MRI, CT scan, and terahertz technology.

The use of mHealth in pharmaceutical monitoring offers myriad methods to address the irrational use of medical. These methods, ranging from simple text message reminder systems to ecological momentary interventions (EMIs), help leveraging interface versatility for eHealth literacy solutions. In Chapter 33 of this section, current scientific evidence will be used to explore the role of mHealth in enhancing pharmacy-related outcomes in varying health-care settings and diverse populations. Current and future developments in mHealth will be explored through the lens of select disease states and patient populations.

Section I

mHealth Preventive and Curative Medicine

2

Placing Prevention in the Pockets: The Role of mHealth in Preventive Medical Services

Alireza Ahmadvand, Robyn Whittaker, and Megan S.C. Lim

CONTENTS

2.1 Introduction

Prevention is an essential part of the continuum of care in medicine and health. The focus of preventive services can be on individuals as well as communities and populations. Traditionally, preventive interventions in medical care have been classified into three hierarchical levels:

1. *Primary prevention*: includes methods that help in avoiding disease occurrence, specifically in healthy people (e.g., health promotion)
2. *Secondary prevention*: includes methods for diagnosing or treating risk factors and/or diseases at early stages with a special focus on at-risk people before they experience morbidity (e.g., screening)
3. *Tertiary prevention*: includes methods that help in reducing the impact of existing diseases in patients, usually by preventing progression of disease or the development of complications and by rehabilitation (Pomeroy and Steiker 2012)

2.1.1 Overall Role of mHealth Interventions in Each Level of Prevention

Examples such as a depression prevention intervention in adolescents using mobile phones (Whittaker et al. 2012d) and a weight management intervention for overweight and obese adults using short message service (SMS) (Donaldson et al. 2013) can be categorized under primary prevention.

Efforts such as sending SMS to men who have sex with men to motivate them undergo HIV testing (Menacho et al. 2013) and SMS reminders for postpartum women to test for type 2 diabetes after they had been diagnosed with gestational diabetes mellitus (Heatley et al. 2013) are some forms of secondary preventive interventions.

Examples such as developing a care model for cardiac rehabilitation (CR) based on mobile phones in outpatient settings (Walters et al. 2010) and sending SMS to increase patient retention in care after HIV diagnosis (van der Kop et al. 2013) can be classified under tertiary preventive measures.

2.2 Modalities for mHealth Preventive Interventions

2.2.1 Mobile Devices and Mobile Phone Services

Mobile devices include smartphones, tablet computers, ordinary (nonsmart) cell phones, personal digital assistants (PDAs), medically specific devices such as wearable devices, and game consoles. Smartphones, tablets, and ordinary (nonsmart) cell phones comprise the largest share of mobile devices available in the market, so this chapter focuses on these three types of mobile devices, unless stated otherwise.

Mobile services are very diverse in nature; popular services include SMS, multimedia messaging service (MMS), direct phone calls (to care providers or to call centers), and mobile Internet access. The published evidence shows that SMS and MMS are two of the most popular services that have undergone vigorous and extensive assessments in different mobile health (mHealth) contexts and well-designed projects. In this chapter, the authors will be focused on these types of mobile services, unless stated otherwise.

Another important service is smartphone applications (or *apps*) developed for different operating system platforms. Apps are programs or pieces of software designed in a self-contained way for a specific purpose that can be downloaded or installed by the smartphone user onto their mobile device. There has been a tremendous growth in the number of apps for health and medical purposes in recent years.

2.2.2 Characteristics of High-Quality mHealth Modalities for Preventive Interventions

Countless mHealth modalities for preventive interventions have been designed and tested all around the world. Although they are all mHealth programs, they differ with respect to their quality of design, comprehensiveness of functionality, quality of development process, adaptation for the needs and requirements of customers, proven efficacy and effectiveness in scientifically sound research, and their capacity to meet the standards of a high-quality service or intervention (Bastawrous and Armstrong 2013).

Norris et al. provided a strategic framework for sustainable mHealth and highlighted phases and tasks for a sustainable strategy identified from the survey of senior executives, strategists, planners, and managers in New Zealand. They introduced three phases: identifying suitable applications, channel development activity, and confirm activity for sustainability. Norris et al. insisted on the fact that designing mHealth solutions for each part of the continuum of care (i.e., prevention, monitoring, treatment, and support) should follow through the proposed phases in order to produce high-quality modalities. Different tasks were suggested under each phase, which is beyond the scope of this chapter (Norris et al. 2009).

Many researchers and evaluators have tried to define the requirements and standards of mHealth interventions within the *design–evaluation spectrum* with more or less comprehensive approaches. One of the most important aspects of this spectrum has been the use of theoretical models of behavior change within the specific study context. In 2011, Riley et al. tried to determine the application of health behavior theories to mobile interventions in smoking cessation, weight loss, adherence to medical regimens, and also disease management. They showed that self-regulation, transtheoretical model, social–cognitive theory, social marketing, theory of planned behavior, self-efficacy theory, systems contingency approach, and cognitive–behavioral model had been widely (although not universally) used in smoking cessation and weight loss mobile interventions to enrich the theoretical basis of the corresponding studies. This has not been the case in mHealth projects targeting other health areas (Riley et al. 2011).

Many studies have tried to publish their theoretical basis of behavior change. One example is presented in Box 2.1, which summarizes a case study from New Zealand in which their development and evaluation process for mHealth interventions has been presented within the specific experience of behavior change in mHealth smoking cessation and depression prevention programs (Whittaker et al. 2008; Free et al. 2011; Whittaker et al. 2011; Whittaker et al. 2012c).

BOX 2.1   DEVELOPMENT AND EVALUATION PROCESS FOR mHEALTH INTERVENTIONS: A CASE STUDY FROM NEW ZEALAND

Smoking cessation interventions based on mobile phones have long been the focus of development and testing at the University of Auckland in New Zealand. Their process for the development and testing of mHealth interventions has gone through stages of advancement over many years. This process involves a series of steps based on research that emphasizes an evidence-based or guideline-oriented process of growth. This may help other developers and evaluators consider formative research, effectiveness trials, or impact evaluations according to their interests and missions.

The 7-step development and evaluation process (described using the examples of a mobile-based video messaging intervention for smoking cessation and a similar intervention for depression prevention) includes:

1. Conceptualization
2. Formative research to inform the development (using focus groups and online surveys)
3. Pretesting content (using focus groups, online surveys, and individual interviews)
4. Pilot study (in a small, nonrandomized approach)
5. Pragmatic community-based randomized controlled trials
6. Further qualitative follow-up to inform improvement (using semistructured interviews)
7. Evaluation of implementation impact (using phone/online surveys and semi-structured interviews)

Involvement of the target audience right from the start of development and using research-based methods for effectiveness assessment are considered as strengths of this process.

But it has also its inherent limitations; for example, it may be a less formalized and randomized approach than other proposed processes such as the multiphase optimization strategy (MOST) proposed by Collins et al. in 2007. In MOST, different components of an intervention are examined randomly before conducting a full trial. This three-phase strategy (screening, refining, and confirming phases) may lead to optimization of the intervention, especially for less complex interventions (Collins et al. 2007; Whittaker et al. 2012c).

2.3 Successful mHealth Projects in Preventive Medicine

Numerous projects have explored the role of mHealth in the prevention spectrum. The results of many are available in online databases and reports or through commercial services (Figure 2.1).

In this chapter, the authors have preferentially included studies published in peer-reviewed scientific journals. Overall, a review of peer-reviewed evidence shows that mHealth for prevention can be categorized as either of the following:

1. Changing consumer behavior for the prevention of disease (with a focus on risky components of individual behavior of healthy people and also on specific disease risk factors)

FIGURE 2.1
Launched by the mHealth Alliance in 2010, Health Unbound (HUB) is the interactive network and online knowledge resource center for the mobile health (mHealth) community. (Last accessed, June 30, 2014. http://www.healthunbound.org.)

2. Screening and early diagnosis and managing disease in order to prevent progression and sequelae (with a focus on patients)

Managing health information, controlling costs of care, and reaching vulnerable and marginalized populations have also been other areas of assessment and development in mHealth projects. Projects have targeted a range of populations: elderly, pregnant women, children, young people, geographically remote residents, rural populations, or hidden populations (such as female sex workers or intravenous drug abusers). In the following sections, the authors have summarized selected pieces of evidence with regard to these focus themes.

2.3.1 Changing Consumer Behaviors

2.3.1.1 Quitting Smoking

Quitting smoking has been one of the earliest behaviors to be targeted by mHealth researchers and developers. There are some possible reasons for this interest:

1. The act of quitting is a tangible primary outcome that can be assessed with confidence and low risk of bias.

2. A date can be set for quitting and the behavior change (which can be supported by mHealth interventions) can be focused on that specific date.

3. Smoking is a global risk factor for many noncommunicable diseases; so the positive effects of cessation can be assessed not only immediately for at-risk individuals but also later for secondary health outcomes.

4. Behavioral change theories have been historically a fundamental part of research on smoking cessation. So, the inclusion of mHealth interventions has been assessed with these theories in mind.

In this section, the authors summarize the course of evolution of evidence, primarily from earlier low- to middle-quality primary studies showing inconsistencies in effectiveness into later high-quality primary and secondary studies of effectiveness.

One of the earliest projects in this area was done by Rodgers et al. who assessed the effectiveness of sending SMS to support smoking cessation in New Zealand. In a single-blind randomized trial, personalized SMS containing advice, support, and distraction messages were sent regularly for 26 weeks. The main outcome measure was current non-smoking (no smoking in the previous week) at 6 weeks after randomization. Secondary outcome measures were current nonsmoking at 12 and 26 weeks. After assessment of 355 Maori (the indigenous population of New Zealand) and 1350 non-Maori participants, the proportion of reported quitting in the intervention group was 28% versus 13% in the control group with a relative risk of 2.20 (95% CI, 1.79–2.70; $p < 0.0001$). Subgroups defined by age, sex, income level, or geographic location did not change the treatment effect and the effect was consistent. At 12 weeks, the treatment effect was high as the proportion of reported quitting in the intervention group was 29% versus 19% in the control group with a relative risk of 1.55 (1.30–1.84; $p < 0.0001$). At 6 months, reported quit rates were still high but less clear in the control group; the reported quit rates increased from 13% at 6 weeks to 24% at 26 weeks in the control group. The proportion of participants who felt extremely confident in staying abstinent was 33% and 20% in the intervention and control groups, respectively ($P = 0.04$) at 26 weeks. Rodgers et al. concluded that their SMS-based smoking cessation program nearly doubled quit rates at 6 weeks, and it could be a novel way to help smokers (especially young people) quit with an inexpensive, customized, and age-appropriate intervention. In another complimentary publication based on the same trial, Bramley et al. showed that this new intervention could be an innovative public health initiative for smoking cessation and was similarly effective in both Maori and non-Maori populations (Bramley et al. 2005; Rodgers et al. 2005).

This group went on to develop a multimedia mobile phone smoking cessation intervention for young people after the success of their previous SMS-based program. An expert group oversaw the development process, with experience in youth development principles, social cognitive theory, effective interventions for smoking cessation, and social marketing for behavior change. A three-phase content development process included (1) focus groups and online survey for consultation, (2) pretesting the content, and (3) role model selection. With the participation of 180 young people in the consultation phase, they found that communicating believable stories from selected, real, and honest role models was important. Ex-smoker role models were selected by the young adult participants. The role models, own quitting experiences were used and turned into short (less than 30 s) video messages. Most participants in the pilot study (12 out of 15) found the program likeable, nearly 60% of them were happy with the frequency of two messages per day, and nine participants (60%) managed to stop smoking. Whittaker et al. concluded that smoking cessation based on a video-message-based multimedia mobile phone program was feasible and acceptable to the target population. A full trial of the effectiveness of the 6-month *STUB IT* intervention was conducted with young adults in New Zealand; the intervention group received a customized package of video and text messages for 6 months. They were able to select a quit date, the timing of messages, and their role models by themselves. They also received additional messages based on their cravings symptoms for the difficult times. For the control group, a quit date was selected but they received only biweekly video messages containing general health information. The study suffered from difficulty in the recruitment of young adults who wanted to quit, and practically, the study could not show the effectiveness of the intervention. At the end of the 6-month follow-up, 26.4% in the intervention group and 27.6% in the control group continued to quit. Qualitative feedback on the intervention was positive, however (Whittaker et al. 2011).

In the United Kingdom, Free et al. aimed to investigate if the New Zealand automated SMS-based smoking cessation program could be adapted and effective in the United Kingdom. They conducted a single-blind, randomized trial in which the intervention arm received motivational SMS and support for behavior change and the control arm received unrelated SMS messages. Within 2 weeks of randomization, the intervention group set a quit date, and thence, they received customized text messages for the next 31 weeks (5 per day for the first 5 weeks and 3 per week for the next 26 weeks). This was a very-high-quality trial, with a primary outcome of 6-month continuous abstinence gathered via self-report, which then was biochemically verified. The follow-up completion was very high in this trial as primary outcome data were available for 5524 participants (2735 [94%] in the intervention group and 2789 [97%] in the control group) at 6 months. The primary outcome data showed that continuous abstinence (biochemical) had been increased in the intervention group (10.7% vs. 4.9% in the control group) with a relative risk of 2.20 (95% CI, 1.80–2.68). At 4 weeks, self-reported no smoking in the past 7 days was 28.7% in the intervention group versus 12.1% in the control group ($P < 0.0001$). This changed to 24.2% and 18.3% at 6 months but was still statistically significant ($P < 0.0001$). Free et al. also concluded that the SMS-based program improved smoking cessation rates at 6 months and that it should be clearly integrated within health-care services in the United Kingdom (Free et al. 2011).

A Cochrane systematic review and meta-analysis was first conducted in 2009 to determine the effectiveness of mobile phone–based interventions in helping people quit smoking. Whittaker et al. included randomized or quasi-randomized trials on any type of mobile-phone-based intervention in their systematic review. The SMS-based programs from New Zealand and from the United Kingdom were included, plus an Internet and mobile phone program from Norway trialed on two different groups. Meta-analysis of the SMS programs showed the effectiveness of the intervention in increasing self-reported quitting in the short term (4 weeks) with a relative risk of 2.18 (95% CI, 1.80–2.65). Meta-analysis of the Internet and mobile phone programs showed the effectiveness of the intervention in increasing self-reported quitting, both in the short and the long term (12 months) with a relative risk of 2.03 (95% CI, 1.40–2.94). They concluded that, while short-term results were promising, more rigorous studies on long-term effects were necessary (Whittaker et al. 2009).

In 2012, the Cochrane systematic review was updated. This time, the Norwegian studies were not included as the criteria were mobile-only interventions. The new studies included (only those with 6-month outcomes) were the SMS arm of a study with multiple arms

(SMS, Internet-based service, and combination of both), and the video messaging intervention described earlier. Pooling the results of all five studies with a total of over 9000 participants proved that mobile phone interventions increased long-term quit rates with a relative risk of 1.71 (95% CI, 1.47–1.99). Whittaker et al. concluded that mobile-phone-based smoking cessation interventions were beneficial on modifying long-term outcomes. Because SMS was the predominant type of intervention, they proposed assessment of other types of interventions for smoking cessation based on mobile phones as well. They also stated the need for conducting studies in low-income countries and assessing the cost-effectiveness of different interventions (Whittaker et al. 2012b).

More recently, Free et al. conducted a systematic review to assess the effectiveness of mHealth interventions on a range of consumer health behaviors. They reviewed 26 trials and showed that nearly all trials were designed and conducted in high-income countries. Text messaging (SMS) support was effective on biochemically verified smoking cessation with a relative risk of 2.16 (95% CI, 1.77–2.62). mHealth interventions showed some benefits for behaviors other than smoking cessation, but their results were inconsistent. Free et al. concluded that SMS-based interventions were efficacious in smoking cessation and should be integrated into routine health-care services (Free et al. 2013).

In summary, there is substantial published evidence on the effectiveness of mHealth interventions on smoking cessation (Table 2.1). Earlier studies were focused on short-term outcomes and were conducted with shorter follow-ups. This may be one of the reasons that earlier systematic reviews were not able to prove the effectiveness of mHealth interventions on long-term smoking cessation outcomes. But when randomized clinical trials continued to develop further and become more focused on long-term (specifically 6-month) outcomes with more tangible clinical outcomes, updated systematic reviews reflected their effectiveness.

2.3.1.2 Sexual Health

The effect of mobile phones in sexual health promotion has been well explored.

Levine et al. developed a SMS service in San Francisco in 2008 named SEXINFO, which was focused on sexual health. SEXINFO was the result of partnership between Internet Sexuality Information Services, Inc. and the San Francisco Department of Public Health. As an information (in a frequently-asked-questions [FAQs] format) and referral service for African-American youth to respond to rising gonorrhea rates, appropriate referral services were planned by collaboration of community organizations, religious groups, and health agencies. Young people aged 15–19 years participated in focused group discussions about viability of referral services. The service showed great success as the usage was higher than expected; more than 180 SMS queries per week were sent to the central phone number to ask about information and advice; the most frequent inquiries were about condom break (532 queries), information about sexually transmitted diseases (486 queries), and unwanted pregnancies (372 queries). Levine et al. concluded that SEXINFO promisingly increased access to sexual health services among adolescents (especially the at-risk ones) (Levine et al. 2008).

One of the first reviews of the literature was conducted by Lim et al. in 2008 to assess the uses of SMS as a popular, fast, and low-cost communication method in the field of sexual health. The authors described the benefits of SMS in comparison to other modes of communication (e-mail, phone call, or postal letter) such as high mobility, instant speed, low cost, less time required for communication with multiple recipients, and also popularity and accessibility. In this review, it was reported that SMS has been used for clinical management for appointment reminders, providing results of sexually transmitted infection (STI)

TABLE 2.1

Summary of Selected Studies on the Effectiveness of mHealth on Smoking Cessation

Authors	Country	Year	Study Type	Intervention(s)	Main Outcome Measures	Effect	Authors' Conclusion
Rogers et al.	New Zealand	2005	Single-blind RCT	Personalized SMS containing advice, support, and distraction message for 26 weeks	Proportion of current nonsmoking (no smoking in the previous week) at 6 weeks	28% in the intervention group versus 13% in the control group.	SMS-based smoking cessation program nearly doubled quit rates at 6 weeks.
Free et al.	United Kingdom	2011	Single-blind RCT	Motivational SMS and support for behavior change	6-month continuous abstinence via self-report, then biochemically verified	Continuous abstinence of 10.7% in the intervention group versus 4.9% in the control group.	SMS-based program improved smoking cessation rates at 6 months and that it should be clearly integrated within health-care services in the United Kingdom.
Whittaker et al.	Studies from New Zealand. United Kingdom, Norway, and other countries	2009	Cochrane systematic review and meta-analysis	—	Self-reported quitting in the short term (4 weeks) and long term (12 months)	Effectiveness of the intervention in increasing self-reported quitting, both in the short and the long term (12 months).	Short-term results were promising; more rigorous studies on long-term effects were necessary.
Whittaker et al.	Studies from New Zealand, United Kingdom, and other countries	2012	Cochrane systematic review and meta-analysis	—	Long-term quit rates	Mobile phone interventions increased long-term quit rates with a relative risk of 1.71.	Mobile phone-based smoking cessation interventions were beneficial on modifying long-term smoking cessation outcomes.

Notes: RCT, randomized controlled trial.

tests, exchanging sexual health information, and tracing contacts following STI diagnosis. Introducing novel ways of access to sexual health services was another area of focus in this review. The authors discussed practical applications such as using SMS for requesting condoms to be delivered to a home address in unmarked packages, sending reminders to women to take their oral contraceptive pills, and registration for alerts to help couples conceive at the most fertile time. Health promotion, including general messages to broad populations and specific messages in response to questions by specific populations (e.g., sexual health questions from young people), was another area of the effectiveness of SMS in sexual health. General messages can inform groups of people about aspects of sexual health but specific messages tailored to queries from costumers require more established systems. Finally, Lim et al. concluded that there is the need for further evaluation of the benefits and also the effectiveness of SMS for sexual health purposes (Lim et al. 2008).

In a qualitative study, Gold et al. assessed the acceptability, utility, and efficacy of health promotion by SMS. They conducted focus group discussions with young participants of a study that tried to examine the effect of SMS for sexual health promotion in Victoria, Australia. Forty-three participants viewed SMS as an acceptable method of health promotion and insisted on informal language for the messages with positivity, shortness, and relevance as the strengths of this communication method. Funny messages were more likely to be remembered and shared, as were those sent on specific annual event dates. Participants found that providing new information, receiving reminders enforcing existing knowledge, and reducing anxiety about testing for STIs were beneficial. Gold et al. concluded that message style, language, and broadcast plan were important aspects to consider in future studies of health promotion by SMS, as well as other mediums such as social networking sites (Gold et al. 2010).

Gold et al. conducted a randomized controlled trial to evaluate the effectiveness of SMS advertising for health promotion with specific focus on safer sex and sun safety in people aged 16–29 years in Victoria, Australia. Participants received eight SMSs during summer in 2008–2009. Knowledge and behavior of the participants regarding safe sex and sun protection were measured by questionnaires customized to be filled using mobile phones. At the end of follow-up, sexual health knowledge was higher and sexual partners were lower in the safe sex group than the sun group. Mobile advertising for health promotion in young people was considered successful by the investigators (Gold et al. 2011).

Gold et al. also conducted another related study in Australia on participants with 16–29 years of age attending a music festival. After completing a short survey, the participants received fortnightly sexual health SMSs for 4 months. At the end of follow-up, they completed an online survey again. The majority of the participants found the SMSs entertaining and informative. Increase in knowledge and STI testing over the study period was significant in both sexes. The authors concluded that SMS was "a feasible, popular, and effective method of sexual health promotion to young people with a relatively low withdrawal rate, positive feedback, and an observed improvement in sexual health knowledge and STI testing" (Gold et al. 2011).

In another expanded version of their previous studies, Lim et al. designed and conducted a randomized controlled trial to determine the effectiveness of e-mail and SMS on promoting young people's sexual health. The intervention (lasting for 12 months) utilized prevention slogans about STIs sent over SMS and e-mail. At the start of the study and after 3, 6, and 12 months, participants completed an online questionnaire. Health-seeking behavior, condom use with risky partners, and STI knowledge were the main outcome measures. STI knowledge was higher for both sexes in the intervention group. Women in the intervention group were more likely to undergo an STI test or discuss with a clinician about their sexual health, compared with the control group. Condom use showed no change. Lim et al. concluded that SMS and e-mail improved STI knowledge and STI testing while having no positive effect on condom use (Lim et al. 2012).

2.3.1.3 HIV/AIDS Screening and Transmission Prevention

Many mHealth projects and research efforts have been conducted with a focus on HIV/ AIDS as a millennium development goal. Different aspects of HIV/AIDS have been explored including adherence to antiretroviral therapy (ART) (Pellowski and Kalichman 2012), HIV prevention for men who have sex with men (Muessig et al. 2013), disseminating HIV information (Leite et al. 2013), and patient retention in care (van der Kop et al. 2013). Pellowski et al. systematically reviewed the literature to assess advances in technology-based interventions for HIV-positive people. They found that most studies have been focused on medication adherence and have used different technologies (such as SMS, cell phones, or computers) for providing specific interventions. They also identified several gaps in the literature, specifically lack of enough evidence on engagement of HIV-positive people and sexual risk reduction. So in this section, the authors will focus on mHealth interventions for screening or testing of HIV and also on preventing its transmission.

Dean et al. conducted a pilot study in an urban setting in South Africa to determine the feasibility of using interactive SMS on preventing mother-to-child HIV transmission. They formed a facilitator-based support group of a few HIV-positive pregnant women (6–10 people) connected to an obstetrician and gynecologist for answering questions, providing accurate information, and advancing the conversations. The group communicated through SMS to ask for medical information (questions, concerns, and ideas about pregnancy and HIV, and also well-being), addressing psychological matters, and providing social comments containing support and advice. After completion of the study, participants were satisfied generally, and they recommended that the program continue to be offered in the future (Dean et al. 2012).

Menacho et al. conducted a qualitative study in Peru about sending SMS to motivate men who have sex with men to test for HIV. They used the information–motivation–behavioral skills model and social support theory as their theoretical framework to guide their focus group discussions and develop the content and tailor the messages. They found that SMS can be used for promoting HIV testing in this group of homosexual men depending on message frequencies, content, and costs (Menacho et al. 2013). Similarly, Muessig et al. conducted a qualitative study in North Carolina to develop mHealth HIV interventions for men who have sex with men. The participants stated that mHealth HIV interventions were acceptable and an HIV app should provide user-friendly content about "test site locators, sexually transmitted diseases, symptom evaluation, drug and alcohol risk, safe sex, sexuality and relationships, gay-friendly health providers, and connection to other gay/HIV-positive men" (Muessig et al. 2013).

In a clinical trial, Odeny et al. assessed the effectiveness of SMS on increasing appointment attendance of circumcised adult males for HIV testing after their operation as high proportions of men were missing their appointment 1 week of their circumcision. By analyzing data from 592 and 596 participants in the intervention and control groups, respectively (follow-up completion of between 98.7% and 99.3%), they showed that SMS modestly improved the attendance (65.4% vs. 59.7%; RR, 1.09; $p = 0.04$) and higher effects were seen for people with lower transportation costs ($P < 0.001$) and higher educational level ($P = 0.07$) (Odeny et al. 2012).

Box 2.2 summarizes the recent experience by Suwamaru et al. in Papua New Guinea (PNG) for developing an education and awareness model in rural areas for HIV/AIDS based on SMS (Suwamaru 2012).

**BOX 2.2   SMS-BASED HIV/AIDS EDUCATION AND AWARENESS
MODEL FOR RURAL AREAS IN PAPUA NEW GUINEA**

Papua New Guinea (PNG) is a geographically rugged country resulting in major challenges in providing access to basic health-care needs of the population in many areas. HIV/AIDS is a major public health concern in PNG and adds to the burden of TB, malaria, and diarrheal diseases as PNG's major health challenges.

Policies to promote competition in mobile phone networks were introduced in mid-2007. This improved the penetration of mobile phone technologies across the country, especially in rural areas where mobile phones have become the preferred medium of communication.

For the purpose of establishing an SMS-based model to educate and disseminate information on HIV/AIDS in rural PNG, these major steps were taken:

1. Conducting a mobile phone usage survey: to evaluate the experiences, attitudes, and perceptions of people and identify variables of interest that may have effects on them
2. Performing a factor analysis: to identify the most important factors that may help in understanding and analyzing participants' experiences, attitudes, and perceptions
3. Interpretation of factors to prepare the inputs for development of an education model
4. Identifying the major issues to be targeted by the educational model
5. Designing an SMS-based HIV/AIDS education model, consisting of a database and an application server integrated into mobile phone networks
6. Developing SMS-based HIV/AIDS education modules: to provide content to closed user groups and to disseminate content (e.g., providing multiple-choice questions to assess knowledge)

Suwamaru proposed that this model could be implemented on a large scale by collaborative partnerships between different stakeholders, mainly mobile network operators (MNOs), departments of health, etc. (Suwamaru 2012).

2.3.1.4 Controlling or Losing Weight

Donaldson et al. conducted a controlled trial on adults to assess if SMS helped overweight and obese people maintain or even lose weight after undergoing a weight loss program. A 12-week SMS intervention was designed in which the participants sent progress messages and received feedback from practitioners regularly. The intervention group showed significant decline in body weight, waist circumference, and body mass index (BMI). Quality of life and depression scores, but not anxiety scores, were also improved. The rate of follow-up attendance also showed significant improvement. Donaldson et al. concluded that SMS interventions after initial weight loss would have positive weight control outcomes (Donaldson et al. 2013).

Brindal et al. provided support using a smartphone app for overweight and obese women on a meal replacement program (MRP) in an 8-week randomized clinical trial. The intervention group received support from an interactive MRP app that "provided information, simplified food intake recording, rewarded positive behavior and prompted regular interaction through reminders and self-monitoring of weight and diet." The application was linked to a web service to "record user data, perform processing, log events and deliver content." The control group was provided with a static app containing information about the MRP. Assessment of usage data showed that engagement with the MRP app was significantly higher among the intervention group in the study period. The intervention group completed 54% of their prompts to enter meal data into the app prior to the next prompt on the evening on the same day. Weight loss was not significantly different between groups ($P = 0.08$). Mood and affect improved in women in the intervention group. The authors concluded that the MRP support app could be a helpful compliment to available MRPs, especially regarding psychological outcomes (Brindal et al. 2013).

Napolitano et al. focused on overweight or obese college students. Their weight loss intervention was an 8-week program and they aimed to assess its feasibility, acceptability, and efficacy. The study had three arms: Facebook, Facebook plus SMS and personalized feedback, and a wait list control. Two assessments were conducted at 4 and 8 weeks. Weight loss was greatest in the Facebook plus SMS arm than the other two groups ($P < 0.05$) after 8 weeks. The Facebook and waiting list groups showed no difference in weight change. The investigators stated the initial efficacy and acceptability of the combined interventions showed potential for developing innovative intervention based on popular technology platforms for college students (Facebook and SMS) (Napolitano et al. 2013). Although this trial has not differentiated the share of weight loss from SMS and Facebook separately, it seems promising to combine different technologies with mHealth interventions to produce outcomes.

A summary of the studies can be found in Table 2.2.

2.3.1.5 Increasing Physical Activity and Fitness

Physical inactivity is a risk factor for many noncommunicable diseases and conditions. The authors retrieved nearly 35 published studies from different biomedical databases (including commentaries, clinical trial protocols, randomized controlled trials, systematic reviews, and meta-analyses), which had been partly or fully focused on physical activity and/or fitness and mobile technologies. Some of the studies had tried to become more focused on a subset of individuals or patients with chronic conditions such as cardiovascular diseases or diabetes. Fanning et al. synthesized the evidence on using mobile devices for influencing physical activity behavior. They aimed to assess (1) whether mobile devices were efficacious in physical activity settings, (2) how device features were implemented in different studies, and (3) future directions for the development of interventions. Summarizing data from 9 out of 11 studies showed that mobile phones (7 studies), SMS (6), native applications (3), and PDAs (2) were the most popular modalities of intervention for increasing physical activity. The authors extracted data for different outcomes including duration and frequency of moderate to vigorous physical activity, percentage of active time spent in such activity, pedometer step counts, number of days of exercise per week, and days per week walking for exercise. They concluded that using mobile devices in research was becoming popular and that mobile platforms can positively influence physical activity in terms of increasing pedometer steps. The effects were more significant if mobile phones were used for delivery of the intervention in comparison with using PDAs. Duration of moderate to vigorous physical activity did not show significant differences with mobile-phone-based interventions in this review, possibly because of large heterogeneity of included studies. Fanning et al. concluded that mobile devices could be used

TABLE 2.2

Summary of Selected Studies on the Effectiveness of mHealth on Controlling or Losing Weight

Authors	Country	Year	Study Type	Intervention(s)	Main Outcome Measures	Effect	Authors' Conclusion
Donaldson et al.	United Kingdom	2013	RCT	12-week SMS intervention with progress messages and feedback from practitioners	Body weight, waist circumference, and body mass index (BMI)	Significant decline in body weight, waist circumference, and BMI.	SMS interventions after initial weight loss would have positive weight control outcomes.
Brindal et al.	Australia	2013	RCT	8-week support from an interactive meal replacement program app	Weight loss	Weight loss was not significantly different; but mood and affect improved in women.	Meal replacement program support app could be a helpful compliment to available meal replacement programs (MRPs) especially regarding psychological outcomes.
Napolitno et al.	United State	2013	RCT	8-week program with Facebook or Facebook plus SMS plus personalized feedback and a wait list control	Weight loss at 4 and 8 weeks	Weight loss was greatest in the Facebook plus SMS arm after 8 weeks.	

Notes: RCT, randomized controlled trial.

as tools to "measure and understand behavior based on theoretically grounded behavior change interventions" and as "an effective tool for increasing physical activity" (Fanning et al. 2012).

Buchholz et al. conducted a systematic review to assess the effects of SMS on improvement of physical activity in adults. They reviewed 10 studies, more than half of which were randomized controlled trials, mostly focused on young to middle aged women.

Physical activity was the primary outcome in six studies, while in the others, it had been assessed as one of multiple healthy behaviors for weight management. Different studies had provided various measures such as self-reported physical activity (in minutes) or frequency of physical activity. Recording steps with a pedometer or an accelerometer was also the measure in a few studies.

Moreover, diverse theoretical frameworks had been used alone or in combination in the included studies such as (but not limited to) social cognitive theory, self-efficacy, transtheoretical model, protection motivation theory, or health belief model. Interventions differed greatly in terms of content and combination with other technologies, ranging from SMS only to combining them with educational materials, counseling sessions, or Internet technology. The effect size of each study was reported in terms of Cohen's *d*, which ranged between 0.26 and 2.22; the median effect size (*d*) was 0.50 for the included studies, which was considered as *medium* effect; that is, the mean in the intervention group was 0.50 unit (standard deviation units) larger than the control group (*d* is the difference between the means divided by standard deviation of either group—technically, *d* = 0.5 means that the mean of the intervention group is at the 69th percentile of that of controls). Buchholz et al. concluded that further research was necessary for expanding the role of mobile technology to increasing physical activity (Buchholz et al. 2013).

In a pilot randomized controlled trial, Kim and Glanz focused on the efficacy of SMS interventions for physical activity promotion in elderly people. They designed a 6-week program for older African-Americans between 60 and 85 years of age living in an urban area. Motivational SMSs were sent three times a day, 3 days a week, for 6 weeks to the intervention group. The primary outcome measure was step count (recorded by pedometers and walking manuals) and perceived activity level was the secondary outcome measure (assessed by questionnaire). Motivational SMS greatly improved step count and perceived activity levels in the intervention group. Kim and Glanz concluded that their 6-week program was effective in increasing physical activity among older African-Americans (Kim and Glanz 2013).

2.3.1.6 Controlling Alcohol Drinking

The amount of evidence on alcohol prevention is very limited in comparison to other sections. Weaver et al. recently evaluated the characteristics, acceptability, and use of smartphone applications related to alcohol. They assessed the correctness and quality of information provided in alcohol-related apps as well as young people's opinions of alcohol apps. After content analysis of 500 apps on iTunes and Google Play, the authors included 384 apps, of which just 11% were related to health promotion or reducing alcohol consumption. The majority of apps encouraged alcohol consumption. Many apps measured blood alcohol concentration, and although these were widely available, they were largely unreliable and inaccurate. Weaver et al. stated the need for regulating the quality of health-related apps to give them credibility (Weaver et al. 2013).

2.3.1.7 Antenatal and Pediatric Health Promotion

An example from the United State was *Text4baby*—an SMS service aimed at providing information to pregnant women and new mothers to improve their health and that of their babies. This national program was launched by a large public–private partnership and saw rapid uptake by more than 320,000 people in just 2 years. The process of development and launch of such a large program were described in 2012 (Whittaker et al. 2012a). Later commentaries by Remick and Kendrick showed positive and promising feedback from program participants, although the feedback came from a small sample of enrollees within the whole program. Remick included Text4baby's promotional strategy, and broad public private partnerships at different levels, as its unique characteristics. Carefully developed, frequently reviewed, and accurate content were considered principles of success in this program (Remick and Kendrick 2013).

2.3.1.8 Preventing Depression in Adolescents

Whittaker et al. developed and tested a mobile-phone-based intervention for preventing depression in high-school adolescents in a double-blind randomized controlled trial in

New Zealand and assessed satisfaction of the participants. Development of the program and its messages was based on cognitive–behavioral therapy (CBT). Combinations of text messages (two messages per day for 9 weeks), video, and cartoon messages, in addition to a mobile website, were used in the program and compared with an attention control program, which had different messages. Over 90% in the intervention group stated they would introduce the program to a friend. Participants reported the program helped them to be more positive and to have less negative thoughts—this was significantly higher than in the intervention group. The authors concluded that mobile phones have the potential to be a helpful channel to provide key messages from CBT. Objective outcomes, such as clinician-rated depression symptom scores, are awaited (Whittaker et al. 2012d).

2.3.1.9 Vaccination

Evidence about the effectiveness of mHealth solutions in promoting vaccinations has been, to some extent, inconsistent. Hofstetter et al. gathered opinions of parents and clinicians about using SMS messages as reminders for early childhood vaccination. They conducted a survey among 200 parents, 26 care providers, and 20 medical staff in 2011 at four academic pediatric centers in New York. Nearly 88% of parents were comfortable receiving health-related SMS, although 84% had never received such health-related SMSs. Receiving reminder/recall SMS was acceptable to almost all parents and was considered more appealing than phone calls or letters. Vaccine-related SMSs were also supported by providers and staff, although some concerns were noted including correct cell phone numbers, appointment availability, and increased call volume. Hofstetter et al. concluded that in order to maximize the effectiveness of SMS vaccination reminders, researchers and providers should identify barriers to implementation (Hofstetter et al. 2013).

Moniz et al. conducted a randomized controlled trial with an aim to assess the effect of SMS on influenza vaccine uptake among pregnant women. Pregnant women at less than 28 weeks of gestation were randomized to receive the intervention (weekly general pregnancy health plus influenza vaccination plus influenza vaccination information for 12 weeks) or the usual care (the same intervention without influenza information). Two surveys about preventive health beliefs were done before and after intervention. The authors reviewed prenatal records to check for influenza vaccine uptake by pregnant mothers. Of the 204 participants available for analysis, most were African-American, had low educational level, and had mainly public or no insurance coverage. The overall rate of influenza vaccination showed no difference between the intervention and control groups. Moniz et al. concluded that sending SMS to low-income, urban, ambulatory pregnant women had no effect in increasing influenza vaccination rates (Moniz et al. 2013).

2.3.1.10 Family Planning

Providing family planning services using mHealth solutions has been the focus of different studies. We previously mentioned a review by Lim et al. (2008) in which registration for alerts to help couples conceive at the most fertile time of the female partner had been mentioned as a possible family planning service. L'Engle et al. assessed if providing family planning information to people in Tanzania in an automated way might be feasible or effective in behavior change. They used 10-month user data from the Mobile for Reproductive Health (m4RH) program. Queries from users were logged and then four questions about gender, age, promotion point, and potential family planning impact were sent to users by SMS. A total of 2870 unique users (56% female and 60% younger than

30 years of age) produced 4813 queries during the 10-month pilot period. The most frequent contraceptive-specific information accessed by users was natural family planning (26%), followed by emergency contraception (21%), implants (16%), condoms (16%), and injectables (14%). Many changes in family planning behaviors and outcomes were reported by 509 participants due to using m4RH; changes were consistent with users' stage of reproductive life cycle. The most frequent changes were switching to short-acting family planning methods including injectables (19%) and oral contraceptives (10%), switching to long-acting methods such as implants (16%) and IUDs (12%), and using condoms (16%). L'Engle et al. recommended that mobile phones could be used for reaching younger people for delivering family planning information, especially during reproductive age (L'Engle et al. 2013).

2.3.2 Early Diagnosis and Management of Diseases

Leaving the area of *at risk* and entering the spectrum of *disease*, preventive strategies focus more on tertiary prevention. Many mHealth projects and programs have been designed to help in diagnosis and management of different diseases such as diabetes, cardiovascular diseases, cancers, asthma, mental and psychiatric conditions, maternal and child diseases, chronic obstructive pulmonary disease, tuberculosis (TB), malaria, and others. In this section, the authors briefly review selected pieces of evidence related to tertiary preventive mHealth interventions with a focus on diseases and an individual approach to prevention.

2.3.2.1 Self-Management of Noncommunicable Diseases

de Jongh et al. conducted a systematic review to assess the effects of SMS and MMS in helping people self-manage their long-term illnesses. They also aimed to assess the users' opinion about the mHealth interventions, utilization and costs of health-care services, and also safety of interventions. After reviewing four randomized controlled trials (all classified as of moderate quality), synthesis of the evidence showed no significant effects for SMS or MMS in controlling glycosylated hemoglobin (HbA1c) (and index of diabetes control over prolonged period of time), complications, and body weight in diabetic patients in comparison to usual care or e-mail reminders; no significant effect in hypertensive patients for mean blood pressure or frequency of achieving controlled blood pressure; but significant improvement in asthma patients in peak expiratory flow variability and symptoms score. Self-management capacity scores were improved in diabetic patients. The authors concluded that messaging interventions might be beneficial in self-management of long-term illnesses. But long-term effects, acceptability, costs, and risks of these interventions should be assessed accordingly, considering the enthusiasm with expansion of mHealth interventions (de Jongh et al. 2012).

2.3.2.2 Acceptance of Health Care, Reminders, and Medication Adherence

Acceptance of health care and compliance with medical advice and treatment has been one of the major challenges for today's health-care systems worldwide. Mobile phone–based interventions have gathered interest and are considered promising in this area of health systems.

2.3.2.2.1 Reminders

It has been shown that SMS and telephone reminders are effective in increasing the rate of attendance in clinics in diverse settings. Wide penetration of mobile phones in addition to directness, convenience, immediacy, confidentiality, and low-cost characteristic of SMS makes it an appealing way of increasing attendance rates.

Guy et al. conducted a systematic review in 2012 of 8 randomized controlled trials (RCTs) (6 of them nonblinded) and 10 observational studies with concurrent or historical control groups. Most studies were done in the United Kingdom and Australia, primarily in hospital outpatient or primary care clinics. Meta-analysis showed that SMS reminders increased attendance rates with a summary odds ratio of 1.48 (1.33–1.72). Type of clinic, target age group (pediatric, older), and message timing before the scheduled appointment did not change the overall odds ratio in subgroup analyses. Guy et al. concluded that SMS reminders were a simple and efficient option in reducing nonattendance rates in a wide variety of health services settings (Guy et al. 2012).

A more recent noninferiority randomized controlled trial in an academic primary care setting in Switzerland demonstrated that SMS reminders were considered cost effective in comparison with phone reminders (Junod Perron et al. 2013). However, the same group of investigators assessed the effectiveness of SMS in attendance of young people at a university hospital clinic. This intervention showed no significant impact on attendance rates as the people who received SMS had 16.4% missed appointments versus 20.0% in the control group ($P = 0.346$). Narring et al. concluded that SMS reminders did not reduce missed appointments because most patients did not book appointments themselves but were usually referred by professionals or parents (Narring et al. 2013).

One of the examples of future projects in this area is the efforts by Heatley et al. to evaluate the effectiveness of SMS reminders on increasing the attendance of women with gestational diabetes to test for oral glucose tolerance testing postpartum within 6 months after delivery. They are conducting a parallel randomized trial in which the intervention group receives SMS reminders at 6 weeks, 3 months, and 6 months postpartum until they complete doing the tolerance test. The control group will just receive a single SMS at 6 months postpartum. Heatley et al. hope to increase the rate of attendance testing, in order to diagnose and intervene earlier, if necessary, for type 2 diabetes (Heatley et al. 2013).

2.3.2.2.2 *Improving Medication Adherence*

An important subgroup of acceptance of care and compliance that has been explored much further is medication adherence.

Mbuagbaw et al. assessed the effects of SMS on improving adherence to ART. They conducted a parallel double-blind randomized clinical trial on 200 HIV-positive adults to determine whether weekly motivational SMSs were effective in comparison to usual care alone. They assessed adherence with a visual analogue scale; but they also measured number of doses missed and pharmacy refill data to validate this primary outcome measure at 3 and 6 months. The overall retention of participants was just over 81% at 6 months. They found that motivational SMSs were not effective in improving adherence to ART. They proposed development of other kinds of messages and trials with longer follow-ups (Mbuagbaw et al. 2012).

In contrast to the findings by Mbuagbaw et al. in 2012, Free et al. conducted a systematic review to assess the effectiveness of mHealth interventions on health behaviors of consumers in the United Kingdom. One of their aims in reviewing 26 trials was to assess if SMS was effective in increasing adherence to ART (Free et al. 2013). They showed that communicating with SMS decreased the viral load with a relative risk of 0.85 (95% CI, 0.72–0.99). However, they concluded that although this intervention was effective, it should be tested in other contexts and also its cost-effectiveness should be investigated (Free et al. 2013). Later, Blake reviewed this piece of evidence and concluded that we have promising evidence that shows technology-based approaches will be helpful and effective for changing health behavior and managing diseases, considering their feasibility, acceptability, and cost-effectiveness in mind (Blake 2013).

In Kenya, Lester et al. assessed the effects of communication by SMS between health-care providers and patients who started antiretroviral therapy (ART) on adherence to treatment and plasma HIV-1 RNA load. They conducted a randomized controlled trial to compare the effects of SMS intervention in addition to standard care. Patients received weekly messages from a nurse and they were asked to respond within 48 h. Primary outcomes were self-reported adherence to ART at 6 and 12 months of follow-up visits (i.e., consuming over 95% of prescribed doses in the past month) and suppression of plasma HIV-viral RNA load at 12 months (i.e., under 400 copies/mL). Adherence to ART was nearly 62% in the intervention group in comparison to 50% in the control group. They reported a relative risk for nonadherence of 0.81 (95% CI, 0.69–0.94; $p = 0.006$). Suppressed viral loads were reported in 57% and 48% of patients in the intervention and control groups, respectively (relative risk for virologic failure, 0.84; 0.71–0.99; $p = 0.04$). They concluded that SMS support significantly improved ART adherence and rates of viral suppression and was effective in improving patient outcomes especially in resource-limited settings. They communicated the low-cost status of SMS as an appealing factor in its adoption: "the SMS intervention is inexpensive (each SMS costs about U.S.$0.05, equivalent to U.S.$20 per 100 patients per month, and follow-up voice calls averaged U.S.$3.75 per nurse per month) and the mobile phone protocol uses existing infrastructure. This protocol is also probably less expensive than in-person community adherence interventions" (Lester et al. 2010).

One innovative project to assess and improve medication adherence by mHealth interventions was conducted by Morak et al. in diabetic patients. They assessed the feasibility of mHealth and near field communication (NFC) technology in medication adherence as a telemonitoring solution. In their solutions, smartphones with NFC functionality were used along with smart blisters. They handed out smart blisters (i.e., pharmaceutical packages that monitored when patients took the pills out of packages) to 59 patients and followed them up for just over 1 year to monitor their takeout events. The events were recorded and transmitted via mobile phones. The findings showed the feasibility of this system in adherence monitoring (Morak et al. 2012).

Results of studies on mHealth interventions to improve adherence to TB treatment are awaited. The usual duration of TB treatment is 6 months and it is composed of multiple drug therapies. Nglazi et al. are conducting a systematic review on randomized and nonrandomized trials to assess the effectiveness of SMS in adherence to TB treatment to inform researchers, practitioners, and also policy makers (Nglazi et al. 2013).

2.3.2.3 *Better Control of Diabetes*

Herbert et al. conducted a systematic review of SMS-based interventions for children and adolescents with type 1 diabetes. Changes in glycosylated hemoglobin (HbA1c) levels as one of the most important clinical outcomes in diabetes control were mixed with positive, neutral, and negative results. Participant satisfaction also was mixed and the retention rates were varied. Herbert et al. recommended that the format, frequency, and timing of SMS interventions should be planned and the solution should be tested before implementation. They recommended future randomized controlled trials to further evaluate the effect of SMS on specific medical outcomes (such as for blood glucose monitoring, insulin use, physical activity, nutrition, and sick day management) or psychosocial outcomes of diabetes (such as higher self-efficacy, self-reported adherence, and social support) (Herbert et al. 2013).

Sieverdes et al. reviewed the guidelines for diabetic care and assessed the situations in which mHealth technology can be of support to guideline recommendations. They identified seven major topics to support behavior change in diabetic patients: "(1) glycemic control and self-monitoring of blood glucose, (2) pharmacological approaches and medication management, (3) medical nutrition therapy, (4) physical activity and resistance training, (5) weight loss, (6) diabetes self-management education, and (7) blood pressure control and hypertension." The authors stated that patients and care providers should explore innovative ways to incorporate these seven major concepts with mHealth solutions (Sieverdes et al. 2013).

2.3.2.4 *Management of Cardiovascular Diseases and Improving Exercise Capacity*

Nundy et al. evaluated the effects of SMS in terms of feasibility and acceptability on self-management of acute decompensated heart failure in a low-income largely African-American population. They conducted a pre–post study, in which hospitalized patients received automated SMS-based program for 30 days following discharge from heart failure admission. Self-care reminders and patient education about diet, symptoms, and how to access health-care services comprised the content of messages. The main outcome measure was the self-care of heart failure index. A total of six patients completed the postintervention survey and majority of them considered the program easy to use (83%), helpful in reminding drug consumption (66%), and helpful in decreasing salt intake (66%). The participants showed high degrees of satisfaction and improvement in self-management of heart failure (Nundy et al. 2013).

Huang et al. developed a mobile wearable efficient telecardiology system (WE-CARE) at Peking University, Beijing, China. By using seven lead mobile ECG devices along with various wireless connectivity networks, they developed an intelligent monitoring system for capturing, transmitting, and interpreting the electrocardiograms for cardiovascular disease patients. WE-CARE showed a high detection rate of over 95% for abnormal cardiac rhythms, and the authors concluded that it could be efficiently used for cardiovascular disease diagnosis and management in medical settings (Huang et al. 2014).

Pfaeffli et al. presented the process of content development for an mHealth exercise intervention focused on CR. They gained patients' input in the development steps: (1) conceptualization under the supervision of experts, (2) formative research using focus group and interview sessions with patients, (3) pretesting, and (4) pilot testing. Participants stated positive impressions of the mHealth program in motivating them to exercise. The idea was considered good by all participants; nearly 85% of patients stated they would sign up for video messages on a website and 60% for a SMS program to receive information. Older people thought that the technology could be a potential barrier because of their unfamiliarity with SMS and not having a mobile phone. Pfaeffli et al. concluded that for delivering exercise-based CR, the mHealth multimedia exercise program could be effective (Pfaeffli et al. 2012).

2.4 Gaps in mHealth Resources or Evidence for Preventive Medical Services

It seems that more comprehensive approaches to providing evidence about mHealth preventive medical services are yet to come. Kumar et al. summarized the results of mHealth evidence workshop held by the US National Institutes of Health in 2011. Evaluating assessments, evaluating interventions, and reshaping evidence generation using mHealth were the main broad categories of the workshop, and the authors tried to describe evaluation standards along with future possibilities and goals for mHealth research as an evolving field. Kumar et al. stated that mHealth applications have been developed and evaluated in the context of many diseases and risk factors, including but not limited to diabetes, asthma, obesity, stress management, depression, and smoking cessation. They insisted upon the fact that the potentials and challenges of mHealth to improve health outcomes should be explored using comprehensive research, especially in a health-care system that already was threatened by suboptimal outcomes and high costs. They concluded that premature adoption of untested mHealth technologies and those that had not proven efficacious might not contribute to health improvement, if not cause harm (Kumar et al. 2013).

Martinez-Perez et al. compared the availability of existing evidence on mHealth apps dedicated to major and high-burden diseases according to the Global Burden of Disease (GBD) study. The top eight health conditions in the world according to the 2004 update of GBD report were iron-deficiency anemia (with 1159.3 million affected individuals), hearing loss (636.5 million), migraine (324.1 million), low vision (272.4 million), asthma (234.9 million), diabetes mellitus (220.5 million), osteoarthritis (151.4 million), and unipolar depressive disorders (151.2 million).

They concluded that work on mobile apps for the eight most prevalent conditions had been fairly unequal in a way that diabetes and depression outnumber other diseases in number of specific apps available and research published. Martinez-Perez calls for filling the gap for research on mHealth to prevent or manage anemia, hearing loss, or low vision (Martinez-Perez et al. 2013).

2.5 Business Case for mHealth Preventive Services

Many preventive measures have in general been proved cost effective, including but not limited to reducing tobacco use, promoting physical activity, reducing harmful alcohol use, and promoting healthy diets and also specific strategies to prevent cardiovascular diseases, cancer, diabetes, and other major noncommunicable disease (WHO 2011). These preventive measures are the exact types of solutions in which mHealth has shown promise as mHealth interventions are likely to be cheaper than existing or traditional methods for delivering preventive services. The *at-risk* populations (especially from low- to middle-income developing countries) are of great interest for different stakeholders such as sponsors, health service funders, researchers, regulators, and policy makers. However, economic analysis or discussions of the costs have been sparse in the published literature.

2.5.1 Returns on Investments

There is little specific evidence available for returns on investments in mHealth preventive medical interventions. Guerriero et al. recently published results of their cost-effectiveness analysis on the previously published TXT2Stop trial. The effectiveness of SMS smoking cessation support had been proved so they considered the U.K. National Health Service's perspective to assess incremental costs and benefits of adding text-based support to routine smoking cessation services. Cost per quitter, cost per life year gained, and cost per quality-adjusted life years (QALYs) gained were the main outcomes of interest in the analysis. By assessing treatment costs and using health state values of lung cancer, stroke, myocardial infarction, chronic obstructive pulmonary disease, and coronary heart disease, they showed that text-based smoking cessation support would be cost saving; that is, with a cost of 16,120 pound sterling per 1,000 smokers, 18 life years and 29 QALYs would be gained (Guerriero et al. 2013).

In another example, Junod Perron showed that SMS reminders were cost effective in reducing missed appointments in academic clinics (Junod Perron et al. 2013). From the health service perspective, this is seen as a significant saving in terms of the costs of missed appointments on clinician time and administration time and cost.

Other stakeholders with direct profit from mHealth preventive solutions are mobile network operators (MNOs). Increased SMS or MMS traffic, mobile phone calls, and data service usage by the involved parties in mHealth interventions are profitable to MNOs, which may validate their investment in expansion of mHealth services. Mobile handset manufacturers are also seeing the benefit of adding mHealth applications to distinguish and expand their consumer offerings.

The main benefit of mHealth preventive measures in the long run is the prevention of the onset or progression of disease to costly medical care in the future.

2.6 Future Perspective

The way health care is currently delivered is recognized as being unsustainable in many countries across the globe. Prevention is a vital focus in attempting to address the exponential increase in noncommunicable disease expected in the future. Transformation of preventive medical services in health-care delivery is, at the same time, a major challenge and a promising breakthrough that is already present in and inherently expected from many mHealth interventions. An mHealth intervention can only change the health-care service delivery system if it meets several important requirements. An mHealth intervention must be based on a practical needs assessment about prioritized preventive medical services in different communities of interest:

1. It should be customized according to wants and demands of individuals in the community.

2. It should be designed in a collaborative partnership between different stakeholders.

3. The design of the intervention should be based on theoretically acceptable frameworks.

4. It should be pretested and pilot tested with practical steps taken to gather stakeholders' feedback (especially from users).

5. It should be evaluated with respect to efficacy, effectiveness, and cost-effectiveness in appropriately designed primary studies (ideally in randomized controlled trials).

6. It should be designed with implementation or commercialization in mind; integration into the routine care process, into other solutions to provide packages of care, into other technologies to develop combined interventions, or into the daily lives of people.

7. It should be evaluated regarding changes in appropriate preventive outcomes and impacts.

Unfortunately, many mHealth solutions will not go through structured stages of development in order to address these requirements (Labrique et al. 2013).

This chapter has shown that there is substantial promise for mHealth in prevention. Many areas have demonstrated effectiveness (e.g., smoking cessation, sexual health). Key features of mobile technologies give advantages over other technologies for preventive measures. These include being able to deliver messages proactively to individuals wherever they may be, at appropriate times, on a mass scale. The future for mHealth in preventive services may be in the expansion of the role of mHealth in underexplored needs of communities, an increase in randomized clinical trials for evaluation, and the integration of mHealth with social networking and other technologies (Househ 2012). mHealth preventive medical services should also be included in multidisciplinary university curricula to foster the development of future researchers, academicians, developers, and care providers who are empowered enough to develop and use mHealth in the future.

As the digital divide has narrowed most for mobile phone use on a global scale (Free et al. 2010), it will be promising to see the expansion of mHealth preventive health research and practice in low-to-middle-income countries as well as high-income countries.

Questions

1. Put these mHealth interventions in correct order regarding their relevance to each level of prevention (primary, secondary, or tertiary):
   1. SMS reminders for postpartum women to test for type 2 diabetes after they had been diagnosed with gestational diabetes mellitus
   2. Weight management intervention for overweight and obese adults using SMS
   3. Sending SMS to increase patient retention in care after HIV diagnosis
   4. Depression prevention intervention in adolescents
   5. Sending SMS to men who have sex with men to motivate them undergo HIV testing
2. Based on Norris et al. (2009), what are the three phases for a sustainable strategy toward mHealth? What would be the objective of this strategy?
3. In the New Zealand case study about the development and evaluation process for mHealth interventions, what are the objectives of the second and sixth steps, that is, formative research after conceptualization and qualitative follow-up after community-based randomized controlled trials?
4. Discuss the reasons why quitting smoking has been one of the earliest behaviors to be targeted by mHealth researchers and developers.
5. Identify two gaps in the literature on advances in technology-based interventions for HIV-positive people, based on the systematic review by Pellowski et al.
6. Discuss an example of mHealth intervention for self-management of noncommunicable diseases with the focus on long-term effects of the intervention.
7. How will you answer to Question 6, if you want to discuss based on the systematic review by Free et al. to assess if SMS is effective in increasing adherence to anti-retroviral therapy (ART) or not?
8. Discuss two examples about the returns on investment of mHealth preventive medical interventions.

References

Bastawrous, A., M. J. Armstrong (2013). Mobile health use in low-and high-income countries: An overview of the peer-reviewed literature. *J R Soc Med* **106**(4): 130–142.

Blake, H. (2013). Text messaging interventions increase adherence to antiretroviral therapy and smoking cessation. *Evid Based Med* **19**: 35–36.

Bramley, D., T. Riddell et al. (2005). Smoking cessation using mobile phone text messaging is as effective in Maori as non-Maori. *N Z Med J* **118**(1216): U1494.

Brindal, E., G. Hendrie et al. (2013). Design and pilot results of a mobile phone weight-loss application for women starting a meal replacement programme. *J Telemed Telecare*, **19**(3): 166–174.

Buchholz, S. W., J. Wilbur et al. (2013). Physical activity text messaging interventions in adults: A systematic review. *Worldviews Evid Based Nurs* **10**(3): 163–173.

Collins, L. M., S. A. Murphy et al. (2007). The multiphase optimization strategy (MOST) and the sequential multiple assignment randomized trial (SMART): New methods for more potent eHealth interventions. *Am J Prev Med* **32**(5 Suppl): S112–S118.

de Jongh, T., I. Gurol-Urganci et al. (2012). Mobile phone messaging for facilitating self-management of long-term illnesses. *Cochrane Database Syst Rev* **12**: CD007459.

Dean, A. L., J. D. Makin et al. (2012). A pilot study using interactive SMS support groups to prevent mother-to-child HIV transmission in South Africa. *J Telemed Telecare* **18**(7): 399–403.

Donaldson, E. L., S. Fallows et al. (2013). A text message based weight management intervention for overweight adults. *J Hum Nutr Diet* **27**: 90–97.

Fanning, J., S. P. Mullen et al. (2012). Increasing physical activity with mobile devices: A meta-analysis. *J Med Internet Res* **14**(6): e161.

Free, C., R. Knight et al. (2011). Smoking cessation support delivered via mobile phone text messaging (txt2stop): A single-blind, randomised trial. *Lancet* **378**(9785): 49–55.

Free, C., G. Phillips et al. (2010). The effectiveness of M-health technologies for improving health and health services: A systematic review protocol. *BMC Res Notes* **3**: 250.

Free, C., G. Phillips et al. (2013). The effectiveness of mobile-health technology-based health behaviour change or disease management interventions for health care consumers: A systematic review. *PLoS Med* **10**(1): e1001362.

Gold, J., C. K. Aitken et al. (2011). A randomised controlled trial using mobile advertising to promote safer sex and sun safety to young people. *Health Educ Res* **26**(5): 782–794.

Gold, J., M. S. Lim et al. (2010). What's in a message? Delivering sexual health promotion to young people in Australia via text messaging. *BMC Public Health* **10**: 792.

Gold, J., M. S. Lim et al. (2011). Determining the impact of text messaging for sexual health promotion to young people. *Sex Transm Dis* **38**(4): 247–252.

Guerriero, C., J. Cairns et al. (2013). The cost-effectiveness of smoking cessation support delivered by mobile phone text messaging: Txt2stop. *Eur J Health Econ* **14**(5): 789–797.

Guy, R., J. Hocking et al. (2012). How effective are short message service reminders at increasing clinic attendance? A meta-analysis and systematic review. *Health Serv Res* **47**(2): 614–632.

Heatley, E., P. Middleton et al. (2013). The DIAMIND study: Postpartum SMS reminders to women who have had gestational diabetes mellitus to test for type 2 diabetes: A randomised controlled trial—Study protocol. *BMC Pregnancy Childbirth* **13**: 92.

Herbert, L., V. Owen et al. (2013). Text message interventions for children and adolescents with Type 1 diabetes: A systematic review. *Diabetes Technol Ther* **15**(5): 362–370.

Hofstetter, A. M., C. Y. Vargas et al. (2013). Parental and provider preferences and concerns regarding text message reminder/recall for early childhood vaccinations. *Prev Med* **57**(2): 75–80.

Househ, M. (2012). Mobile social networking health (MSNet-health): Beyond the mHealth frontier. *Stud Health Technol Inform* **180**: 808–812.

Huang, A., C. Chen et al. (2014). WE-CARE: An intelligent mobile telecardiology system to enable mHealth applications **18**(2): 693–702.

Junod Perron, N., M. D. Dao et al. (2013). Text-messaging versus telephone reminders to reduce missed appointments in an academic primary care clinic: A randomized controlled trial. *BMC Health Serv Res* **13**: 125.

Kim, B. H., K. Glanz (2013). Text messaging to motivate walking in older African Americans: A randomized controlled trial. *Am J Prev Med* **44**(1): 71–75.

Kumar, S., W. J. Nilsen et al. (2013). Mobile health technology evaluation: The mHealth evidence workshop. *Am J Prev Med* **45**(2): 228–236.

L'Engle, K. L., H. L. Vahdat et al. (2013). Evaluating feasibility, reach and potential impact of a text message family planning information service in Tanzania. *Contraception* **87**(2): 251–256.

Labrique, A., L. Vasudevan et al. (2013). H_pe for mHealth: More "y" or "o" on the horizon? *Int J Med Inform* **82**(5): 467–469.

Leite, L., M. Buresh et al. (2013). Cell phone utilization among foreign-born Latinos: A promising tool for dissemination of health and HIV information. *J Immigr Minor Health*.

Lester, R. T., P. Ritvo et al. (2010). Effects of a mobile phone short message service on antiretroviral treatment adherence in Kenya (WelTel Kenya1): A randomised trial. *Lancet* **376**(9755): 1838–1845.

Levine, D., J. McCright et al. (2008). SEXINFO: A sexual health text messaging service for San Francisco youth. *Am J Public Health* **98**(3): 393–395.

Lim, M. S., J. S. Hocking et al. (2008). SMS STI: A review of the uses of mobile phone text messaging in sexual health. *Int J STD AIDS* **19**(5): 287–290.

Lim, M. S., J. S. Hocking et al. (2012). Impact of text and email messaging on the sexual health of young people: A randomised controlled trial. *J Epidemiol Community Health* **66**(1): 69–74.

Martinez-Perez, B., I. de la Torre-Diez et al. (2013). Mobile health applications for the most prevalent conditions by the world health organization: Review and analysis. *J Med Internet Res* **15**(6): e120.

Mbuagbaw, L., L. Thabane et al. (2012). The Cameroon Mobile Phone SMS (CAMPS) trial: A randomized trial of text messaging versus usual care for adherence to antiretroviral therapy. *PLoS ONE* **7**(12): e46909.

Menacho, L. A., M. M. Blas et al. (2013). Short text messages to motivate HIV testing among men who have sex with men: A qualitative study in Lima, Peru. *Open AIDS J* **7**: 1–6.

Moniz, M. H., S. Hasley et al. (2013). Improving influenza vaccination rates in pregnancy through text messaging: A randomized controlled trial. *Obstet Gynecol* **121**(4): 734–740.

Morak, J., M. Schwarz et al. (2012). Feasibility of mHealth and near field communication technology based medication adherence monitoring. In: *2012 Annual International Conference of the IEEE Engineering in Medicine and Biology Society*, San Diego, CA, pp. 272–275.

Muessig, K. E., E. C. Pike et al. (2013). Putting prevention in their pockets: Developing mobile phone-based HIV interventions for black men who have sex with men. *AIDS Patient Care STDS* **27**(4): 211–222.

Napolitano, M. A., S. Hayes et al. (2013). Using Facebook and text messaging to deliver a weight loss program to college students. *Obesity (Silver Spring)* **21**(1): 25–31.

Narring, F., N. Junod Perron et al. (2013). Text-messaging to reduce missed appointment in a youth clinic: A randomised controlled trial. *J Epidemiol Community Health* **67**(10): 888–891.

Nglazi, M. D., L. G. Bekker et al. (2013). Mobile phone text messaging for promoting adherence to anti-tuberculosis treatment: A systematic review protocol. *Syst Rev* **2**: 6.

Norris, A. C., R. S. Stockdale et al. (2009). A strategic approach to m-health. *Health Inform J* **15**(3): 244–253.

Nundy, S., R. R. Razi et al. (2013). A text messaging intervention to improve heart failure self-management after hospital discharge in a largely African-American population: Before-after study. *J Med Internet Res* **15**(3): e53.

Odeny, T. A., R. C. Bailey et al. (2012). Text messaging to improve attendance at post-operative clinic visits after adult male circumcision for HIV prevention: A randomized controlled trial. *PLoS ONE* **7**(9): e43832.

Pellowski, J. A., S. C. Kalichman (2012). Recent advances (2011–2012) in technology-delivered interventions for people living with HIV. *Curr HIV/AIDS Rep* **9**(4): 326–334.

Pfaeffli, L., R. Maddison et al. (2012). A mHealth cardiac rehabilitation exercise intervention: Findings from content development studies. *BMC Cardiovasc Disord* **12**: 36.

Pomeroy, E. C., L. H. Steiker (2012). Prevention and intervention on the care continuum. *Soc Work* **57**(2): 102–105.

Remick, A. P., J. S. Kendrick (2013). Breaking new ground: The Text4baby program. *Am J Health Promot* **27**(3 Suppl): S4–S6.

Riley, W. T., D. E. Rivera et al. (2011). Health behavior models in the age of mobile interventions: Are our theories up to the task? *Transl Behav Med* **1**(1): 53–71.

Rodgers, A., T. Corbett et al. (2005). Do u smoke after txt? Results of a randomised trial of smoking cessation using mobile phone text messaging. *Tob Control* **14**(4): 255–261.

Sieverdes, J. C., F. Treiber et al. (2013). Improving diabetes management with mobile health technology. *Am J Med Sci* **345**(4): 289–295.

Suwamaru, J. K. (2012). An SMS-based HIV/AIDS education and awareness model for rural areas in Papua New Guinea. *Stud Health Technol Inform* **182**: 161–169.

van der Kop, M. L., D. I. Ojakaa et al. (2013). The effect of weekly short message service communication on patient retention in care in the first year after HIV diagnosis: Study protocol for a randomised controlled trial (WelTel Retain). *BMJ Open* **3**(6): e003155.

Walters, D. L., A. Sarela et al. (2010). A mobile phone-based care model for outpatient cardiac rehabilitation: The care assessment platform (CAP). *BMC Cardiovasc Disord* **10**: 5.

Weaver, E. R., D. R. Horyniak et al. (2013). "Let's get Wasted!" and other apps: Characteristics, acceptability, and use of alcohol-related smartphone applications. *JMIR mHealth uHealth* **1**(1): e9.

Whittaker, R., R. Borland et al. (2009). Mobile phone-based interventions for smoking cessation. *Cochrane Database Syst Rev* **4**: CD006611.

Whittaker, R., E. Dorey et al. (2011). A theory-based video messaging mobile phone intervention for smoking cessation: Randomized controlled trial. *J Med Internet Res* **13**(1): e10.

Whittaker, R., R. Maddison et al. (2008). A multimedia mobile phone-based youth smoking cessation intervention: Findings from content development and piloting studies. *J Med Internet Res* **10**(5): e49.

Whittaker, R., S. Matoff-Stepp et al. (2012a). Text4baby: Development and implementation of a national text messaging health information service. *Am J Public Health* **102**(12): 2207–2213.

Whittaker, R., H. McRobbie et al. (2012b). Mobile phone-based interventions for smoking cessation. *Cochrane Database Syst Rev* **11**: CD006611.

Whittaker, R., S. Merry et al. (2012c). A development and evaluation process for mHealth interventions: Examples from New Zealand. *J Health Commun* **17**(Suppl 1): 11–21.

Whittaker, R., S. Merry et al. (2012d). MEMO—A mobile phone depression prevention intervention for adolescents: Development process and postprogram findings on acceptability from a randomized controlled trial. *J Med Internet Res* **14**(1): e13.

WHO (2011). *Global Status Report on Noncommunicable Diseases 2010*. World Health Organization, Geneva, Switzerland.

3

Going Beyond:
Challenges for Using mHealth Applications for Preventive Medicine

Joerg Brunstein, Angela Brunstein, and Michael K. Martin

CONTENTS

3.1 Introduction

In 2008, Patrick and colleagues provided an optimistic outlook for the potential of mobile phone technology "… to impact delivery of healthcare services and promotion of personal health" (Patrick 2008, p. 181). Four years later, Collins repeats stressing the potential of "… mobile devices to become powerful medical tools" (Collins 2012, p. 16). While the numbers of sold smartphones, available sensors, and mobile health and fitness applications have skyrocketed, truly intelligent mHealth applications for a majority of citizens at risk for noncommunicable diseases (NCDs) are still hard to find.

In this chapter, we describe current challenges for future mHealth applications in preventive medicine posed by type of users who are neither athletes nor patients, by increasing amounts of available sensor data, and by currently missing guidance for health behavior change. For mastering these challenges, we propose a modular framework with theory, modeling, and intelligent tutoring as practiced in cognitive science. This approach clearly discriminates between theoretical knowledge acquired in the field and specific mobile applications designed to foster behavior change. That split allows reducing effort and costs. There is a need for modeling to bridge the theory–application gap. Without modeling, theories tend to be too unspecific to be implemented in applications. With models implementing the theory for specific tasks, there is a structured translation between theory and applications that allows transfer of knowledge in both directions.

3.1.1 Different Users with Different Needs as a Call for a Framework

Current mHealth applications cover a variety of services for management of *patients* (e.g., Tomlinson et al. 2013) with most evidence coming from management of chronic conditions programs (de Jongh et al. 2012; Fiordelli et al. 2013). Another strong domain on the market related to mHealth is advanced mobile sport applications under the Health & Fitness category of iTunes and android app stores. These programs allow athletes to track performance, similar to monitoring clinical conditions, and to plan workout, similar to management of chronic conditions. Obviously, both kinds of applications aim for different users, fulfill different functions, and serve a different purpose.

In between these two extremes of functioning support, huge numbers of mobile health and fitness applications serve the majority of citizens who are neither athletes nor patients and try to lose weight or aim to become a little more physically active for preventing chronic illness. This segment of users is expected to grow significantly over the next decades given the current obesity and diabetes epidemic (WHO 2010), aging societies in developed and developing countries (WHO 2012), and huge prevalence of NCDs worldwide (WHO 2010).

These users are not patients with chronic conditions yet. They have not had a heart attack yet, but they might be at risk for cardiovascular diseases. This makes a huge difference in motivation: patients after heart attack are likely to be motivated to manage their condition well to prevent the next heart attack based on their own experience. People at risk for cardiovascular diseases fight a potential heart attack that might or might not come in a few years or decades time. For those, it is quite rational to choose immediate benefits from a sedative lifestyle for sure over only potential hazards in the far future. According to a McKinsey Retail Healthcare Consumer Survey (Dixon-Fyle et al. 2012), 57% of obese participants overestimate their health status as *excellent* or *very good*. In addition, majority of U.S. adults have insufficient health literacy with more than half finding it difficult to follow directions on a prescription drug label (White and Dillow 2005). These biases impact people willingness to engage in health behaviors.

People targeted by preventive medicine are not athletes either. They are often not as young and skilled as users targeted by health and fitness applications. And they might come with clinical conditions that impact their exercise program. They have a longer learning history of not living healthy than athletes. Their age and social and work environment might make it hard to exercise or eat healthy. For older adults at risk, there might also be technological barriers (Parker et al. 2013): they might not possess a smartphone or know how to operate one.

To summarize, people at risk for disease or disability differ significantly from managed patients with chronic conditions and from athletes in terms of motivation, skills, habits, and clinical conditions. When providing mobile health and fitness applications, these issues need to be addressed for being useful for this group of users who will be soon the majority of citizens for many societies.

3.1.2 Data Flooding as a Call for Modeling

Both mHealth applications for the management of chronic conditions and mobile sports applications track users' performance and physiological states for monitoring and progress analysis. Similarly, many health and fitness applications count steps, consumed and burned calories, etc. This is matched by the highly active and highly productive technological research and development, for example, in the area of the *Internet of things* (e.g., Reina et al. 2013; Rodriguez-Molina et al. 2013) or body sensor networks (e.g., Lai et al. 2013). Based on its market analysis, PRWEB (2013) predicts an increase of 11% of microelectromechanical systems between 2013 and 2018. In tendency, physiological sensors become cheaper, more accurate, more convenient, better integrated in existing infrastructure, and smarter for analyzing users' health status (e.g., Sarasohn-Kahn 2013). For example, iRhythm Technology offers a small Band Aid–like sensor that records cardio data and transmits data wireless to a smartphone. Other sensors are integrated in shoes, wristwatches, shirts, sports bras, or smartphones.

Users welcome this development. According to a survey of the Pew Research Center (Fox and Duggan 2012), 85% of U.S. adults own a cell phone and half of those own smartphones. According to Gartner's market analysis (van der Meulen and Rivera 2013), the second quarter of 2013 has seen an increase of 47% in smartphone sales. For the first time, sale of smartphones was bigger than sale of feature phones with 225 million sold smartphones worldwide during last year. The Pew survey (Fox and Duggan 2012) adds that 31% of cell phone owners have used their phone for gathering health information. Eighty-two percent of people surveyed in the United States and United Kingdom would be willing to pay for health and fitness sensors that sync with their smartphones and would spend up to $140 or £90 (Arrowsmith 2013, IHS Inc.).

These data paint an optimistic outlook for serving high numbers of customers with reliable, health-relevant information. Potentially sensor data can be used for monitoring users' existing health conditions and for warning them or their health-care provider in case they reach predefined thresholds. They could be also used for developing intelligent mobile health and fitness applications that learn from the users and accordingly adapt their support to users' needs and preferences.

At the same time, an increase in the number of users with increasing numbers of health conditions to be monitored multiplied by the increasing number of sensors will result in an exponential growth of the generated data. This holds especially for continuously monitored variables, like heart rate or blood oxygen levels. In critical care monitoring, up to 90% of alarms are false positives (Imhoff and Kuhls 2006). Comparable numbers for preventive care would overtax health-care providers' capacities to respond to patients' needs given the expected *data tsunami* (e.g., Richardson and Reid 2003, for pain management; Sarasohn-Kahn 2013). In OECD states, there are on average 3 physicians per 1000 patients (OECD Health Data 2012). Who among these three will respond to thousands of warnings for their patients on a daily basis?

The information problem does not only concern the amount of data but also the complexity of data. Physiological systems have many parameters that interact with each other, develop dynamically over time, tend to be intransparent, have delayed response, and have accumulation and homeostasis associated. There is evidence that even well-educated young adults have difficulties to understand complex systems, including their bank account, calorie budget, and the weather (Booth et al. 2000; Cronin and Gonzalez 2007; Cronin et al. 2009; Sterman and Booth Sweeney 2002). For those domains, research participants perform usually very well reading system inputs and outputs in a diagram, but they perform alarmingly poorly when judging the system's state from inputs and outputs. For fluid management, 90% of medical students in our research could tell when a patient in intensive care had received and excreted most fluid. But only 7% could tell when the patient was most at risk of volume overload and 0% when he was most at risk of dehydration (Brunstein et al. 2010). In system dynamics research, participants are confronted with diagrams plotting two variables over a period of time. With increasing numbers of sensors, we would expect increasing difficulties to decide which readings are important and which are not. Given that evidence, why should users understand complex sensor readings? Issues with huge amounts of sensor data become even more urgent with rising numbers of potential users. It is comparably easy to serve a few world-class athletes or a few hundreds of patients during a clinical trial. It will be extremely challenging to serve the majority of population with diverse backgrounds, educational levels, clinical conditions, goals, etc.

To summarize, current development of physiological sensors that can interact with smartphones promises to cover continuously and completely users' physiological and behavioral measures. This will be used for monitoring users' health conditions, for example, during exercise. At the same time, the fast-growing amount of sensor data for a user and rapidly rising numbers of potential users create a huge amount of potentially available information. Most users will not be able to understand this information, and professionals will not have the capacity for processing that much information for that many patients.

3.1.3 Guidance Desert as a Call for Intelligent Tutoring

Even if users would understand the data, it is unlikely that they can apply the implications for their behavior and health. Research on health behavior has investigated for decades the *motivation–action gap* (Sheeran 2002) within the field of preventive medicine: many people know that they should change their lifestyle, but only a few succeed. One result from that research is that people need metacognitive support for translating their goals into specific behaviors. For example, people often start with general goals, like losing 4% body weight in 4 weeks (*DietBet*, version 1.2, DietBet, Inc.). For reaching that goal, they need to plan actions that are expected to result in that outcome. Not knowing which action will result in how much weight loss contributes to the motivation–action gap. The second hurdle for reaching health aims is the difficulty to backtrack success or failure to causing actions. Because physiological systems are complex, it is impossible to decide what exactly caused the weight loss when it shows on the scale after a week's time. Sensor readings cannot solve that problem. They can protocol every cookie eaten and every step made, but they cannot bridge the utility assignment gap that guides rational skill acquisition (e.g., Fu and Anderson 2008). For example, what if calories look good during workout, but heart rate looks bad? Does it matter when to eat something? All these questions cannot be answered even with an accurate instant weight measurement. A current trend for mobile health and fitness applications is to upload sensor data to a web-based system and receive performance and progress reports on the website or via e-mails on monthly subscription basis (e.g., MapMyFitness, Runtastic). Most of these reports plot one variable at the time. They do not explain relationships and they do not conclude advice based on the data despite having the technological potential to do so.

To summarize, given expected increase in user numbers and sensor data for mobile health and fitness applications as implementation of preventive medicine, there are issues to address concerning users' motivation, needs, and preferences. For our current review of 61 mobile health and fitness applications, only 17 applications asked for user's goals during setup (see Chapter 9). These users do not match patients or athletes. It is unlikely that increasing the amount of potentially available data will increase users' understanding or progress toward their personal health goals. First, they have difficulties with reading, understanding, and integrating the data. Second, they will encounter problems when concluding implications of data for their behavior.

3.2 Intelligent Solutions

For solving the challenges concerning users' characteristics, increasing numbers of users and tracked data, and understanding data for deriving implications for users' behavior, we propose a modular approach of architecture, modeling, and tutoring as successfully practiced in cognitive modeling of human cognition and behavior.

Cognitive architectures mirror our current understanding of the human mind (Anderson 2007) and provide the theoretical background for interventions by defining parameters of attention, memory, reasoning, problem solving, learning, and skill acquisition. For mobile prevention applications, health behavior theories would fill that position.

Cognitive models for specific tasks complement the theory with task-specific knowledge and skills while adhering to the structural constraints of the cognitive architecture for a variety of task domains. For example, Salvucci (2006) models driver performance. Lebiere et al. (2001) model performance for an air traffic control task. Great advantage of cognitive models is that they simulate an individual user performing a specific task and produce exactly the same data as human participants do in terms of response time, mistakes, learning curve, and brain activity, for example, when learning algebra using the algebra tutor (Anderson et al. 2010). This makes it easy to judge whether we have correctly understood human performance or not. It is important to note that cognitive models *simulate* behavior. Software and an interface facilitate behavior. This allows modelers to focus on the aspect of interest for that task without need to model the whole body around. This means behavior change models do not need to go jogging. All they have to do is to produce the same simulated sensor data via interface as users would base on model states. Finally, *cognitive tutoring systems* build instruction for a domain on the cognitive architecture, on knowledge about the task domain, and on educational principles.

Cognitive tutoring systems (e.g., Anderson et al. 1995) are able to provide in-time, personalized support because they have an internal model of the user integrated that mirrors a user's current understanding and updates constantly while the user learns. For behavior change, mobile applications supported by remote servers would do the job of cognitive tutoring systems for guiding users toward their health goals.

The modular approach of architecture, modeling, and tutoring permits channeling of knowledge as it accumulates in the field for behavior change in general, for specific behaviors, and for guiding instruction. It allows to split general parameters from task- or situation-specific parameters. It supports discriminating between theory and implementation via modeling or simulation. Updating one conceptual level can be done without modifying all other levels as well but impacting those downward from theory through model to tutoring system. Finally, it supports sharing the workload between mHealth community for the architecture, work or research groups for modeling, and health-care providers or companies for tutoring. It also allows the "blooming of a thousand flowers": if app designers can easily rely on evidence and can use a common, open framework for development, applications will become more intelligent without replicating developing costs a thousand times.

3.2.1 Theoretical Framework on Behavior Change: Health Behavior Theories as Architecture

Users at risk for NCDs are typical targets of preventive medicine. For this type of users, health behavior theories have investigated the factors that impact health behavior change: health behaviors are defined as aiming for good health by preventing illness when healthy or restoring health when ill (Kasl and Cobb 1966). Theories agree that there exist several factors impacting behavior change and/or that change happens along different states. The *health action process approach* (HAPA, Schwarzer 2008) integrates concepts from most health behavior theories and describes behavior change as a process along several states, from motivation through planning to action. HAPA also spells out factors impacting the transition from one state to the next: a person will progress only from an unspecific motivation to lose weight to making specific workout and diet plans, if he perceives a health risk, believes there is a solution to the problem, and feels capable to perform the associated action. A user will cope well with challenges when practicing new health behaviors if she feels capable of managing challenges and coming back to normal after relapse. Based on states and associated factors, HAPA allows to assess a user's current state along the behavior change process and to provide support tailored for that state. These interventions have been demonstrated to be effective for several kinds of health behaviors and for participants in different states of behavior change (Craciun et al. 2012; Scholz et al. 2005; Wiedemann et al. 2012).

Interventions can be potentially also implemented in mobile health and fitness applications. For adjusting support to the current state of a user concerning behavior change, an application needs to display at least a basic version of system intelligence. Schulz and colleagues suggest four dimensions for designing and evaluating *Quality of Life Technology* systems (QoLT; Schulz et al. 2012): system intelligence, functional domain, kind of functioning support, and mode of operating the system. System intelligence refers to the system's ability to learn from the user and to adjust support correspondingly to the user's needs and preferences. Schulz and colleagues argue that an intelligent QoLT system can cover the complete scale of functioning support from compensating for diminished functioning over preventing decline of functioning to enhancing normal functioning. For vision, glasses compensate for diminished functioning, while night vision systems enhance normal functioning. Current health and fitness applications tend to cover primarily one functional domain, for example, diet or exercise, and one part of the functioning continuum. Finally, a QoLT system can be operated passively, requiring no user input for operation, for example, passive tracking of performance, or it can be operated interactively, for example, requiring manual entry for consumed and burned calories.

Obviously, recognizing which state a user has reached concerning health behavior change requires system intelligence, best paired with passively tracking a user's status and interactively providing guidance.

There are a few interesting implications from matching HAPA with the QoLT evaluation dimensions for this kind of user modeling with mobile technology: both frameworks describe a process of change. HAPA's behavior change participants move from a stable state of not eating healthy, for example, *uphill* to adopting new healthy eating habits. Interventions help them to boost their self-efficacy and to learn required skills. The QoLT dimensions refer to constant decline of function due to aging and disability and support patients with technical means for resisting the *downhill* decline as long as possible. Both the HAPA theory and the QoLT evaluation criteria emphasize specific support according to a specific state of users. Both request adaptation to users' needs and preferences. Both focus on coping with a current setting.

There are also important differences fostering development of both frameworks for supporting people addressed by preventive medicine. HAPA focuses on cognitive coping and skill acquisition and ignores clinical conditions of users that might impact behavior change as well as health outcomes as rewards for behavior. Integrating those in the theory would prepare HAPA to serve aging users or users with disabilities in addition to mostly healthy young adults. Physiological theories might complement health behavior theories by contributing physiological parameters for behavior change. The QoLT evaluation dimensions focus on physical decline of functioning and technical support, neglecting patients' cognitive coping strategies. Adjusting to a patient's current physical functioning level displays global system intelligence. Taking into consideration day-to-day performance adds a layer of local system intelligence of adjusting support to current needs.

3.2.2 Modeling Health Behavior Change and Physiological Implications for Intelligent Guidance

Taking together, HAPA and QoLT dimensions can guide the development of intelligent mobile health and fitness applications that support users' efforts for changing health behaviors to reduce health risks and prevent illness and disability. It indicates modeling users on a behavioral level for anticipating needs for support. Currently, neither HAPA nor the QoLT dimensions are suitable yet to model a *single user*'s performance over time and provide appropriate, in-time support. The QoLT dimensions have not been used for modeling patients' performance. HAPA models typically use path analysis (e.g., Chiu et al. 2011) for predicting who will change on level of a patient cohort. Similarly, intervention efficiency has been studied on level of patient cohorts (e.g., Wiedemann et al. 2012). The path analytic method has been vital for spelling out direct and indirect links between impacting factors (e.g., Edwards and Lambert 2007), like participants' action self-efficacy, and resulting behavior change during preparation stage. This method is also suitable for discriminating between relevant and irrelevant sensor data sets on the level of a patient or user cohort. However, several authors (Boorsboom et al. 2003; Dunton and Atienza 2009; Ogden 2003) have argued that health behavior theories with their current modeling approach are not suitable for providing intelligent support for behavior change on an *individual user* level. Riley et al. (2011) propose using dynamic system models as modeling frameworks for health behavior theories. They also provide evidence for effectiveness for preventing children from developing conduct disorder with family counseling using this approach (Navarro-Barrientos et al. 2010). In addition, HAPA does not cover health implications and barriers based on clinical conditions or injury. Gortmaker et al. (2011) propose using quantitative modeling methods for understanding factors contributing to the obesity epidemic from individual level up to population level, including energy gap models. This modeling approach can be used for understanding the implications of behavior on body weight given complex systems with intransparency, accumulation, and delayed response.

Modeling users' *behavior* for intelligent behavior change support is only half of the story. Cognitive modeling can focus on cognition and behavior only, because concerned behavior typically results in acquired skills and learned knowledge. Health behavior change results in learning only as a side effect. The main outcome of health behavior is impacting a person's health. Also, users are not primarily interested to get learning support. They need guidance for reaching their health goals. Therefore, modeling health behavior change needs to integrate social, cognitive, behavioral, and physiological aspects of this process.

3.2.3 Adaptive Visualization and Simulation for Providing Intelligent Guidance for Health Behavior Change

System dynamics research shows that people have difficulties understanding complex systems, including calorie balance or health in general (e.g., Cronin et al. 2009). Diagrammatic reasoning research shows that people have difficulties understanding even simple graphs (e.g., Mazur and Hickam 1993). This implies users of mobile prevention applications will not be able to read, understand, and integrate physiological sensor data and the physiological system behind those readings.

Simulation has been demonstrated to be an efficient method for teaching complex systems to professionals in several domains, including aviation (e.g., Hays et al. 1992), operating power plants, finance, policymaking, and surgery (e.g., Milburn et al. 2012). Simulation allows exploring complex systems in a safe environment, without time constraints, if needed, transparent and without consequences.

There is also evidence that simulation helps novices to understand complex systems (e.g., Gonzalez and Dutt 2011; Vollmeyer et al. 1996). Therefore, modeling the physiological system is important to link health behaviors to health outcomes and for allowing users to explore consequences of their behavior via simulation. Seeing an avatar aging faster or slower depending on health behavior choices might be a strong motivator for implementing a healthier lifestyle even if health outcomes show only much later, like for weight loss in obese patients. There exist several promising proposals for simulating users (e.g., see Jiang et al. 2012 or Macal and North 2007 for an overview on agent-based modeling).

As research participants can learn to understand complex systems, participants in diagrammatic reasoning research fail systematically for some aspects of diagrams and systematically excel for others indicating that there is some hope for users understanding their physiological sensor readings. First, knowing the task domain can partially compensate for missing diagrammatic literacy (e.g., Mazur and Hickam 1993). Second, starting with Larkin and Simon (1987), researchers have discriminated between perceptual and conceptual processing of diagrams (e.g., Kim et al. 2000). *Perceptual processing* covers information that can be directly read out of a diagram. For our research (Brunstein et al. 2010), that was calories consumed and expended during a time period. Indeed, participants performed very well for that information directly displayed in a diagram. *Conceptual processing* covers information that cannot be directly read out of a diagram but can be inferred from what is given. If there are no other impacting variables, weight directly results from current weight plus calories consumed minus calories expended. This kind of inference did not work well for novices in our study. The US Department of Agriculture (USDA) has retired the famous food pyramid in their campaigns in 2011 and has replaced it with a plate displaying the proportions of nutrition groups for a healthy diet (USDA Center for Nutrition Policy and Promotion 2013) consumers associate a plate more with eating than a pyramid predicting more transfer to America's breakfast tables than the pyramid has had. The difference between perceptual and conceptual processing compares to foreground and background in a picture. People perform very well for information in the foreground and less so for information in the background. No artist would portrait a king *behind* his castle. The other way around, only who knows the location well can describe the castle behind the king in detail. If users are interested in health outcomes, why would we display sensor readers in the foreground and let them infer the health implications themselves?

Therefore, progress or monitoring displays should make relevant information salient, explicit, and in the foreground and hide not relevant information in the background. What is relevant depends again on the users' needs. If they want to track their progress in weight loss, current weight plots might be sufficient. In contrast, for planning behavior and backtracking perceived health outcomes to actions, the relationship between behavior and expected or experienced outcomes is important and needs to be displayed. "How much difference in body weight can I expect when jogging for 30 min instead 20 min?" "The scale looks still as bad as in the beginning of the program. When can I expect to harvest results?" HAPA understands action planning as mental simulation of the concerned behavior (Schwarzer 2008). Visualizations that support this kind of mental simulation should display related parameters. Most likely relevant parameter will integrate several sensor readings over time in higher-order concepts and not present individual readings given the complexity of the physiological, behavioral, and social systems. Technical solutions for integrating sensor data in a meaningful way start to emerge (e.g., Rodriguez-Molina et al. 2013).

Larrik and Soll (2008) describe a successful implementation of making relevant information salient in diagrams. When presenting fuel efficiency of cars in miles per gallons, most college students did not understand that greatest improvement comes from retiring least efficient cars. This misconception disappeared when presenting fuel efficiency in gallons per mile.

Displaying relevant information that enables users to understand the complex system of behavior, health, and context is the first part of intelligent QoLT systems in general and mobile health and fitness applications in particular. As important is guiding the user toward the most efficient way for reaching his or her health goals. Intelligent health tutoring system can generate this guidance based on the theoretical framework or architectures and from models on relevant health behaviors. Simulations with the current user parameters can identify one or a few optimal paths through the problem space of changing health behavior and can provide users with corresponding suggestions and motivation support.

3.3 Conclusions

Building an architecture reflecting our current understanding of users' health behavior change and physiological processes needs to be a mHealth community-wide project. Platforms like Open mHealth (Chen et al. 2012) might facilitate that project. Modeling specific aspects of health behaviors and/or physiological processes can be a task for research groups who implement evidence from empirical studies and generate specific and detailed hypotheses for upcoming studies. Finally, app design concerns both, all-around talent as virtual personal trainers on a healthy lifestyle and special purpose applications, for example for counting steps or spelling out nutrition values. These have to be delivered by healthcare providers or commercial suppliers of the mobile app stores.

Patrick et al. (2008) and Collins (2012) advertised the potential of mobile technology for intelligent support of patients and citizens. By guiding technological process with theory and evidence-driven concepts, we hopefully do not need 5 more years to get there.

CASE SCENARIO 1

Rodriguez et al. (2013) describe a wireless software network that integrates sensor data from users during workout for producing warnings, for example, if body temperature rises above the threshold given the current environment temperature.

QUESTION:

Why is this application better than simply displaying body temperature readings next to graphs on heart rate, oxygen level, and burned calories?

The answer is available at the end of the book in "Answers to the End-of-Chapter Questions."

CASE SCENARIO 2

Mr. Smith, 50 years old, describes himself as a couch potato who plans to become a little more active to keep up with his grandchildren. He recently bought a mobile phone application *WALK!* that tracks every step he makes and syncs in several physiological sensors, including heart rate, and performance measures, including distance, pace, and time. After his walk through the park, *WALK!* presents an impressive analysis of his performance with map, color-coded fast- and slow-track aspects of his workout. Based on his readings, the program recommends to spend 5 min cooling down and shows a male in his 40s performing the stretching exercises as a model.

QUESTION:

Concerning guiding Mr. Smith's through his fitness program, which aspects should be covered by the theory, by modeling, and by the tutoring system?

The answer is available at the end of the book in "Answers to the End-of-Chapter Questions."

3.4 Future Directions

Users, who are neither athletes nor patients, increasing amounts of available sensor data and currently missing guidance for health behavior change, pose challenges for future mHealth applications. For mastering these challenges, we propose a modular framework with theory, modeling, and intelligent tutoring as practiced in cognitive science. This approach clearly discriminates between theoretical knowledge acquired in the field and specific mobile applications designed to foster behavior change. That split allows reducing effort and costs and discrimination between general theoretical knowledge and task-specific evidence. There is a need for modeling for bridging the theory–application gap. Without modeling, theories tend to be too unspecific to be implemented in applications. With models implementing the theory for specific tasks, there is a structured translation tool between theory and applications that allows transfer of knowledge in both directions.

Questions

1. How do users of mobile health and fitness applications differ from patients with chronic conditions or athletes?
2. Why do some researchers warn about a potential *data tsunami* in respect to mHealth sensor readings?
3. Why is it not good enough for mHealth applications to display information, like sensor readings or nutrition values for food, but also provide guidance on action?
4. Why do we need a theory on behavior change and physiological processes when designing mHealth applications?

5. How does a user model for a specific health behavior complement health behavior theories?

6. Why can simulation of health behavior and physiological processes help users to take action?

References

Anderson, J. R. 2007. *How Can the Human Mind Occur in the Physical Universe?* New York: Oxford University Press.

Anderson, J. R., Betts, S. A., Ferris, J. L., and Fincham, J. M. 2010. Neural imaging to track mental states while using an intelligent tutoring system. *Proceedings of the National Academy of Science* 107:7018–7023.

Anderson, J. R., Corbett, A. T., Koedinger, K., and Pelletier, R. 1995. Cognitive tutors: Lessons learned. *Journal of Learning Sciences* 4:167–207.

Arrowsmith, L. 2013. *Are Dedicated Sports and Fitness Monitors Still in the Running?* Press release July 02, 2012. Englewood, CA: IHS Inc.

Boorsboom, D., Mellenbergh, G. J., and van Heerden, J. 2003. The theoretical status of latent variables. *Psychological Review* 110:202–219.

Booth Sweeney, L. and Sterman, J. D. 2000. Bathtub dynamics: Initial results of a systems thinking inventory. *System Dynamics Review* 16:249–286.

Brunstein, A., Brunstein, J., and Limam Mansar, S. 2012. Integrating health theories in health and fitness applications for sustainable behavior change: Current state of the art. *Creative Education* 3(Suppl):1–6.

Brunstein, A., Gonzalez, C., and Kanter, S. 2010. Effects of domain experience in the stock-flow failure. *System Dynamics Review* 26:347–354.

Chen, C., Haddad, D., Selsky, J. et al. 2012. Making sense of mobile health data: An open architecture to improve individual- and population-level health. *Journal of Medical Internet Research* 14:e112. doi: 10.2196/jmir.2152.

Chui, C.-Y., Lynch, R. T., Chan, F., and Berven, N. L. 2011. The health action process approach as a motivational model for physical activity self-management for people with multiple sclerosis: A path analysis. *Rehabilitation Psychology* 56:171–181.

Collins, F. 2012. The real promise of mobile health apps. *Scientific American* 307:16.

Craciun, C., Schuez, N., Lippke, S., and Schwarzer, R. 2012. A mediator model of sunscreen use: A longitudinal analysis of social-cognitive predictors and mediators. *Behavioral Medicine* 19:65–72.

Cronin, M. and Gonzalez, C. 2007. Understanding the building blocks of system dynamics. *System Dynamics Review* 23:1–17.

Cronin, M. A., Gonzalez, C., and Sterman, J. D. 2009. Why don't well-educated adults understand accumulation? A challenge to researchers, educators, and citizens. *Organizational Behavior and Human Decision Processes* 108:116–130.

de Jongh, T., Gurol-Urganci, I., Vodopivec-Jamsek, V., Car, J., and Atun, R. 2012. Mobile phone messaging for facilitating self-management of long-term illnesses. *Cochrane Database of Systematic Reviews* 12. Art. No.: CD007459. doi: 10.1002/14651858.CD007459.pub2.

Dixon-Fyle, S., Gandhi, S., Pellathy, T., and Spatharou, A. 2012. Changing patient behavior: The next frontier in healthcare value. *Health International* 12:65–73. McKinsey's Healthcare Systems and Services Practice.

Dunton, G. F. and Atienza, A. A. 2008. The need for time-intensive information in healthful eating and physical activity research: A timely topic. *Journal of the American Dietetic Association* 109:30–35.

Edwards, J. R. and Schurer Lambert, L. 2007. Methods for integrating moderation and mediation: A general analytic framework using moderated path analysis. *Psychological Methods* 12:1–22.

Fiordelli, M., Diviani, N., and Schulz, P. J. 2013. Mapping mHealth research: A decade of evolution. *Journal of Medical Internet Research* 15:e95. doi: 10.2196/jmir.2430.

Fox, S. and Duggan, M. 2012. Mobile health 2012. *Half of Smartphone Owners Use Their Devices to Get Health Information and One-Fifth of Smartphone Owners Have Health Apps.* Washington, DC: Pew Research Center's Internet & American Life Project.

Fu, W.-T. and Anderson, J. R. 2008. Solving the credit assignment problem: Explicit and implicit learning of action sequences with probabilistic outcomes. *Psychological Research* 72:321–330.

Gonzalez, C. and Dutt, V. 2011. A generic dynamic control task for behavioral research and education. *Computers in Human Behavior* 27:1904–1914.

Gortmaker, S. L., Swinburn, B., Levy, D. et al. 2011. Changing the future of obesity: Science, policy and action. *Lancet* 378:838–847.

Hays, R. T., Jacobs, J. W., Prince, C., and Salas, E. 1992. Flight simulator training effectiveness: A meta-analysis. *Military Psychology* 4:63–74.

Imhoff, M. and Kuhls, S. 2006. Alarm algorithms in critical care monitoring. *International Anesthesia Research Society* 102:1525–1537.

Jiang, H., Karwowski, W., and Ahram, T. Z. 2012. Applications of agent-based simulations for human socio-cultural behavior modeling. *Work* 41:2274–2278.

Kasl, S. V. and Cobb, S. 1966. Health behavior, illness behavior, and sick role behavior. I. Health and illness behavior. *Archives of Environmental Health* 12:246–266.

Kim, J., Hahn, J., and Hahn, H. 2000. How do we understand a system with (so) many diagrams? Cognitive integration processes in diagrammatic reasoning. *Information Systems Research* 11:284–303.

Lai, X., Liu, Q., Wei, X., Wang, W., Zhou, G., and Han, G. 2013. A survey of body sensor networks. *Sensors* 13:5406–447.

Larkin, J. and Simon, H. A. 1987. Why a diagram is (sometimes) worth ten thousand words. *Cognitive Science* 11:65–99.

Larrik, R. P. and Soll, J. B. 2008. The MPG illusion. *Science* 320:1593–1594.

Lebiere, C., Anderson, J. R., and Bothell, D. 2001. Multi-tasking and cognitive workload in an ACT-R model of a simplified air traffic control task. In *Proceedings of the 10th Conference on Computer Generated Forces and Behavior Representation*. Norfolk, VA.

Macal, C. M. and North, M. J. 2007. Agent-based modeling and simulation: Desktop ABMS. In *Proceedings of the 2007 Winter Simulation Conference*, eds. S. G. Henderson, B. Biller, M.-H. Hsie et al., pp. 95–106. Washington, DC.

Mazur, D. J. and Hickam, D. H. 1993. Patients' and physicians' interpretations of graphic data displays. *Medical Decision Making* 13:59–63.

Milburn, J. A., Khera, G., Hornby, S. T., Malone, P. S. C, and Fitzgerald, J. E. F. 2012. Introduction, availability and role of simulation in surgical education and training: Review of current evidence and recommendations from the Association of Surgeons in Training. *International Journal of Surgery* 10:393–398.

Navarro-Barrientos, J. E., Rivera, D. E., and Collins, L. M. 2010. A dynamical systems model for understanding behavioral interventions for weight loss. In *2010 International Conference on Social Computing, Behavioral Modeling, and Prediction (SBP 2010)*, eds. S. K. Chai, J. J. Salerno, P. L. Mabry, pp. 170–179. Heidelberg, Germany: Springer.

OECD.org 2013. *OECD Health Data 2013.* Paris, France: OECD.

Ogden, J. 2003. Some problems with social cognition models: A pragmatic and conceptual analysis. *Health Psychology* 22:424–428.

Parker, S. J., Jessel, S., Richardson, J. E., and Reid, M. C. 2013. Older adults are mobile too! Identifying the barriers and facilitators to older adults' use of mHealth for pain management. *BMC Geriatrics* 13:43. http://www.biomeddentral.com/1471-2318/13/43. Accessed on July 19, 2014.

Patrick, K., Griswold, W. G., Raab, F., Intille, S. S. 2008. Health and the mobile phone. *American Journal of Preventive Medicine* 35:177–181.

PRWEB. 2013. *MEMS Market Biosensors and Nanosensors 2013 Analysis in New Research Report at ReportsnReports.com.* Press release August 28, 2013. Dallas, TX: PRWEB.

Reina, D. G., Toral, S. L., Barrero, F., Bessis, N., and Asimakopoulou. 2013. The role of ad hoc networks in the internet of things: A case scenario for smart environments. In *Internet of Things and Inter-Cooperative Computational Technology* SCI 460, eds. N. Bessis et al., pp. 89–113. Berlin, Germany: Springer.

Richardson, J. E. and Reid, M. C. 2013. The promises and pitfalls of leveraging mobile health technology of pain care. *Pain Medicine.* doi: 10.1111/pme.12206.

Riley, W. L., Rivera, D. E., Atienza, A. A., Nilsen, W., Allison, S. M., and Mermelstein, R. 2011. Health behavior models in the age of mobile interventions: Are our theories up top the task? *Translational Behavioral Medicine* 1:53–71.

Rodriguez-Molina, J., Martinez, J.-F., Castillejo, P., and Lopez, L. 2013. Combining wireless sensor networks and semantic middleware for an internet of things-based sportsman/woman monitoring application. *Sensors* 13:1787–1835.

Salvucci, D. D. 2006. Modeling driver behavior in a cognitive architecture. *Human Factors* 48:362–380.

Sarasohn-Kahn, J. 2013. *Making Sense of Sensors: How New Technology Can Change Patient Care.* Oakland, CA: California HealthCare Foundation.

Scholz, U., Sniehotta, F. F., and Schwarzer, R. 2005. Predicting physical exercise in cardiac rehabilitation: The role of phase-specific self-efficacy beliefs. *Journal of Sport and Exercise Psychology* 27:135–151.

Schulz, R., Beach, S. R., Matthews, J. T., Courtney, K. L., and De Vito Dabbs, A. J. 2012. Designing and evaluating quality of life technologies: An interdisciplinary approach. *Proceedings of the IEEE* 100:2397–2409.

Schwarzer, R. 2008. Modeling health behavior change: How to predict and modify the adoption and maintenance of health behaviors. *Applied Psychology* 57:1–29.

Sheeran, P. (2002). Intention–behavior relations: A conceptual and empirical review. *European Review of Social Psychology* 12:1–36.

Sterman, J. D. and Booth Sweeney, L. 2002. Cloudy skies: Assessing public understanding of global warming. *System Dynamics Review* 18:207–240.

Tomlinson, M., Rotherarm-Borus, M. J., Swartz, L., and Tsai, A. C. 2013. Scaling up mHealth: Where is the evidence? *PLoS Medicine* 10(2):e1001382. doi: 10.1371/journal.pmed.1001382.

USDA. 2011. MyPlate. http://fnic.nal.usda.gov/dietary-guidance/myplatefood-pyramid-resources/usda-myplate-food-pyramid-resources. Accessed on June 29, 2014.

USDA Center for Nutrition Policy and Promotion [Internet]. 2013. Updated December 11, 2013; cited July 19, 2014. http://www.cnpp.usda.gov/MyPlate.htm

Van der Meulen, R. and Rivera, J. 2013. *Gartner Says Smart Phone Sales Grew 46.5 Percent in Second Quarter of 2013 and Exceeded Feature Phone Sale for First Time.* Press Release. Egham, U.K., August 14, 2013. Stamford, CO: Gartner, Inc.

Vollmeyer, R., Burns, R., and Holyoak, K. J. 1996. The impact of goal specificity on strategy use and the acquisition of the problem structure. *Cognitive Science* 20:75–100.

White, S. and Dillow, S. 2005. *Key Concepts and Features of the 2003 National Assessment of Adult Literacy.* Washington, DC: National Center for Education Statistics.

Wiedemann, A. U., Lippke, S., and Schwarzer, R. 2012. Multiple plans and memory performance: Results of a randomized controlled trial targeting fruit and vegetable intake. *Journal of Behavioral Medicine* 35:387–392.

World Health Organization. 2010. *Global Status Report on Non-Communicable Diseases.* Geneva, Switzerland: World Health Organization.

World Health Organization. 2012. *Bulletin of the WHO.* Vol. 90, pp. 77–156. Geneva, Switzerland: World Health Organization.

4

Mobile Services for Diabetic Patients' Sustainable Lifestyle

Shakira Shaheen, Abbas Mohammed, and Nedelko Grbic

CONTENTS

4.1 Introduction

Diabetes is an illness in which there is an abnormally high level of glucose in the blood. It is caused by our genes and lifestyle (*environmental risk*) factors. The number of people suffering from diabetes, and related conditions, has skyrocketed over the past few decades. In addition, more people have blood sugar levels that are not high enough to be classified as diabetes, or are too high for good health. This condition is known as glucose intolerance, or prediabetes. According to a recent report by the World Health Organization (WHO), more than 347 million people worldwide have diabetes [20], and it is predicted that this number would double by 2030 [1] and that diabetes death will double between 2005 and 2030 [20]. This alarming increase in diabetic patients means increased costs from an already strained health-care resource, which would impact adversely on a global level.

For diabetic patients, it is very important for the health-care professionals to constantly monitor their blood glucose trends. If the glucose level is not controlled, it can lead to some major complications like cardiovascular diseases, blindness, nerve disorders, and kidney failures. In order to maintain a proper glucose level, regular interaction between patient and doctor is very essential; however, this will increase the health-care expenditure incurred by diabetic patients. In addition, travel to the hospital or doctor's clinic for regular health-care checks will increase the already strained health-care expenditure in many countries and impact adversely on the carbon footprint (carbon footprints are measurement of all greenhouse gases we individually produce [16]), and there might also be more risk of picking colds, flu, etc., from other patients. Thus, it is beneficial to provide diabetic patients with additional, effective, and efficient self-management health-care monitoring tools in their own environment to determine the amount of exercise and diet they should follow and simultaneously mitigating the effects of the previously mentioned concerns. Regular exercise and proper diet to control the glucose level are considered as two main cornerstones for self-management.

Current diabetes monitoring methods do not meet patients' needs and are not effective in helping people control the disease [1]. Results performed for e-health in Helsinki University of Technology, Finland, by Ekroos and Jalonen [17] concluded that "There is a potential to provide interactive tools to empower diabetic patients self-management as well as the communications between the patients and their healthcare providers." So far, there have not been many mobile services and biosensors to encourage community-based physical activities for diabetic patients who are eager to improve their health and achieve a sustainable lifestyle [1]. What is needed is an effective self-management wireless solution that collects, stores, and transmits these important parameters in real time to the health professionals and receives a feedback correspondingly.

Considering the previously mentioned facts, telemedicine and wireless technologies can be exploited to overcome some of the difficulties encountered by diabetic patients. The introduction of mobile phone applications on an open-source platform and employing biosensor systems such as wireless body area network (WBAN) can be used as efficient self-management tools for diabetic patients. WBAN is an emerging technology consisting of sensors placed on various parts of the body, for example, wearable devices such as glucose sensors that collect the information of the glucose level in the body periodically. WBANs also have advantages in the mobility of patients due to their portable devices and the location-independent monitoring capabilities where the doctor can monitor the patient's health remotely from a clinic [5,9,10,12,15,16,21]. The body area network (BAN) requires devices that are expected to work over a prolonged period of time; since they are implanted medical devices, replacement or charging of the battery is not feasible [19]. IEEE 802.15.6 is a WBAN standard designed with high priority to low-power consumption with a duty cycle of less than 1% [19].

Mobile phones have made communications much easier, while smartphones have brought a great revolution in changing people's lifestyle. Previously where phones were a means of conventional communication, smartphones have enhanced mobile phones with varied and better features. Using smartphones for biomedical purposes has made people's life much easier, and employing technologies like WBAN has made considerable evolution, where patients can monitor their health from their own premises. Google has developed an innovative technology to connect people with a mobile operating system using a new, open and comprehensive platform for mobile devices called Android. It is a user-friendly and yet a versatile operating system. As Google offers Android as an open source, any handset manufacture can use it as a software development platform [2–4,7].

Mougiakakou has suggested a prototype mobile phone application (MPA) that helps for the self-management of diabetic patients. The application consists of blood glucose measurements, blood pressure management, insulin dosage, food intake, physical activity, and an alarm in case of emergency [16].

The self-management application we propose runs on an Android platform, where the patient's daily exercise is recorded and sent to the doctor through a web server located at the doctor's clinic. With the updates received from the web server, the doctor can remotely monitor the patient's activities. This MPA is user friendly and cost efficient and helps to reduce the carbon footprint. In this chapter, we show how this application can be applied to improve the health of diabetic patients. This chapter is divided into three sections: Section 4.2 highlights the user needs and survey results from diabetic patients, Section 4.3 shows WBAN working architecture, and Section 4.4 presents the patient, physician, and web server view of the fully implemented system.

4.2 User Finding and Survey

4.2.1 User Finding

The proposed application is targeted to specific users with the following profile:

- A diabetic patient with type 2 diabetes
- Having trouble exercising alone and wishing to exercise in a group
- Difficulty in meeting the doctor often
- Familiarity in using Android phone applications
- Comfortable wearing sensors on the body
- Uses social networking sites such as Twitter or Facebook
- Having an appropriate knowledge in using smartphones and applications

4.2.2 Survey

In order to understand the need of self-management tools and the related technology for diabetic patients, a survey was conducted to get better understanding of the problems faced by diabetic patients and their reaction for the technology and use of body sensors. The survey was performed on 200 diabetic patients with an age group of 35–55 years, and their feedback was registered and analyzed. Seventy-five percent of the patients' doctors suggested that their patients need to exercise daily of which only thirty percent did. Seventy percent of the participants were interested in walking in a group, eighty percent were participants of social networking, and eighty-five percent did not have objections to having sensors attached on their body.

The results of the survey showed that over two-thirds of patients are interested in walking in a group and most of them were a part of a social network. In addition, the patients acknowledged that they might achieve greater benefits by participating in an active lifestyle in groups (rather than individually) and to network socially with other patients and be supportive to one another. Thus, in this chapter, we present a framework that helps diabetic patients to interact with new patients who have similar desires to network and exercise in groups. Observing many users are acquainted with social networking sites, this chapter utilizes the existing mobile services and the well-known social networking sites to achieve the desired objectives.

4.3 WBAN

WBAN is a special wireless network that consists of small, intelligent devices attached to clothing or on the body or even implanted under the skin that are capable of establishing a wireless communication link [9]. The wireless nature of these networks, and the increasing sophistication of implantable and wearable medical devices and their integration with wireless sensors, provides promising applications in medical monitoring systems and real-time feedback to the user or medical personnel to improve health care. They allow patients to experience greater physical mobility, where they are able to conduct their daily activities in a normal living environment. Consequently, the patients are no longer compelled to stay in the hospital and their quality of life improves considerably. The sensors of a WBAN are lightweight, low powered, and able to measure important parameters such as body temperature, heartbeat, and glucose level [5,6,16]. Location-based information can also be provided if needed.

The sensors transmit the measured data from the human body to the gateway through a wireless personal area network (WPAN) implemented using ZigBee (802.15.4) or Bluetooth (802.15.1) [8,9,13] and collected by a mobile phone. These sensors receive initialization commands and they respond to queries from the server through a gateway node. The gateway node is used to connect the sensors to different communication systems such as mobile phones of the patients, physicians, health-care professionals, or emergency services [10,11,14]. When something abnormal is detected, the sensors send the detected data to the gateway, which in turn delivers it to the doctor's clinic, emergency center, or patient's relatives on their cellular phones or via the Internet, and in critical conditions, it can also take appropriate decisions [8,9].

A body-worn glucose sensor that can periodically measure the glucose level in the body is analyzed in this chapter. When the glucose level is increased beyond normal, an injection controller in the gateway decides on the amount of insulin needed to be injected. Then a pump injects the required amount of insulin based on the instructions received from the controller.

A WBAN system (Figure 4.1) has three components:

1. The first component is the WBAN with medical sensors implanted on the human body.
2. The second component is the server providing Internet access and sending patient's information to the doctor and health-care centers in case of emergency.
3. The third component is the health center that keeps the record of the patient database and monitors the patient's health.

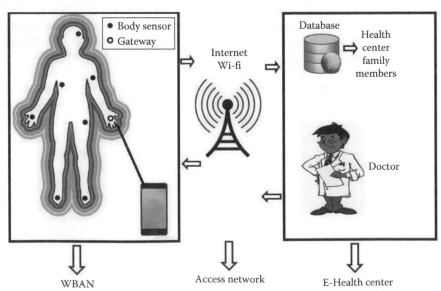

FIGURE 4.1
WBAN system.

4.3.1 WBAN Architecture

The proposed MPA in this chapter is a self-management tool for diabetic patients. It is a client–server model as shown in Figure 4.2. The MPA is comprised of two units: the client unit and server unit. The client unit is the patient's Android smartphone and the physician's smartphone or workstation. The server unit is the physician's database and a web server.

Biosensors are attached on the patient's body and communicate with the patient's smartphone via Bluetooth. The self-management application, installed in the smartphone, collects the information periodically from the sensors and sends it to the web server. This information will be stored on the physician workstation in the patient's database, and the physician can regularly monitor the patient's condition by logging into a dedicated website or directly from a smartphone.

By using the proposed system and application, the diabetic patient can check his or her glucose level regularly on a smartphone and take proper precautions to maintain it at an appropriate level. In addition, the physician can check the patient's data regularly and provide feedback directly to the patient when needed. This interactive application will help to improve the patient's health accordingly.

The detailed implementation of the whole system is out of scope for this chapter, so we focus mainly on a major service of the system, that is, the patient's daily exercise monitoring activity and patient-to-patient interaction using mobile services and social networking sites such as Twitter. A simple view of the application procedure is shown in Figure 4.3. The Android application collects Global Positioning System (GPS) data in 30 s intervals and sends it to the server. The server calculates the distance from two consecutive coordinates using the Haversine formula [16]:

$$d = 2r\sin^{-1}\left(\sqrt{\sin^2\big((\phi_2)-(\phi_1)\big)/2 + \cos\varphi_1\,\cos\varphi_2\,\sin^2\left(\frac{(\varphi_2-\varphi_1)}{2}\right)}\right)$$

where
　　d is the distance between two coordinates
　　r is the earth radius
　　ϕ is latitude
　　φ is longitude

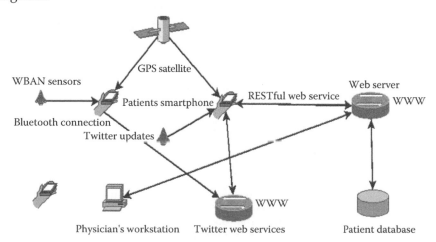

FIGURE 4.2
WBAN architecture.

This information is added to the previous distance and the total walking distance, and current coordinates are stored in the database as shown in Figure 4.3.

If the patient wishes to walk in a group, then the patient sends his or her id and coordinates to the requesting patient and the server will calculate the position of other patients nearest to the user. If the other patient wishes to join, then the two patients can form a group and exercise. In a similar way, all patients who wish to walk in a group can form a community and exercise together. The application will send Twitter/Facebook updates to other patients. This procedure is shown in Figure 4.4.

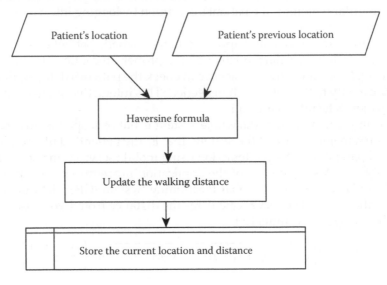

FIGURE 4.3
Distance calculation.

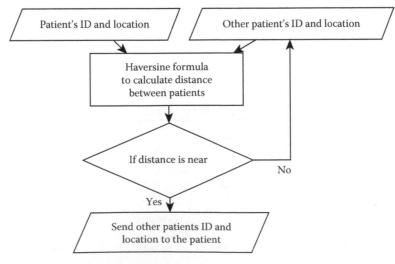

FIGURE 4.4
Finding the nearest patient.

4.4 Implementation

The implementation of the system is comprised of three views: the patient's view, the physician's view, and the server's view. The patient's view consists of one easy-to-navigate main menu (Figure 4.5) from which the patient can go to the subservices; for example, if the patient wishes to see the doctor's feedback, he can go to the main menu and select the corresponding option. The main menu consists of the total daily exercise requirements, doctor's feedback, finding a partner option, checking updates in Twitter, and a button for emergency situations.

The physician's side is implemented as a web view that the physician can access from his or her workstation by any web browser as shown in Figure 4.6. The website has a login page to authenticate the physician, and after authentication, the physician can check the patient's daily activities through a graphical representation of the patient's exercise. In addition, there is also an option to send direct feedback to the patients.

The server side is implemented as a REpresentational State Transfer (RESTful) that is often used in mobile applications and social networking websites. It defines a set of architectural principles by which one can design web services that focus on a system's resources, including how resource states are addressed and transferred over a web protocol by a wide range of clients written in different languages. Systems that conform to REST principles are referred to as RESTful. In our application, this system implementation provides functionalities such as updating current location and passed distance, finding neighboring patients and sending their current location, and broadcasting emergency messages to the patient's family and physician. Figure 4.7 shows a simple graphical user interface using

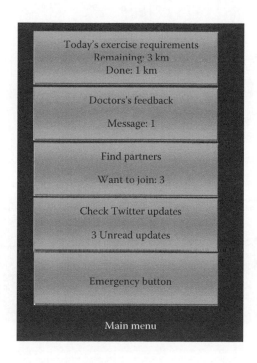

FIGURE 4.5
Patient's view.

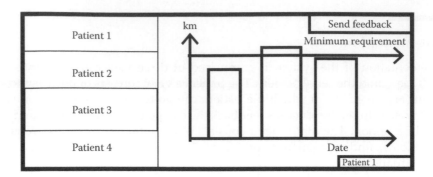

FIGURE 4.6
Physician view.

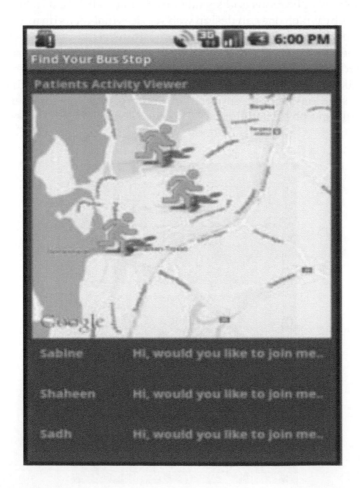

FIGURE 4.7
User interface.

the mobile services of the Android platform, where three users at different locations are interacting with each other using the GPS facility in their smartphones.

4.5 Future Directions

Future WBANs should focus on better hardware design; for example, the sensors should be designed as small as possible to provide more comfort to patients and more power-efficient sensors should also be developed. A new, improved Android application should be developed with the following features:

- The patient gets a diet update everytime the glucose level is tested. This update should come from the inbuilt data in the mobile phone as it is difficult for the doctor to attend to the patient too often.
- Wireless security should be provided by developing a secure wireless transmission to avoid eavesdropping or tampering of the patient's data.
- An alarm signal should be activated so that when the sensor's battery is low, it can send a message to alert the user.

4.6 Conclusions

In this chapter, we presented a self-monitoring application for diabetic patients on an Android mobile platform and have shown how using the current mobile services on smartphones can be exploited to improve diabetic patients' health. We conducted surveys on diabetic patients, and from the results, we presented a framework to improve the self-management of individuals and ways of interaction with other patients to exercise in groups. In addition, this application offers the doctors with a clearer idea about the patient's health history through an updated database showing the glucose level and amount of exercise accomplished by patients. Simultaneously, the doctors can use the application to send direct feedback to their patients if needed.

Questions

1. Briefly define diabetes and its main causes.
2. What is the most important parameter that must be monitored in diabetic patients?
3. What are the main complications if the blood glucose level is not controlled in diabetic patients?
4. What is the effectiveness status of current diabetic monitoring methods?
5. What are the most important features in diabetic self-monitoring tools?
6. What is wireless body area network (WBAN)?

7. Why are WBANs attractive for medical monitoring applications?
8. What is global positioning systems (GPS)?
9. What is Android?
10. What is REpresentational State Transfer (REST)?

References

1. Epinex Diagnostics. 2011. Diabetic problem. http://www.epinex.com/the-diabetes-problem.php (accessed January 22, 2014).
2. Butler, M. Android: Changing the mobile landscape, *IEEE Pervasive Computing*, 10(1), 4–7, January–March 2011.
3. Cheng, Y.-H., Kuo, W.-K., and Su, S.-L. An android system design and implementation for telematics services, *IEEE International Conference on Intelligent Computing and Intelligent Systems (ICIS)*, Xiamen, pp. 206–210, October 29–31, 2010.
4. Gozalvez, J. First Google's android phone launched, *IEEE Vehicular Technology Magazine*, 3, 3–69, September 2008.
5. Jovanov, E. Wireless technology and system integration in body area networks for m-health applications, *Annual International Conference of the Engineering in Medicine and Biology Society*, Beijing, pp. 7158–7160, January 17–18, 2006.
6. Raazi, S.M.K.-u.-R., Lee, S., Lee, Y.-K., and Lee, H. BARI: A distributed key management approach for wireless body area networks, *International Conference on Computational Intelligence and Security, CIS '09*, vol. 2, pp. 324–329, December 11–14, 2009.
7. Conti, J.P. The androids are coming, *Engineering & Technology*, 3(9), 72–75, May–June 24, 2008.
8. Rehunathan, D. and Bhatti, S. Application of virtual mobile networking to real-time patient monitoring, *Australasian Telecommunication Networks and Applications Conference (ATNAC)*, Auckland, pp. 124–129, October 31–November 3, 2010.
9. Abousharkh, M. and Mouftah, H. A SOA-based middleware for WBAN, *IEEE International Workshop on Medical Measurements and Applications Proceedings (MeMeA)*, pp. 257–260, May 30–31, 2011.
10. Ivanov, S., Foley, C., Balasubramaniam, S., and Botvich, D. Virtual groups for patient WBAN monitoring in medical environments, *IEEE Transactions on Biomedical Engineering*, 59(11), 3238–3246, November 2012.
11. Kim, D.-Y. and Cho, J. WBAN meets WBAN: Smart mobile space over wireless body area networks, *IEEE 70th Vehicular Technology Conference Fall (VTC 2009-Fall)*, Anchorage, pp. 1–5, September 20–23, 2009.
12. Jobs, M., Lantz, F., Lewin, B., Jansson, E., Antoni, J., Brunberg, K., Hallbjorner, P., and Rydberg, A. WBAN mass: A WBAN-based monitoring application system, *Second IET Seminar on Antennas and Propagation for Body-Centric Wireless Communications*, pp. 1–5, April 20–20, 2009.
13. Choi, I.M., Won, K.H., and Choi, H.-J. Group manchester code modulation for medical in-body WBAN systems, *17th Asia-Pacific Conference on Communications (APCC)*, Sabah, pp. 867–871, October 2–5, 2011.
14. Narazaki, H., Tani, S., Ishizaki, J., Nakazawa, K., Inada, H., and Nishimura, H., Development of a support system for diabetic patients at home using a smartphone, *Joint 6th International Conference on Soft Computing and Intelligent Systems (SCIS) and 13th International Symposium on Advanced Intelligent Systems (ISIS)*, Kobe, pp. 1134–1137, November 20–24, 2012.
15. Nordin, N.A.M., Zaharudin, Z.A., Maasar, M.A., and Nordin, N.A., Finding shortest path of the ambulance routing: Interface of A* algorithm using C# programming, *IEEE Symposium on Humanities, Science and Engineering Research (SHUSER)*, Kuala Lumpur, pp. 1569–1573, June 24–27, 2012.

16. Mougiakakou, S.G., Kouris, I., Iliopoulos, D., Vazeou, A., and Koutsouris, D., Mobile technology to empower people with Diabetes Mellitus: Design and development of a mobile application, *Ninth International Conference on Information Technology and Applications in Biomedicine*, Larnaca, p. 1, November 4–7, 2009.

17. Ekroos, N. and Jalonen, K. E-Health and diabetes care, *Journal of Telemedicine and Telecare*, 13(1), 22–23, July 1, 2007.

18. Hwang, T.-H., Kim, D.-S., and Kim, J.-G. An on-time power-aware scheduling scheme for medical sensor SoC-based WBAN systems, *Journal of Sensors*, 13(1), 375–392, December 27, 2012.

19. Tachtatzis, C., Di Franco, F., Tracey, D.C., Timmons, N.F., and Morrison, J., An energy analysis of IEEE 802.15.6 scheduled access modes, *2010 IEEE GLOBECOM Workshops*, Miami, pp. 1270–1275, December 6–10, 2010.

20. World Health Organization, Fact Sheet No. 312, October 2013. http://www.who.int/mediacentre/factsheets/fs312/en/index.html (accessed January 22, 2014).

21. Khan, S., Khan, A., and Khan, Z.H. Artificial pancreas coupled vital signs monitoring for improved patient safety, *Arabian Journal for Science and Engineering*, 38, 3093–3102, Springer, 2012.

Section II

Consumer and Patient-Centric Approaches in mHealth

5

mHealth Diet and Nutrition Guidance

Yasmine Probst, Duc Thanh Nguyen, Megan Rollo, and Wanqing Li

CONTENTS

5.1 Introduction

The risk factors for many chronic diseases such as cardiovascular disease, type 2 diabetes mellitus, and some cancers are associated with poor diet and/or weight gain. Preventing diseases through improved nutrition is now a global health priority. However, before adults can improve the quality of their diet, they need to have a greater awareness of what constitutes healthy eating. Many adults are unaware of what they are eating and believe their diet is healthy when it isn't. Therefore, accurate and timely assessment of food intake is necessary so that consumers can be better informed on what constitutes a healthy diet consistent with a country's dietary guidelines.

Health informatics has slowly progressed in the field of nutrition, with nutrition informatics now a recognized area of dietetic practice [1] in some countries. In past decades, there was increased resistance to technological approaches to nutrition, in particular dietetics, though societal change has seen nutrition professionals embrace the efficiencies that technology may create. Automation of such methods now provides the added advantage of other health professions being able to embrace nutrition concepts in their client interactions. In particular, capture of accurate food intake information has long been a challenge for dieticians and nutritionists, though with the progress from paper and pen to automated intake assessments, much of the previously onerous tasks for both the individual and health professional may be minimized.

Recall methods to assess diet are often affected by an individual's ability to remember food descriptions and portion sizes. Further to this, participant burden of handwritten dietary records is high, resulting in a low level of accuracy and in turn limiting the tailored nutrition advice that may be provided to an individual to help them manage their nutrition-related health concerns. Technological innovations offer promise with respect to the design of new dietary assessment tools to overcome some of the limitations outlined earlier. This has been recognized within the field of dietary assessment since the 1970s with the first automation of diet history interviews using computers [2]. Since then, many other forms of technology, such as digital cameras and personal digital assistants, have been employed in an attempt to improve upon paper and pen method [3,4]; however, few have been able to overcome weaknesses of traditional tools, in particular the high levels of participant burden and portion size estimation errors [3,5,6].

As the capabilities of technology have continued to progress, engineers and software developers focused on end-user satisfaction, simplifying interactions with devices and producing applications without *bugs*; nutrition experts continue to seek an application that provides accurate and reliable dietary information, while consumers are looking for low-cost real-time feedback from a credible source. In addition, consumers also want information about their diet that is personally relevant to them and based on the best available evidence with feedback that is achievable and easy to incorporate into their daily food patterns. mHealth is seen as a suitable platform for achieving each of these stakeholder needs, and ideally, application development for nutrition needs to consider all of these aspects as showcased throughout the later sections of this chapter.

An overview of the technology required as well as current developments and future challenges and opportunities for the use of mobile devices for informing, guiding, monitoring, and improving nutritional intake of individuals, groups, and populations will be outlined in this chapter. To begin with, the assessment methods employed by nutrition professionals are explored emphasizing the importance of nutrition training when food intake information is to be collected.

5.2 Traditional Methods for the Assessment of Diet

Assessment of food intake is a difficult task as it requires individuals to describe, in detail, the food or drinks consumed. Most consumers have little knowledge of the composition or amount of food or drinks they commonly consume, increasing the complexity of this task. Further, as a large proportion of food is now eaten away from home, the task of recording food intake is even more challenging. Methods for the assessment of dietary intake in both practice and research have been used since the early 1940s with four methods traditionally employed. These methods can be categorized as prospective or retrospective based on when the dietary intake information is collected. Retrospective methods are collected after the food has been consumed and include food frequency questionnaires, diet history interviews, and 24-h recalls, while prospective methods include food records and are collected prior to and/or at the time the food is consumed. The reference time period for which food information is to be either collected or recalled over is dependent on the nutrient of interest, with energy (calories) and macronutrients (e.g., carbohydrate) requiring fewer days compared to micronutrients (e.g., zinc, vitamin E). Each method has inherent strengths and weaknesses and, when applied to mHealth, creates unique challenges and opportunities. A brief outline of each of the four traditional dietary assessment methods is provided below though is not meant to serve as a complete review. Readers are encouraged to source additional information on these methods to inform decisions on the appropriateness of their use beyond the work of this chapter.

5.2.1 Food Frequency Questionnaire

Food frequency questionnaire is the recall of the consumption frequency of selected group(s) of food items over a specified time period to estimate intake of single or multiple nutrient(s), food groups, and eating patterns. Food frequency questionnaires were designed to gather qualitative data over a defined period of time to determine links between intake and the etiology of disease [7]. Therefore, this method aims to measure exposure in terms of weeks, months, and/or years of the individual to a nutrient(s) and/or particular foods [8].

5.2.2 Diet History Interview

Diet history interview is the recall of eating patterns and foods consumed over a usual period of intake, with items quantified in household measures and with a *cross-check* process used to clarify recorded information and probe for consumption of additional foods. Typically, the diet history interview is used to gather detailed information on habitual diet, including eating behaviors, cooking, and preparation methods. The interview is performed by a trained interviewer and is labor and time intensive [9], although self-administered forms using computer-based versions have been trialed [10,11].

5.2.3 24 Hour Recall

A 24-h recall is the recall of all foods (description and quantity) consumed over the previous 24-h period; this method is typically used to estimate the average intake of a group or population; however, usual intake at the individual level can be estimated with multiple days of recording [9]. The relatively quick and easy administration of this method results in lower individual burden [9] and has been used in large population surveys such as the National Health and Nutrition Examination Survey in the United States and the Australian Health Survey. The 24-h recall may be administered by a trained interviewer, while more recently self-administered versions have been tested [12,13]. Multiple-pass approaches (standardized interview structure) are often recommended for the collection of intake information as it has been shown to assist individuals to recall accurately and to minimize bias [14].

5.2.4 Food Record/Food Diary

Food record/food diary is the real-time recording of food intake prior to consumption, with quantities either measured/weighed or estimated using household measures (e.g., cups and spoons). Food record information is typically documented via paper and

pen record and requires strong literacy and numeracy skills, in addition to highly motivated individuals [15,16]. Although considered burdensome, the food record is acknowledged to be the most precise technique available for estimating usual food and nutrient
intakes of individuals [9] and has also been used in large population surveys such as the
British National Diet and Nutrition Survey. Recent work found that 3-day weighed food
records produce comparable measures of usual 7-day intake [17], suggesting that shorter
recording periods can produce accurate estimates of intake.

The traditional dietary assessment methods outlined earlier have evolved with advances
in technology and associated capabilities. Web-based versions in which the general processes have been automated were the initial advances and have evolved significantly with
time as the capabilities of technology have progressed. In particular, the increased uptake
and capabilities of cellular phones and smartphones has seen a marked increase in the
use of these devices to collect prospective or *real-time* dietary intake information in more
recent times.

5.3 Advances in the Automation of Dietary Intake Assessment Methods

5.3.1 Cellular and Smartphone Devices

Cellular phones, and in particular smartphones, offer a new platform that has the potential to improve both accuracy and efficiency of dietary assessment. Mobile devices have
increasingly been used to guide individuals in the selection of foods, self-monitoring of
dietary intake in relation to the management of diseases, monitoring of intake to inform
assessment, and assisting in the provision of nutrition intervention by health professionals.
Mobile devices, such as smartphones and tablets, can be used to capture food intake and
deliver dietary feedback, due to the portability of these devices that allows individuals to
record intake in an increasing number of situations. Despite these many advantages, the
field of mHealth in nutrition has seen applications that provide weight loss tips and nutrition and calorie trackers as one of the fastest-growing application genres. This vast influx
has challenged the credibility of mHealth nutrition applications with users advised that
they should be used with caution often due to limited information on the content source
and/or relevance to native food supply of the user, both of which are important to the
accurate translation of dietary advice.

Unique advantages of smartphones as a mobile device include their ubiquitous nature,
incorporation of cameras, and the capacity to connect and send information to a remote
server [18,19]. Additionally, the use of images rather than text to report intake offers the
opportunity to create tools suitable for low literacy groups [20], a population in which estimation of dietary intake is particularly problematic. Photographic dietary records (PhDRs)
collected via mobile devices have been used in a diverse range of groups to assess intake
including adolescents [21], college students [22–24], overweight and obese adults [25], and
older adults with a chronic disease [22,26].

Emerging technologies are also facilitating automatic visual recognition and volume
computation enabling, for example, food identification and portion size estimation from
cellular phone photographic food records [5,27]. Table 5.1 summarizes PhDR methods collected using smartphones, cellular phones, and other similar devices for the estimation of
nutrient intake. Although these applications have the ability to track overall food intake
and/or analyze intake of specific nutrients (e.g., calories, fat), most require individuals to

TABLE 5.1

Characteristics of PhDR Methods Collected by a Mobile Device and Used to Assess Nutrient Intake

Name of PhDR Method/Project (If Available) [Studies That Have Used a PhDR Method and/or Have Described It's Development]	Method of Collection and Analysis of PhDR
Wellnavi [30–32]	• PDA with camera function, description written using stylus onto photograph of food displayed on screen. • Reference object: stylus was placed next to food items. • Free-living subjects collected PhDR of own intake; PhDR transferred automatically for analysis. • Manual analysis—dietitian only.
Remote food photography method [25,33]	• Mobile phone with camera function + prompts to remind subject to record + foods consumed but not captured using PhDR collected using written record or voice record. • Telescopic pen used to standardize distance photograph captured in early trials; reference card in later trials. • Subject collected PhDR of own intake (preprepared food items provided and eaten in controlled/free-living environments); PhDR transferred automatically for analysis. • Semiautomated analysis—dietitian in combination with searchable database of reference food photographs.
Technology-assisted dietary assessment project [21,23,24,27,34]	• Mobile phone with camera function. Early prototypes tested required the subject to identify food items in PhDR by writing on screen or confirm food items identified. • Reference object: fiduciary marker. • Intended for free-living subjects to collect PhDR of own intake; PhDR transferred automatically for analysis. • Automatic analysis—image analysis algorithms to identify and quantify food items contained in PhDR.
Food intake visual and voice recognizer [35]	• Mobile phone with camera function + voice recorded on phone used to clarify items unable to be automatically identified in PhDR. • Reference object: fiduciary marker. • Intended for free-living subjects to collect PhDR of own intake; PhDR transferred automatically for analysis. • Automatic analysis—image analysis algorithms to identify and quantify food items contained in PhDR.
DietCam [38]	• Mobile phone with camera function. • Reference object: credit card (camera calibrated prior to use). • Intended for free-living subjects to collect PhDR of own intake; PhDR transferred automatically for analysis. • Automatic analysis—image analysis algorithms to identify and quantify food items contained in PhDR.

(Continued)

TABLE 5.1 (*Continued*)

Characteristics of PhDR Methods Collected by a Mobile Device and Used to Assess Nutrient Intake

Name of PhDR Method/Project (If Available) [Studies That Have Used a PhDR Method and/or Have Described It's Development]	Method of Collection and Analysis of PhDR
Nutricam dietary assessment method [22,26]	• Mobile phone with camera function and voice recorder. In a later version, a brief phone call to users the day following recording was incorporated to clarify contents of the PhDR and probe for forgotten foods.
	• Reference object: Card with recording instructions.
	• Free-living subjects collected PhDR of own intake; PhDR transferred automatically to a website for analysis.
	• Manual analysis—dietitian + portion size estimation aid (photographs and graphics).

Source: Adapted from Rollo, M.E., An innovative approach to the assessment of nutrient intake in adults with type 2 diabetes: The development, trial and evaluation of a mobile phone photo/voice dietary record, Faculty of Health, School of Exercise and Nutrition Sciences, Queensland University of Technology, Brisbane, Queensland, Australia, 2012.

Notes: PhDR, photographic dietary record; PDA, personal digital assistant.

select each food item from a list of foods. This type of application may appeal to highly motivated individuals interested in calorie tracking, but digital entry of food and drink names has many of the same barriers to completion as paper-based methods.

Image-based approaches using the integrated cameras on mobile devices for monitoring diet are gaining appeal as they reduce the burden on the participant by simplifying the recording process. Accuracy of data capture has been trialed by some with individuals also recording an audio clip describing the food featured in the photograph, a process that becomes void with increasingly refined photographic applications, though the added information may still improve the speed of differentiation between like foods. These types of applications remove the need for individuals to log their food intake on a website or keep a paper-based record and reliance on cognitive capacity and memory. However, improving the automation of image analysis in order to reduce the server-side requirements is an important step in the development of these applications.

5.3.2 Wearable Devices and Sensor-Based Technologies

Although portable, cellular and smartphones have inherent disadvantages for recording, from a computation perspective, while still outweighing the time taken to obtain such information by comparison with a purely paper-based format. New methods that incorporate digital camera technologies into wearable, wireless data transmission devices are also being drawn on for food intake information [28,29,36,37]. Many such devices use various forms of visual tracking such as single-frame food images or continuous-frame footage at a specified period of time, allowing the eating or purchasing situations related to food to be contextualized for the user of the data. In a pilot trial of the Microsoft SenseCam device used to capture information in a 24-h period, it was found that recall of food information by an individual could be assisted with the use of the sensor technology over a recall of the period's food intake alone. Limitations include the need to actively turn the device on and off in situations that do not warrant recording, leaving room for bias in the scenarios that are actually recorded by the individual. The early developmental challenges faced included the narrow field of view and poor battery life, each of which was independent of the food recall itself [29]. The sensor was also worn on a lanyard around the neck capturing both passive and objective information about physical activity and food intake. The placement of the sensor on a lanyard also creates practical challenges in terms of the field of vision when seated for a meal, for example, suggesting that a sensor placed at a higher focal point may warrant better imagery for analysis. Further developments in the field include electronic chest buttons to specify placement of the sensor. These early platforms still require manual analysis of the food information to obtain nutrient information showing a clear link between this assessment form and the image processing techniques outlined later.

5.4 Determining Food Portion Size Using Image Processing Techniques

During the last decade, advances in mobile technology have been developed significantly to provide a range of facilities. As part of this development, imaging devices such digital cameras have also been integrated into the cellular phone. This has made image capture easy and accessible. Exploiting this advantage, various applications using image processing techniques have been proposed, developed, and tested in laboratory settings to determine the portion size of foods present in a PhDR; however, application within the free-living environment with a diverse food supply is still to be established.

The food image in this instance is then sent to a server for automated recognition of food type, for example, burger and salads. The server recognizes the food type that is predefined offline using pattern recognition algorithms and sends the recognized food type back to the cellular phone. Figure 5.1 illustrates the process of this application. As can be seen, image classification is the core part of this application.

Research effort in food image classification for the estimation of food intake has until recently been scant. It is vital to recognize that the challenges of the food image classification problem [39–42] are largely due to the variations in appearance (color, texture, and shape), angles to take the images from, and recording environment (complex background, presence of other objects, uncontrolled photographing conditions, and illumination) in practice.

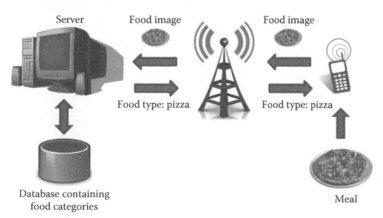

FIGURE 5.1
Nutrition guidance on mobile phones using image-based food recognition.

Generally speaking, state-of-the-art methods for food image recognition and classification have used various descriptors that exploit appearance features including color [40], texture [41], shape features [39,42], and a suitable model to represent a food item.

Color is considered as one of the important features for food image classification [40]. However, color is sensitive to illumination conditions. For texture information, Gabor texture features are often employed [41]. Shape information [39–42] exploits the geometric characteristics of the food item and can be classified as global shape and local shape information. For global shape [42], for example, the size (counted as the area) and the aspect ratio of the food image can be used as features if prior knowledge of the size and shape of the food object is available. This is rarely the case. Therefore, more recently, the scale-invariant feature transform (SIFT) descriptor [43] has been employed to describe the local shape of food objects [39]. Using a set of points at which the appearance of the food tends to be more descriptive may be applied to the food images to detect a set of key points, for example, SIFT [43] key points, unique to the image. The scale of the local image region centered at key points is subsequently calculated from these key points. Information relating to the scale and size of the local image regions obtained for food images are held in a training dataset. This training dataset is a collection of images whereby the algorithm *learns* how to recognize and differentiate foods from one another and is used to construct a *probability density function* (*pdf*) of the food portion sizes. In practice, when a person takes a photo of the food that they eat, key points and their scale on the food image are first computed and the volumetric portion size is then estimated based on the precomputed *pdf* of the scale and size. Generally, the advantage of the SIFT feature is that it is highly informative and insensitive to local deformations such as scale, rotation, and distortion due to the variation of viewpoints when a number of clients are asked to take photographs of their foods. The combination of various feature combinations has also been investigated [41].

While existing studies on food image classification using multiple features [39–41] have shown promising results, structural information of food objects has largely been ignored. It is anticipated that a combination of appearance features and structural features would enhance the overall classification performance.

Following the estimation of the portion size of food items contained within PhDRs, (specifically volume), this information must then be converted to a measurement unit, more often a weight, before analysis of nutrient intake can be undertaken. A lack of food density data has been one restriction in the past [44]; however, a more comprehensive database is now available for many foods [45]. Due to the complexity of these processes, such features are housed behind the scenes of developed applications requiring the individual only to interact with the camera function of the cellular or smartphone device and receive output often in the form of nutrition education messages tailored to the information they captured about their food intake.

5.5 Providing Mobile Nutrition Education

Computer-tailored interventions are a relatively new health approach [46] and may be useful for providing dietary advice to a large population of people. Computer-tailored interventions are characterized by the fact that the content of the materials is tailored to the specific needs of an individual [46], meaning that the messages generated are likely to be of more personal relevance and may have stronger motivational effects [47]. According to Brug and colleagues, computer-tailored nutrition education is more likely to be read, remembered, and experienced as personally relevant compared to general nutrition materials [47]. The form in which such education is delivered, namely, using a traditional personal computer or laptop versus using a tablet, smartphone, or even digital audio device, remains to be determined. As a point of interest, studies on computer-tailored feedback have indicated that even people who think their diets are in accordance with nutrition recommendations are still interested in personal feedback and do volunteer to receive tailored messages [48]. Such tailored messages from an mHealth perspective can include the use of short message service (SMS) in which a library of motivational messages are developed and drawn on at particular time points depending on the stage of change and level of interest of the receiver [49]. They may be used to encourage engagement with a particular program or send reminders for health practitioner appointments. Smartphone applications offer similar functions but also allow for a more two-way interaction between the device and users. Generally, such technologies are used to help them keep recipients on the right track or as a self-monitoring tool to check that their food intake, eating behaviors, and nutritional status (e.g., body weight) are in accordance with the recommendations. The main advantage of computer tailoring over traditional counseling is that it provides health professionals with the opportunity to reach more people, at a lower cost, than would be possible with interpersonal counseling [50]. When developed by a credible source, the use of mHealth in education also allows for the inclusion of nutrition in multidisciplinary projects when resources do not allow for an employed nutritionist or dietician. However, costs associated with the development of these systems are likely to be higher in the short term and there is a need for regular maintenance, and therefore, although there are advantages over the traditional paper-based approach, caution should be applied.

Tailored nutrition education encompasses three underlying factors:

1. A combination of nutrition information or dietary change strategies.
2. It is aimed at a specific person based on this person's dietary habits and/or stage of change and the determinants of these habits.
3. It is developed for the specific person's needs [51].

This is what most dieticians and nutrition counselors aim to use each day [51]; however, personal counseling can be very time-consuming, difficult to access, and often costly to the individual [51]. An alternative is to apply tailored nutrition education by using interactive technology, allowing for personalization of key messages to large groups of people with lower resource needs [51].

5.5.1 Does Computer-Tailored Nutrition Education Work?

Several reviews have demonstrated that tailored print materials generally outperform standard health education messages [47,52]. Brug and colleagues reviewed the literature on the impact of computer-tailored nutrition education, specifically for dietary change, and concluded that tailored nutrition education is more effective than generalized nutrition education, especially fat reduction [47]. An additional analysis of the pooled results of three trials of computer-tailored feedback showed that computer-tailored feedback resulted in a 5.4% lower fat intake, compared with a 1.4% decrease in a general nutrition information control group about 4 weeks after the intervention [53]. Thus, it can be presumed that computer-tailored nutrition education is effective in improving people's eating habits toward dietary recommendations. Specific focus on the mobile platforms requires further investigation, though the added convenience created suggests it to be of further benefit. It is likely that such avenues will be increasingly pursued once the assessment of intake can be accurately captured and a logical flow through of food information can be translated to appropriate nutrition education messages.

5.6 mHealth in Practice

5.6.1 Nutrition Perspective

Case Study 1:   Passive Image Capture of Food Intake through Portable Life-Logging Sensors

Digitizing and creating a log of a person's daily activities is a novel approach in the field of nutrition drawing on global positioning systems and user-aided video recordings to potentially capture food intake in the location and at the time of consumption. More often seen in other areas of daily life, the Microsoft SenseCam is considered a portable life-logging device. Incorporating a real-time logging technology and digital camera, movement sensor (triaxial), as well as infrared heat and light level sensors, the device is worn around the neck capturing images at specified intervals and in response to environmental triggers such as shifts in temperature, light, and motion. Minimal user input is required, though a privacy button minimizes inappropriate recordings during long-term wear. Initially trialed to assist with recall in persons with severe memory impairment, this technology was flagged as an ideal platform to use in food recall. Examined in the United Kingdom, adults wore the device from wake until sleep for a period of 2 days. For comparison, a 24-h recall assessment was also conducted. Forgotten food items were captured via this process, though often forgotten in traditional assessment methods. Image quality remained a challenge due to low-light environments in which foods are commonly consumed. Between the two assessment methods, a 12.46% increase in energy intake was found and 42 missing foods identified in a pilot sample of 10 users. The impact of incorrect portion size was found to be less significant on the overall recall of the diet than the missing foods. Misreported foods, however, can be corrected with the sensor technology allowing the researcher to replace the reported food with the correct food that was recorded. The tool provides promise for food intake assessment, though limitations such as positioning on the user and short battery life presently limit its uptake. The research groups are presently experimenting with a higher placement of the sensor device in relation to common eating situations. Think about your own eating environment and how much of your intake would be captured if you wore a lanyard around your neck. Which items in particular would be missing from your food intake due to placement of the sensor alone? Is there a pattern to these missing foods based on meal or food type?

Case Study 2: Connecting Health and Technology (CHAT) Protocol

Drawing on iPod touch technology, used socially by a vast number of adolescents, Connecting Health and Technology (CHAT) draws on the food record method of assessing a person's food intake [54]. Using mHealth platforms, this work draws on both a mobile device and text messaging to communicate with its users. The messages relate to key food groups to create an individualized approach while the iPod (mobile device) requires the user to photograph all foods consumed. The iPod device was selected due to ease of use and feasibility. Following training in the use of a fiducial marker (standardized 10 × 10 cm checked square), to be placed in all photographs to assess the platform with image recognition, the users were asked to record their food intake for a period of 3 days including 2 week days and 1 weekend day. Targeting young adults below the age of 30 years, the program aimed to increase fruit and vegetable intake and also decrease junk food intake through the provision of regular text messages to the participants. All messages had motivationally focused wording determined through focus groups research prior to implementation. A library of over 50 messages was subsequently developed. User testing was also employed with the term *junk food*, for example, seen to match well with the user group as opposed to discretional foods as labeled in the nutrition guidelines. CHAT drew on a free-living community-based approach with a specific focus on increasing fruit and vegetable intake during user testing. The users were required to photograph their foods before and after eating accessed by a dietician via remote server. The dieticians also provided the SMS based on the food information received. The food image recognition approach used for CHAT was based on that of the technology assisted dietary assessment (TADA) project (http://www.tadaproject.org/) with food recognition for nutrient processing performed manually by the dietician due to the U.S. focus of TADA and the need for regionally specific food information. The investigation similar to the sensor-based intake assessment also found the food environment to be of particular interest seeing young adults rarely using traditional sit-down eating scenarios when photographing their foods. This approach, however, is well suited to a range of other domains beyond nutrition, pending a user-focused setup and targeted approach to change. Furthermore, the user group needs to be well aligned with the mobile platform being used.

5.6.2 Information Technology Perspective

Case Study 3: Application of Food Image Recognition

In the context of automated image analysis of PhDRs, there are three main factors for consideration:

1. Shape
2. Color
3. Texture

The food image classification method proposed extends the bag-of-words (BoW) paradigm [45] applied for food image classification [40,41]. In particular, the local appearance of food objects is encoded as code words, while the spatial distribution of code

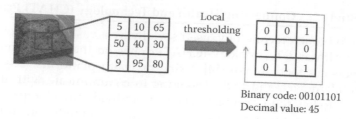

Binary code: 00101101
Decimal value: 45

FIGURE 5.2

Illustration of LBP descriptor.

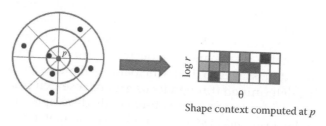

Shape context computed at p

FIGURE 5.3

Illustration of shape context descriptor extracted at a point p. Dark intensity represents high value of the shape context.

words is considered in describing food items. A process of local binary pattern (LBP) recognition is employed to describe the appearance of code words [55]. Intuitively, LBP describes the image texture based on the changes of intensity at a local image region. LBP has been widely applied for face recognition [56] and object detection [57,58]. The concept of an LBP code for a pixel is illustrated in Figure 5.2, in which the central pixel's intensity is compared with the intensities of neighboring pixels. The neighboring pixels whose intensities are equal to or higher than the central one are labeled as *1*; otherwise they are labeled as *0*.

Shape context as a descriptor was proposed by Belongie and colleagues [59] and has been successfully used in various shape matching and object recognition tasks. Given a set of points, the shape context of a point is represented by the histogram of the relative coordinates of this point to other points in the set. Figure 5.3 illustrates an example of the shape context descriptor.

5.6.3 Food Image Classification Using LBP and Shape Context

Given a food image containing an instance of a food item, the SIFT detector [44] is invoked to detect a set of key points. The LBP is used to describe the local appearance at key points. These key points are then matched with a predefined codebook and a histogram of matched code words is created.

Considering every key point as a point on a shape, the shape context descriptor [59] is applied to analyze the spatial relationship between the best matching code words found at the key points [60].

Using the BoW model with both the appearance and structural information, a food image can be described by integrating the appearance and structural information and classified using a support vector machine [43]. By using shape context to represent the spatial relationship between code words, the proposed method can accommodate with the global deformations and transformations in the shape of food objects.

TABLE 5.2

Comparison of Using Color, SIFT, and Appearance Feature (LBP) with and without Structural Feature Represented by Shape Context

	Color	SIFT	Appearance (LBP)	Appearance + Structure (LBP + Shape Context)
Sandwiches	0.74	0.51	0.54	0.67
Meat	0.47	0.56	0.61	0.68
Salads	0.66	0.90	0.75	0.90
Donuts	0.21	0.43	0.46	0.48
Hamburger	0.61	0.49	0.54	0.63
Miscellaneous	0.31	0.64	0.67	0.72

Notes: SIFT, scale-invariant feature transformation; LBP, local binary pattern.

5.6.4 Experimental Results

The LBP with eight neighboring pixels (as shown in Figure 5.2) was employed to encode the appearance of food items. Both the training and testing were performed on the Pittsburgh Fast-Food Image (PFI) dataset [40]. This dataset contains a total of 4545 still images, 606 stereo pairs, 303 360° videos, and 27 privacy-preserving videos of eating events of volunteers. Food images in the PFI dataset were labeled as six different categories: *sandwiches, meat, salads, donuts, hamburger,* and *miscellaneous.* The dataset was separated into the training and test sets so that no instance of a food item appeared in both the sets. The dataset was also accompanied by a set of baseline experiments.

Experimental results show that the category *salads* achieved the highest performance accuracy, while the category *donuts* had the lowest accuracy. One possible reason might be the low resolution of the donut images and their less diverse surfaces, which could result in fewer key points compared with other categories of food.

As shown in the experimental results, the LBP slightly outperformed SIFT when the overall performance was considered. The importance of the structural information represented by the shape context descriptor [59] was also verified by employing only the appearance features (LBP) and combining appearance features with the shape context in describing food items. As shown in Table 5.2, the use of structural information improved the classification accuracy for the food items. Table 5.2 also shows the impact of color. It is, therefore, important to consider color information to enhance appearance features. For instance, the category *sandwiches* often had a white surface, while other categories may not have any white surfaces at all.

5.7 Conclusion

Accurate and reliable assessments of dietary intake are challenging to obtain, in particular, given the diversity of our food supply and variability in access to appropriate food-nutrient databases throughout the world. Inherent methodological limitations with traditional approaches combined with day-to-day natural variations in intake also compound the task. The enhanced functionality and focus on user-centered design have seen a rapid rise in the uptake of mobile devices and mHealth. The portability of these devices allows for some of the challenges associated with traditional dietary assessment methods to be addressed and also offers new modes for the collection of dietary intake information via PhDRs. This aspect of mHealth is relatively novel and, through continued evaluation of image analysis techniques, is continuing to advance. Although PhDRs are often preferred by the user due to their simplicity for collecting dietary intake information, new challenges arise relating to developing automated algorithms that not only efficiently analyze this type of dietary record but are also adaptable and can account for the complexity and variety that is present within the diets of individuals, groups, and populations. As this area of mHealth becomes refined, it is envisaged that mobile applications will more effectively use PhDRs (in combination with other data collected via the device) to provide tailored user feedback relating to dietary intake, eating behaviors, and nutritional status. As a result, a more ubiquitous two-way communication stream for diet and nutrition guidance will emerge.

5.8 Future Directions for mHealth in Nutrition

One of the main benefits of technology in nutrition as with other fields of healthcare is the reduction in the paper usage and resultant storage space needed. As smartphones and cellular phones become more sophisticated and compact, it is likely that they may be incorporated into wearable devices in the future. Through combination with a greater number of mHealth platforms and increased capabilities in the automation of analysis, especially with regard to photographic dietary records, there will be more opportunities to provide the user with real-time, in-depth dietary self-monitoring and feedback of nutrient intake. Cloud computing also allows the dietitian or nutritionist the benefits of increased portability while also creating a virtual storage environment for educational materials and resources. In countries such as Australia, *cloud* can also provide benefits to those seeing remote and rural patients. The primary present day shift is toward electronic health is being closely followed by mobile health as outlined in this chapter. Sensor technologies are progressing slowly into the field of nutrition allowing increasingly accurate assessments to be conducted and less reliance on self-report about a person's food intake. Looking toward the future though, the concept of remote and real-time ubiquitous health (uHealth) platforms of wireless technologies will allow patients to be monitored in home and with wireless messages transmitted to the dietitian. Likely commencing as wearable devices, such devices will enable more ubiquitous use and facilitate the collection of extensive information on diet and nutrient intake, the existing challenges of ensuring that this information is accurate and reliable will remain. UHealth so far has been trialled for the assessment of intake though.

What Did You Learn? Discussion Points

1. Mobile devices have simplified the recording of dietary intake information for the user; however, caution needs to be used when moving from a traditional paper-based approach to various platforms. Consider the impact of moving a diet history interview to a laptop-based platform, a tablet, or a smartphone. What are the advantages and disadvantages of each?

2. Image recognition has shown its potential in nutrition guidance, in particular in the area of portion size identification in the analysis of PhDRs. Considering the complexity of the food supply, what are three challenges related to the use of image recognition systems in this context?

3. Think about some user characteristics that need to be considered when using mobile-based dietary assessment methods. For example, consider how you would use this technology to assess the diets of children compared to adults? Hint: Ensure the age, social, economic, demographic, and location of the users are suited to the approach you choose.

Questions

1. How can key points be determined from an input image?
2. How is a food image encoded in the bag-of-words model?
3. What information does the LBP represent?
4. In which decade were dietary assessment methods first automated?
5. Name two of the most commonly used methods for dietary assessment?
6. Which component of a dietary assessment is mHealth well suited to?

References

1. Ayres, E.J., Greer-Carney, J.L., Fatzinger McShane, P.E. et al., Nutrition informatics competencies across all levels of practice: A national Delphi study. *Journal of the Academy of Nutrition and Dietetics*, 2012. 112(12): 2042–2053.
2. Medlin, C. and Skinner, J., Individual dietary intake methodology: A 50-year review of progress. *Journal of the American Dietetic Association*, 1988. 88(10): 1250–1257.
3. Shriver, B.J., Roman-Shriver, C.R., and Long, J.D., Technology-based methods of dietary assessment: Recent developments and considerations for clinical practice. *Current Opinion in Clinical Nutrition and Metabolic Care*, 2010. 13(5): 1363–1950.
4. Thompson, F.E., Subar, A.F., and Baranowski, J., Need for technological innovation in dietary assessment. *Journal of the American Dietetic Association*, 2010. 110: 48–51.
5. Hongu, N., Hingle, M., Merchant, N. et al., Dietary assessment tools using mobile technology. *Topics in Clinical Nutrition*, 2011. 26(4): 300–311.
6. Illner, A.K., Freisling, H., Boeing, H. et al., Review and evaluation of innovative technologies for measuring diet in nutritional epidemiology. *International Journal of Epidemiology*, 2012. 41(4): 1187–1203.
7. Bingham, S.A., The dietary assessment of individuals; methods, accuracy, new techniques and recommendations. *Nutrition Abstracts and Reviews (Series A)*, 1987. 57: 705–742.
8. Willett, W.C., *Nutritional Epidemiology*, 2nd edn., 1998, New York: Oxford University Press.
9. Gibson, R.S., *Principles of Nutritional Assessment*. 2nd edn., 2005, New York: Oxford University Press.
10. Probst, Y.C., Faraji, S., Batterham, M. et al., Computerized dietary assessments compare well with interviewer administered diet histories for patients with type 2 diabetes mellitus in the primary healthcare setting. *Patient Education and Counseling*, July 2008. 72(1): 49–55. DOI: 10.1016/j.pec.2008.01.019.

11. Slattery, M.L., Murtaugh, M.A., Schumacher, M.C. et al., Development, implementation, and evaluation of a computerized self-administered diet history questionnaire for use in studies of American Indian and Alaskan native people. *Journal of the American Dietetic Association*, 2008. 108(1): 101–109.

12. Zoellner, J., Anderson, J., and Gould, S.M., Comparative validation of a bilingual interactive multimedia dietary assessment tool. *Journal of the American Dietetic Association*, 2005. 105(8): 1206–1214.

13. Arab, L., Wesseling-Perry, K., Jardack, P. et al., Eight self-administered 24-hour dietary recalls using the Internet are feasible in African Americans and Whites: The energetics study. *Journal of the American Dietetic Association*, 2010. 110(6): 857–864.

14. Moshfegh, A.J., Rhodes, D.G., Baer, D.J. et al., The US Department of Agriculture Automated Multiple-Pass Method reduces bias in the collection of energy intakes. *American Journal of Clinical Nutrition*, 2008. 88(2): 324–332.

15. Rebro, S.M., Patterson, R.E., Kristal, A.R. et al., The effect of keeping food records on eating patterns. *Journal of the American Dietetic Association*, 1998. 98(10): 1163–1165.

16. Vuckovic, N., Ritenbaugh, C., Taren, D.L. et al., A qualitative study of participants' experiences with dietary assessment. *Journal of the American Dietetic Association*, 2000. 100(9): 1023–1028.

17. Fyfe, C.L., Stewart, J., Murison, S.D. et al., Evaluating energy intake measurement in free-living subjects: When to record and for how long? *Public Health Nutrition*, 2010. 13(02): 172–180.

18. Ozdalga, E., Ozdalga, A., and Ahuja, N., The smartphone in medicine: A review of current and potential use among physicians and students. *Journal of Medical Internet Research*, 2012. 15(5): 128–143.

19. Klasnja, P. and Pratt, W., Healthcare in the pocket: Mapping the space of mobile-phone health interventions. *Journal of Biomedical Informatics*, 2012. 45: 184–198.

20. Ngo, J., Engelen, A., Molag, M. et al., A review of the use of information and communication technologies for dietary assessment. *British Journal of Nutrition*, 2009. 101: S102–S112.

21. Six, B.L., Schap, T.E., Zhu, F.M. et al., Evidence-based development of a mobile telephone food record. *Journal of the American Dietetic Association*, 2010. 110(1): 74–79.

22. Rollo, M.E., An innovative approach to the assessment of nutrient intake in adults with type 2 diabetes: The development, trial and evaluation of a mobile phone photo/voice dietary record. Faculty of Health, School of Exercise and Nutrition Sciences, Queensland University of Technology: Brisbane, Queensland, Australia, 2012.

23. Mariappan, A., Bosch, M., Zhu, F. et al., Personal dietary assessment using mobile devices. In *Proceedings of the IS&T/SPIE Conference on Computational Imaging VII*, San Jose, CA, 2009.

24. Woo, I., Otsmo, K., Kim, S. et al., Automatic portion estimation and visual refinement in mobile dietary assessment. In *Proceedings of the Electrical Imaging Science and Technology, Computational Image VIII*, San Jose, CA, 2010.

25. Martin, C.K., Correa, J.B., Han, H. et al., Validity of the remote food photography method (RFPM) for estimating energy and nutrient intake in near real-time. *Obesity*, 2012. 20(4): 891–899.

26. Rollo, M.E., Ash, S., Lyons-Wall, P. et al., Trial of a mobile phone method for recording dietary intake in adults with type 2 diabetes: Evaluation and implications for future applications. *Journal of Telemedicine and Telecare*, 2011. 17(6): 318–323.

27. Zhu, F., Mariappan, A., Boushey, C.J. et al., Technology-assisted dietary assessment. In *Proceedings of the IS&T/SPIE Conference on Computational Imaging VI*, San Jose, CA, 2008.

28. Doherty, A.R., Hodges, S.E., King, A.C. et al., Wearable cameras in health: The state of the art and future possibilities. *American Journal of Preventive Medicine*, 2013. 44(3): 320–323.

29. Gemming, L., Doherty, A., Kelly, P. et al., Feasibility of a SenseCam-assisted 24 h recall to reduce under-reporting of energy intake. *European Journal of Clinical Nutrition*, 2013. 67(10): 1095–1099. DOI:10.1038/ejcn.2013.156.

30. Wang, D.-H., Kogashiwa, M., Ohta, S. et al., Validity and reliability of a dietary assessment method: The application of a digital camera with a mobile phone card attachment. *Journal of Nutritional Science and Vitaminology*, 2002. 48(6): 498–504.

31. Wang, D.-H., Kogashiwa, M., and Kira, S., Development of a new instrument for evaluating individuals' dietary intakes. *Journal of the American Dietetic Association*, 2006. 106(10): 1588–1593.

32. Kikunaga, S., Tin, T., Ishibashi, G. et al., The application of a handheld personal digital assistant with camera and mobile phone card (Wellnavi) to the general population in a dietary survey. *Journal of Nutritional Science and Vitaminology*, 2007. 53(2): 109–116.

33. Martin, C.K., Han, H., Coulon, S.M. et al., A novel method to remotely measure food intake of free-living individuals in real time: The remote food photography method. *British Journal of Nutrition*, 2009. 101(3): 446–456.

34. Boushey, C.J., Kerr, D.A., Wright, J. et al., Use of technology in children's dietary assessment. *European Journal of Clinical Nutrition*, 2009. 63(Suppl 1): S50–S57.

35. Weiss, R., Stumbo, P.J., and Divakaran, A., Automatic food documentation and volume computation using digital imaging and electronic transmission. *Journal of the American Dietetic Association*, 2010. 110(1): 42–44.

36. Sun, M., Fernstrom, J.D., Jia, W. et al., A wearable electronic system for objective dietary assessment. *Journal of the American Dietetic Association*, 2010. 110(1): 45–47.

37. Jia, W., Wenyan, J., Fernstrom, J.D. et al., Imaged based estimation of food volume using circular referents in dietary assessment. *Journal of Food Engineering*, 2012. 109(1): 76–86.

38. Kong, F. and Tan, J., DietCam: Automatic dietary assessment with mobile camera phones. *Pervasive and Mobile Computing*, 2012. 8(1): 147–163.

39. Wu, W. and Yang, J., Fast food recognition from videos of eating for calorie estimation. In *IEEE International Conference on Multimedia and Expo (ICME'09)*, New York, 2009, pp. 1210–1213.

40. Chen, M., Dhingra, D., Wu, W. et al., PFID: Pittsburgh fast-food image dataset. In *IEEE International Conference on Image Processing (ICIP'09)*, Piscataway, NJ, 2009, pp. 289–292. doi: 10.1109/ICIP.2009.5413511.

41. Joutou, T. and Yanai, K., A food image recognition system with multiple kernel learning. In *IEEE International Conference on Image Processing (ICIP'09)*, Tokyo, Japan, 2009, pp. 285–288.

42. Pishva, D., Kawai, A., and Shiino, T., Shape based segmentation and color distribution analysis with application to bread recognition. In *IAPR Workshop on Machine Vision Applications (MVA'00)*, Tokyo, Japan, 2000, pp. 193–196.

43. Lowe, D., Distinctive image features from scale invariant key points. *International Journal of Computer Vision*, 2004. 60(2): 91–110.

44. Food and Agriculture Organization of the United Nations. 2012. FAO/INFOODS density database (Version 2.0). http://www.fao.org/docrep/017/ap815e/ap815e.pdf. Accessed on September 5, 2013.

45. Stumbo, P. and Weiss, R., Using database values to determine food density. *Journal of Food Composition and Analysis*, 2011. 24: 1174–1176.

46. De Vries, H. and Brug, J., Computer-tailored interventions motivating people to adopt health promoting behaviours: Introduction to a new approach. *Patient Education and Counselling*, 1999. 36: 99–105.

47. Brug, J., Campbell, M., and van Assema, P., The application and impact of computer-generated personalized nutrition education: A review of the literature. *Patient Education and Counseling*, 1999. 36: 145–156.

48. Brug, J. and van Assema, P., Differences in use and impact of computer-tailored dietary feedback according to stage of change and education. *Appetite*, 2000. 34: 285–293.

49. Whittaker, R., Dorey, E., Bramley, D. et al., A theory based video messaging mobile phone intervention for smoking cessation: A randomized controlled trial. *Journal of Medical Internet Research*, 2011. 13: e10.

50. Oenema, B., Brug, J., and Lechner, L., Web-based tailored nutrition education: Results of a randomized controlled trial. *Health Education Research*, 2001. 16(6): 647–660.

51. Brug, J., Oenema, A., and Campbell, M., Past, present, and future of computer-tailored nutrition education. *American Journal of Clinical Nutrition*, 2003. 77: 1028S–1034S.

52. Skinner, C.S., Campbell, M.K., Rimer, B.K. et al., How effective is tailored print communication? *Annals of Behavioural Medicine*, 1999. 21: 290–298.

53. Brug, J., Computerized smoking cessation program for the worksite: Treatment outcome and feasibility. *Journal of Consulting and Clinical Psychology*, 1999. 57: 619–622.

54. Kerr, D.A., Pollard, C.M., Howat, P. et al., Connecting Health and Technology (CHAT): Protocol of a randomized controlled trial to improve nutrition behaviours using mobile devices and tailored text messaging in young adults. *BMC Public Health*, June 2012. 12: 477.

55. Ojala, T., Pietikäinen, M., and Harwood, D., Distinctive image features from scale-invariant keypoints. *Pattern Recognition*, 1996. 29(1): 51–59.

56. Ahonen, T., Hadid, A., and Pietikäinen, M., Face recognition with local binary patterns. In *European Conference on Computer Vision (ECCV'04)*, Prague, Czech Republic, 2004, pp. 469–481. doi: 10.1007/978-3-540-24670-1_36.

57. Hussain, S. and Triggs, B., Feature sets and dimensionality reduction for visual object detection. In *British Machine Vision Conference (BMVC'10)*, Wales, U.K., 2010, pp. 1–10. doi: 10.5244/C.24.112.

58. Nguyen, D. Zong, Z., Ogunbona, P. et al., Object detection using non-redundant local binary patterns. In *IEEE International Conference on Image Processing (ICIP'10)*, Hong Kong, People's Republic of China, 2010, pp. 4609–4612.

59. Belongie, S., Malik, J., and Puzicha, J., Shape matching and object recognition using shape contexts. *IEEE Transactions on Pattern Analysis and Machine Intelligence*, 2002. 24(4): 509–522. doi: 10.1109/34.993558.

60. Zong, Z., Nguyen, D.T., Ogunbona, P., et al. On the combination of local texture and global structure of food classification, in IEEE International Symposium on Multimedia (ISM'10) 2010. p. 204–211.

6

Step-By-Step Guide to Designing Effective Wellness Apps

Sandy Ng and Lucia Vuong

CONTENTS

6.1 Introduction

Leveraging on the rise of consumer usage of mobile application technology, the health-care industry currently has more than 97,000 health-related apps listed on 62 full catalogue app stores (Research2guidance 2013). The projected revenue of these apps is to reach $26 billion, with 3.4 billion people would have access and 50% would have downloaded these apps by 2017 (Mobihealthnews 2013). Clearly, the health-related app market is on an upward trajectory, transforming the way we consume health-related services, and, importantly, has the potential to achieve both social and economic benefits to communities.

In this chapter, the discussion is focused on wellness apps that are preventative health focus in nature. Wellness apps are purposefully created to help consumers achieve their health behavior goal, to improve their general well-being and lifestyle. It is a cost-effective medium for the health-care industry to disseminate current public health information and services on health prevention and awareness messages and provide an avenue for health service providers to form closer relationships with their consumers (Hennig-Thurau et al. 2010).

Quite simply, prevention is better than cure. "Health is a state of complete physical, mental, and social well-being and not merely the absence of disease or infirmity" (WHO 2013). If health-care providers are able to shift consumers' mindset from a "sickness" to a "wellness" paradigm, the broad implication is a reduction in patient load on the already strained health-care systems that exist in most developed countries and improvements in general health indices (GSMA-PWC 2012).

Despite this significance, wellness apps face challenges that could limit their potential. First, the classification of wellness apps is generally poorly curated (Mobihealthnews 2012). Two of the largest app stores, iTunes and Google Play store, grossly classified all wellness apps into the single category, *Health and Fitness*. This current segmentation is broad and uninformative. Overloaded with a myriad of choices, oftentimes, it is difficult for a consumer to find the health app that can truly meet his or her health needs (van Velsen et al. 2013). App cynics may be skeptical of the ranking system, blogs, and reviews written by others, as they may be biased. Therefore, users that would want the option of choosing an app independent of other's opinions would likely experience consternation with current search options.

Second, there is doubt over whether wellness apps are effective as a preventative health intervention tool. The statistics call into question of wellness apps' effectiveness. Although wellness apps continue to increase, the download rates remain low at 10% and a high dropout rate of around 74% (iHealthBeat 2012). Academically, evidence-based studies on app effectiveness remain scant (Kumar et al. 2013) and prior studies aimed at evaluating the effectiveness of health-related apps have yielded mixed results (McCurdie et al. 2012). Collectively, these two challenges open up opportunities to improve the potential of wellness apps. To this end, this chapter provides solutions to the aforementioned challenges through a four-step guide to designing effective wellness apps.

6.2 Significance of This Chapter

This chapter is an essential read for app creators to design effective wellness apps. Much of the literature discussing the design of mobile apps for the health sector is focused on apps related to weight loss (Hebden et al. 2012), categorization of mobile health apps (Mosa et al. 2012), and evaluation of the effectiveness of clinical health applications (Murfin 2013). While prior literature adds knowledge to understanding health-related mobile apps broadly from a business technology perspective, more work is required to investigate the effectiveness of this product offering in the health industry that only has a relatively short market existence since 2009. To date, there is little research addressing the design of wellness apps specifically using a user-centered design (UCD) approach. Since wellness apps, unlike medical or clinical apps, are primarily used by the general public consumer market, a user-centered approach is necessary. Consumers will only choose apps that they feel address their needs. This chapter provides a practical guide for app creators focused on the design and ongoing maintenance of effective wellness apps.

Mobile wellness apps are calling for the coming together of various disciplines such as health practitioners, marketing, and software engineering to work together in order to address consumer demands and thus remain competitive and relevant. Wellness app creators require practical and flexible advice that can be tailored to the specific consumer needs that their apps are trying to address. To this end, the four-step process outlined in this chapter takes a multidisciplinary approach toward wellness app design.

6.3 Step-By-Step Guide to Designing Effective Wellness Apps

This guide provides a structured plan to design wellness apps that are meaningful, engaging, and effective to consumers. The first step starts with matching the value offered by the health app with the behavioral goal orientation of the target audience. Second, scan the market to identify if there are similar-type apps, and if so, determine how to differentiate your health value offering from others. In step 3, the philosophy of cocreating the app development process using the "user-centered approach" is discussed in detail as it underpins successful design. In addition, five key app elements—namely—usability, aesthetics, personalization, sociality, and trustworthiness—are presented for design consideration. Finally, step 4 proposes periodic evaluation, particularly if you are considering expanding variations of the app to other customer bases; this requires you to revisit the process from step 1. The steps are shown in Figure 6.1 and Sections 6.3.1 through 6.3.4 will discuss each step in detail.

6.3.1 Step 1: (Re)Define the Behavior Goal Orientation Offering of Wellness Apps

As with the development of any innovative service products, health organizations must understand consumer's needs in order to develop an app that fits their purpose. Consumers select wellness apps to achieve positive change in health behavior. There are 15 ways in which human behaviors can change as shown in the Fogg Behavior Grid (FBG) in Figure 6.2 (Fogg 2009).

App creators must determine the app value offering based on the type of behavior goal it can assist consumers to achieve. For instance, it is not sufficient to simply state the

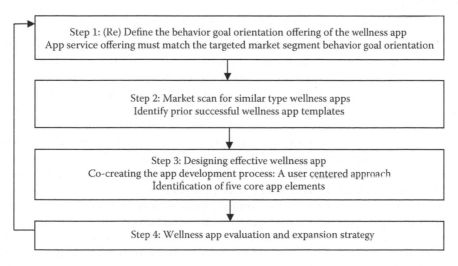

FIGURE 6.1
Wellness app design steps.

	1 Do new behavior	2 Do familiar behavior	3 Increase behavior intensity	4 Decrease behavior intensity	5 Stop existing behavior
Dot one time	Dot 1 *Do a new behavior one time*	Dot 2 *Do familiar behavior one time*	Dot 3 *Increase behavior one time*	Dot 4 *Decrease behavior one time*	Dot 5 *Stop behavior one time*
Span period of time	Span 1 *Do behavior for a period of time*	Span 2 *Maintain behavior for a period of time*	Span 3 *Increase behavior for a period of time*	Span 4 *Decrease behavior for a period of time*	Span 5 *Stop behavior for a period of time*
Path from now on	Path 1 *Do new behavior from now on*	Path 2 *Maintain behavior from now on*	Path 3 *Increase behavior from now on*	Path 4 *Decrease behavior from now on*	Path 5 *Stop behavior from now on*

FIGURE 6.2
Fogg behavioral grid.

functionalities of the app. It is more meaningful for app creators to communicate to its users, for example, how a cardio app can help users achieve their behavioral goal. Consider this. Consumers who wish to select a cardio fitness dominant app may differ in their behavioral goal orientation. Some of us will want to have a cardio app that gets us doing cardio exercises daily and at a regular basis (green path) (e.g., http://dailyworkoutapps.com/dailyworkoutapps/Daily_Cardio_Workout.html), whereas others will want a cardio app that helps them in marathon training (purple span) (e.g., Hal Higdon Marathon Training Program—Novice 1 [http://www.halhigdon.com/]). Therefore, it is crucial to clearly define and communicate the behavior change goal that a specific health app can offer in order to match it with a receptive audience. In Case Study 6.1, an analytical method is applied to cardio apps as a context, to demonstrate that it is possible to further segment the apps based on their functionalities and behavioral goal orientation. This case study highlights the aforementioned point that at present, the existing category that encapsulates all types of wellness apps, health and fitness, is poorly curated. It is recommended that wellness apps should communicate clearly its single or multifunctionalities and importantly the behavioral goal orientation in order to provide a useful, informed product to consumers.

Case Study 6.1: System to "Unpack" Cardio Apps' Behavioral Goal Orientation

Dan's thought bubble

All I want is to find an app that suits what I need! I have read the reviews and look up blog sites and so on … followed their recommendations and downloaded the few apps only to delete it because they are not quite what I am after. I cannot believe it is so labor intensive! This is ridiculous. Why can't I just simply download an app that I WANT? It has to be out there … Arrrghh" A frustrated Dan speaks out loud.

Can you relate to Dan's frustration in searching for an app that suits you? With millions of apps in the marketplace, it is not surprising that most users will defer to reviews (basically an indication of app popularity) that guide their app download decisions. However, what happens when none of the reviewed apps suit the health behavioral goal that you want to achieve? Must the search for the app that suits you be labor intensive?

Using FBG (Fogg 2009), there are 15 ways in which behavior can change. A preliminary study was conducted to demonstrate if it is possible to group the health apps based on their behavioral goal offering. To begin, a small pool of cardio-only fitness apps were identified from Apple's iTunes Store (see Figure 6.3).

To ensure that the identified 80 apps are pure cardio apps only, the researcher examined the description of the apps' functionalities and, after further analysis, determined that only 30 apps are applicable (see Figure 6.4).

Following, using judgmental forecasting, three expert opinions were gathered. Their tasks were to read the description of the 30 cardio fitness apps and to classify the behavioral goal orientations of these apps to the FBG (Fogg 2009). From this analysis, these apps fit into four distinct behavioral goal orientations as shown in Table 6.1.

This analysis demonstrates two points. First, the current grouping of apps in app stores' "health and fitness" category is uninformative and difficult for consumers to easily navigate through the entire list to find the app they want. Second, detailed analysis shows that even within a subcategory, for example, like cardio apps can be grouped more precisely (as shown in Figure 6.2) in accordance with their functionalities. Importantly, the analysis shows that within cardio-only apps, there are four behavioral goals that are offered to consumers. Hence, it is recommended that app

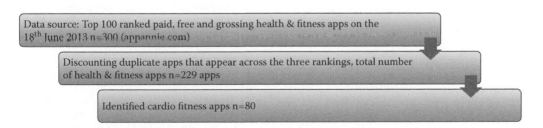

FIGURE 6.3
Cardio fitness apps.

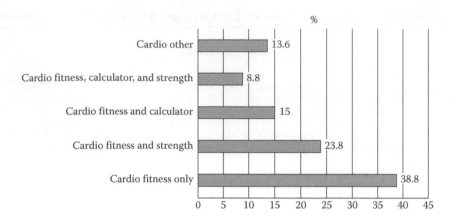

FIGURE 6.4
Cardio fitness apps function breakdown.

TABLE 6.1

Cardio Fitness Dominant App Behavioral Goal Orientation Typology

Behavioral Goal Orientation	User Goal	App Names
Blue path – to maintain behavior from now on	Users download these apps to maintain their running habits	WalkJogRun GPS Running Routes Cruise control: Run Zombies, Run! 5k Training Zombies, Run!
Green path – do new behavior from now on	Users download these apps to get them into the habit of running	Daily cardio workout free Daily cardio workout
Purple path – increase behavior from now on	Users download these apps to help them increase their running intensity from now on	Beep test Beep test trainer Couch to 5k Nike+ Running Moves Beep Test Solo 5K Runner: 0–5k to 10k run training, free Beep test free 10k runner: 0–5k to 10k run training Ease into 5k Cyclemeter GPS: Track cycling running and mountain biking Walkmeter GPS Track walking hiking running cycling Strava Cycling Strava Run: GPS running, training, and cycling workout tracker Get running (Couch to 5k) Pedometer pro runner Team beep test 5k runner: 0–5k run training beep fitness test
Purple span – increase behavior for a period of time	Users download these apps to help them train specifically for a running event within a specific timeframe	Hal higdon marathon training program – novice 1 10k trainer free: run for PINK—couch to 10k 10k trainer pro: run for PINK—couch to 10k C25k: 5k Trainer pro C2K: 5k Trainer free

creators need to be strategic in describing the purpose of their app, going beyond stating the features of the apps, to clearly state the functionalities and the specific health behavior goal that the app will assist the users to achieve.

Once the behavior goal orientation of the app is matched with the target audience, the features of the app should then be scrutinized. In particular, app creators should design appropriate triggers, for example, timely reminder alerts to prompt a user to act (Fogg 2009), to prompt users to perform a healthy behavior in order to achieve their goal. To this end, the next step required is to identify the app functionalities that have been proven successful in triggering behaviors.

6.3.2 Step 2: Gather Market Intelligence to Identify Successful Wellness App Functionalities

Conduct an environmental scan and collate the best three to five wellness apps currently in the market that offer a similar behavior change goal as your app. Pay attention to the market feedback of these selected apps (e.g., ranking, qualitative comments, blogs, forums) and then compile a checklist of the functionalities that consumers like about these apps, features that can be improved and wish list features. By doing so, this provides a basic working template.

6.3.3 Step 3: Designing Effective Wellness Apps

Undermining a health app intervention's effectiveness is commonly reported when consumers do not continue using the app if it is nonengaging (McCurdie et al. 2012). Apps that are effective are those that consumers are engaged with and deliver a solution to a problem. A UCD process, as shown in Figure 6.5, is effective in achieving user engagement to improve app health intervention's effectiveness (McCurdie et al. 2012).

To create a valuable app, creators need to cocreate the design process with consumers. In fact, the World Health Organization advises that user insights and evaluation should be incorporated into any mHealth project life cycle to ensure effective outcomes (WHO 2011). Therefore, the cocreation of the health app is important to ensure that the app is an effective intervention. A recent study found evidence that the involvement of a target group in the development of a mental health app proved beneficial because it went beyond the initial scope of the pilot study and, as a result, has greatly improved the app capability and design (Pelletier et al. 2013).

The cocreation of the app also allows health businesses that build the app to adapt and change their service offering accordingly in order to offer competitive value propositions to their users (e.g., Lusch et al. 2010). Given that cocreating the app requires all parties involved in the production and consumption process to collaborate closely in a network in order to determine the value propositions, mutually beneficial relationships have a greater chance of being formed among all stakeholders (e.g., Vargo 2009).

From the extant literature, there are five key elements that are important to consumers in their evaluation of apps. These are (1) the usability of the app, (2) the aesthetic value, (3) the degree of customization the app provides, (4) app social functions such as links to Facebook and Twitter, and (5) trustworthiness of the app (see Figure 6.6).

Broadly speaking, *usability* can be defined as the ease of use and perceived usefulness of the app (McMahon et al. 2012). Usability perception is affected by the efficacy of the user, which is the ability of the user in trying to accomplish a goal with the application, and the

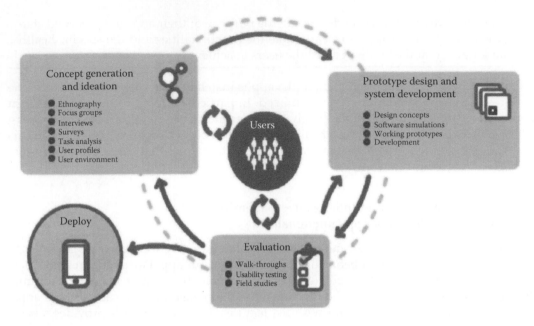

FIGURE 6.5
The user-centered design process.

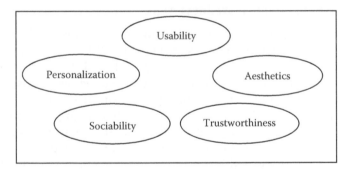

FIGURE 6.6
The five app elements.

context in which the user uses the application (Harrison et al. 2013). Therefore, app design needs to consider user ability and motivation level (low to high) toward performing the task (Fogg 2009), ensuring that the efficacy of the user and the difficulty of the task that the app user needs to perform are matched. Some aspects that enhance the usability of the app are tracking capability and reward features.

Consumers reported that in tracking capability that allows users to track his or her historical performance, past searches are important as they are a source of tangible evidence that reflect how they perform with their change to healthy behavior over time. In addition, they want a "reward" feature to support and motivate users to continue their efforts toward achieving their health goals. However, the "reward" feature must not be excessive or exaggerated (e.g., over the top) relative to the goals that the user achieved. Or else, it will be perceived by users as patronizing or, worse, silly.

App creators need to design appropriate triggers—app notifications—to ensure that users will use their app in the right context (Fogg 2009). App notifications, at present, are repetitive, irrelevant (after some time), and, at times, annoying. To this extent, users will turn off notifications, which defeats the purpose of app creators building this feature to begin with! From the users' perspective, app notifications must take into consideration the context. In other words, the notifications have to be relevant to each individual context; provide interesting, nonrepetitive information; and also, allow users to have the ability to customize their own reminder messages or sound badges as part of the notification feature. As such, users will then be more engaged with using the app. With the abundance of information that an app could potentially offer to users, it is important to provide content, which is also *customisable*. By doing so, this will increase the relevance of the information to consumers, especially if the content is personalized toward an individual's characteristics and context (van Velsen et al. 2013).

Aesthetically, app design can be further improved as the majority of the current apps are simplistic, highly derivative, or not overly compelling from graphical, creative, or usability points of view (Falchuk 2009). A visually appealing app is designed to be interactive in nature, with the core aim to improve the representation and understanding of wellness information (Falchuk 2009). An example is the gamification feature of mobile apps, which is popular and can work well in wellness apps for certain target segments. For example, Zombie, Run! (a popular running app) clearly recognizes the potential of gamification and builds its running app that differentiates from other running apps and, as a result, yields great success and popularity among its users.

Increasingly, many apps incorporate social network functionalities as they allow users to exchange personal experiences and discuss their health issues with others, for example, Nike + Running and My Fitness Pal (Gay and Leijdekkers 2011). However, these must be executed with care. Gay and Leijdekkers (2011) found that there are positive and negative aspects of incorporating social functionalities to apps. Incorporating social network functionality in the app allows users a channel to share their health data and communicate with their peers and draws on health communities for support. Indeed, a recent survey conducted by Infosys suggests that more users are ready to share their personal health information online, with the proviso it is with trusted sources (Foo 2013). However, the threats that users face is that shared health data may be compromised as the security and privacy of the information may not be guaranteed. Further, the health information received from community may include incorrect information from a potentially untrustworthy source and hence be hazardous (Gay and Leijekkers 2011). This highlights the importance of trust and consumer confidence in an app.

Consumer trust is a salient concept that bonds the consumer–health provider relationship and it has a significant impact on consumer decision-making (Chang et al. 2013). Trustworthiness of the app is crucial and needs to be carefully considered in the app design. Research has shown that trustworthiness of an app must demonstrate competency in providing good information, has good intent toward the user, and is reliable and honest and the user is able to access information whenever required (Aker et al. 2011). Hence, users would like to have evidence-based information.

The Food and Drug Administration (FDA) of the U.S. Department of Health and Human Services recognizes that mobile health apps are gaining in popularity and that they can help people manage, prevent, and treat health conditions. Mobile medical applications are defined as "a mobile app that meets the definition of device in section 201(h) of the Federal Food, Drug, and Cosmetic Act (FD&C Act); and either is intended:

- To be used as an accessory to a regulated medical device; or
- To transform a mobile platform into a regulated medical device."

While the FDA has developed a set of standards to regulate high-risk mobile medical applications, currently, the FDA does not regulate wellness apps as they do not fall into this category of mobile medical application. The FDA "intends to apply its regulatory oversight to only those mobile apps that are medical devices and whose functionality could pose a risk to a patient's safety if the mobile app were to not function as intended" (p. 4). The majority of wellness apps would fall into this latter category and therefore are not regulated.

Of interest are some of the FDA's regulatory requirements around establishment registration, medical device listing, labeling requirements, premarket submission for approval or clearance, and reporting. If wellness apps were required to undergo a similar process before being made available to the public, this could increase consumer trust in wellness apps as well as ensure that apps provide the appropriate, regulated health advice to end users. Evaluation and regulation of wellness apps could also help app creators understand the effectiveness of their products in helping users achieve their desired health goals. In our opinion, there should be a regulatory body that governs information on wellness apps too.

6.3.4 Step 4: Evaluation and Expansion Strategy

It is important that once a health app is designed using the UCD process and reaches the market place, it needs to be periodically evaluated. Case Study 6.2 shows a wellness app evaluation—SunSmart app, created by Cancer Council Victoria, SunSmart team.

Case Study 6.2:   Do I Know What Consumers Want from My App? The SunSmart App Assessment

---

Health apps are useful resource to consumers, empowering them to make informed decisions and choices about their health care. From a public health perspective, wellness apps offer an important value proposition—potentially reducing the "sickness" load on the health-care system. For example, Australia has one of the highest skin cancer mortality rates in the world (SunSmart 2013). Yet it is the most preventable form of cancer, with 95% of skin cancers being successfully treated if detected early (SunSmart 2013). In a bid to lower the number of public from getting skin cancer, the SunSmart team at Cancer Council Victoria, Melbourne, Australia, launched the SunSmart app in November 2010, with the purpose of educating the public to be UV savvy at all times (for more information about the app, see http://www.sunsmart.com.au/tools/interactive-tools/free-sunsmart-app; SunSmart 2014).

As the promotion of this app progressed, the download rate steadily increased, suggesting that the app was gaining in popularity. However, increased download rates of an app do not naturally indicate that an app is useful to consumers. This is because many consumers may download an app out of curiosity, only to abandon it shortly after they have experienced using it, having deemed it as "useless." Indeed, the dropout rate of mHealth apps is high (iHealthBeat 2012).

Given that the usefulness of the SunSmart app requires user cooperation to engage with the app frequently and continually, it is necessary to evaluate the SunSmart app. Data were collected over a 6-week period between November and December 2011. After cleansing the data, a total of 368 respondents filled out the survey and each state in the country was represented, with the majority of the survey population from Victoria, New South Wales, and Queensland—the three most densely populated states within Australia. The majority of the respondents were female aged between 25 and 44 years old, with university education and above, and typically earns $50,000–$69,999. The sample characteristics are interesting, given that the SunSmart app was designed with a broad appeal in mind.

The majority of the users reported that they use the SunSmart app either on a daily basis or weekly, 78% of the respondents are happy to tell others about it, and 90% of the respondents believe that the app is important to them. This suggests that users generally find that the app offers a good value proposition. As displayed in Figure 6.7, users are generally satisfied, believe that the app provides good value, and have good intentions to continue using the app. Users were asked to rate 1 = poor and 5 = very good.

Based on prior research/reports, nine app features were considered to assess app user experience. As displayed in Figure 6.7, users rated the nine features positively. However, the results suggest that the aesthetic value, enjoyment factor, and the customizability of the SunSmart app can be further enhanced. This result provides the SunSmart team with specific information on enhancing specific aspects of the app, in an effort to improve its overall interface.

Naturally, as users become more experienced, customization will be necessary to cater for users with different levels of experience (e.g., teenagers). For an app to remain relevant, periodic feedback from users is necessary to ensure that the app's value proposition is in sync with consumer needs. For future direction, cocreating the app development process with users will result in the creation of better apps and thus lower user dropout rate.

Future technological directions of health apps
There is mounting public concerns that a substantial number of wellness apps do not adhere to evidence-based health information and recommendations (Breton et al. 2011). It is potentially hazardous for consumers to follow health information that is not

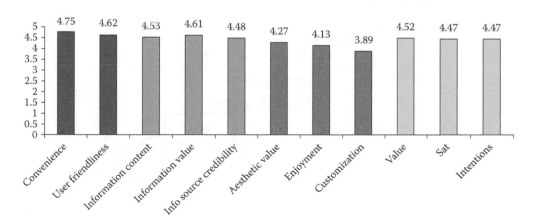

FIGURE 6.7
SunSmart App Features and Evaluation.

evidence-based. Therefore, policies must soon emerge that enforce stringent industry standards that app creators have to comply with, to ensure that the app content is evidence-informed (Breton et al. 2011). From a technological perspective, this means that creators need to ensure that health information and instruction comply with health practices and guidelines in order to minimize confusion among users.

Another issue that app creators must overcome is to reduce the high consumer dropout rate (at 74%) at wellness app usage after 1 month (iHealthBeat 2012). To overcome this, app creators recognized that it is important to include consumers as part of the development process—collaboration between end users—and yet this approach is not practiced consistently (Hoda et al. 2011). From an app development perspective, broadly speaking, most organizations adopt one of two approaches toward software development: waterfall or agile. Waterfall methodologies involve a series of sequential implementation steps from requirements analysis through to implementation and delivery and rely on requirements and scope being thoroughly defined and fixed from the beginning of a project. Dr. Winston Royce (1970), often considered the founder of the waterfall methodology, identifies the following steps as necessary for the management of the development of large software systems: the steps are as shown in Figure 6.8.

Figure 6.8 demonstrates why this approach is called a "waterfall" method, where each phase is self-contained and the outputs from a preceding step become the input to the following step. Often, business units work in isolation from one another, with *minimal feedback* from one phase to another and involving only the transfer of a small set of artifacts such as documents, models, or code (Kruchten 2004). In his article, Dr. Winston Royce (1970) discusses the limitations of this method. This process is described as risky and costly. In spite of this, many organizations around the world have adopted a waterfall approach toward managing software development projects. A key limitation of this approach is that cocreation of wellness apps with consumers is prohibiting.

Agile methodologies, in contrast, promote iterative development where the goal is to deliver small sets of working and valuable functionality iteratively. Scope and requirements are flexible, while testing occurs alongside development. This allows the project

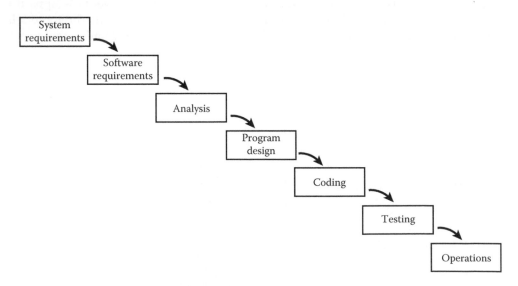

FIGURE 6.8
Winston Royce's waterfall methodology (1970).

TABLE 6.2

Comparison of Waterfall and Agile Methodologies

Waterfall	Agile
Requirements are fixed and thoroughly defined from the beginning of the project.	Requirements may change throughout the development process.
Scope is fixed and delivered in large pieces of work.	Scope is flexible and adjusted depending on feedback from different business units. Focus is to deliver small packets of valuable and usable software.
Projects can take a long time to be delivered.	Projects can be delivered in smaller increments faster to the end user.
Testing occurs when development is complete. Results are used for validation of preestablished use cases.	Testing occurs throughout the development process and provides feedback to both developers and product owners about the value of the end product.
Business units work in isolation from one another and the outputs of one phase become the input to the next.	Business units work collaboratively and concurrently to produce smaller deliverables iteratively.

to be flexible to changing requirements and unexpected use cases found during testing. Agile methodologies also promote collaborative work processes where communication between different business units is key to the success of the project (http://agilemanifesto.org/) (Manifesto for Agile Software Development 2014). This also allows the possibility of cocreating the app with end users. Table 6.2 summarizes the key differences between the waterfall and agile methodologies.

The agile methodology provides app creators with a process that allows for continuous improvement based on user feedback and changing requirements. It also allows app creators to deliver smaller, working components of apps faster to users and assess the value of particular features and functionalities in the market. In short, agile allows app creators to test small feature sets and adjust the course of a project accordingly, while waterfall can, in contrast, result in costly apps that are of little value to users. Hence, we think that the agile methodology (Agile Methodology 2013; visit http://www.agilemethdology.org), which echoes the importance of the cocreation philosophy in app development, is a better approach. Through the agile methodology, app creators will be able to fine-tune the five key app elements stated in this chapter—usability, aesthetics, personalization, sociability, and trustworthiness—with respective stakeholders as stated in this chapter.

Third, if we feel as though we have an app for almost anything, we probably do. The downside is consumers feel overwhelmed and overloaded with the number of apps they have on their phones or tablets (van Velsen et al. 2013). Besides, health information and features are often fragmented over too many apps; it limits usefulness (van Velsen et al. 2013). A future direction that will benefit both consumers and app creators is the development of an integrated app, which provides a gateway for consumers to get the holistic health and/or wellness information required as opposed to downloading multiple apps. Imagine being able to download an integrated wellness app that has your yoga, cardio, weights, and pilates exercises all in one!

Health organizations that have complementary health service offerings and are willing to combine their resources to develop an integrated app will find the step-by-step guide presented in this chapter a good instrument to map out their action plan. Importantly, the development of integrated apps has a number of benefits. They are as follows: (1) The integrated app will be a product that is more appealing compared to a stand-alone app as it is able to help consumers achieve multiple health goals; (2) it will offer a stronger value proposition to consumers and thus differentiate the integrated app from competition; and (3) collaborative health organizations will be better resourced and thus stand a strong chance at designing an engaging app that is also effective in health intervention. From a technological perspective, this means that app creators need to work together and be more cost-efficient and join the open-source movement (van Velsen et al. 2013) to develop an integrated app that will provide relevant high-quality health information to potentially a bigger group of customers. Collectively, these future technological directions points toward the need for app creators to use a collaborative and relational approach to design meaningful, engaging, and effective health apps.

Notwithstanding, as a caveat, there are also benefits for wellness apps that are single focus with their offering. An example being the current popular running app—Zombies, Run! Obviously, the benefit of single-goal apps is obviously to help users focus on what they need to achieve. Hence, whether the future of wellness apps will remain largely single focus or move toward integrated apps will be dictated mainly by consumer demand and needs. App developers, as much as it is required of them to follow the principles of the step-by-step guide to designing effective wellness apps, also need to pay attention to consumer demand and match the wellness app core offering (whether single focus or integrated) to their respective target segments. It is only then that the well-designed wellness app will truly be effective in persuading the appropriate target segments to make a positive healthy change to their lifestyle.

It is easy to fall into the complacency trap, thinking that all the grinding hard work is done and you have designed a great app for your target audience. Just as seasons change, consumer needs will evolve and change. Ongoing evaluation of user feedback can provide ideas for new features as well as insight into changing user goals. Iacob and Harrison (2013) found that 23.3% of user reviews represented feature requests, highlighting the amount of data available for app creators to use to improve and adjust app functionalities to better suit user needs. User feedback can also be used as a gauge to measure the effectiveness of a wellness app.

Changing technologies and new hardware also provide app creators with new opportunities to help users reach their health goals. Wearable technology (such as Google Glass, Fitbit, Pebble) and other hardware innovations must be continually assessed to ensure that app creators do not fall behind fellow competitors and user demand. In addition to this, user health goals will also change. For example, a user may download a wellness app to start a new healthy behavior such as running a marathon for the first time (GreenDot); this may, down the track, become a goal to maintain the ability to run a marathon (thus becoming BluePath). Thus, it is necessary for app creators to continue the conversation with their target audience to measure the effectiveness of their app, to evolve the app as new technologies and hardware become available, and, most importantly, to continue to remain relevant to their target audiences' changing health goals.

6.4 Conclusions

This chapter opens with the premise that wellness apps will continue to increase and have the potential to transform the health-care landscape globally. Wellness apps enable us to make good decisions with regard to our own health-care regime. Despite the significance, statistics showed that it has limited effectiveness. The purpose of this chapter is to highlight more research needs to focus on understanding the relationship between wellness apps and users. After all, it is the end user that these apps are designed for, and as such, this chapter, contributed from the perspective of an academic and a practitioner, aims to bridge the gap toward improving user design of wellness apps, to increase its effectiveness in behavior change.

This chapter forwards a four-step guide to designing effective wellness apps. Underpinning the success factor of each respective step is the need for research and market intelligence. Without proper research and intelligence to support each step, the full potential of this guide will not be realized. The first step is to (re)define the behavior goal orientation offering of wellness apps. App creators need to be specific with regard to the purpose of the app (value offering). Specifically, Fogg (2009) forwards 15 types of behaviors that users can seek to modify/change. For a wellness app creator, this means prior to developing the app, which behavior(s) that the app is seeking to persuade the users to change must be identified and matched with its target segment. Failure to do so will limit the effectiveness of the app as users may find the app unappealing to them and will result in high dropout rates. To ensure that app developers understand the importance of this step, Case Study 6.1 is used, which exemplifies this step.

Step 2 requires app creators to do research and gather market intelligence to identify successful wellness app functionalities. The purpose is to provide a good template for app creators to improve on once the behavior goal (value) offering of the app is decided. The template serves as a guide prior to the app building phase. Step 3 provides guidance on designing effective wellness apps. Based on prior literature and current research, we have provided five elements for you to reflect in your design of wellness apps. However, it is not an exhaustive list. As mentioned earlier, more studies are required in this area.

Finally, step 4 highlights the need for periodic evaluation of the apps (e.g., Case Study 6.2) and also to consider, if it is possible to build on the success of your initial app, and expand the brand to other wellness areas. However, app creators need to weigh up if it is wise to expand its brand to other wellness areas. Potential trappings of doing so are that resources might be spread thinly to manage all wellness apps under the same brand and, as such, the other wellness apps are not as well curated as the original.

In conclusion, the authors have put forward a four-step approach toward designing effective wellness apps. We have stated the purpose, limitations, and results of each step. We argue that for wellness apps to be effective (and not a fad), its design needs to be user-centric. To our knowledge, this chapter is one of the first that offers a practical guide, which is underpinned by research and literature that assist app creators in their app design thinking to focus on the improvement of end user and wellness app relationship.

Questions

1. Why are wellness apps important to the society?
2. What are the market challenges of wellness apps?
3. Detail and explain the four-step guide to designing effective wellness apps.
4. Explain the FBG model in detail. Explain why it is necessary to consider segmenting wellness apps using this model.
5. Why is the user-centered approach important in the design of wellness apps?

Acknowledgments

The authors acknowledge the School of Economics, Finance and Marketing, RMIT University, and SunSmart, Cancer Council Victoria, for providing funding and in-kind support to conduct the SunSmart app evaluation mentioned in Case Study 6.2 and thanks to Mr. David Fitzgerald from Ecomarketing Group for providing editorial support.

References

Akter, S., J. D'Ambra, and P. Ray, Trustworthiness in mHealth information services: An assessment of a hierarchical model with mediating and moderating effects using partial least squares (PLS). *Journal of the American Society for Information Science and Technology* 62(1) (2011): 100–116.

Ben, F., Visual and interaction design themes in mobile healthcare, In *First International Workshop on Ubiquitous Mobile Healthcare Applications: Proceedings of the Mobile Healthcare Conference*, November 10, 2009, Toronto, Ontario, Canada.

Breton, E.R., B.F. Fuemmeler, and L.C. Abroms, Weight loss—There is an app for that! but does it adhere to evidence-informed practices? *Translational Behavioral Medicine* 1(4) (2011): 523–529.

Chang, C.-S., S.-Y. Chen, and Y.-T. Lan, Service quality, trust and patient satisfaction in interpersonal-based medical service encounters. *BMC Health Services Research* 13 (2013): 22–33.

Fogg, B.J., BJ Fogg's behavior model, Stanford University, Stanford, CA, http://www.behaviormodel.org/. (Accessed October 10, 2012).

Fogg, B.J., The behavior grid, Persuasive Technology Lab @ Stanford University, Stanford, CA, http://www.behaviorgrid.org/. (Accessed October 10, 2012).

Fran, F., Apps user ready to share medical history. *The Australian* (Australia), 1st edition, vol. 4, July 3, 2013. Retrieved from http://www.theaustralian.com.au/technology/apps-user-ready-to-share-medical-history/story-e6frgakx-1226673071392?nk=86c3b0c15ccb7176c22128f9183c5612.

Harrison, R., D. Flood, and D. Duce, Usability of mobile application: Literature review and rationale for a new usability model. *Journal of Interaction Science* 1(1) (2013): 1–16.

Hebden, L., A. Cook, H.P. van der Ploeg, and M. Allman-Farinelli, Development of smartphone applications for nutrition and physical activity behavior change. *JMIR Research Protocols* 1(2) (2012;): e9. URL: http://www.researchprotocols.org/2012/2/e9/. (Accessed January 29, 2014).

Hennig-Thurau, T., E.C. Malthouse, C. Friege, S. Gensler, L. Lobschat, A. Rangaswamy, and B. Skiera, The impact of new media on customer relationships. *Journal of Service Research* 13(3) (2010): 311–330.

Hoda, R., J. Noble, and S. Marshall, The impact of inadequate customer collaboration on self-organizing agile teams. *Information and Software Technology* 53(5) (2011): 521–534.

iHealthBeat, Number of health apps rising, but download rates remain low, iHealthBeat, http://www.ihealthbeat.org/articles/2012/7/17/number-of-health-apps-rising-but-download-rates-remain-low.aspx. (Accessed 10 October, 2012).

Jonah, C., 1.7B to download health apps by 2017, Mobihealthnews, http://mobihealthnews.com/20814/report-1-7b-to-download-health-apps-by-2017/. (Accessed August 20, 2013).

Kruchten, P., Going over the waterfall with the rup. *DeveloperWorks*, 26 April 2004, IBM Corporation, http://www.ibm.com/developerworks/rational/library/4626.html. (Accessed January 28, 2014).

Kumar, S., W.J. Nilsen, A. Abernethy, A. Atienza, K. Patrick, M. Pavel, W.T. Riley et al., Mobile health technology evaluation: The mHealth evidence workshop. *American Journal of Preventative Medicine* 45(2) (2013): 228–236.

Lusch, R.F., S.L. Vargo, and M. Tanniru, Service, value networks and learning. *Journal of the Academy of Marketing Science* 38 (2013): 19–31.

Manifesto for Agile Software Development, *Agile Methodology*, Manifesto for Agile Software Development 2014. http://agilemanifesto.org/. (Accessed July 27, 2014).

McCurdie, T., S. Taneva, M. Casselman, M. Yeung, C. McDaniel, W. Ho, and J. Cafazzo, mHealth consumer apps: The case for user-centered design. *Biomedical Instrumentation & Technology* 46(2) (2012): 49–56.

McMahon, S., M. Vankipuram, E.B. Hekler, and J. Fleury, Design and evaluation of theory-informed technology to augment a wellness motivation intervention. *Translational Behavioral Medicine* 4 (2013): 1–13.

MobiHealthNews, Consumer health apps for Apple's Iphone, http://mobihealthnews.com/research/consumer-health-apps-for-apples-iphone/. (Accessed October 10, 2012).

Mosa, A.S.M., I. Yoo, and L. Sheets, A systematic review of healthcare applications for smartphones. *BMC Medical Informatics and Decision Making* 12 (2012): 67.

Murfin, M., Know your apps: An evidence-based approach to evaluation of mobile clinical applications. *The Journal of Physician Assistant Education* 24(3) (2013): 38–40.

Pelletier, J.-F., M. Rowe, N. Francios, J. Bordeleau, and S. Lupien, No personalization without participation: On the active contribution of psychiatric patients on the development of a mobile application for mental health. *BMC Medical Informatics and Decision Making* 13(1) (2013): 78–85.

Price Waterhouse Coopers. Emerging mHealth paths for growth, http://www.pwc.com/en_GX/gx/healthcare/mhealth/assets/pwc-emerging-mhealth-full.pdf. (Accessed August, 20, 2013).

Research2guidance, Mobile health market report 2013–2017: The commercialization of mhealth applications (Vol. 3), Research2guidance 2013, http://www.research2guidance.com/shop/index.php/downloadable/download/sample/sample_id/262/. (Accessed August 20, 2013).

Royce, W.W., Managing the development of large software systems. *Proceedings, IEEE WESCON*, Los Angeles, CA, August 1970, pp. 1–9.

SunSmart 2013, http://www.sunsmart.com.au/skin_cancer. (Accessed September 5, 2013).

SunSmart 2014, http://www.sunsmart.com.au/tools/interactivetools/free-sunsmart-app. (Accessed 27 July, 2014).

U.S. Food and Drug Administration, Mobile medical applications, http://www.fda.gov/medicaldevices/productsandmedicalprocedures/connectedhealth/mobilemedicalapplications/default.htm. (Accessed January, 28, 2014).

Valerie, G. and P. Leijdekkers, The good, the bad and the ugly about social networks for health apps, In *2011 IFIP Ninth International Conference on Embedded and Ubiquitous Computing: Proceedings of IEEE*, Melbourne, Victoria, Australia, October 24, 2011.

Van Velsen, L., D.J. Beaujean, and J.E. van Gemert-Pijnen, Why mobile health app overload drives us crazy, and how to restore the sanity. *BMC Medical Informatics and Decision Making* 13(1) (2013): 23–27.

Vargo, S.L., Toward a transcending conceptualization of relationship: A service-dominant logic perspective. *Journal of Business and Industrial Marketing* 24(5/6) (2009): 373–379.

Wirth, A., Enabling mHealth while assuring compliance: Reliable and secure information access in a mobile world. *Biomedical Instrumentation & Technology* 46 (2012): 91–96.

World Health Organization, *mHealth: New Horizons for Health Through Mobile Technologies*. Geneva, Switzerland: WHO, 2011.

World Health Organization, WHO definition of health, http://www.who.int/about/definition/en/print.html. (Accessed August 20, 2013).

7

Collaborative Health-Care System (COHESY) Model

Vladimir Trajkovik, Saso Koceski, Elena Vlahu-Gjorgievska, and Igor Kulev

CONTENTS

7.1 Introduction

Recent trends in health-care support systems are focused on developing patient-centric pervasive environments and the use of mobile devices and technologies in medical monitoring and health-care systems (Ballegaard et al. 2008). This is of particular importance for patients with chronic diseases who require 24 h medical care.

The COllaborative HEalth-care SYstem (COHESY) model, presented in this chapter, allows monitoring of users' health parameters and their physical activities. The proposed model uses a social network that allows communication between users with the same or similar conditions and exchange of their experiences. This model creates an opportunity for increasing users' health care within their homes—24 h medical monitoring on one hand and increased capacity of health institutions on the other hand—resulting in reduction of overall costs for consumers and health-care institutions (Zimmerman and Chang 2008). The proposed model helps its users to actively participate in their health care and prevention, thereby providing an active life in accordance with their daily responsibilities at work and with family and friends.

The COHESY model gives a new dimension in the usage of novel technologies in health care. This system uses mobile, web, and broadband technologies, so the citizens have ubiquity of support services wherever they may be, rather than becoming bound to their homes or health centers (Khan et al. 2009). The most important benefits of COHESY are the possibility for patient notification in different scenarios, transmissions of the collected biosignals (health parameters such as blood pressure, heart rate, blood-sugar level) automatically to medical personnel, and increased flexibility in collecting medical data. The usage of the social network and its recommendation algorithm are the main components and advantages of COHESY that differentiate it from other health-care systems. These components provide a new perspective in the use of information technologies in pervasive health care and make this system model more accessible to users. COHESY bridges the gap between users, clinical staff, and medical facilities, strengthening the trust between them and providing relevant data from a larger group of users, grouped on the basis of various indicators.

7.2 Related Work

The rapid spread of information and communication technology (computers and information systems) led to a significant transformation of public health and health care, changing the organization rules of health care, health requirements, and services.

The use of information and communication technologies in the health sector is often realized by the implementation of complex information systems. Health-care information systems contribute to the improvement of the health quality and health care through monitoring errors in different phases of patient's care and treatment (Fichman et al. 2011). Health-care information systems in combination with the Internet are expected to contribute to reduce the incidence of medical errors, promote price transparency and performance, improve quality of care by strengthening the relationship between patient and medical worker, and encourage patients to focus on their own health care as well as to enable users to participate in the transformation of the health system.

There are different divisions of information systems used in health care. Blaya et al. (2010) carried out the division of health-care information systems in eight categories: monitoring, evaluation, and patient tracking system; clinical decision support system; patient reminder system; research/data collection system; electronic health record; patient registration or scheduling system; laboratory information management system; and pharmacy information system. In this categorization, the health-care social networks, which today are often used, are not categorized (Chorbev et al. 2011, Fernandez-Luque 2009, Kargioti et al. 2010). Also, it must be noted that a health-care information system cannot always be classified in one category. Namely, in order to provide greater functionality, the structure of health-care information systems becomes more complex. Thus, some components of health information systems may belong to a subcategory, while the whole system can belong to two or more of the aforementioned categories.

Hereinafter, some models of health-care information systems reported in the literature will be discussed and compared to the proposed COHESY model.

The MobiHealth project (van Halteren et al. 2004) has developed a system for monitoring users' health preferences, based on body sensor networks. This project also develops a platform for mobile health-care service (mHealth) based on the use of new-generation wireless networks and mobile devices. MobiHealth allows inclusion of various medical sensors through wireless connections and direct transmission of the measured health parameters via public wireless networks to the health-care centers.

Jung and Hinze (2005) are proposing a concept of a mobile system to support patients with chronic diseases. This system allows monitoring the health of patients and notifying the medical staff if immediate intervention is needed. The main objective of the system proposed in Jung and Hinze (2005) is the active involvement of patients in their own treatment and release of medical workers from bureaucratic work, leading to overall improved outcomes of health treatment. The advantage of this system also is the automatic input of the users' health parameters (blood pressure, blood-sugar level) in their electronic health record.

Shopov et al. (2009) provide a detailed overview of the building blocks and architecture models for web-based personal health systems. The authors distinguish several main blocks: body sensor networks, personal server, data warehouse, medical server, and medical web portal. If we analyze the building blocks of Shopov et al. (2009), we can conclude that the personal server from Shopov et al. (2009) is similar to the mobile application and social network in the proposed model. The personal server from Shopov et al. (2009) can perform logical reasoning and determine the health status of the user based on data obtained from numerous sensors and provide feedback through an interactive user interface in a graphical or audio format. However, in the proposed COHESY model, these parts are more efficient, providing emergency call (based on the health of the user), analysis of previous medical conditions, and client communication between users to exchange experience.

Unlike MobiCare (Chakravorty 2006) and Personal Care Connect (Blount et al. 2007) systems, the proposed COHESY model is more advanced and includes communication between the user, his or her mobile application, the social network, health centers and institutions, and additional services. This communication allows greater complexity of the data, broader analysis, and satisfactory degree of certainty.

The proposed system secure and independent living (SINDI) in Mileo et al. (2010) uses a smart environment (environment-positioned sensor networks), which is an advantage over COHESY. On the other hand, the advantage of the COHESY is its collaborative component (social network) that enables the collection of data for different users. Although both systems generate recommendations, there is a significant difference in the data used to generate these recommendations. Unlike SINDI where recommendations are generated based only on medical data, the proposed model generates recommendations based on data read from the bionetwork, data from physical activity of the user (obtained by the mobile application), user's medical records (obtained from health facilities), data for user's characteristics (obtained from a social network), and other data from similar clients in the social network.

The proposed COHESY model, in addition to monitoring physical activity, also monitors health parameters of the user, which is not the case in the model presented in Cortellese et al. (2010). Despite the recommendation differences that are generated in both systems (e.g., the proposed collaborative model recommends various activities with different intensities and duration), there is a substantial difference in the evaluation of the recommendation. Unlike the model in Cortellese et al. (2010) where the recommendation is validated only if the activity is executed by the user or if the user entered new data, it is not the case in our model. Evaluating the effectiveness of the recommendation in COHESY is done on the basis of improved health parameters of the user after performing physical activity. This is a basic input parameter in the process of generating recommendations by the recommendation algorithm in the proposed model.

The Jog Falls system mostly resembles the COHESY model (Nachman et al. 2010). The Jog Falls system is an end-to-end system to manage diabetes that blends activity and energy expenditure monitoring, diet logging, and analysis of health data for patients and physicians. This is an integrated system for diabetes management providing the patients with continuous awareness of their diet and exercise, automatic capture of physical activity and energy expenditure, simple interface for food logging, and ability to set and monitor goals and reflects on longer-term trends. Its interface gives physicians comprehensive and unbiased visibility into the patients' lifestyles with respect to activity and food intake, as well as enabling them to track their progress towards agreed goals. Authors put their main emphasis on their novel method for fusing heart rate and accelerometer data that improves the accuracy of energy expenditure estimation (a key feature in enabling weight loss).

The advantage of Jog Falls, in relation to the proposed model, is the ability to enter data on the food consumed by the user. But on the other hand, Jog Falls has no social network and collaborative algorithms, which are the main advantages of the proposed model.

There are various examples of mobile systems for sending messages and notifications (Cortellese et al. 2010, Jung and Hinze 2005, Lee et al. 2007, Mohan et al. 2008, Shim et al. 2008). In these systems, the personalization of messages is typically according to the user's health. Although there are examples where generation of messages, despite the health of the user, exploits some prior knowledge of other users, analysis is usually performed by a medical service. On the other hand, there are quite developed health social networks and websites containing information on different diseases and accessible communication with users and medical personnel who have relevant experience in a particular area (Chorbev et al. 2011, Fernandez-Luque 2009, Kargioti et al. 2010). To the best of our knowledge, no prior health-care system model that integrates these different modules, a mobile application, social network, and medical services, has been reported in the literature. That is the innovation offered by the COHESY model. The proposed model offers support of the mobile technology collaborating with the social network and the medical information systems. The three components are further extended with various medical databases and additional services.

The comprehensiveness of the proposed model provides connectivity and communication of the different components of the system. This expands information exchange leading to quantitative data increase for different types of treatments, therapies, and activities for users with the same or similar diagnosis. These data enable further analysis and research that result in improved diagnosis, treatment, and therapies for users.

7.3 COHESY Description

COHESY has simple graphical interfaces that provide easy use and access not only for the young but also for elderly users. The system model has more purpose and includes use by multiple categories of users (patients with different diagnoses). Some of its advantages are scalability and ability of data information storing when communication link fails. COHESY is an interoperable system model that allows data sharing between different systems and databases.

The COHESY architecture enables 24 h monitoring of the health and physical condition of the patients and the possibility of sending an emergency call for the sudden deterioration of their medical condition, ability to send various notifications to users, and increased medical prevention. In addition, the system enables the patient (system user) to contact other people with a similar condition and exchange their experience.

7.3.1 System Layers

COHESY is deployed over three basic usage layers, as shown on Figure 7.1. The first layer consists of the bionetwork (implemented from various body sensors) and mobile application that collects users' bio data during various physical activities (e.g., walking, running, cycling) and users' health parameters (e.g., weight, blood pressure, blood-sugar level, and heart rate). The second layer is presented by the social network implemented as a web portal that enables different collaboration within the end user community. The third layer enables interoperability with the primary/secondary health-care information systems that can be implemented in the clinical centers and different policy-maker institutions.

The communication between the first and the second layer is defined by users' access to the social network where the user can store their own data (e.g., personal records, health-care records, bionetwork records, readings on physical activities). The social network allows communication between users based on collaborative filtering techniques, thus connecting the users with the same or similar diagnoses, sharing their results, and exchanging their opinions about performed activities and received therapy. Users can also receive average results from the other patients that share the same conditions in a form of recommendation or notifications. These notifications can vary from the average levels of certain bio data calculated for certain geographical region, age, and sex to the recommendation for certain activity based on the activities of other users. Collaborative filtering can be used to achieve different recommendations in these contexts. The different levels of the validity of the data used by the collaborative algorithms affect the validity of obtained results.

The communication between the first and the third layer is determined with the communication between patient and health-care centers. The patient has 24 h access to medical personnel and a possibility of sending an emergency call. The medical personnel remotely monitors the patient's medical condition, reviews the medical data (blood pressure, blood-sugar level, heart rate), and responds to the patient by suggesting most suitable therapy (if different from the one that is encoded in the mobile application) as well as sending him or her various notifications (e.g., tips and suggestions) regarding his or her health condition.

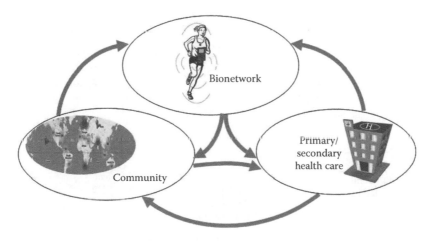

FIGURE 7.1
COHESY layers. (From Fogg, B.J., Fogg behavioral grid, http://www.behaviorgrid.org/, 2013.)

The second and the third layer can exchange data and information regarding a larger group of patients grouped by any significant indicator (region, time period, sex, type of activities) that can later be used for research, policy recommendations, and medical campaign suggestions.

7.3.2 System Architecture

Collaboration between different entities in COHESY, the flow and exchange of data information, is given in Figure 7.2 and Table 7.1.

The data information in the proposed system model are users' personal data (name, age, height, diagnosis), data from users' bionetwork (weight, heart rate, blood pressure, blood-sugar level), realized and recommended activity (type of activity, path length, time interval, average speed), and recommendation and suggestions.

The mobile technologies (devices and applications) in this system are used to support and enable collaboration. The installed mobile application, using various sensors (bionetwork), performs readings regarding users' health during their physical activities (walking, running, cycling) and, based on them, gives appropriate instructions, proposals, and constraints of their execution, in order to improve their own health. In the presented system model, mobile application attempts to categorize events by processing the collected data, from the patient's current state and patterns of biologic and environmental sensors. For example, if the heart rate monitor detects an increase in pulse rate, the accelerometer detects movement, and the GPS detects a frequent change in location, it can be inferred that the user is exercising. Threshold parameters are then adjusted automatically to minimize interaction with the user. In cases where only the heart rate may spike and no accelerometer and GPS activities are detected, the user is quickly prompted to provide simple

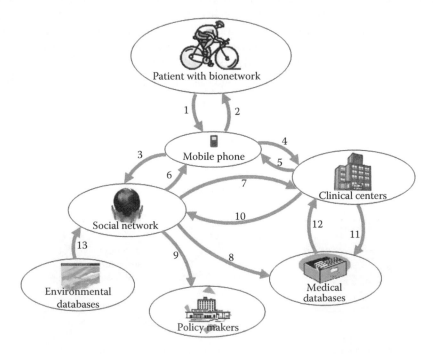

FIGURE 7.2
COHESY architecture and data exchange. (From McCurdie, T. et al., *Biomed. Instrum. Technol.*, Fall, 49, 2013.)

TABLE 7.1

Data Flow

Link	Data	Link	Data
1	Heart rate, blood pressure, blood-sugar level, weight	2, 6, 7, 8	Name and surname of the patient, realized activities: • Type of activity • Weight • Heart rate • Blood pressure • Blood-sugar level • Path length • Time interval • Average speed • Weather temperature • Atmospheric pressure • Air humidity • Wind speed
3	Age diagnosis, height, weight, heart rate, blood pressure, blood-sugar level, realized activity: • Type of activity • Path length • Time interval • Average speed	4	Emergency call, weight, heart rate, blood pressure, blood-sugar level, realized activity: • Type of activity • Path length • Time interval • Average speed
5	Emergency call, diagnosis, recommended therapy, suggestions, recommended activity: • Type of activity • Path length • Time interval • Weather temperature • Atmospheric pressure • Air humidity	9	Realized activities: • Type of activity • Weight • Heart rate • Blood pressure • Blood-sugar level • Path length • Time interval • Average speed • Weather temperature • Atmospheric pressure • Air humidity • Wind speed
10, 11, 12	Diagnosis, recommended therapy, suggestions, recommended activity: • Type of activity • Path length • Time interval • Weather temperature • Atmospheric pressure • Air humidity	13	Weather temperature, atmospheric pressure, air humidity, wind speed

feedback. This feedback is stored to be used in machine learning algorithms for future decisions. Based on such results, the dataset can be flagged for further analysis by medical professionals as a supplement to a more complex analysis of the patient.

The patients are not restricted in their movements or their location. By using their mobile phones (the installed application), they have access to the medical personnel at any time. The medical personnel can remotely monitor the patient's medical condition, reviewing the medical data (blood pressure, blood-sugar level, heart rate) arriving from the mobile application of the patient. At the same time, the patient individual data can be compared with average data obtained using different collaborative filtering techniques. In this way, the medical personnel can quickly respond to the patient by suggesting the most suitable therapy as well as when to receive it, focusing on activities that are necessary for his or her rehabilitation and maintenance of his or her health, sending him or her various tips and suggestions for improving his or her health. Even more important, the social network can learn from this recommendation and generate notifications and recommendations based on the most successful scenarios.

The installed mobile application has access to the social network (web portal) where it can store users' data and read average data readings on bio and physical activities of all users. The social network allows direct communication between users (if approved by the user and stored in the user profile) and sharing of their results. This portal can provide an interface and use data from a variety of medical databases and environmental databases (temperature, wind speed, humidity). In this way, mobile application within the COHESY provides a tool for a complete personal health care.

The conclusions drawn from research data, while exploring medical databases, can be routed back to the clinical centers. These data are used to individually analyze the condition of patients. Clinical centers have access to data on physical activities of patients collected by the application installed on their mobile device. Therapies and recommendation are drawn from the analysis of the overall data obtained by the clinical centers. Those therapies and recommendations now can be easily routed back to the patients' mobile devices.

Simultaneously, clinical centers can exchange data and information with a social network and thus have access to a larger group of patients that can share the research, recommendation, and suggestion of the medical personnel. Received therapies and recommendations can be used by the application to suggest to users when, where, and what action to accomplish in order to improve their health. The social network has incorporated collaborative filtering that allows filtering large amounts of data on concrete condition. So policy makers can get those data, make specific analysis of it, and give recommendations for national action by governments and nongovernment organizations, including programs and strategies.

7.3.3 Security Issues

The fundamental goals of secure health-care systems are safely exchanging the patient's information issued by mobile devices and preventing improper use of illegal devices, such as intercepting transferred data, eavesdropping communicating data, replaying out-of-date information, or revealing the patient's medical conditions.

There is a growing research done to address the privacy and security concerns to mobile and internet health applications (Hung 2005, Raman 2007, Zheng et al. 2007). For example, the development of a privacy-preserving trust negotiation protocol for mobile health-care systems (Dong and Dulay 2006) facilitates trust between user devices in compliance with predefined access control and disclosure policies.

With the emergence of e-health networks and their offer for web-based services, the future success of e-health is a lot more likely to depend on how effectively patients can obtain and manage their information. In the past decade, several leading technology vendors and consumer-oriented enterprises have established the Liberty Alliance project (Peyton et al. 2007). This framework is being adapted to establish "Circle of Trust" (CoT) for cooperating enterprises such as hospitals, labs, pharmacies, and insurers that will enable them to offer web-based services to the patients. In this framework, personally identifiable information is managed by a designated "Identity Provider" who provides pseudonymous identities of patients for transactions among partners. Further, an audit service, provided by an independent organization, logs all transaction requests by the members of CoT that enables validation of privacy compliance by a regulatory agency and verification on how their data are used by individual patients and ability to challenge its accuracy. Users need to feel confident that no personal information will be shared to other users or unauthorized personnel. This is important for the success of the entire system since it relies heavily on a large amount of subscribed users sending (uploading) data so it can have enough information to produce reliable recommendations about one's health.

The safety and protection of the private data are of considered important for the success of the proposed model. The model is based on the vast amount of data, entered by the user, that are further used for generating recommendations for improving users' health.

Based on the potential threats of mobile healthcare, specific security requirements will have a significant influence on the performance of the system:

Data storage and transmission: Local database (in mobile phone) stores data received by sensors, in case there is always back up of data (they will be saved only for some period of time). When there are problems in sending data to the clinical center (e.g., some of data are not sent), all transactions will be rolled back, so when the service will be available, data will be (re)sent. By this, we provide quality of service (QoS) facilities since these clearly demand for high reliability, guaranteed bandwidth, and short delays.

Data confidentiality: Most patients do not want anyone to know their medical information, except their family doctor or medical specialist. The solutions are to use a cryptographic algorithm to encrypt medical information and protect the necessary data.

Authentication: Only an authenticated entity can access the corresponding data that are available for that entity; unauthenticated entities are denied access when they visit data information that they do not have the rights to obtain. For example, asymmetric cryptography (i.e., public key infrastructure [PKI]) is often used, because these private keys are credentials shared only by the communicating parties.

Access control: In traditional network security models, access control determines whether a subject can access an object based on an access control list (ACL).

Privacy concerns: Every user can choose what information can be private or public. The user can choose his or her records to be public: (1) for medical purposes, (2) to all visitors of the social network, (3) to users in his or her category, and (4) to none. In order to have medical support, the user has to agree to share personal information with clinical centers and medical databases, whose data are also protected. According to the user agreement policy, those data information would be exchanged through the system.

Though many health-care researchers are interested in collecting and recording medical sensor data, these data may contain many personal facts, meaning patients are not willing to reveal them. Especially in an open wireless environment, an intruder may observe network traffic and thereby infer the relationships and identities of the communicating nodes.

One possible approach for solving this issue is to apply the theory of trust to identify malicious nodes and after that to exclude them from a presently health-care network. As an emerging technique, trust can be defined as "the degree to which a node should be trustworthy, secure, or reliable during any interaction with the node" (Boukerche and

Ren 2008). This means that if one node trusts another node to perform the intended operation, the trust relationship between these two nodes can be established reliably from the communicating initiator's point of view. So the technique of trust evaluation without a centralized trust management authority can significantly improve the security and reliability of the network while also reducing the complexity of the traditional trust schemes and thus improving efficiency.

7.4 Recommendation Algorithm in COHESY

The recommendation algorithm is part of the second level in COHESY (the social network). It is implemented as a web service and its purpose is to recommend the physical activities that the users should carry out in order to improve their health. The algorithm uses the data read by the bionetwork, the data about the user's physical activities (gathered by the mobile application), the user's medical record (obtained from a clinical center), and the data contained in the user profile on the social network (so far based on the knowledge of the social network).

The main purpose of this algorithm is to find the dependency of the users' health condition and the physical activities they perform. The algorithm incorporates collaboration and classification techniques in order to generate recommendations and suggestions for preventive intervention. To achieve this, we consider datasets from the health history of the users and we use classification algorithms on these datasets to group the users by their similarity. The usage of classified data when generating the recommendation provides more relevant recommendations because they are enacted on knowledge from users with similar medical conditions and reference parameters.

The steps of the proposed algorithm in the form of an activity diagram are shown in Figure 7.3.

7.5 Simulation Results and Discussion

In order to make the validation of the proposed model, we made few simulations using generic data. Three simulations are made, and in all of them, two types of activities are generated: a positive activity (activity whose performance increases the value of a given parameter) and a negative activity (activity whose performance reduces the value of a given parameter). Each activity has individual influence on the global parameter change and it is presented by a function whose shape is similar to a Poisson probability mass function. Simulation results are presented in Table 7.2.

The analysis of data obtained from the simulations of the recommendation algorithm on generic data shows that the algorithm generates appropriate recommendations with an accuracy of 82%–92%. As the time period and the number of activities extend, so does the percentage of appropriate recommendations generated by the algorithm increase. However, the analysis showed that the proposed model has deficiencies, as the *cold start* and the extension of the initial period generating inappropriate recommendations, which should be treated with more attention in the future. A possible solution to this problem

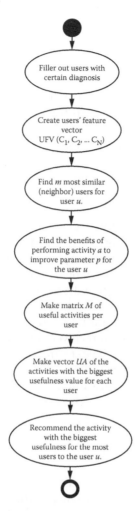

FIGURE 7.3
Activity diagram of the COHESY recommendation algorithm.

TABLE 7.2

Simulation Results

	First Simulation	Second Simulation	Third Simulation
Number of generated recommendation	28	45	58
Accuracy of the generated appropriate recommendations	82.14	84.44	91.38
Time period of inappropriate recommendations	Till ~8th day	Till ~15th day	Till ~9th day

is to generate prior knowledge before the following simulations. This will also avoid the elongation of the period that generates inappropriate recommendations as well as the issue of a cold start.

Executed simulations are the first step to evaluate the effectiveness of the recommendation algorithm and proposed model. The effectiveness of the proposed model has been (could be) evaluated using other methodological approaches as well (Kulev et al. 2013).

7.6 Case Studies

7.6.1 Use Case Scenario No. 1

The user (with diagnosed diabetes) switches on the application on his or her mobile phone. The application using Bluetooth connects to the device that measures the user's weight, blood pressure, and blood-sugar level and reads the measured parameters. These values are sent to the user's profile on the social network.

The algorithm deployed on the social network generates recommendation with activity that is best for user's health to be performed. The recommendation is based on the user's data, average data, and previously generated successful recommendations from the social network, previous clinically originated recommendations, and patient history. The application sends this recommendation to the user.

The user starts the recommended activity. The application reads the user's heart rate from his or her bionetwork. During running, an irregularity occurs. While reading the data, the application detects that the user's heart rate is quite higher than his or her average and the application sends a message with these data to the medical center and social network and signals the user that there is some irregularity happening.

Medical personnel review the submitted data and the user's previous medical records. Based on the user's diagnosis, treatment received, and his or her activity currently carried out, along with the medical data received from the application, the medical personnel decides that the user should stop the activity and pause for 15 min. This recommendation is issued back to the user's mobile application. The application signals the user that a message from the medical center has arrived. The user applies the recommendation from the medical center.

7.6.2 Use Case Scenario No. 2

The user switches on the application on his or her mobile phone and connects to the social network. The user is suspecting that he or she has diabetes. The algorithms deployed on the social network checks his or her profile and compares his or her health and physical parameters with the average results of the other users with diagnosed diabetes (obtained from the social network). The social network gives notification to the user that his or her health data matches with the average data gathered from people having this condition (diabetes), as stated by medical institutions and other users with a similar profile, which can confirm, diagnose, and advise him or her to talk to a physician.

7.7 Future Directions

Each year, millions of people are diagnosed with chronic disease, and millions more die from their condition. Despite great improvements in therapies and treatment, disease rates have risen dramatically. The metabolic syndrome, which is a synonym for a group of symptoms (e.g., obesity, high blood pressure, insulin resistance), is emerging as a major public health issue across the world. The rising rate of chronically ill people will become a crucial contributor to rising medical expenditures over time.

The importance of health care to individuals and governments and its growing costs to the economy have contributed to the emergence of health care as an important area of research for business and other disciplines. The solution to decrease the cost of health-care services requires necessary changes: moving from reactive to preventive medicine and concentrating on the long-term care rather than only acute care and patient-centered care rather than hospital-centered care, including remote care delivery mechanisms.

In addition to the embedded role of information technology in clinical and diagnostics equipment, the advances in communication and computer technologies have revolutionized the way health information is gathered, stored, processed, and communicated to decision makers for better coordination of health care at both the individual and population levels.

Therefore, the future trends in health-care support systems are based on the development of patient-centric pervasive environments in addition to the hospital-centric ones. Such systems should enable health-care personnel to be able to timely access, review, update, and send patient information from wherever they are, whenever they want. In that way, pervasive health care will contribute to design, develop, and evaluate computer technologies that will help citizens participate more closely in their own health care, on one hand, and will provide flexibility in the life of patients who lead an active everyday life with work, family, and friends, on the other.

The future health-care systems will have a sense-making component in order to better understand patient personal condition and clinical or living environment and, in that way, increase health-care system efficiency by increasing its availability. It is very important to discuss ways in which information systems that support health care can make that kind of sense making.

The COHESY model, presented in this work, will give a new dimension in the usage of novel technologies in health care. By using mobile, web, and broadband technologies, the citizens will have ubiquity of support services wherever they may be, rather than becoming bound to their homes or health centers. COHESY allows connecting users with the same or similar diagnoses, sharing their results, and exchanging their opinions about performed activities and received therapy. At the same time, the system generates average values based on filtering large amounts of data about concrete conditions such as are geographical region, age, sex, and diagnosis.

In the future, we are planning an experiment with real data to be done. These experimental data will allow us to determine the effectiveness (accuracy) of the system in the improvement of health parameters. For instance, based on data from different groups of participants, we will be able to make a qualitative analysis for improving the health parameters of participants who use the recommendations from the system, compared with the improvement of health parameters of users who do not use recommendations from the system. In this way, we will be able to confirm our assumptions, made on generic data, that the recommendations generated by COHESY are of great importance to users.

7.8 Conclusions

The presented system model is a complex system composed of mobile application, social network, information systems that are used by medical personnel, medical databases, and additional services. The proposed model provides monitoring of health parameters and tracking of the user's physical activities, communication between users, automatic data transfer, data exchange between medical centers, and databases. But what distinguishes the proposed model from the rest, meaning its main advantage, is the communication and exchange of data between the various components. The proposed model offers a collaboration that is important not only in the patients' treatment and rehabilitation but also to generate average values based on filtering large amounts of data for specific attributes such as geographic region, age, sex, and diagnosis.

From the aforementioned, it can be concluded that the proposed model offers the following: increased medical prevention, faster response time of emergency medical staff, 24 h monitoring of the user's health condition and performed physical activities, the ability to send different notifications to users, automatic transmission of the measured health parameters to the medical staff, as well as increased flexibility in data collection. But in addition to these services for users, one of the objectives of the proposed model is a collection of different data types and combining them into complex data structures. Analysis and research of such structures make a possibility to recognize the influence of the physical activity, applied therapy, time parameters, and other factors on the development of the patients' health condition. Such analyses can be further used by medical facilities for the diagnosis, treatment, and therapy of patients. On the other hand, data obtained from health facilities and social network, sent as recommendations through mobile application, allow the user to adapt and direct their physical activities in order to improve their health condition and rehabilitation and thus enable self-care regarding their health.

Questions

1. What is the main advantage of the presented health-care system model COHESY?
2. How many layers does COHESY have?
3. Explain the role of the mobile application in COHESY?
4. What are the specific security requirements that a health-care system should have?
5. What is the purpose of the recommendation algorithm in COHESY?
6. What data do the recommendation algorithm use?

References

Ballegaard, S.A., Hansen, T.R., and Kyng, M. 2008. Healthcare in everyday life: Designing healthcare services for daily life. In *Proceedings of the Conference on Human Factors in Computing Systems*, Florence, Italy, pp. 1807–1816.

Blaya, J.A., Fraser, H.S., and Holt, B. 2010. E-health technologies show promise in developing countries. *Health Affairs* 29:244–251.

Blount, M., Batra, V.M., Capella, A.N. et al. 2007. Remote health-care monitoring using personal care connect. *IBM Systems Journal* 46(1):95–113.

Boukerche, A. and Ren, Y. 2008. A trust-based security system for ubiquitous and pervasive computing environments. *Computer Communications* 31:4343–4351.

Chakravorty, R. 2006. MobiCare: A programmable service architecture for mobile medical care. In *Proceedings of Fourth IEEE Conference on Pervasive Computing and Communications Workshops*, Pisa, Italy, pp. 532–536.

Chorbev, I., Sotirovska, M., and Mihajlov, D. 2011. Virtual communities for diabetes chronic disease healthcare. *International Journal of Telemedicine and Applications* 11:1–7.

Cortellese, F., Nalin, M., Morandi, A., Sanna, A., and Grasso, F. 2010. Personality diagnosis for personalized eHealth services. In *Electronic Healthcare, Lecture Notes of the Institute for Computer Sciences, Social Informatics and Telecommunications Engineering*, vol. 27, ed. P. Kostkova, pp. 173–180, Berlin, Heidelberg, Germany: Springer.

Dong, C. and Dulay, D. 2006. Privacy preserving trust negotiation for pervasive healthcare. In *Proceedings of Pervasive Conferences and Workshops*, Austria, pp. 1–9.

Fernandez-Luque, L. 2009. My health educator: Personalization in the age of health 2.0. In *Proceedings of the International Workshop on Adaptation and Personalization for Web 2.0*, Trento, Italy, pp. 139–142.

Fichman, R.G., Kohli, R., and Krishnan, R. 2011. Editorial overview-the role of information systems in healthcare: Current research and future trends. *Information Systems Research* 22(3):419–428.

Fogg, B.J., Fogg behavioral grid, http://www.behaviorgrid.org/, 2013.

Hung, P.C.K. 2005. Towards a privacy access control model for e-healthcare services. In *Proceedings of 3rd Annual Conference on Privacy, Security and Trust*, New Brunswick, Canada.

Jung, D. and Hinze, A. 2005. A mobile alerting system for the support of patients with chronic conditions. In *Proceedings of the 1st Euro Conference on Mobile Government*, Brighton, U.K.

Kargioti, E., Kourtesis, D., Bibikas, D., Paraskakis, I., and Boes, U. 2010. MORMED: Towards a multilingual social networking platform facilitating medicine 2.0. In *Proceedings XII Mediterranean Conference on Medical and Biological Engineering and Computing*, Chalkidiki, Greece, pp. 971–974.

Khan, P., Hussain, A., and Kwak, K.S. 2009. Medical applications of wireless body area networks. *International Journal of Digital Content Technology and its Applications* 3(3):185–193.

Kulev, I., Vlahu-Gjorgievska, E., Trajkovik, V., and Koceski, S. 2013. Development of a novel recommendation algorithm for collaborative health-care system model. *Computer Science and Information Systems* 10(3):1455–1471.

Lee, R.G., Chen, K.C., Hsiao, C.C., and Tseng, C.L. 2007. A mobile care system with alert mechanism. *Transaction on Information Technology in Biomedicine* 11(5):507–517.

McCurdie, T., Taneva, S., Casselman, M., Yeung, M., McDaniel, C., Ho, W., and Cafazzo, J. 2013. mHealth consumer apps: The case for user-centered design. *Biomedical Instrumentation & Technology*, Fall:49–56, 2013.

Mileo, A., Merico, D., and Bisiani, R. 2010. Support for context-aware monitoring in home healthcare. *Journal of Ambient Intelligence and Smart Environments* 2(1):49–66.

Mohan, P., Marin, D., Sultan, S., and Deen, A. 2008. MediNet: Personalizing the self-care process for patients with diabetes and cardiovascular disease using mobile telephony. In *Proceedings of 30th Annual International Conference of the IEEE Engineering in Medicine and Biology Society*, Vancouver, British Columbia, Canada, pp. 755–758.

Nachman, L., Baxi, A., Bhattacharya, S., and Darera, V. 2010. Jog falls: A pervasive healthcare platform for diabetes management. In *Proceedings of the 8th international conference on Pervasive Computing*, Helsinki, Finland, pp. 94–111.

Peyton, L., Hu, J., Doshi, C., and Seguin, P. 2007. Addressing privacy on a federated management network for e-health. In *Proceedings of the 8th World Congress on the Management of eBusiness*, Toronto, Canada, pp. 12–12.

Raman, A. 2007. Enforcing privacy through security in remote patient monitoring ecosystems. In *Proceedings of 6th International Special Topic Conference on Information Technology Applications in Biomedicine*, Tokyo, Japan, pp. 298–301.

Shim, H., Kim, H.M., Song, S.H., Lee, J.H., and Yoon, Y.R. 2008. Personalized healthcare comment service for hypertension patients using mobile device. In *Proceedings of 30th Annual International Conference of the IEEE Engineering in Medicine and Biology Society*, Vancouver, Canada, pp. 1521–1524.

Shopov, M., Spasov, G., and Petrova, G. 2009. Architectural models for realization of web-based personal health systems. In *Proceedings of International Conference on Computer Systems and Technologies and Workshop for PhD Students in Computing*, Ruse, Bulgaria, pp. 1–6.

van Halteren, A., Bults, R., Wac, K. et al. 2004. Wireless body area networks for healthcare: The MobiHealth project. In *Wearable eHealth Systems for Personalised Health Management, volume 108 of Studies in Health Technology and Informatics*, eds. A. Lymberis and D. De Rossi, pp. 121–126. Amsterdam, the Netherlands: IOS Press.

Zheng, Y., Cheng, Y., and Hung, P.C.K. 2007. Privacy access control model with location constraints for XML services. In *Proceedings of the 23rd International Conference on Data Engineering Workshop*, Istanbul, Turkey, pp. 371–378.

Zimmerman, T. and Chang, K. 2008. Simplifying home health monitoring by incorporating a cell phone in a weight scale. In *Proceedings of the 1st International Conference on PErvasive Technologies Related to Assistive Environments*, Athens, Greece, pp. 1–4.

Section III

Psychological, Behavioral, and Mental Verticals of mHealth

8

mHealth for Behavior Change and Monitoring

Donna Spruijt-Metz, Wendy Nilsen, and Misha Pavel

CONTENTS

Mobile and wireless health (mHealth) technologies have developed at an exponential pace in recent years; however, the integration and translation of these cutting-edge technologies into rigorously evaluated health research and healthcare tools have lagged behind. Remarkable advances have been made in the last decade, for example, in low-cost, real-time technologies to assess and/or intervene on disease, movement, images, behavior, social interactions, environmental toxins, hormones, and other physiological variables. These advances are due to increased computational sophistication, as well as reductions in size and power requirements.[1] These technologies provide the potential to advance research, prevent disease, enhance diagnostics, improve treatment, reduce health disparities, increase access to health services, and lower healthcare costs in ways previously unimaginable.[2]

However, although the basic engineering and computer science knowledge exists to develop technologies, the translation of these devices into effective behavior change and monitoring tools has not kept pace. Recent reviews of the mHealth literature highlight weak empirical support for the use of mobile technologies to affect behavior change.[3–5] Some argue that this may be because the field is transitioning towards rigorous evaluation, as evidenced by the proliferation of new studies that have been recently funded.[6] The fact remains that, to date, the hype of mHealth has far exceeded the actual scientific justification,[7] leaving many concerned that mHealth may not meet its expected potential.[8]

The promise of mHealth is based on the premise that mobile technologies can allow us to develop behavioral monitoring and change systems that are personalized and delivered in real time.[9–11] Available research suggests that immediate and contextualized feedback to promote behavior change[12] as well as improve learning[13,14] is superior to traditional approaches. These technologies also enhance the possibility of scaling surveillance, and prevention and intervention efforts in ways that have been unthinkable with conventional face-to-face programs. Further, these devices have the ability to capitalize on small, but frequent, intervention doses at a timing that is more optimal for the user. Wearable and deployable sensors and smartphones can provide rich, temporally dense contextualized data that allow for the personalization of feedback on the fly, ensuring appropriate feedback, messaging, timing, and dose that is adaptive.[15] Adaptive technologies can be set to prompt for answers to questions or deliver intervention content based on current behaviors, behavior changes, and locations, at random or in response to particular events.[11,16] With all of these advantages, why have we not yet seen the potential of mHealth realized? Although the question is ultimately empirical in nature, this chapter will examine some of the factors believed to have led to some of the findings reported in the field thus far.

8.1 State of the Science

Adverse and suboptimal health behaviors and habits are responsible for approximately 40% of preventable deaths, in addition to their adverse effects on quality of life and economics.[17–20] Although many of these behaviors and habits are driven by personal decisions that can be influenced by effective programs, behavior change programs have not been well integrated into public health or the healthcare system.[21] Furthermore, while a solid empirical base exists for the effectiveness of interventions in initiating some behavior change,[22–25] much less support exists for interventions that sustain behavior change or change some of the most challenging and prevalent behaviors (e.g., smoking,[26] poor diet, and lack of physical activity[27,28]). Scaling up much of the available work on interventions is a challenge because the interventions require considerable time and expense[29] and are not easily adapted to different populations or settings. Because of these issues, using mHealth for behavior change is a logical step and one that has been the focus of considerable scientific discussion.

Recent reviews in the area of behavior change in mHealth have highlighted the challenges within this field. These reviews of the published research[3–5,30] suggest areas of strength and needed enhancements. For example, Kaplan and Stone[5] found that of the 20 randomized control trials available in mHealth, only six showed significant main effects for the use of mobile technology to promote behavior change. Of the others, 55% showed no effect of the intervention on the targeted behavior and three others had mixed outcomes. In this review and others,[4,30] the clearest benefits have been demonstrated for both smoking cessation and medication adherence, but these findings mask the diverse efforts to enhance behavior change across the spectrum of behaviors and health conditions and the complexities of the findings. The following section highlights major findings in smoking cessation, weight loss, chronic disease management, and behavioral health (i.e., mental health and addictions).

8.1.1 Smoking Cessation

Despite the well-documented pernicious effects of tobacco on health, smoking rates are increasing around the developing world and have not continued to decrease in developed countries.[31] Smoking cessation is one of the most developed areas of mHealth research. A recent Cochrane review[32] of this research area highlights the success of mHealth in increasing quit rates in five studies with a total of 9000 participants. Much of the work in this area has begun in New Zealand and then exported to other countries.[33–35] The original intervention[33] recruited via advertising people who wanted to quit smoking and who owned a mobile phone. The program included having people set a quit day and then receiving automated personalized but standard text messages (SMS) to help them achieve cessation. The messages were selected from a database according to participant characteristics and time from quit day—with daily messages leading up to quit day, an intensive month of 5–6 messages per day, followed by a maintenance phase of one message every 2 weeks. Messages included quitting advice and motivational messages to encourage abstinence mixed with some distraction/general interest messages. Control groups received access to the automated quitline and a subsidy program for nicotine replacement therapy. Overall, the Cochrane review reported a significant effect for mHealth interventions to increase smoking cessation (RR 1.71, 95% CI 1.47–1.99, P = 0.001), using a definition of abstinence of no smoking at 6 months since quit day but allowing up to three lapses or up to five cigarettes. These findings include trials with young adults through seniors and with self-report and biologically verified quit rates. Interestingly, of the studies included for analyses, three included text message only.[33–35] More recent research has shown little added effect for components beyond text message in reducing smoking.[30] Future research will begin using adaptive interventions to personalize treatments and enhance cessation across populations and cultures.

8.1.2 Obesity Prevention and Treatment

Obesity has reached epidemic proportions in the United States[36,37] and internationally[38] and is recognized as one of the most important public health challenges to date. Obesity is linked to a host of chronic diseases. These include diabetes,[39] hypertension and cardiovascular disease,[40] and several types of cancers (endometrial, postmenopausal breast, kidney, and colon cancers).[41] In the United States, 35.9% of all adults aged 20 and above are obese, while 69.2% of adults are overweight.[37] In children, obesity in the United States has tripled in the last decade.[28] Although increases in physical activity and improved diet are related to weight loss and improved metabolic health,[23] interventions to prevent and treat obesity have, to date, met with limited success. It is thus not surprising that efforts to use mobile technologies to increase physical activity, decrease sedentary behavior, and improve dietary intake have proliferated in recent years. Weight loss behaviors that lend themselves to mHealth interventions and have consistently been associated with weight loss include self-monitoring (diet, physical activity, and weight), goal setting (to reduce calorie intake, increase physical activity, or decrease sedentary time), feedback on weight loss behaviors, and social support from coaches or peers.[42] Although much work remains to be done, mHealth weight loss interventions are showing promise. In a recent review of interventions using mobile technologies to improve physical activity, 75% of the 12 studies included in the review reported significant increases in physical activity and/or decreases in sedentary behavior.[43] Components included tailored messaging, displays of personal physical activity data and progress towards physical activity goals, and personalized feedback based on self-reported data. Although a few of the studies tested the feasibility of real-time feedback using ubiquitous measures, none tested the impact of real-time, tailored, adaptive interventions on physical activity outcomes using these technologies. A recent randomized controlled trial that intervened on multiple obesity-related behaviors used remote coaching supported by mobile decision support technology. In one of the four groups, daily fruit/vegetable intake increased from 1.2 servings to 5.5 servings, sedentary leisure decreased from 219.2 to 89.3 min, and saturated fat decreased from 12.0% to 9.5% of calories consumed.[44] Thus, tailored, real-time mHealth interventions for obesity hold much promise, particularly as interventionists begin to utilize mobile technologies to their full potential to support just-in-time, adaptive interventions.

8.1.3 Chronic Disease Management

There has been considerable interest in harnessing mobile technologies to enhance chronic disease self-management. As populations age, the prevalence of chronic diseases have increased, putting increased burden on patients and healthcare systems in the United States and across the world.[45] Chronic disease self-management generally includes patient education, symptom monitoring, medication adherence, diet, and physical activity requirements.[21] The three areas with the greatest focus for mHealth have been in HIV/AIDS, diabetes, and behavioral health. These areas are described in more depth in the following.

8.1.4 HIV/AIDS

The treatment of HIV/AIDS requires vigilance in medication adherence. Adherence to the antiretroviral regimen is associated with a reduced viral load, less symptoms, and a reduced probability of disease transmission. Thus, taking prescribed medications in HIV/AIDS has both a treatment and a preventive effect in a disease that was fatal only a decade ago.[46] The ability to use mHealth to enhance adherence in populations around the world, for which there is often limited healthcare services and in where mobile services have leapfrogged wired capabilities, was embraced early in the field. An example of this is Lester et al.'s[47] WelTel trial. This randomized clinical trial examined the effects of weekly SMS messaging on adherence to antiretroviral therapy and viral loads in people with HIV/AIDS in Kenya. This study found a significant reduction in virological failure in patients who received a weekly message designed to assess their adherence and other problems with treatment. Other studies have had similar results, suggesting that SMS is an efficacious method for enhancing medication adherence in HIV/AIDS in the developing world.[48] More recent work has sought to expand this work by utilizing SMS to improve awareness of HIV and testing.[48]

8.1.5 Diabetes

Management of diabetes requires daily monitoring of blood glucose, a specific diet and physical activity regimen, and medicines to manage not only glucose but also common commodities such as high blood pressure and cholesterol. mHealth seems to have unique strengths in this area, including the ability to wireless sync sensor data from glucometers, weight scales, and accelerometers with patient self-report to assess and provide feedback in all of the critical areas of diabetes disease management.[49] A recent review of diabetes self-management using both computers and mobile technologies found that interventions using mobile demonstrated an edge over traditional web-based (i.e., eHealth) applications.[50] The effect was limited to blood glucose control and did not extend to weight management and emotion regulation, which were also common targets of the interventions. Quinn and colleagues[51] found that the use of mHealth for diabetes self-management also helped patients to better organize their information and provided data to care providers that facilitated treatment decisions.

8.1.6 Behavioral Health

Given the high presence of mental health and substance abuse disorders in the United States, it is not surprising that mobile technologies have made their way into this area. Behavioral health seems an especially promising area of research in mHealth as relapses and failures in functioning happen in real time when people are in their natural environment, and not when they are in the structured treatment setting. For example, using a multimedia smartphone intervention that utilizes standard text messaging, as well as psychoeducation, social media, videoconferencing with counselors, and GPS for identification of environmental cues, Gustafson and colleagues have shown that mHealth can reduce relapse days and rehospitalizations for people being discharged from in-patient treatment for alcoholism.[52] Other studies are examining the value of mHealth in treating drug abuse,[53] based on the premise that just-in-time behavior change tools and cues for healthy behavior should decrease addictive behaviors and improve health.

Ideally, to be most effective, mHealth should increase access to mental health services, reduce the stigma associated with care, and allow people to utilize information and prompts to action when they are in real-world settings far from traditional therapeutic supports.[54] Early work has targeted a range of mental illness, such as schizophrenia and depression.[54] For example, Ben-Zeev and colleagues[55] have shown that patients with schizophrenia perceive value in mHealth interventions designed to track and manage their symptoms and provide psychoeducation in real time. For disorders such as depression, researchers have begun working with SMS as between sessions prompts to complete scheduled *homework*, such as tracking symptoms and doing mood enhancing activities.[56] In addition, projects such as the Veterans Administration's PTSD Coach[57] provide therapeutic supports in a mobile form that is based on well-established evidenced-based behavioral treatment.[58] Current trials of the program will show the relative strengths and weaknesses of the treatment compared to in-person therapy in the treatment of posttraumatic stress disorder (PTSD). Given the majority of people with mental health symptoms never get to treatment, the potential to increase access to services in this area is huge. These projects highlight the possibilities of mHealth for behavioral health, as well as the early stage of research in this area.

8.1.7 State of the Science Conclusion

These areas of behavior change research highlight some of the successes and failures of current mobile technologies. They also highlight the opportunities in this area. For example, reviews of the scientific underpinnings of the interventions make it clear that we are in need of research on the effects of differing mobile intervention modalities (e.g., SMS vs an app vs videoconferencing vs sensor-based tracking and feedback), tailoring and targeting of interventions, and the frequency and directionality of the intervention/messaging. This is critical, as there have been variables shown to affect the outcomes in nontechnological behavior change research but have not yet been adopted into as a standard of mHealth research. For example, targeting and tailoring messages based on participant characteristics or situations specific to the intervention (e.g., a craving sensation in smoking cessation treatment) have long been shown to have an effect on behavior change in nonmobile interventions (see Noar and Harrington[59]). In a recent review of these factors in SMS interventions, Head and colleague[30] found that treatments were more efficacious when they employed tailoring on psychosocial or demographic characteristics, as well as those that personalized the intervention with names than studies in which tailoring or targeting was not included. Similar issues arise in the frequency of communications. Lester et al.'s[60] successful SMS research in HIV adherence has been used as an exemplar for the use of two-way messaging for enhancing behavior change, but the study did not test the role of interactivity. Thus, it cannot be used as evidence for the value of two-way communications. This example highlights how ripe this area is for research. The same is true of the utilization of multiple modalities (e.g., SMS with a mobile phone app) in mHealth treatment. None of the recent reviews have found that multiple modalities increased the effect of interventions,[30,32] but the tests have been of limited scope and a restricted number of modalities. Given the move in consumer electronics to offer smartphone applications that include multiple modalities, it may be that we need a stronger research portfolio in mHealth to understand the role of modalities in the behavior change arena.

8.2 Next Steps: Challenges and Opportunities

The empirical findings from the literature suggest that the move to *mobilize* behavior change has begun to show success in some of the areas in which it has been applied. Many areas remain (e.g., diabetes and behavioral health) in which the number of trials or the conflicts in findings are so great (e.g., cardiovascular) that it is far too early to declare victory.

mHealth research is growing, but the development of the scientific underpinnings is slower than might have been expected given the initial hype. Currently, there are some pressing issues that likely have hampered mHealth research. Those that have received the most attention are the need for transdisciplinary teams, creating shared behavior change vocabularies, the integration of complex sensing into mHealth, and the development of new theories of behavior change that utilize existing knowledge but develop it into quantifiable models of behavior. These factors are described in more detail in Section 8.3, below.

8.3 Challenges

8.3.1 Transdisciplinary Teams

One of the factors hampering the current field of mHealth is that much of the work currently being done arises from single disciplines without integration of the behavioral, social sciences, and medical research fields. Reviews of much of the developing mHealth technology suggest some of the work being done is driven by the technology industry and communication/marketing fields with a focus on the creation of high-end products with little reference to behavior theory or to potential application in biomedical settings.[61] Products uninformed by the corpus of clinical, behavioral, and measurement methodology literature will not be maximally effective. mHealth research requires an integration of expertise from a range of various health sciences (e.g., behavioral, social, and biomedical sciences), technological fields (e.g., engineering and computer science), and related disciplines (human–computer interaction, artificial intelligence, and game design) that would ensure that the appropriate technology was being utilized in a way most conducive to health research[62] and healthcare delivery. This will allow us to ask questions from a traditional health research perspective, as well as to ask new questions by exploiting the synergies of working across the technology disciplines.[1] By creating transdisciplinary teams we can draw upon the synergies between disciplines to ask new questions, explore old problems in novel ways, and realize the potential of mHealth as well as develop the underlying interdisciplinary science. Moreover, teams in mHealth provide an opportunity for current health scientists to include technology in their research even when the focus is not on mobile. This cross-fertilization will strengthen and support ongoing efforts.

Another transdisciplinary opportunity that mHealth offers is the utilization of knowledge from the consumer technology industry to improve health. Design elements such as gaming, interactivity, personalization, videos, and rewards are all tools used by technology design professionals to change consumer behavior. Although there has been some use of these tools in health behavior change and monitoring research,[65] to date, they have not yet been fully embraced. Companies that have partnered with scientists have seen the development of empirically evaluated products (e.g., the BlueStar diabetes management system[66]). These not only benefit the companies that make them but also disseminate evidenced-based treatment to those in need in ways not traditionally possible. Thus, by working together and capitalizing on their respective disciplinary strengths, experts from across the behavioral and social sciences, biomedical research, engineering, human–computer interaction, and computer sciences will be able to optimize the design of mHealth tools that can effectively measure and improve health.

Another important challenge relates to the quantification and measurement of psychological health and behavior-related aspects. These have generally been the domain of psychological or neuropsychological assessment and performed by instruments comprised of questionnaires, subjective responses to carefully crafted questionnaires. These instruments have generally been applied relatively infrequently and in specialized settings; that is, at intake and perhaps after treatment, and in medical settings requiring clinically trained personnel. In contrast, mHealth opens the opportunity to frequent and often completely unobtrusive objective assessment of behavioral and mental characteristics and/or their directly measurable correlates. To render this capability of mHealth for quantification effective, the development of new scientific areas comprising mathematical models that characterize the relationship between the measurable quantities and the behaviorally relevant aspects is necessary. For example, a fast sampling accelerometer output (of little interest in itself) must be calibrated and processed by appropriate algorithms to reflect the activity or energy expenditure of the individual in order to be useful in the assessment and behavioral modification. The development of such algorithms, so that their outputs are relevant and valid, is a considerable challenge to the interdisciplinary teams. Furthermore, the vast battery of trusted and highly utilized questionnaires and other measures in health has heretofore been developed to be stable across time and/or to detect stable changes, and can frequently take quite a bit of time to complete. However, new technologies and frequent measurement capabilities demand momentary assessments that are sensitive to momentary fluctuations in state.

A final opportunity in teams is to develop a connection between behavior change best practice guidelines and the mHealth development world. The rationale is most obvious in the commercial app world where a current review of mobile apps for diabetes highlights discrepancies between what is featured in a consumer app and the current medical guidelines.[67] Thus, while medication adherence is central in both the guidelines and the majority of apps (62%), patient education (another central feature of the guidelines) is largely absent from the apps. The gaps between the health content available to help consumers change behavior through commercial apps and what is recommended in the best practice guidelines have also been shown both in smoking cessation[61] and in weight loss.[68] Thus, there is also considerable room for enhancing the content of commercial apps to better match the best practices developed through decades of behavior change research. This also has the positive benefit of disseminating best practices in behavior change to a larger population than previously possible.

8.3.2 Finding Shared Vocabularies and Goals

Transdisciplinary teams bring a wealth of knowledge but also discipline-specific terminologies and jargon that can form barriers to collaboration. This has been referred to as the *Tower of Babel* problem.[11] Across disciplines, the same term can mean different things depending upon discipline. For instance, a human–computer interface designer and an anthropologist may mean very different things when they use the term *participation*. Different disciplines may also attach very different normative valences and values to particular terms. Furthermore, different disciplines often study the same problem in different ways, with different emphases and different goals. For example, engineers are often focused on developing sensors that *work*, which means that they respond to the physical signals with appropriate sensitivity and do not malfunction while in use. For behavior change specialists, *works* would include all of that, plus ensuring that sensors consistently measure behavior-relevant aspects (reliability and validity). Communicating these issues within the teams and including them in the requirements before engineering design and development may be complicated by geographic dispersion on campus in addition to the differing uses of vocabulary. Even between relatively colocated disciplines, such as health promotion and health psychology, major efforts have been undertaken to identify and agree upon unifying concepts and terminology.[69–71] Finally, academic disciplines needed to develop mHealth technologies and interventions might not share the same academic objectives. One example is differences in publication practices that impact progression to tenure. Publication through prestigious, peer-reviewed conferences is an accepted and valued practice in engineering, while health professionals are expected to publish in journals. Standards for acceptable evidence that something *works* or is *valid* might differ across disciplines, and when industry is brought into the mix, differences in these standards become even broader.

8.4 Opportunities

8.4.1 Sensing and Other Data

mHealth also offers a range of sensing opportunities that research has not yet fully exploited. Even a standard mobile phone is full of complex sensors, many of which can be employed for behavioral monitoring and change. For example, researchers are currently testing mobile phones to be used as balance sensors for the elderly and as accelerometers in exercise trials. But mHealth is not limited to mobile phones, and the wide range of sensors available means that researchers can now have access to continuous data across a range of variables. Recent mobile and wireless sensors have been developed to measure almost any behavior imaginable. From sensors that can monitor precise movements in patients after a stroke[72] to toe movements with foot ulcers[73] to those designed to track hand movements in eating[74] to cameras that track daily activity[75] to multiple sensors whose data can be fused to infer stress,[76] the potential for wireless measurement is great. Many of the sensors being developed now and complemented by appropriate algorithms that convert the raw measurements to useful metrics can be used to monitor behavior directly or infer it from related behaviors, and then use this information to provide feedback or track factors thought to mediate or moderate behavior change.

In addition, mHealth can also take advantage of stationary data that can be correlated with movement through global positioning systems (GPS). Thus, with proper consents, data about food purchases can be culled from loyalty cards and air quality from fixed environmental sensors. These data will allow us to better measure phenomena, but also explore complex interactions that were previously inaccessible because the data could not be collected.

8.4.2 New Theories of Behavior Change

Existing health behavior theories have served for many years as guides to intervention development and delivery, and these theories have been the basis for many face-to-face behavior change interventions. As mHealth speeds into the twenty-first century, so must behavior change theories.[77] A recent review found no effect for theory on behavior change outcomes,[30] although this may be because current theory does not capture and utilize the measurements and the metrics that would have the capabilities in timing, personalization, and timing that are inherent in modern technology. To fully leverage the potential of mobile technologies for health behavior interventions, health behavior theories need to be able to guide the development of complex interventions that adapt rapidly over time in response to real-time and real-world inputs.[11,62]

As intervention developers take full advantage of mobile technologies, health behavior models will be required to guide tailored adjustments not only at the start of an intervention but also through the dynamic process of frequent iterative adjustments during the course of intervention. The content and timing of a specific intervention can be driven by a range of variables including (1) the target behavior frequency, duration, or intensity; (2) the effect of prior intervention effects on the target behavior; and (3) the current context of the individual. Such interventions require health behavior models that have dynamic, regulatory system components to guide rapid intervention adaptation based on the individual's current and past behavior and situational context.[16] The predominately linear and static nature of current models severely limits their ability to guide the dynamic, adaptive interventions possible via mobile technologies.

8.5 Conclusion

This chapter highlights some of the current trends in behavior change research using mobile technologies. Despite tremendous enthusiasm and activity, we currently know very little about mHealth's ability to change behavior. Large numbers of clinical trials are currently underway[6] and the next few years should see an explosion of published work in this area. Provided that the field will address the challenges outlined earlier, there is every reason to expect that real-time sensing and feedback capabilities of mobile and wireless technologies will have a huge impact on behavior change research in the future.

CASE SCENARIO

Dr. Zemi, a noted exercise physiologist, is pulling together a team of researchers to develop a mHealth intervention to treat obesity through increasing physical activity in African American women. This population has a high rate of sedentary and is at increased risk for obesity and related diseases. His team includes a computer scientist, a biomedical engineer, and an electrical engineer. They have developed a mHealth application for an intervention that delivers exercise advice tailored to African American populations at set times during the day. The application has been developed by distilling information from an evidenced-based intensive 16-week face-to-face interventions.

Questions

1. The system's first trial in the field is disappointing. Name at least one discipline that Dr. Zemi is missing and explain what that person could bring to the endeavor.
2. The regularly timed messages do not seem to be working. Why might that be?
3. What can Dr. Zemi do to improve the impact of the program?
4. Dr. Zemi's intervention is based on messages being delivered at regular intervals and does not take full advantage of mHealth technologies. How might Dr. Zemi augment the intervention?
5. What is a tailored intervention?
6. Is Dr. Zemi's intervention as it now stands tailored? If not, please suggest ways of tailoring the intervention.

References

1. Kumar S, Nilsen WJ, Abernethy A, Atienza A, Patrick K, Pavel M, Riley WT et al. Mobile health technology evaluation: The mHealth Evidence Workshop. *Am J Prev Med*. 2013; **45**(2): 228–236.
2. Collins F. How to fulfill the true promise of "mHealth." *Sci Am*. 2012; **307**(1): 16.
3. Déglise C, Suggs LS, Odermatt P. Short message service (SMS) applications for disease prevention in developing countries. *J Med Internet Res*. 2012; **14**(1): e3.
4. Free C, Phillips G, Galli L, Watson L, Felix L, Edwards P, Patel V, Haines A. The effectiveness of mobile-health technology-based health behaviour change or disease management interventions for health care consumers: A systematic review. *PLoS Med*. 2013; **10**(1): e1001362.
5. Kaplan RM, Stone AA. Bringing the laboratory and clinic to the community: Mobile technologies for health promotion and disease prevention A. *Annu Rev Psychol*. 2013; **64**: 471–498.
6. Labrique A, Vasudevan L, Chang LW, Mehl G. H pe for mHealth: More "y" or "o" on the horizon? *Int J Med Inf*. 2013; **82**(5): 467–469.
7. PLOS Medicine Editors. A reality checkpoint for mobile health: three challenges to overcome. *PLoS Med*. 2013; **10**(2): e1001395.

8. Tomlinson M, Rotheram-Borus MJ, Swartz L, Tsai AC. Scaling up mHealth: Where is the evidence? *PLoS Med*. 2013; **10**(2): e1001382.

9. Saranummi N, Spruijt-Metz D, Intille S, Korhonen I, Nilsen WJ, Pavel M. Moving the science of behavioral change into the 21st century. *IEEE Pulse*. 2013; **4**(5): 23–24.

10. Nilsen WJ, Pavel M. Moving behavioral theories into the 21st century. *IEEE Pulse*. 2013; **4**(5): 25–28.

11. Hekler EB, Klasnja P, Traver V, Hendriks M. Reailizing effective behavioral management of health. *IEEE Pulse*. 2013; **4**(5): 25–28.

12. Petersen JE, Shunturov V, Janda K, Platt G, Weinberger K. Dormitory residents reduce electricity consumption when exposed to real-time visual feedback and incentives. *Int J Sustainability High Educ*. 2007; **8**(1): 16–33.

13. Kehrer P, Kelly K, Heffernan N. Does immediate feedback while doing homework improve learning? *The Twenty-Sixth International FLAIRS Conference*, St. Pete Beach, Florida. May 22–24, 2013.

14. Peck SD, Raleigh DM, Stehle Werner JL. Improved class preparation and learning through immediate feedback in group testing for undergraduate nursing students. *Nurs Educ Perspect*. 2013; **34**(6): 400.

15. Collins LM, Murphy SA, Bierman KL. A conceptual framework for adaptive preventive interventions. *Prev Sci*. 2004; **5**(3): 185–196.

16. Spruijt-Metz D, Saranummi N, Intille S, Korhonen I, Nilsen W, Spring B, Hekler EB et al. Building new computational models to support health behavior change and maintenance: New opportunities in behavioral research under review.

17. Schroeder SA. Shattuck lecture. We can do better—Improving the health of the American people. *N Engl J Med*. 2007; **357**(12): 1221–1228.

18. Mokdad AH, Marks JS, Stroup DF, Gerberding JL. Correction: Actual causes of death in the united states, 2000. *J Am Med Assoc*. 2005; **293**(3): 293–298.

19. Mokdad AH, Marks JS, Stroup DF, Gerberding JL. Actual causes of death in the united states, 2000. *J Am Med Assoc*. 2004; **291**(10): 1238–1245.

20. Keeney RL. Personal decisions are the leading cause of death. *Oper Res*. 2008; **56**: 1335–1347.

21. Lorig K. Patient-centered care: Depends on the point of view. *Health Educ Behav*. 2012; **39**(5): 523–525.

22. Waters E, de Silva-Sanigorski A, Hall Belinda J, Brown T, Campbell Karen J, Gao Y, Armstrong R, Prosser L, Summerbell Carolyn D. Interventions for preventing obesity in children. *Cochrane Database Syst Rev*. 2011; **2011**(12): CD001871.

23. Shaw Kelly, A., Gennat Hanni, C., O'Rourke, P., & Del Mar, C. Exercise for overweight or obesity. *Cochrane Database of Systematic Reviews*, (4).2006. http://onlinelibrary.wiley.com/doi/10.1002/14651858.CD003817.pub3/abstract. doi:10.1002/14651858.CD003817.pub3.

24. Stead Lindsay, F., and Lancaster, T. Group behaviour therapy programmes for smoking cessation. *Cochrane Database of Systematic Reviews*, (2). 2005. http://onlinelibrary.wiley.com/doi/10.1002/14651858.CD001007.pub2/abstract. doi:10.1002/14651858.CD001007.pub2.

25. Niederdeppe J, Farrelly MC, Haviland ML. Confirming "Truth": More evidence of a successful tobacco countermarketing campaign in Florida. *Am J Public Health*. 2004; **94**(2): 255–257.

26. Grimshaw, G., and Stanton, A. Tobacco cessation interventions for young people. *Cochrane Database of Systematic Reviews*, (4).2006. http://onlinelibrary.wiley.com/doi/10.1002/14651858.CD003289.pub4/abstract. doi:10.1002/14651858.CD003289.pub4.

27. Stamatakis KA, Leatherdale ST, Marx CM, Yan Y, Colditz GA, Brownson RC. Where is obesity prevention on the map? Distribution and predictors of local health department prevention activities in relation to county-level obesity prevalence in the United States. *J Public Health Manage Pract*. 2012; **18**(5): 402–411.

28. Spruijt-Metz D. Etiology, treatment, and prevention of obesity in childhood and adolescence: A decade in review. *J Res Adolescence*. 2011; **21**(1): 129–152.

29. Rychetnik L, Bauman A, Laws R, King L, Rissel C, Nutbeam D, Colagiuri S, Caterson I. Translating research for evidence-based public health: Key concepts and future directions. *J Epidemiol Community Health*. 2012; **66**(12): 1187–1192.

30. Head KJ, Noar SM, Iannarino NT, Grant Harrington N. Efficacy of text messaging-based interventions for health promotion: A meta-analysis. *Soc Sci Med*. 2013; **97**: 41–48.

31. World Health Organization. Prevalence of tobacco use. 2009 [cited 2013]; Available from: http://www.who.int/gho/tobacco/use/en/. Accessed on January 19, 2014.

32. Whittaker R, McRobbie H, Bullen C, Borland R, Rodgers A, Gu Y. Mobile phone-based interventions for smoking cessation. *Cochrane Database Syst Rev*. 2012; **11**: CD006611.

33. Rodgers A, Corbett T, Bramley D, Riddell T, Wills M, Lin R-B, Jones M. Do u smoke after txt? Results of a randomised trial of smoking cessation using mobile phone text messaging. *Tob Control*. 2005; **14**(4): 255–261.

34. Free C, Whittaker R, Knight R, Abramsky T, Rodgers A, Roberts I. Txt2stop: A pilot randomised controlled trial of mobile phone-based smoking cessation support. *Tob Control*. 2009; **18**(2): 88–91.

35. Free C, Knight R, Robertson S, Whittaker R, Edwards P, Zhou W, Rodgers A, Cairns J, Kenward MG, Roberts I. Smoking cessation support delivered via mobile phone text messaging (txt2stop): A single-blind, randomised trial. *Lancet*. 2011; **378**(9785): 49–55.

36. Ogden CL, Carroll MD, Kit BK, Flegal KM. Prevalence of obesity and trends in body mass index among US children and adolescents, 1999–2010. *J Am Med Assoc*. 2012; **307**(5): 483–490.

37. Flegal KM, Carroll MD, Kit BK, Ogden CL. Prevalence of obesity and trends in the distribution of body mass index among US adults, 1999–2010. *J Am Med Assoc*. 2012; **307**(5): 483–490.

38. James PT, Leach R, Kalamara E, Shayeghi M. The worldwide obesity epidemic. *Obes Res*. 2001; **9**(Suppl 4): 228S–233S.

39. Lazar MA. How obesity causes diabetes: Not a tall tale. *Science*. 2005; **307**(5708): 373–375.

40. Reaven GM. Insulin resistance: The link between obesity and cardiovascular disease. *Med Clin North Am*. 2011; **95**(5): 875–892.

41. Taubes G. Unraveling the obesity-cancer connection. *Science*. 2012; **335**(6064): 28–32.

42. Coons MJ, Demott A, Buscemi J, Duncan JM, Pellegrini CA, Steglitz J, Pictor A, Spring B. Technology interventions to curb obesity: A systematic review of the current literature. *Curr Cardiovasc Risk Rep*. 2012; **6**(2): 120–134.

43. O'Reilly GA, Spruijt-Metz D. Current mHealth technologies for physical activity assessment and promotion. *Am J Prev Med*. 2013; **45**(4): 501–507.

44. Spring B, Schneider K, McFadden HG, Vaughn J, Kozak AT, Smith M, Moller AC et al. Multiple behavior changes in diet and activity: A randomized controlled trial using mobile technology. *Arch Intern Med*. 2012; **172**(10): 789–796.

45. Tunstall-Pedoe H. *Preventing Chronic Diseases. A Vital Investment: WHO Global Report*. Geneva, Switzerland: World Health Organization, 2005, p. 200. CHF 30.00. ISBN 92 4 1563001. Also published on http://www. who. int/chp/chronic_disease_report/en. *Int J Epidemiol*. 2006; **35**(4): 1107. Accessed on January 19, 2014.

46. AIDS.gov. A timeline of AIDS. 2011 [cited 2013]; Available from: http://aids.gov/hiv-aids-basics/hiv-aids-101/aids-timeline/. Accessed on January 19, 2014.

47. Lester RT, Ritvo P, Mills EJ, Kariri A, Karanja S, Chung MH, Jack W, Habyarimana J, Sadatsafavi M, Najafzadeh M. Effects of a mobile phone short message service on antiretroviral treatment adherence in Kenya (WelTel Kenya1): A randomised trial. *Lancet*. 2010; **376**(9755): 1838–1845.

48. Mukund BK, Murray P. Cell phone short messaging service (SMS) for HIV/AIDS in South Africa: A literature review. *Stud Health Technol Inform*. 2010; **160**(Pt 1): 530.

49. Cafazzo JA, Casselman M, Hamming N, Katzman DK, Palmert MR. Design of an mHealth app for the self-management of adolescent type 1 diabetes: A pilot study. *J Med Internet Res*. 2012; **14**(3): e70.

50. Pal K, Eastwood Sophie V, Michie S, Farmer Andrew J, Barnard Maria L, Peacock R, Wood B, Inniss Joni D, Murray E. Computer-based diabetes self-management interventions for adults with type 2 diabetes mellitus. *Cochrane Database Syst Rev*. 2013; **3**: CD008776.

51. Quinn CC, Gruber-Baldini AL, Shardell M, Weed K, Clough SS, Peeples M, Terrin M, Bronich-Hall L, Barr E, Lender D. Mobile diabetes intervention study: Testing a personalized treatment/behavioral communication intervention for blood glucose control. *Contemp Clin Trials*. 2009; **30**(4): 334–346.

52. McCarty D, Gustafson D, Capoccia V, Cotter F. Improving care for the treatment of alcohol and drug disorders. *J Behav Health Serv Res*. 2009; **36**(1): 52–60.

53. Marsch LA. Technology-based interventions targeting substance use disorders and related issues: An editorial. *Subst Use Misuse*. 2011; **46**(1): 1–3.

54. Ben-Zeev D, Drake RE, Corrigan PW, Rotondi AJ, Nilsen W, Depp C. Using contemporary technologies in the assessment and treatment of serious mental illness. *Am J Psychiatric Rehabil*. 2012; **15**(4): 357–376.

55. Ben-Zeev D, Kaiser SM, Brenner CJ, Begale M, Duffecy J, Mohr DC. Development and usability testing of FOCUS: A smartphone system for self-management of schizophrenia. *Psychiatric Rehabi J*. 2013; **36**(4): 289–296.

56. Aguilera A, Mu√±oz RF. Text messaging as an adjunct to CBT in low-income populations: A usability and feasibility pilot study. *Prof Psychol Res Pract*. 2011; **42**(6): 472.

57. United States Department of Veteran's Affairs. Mobile App: PTSD Coach. 2013 [cited 2013]; Available from: http://www.ptsd.va.gov/public/pages/ptsdcoach.asp. Accessed on January 19, 2014.

58. Foa EB, Kozak MJ. Emotional processing of fear: Exposure to corrective information. *Psychol Bull*. 1986; **99**(1): 20.

59. Noar SM, Harrington NG, Aldrich RS. The role of message tailoring in the development of persuasive health communication messages. *Commun Yearbook*. 2009; **33**: 73–133.

60. Lester RT, Ritvo P, Mills EJ, Kariri A, Karanja S, Chung MH, Jack W et al. Effects of a mobile phone short message service on antiretroviral treatment adherence in Kenya (WelTel Kenya1): A randomised trial. *Lancet*. 2010; **376**(9755): 1838–1845.

61. Abroms LC, Padmanabhan N, Thaweethai L, Phillips T. iPhone apps for smoking cessation: A content analysis. *Am J Prev Med*. 2011; **40**(3): 279–285.

62. Spring, B., Gotsis, M., Paiva, A., and Spruijt-Metz, D. Healthy apps: mobile devices for continuous monitoring and intervention. *IEEE PULSE*; 2013. **4**(6), 34–40. doi: 10.1109/mpul.2013.2279620.

63. WellDoc. BlueStar. 2013; Available from: http://www.bluestardiabetes.com. Accessed on January 19, 2014.

64. Chomutare T, Fernandez-Luque L, Årsand E, Hartvigsen G. Features of mobile diabetes applications: Review of the literature and analysis of current applications compared against evidence-based guidelines. *J Med Internet Res*. 2011; **13**(3): e65.

65. Pagoto S, Schneider K, Jojic M, DeBiasse M, Mann M. Evidence-based strategies in weight loss mobile apps. *Am J Prevent Med*. 2013; **45**(5): 576–582.

66. Michie S, van Stralen MM, West R. The behaviour change wheel: A new method for characterising and designing behaviour change interventions. *Implement Sci*. 2011; **6**: 42.

67. Michie S, Johnston M, Abraham C, Lawton R, Parker D, Walker A. Making psychological theory useful for implementing evidence based practice: A consensus approach. *Qual Saf Health Care*. 2005; **14**(1): 26–33.

68. Abraham C, Michie S. A taxonomy of behavior change techniques used in interventions. *Health Psychol*. 2008; **27**(3): 379.

69. Dobkin BH, Dorsch A. The promise of mhealth daily activity monitoring and outcome assessments by wearable sensors. *Neurorehabil Neural Repair*. 2011; **25**(9): 788–798.

70. Shaporev A, Gregoski M, Reukov V, Kelechi T, Kwartowitz DM, Treiber F, Vertegel A. Bluetooth TM enabled acceleration tracking (BEAT) mHealth system: Validation and proof of concept for real-time monitoring of physical activity. *E-Health Telecommun Syst Networks*. 2013; **2**(3): 49–57.

71. Salley JN, Scisco JL, Muth ER, Hoover A. A comparison of user preferences and reported compliance with the bite counter and the 24-hour dietary recall. *Proceedings of the Human Factors and Ergonomics Society Annual Meeting, 2012*: SAGE Publications, Boston, Mass, 2012, pp. 2177–2180.

72. Doherty AR, Hodges SE, King AC, Smeaton AF, Berry E, Moulin C, Lindley Sn, Kelly P, Foster C. Wearable cameras in health: The state of the art and future possibilities. *Am J Prev Med.* 2013; **44**(3): 320.
73. Ertin E, Stohs N, Kumar S, Raij A, al'Absi M, Shah S. AutoSense: Unobtrusively wearable sensor suite for inferring the onset, causality, and consequences of stress in the field. *Proceedings of the 9th ACM Conference on Embedded Networked Sensor Systems, 2011*: ACM. Seattle, Washington, 2011, pp. 274–287.
74. Riley WT, Rivera DE, Atienza AA, Nilsen W, Allison SM, Mermelstein R. Health behavior models in the age of mobile interventions: Are our theories up to the task? *Transl Behav Med.* 2011; **1**(1): 53–71.

9

Implementing Behavior Change: Evaluation Criteria and Recommendations for mHealth Applications Based on the Health Action Process Approach and the Quality of Life Technology Framework in a Systematic Review

Angela Brunstein, Joerg Brunstein, and Michael K. Martin

CONTENTS

9.1 Introduction

In this review of current mobile health and fitness applications, we focus on mHealth applications' potential for preventive medicine interventions. According to the American College of Preventive Medicine (ACPM), this discipline aims to "… protect, promote, and maintain health and well-being and to prevent disease, disability and death … [in] individuals, communities, and defined populations" (American College of Preventive Medicine 1998). Currently, the biggest application area for preventive medicine is noncommunicable diseases (NCD). In 2008, NCD accounted for 63% of global death (WHO 2010) and prevalence of NCD is predicted to increase significantly over the next years. According to the global status report on NCD (WHO 2010), the top four killers are cardiovascular disease, cancers, diabetes, and chronic lung diseases. Risk factors for these conditions are strongly related to lifestyle and health behavior, including unhealthy diet, physical inactivity, and tobacco use (WHO 2009). Increasing protective health behaviors in patients is related to reducing health risks, improving health, and prolonging expected life span (e.g., Ahmed et al. 2013). Therefore, the World Health Organization (WHO 2010) has published recommendations for different health behaviors, including healthy diet and exercise, for reducing risk factors for NCD. For children, the WHO (2010) recommends at least 60 min of at least moderate physical activity a day and for adults 150 min of moderate or 75 min of vigorous physical activity during the week with aerobic activity sessions lasting at least 10 min.

Despite consensus and evidence, majority of adults do not adhere to recommended levels of health behaviors, for example, of physical activity (Guthold et al. 2008; WHO 2002). In 2009, the WHO published a review of interventions that have been demonstrated to increase physical activity and concluded based on 937 diet studies and 776 physical activity studies that best interventions addressed both physical activity and healthy diet, and not just either of the two. In addition, involving users is more promising than just prescribing healthy levels. On a conceptual level, health behavior theories have been shown, first, to predict who is likely to succeed in improving their lifestyle and who will not (e.g., Armitage and Conner 2001). Therefore, these theories can be used for assessing potential users of mHealth applications. Second, they have successfully identified relevant dimensions contributing to behavior change. These dimensions need to be addressed in interventions when fostering behavior change. Mobile health and fitness applications could tap into this potential of health theories for increasing health behavior and reducing risk of NCD.

When implementing interventions using mobile technology, mobile phone applications can use intrinsic properties of mobile technology to support behavior change, namely, that mobile phones are easily accessible, online, and interactive. These properties allow mobile technology to bridge spatial and temporal distance. Instead seeing family physician every other month, mobile phone applications can provide guidance for every meal and every workout. This means that advice becomes more specific: Instead of "Reduce fat intake!" it can say "Eat this, not that!" At the same time, reported health behavior becomes more specific and more accurate. Instead of a patient reporting back: "… the last month went mostly ok," a mobile application can report every step and every consumed calorie with a timestamp. This kind of data permits an accurate analysis of potential challenges and correspondingly designing better interventions. Evidence for effective mHealth interventions starts to accumulate (e.g., de Jongh et al. 2012; Tomlinson et al. 2013). At the same time, there is increasing awareness of the need for safe and valid applications given the vast amount of applications available on the commercial market resulting, for example, in the U.S. Food and Drug Administration's guidance for mobile medical applications (2013).

As mentioned earlier, health behavior theories can assist creating efficient mobile applications by providing means for assessment, relevant dimensions to be covered, and evidence-based interventions. There are several health behavior theories in the field, each associated with evidence and all of them with partially overlapping concepts. The most prominent theories are the *health action process approach* (HAPA, Schwarzer 2008), the *health belief model* (HBM, Becker and Rosenstock 1984), the *theory of planned behavior* (TPB, Ajzen and Fishbein 1980), and the *transtheoretical model* (TTM, Prochasta and Velicer 1997). They all describe the intention–behavior gap (Sheeran 2002): not all people who intend to improve their health behavior actually succeed. Theories also agree that different factors determine who will succeed and who will not, and most of them have been used to create successful interventions for improving diverse health behaviors in patients.

For this evaluation of current mHealth applications, we chose the HAPA as a hybrid model that unifies two lines of health behavior models, models covering factors impacting health behaviors and models focusing on the process of change. HAPA has been designed to cover most relevant factors from earlier theories. As a hybrid model, it stresses the process of changing behavior as well as the factors contributing to different states during that process. HAPA has been used for predicting and impacting diverse health behaviors, including dental flossing (Schuez et al. 2006, 2009), breast self-examination (Luszcynska and Schwarzer 2003), physical activity after cardiac rehabilitation (Scholz et al. 2005), and weight loss (Renner and Schwarzer 2005). For these application areas, HAPA's framework and interventions have been extensively tested using statistical modeling methods on empirical data. When applying HAPA components for evaluation of mobile health and fitness applications, these theoretical concepts need to be translated into functionality of applications. In 2012, we performed this exercise of translating components of HBM, TPB, and TTM into evaluation criteria and used those for a systematic review of 14 out of 100 iTunes' *Health & Fitness* applications (Brunstein et al. 2012). Similarly, Azar and colleagues (2013) used criteria based on HBM, TTM, TPB, and social cognitive theory (Bandura 2001) for evaluating 23 free, stand-alone diet applications out of 200 iTunes Health & Fitness applications. Here, we have chosen HAPA for our evaluation because it unifies theoretical constructs from different health behavior models in one approach. Compared with the eclectic pooling of single components from different theories, a unified framework is more powerful because it allows investigating interactions between factors over time. Earlier reviews have identified very little use of health behavior theories in current applications. Using a unified framework like HAPA goes beyond detecting missing coverage by permitting recommendations on how to implement relevant factors in mobile applications for different user groups and for different time points along the process of adopting healthier habits.

Even if a mobile application would include all relevant HAPA criteria, it would be useless without minimal standards for usability. Probably some people would stop eating cookies if they had to monitor their diet, and entering a cookie would take 20 clicks and searching several databases. There exist several approaches for designing successful mobile applications (e.g., DeLone and McLean 2003). We decided to use Schulz and colleagues' interdisciplinary approach of designing and evaluating *Quality of Life Technologies* (QoLTs; Schulz et al. 2012) because this approach is patient centered and can be used for design and evaluation of patient support systems along four dimensions. In the following, we describe the HAPA model and the QoLT dimensions as used in our review more in detail.

9.1.1 Health Action Process Approach

HAPA discriminates between different groups of users based on phases during the process of behavior change. This is important because users' needs for support change during that process. In addition, HAPA defines several factors impacting the process. Each factor is relevant for user's behavior change during a specific phase or state of the process, but not during other phases. For mobile health and fitness applications, the HAPA phases can be used for assessing users, and the factors can be used for providing appropriate support for users given their current state.

9.1.1.1 Assessing Phases of Change and Associated User Groups

HAPA discriminates two broad phases of change, a *motivation* or *goal setting phase* and a *volition* or *goal pursuit phase*. Users during the motivation phase might know that they need to lose weight and therefore have to become more physically active. As long as they have not started planning behavior change, HAPA calls these users *nonintenders.* This label is a little misleading because the motivation phase includes a state of intention. Basically, nonintenders move from a preintentional state to an intention state within the motivation phase resulting in a general but unspecific aim to change their health behavior.

Once users have decided to actually change their lifestyle correspondingly, they enter the volition phase that includes a planning state and several action states. Users during the planning state are called *intenders*. During this state, intenders define the *when*, *where*, and *how* of behavior for *action planning.* They might subscribe to a fitness club or buy running shoes. In contrast to intention state, intenders plan to start workout next Monday, not sometime soon. In addition to spelling out action plans for workout and diet, intenders should think about potential barriers when changing their lifestyle during *coping planning.* Maybe it is easier to keep a healthy diet when eating alone, but not during family meals or eating out with friends. They should also identify strategies for coping with these challenges in advance. One option to combat grandma's pies is eating a light meal before joining the feast for feeling less hungry.

Once users have started to enact their plans, they enter the action state and are called *actors*. Within the action phase, HAPA discriminates between early implementation of new behaviors, the *initiative state*, later building of habits during *maintenance state*, and maybe *recovery state* after relapse. The initiative state covers the first weeks in the gym. During this state, actors acquire necessary skills. Once actors have become accustomed to new behaviors and start to build and maintain a habit of exercising over a period of months or years, they are in a *maintenance state*. From there, they might remain in maintenance state; they might leave once they have reached their personal goal, for example, fitting into the wedding dress; or they might suffer a relapse. After relapse, they might pass through a recovery state and kick off another planning, initiative, or maintenance state. According to HAPA, states are not fixed and users can move forward and backward between states several times. It is important for mobile health and fitness applications to assess a user's current state within the process of behavior change for providing appropriate support for that state.

9.1.1.2 Identifying Relevant Factors Impacting the Process of Behavior Change

Users in different states have different needs for support: nonintenders might profit from motivational support for setting personal weight loss goals. That support would be wasted on intenders who have already set their goals. Intenders and actors during initiative state might profit from detailed instruction on how to perform exercise without injury. This would be useless for users in maintenance state who are already exercising for months. HAPA has identified five factors impacting transitions between states. Nonintenders will come up with specific weight loss goals only if they believe that they can perform exercise (i.e., *action self-efficacy*), if they perceive a health risk if not exercising (i.e., *risk perception*), and if they believe that exercising could reduce that health risk (i.e., *positive outcome expectancies*). According to HAPA, risk perception and outcome expectancies impact only intention, but not planning or action states. In contrast, action self-efficacy also impacts planning. During action state, action self-efficacy loses impact as predicting variable, probably because everybody who is exercising knows now by experience that they can do. From planning state on, *coping self-efficacy*, as the belief to be able to overcome barriers, becomes important. Users who score high on coping self-efficacy are more likely to engage in coping planning. They might decide going jogging on Monday, which is their original rest day, if they have an important business meeting on Tuesday and go back to normal on Wednesday. Not surprisingly, these users will also cope better with barriers when they materialize. Therefore, coping self-efficacy also impacts the action phase. Finally, *recovery self-efficacy* concerns users' belief whether they can come back to action after relapse. People with high recovery self-efficacy tend to attribute relapse to a difficult situation or setting. They perceive relapse less dramatically than people who score low on recovery self-efficacy and attribute relapse to their own failure. They are also more likely to master relapse and return quickly to normal.

Integrating phases and impact factors, HAPA has been successfully used to design interventions. Nonintenders have profited from *risk and resource communication* (e.g., Cranium et al. 2012). A mobile application might calculate the body mass index (BMI) for a user and list the associated health risks. It might also suggest small changes in behavior and list associated health benefits. Intenders have benefited from an exercise for action and coping planning that provides skills for translating intentions into actions (e.g., Wiedemann et al. 2012): they were asked to write down a few specific goals with spelling out the *when, where, and how* for action planning. They were also asked to envision a few specific challenges and to identify solutions for those challenges for coping planning. A mobile app might ask for specific goals during setup and adjust suggested workout to those goals. In addition, the app might suggest automatic reminders during setup, for example, for drinking water. Actors usually need very little support, but might profit from relapse prevention and management when translating newly acquired behaviors into habits. A mobile application might present frequently encountered barriers and strategies for overcoming those barriers. These will be particularly important in the context of quality of life dimensions.

9.1.2 Quality of Life Technology Evaluation Dimensions

Schulz and colleagues' framework (2012) has been used for designing QoLT systems and for evaluating existing QoLT systems. The framework suggests four dimensions for defining outcome measures and for evaluating ease of operation in general.

The first dimension is the *functional domain* that an application is designed for. That domain could be as general as weight loss or as specific as yoga instruction. Specifying the functional domain is necessary for defining outcome measures. It also prevents comparing apples with oranges. Therefore, we categorized health and fitness applications by functional domain before comparing their functionality.

Second, Schulz and colleagues differentiate *kinds of functioning support* along a continuum between compensating diminished functioning, like a wheelchair for restoring mobility, and enhancing normal functioning, like a bicycle for going beyond a person's normative physical mobility. In between these two endpoints, systems can help to maintain normative functioning or prevent decline, like regular workout or physiotherapy. Most mobile applications support maintaining functioning and preventing decline. Some athletic applications support enhancement of normative functioning. Applications for management of clinical conditions help to compensate for diminished functioning. Schulz and colleagues state that highly intelligent systems address needs of users throughout the continuum. Closest to that ability to adapt to users' functioning level are series of applications, like Couch to 5K, *5K to 10K, 10K to Half Marathon,* and *Half Marathon to Marathon* running applications that span a range from prevention to enhanced functioning, but do not include compensation.

Third, Schulz and colleagues address the *mode of operating the system.* QoLT systems can be operated passively without requiring user input or interactively requesting user input for operation. Passive calorie tracking is obviously more convenient than manually entering every cookie. On the other hand, it can be very frustrating to discover that only 200 steps have been tracked after hiking for hours in the woods because of poor GPS coverage. Therefore, a hybrid model with options for passive or (inter-)active operating technology might be optimal for mobile health and fitness applications.

Fourth and finally, Schulz and colleagues address the *system intelligence* as the ability to learn from a user and to adapt to a user's needs and preferences. For example, some workout planning tools (e.g., *Gymprovise,* Gymprovise) adjust a suggested workout to available equipment and suggest alternatives if currently not available. We found it useful to discriminate between *global* and *local system intelligence.* Many mobile calorie tracking applications globally adjust their calculations to a person's body weight, for example, when displaying burned calories for a workout. Applications with local system intelligence can adapt to current performance during workout or diet. This is essential for relapse prevention and management.

Targeted functional domain, kind of functioning support, passive or interactive mode of operating the system, and system intelligence do not assess the complete spectrum of mobile applications' usability, but they provide a useful proxy for assessing what is currently on the market and what is not yet.

9.2 Methods

9.2.1 Selecting mHealth Applications

We have reviewed application descriptions for mobile applications at the iTunes Store and the Google Play App Store that are listed under the *Health & Fitness* categories as of June 21, 2013. I wish we could update the analysis. But this would mean to redo the complete research. It might be that some of the apps that we analyzed cannot be found anymore and that others that we did not consider are relevant now or new on the market. Nevertheless, the framework would apply to the recent apps as well and the examples should sufficiently illustrate how to use the framework for evaluating recent applications. Within those categories, we selected the list of top free and paid applications. This resulted in 571 health and fitness applications for Mac and 400 medical and 400 health and fitness applications for iPhones. For Androids, we gathered 480 medical and 480 health and fitness applications. For both platforms, we chose the U.S. stores resulting in total 2331 applications in the first selection list.

We applied several rounds of filtering before generating the final list based on titles, publishers, application description, and screenshots on their store website. Mostly, we excluded duplicates and applications that were either unrelated, did not utilize mobile technology, were not directly related to relevant health behavior of humans, or did not work on our devices.

During the first filtering, we included applications that are related to health behavior, have an English title and description, and are designed for humans. We excluded non-English titles or descriptions, medical education materials for exam preparation, medical reference and translation tools, applications for animals, games without health content, beauty tips, and applications that seem to be misclassified, like egg timer. After the first filtering, there were 1385 applications left.

During a second filtering, we eliminated applications that had several versions across and within platforms. For example, we kept only one copy of applications that have a version for iPhone and Android. We also kept only one version for applications that have several siblings. For example, *Daily Workout* (Daily Workout Apps) exists for arm, butt, leg, cardio, and yoga. Similarly, several running apps have subversions for couch to 5K, 10K, Half Marathon, and Marathon. If possible, we preferred general applications (e.g., *Daily Workout*, Daily Workout Apps) before specific applications (e.g., *Daily Leg Workout*, Daily Workout Apps) and applications that had different programs integrated within the application (e.g., from *couch to 5K* up to *Half Marathon to Marathon*) before applications for a single program. We categorized applications based on the QoLT evaluation dimensions according to their targeted functional domain. Thereafter, we excluded functional domains that are not central to health behavior, for example, hypnosis and meditation, or that target specific situations in a user's life, like pregnancy or baby applications. After filtering, we had 435 applications left with a majority from exercise ($N = 187$) and diet ($N = 132$) but also health management, including monitoring physiological measures ($N = 47$), for example, blood pressure; adherence ($N = 18$); diabetes management ($N = 14$); sleep, relaxation, and stress management support ($N = 14$); substance cessation ($N = 12$), mainly alcohol and smoking; patient education ($N = 6$); and rehabilitation support ($N = 5$).

We decided to focus on diet and exercise as the most central health behaviors impacting people's weight loss and health, as covered best in the literature and based on the biggest numbers of applications covered in the Health & Fitness applications stores.

Within the exercise category, we discriminated between general workout ($N = 85$), including short-duration workout, HIIT, and workouts of the day (WOD); distance sports programs ($N = 50$), like running and biking; activity tracking ($N = 13$); bodybuilding ($N = 11$); yoga and Tai Chi ($N = 13$); warm-up and stretching ($N = 4$); and lifestyle ($N = 5$), including weight loss and fitness assessment. We excluded six applications that were either unrelated after detailed inspection, like music selection when jogging, or had too few applications for creating a subcategory.

Within the diet category, we discriminated between specific diet applications ($N = 26$), like Paleo diet, food databases ($N = 43$), calorie trackers ($N = 60$), and comprehensive weight loss programs ($N = 3$). For both exercise and diet, we found an impressive range of functionality and creative use of mobile technology potential among applications. For example, for measuring heart rate at rest and during exercise, there exist comparably accurate sensor measures next to iPhone camera estimates. We also found significant overlap in functions across applications. For example, there exist several GPS trackers with only marginal differences that would look all the same in our analysis. Therefore, we applied an approach of *fishing for features* for representing the range of functions and features that are currently on the market for presentation in this chapter. For each subcategory, we identified one or two candidates with basic functionality, for example, a manual entry running log. In addition, we sampled advanced options, like pedometers, GPS trackers, integration of sensor data, optional life tracking and sharing, color-coded performance analysis, and race against earlier runs or opponents.

Tables 9.1 through 9.3 include evaluation results for 41 exercise applications and 21 diet applications. We downloaded these applications from the iTunes or Google Play shop and installed them onto an iPhone 4s, iPod touch 32 GB, or onto a Nexus 4 (16 GB) phone or an Acer Iconia B1 tablet. After inspecting the applications and using it for its designed purpose for a few days, we classified them concerning HAPA coverage and QoLT dimensions.

9.2.2 Classifications for HAPA Components and QoLT Dimensions

Based on Schwarzer (2008) for the HAPA components and on Schulz and colleagues (2012) for the QoLT dimensions, we developed the following operational definitions for evaluating current health and fitness applications. We revised these definitions within the workgroup and finally reached an interrater reliability of 97% for a subgroup of 10 applications rated independently.

9.2.2.1 Assessing Health Status

Assessing a person's health status is relevant for communicating risks to nonintenders (HAPA) and is a prerequisite for global system intelligence (QoLT dimensions). We classified an application as assessing initial health measures if asking at least for a user's age, gender, weight and height during setup, and better also for clinical conditions.

9.2.2.2 Assessing Status of Behavior Change

HAPA's strongest claim is tailoring behavior change support according to a user's needs based on their state along the process of behavior change.

We classified an application as assessing behavior change states (BCSs) if at least asking for users' goals. During this stage, we did not investigate whether applications use that information for tailoring support. This feature was evaluated under system intelligence.

9.2.2.3 Intention State Support

HAPA recommends for intention state risk and resource communication. Most health and fitness applications do not serve nonintenders by definition, because users start to search for appropriate applications during planning state *after* setting a goal.

9.2.2.4 Risk Communication

When discussing pros and cons of a critical health behavior, risks are associated with *not* performing that health behavior. Communicating injury risk associated with performing exercise does not qualify for risk communication.

We classified applications as communicating risks if they addressed health risks associated with poor diet or physical inactivity.

TABLE 9.1

Evaluated Performance Tracking and Walking Program Applications, with Supported Platforms

Applications		Platform		Assessment		Intervention							Functioning Support		Mode of Operating		Intelligence
						I		II		III							
Application	Publisher	iT	An	H	BCS	Ri	Re	AP	CP	IS	MS	RP	Pr	En	I	T	
(a) Accupedo-Pro	Corusen LLC	1	1	1	1	0	0	0	0	1	1	0	1	0	0	P	L
BiCycle	Valley Development	1	1	1	0	0	0	0	0	0	0	0	1	0	0	P	0
Endomondo	Endomondo	1	1	1	0	0	0	0	0	1	1	0	1	0	0	P	0
Heart Rate Pro	Runtastic	1	1	1	0	0	0	0.5	0	1	1	0	1	0	0	A	0
MapMyFitness+	MapMyFitness Inc.	1	1	1	1	0	0	1	0	0	1	0	1	0	0	P+A	0
MyFitnessCompanion	myFitnessCompanion	1	0	1	0	0	0	0	0	0	1	0	1	0	0	P+A	0
Nexercise	Nexercise	0	1	0	0	0	0	1	0	1	1	0	1	0	A	P	L
Road Bike Pro	Runtastic	1	1	0	0	0	0	1	0	0	1	0	1	0	0	P	G+L
SmartRunner Pro	APPSfactory	1	1	1	0	0	0	0	0	0	1	0	1	0	0	P	0
Walkmeter	Abvio Inc.	1	1	1	0	0	0	1	0	1	1	0	1	0	P	P	G
WOD Tracker Pro	Pete Wood	1	0	0	0	0	0	0.5	0	0	1	0	0	1	0	A	0
(b) 5K Runner	Clear Sky Apps LTD	1	0	0	0	0	1	1	0	0	1	0	1	0	P	0	0
B210K Pro	Guy Hoffman	0	1	0	0	0	0	0.5	0	0	1	0	1	0	P	0	0
FIRST	Bryan Catron	1	0	0	1	0	0	1	0	0	1	0	1	1	P	0	G
Get Running	Splendid Things	1	1	0	0	0	0	0	0	1	0	0	1	0	P	0	0
Higdon Novice1 26.2	Bluefin Software, LLC	1	1	0	1	1	1	1	1	1	0	0	1	1	P	P	0
Run a 10K PRO!	Red Rock Apps	1	1	0	0	1	1	0.5	1	1	1	1	1	0	P	P	0
Zombies, Run!	Six to Start	1	1	0	0	0	0	0	0	1	1	1	1	1	P	P	0

Notes: iT, iTunes; An, Android; initial health (H) and BSC evaluation; offered interventions (I or the intention state: Ri, Re, risk and resource communication; II or the planning state: AP, CP, action and coping planning support; III or action states: IS, initiative support; MS, maintenance support; RP, relapse prevention); kind of functioning support (Pr, prevention; En, enhancing normal functioning); active versus passive mode of operating the system (I, instruction; T, tracking); system intelligence (G, global; L, local).

TABLE 9.2

Evaluated Workout Applications, with Supported Platforms

Application	Publisher	Platform iT	An	Assessment H	BCS	Intervention I Ri	Re	II AP	CP	IS	III MS	RP	Functioning Support Pr	En	Mode of Operating I	T	Intelligence G
(a) 7 min Workout	UOVO	1	0	0	0	0	0	0.5	0	1	0	0	1	0	P	0	0
Daily Workouts	Daily Workout Apps	1	1	0	0	0	0	0.5	0	1	0	0	1	0	P	0	0
Gorilla	Heckr LLC	1	1	1	0	0	0	0.5	0	1	0	0	1	0	P	0	G
(b) BYFWM!!	Creative Freaks	1	1	0	0	0	1	0.5	0	1	0	1	1	0	P	A	L
Just 6 Weeks	Just Do Inc.	1	1	1	0	0	0	0.5	0	1	0	0	1	0	P	A	G
Pushups	Clear Sky Apps LTD	1	0	0	0	0	0	1	0	1	0	0	1	0	P	A	0
PushUps PRO	Runtastic	1	1	0	0	0	0	0.5	0	1	0	0	1	0	P	P	0
(c) Daily Yoga	DailyYoga.com	1	1	0	0	1	1	1	0	1	1	0	1	0	P	0	0
Stretch Guru: Run	Lori Bryan Living	0	1	0	0	0	1	1	0	1	1	0	1	0	P	0	0
Taichi8+16	ChuFengChin	1	1	1	0	0	0	0	0	1	0	0	1	0	P	0	0
(d) Fitness	Plus Sports	1	0	1	1	1	1	1	1	1	1	1	1	0	P	P+A	0
Caynax HIIT PRO	Caynax	0	1	0	0	0	1	1	0	0	1	0	1	0	0	0	0
Nike Training	Nike	1	1	0	1	0	0	0.5	0	1	0	0	1	0	P	0	0
(e) BBB	WorkoutRoutines	0	1	0	0	0	0	1	0	0	1	0	0	1	P	A	0
Gym Genius	AppBrains	1	0	1	0	0	0	1	0	0	1	0	0	1	0	A	0
Gymme	Matteo Pizzorni	0	1	1	1	0	0	1	0	0	1	0	0	1	A	A	G
Gymprovise	Gymprovise	0	1	1	0	0	0	1	1	1	0	1	1	0	A	P	L
myWOD	theDreamWorkshop	1	1	1	0	0	0	1	0	0	0	1	1	0	P	P	0
WOD Programmer	CrossfitHardcore	1	1	0	0	0	1	0	0	0	0	0	1	0	0	A	0
Workout Hero	Storeboughtmilk LLC	1	0	1	0	0	1	1	0	0	0	1	1	0	P	A	0
(f) U.S. Navy PFA Calculator	Crash Test Dummy	0	1	1	0	0	0	0	0	0	0	0	0	0	0	A	0
SuperBetter	SuperBetter Labs	1	0	0	1	1	1	1	1	1	1	1	1	0	A	A	G+L

Notes: iT, iTunes; An, Android; initial health (H) and BSC evaluation; offered interventions (I or the intention state: Ri, Re, risk and resource communication; II or the planning state: AP, CP, action and coping planning support; III or action states: IS, initiative support; MS, maintenance support; RP, relapse prevention); kind of functioning support (Pr, prevention; En, enhancing normal functioning); active versus passive mode of operating the system (I, instruction; T, tracking); system intelligence (G, global; L, local).

TABLE 9.3

Evaluated Diet Applications, with Supported Platforms

Applications		Platform		Assessment		Intervention							Functioning Support		Mode of Operating		Intelligence
						I		II			III						
Application	Publisher	iT	An	H	BCS	Ri	Re	AP	CP	IS	MS	RP	Pr	En	I	T	
(a) 3 Day Diet	Realized Mobile LLC	1	1	0	0	0	0	1	0	1	0	0	1	0	P	P	0
17 Day Diet Complete	Realized Mobile LLC	1	1	1	1	0	0	1	0	1	0	0	1	0	P	A	0
Fast Metabolism Diet	Random Digital Inc.	1	0	1	0	1	1	1	1	1	0	1	1	0	P	A	0
Low Fodmap Diet	CEMA	1	0	1	0	1	1	1	1	1	1	0	1	0	P	A+P	L
HCG	Appilicious	1	1	1	1	1	1	1	0	1	0	0	1	0	P	A	L
FiiT Chick	Brett Campbell	1	1	0	0	1	1	1	0	1	0	0	1	0	P	0	0
(b) Eat This!	Rodale Inc.	1	1	1	0	0	1	1	1	1	1	1	1	0	P	0	G+L
Green Smoothies	Raw Family	1	1	0	0	1	1	1	0	1	1	0	1	0	P	0	0
NxtNutrio	Nxtranet Inc.	1	0	1	0	1	1	0	1	1	1	0	1	0	P	0	G+L
Paleo Central	Nerd Fitness	1	1	0	0	0	0	0.5	0	0	1	0	1	0	P	0	0
Points Calculator	Ellisapps Inc.	1	1	0	0	0	0	0.5	0	1	1	0	1	0	0	A	0
(c) CaloryGuard Pro	BlueBamboo	1	1	1	0	0	0	0.5	0	1	1	0	1	0	P	A	L
Carb Master	Deltaworks	1	0	1	1	0	0	0.5	0	1	1	0	0	1	0	A	0
CarbsControl	Coheso Inc.	1	0	0	1	0	0	1	0	1	1	0	0	1	P	A	0
Cronometer	BigCrunch Consulting	1	1	1	1	0	0	0.5	0	1	1	0	1	0	P	A	L
DietBet	DietBet	1	0	1	0	0	0	0	0	0	1	0	0	1	P	A	0
Nutrition Menu	Shroomies	1	0	0	0	0	0	0.5	0	0	1	0	1	0	0	A	0
Nutritionist	Nutritionist	1	1	1	1	1	1	1	1	1	1	1	1	0	P	A+P	G
(d) Bodybuilding Diet	Bodybuilding Diet - Pro	0	1	1	1	0	1	1	1	0	1	0	1	0	P	A+P	G
My Diet Coach - Pro	InspiredApps	1	1	1	1	0	1	1	1	1	0	1	1	0	P	A	0
Noom	Noom Inc.	1	1	1	1	0	1	1	1	1	1	1	1	0	A	A+P	G+L

Notes: iT, iTunes; An, Android; initial health (H) and BSC evaluation; offered interventions (I or the intention state: Ri, Re, risk and resource communication; II or the planning state: AP, CP, action and coping planning support; III or action states: IS, initiative support; MS, maintenance support; RP, relapse prevention); kind of functioning support (Pr, prevention; En, enhancing normal functioning); active versus passive mode of operating the system (I, instruction; T, tracking); system intelligence (G, global; L, local).

9.2.2.5 Resource Communication

Benefits associated with physical activity and healthy diet complement health risks associated with inactivity and poor diet. According to HAPA, positive outcome expectancies include health benefits, but also expected positive social or emotional outcomes.

We classified applications as addressing resource communication if they addressed benefits of performing exercise or eating healthy.

9.2.2.6 Planning State Support

HAPA recommends for planning state to provide metacognitive support for action planning and coping planning by translating goals into action.

9.2.2.7 Action Planning Support

Action planning translates goals into behavior by specifying the *when, where, and how* of the concerned behavior. Action planning support can either provide structure for this process or suggest options for this mapping.

We classified applications as supporting action planning if they can be used to envision exercise or diet as specific as possible before actually doing it. As a special case, we coded applications that can be *misused* for planning by fast-forward simulations as action planning support even if the application was not designed to support planning.

9.2.2.8 Coping Planning Support

While action planning spells out parameter for critical behavior, coping planning concerns expected barriers. Coping planning support can either provide structure for this process or suggest typical barriers and solutions.

We classified an application as supporting coping planning if mentioning or asking for potential challenges *before* performing the concerned health behavior.

9.2.2.9 Action State Support

According to HAPA, users in action phase do not need much action support, but might profit from relapse prevention and management support. Because actors solve different tasks during initiative, maintenance, and recovery state, they are expected to change in respect to their needs and preferences during these states.

9.2.2.10 Action Initiative Support

During the action initiative state, users start to perform the desired behavior for a few days or weeks. For the user who planned to join the health club, this is the first few weeks in the gym. During this state, users acquire required new skills. Correspondingly, users will profit from detailed and specific instruction, demonstration, and feedback.

We classified an application as suitable for initiative action state if it provided detailed instruction or demonstration on what to do, for how long to do, and how to do. In contrast to action planning, this instruction should be in time and allow to perform an exercise along with the model.

9.2.2.11 Action Maintenance Support

During the *maintenance state,* users continue to perform the desired behavior over a longer period of time in terms of months to years. During this state, users form and maintain new habits. Users do not need detailed instruction anymore because they have sufficiently learned the procedures already.

We classified an application as suitable for maintenance state if it provided support for performance over longer periods of time, including tracking current performance and logging performance over time, analyzing past workouts, and providing detailed feedback for increasing performance. Applications that fight boredom in the gym might also provide maintenance support.

9.2.2.12 Relapse Prevention

Several factors impact how well a user will master challenges. As for coping planning, users can profit from metacognitive support for anticipating and mastering barriers.

As for the intention state, it is very difficult to support users during a current crisis because it is very unlikely that they will use the application for entering a disastrous over-eating day. It is also unlikely to experience an application's relapse management support when testing an application only for a few days. Therefore, we evaluated applications' *relapse prevention* support, but not *relapse management and recovery* support.

We classified an application as providing relapse prevention support if it encouraged users to analyze current or past challenges in a positive way, provided support for mastering frequently encountered challenges, or supported restoring self-efficacy and motivation after a crisis.

9.2.2.13 Targeted Functional Domain

In addition to HAPA concepts, we classified applications concerning the four QoLT dimensions (Schulz et al. 2012): functional domain, kind of functioning support, passive or active mode of operating the system, and system intelligence.

As described earlier, we discriminated during filtering and evaluating applications between functional domains. At the broadest level, we discriminated between exercise and diet applications. Within those categories, we discriminated different sports and diet approaches.

9.2.2.14 Kind of Functioning Support

By definition, most health and fitness applications address prevention and maintenance. We categorized applications regarding kind of functioning support based on aimed final performance. Running a 5K distance is within normal functioning and aims to maintain functioning. In contrast, running a marathon is out of range for most users and therefore enhancing normal functioning. Managing clinical conditions, like diabetes or allergies, compensates for diminished functioning.

9.2.2.15 Mode of Operating the System

We analyzed mode of operation for instruction and tracking. Passive *instruction* does not consider user's current performance for guidance. As an extreme case, instruction could be prerecorded on a tape. In contrast, interactive instruction would adjust guidance to current levels of performance like a good personal trainer.

Passive *tracking* of current performance does not require users' input, for example, GPS, pedometers, or sensors. In contrast, interactive tracking requires users' input, for example, manual entry diaries.

9.2.2.16 System Intelligence

We considered an application as *globally intelligent* if it assesses a user's current performance level and health status during setup and adjusts the program correspondingly. For intelligent tutoring systems, this feature of curriculum development is referred to as the *outer loop*. We considered an application to be *locally intelligent* if it monitors current performance during a session and adjusts today's program accordingly. For intelligent tutoring systems, this dimension is referred to as the *inner loop*.

9.3 Results

We evaluated exercise applications in subcategories for tracking performance ($N = 11$; see Table 9.1a), running programs ($N = 7$; see Table 9.1b), and general workout programs ($N = 23$; see Table 9.2), including brief workout sessions (N = 3; see Table 9.2a), specific exercises (N = 4; see Table 9.2b), warm-up (N = 3; see Table 9.2c), diverse exercises (N = 3; see Table 9.2d), bodybuilding (N = 7; see Table 9.2d), and others (N = 2; see Table 9.2e). We evaluated diet applications (see Table 9.3) in subcategories for specific diet regimes ($N = 6$; see Table 9.3a), food databases ($N = 5$; see Table 9.3b), nutrition trackers ($N = 7$; see Table 9.3c), and comprehensive weight loss programs ($N = 3$; see Table 9.3d).

9.3.1 Exercise Performance Tracking

We evaluated 11 applications that track performance during exercise (see Table 9.1a). Most advanced tracking applications originate from bike computers and have started to integrate tracking health measures along traditional athletic values. We included *B.iCycle* (Valley Development GmbH) and *Road Bike Pro* (Runtastic) as bike applications. For tracking walking performance, we analyzed *Accupedo-Pro* (Corusen LLC), *SmartRunner Pro* (APPSfactory), and *Walkmeter* (Abvio Inc.). For general workout and activity, we included *Endomondo* (Endomondo.com), *Map My Fitness+* (MapMyFitness Inc.), *MyFitnessCompanion* (MyFitnessCompanion), *Nexercise* (Nexercise), and *WOD Tracker Pro* (Pete Wood). Finally, we included *Heart Rate Pro* (Runtastic) representing a line of development that aims to use mobile technology instead of additional sensors for measuring health parameters. *Heart Rate Pro* uses the iPhone camera for estimating resting heart rate.

Exercise tracking applications mostly target advanced users during maintenance state and correspondingly provide detailed performance monitoring, but with very little instruction. As Table 9.1a shows, 8 of 11 applications assess demographics for estimating burned calories as one frequently displayed measure. Only two applications ask for users' goals. For example, *Accupedo-Pro* asks during setup to set a daily goal and suggests 10,000 steps as default.

If the interface is intuitive and performance does not require specific skills, like walking, applications can be also used for initiative state ($N = 6$). In tendency, tracking applications for beginners can be also used for action planning. *Road Bike Pro* provides trail information in advance including distance and difficulty level. No application communicates risks and resources to nonintenders or supports relapse prevention and management.

Walking and biking can prevent decline of functioning. In contrast, bodybuilding aims to build capacity to lift weights beyond normal functioning. Bike computers and running trackers necessarily track performance during workout passively and accept manual entry for comments after workout. This is a prerequisite for local system intelligence that is still displayed by only a minority of health and fitness applications.

9.3.2 Running Programs

Table 9.1b displays evaluation results for seven running program applications. All applications provide detailed instruction when to run and when to walk during each session. Applications vary in interface design and in tracking and analyzing performance capacities. *5K Runner* (Clear Sky Apps LTD) and *Get Running (Couch to 5K*, Splendid Things) enable beginners during initiative state to run 5K in 8 or 9 weeks time. *B210K Pro* (Bridge to 10K, Guy Hoffman) and *Run a 10K Pro!* (Red Rocks Apps) target advanced users during maintenance state and aim for running 10K in 6 or 10 weeks time for enhancing normal functioning. *Hal Higdon Novice1 26.2 Marathon* (Bluefin Software, LLC) allows choosing a target distance between 1 and 42 km for an 18-week program. *FIRST iPhone Companion* (Bryan Catron) offers 12-week programs for 5K and 10K or a 16-week program for Half Marathon and Marathon. In contrast, *Zombies, Run!* (Six to Start) does not aim for a specific time or distance, but offers 30 or 60 min running sessions integrated in playing a game.

Given their aim, all reviewed running program applications provide detailed instruction during a session. Beginners' applications do not track distance or whether the user is running at all. This makes guidance very passive and not locally intelligent. In contrast, *Higdon Novice1 26.2* and *Run a 10K Pro!* target advanced users and use GPS tracking. *Zombies, Run!* tracks distance, time, and pace during a session. All programs either allow to preview sessions for action planning or could be *misused* for planning by simulating sessions.

None of the running program applications assesses users' initial health parameters. Applications that provide several programs to choose from also ask for users' target (*FIRST* and *Higdon Novice1 26.2*). Those two applications can also be used as race preparation tools and ask for the race day and name during setup. More than half of applications do not provide risk and resource communication. This is matched with missing coping support and absent relapse prevention for most applications. Interestingly, the long-distance applications are more likely to provide coping support. As for performance tracking applications, only minority of running program applications display global or local intelligence given missing assessment and passive, nonadaptive guidance.

9.3.3 General Workout Tools

Table 9.2 shows evaluation results for 23 general workout applications. These are more diverse than performance tracking or running program applications and cover a variety of areas: Three applications (see Table 9.2a) are designed for brief full body workout sessions—*7 Minute Workout* (UOVO), *Daily Workouts* (Daily Workouts Apps, LLC), and *Gorilla Workout* (Heckr LLC). Four applications (see Table 9.2b) cover specific exercises for beginners: *Burn Your Fat With ME!!* (BYFWM!!, Creative Freaks) targets sit-ups and motivates users with a female Japanese cartoon character as personal trainer. *Just 6 Weeks* (Just Do Inc.) offers

6-week programs for different exercises, including push-ups and sit-ups. *Pushups 0 to 100 Exercise Workout Trainer* (Clear Sky App LTD) and *Runtastic Push-Ups Pro* (Runtastic) provide programs aiming for 100 push-ups.

Three applications (see Table 9.2c) represent a collection of warm-up and cooldown exercises or relaxation. *3D Tai Chi 8 + 16 Forms* (ChuFengChin) offers a 3D simulated character that can be rotated to view any perspective during exercise demonstration. *Daily Yoga* (*DailyYoga.com*) provides yoga programs for different body parts, for example, back. *Stretch Guru: Run* (Lori Bryan Living) offers stretch instructions for different cooldown durations.

Three applications (see Table 9.2d) have a collection of diverse exercises with diverse measures tracked and analyzed during workout. *All-in Fitness* (Fitness, Plus Sports) provides a variety of fitness and yoga exercises and passively tracks several measures during workout. *HIIT—Interval Training* Pro (Caynax) offers a high-intensity interval training program as cardiovascular workout for weight loss. *Nike Training Club* (Nike Inc.) is a workout guide including tracking and options for earning virtual rewards.

Seven applications (see Table 9.2e) for advanced users on weight lifting or bodybuilding provide workout planners and trackers. The *Bible of BodyBuilding* (WorkoutRoutines) reads like a manual for bodybuilding with a 12-month workout guide. *Crossfit WOD Programmer* (CrossfitHardcore) supports workout planning and considers preferred equipment and preferred exercises. *Gym Genius—Workout Tracker* (AppBrains) offers six core weight lifting programs with comprehensive workout tracking. *Gymme—Gym Personal Trainer* (Matteo Pizzorni) and *Gymprovise Gym Workout Tracker* (Gymprovise) are exercise-planning tools. *myWOD* (theDreamWorkshop) presents WOD linked with YouTube instruction and benchmark models. *Workout Hero* (Storeboughtmilk LLC) is a weight lifting workout planner and tracker.

Two reviewed applications (see Table 9.2f) stick out for different reasons. The *PFA Calculator For U.S. Navy* (Crash Test Dummy Limited, LLC) is a fitness assessment tool tailored for the U.S. navy applicant rating criteria. *SuperBetter* (SuperBetter Labs) is a resilience support game that almost looks like designed with HAPA in mind. It is not specific for exercise and weight loss, but can be used for those aims and includes tips and instructions on coping that is missing in majority of applications.

Most general workout applications address either beginners during initiative state or advanced users during maintenance state. Beginners' applications for brief workout or specific exercise programs mostly imitate running programs in their evaluation patterns: they provide passive instruction and action planning support. They also have very little initial assessment or risk communication. Only *Daily Yoga* mentions adverse effects of a sedative lifestyle as risk communication. As for running programs, there is a little more resource communication. *Daily Yoga* and *Stretch Guru* emphasize health benefits of exercise to support positive outcome expectancy. *Fitness* provides coping planning support by describing typical mistakes of beginners and how to master those challenges. As for running programs, beginners' workout applications do not provide relapse prevention and management.

Applications for brief workout sessions usually do not track performance as basic running programs do. In contrast, applications on specific exercise programs track repetitions. Most beginners' workout applications do not display global or local intelligence as expected given missing assessment and passive instruction.

Advanced weight lifting and gym workout applications target advanced users during maintenance state, but do not mimic distance tracking applications. They do provide guidance and some instructions are interactive: *Gymme* does not provide information on how to perform an exercise, but predicts how many repetitions are necessary for reaching personal bodybuilding goals given current performance. *Gymprovise* assesses in the beginning preferred exercises and equipment and suggests alternatives if needed. Most programs can be used for planning workout or they import WODs from RSS feeds eliminating the need to plan ahead (e.g., *WOD programmer*). All these applications track performance for displaying progress after assessing initial performance during setup.

Advanced workout applications provide very little coping support or relapse prevention and focus exclusively on action planning and performance support. They mostly do not assess goals and provide no risk and resource communication. Predicting necessary workout for *Gymme* or compensating for occupied gym equipment in *Gymprovise* can be interpreted as indicators for system intelligence. While general workout can contribute to maintain functioning, advanced weight lifting and bodybuilding are more likely to enhance performance beyond normative functioning.

There are two applications that did not fit either in the beginners' applications or in the advanced athletes' applications. *U.S. Navy PFA Calculator* is designed as a fitness assessment tool for one-time use. It asks for a couple of performance measures to be manually entered and calculates a corresponding pass/fail value. It does not track progress and does not provide instruction on how to get in shape for passing the test either. Therefore, *U.S. Navy PFA Calculator* misses by design most evaluation criteria. The other exception, *SuperBetter*, is also not a typical exercise application. It aims to increase users' coping skills. Correspondingly, *SuperBetter* looks very well in our evaluation, but is not necessarily a very good exercise program. It tells users what to do if they do not have sufficient time to go to the gym. It does not tell what to do in the gym in case users have the time to go there.

Similarly, tracking performance is much more important for biking or running than for yoga or Tai Chi. Therefore, missing tracking would be a disadvantage for advanced running programs, but not for yoga programs. This demonstrates the importance of the QoLT dimension of targeted functional domain when evaluating mobile health and fitness applications.

9.3.4 Diet Tools

Table 9.3 presents evaluation results for 21 diet applications with great overlap of functionality. Six of these applications (see Table 9.3a) advertise a specific diet program for a limited period of time: *3 Day Military Diet* (Realized Mobile LLC) suggests keeping a diet regime for 3 days followed by 2 days off each. *17 Day Diet Complete* (Realized Mobile LLC) proposes a 51-day program for weight loss with three stages. *Fast Metabolism Diet App ...* (Random House Digital Inc.) presents a 28-day diet without counting nutrition values. The *Monash University Low FODMAP Diet* (CEMA, Monash University) aims to assist in management of irritable bowel syndrome (IBS) by restricting some carbohydrates for a weeklong challenge with symptom tracking. The *hCG Diet App* (Appilicious) supports hCG treatment for weight loss with a strict eating regime complementing the hormone therapy. The *Fit Chick Meal Planner & Recipe Sharer* (Brett Campbell) comes with a 7-day clean eating meal plan.

Five applications (see Table 9.3b) support healthy eating choices by providing a food database with essential nutrition information: *Eat This, Not That! Restaurants* (Rodale Inc.) provides best and worst options for each restaurant. *Green Smoothies* (Raw Family) presents a recipe collection including nutrition information. *Healthy Food, Allergens, GMOs & Nutrition Scanner* (NxtNutrio, Nxtranet Inc.) analyzes food ingredients and helps spotting allergens. *Paleo Central* (Nerd Fitness) classifies food items as either adhering the Paleo philosophy or not. *Points Calculator for Weight Loss ...* (iTrackBites, Ellisapps Inc.) supports tracking of food score values.

Seven applications (see Table 9.3c) support logging over time calories, body weight, and sometimes exercise: *CaloryGuard Pro* (BlueBamboo) is a diet and exercise tracker. *Carb Master—Daily Carbohydrate Tracker* (Deltaworks) and *CarbsControl—Carb Counter and Tracker* (Coheso Inc.) are diet trackers supporting low-carb diets or diabetes management. *Cronometer* (BigCrunch Consulting Ltd.) is an online diet and exercise tracker. *DietBet—Loose Weight. Make Money* (DietBet) is a social weight loss game that rewards players who lost 4% of their body weight in 4 weeks on cost of players who did not. *Nutrition Menu* (Shroomies) tracks several nutrition and exercise measures and provides nutrition information based on a food database. *Nutritionist* (OTK) is a nutrition and exercise tracker with meal planning.

Finally, three comprehensive weight loss programs (see Table 9.3d) integrate meal planning, nutrition information, and tracking. *Bodybuilding Diet Pro* (Zen Software, LLC) provides a long-term program for building muscle. *My Diet Coach—Weight Loss for Women* (Inspired Apps) visualizes users' weight loss goal with before-/after-animated characters and provides a variety of tips for coping with frequent challenges. *Noom Weight Loss Coach* (Noom Inc.) is the most comprehensive weight loss program in our selection and offers a variety of weight loss support tools.

By definition, weight loss applications address mainly users in the action planning and initiative state (N = 20 of 21), and weight management applications address mainly users during maintenance state (N = 10) with partial overlapping. While *DietBet* and explicit diet trackers provide no instruction, all other applications provide passive information. It is worth noting that database tools provide *information*, but no or very little *guidance*. In contrast, weight loss and management programs provide guidance and action planning support as well. As for exercise applications, majority does not address relapse prevention or management, but more applications provide basic coping support. *My Diet Coach* presents a collection of typical challenges and how to master these and has a *food cravings panic* button. *Nutritionist* explains the basics of dieting, including recommendations to restructure the environment for preparing for challenges and resisting cravings. *Eat This!* suggests the healthiest option even for fast-food restaurants allowing to manage a crisis.

While it is very easy for distance tracking tools to passively monitor performance, passive nutrition tracking is challenging for diet applications. Entering breakfast can easily take 15 min for the first time with significantly less effort for repetitions. A nice exception is *Noom* with an option for fast versus accurate tracking. The tracking challenge also becomes easier for barcode scanners and restaurant trackers that track complete meals. The challenge disappears for diet programs that do not believe in calorie tracking, like *Paleo* or *Fast Metabolism Diet*, or programs that prescribe a menu, like *Fit Chick*. For weight loss programs, current and target weights are usually assessed during setup (e.g., *My Diet Coach* or *Noom*).

About half of the application communicates health risks and/or expected benefits. This holds especially for applications that support management of clinical conditions, like *Low Fodmap Diet* and *NxtNutrio*. In general, weight loss and weight management programs aim to prevent decline of functioning, while management of clinical conditions (*Carb Master, Carbcontrol, Low Fodmap Diet, NxtNutrio*) aims to compensate for diminished functioning.

As for exercise applications, global and local intelligence is still rare among diet applications resulting in passive instruction and missing advanced coping and recovery support.

9.4 Discussion

There are currently huge numbers of mobile health and fitness applications on the market. We started our evaluation project with 2331 applications from iTunes and Google Play stores. These applications offer a great variety of features and variance for implementing the same functions. Therefore, we presented our evaluation results for only 61 applications covering fast majority of features from the originally 435 after filtering. When filtering by functionality, we necessarily introduced bias toward prevention applications targeting health behavior of users. At the same time, our evaluation criteria would not have done justice to some excellent patient management applications.

Our evaluation of covered HAPA components revealed that majority of mobile health and fitness applications are designed for a specific state of behavior change state, but do not assess whether potential users are currently in that state or support users to reach that state. Still, some running programs provide siblings of applications that cover different distances and correspondingly cover action planning, initiative, and action maintenance states. This specialization toward a specific BCS shows best when discriminating between beginners' (or action initiative state) applications and experts' (or action maintenance state) applications that exist for all subcategories. In tendency, applications instruct beginners and track experts.

Most applications have in common that they provide much more action support than coping and relapse prevention support, especially for exercise applications. However, there are systematic variations between and within categories: for all categories, there are implementations of very specific support, like bike computers and food databases, but also all-around talents, like general fitness applications or comprehensive weight loss programs. In tendency, we have seen a greater overlap in functions between subcategories among diet applications than among exercise applications. There is also in tendency more coping support within diet applications than within exercise applications. This seems intuitive given frequently reported food cravings, but not workout cravings. Still, exercise applications would greatly profit from guidance for *difficult* users with clinical conditions or for injury management. For both exercise and diet applications, there are promising exceptions from the rule providing extensive coping support via games, bets, challenges, social support, or virtual love.

From our evaluation of implemented QoLT dimensions, we took, first, the discrimination between functional domains that highlighted differences and similarities between and within categories. It is important to note that our discrimination of functional domains came from a perspective of functions covered by applications. A patient looking for support while recovering from heart attack might have drawn the lines differently.

Given that focus, our selection of mobile health and fitness applications mostly covered the segment of preventing decline of functioning than compensating for diminished functioning or enhancing normal functioning. As for the HAPA components, most applications cover a specific segment along the continuum of functioning support. That matches Schulz and colleagues' (2012) observation for QoLT systems in general.

Also matching Schulz and colleagues' analysis of current QoLT systems, majority of mobile health and fitness applications displays very little system intelligence. Again, there are a couple of basic versions of global system intelligence, like adjusting calories burned during exercise to a user's BMI, and a few promising exceptions. For example, *Bodybuilding Diet* suggests considering an 85+-week program for competing in a bodybuilding competition for a desperate male user with 200 lb body weight and low activity level. Related to low system intelligence, most applications provide passive and not-interactive guidance. Since the advent of GPS tracking, barcode scanners, and sensors integrated in mobile phones, tracking is increasingly passive for monitoring performance. But there remains a challenge to track meals that do not come out of a box or from a restaurant.

Taken together, we found that HAPA theory components and QoLT evaluation criteria nicely complement each other. Without assessment of states as suggested by HAPA, there would be no option for system intelligence. Without adapting to users' preferences and needs, an application cannot provide meaningful coping and relapse support.

9.5 Limitations

The evaluation of mobile health and fitness applications that we have conducted has several limitations. First, we have developed and applied an evaluation schema based on the HAPA theory and QoLT dimensions. Within our workgroup, we have reached a reasonable inter-rater reliability. However, this schema is not fully tested yet and might need further modifications for being sufficiently reliable and valid for evaluating mHealth applications on a regular basis.

Second, here, we applied this schema to diet and exercise applications, but not to other Health & Fitness applications in iTunes and Google Play stores. For some of those, for example, smoking cessation, HAPA has been successfully used for modeling behavior change and designing interventions (e.g., Ochsner et al. 2014). However, in our sample, we had only 12 smoking cessation applications. Therefore, we would not have fairly evaluated those compared to the huge collection of diet and exercise applications. We also did not evaluate patient management applications or sports applications because they either do not implement stand-alone applications or do not aim to reduce health risks.

Third, we sampled the U.S. market and partially the German market for mobile health and fitness applications, but not other regions. We also did not explicitly address cultural issues (see, e.g., Limam Mansar et al. [2014] for an adaptation of weight loss applications to the Middle East).

Fourth, this systematic review mirrors the state of iTunes and Google Play stores Health & Fitness applications in June 2013. Some of the reviewed applications might have been updated since that time or have disappeared from the stores.

Fifth, to our knowledge, there does not exist evidence yet on mobile applications that implement HAPA and QoLT dimensions effectively improving users' health and well-being. This taps into the issues of life cycles for mobile applications that are much shorter than the average duration of clinical trials. However, clustering applications according to HAPA and QoLT dimensions and conducting clinical trials on groups of applications instead of single implementations might help to bypass this limitation of clinical trial methodology.

Finally, this schema is designed for experts who design and evaluate mHealth applications. It is not intended for users of health applications. The complexity of health, interventions, and mobile technology is too high to be easily understood by laypeople and applied to their own life. Potentially, this schema can be developed into a rating system similar to the five-star popularity system for displaying compliance with evidence-based recommendations for efficient and user-friendly mHealth applications.

9.6 Case Study

The operational definitions described in the methods section can be also used for designing mHealth applications. For designing a mobile application to complement a local weight loss program, we (Limam Mansar et al. 2012) first defined the intended QoLT dimensions. The application was designed as add-on for managing health behavior goals, motivational support, and social networking integrated in a weight loss program at a local health clinic. Therefore, it primarily provided cognitive support in semiautomated mode of operation with a basic version of system intelligence. According to HAPA, intended users had already passed intention state. During the 4-week program, users were expected to pass through preparation and action initiative state with in-person support for planning and application-based support for action initiative. During setup, the application asked users to enter personal goals for the coming week and to choose frequency and time of reminders and text messages. During a week, the application provided detailed instruction on performance and feedback on achieved goals supporting users' action and coping self-efficacy (for more details, see Limam Mansar et al. 2012). This is an example for very limited add-on support for users during planning and action initiative state.

9.7 Future Directions

For future development, currently available mobile health and fitness applications demonstrate the departure from clinical patient management programs on one side and from athletic expert systems on the other side and start to approach the area of preventive medicine for fostering citizens' health behavior concerning exercise and diet. Current mobile applications provide more action support than coping support and miss the capacity of required system intelligence to adapt to user's needs and challenges. They do well for young users without medical conditions, but they are not well equipped to serve *difficult* users, for example, a patient recovering from heart attack or a patient trying to stabilize his or her blood glucose readings. Current applications also started to connect diet with exercise and integrated tracking health measures for both areas, but they are still stronger on one side than the other. Comprehensive weight loss and management applications should excel on both for increasing usability, user satisfaction, and effectiveness.

9.8 Conclusions

For improving current mobile health and fitness applications, the greatest challenges are not more advanced sensors. The greatest challenge is to provide adaptive and intelligent guidance for good and bad times. It is naïve to assume that a couch potato masters a 5K program successfully from day 1 on. Therefore, it is not good enough to *tell* that person when to walk and when to run. It is necessary to *track* whether he or she is actually running and providing feedback and motivational support accordingly. There is a need for current mobile health and fitness to become more intelligent to provide users with what they are looking for.

Case Scenario

Mrs. Smith has been told by her family doctor to significantly lose weight after receiving test results indicating prediabetic status. Mrs. Smith signals willingness, but she needs guidance. The doctor recommends trying *Fit!*, a new weight loss application developed under guidance of the local teaching hospital. When opening the application first time, Mrs. Smith is welcomed, is asked a few questions on demographics, and is prompted to choose a fitness goal from a list. After that, the application updates for a moment and invites her to explore different features. All starts very easy with taking a picture of her meals first day and just carrying around the phone in her pocket. During the next days, things become more demanding with signaling that she has reached her daily calorie budget already before lunch. The application suggests walking in the park for half an hour, but it is raining heavily outside. After trying hard for 2 weeks, Mrs. Smith feels exhausted. The application is bossing her around and leaves very little wiggle space. When a coworker invites her for an ice cream, she cannot resist. Mentally counting calories, she switches off her phone and does not turn it on until recovering from her *blackout* 2 days later. She feels ashamed and asks herself whether she is the only weak person on the planet. Her cell phone says "Great job! You sticked to the budget for a week in a row."

Questions

1. According to HAPA, in which state is Mrs. Smith in the end of this episode?
2. Given the description, which kinds of assessment did the *Fit!* application implement?
3. Given the description, which kind of intervention did the *Fit!* application provide?
4. Which kinds of self-efficacy did Mrs. Smith display during her first week of diet?
5. Which modes of operating the system did the *Fit!* application offer according to QoLT evaluation dimensions?
6. What do you think: Does the *Fit!* application display system intelligence?

References

Ahmed, H. M., Blaha, M. J., Nasir, K. et al. 2013. Low-risk lifestyle, coronary calcium, cardiovascular events, and mortality: Results from MESA. *American Journal of Epidemiology*. doi:10.1093/aje/kws453.

Ajzen, I. and Fishbein, M. 1980. *Understanding Attitudes and Predicting Social Behavior*. Englewood Cliffs, NJ: Prentice Hall.

American College of Preventive Medicine (Internet). 1998. Available from: http://www.acpm.org/?page=WhatisPM# (updated and cited July 30, 2014).

Armitage, C. J. and Conner, M. 2001. Efficacy of the theory of planned behaviour: A meta-analytic review. *British Journal of Social Psychology* 40:471–499.

Azar, K. M. J., Lesser, L. I., Laing, B. Y. et al. 2013. Mobile applications for weight management. Theory-based content analysis. *American Journal of Preventive Medicine* 45:583–589.

Bandura, A. 2001. Social cognitive theory: An agentive perspective. *Annual Review of Psychology* 51:1–26.

Becker, M. H. and Rosenstock, I. M. 1984. Compliance with medical advice. In *Health Care and Human Behavior*, eds. A. Steptoe and A. Matthews. pp. 135–152. London, U.K.: Academic Press.

Brunstein, A., Brunstein, J., and Limam Mansar, S. 2012. Integrating health theories in health and fitness applications for sustained behavior change: Current state of art. *Creative Education* 3(Suppl):1–5.

Craciun, C., Schüz, N., Lippke, S., and Schwarzer, R. 2012. A mediator model of sunscreen use: A longitudinal analysis of social-cognitive predictors and mediators. *International Journal of Behavioral Medicine* 19:65–72.

de Jongh, T., Gurol-Urganci, I., Vodopivec-Jamsek, V., Car, J., and Atun, R. 2012. Mobile phone messaging for facilitating self-management of long-term illnesses (Review). *Cochrane Database of Systematic Reviews* 2012(12):CD007459. doi: 10.1002/14651858.CD007459. pub2.

DeLone, W. H. and McLean, E. R. 2003. The DeLone and McLean model of information system success: A ten-year update. *Journal of Management Information Systems* 19:9–30.

Guthold, R., Ono, T., Strong, K. L., Chatterji, S., and Morabia, A. 2008. Worldwide variability in physical inactivity. A 51-Country survey. *American Journal of Preventive Medicine* 34:487–494.

Limam Mansar, S., Brunstein, A., Jariwala, S., and Brunstein, J. 2012. Addressing obesity using a mobile application: An experimental design. *Proceedings of the IADIS Multi Conference on Computer Science and Information Systems (MCCSIS), e-Health (EH 2012)*, Lisbon, Portugal. Accessed on July 19, 2014.

Limam Mansar, S., Jariwala, S., Behih, N. et al. 2014. Adapting a database of text messages to a mobile-based weight loss program: The case of the Middle East. *International Journal of Telemedicine and Applications*, doi: http://dx.doi.org/10.1155/2014/658149. Accessed on July 19, 2014.

Luszczynska, A. and Schwarzer, R. 2003. Planning and self-efficacy in the adoption and maintenance of breast self-examination: A longitudinal study on self-regulatory cognitions. *Psychology and Health* 18:93–108.

Ochsner, S., Luszczynska, A., Stadler, G. et al. 2014. The interplay of received social support and self-regulatory factors in smoking cessation. *Psychology & Health* 29:16–31.

Prochaska, L. O. and Velicer, W. F. 1997. The transtheoretical model of health behavior change. *American Journal of Health Promotion* 12:38–48.

Renner, B. and Schwarzer, R. 2005. The motivation to eat a healthy diet: How intenders and non-intenders differ in terms of risk perception, outcome expectancies, self-efficacy, and nutrition behavior. *Polish Psychological Bulletin* 36:7–15.

Scholz, U., Sniehotta, F. F., and Schwarzer, R. 2005. Predicting physical exercise in cardiac rehabilitation: The role of phase-specific self-efficacy beliefs. *Journal of Sport and Exercise Psychology* 27:135–151.

Schuez, B., Sniehotta, F. F., Mallach, N., Wiedemann, A., and Schwarzer, R. 2009. Prediction of stage transitions in dental hygiene: Non-intenders, intenders, and actors. *Health Education Research* 24:64–75.

Schuez, B., Sniehotta, F. F., Wiedemann, A., and Seemann, R. 2006. Adherence to a daily flossing regimen in university students: Effects of planning when, where, how, and what to do in the face of barriers. *Journal of Clinical Periodontology* 33:612–619.

Schulz, R., Beach, S. R., Matthews, J. T., Courtney, K. L., and De Vito Dabbs, A. J. 2012. Designing and evaluating quality of life technologies: An interdisciplinary approach. *Proceedings of the IEEE* 100:2397–2409.

Schwarzer, R. 2008. Modeling health behavior change: How to predict and modify the adoption and maintenance of health behaviors. *Applied Psychology* 57:1–29.

Sheeran, P. 2002. Intention–behavior relations: A conceptual and empirical review. *European Review of Social Psychology* 12:1–36.

Tomlinson, M., Rotherarm-Borus, M. J., Swartz, L., and Tsai, A. C. 2013. Scaling up mHealth: Where is the evidence? *PLoS* 10:e1001382. doi: 10.1371/journal.pmed.1001382.

U.S. Food and Drug Administration 2013. Mobile medical applications. Guidance for industry and food and drug administrative staff. Available at http://www.fda.gov/downloads/Medical Devices/DeviceRegulationandGuidance/GuidanceDocuments/UCM263366.pdf (retrieved October 23, 2013).

Wiedemann, A. U., Lippke, S., and Schwarzer, R. 2012. Multiple plans and memory performance: Results of a randomized controlled trial targeting fruit and vegetable intake. *Journal of Behavioral Medicine* 35:387–392.

World Health Organization. 2002. *World Health Report 2002: Reducing Risks, Promoting Healthy Life*. Geneva, Switzerland: World Health Organization.

World Health Organization. 2009. *Global Health Risks: Mortality and Burden of Disease Attributable to Selected Major Risks*. Geneva, Switzerland: World Health Organization.

World Health Organization. 2010. *Global Status Report on Noncommunicable Diseases*. Geneva, Switzerland: World Health Organization.

10

mHealth and Population Management

Dirk Boecker, Bernhard Mikoleit, and David Scheffer

CONTENTS

10.1 Challenges of Population Management

Spreading unhealthy behavior of individuals and an aging population drive increasing prevalence of chronic diseases and related comorbidities, which make up 80% of the exploding health-care cost. Historic and current approaches of disease management and general population motivation via prevention and wellness programs have not shown satisfying impact on required personal behavior change. One-size-fits-all programs do not move the needle enough. People need to be inspired using the unique language and channels that meet their individual needs, because change is highly personal.

Current approaches to alleviate the situation through a broad number of direct-to-consumer health programs (health-related apps and web-based programs) provide at best point solutions and feel very piecemeal from the perspective of the consumer. What health plans and care organizations are missing is the opportunity to centralize data capture, consolidate analysis, and perform in-depth analytics, which provide the necessary platform for a new level of comprehensive and highly personalized intervention and outreach programs.

Consumers want more seamless interactions; however, relying on discipline and willpower of the consumer alone will fall short. In order to get health-care cost under control, health plans should pursue an integrated approach of analytics and related personalized intervention design.

Key challenges of population management today are as follows:

- Dramatic increase of prevalence and incidence of chronic diseases often influenced by lifestyle choices and bad health habits; related cost for health care are exploding.
- Limited reach into affected population due to high-cost communication models (mainly call centers staffed with well-trained but high-cost clinicians and nurses).
- Limited effectiveness of intervention models due to the complexity and variety of individual issues.

Traditional disease management programs typically reach and impact only a small percentage of the target population. Even for this small segment, the contact frequency is like several weeks or months, which yields only low effectiveness and only moderately alters health behavior or lifestyle-related habits. Despite the *personal* outreach, the call center agents mostly address *external* behavioral management and do not have the means to focus on stimulating self-motivation.

The current disease management system is set in a legacy communication pattern, which is very difficult to escape in a stepwise way. What is needed is a much more fundamental change: a rollout of low-cost and automated communication processes, which allow addressing broad population segments with high frequency yet in an individualized approach. The advent of mobile health (mHealth)-related technologies and communication systems provides an opportunity to initiate such a paradigm shift in population management.

10.2 Our Solution: Personalization and NeuroIPS™

10.2.1 Overview

What is missing today is the ability to communicate with large numbers of patients in a low-cost yet personalized way. Despite the availability of proven low-cost technologies, which facilitate efficient and broad-scale communication, automated outreach methods combined with individualized member communication are lacking.

Conventional interventions and instruments are often less effective than desired, often because patients are not being addressed in a way that chimes with their emotional state and motivations. However, a high level of patient compliance is the cornerstone of long-term, stable therapy success. In order to achieve this kind of sustained cooperative behavior, patients need to understand the prescribed therapeutic measures and recommended behavioral changes and be intrinsically motivated to take on responsibility.

Our system provides such combination of customized, automated mass communication tuned to patients' individual personalities. Using our NeuroIPS method in combination with our automated outreach system, MassineBoecker identifies the inner feelings of each individual, which allows to effectively communicate with them. People are in different situations and our outreach methods reflect that.

We have developed a thorough understanding of many of the weaknesses in current disease management programs. Our system is designed to overcome these weaknesses and optimize the effectiveness of planned interventions. Our goal is to provide a reliable method for personalizing interventions and communication in a completely new way.

10.2.2 What Makes Our Method Unique

MassineBoecker has developed a new and more effective way for communicating with large populations such as those with health risks or chronic diseases. This new approach starts by determining the psychographic and personal characteristics of individual patients and representative population segments. We create a tailored communication profile that stimulates the patient's internal motivation guaranteeing long-lasting compliance that enables sustained self-management. We integrate all major dimensions: demographic, biometric, educational, sociographic, psychographic, and emotional information of the individual.

The personalized intervention model includes proven proprietary survey and training tools and scalable peer-to-peer social network technology to ensure long-term sustainability of the impact on member behavior. The approach provides personalized coaching through automated outreach via low-cost modalities (e.g., SMS) designed to address the member's psychological barriers to compliance behavior. This individual contact allows providing a platform for building social peer-to-peer networks that will promote member participation. The expected cost savings and related health benefits are substantial.

Our knowledge database is rule bound and dynamically adaptable. It configures automatically to new information ensuring our outreach is fully current with respect to our target groups' situation.

10.2.3 Modular Solution

The intervention concept is constructed on a modular basis and can be utilized as such (Figure 10.1).

10.2.3.1 *Profiling and Target Group Segmentation*

MassineBoecker analyzes groups and individuals in unprecedented depth: from the stratification of individual candidates for clinical studies to profile analysis and segmentation of larger populations. We consider current emotional disposition and personality type, as well as personal abilities and preferences in terms of content, design, tone, and choice of communication channel.

With this segmentation and detailed profiling, we provide companies with a whole new way of selecting and managing communications outreach to individuals. The method enables decisive evaluation and effective stimulation of individuals' intrinsic motivation.

10.2.3.2 *Customer Communication Analysis*

We review our customer communication and loyalty measures and analyze the effectiveness of both current and planned initiatives. We compare current outreach activities

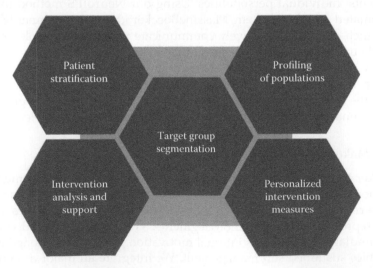

FIGURE 10.1
Overview of our solutions.

(design, tone, choice of channel) with the psychological characteristics of the various target groups and show where there is room for improvement. With these insights, our customers can evaluate and optimize the efficiency of their measures with increased precision.

10.2.4 Key Success Factors for Compliance Support

In order to achieve high rates of compliance, one needs an in-depth insight into the person. We consider the following essential factors:

- Personalization
- Ongoing support
- Training
- Personal data management

The key questions are: *Where* are you, *who* are you, and *what* do you need to improve?

We take into account the patients' emotional disposition (the *where*) and their personality type (the *who*) in order to help them with their plan forward (*the what*).

The core elements of our compliance support system are as follows:

- Contact: frequent contact using the patient's preferred communication channels and all mHealth-related tools and systems.
- Dynamic adaptation: our measures are tailored and continuously tuned to the patient's current situation.
- ESCALATE™: support for patients through their existing family and social networks via an actively managed escalation system.
- microGROUPS™: effective small online groups whose members are selected by their personality type.
- Coaching: interactive coaching via online chat and online blogs on our web portal.
- Training: behavioral and/or subject-specific online training courses.

10.2.5 Four Key Goals of Our Measures

Our approach is integrating four goals:

1. Personalized support and new insights for the patients, with a view to changing the course of their personal health management
2. Awareness of the change in behavior and adjustment to the new circumstances surrounding the health risk or condition
3. Avoidance of emergency situations and their associated costs and aftercare problems
4. Integration of various components that make up individual health management (Figure 10.2)

10.2.6 Unique MassineBoecker Profiling Method

Our profiling method integrates various aspects of the patient's personality. An individual profile is based on the determination of intrinsic motivators (e.g., measured via the

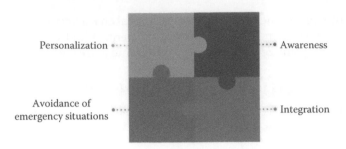

FIGURE 10.2
The four key goals of our measures.

ViQ™ method, which is based on the personality systems interaction (PSI) theory, which is described below), extrinsic tests (e.g., a survey on the subject's educational background), and indirectly accessible data (e.g., via query of databases of health insurance companies) (Figure 10.3). Our system can be integrated with all standard intervention measures (see Table 10.A.1).

MassineBoecker personalizes communication in four ways (Figure 10.4):

1. Frequency: optimization of frequency and duration of all outreach communication
2. Tone: personalization of tone and wording
3. Content: tailoring of content and presentation style to the needs and capabilities of each recipient
4. Modality: the use of the preferred channel of the individual

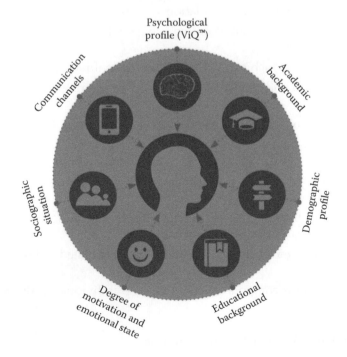

FIGURE 10.3
Features of MassineBoecker's profiling method.

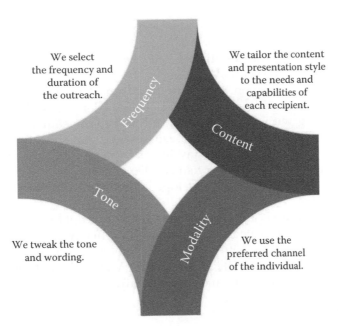

We select
the frequency and
duration of
the outreach.

Frequency

We tailor the content
and presentation style
to the needs and
capabilities of
each recipient.

Content

Tone

Modality

We tweak the tone
and wording.

We use the
preferred channel
of the individual.

FIGURE 10.4
Optimization of communication in four ways.

10.2.7 Benefit from Our Support

- Direct communication (bidirectional): Medication and appointments, reminders, personal goal setting, preparation and follow-up of doctor's appointment, and hospital discharge support.

- Online groups: MassineBoecker has developed a whole new concept for the formation of online groups, based on individuals' psychological characteristics. As a result, we assist chronically ill patients to join a harmonious group to receive tailored advice and assistance (directly or via online-group providers).

- Segmentation: Selection of small groups based on their psychological characteristics; groups pursue activities and challenges tuned to their abilities.

- Mobile communication: Mobile-device-based communication is the ultimate way of successfully addressing a large number of recipients. It is always available and ready to handle communication in both directions. What is more, it is easy to update in order to reflect new insights and changes. Our *mobile coach* can be integrated into our small-group-based support offering an array of tools: online training sessions, group games, snap polls, support lines, and important news and developments.

- Dynamic adaptation of communication: Our fully automated communication with the target group will be managed through our communication engine, which generates short messages, drives quick surveys, or issues small pieces of advice as appropriate for each individual patient. The system allows for timely communication of specific situations with broad segments of your population.

10.2.8 Behavioral and Psychological Triaging System

The PSI theory of Kuhl (2000) provides insights for the dynamic interaction of personality systems, which helps us to understand the behavior of ill and chronically ill patients.

Based on the PSI theory, we were able to develop a full array of 1200 rules to motivate people with chronic diseases to activate relevant health behavior. The rules have three factors that influence the behavior and are activated after a measured discrepancy to a goal state (goal, barrier, intervention).

The NeuroIPS data collection procedure is key part of our patient characterization system. This innovative method was developed by our Chief Scientist Prof. David Scheffer, along with Bernhard Mikoleit and their team. Based at Hamburg's Nordakademie, Professor Scheffer has been researching subconscious motivations on the basis of psychologist Carl Gustav Jung's typology and personality psychologist Professor Julius Kuhl's PSI theory since 1996. Built on this work, David Scheffer, Bernhard Mikoleit, and their team developed the NeuroIPS method and the visual questionnaire (ViQ).

The NeuroIPS procedure is a sophisticated scientific concept that has captured a strong market position. We have repeatedly demonstrated the effectiveness of our psychographic approach in a series of studies with renowned market research institutes, such as GfK,* TNS Infratest,† and ACNielsen.‡ The NeuroIPS and ViQ methods have been extensively researched for more than 15 years and have already been successfully deployed in the health-care and marketing industries.

10.2.9 Our Method in Brief

NeuroIPS stands for *Neuropsychological Implicit Personality System*. It is using a purely visual testing procedure (visual questionnaire or ViQ for short); the method assesses the subconscious (implicit) personality structures of the subject (Figure 10.5).

FIGURE 10.5
A sample question on visual perception.

---

* GfK is one of the world's leading research companies, with more than 12,000 experts working daily in more than 100 markets. http://www.gfk.com/Pages/default.aspx.
† http://www.tns-infratest.com
‡ http://www.nielsen.com/content/corporate/us/en.html

Even if you know that the lines shown are perfectly parallel, you may not see any straight or parallel lines. The degree to which the viewers perceive these lines as slanted depends on their psychological type, in this instance, on the extent to which they are introverted or extraverted.

Depending on how the viewer perceives visual symbols, the ViQ identifies fundamental behavioral patterns and personality traits. *We are what we see*, and the ViQ makes use of this phenomenon, as it links these implicit processes via analysis of our perception to our decision making and ultimate behavior. The test results classify the subject's personality to a high degree of accuracy, which then can be used to construct personalized effective interventions.

10.2.10 Advantages of Personalized Communication

Many economic and psychological theories for human behavior are based on the assumptions of expected utility maximization, that is, people are believed to consciously deliberate the balancing of the costs and benefits of different options like a *homo economicus*. Today, we know from the Nobel Prize–winning work of Daniel Kahnemann (2010) that this assumption is wrong (for a review, see Camerer et al., 2005).

People differ in their needs. They differ in what drives and motivates them how they act and make decisions. People differ in the way they face life. Desires and motives interact in complex, subconscious patterns of motivation, which influence perception, decision making, and behavior (Kuhl et al., 2010). Only a relatively small number of these subconscious activities are realized and can be set proactively as personal goals. Interestingly, these subconscious goals sometimes point in a different direction than the conscious motives and make them appear irrational (Scheffer and Mikoleit, 2013).

While not denying that conscious and rational choice is part of human perception and decision making, Zaltman (2003) estimates that 95% of all human decisions in daily life are actually made by intuitive and/or affective processing. Today, there is a large consensus in psychology, neuroscience, and behavioral sciences that human behavior is guided to a large extent by subconscious and implicit processes deeply imbedded in the personality systems of an individual shaped by genetic and environmental influences.

For example, sympathy, as the saying goes, is the sum of similarities. The use of images and the language that matches the motivations of a person or a target group causes a positive response from the recipient. The result is *unison* between the sender and the receiver of the message.

The target audience is much more open to what has been said, because the word and motives are familiar and put the fulfillment of a deep need within reach. So it is easier to gain the attention of the target audience and generate understanding of your own requests. This applies equally to the customer approach in marketing, the doctor–patient communication, or the motivation of people with a chronic disease to be stimulated to participate in a health program (Scheffer et al., 2006a,b).

10.2.11 Royal Road to Understand Implicit Personality

Personality systems are motivational to an extent so large that some personality researchers began to view the study of perception as the royal road to the understanding of personality (Murray, 1938, 1943). The perceiver's motivation is a top-down process, which strongly and preconsciously influences visual perception (Henderson and Hollingworth, 1999; Balcetis and Dunning, 2006).

10.2.11.1 *The connection between personality systems and design elements*

Advertising professionals, designers, and artists have always had an intuitive understanding in grafting particular personality characteristics into their objects (visuals, products, and brands). Recently, Scheffer and Loerwald (2008) postulated the implicit rules sometimes used by practitioners. Part of the theoretical basis was provided by Jung's personality typology (Jung, 1994), which is stating four orthogonal dimensions:

1. Sensing
2. Intuition
3. Thinking
4. Feeling

and two additional emotional functions:

1. Extraversion
2. Judging

Another component contributing to the approach is the PSI theory (Kuhl, 2000, 2001). The PSI theory has been developed on the basis of neuropsychological research on motivation and self-regulation. According to this theory, motivation and behavior can be seen as a function of systems interaction, which can be modulated by a variety of stimuli.

The PSI theory describes the interaction among two affective and four cognitive systems. An important assumption of this theory refers to the independence of all six systems and their implicit, subconscious nature.

The two affective systems describe positive and negative effect, which are related to stimulation and security motives, respectively (Bischof, 1985; Scheffer and Heckhausen, 2008).

These affective systems modulate four core functions (Kuhl, 2001):

1. Object-recognition system
2. Intuitive behavioral control
3. Intention memory
4. Extension memory

In PSI theory terminology, sensing [S] originates from neuropsychological object-recognition systems activity, intuition [N] originates from intuitive-behavior systems activity, thinking [T] is linked with the intention memory, and feeling [F] is linked with the extension memory.

We have formed the NeuroIPS method by combining key elements of Jung and the PSI theory. This combination provides a system to describe characteristics and behavior patterns of individuals as a function of the actual situation.

The desire for stimulation through action represses intention memory (thinking) and activates intuitive-behavior control (intuition) (Figures 10.6 and 10.7).

The desire for security represses the extension memory (feeling) and activates the object-recognition system (sensing).

Two dimensions can define the structure of communication: information and order. Figure 10.8 outlines the various shapes resulting from modifying the relative value of a shape along these parameters.

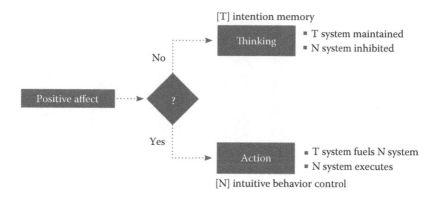

FIGURE 10.6
The influence of positive affect.

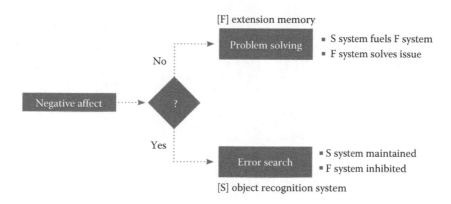

FIGURE 10.7
The influence of negative affect.

Following Jung's theory, each personality type appeals to certain shapes and structures. For example, the sensing-type [S] person would be attracted by design elements containing a lot of details and information.

Kuhl explains this behavior in his PSI theory (Kuhl, 2000; Scheffer and Kuhl, 2005, 2008): Sensing types have a stronger tendency towards using the object-recognition system, which is responsible for discovery of dangerous and improbable stimuli and discrepancies. Contrary to this are design elements containing only little details and information, which are implicitly perceived rather than explicitly recognized and appeal more to intuitive-types [N].

Thinking types [T] prefer an abstract order based on general principles and look for definable figures that are marked off from each other with sharp and straight shapes.

Feeling types [F] like an emotional, tolerant, adaptable order based on personal values and look for connecting, organic figures with diffused and soft shapes.

The two axes construction of *information* and *order* can also be utilized to derive communication preferences of certain personality types on a higher order. Figure 10.9 shows

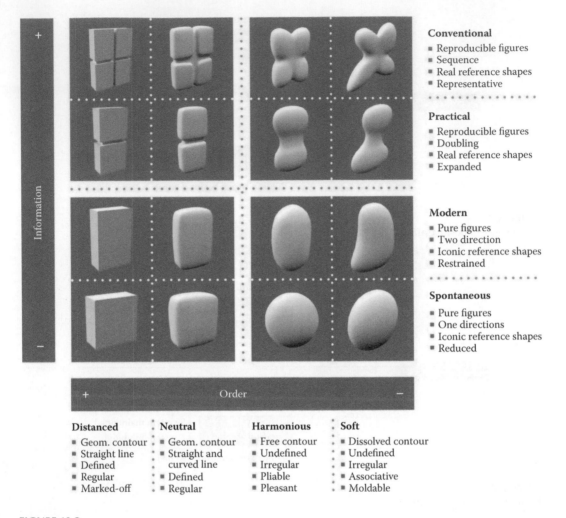

Distanced	Neutral	Harmonious	Soft
▪ Geom. contour	▪ Geom. contour	▪ Free contour	▪ Dissolved contour
▪ Straight line	▪ Straight and	▪ Undefined	▪ Undefined
▪ Defined	curved line	▪ Irregular	▪ Irregular
▪ Regular	▪ Defined	▪ Pliable	▪ Associative
▪ Marked-off	▪ Regular	▪ Pleasant	▪ Moldable

FIGURE 10.8

Shape structure along values of information and order. (From Scheffer, D. and Loerwald, D., Messung von Persönlichkeitseigenschaften mit dem Visual Questionnaire (ViQ)—Attraktivität als Nebengütekriterium, In W. Sarges and D. Scheffer (Hrsg.), *Innovative Ansätze für die Eignungsdiagnostik* [*Innovative Approaches in Personnel Selection*], Hogrefe, Göttingen, Germany, pp. 51–63, 2008.)

examples of communication, differing in emotional content and specificity, for four pairs of personality types.

10.2.12 Psychological Characterization with the Visual Questionnaire

The ViQ is a unique method to measure implicit personality characteristics by using visual items (Figure 10.10).

Based on the theory of C.G. Jung and J. Kuhl, the ViQ gives an insight about the preference for each of the four cognitive dimensions (sensing, intuition, thinking, feeling) and the two affective dimensions (extraversion and judging):

The ViQ measures the intensity of the characteristics in both directions of the cognitive functions that normally reveals a clear tendency. A distinction is drawn between

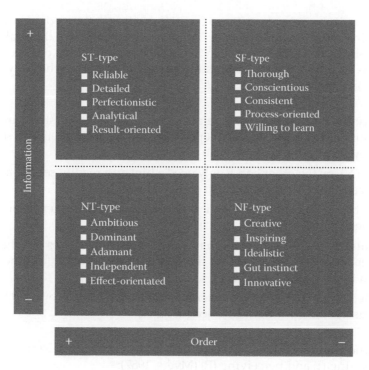

FIGURE 10.9
Characteristics of personality types.

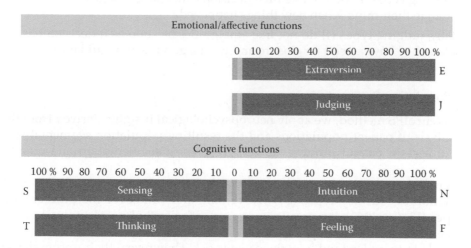

FIGURE 10.10
Dimensions of personality characterization.

perception through the physiological senses, that is, sensing [S], and the more intuitive, that is, intuitive [N], and between a logical–analytical style and a subjective and value-oriented style, that is, thinking [T] and feeling [F].

The procedure measures the level of extraversion [E] and judging [J]. The levels of the functions introversion [I] and perceiving [P] are computed from their corresponding functions extraversion and judging, respectively.

In order to classify people along these functions, a four-letter code is used. The dominant type within each of the four pairs of functions (E/I, J/P, S/N, T/F) is taken as a classifier in this four-letter code. This results in 16 possible combinations or types.

Each type has its individual characteristics, already observed in the early 1920s by Jung (1923/1971). He noticed the following:

- Sensing types (S) perceive very detailed and need a lot of structure and clear information and objectives. They depend exactly on what their five senses show them. They see the detail rather than the whole picture.

- Intuitives (N), on the other hand, only need a hint and readily see the *whole picture*. Their perception is far more abstract and symbolic.

- Thinking types (T) need a very logical and systematic argumentation in order to be convinced.

- Quite contrary, feeling types (F) do not trust facts and figures but their gut feelings—their decisions are based on the nature of relationship to other people as well as on their deeply rooted values.

- Extraverted types (E) need more external attention, while introverted types (I) are more focused on their internal, subjective states.

In addition to the aforementioned functions defined by C.G. Jung, Myers-Briggs added the functions of judging (J) and perceiving (P) (Myers, 1987):

- Perceiving types (P) do not like final decisions; they love to learn new things and to change their mind when new things happened.

- Finally, judging types (J) do not like ambiguities as soon as a decision has been made—they have a need for closure and want a good deal of guidance.

10.2.13 Population Segmentation

With the NeuroIPS method, we apply neuropsychological insights derived from the ViQ to the analysis of patient populations and the resulting population segmentation/stratification of certain patient groups. We achieve precise segmentation and quantifiable success.

The NeuroIPS method ascertains personal behavioral patterns. It starts with the differentiation of four key types (ST, SF, NT, NF) and can cover the differentiation of 16 subtypes (Figure 10.11); it also allows an in-depth psychological analysis of an individual or an entire target group.

Our tests help to understand how the patients *tick*. They reveal their unconscious motivations and drivers influencing compliance with applied therapies and services.

We take advantage of the neuropsychological insights stemming from this methodology and enhance population management in several ways:

- Classify patients psychologically and determine communication preferences
- Reliably connect with patients and significantly boost patient compliance
- Address patients' real needs and optimize product and service acceptance

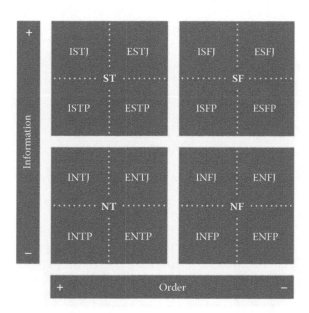

FIGURE 10.11
Categorization of personality types with NeuroIPS.

10.2.14 Emotional Disposition

The attitude of chronically ill patients towards their condition is mainly driven by their emotional disposition, which can be determined along two main levers: responsibility and assertiveness. By spanning these factors from low to high, one can construct four quadrants reflecting four distinct emotional types. In order to design proper communication, we also consider the patient's particular position (Figure 10.12):

- Quadrant I: Positively motivated—The positively motivated patients take responsibility and see themselves as proactive managers of their own health. The aim of health management is invariably to communicate with other patient types in such a way that they achieve this ideal disposition, which is—sadly enough—not the norm.

- Quadrant II: Defensive—The defensive patients (fearful, aggressive) is often highly competent. However, they frequently fail to take responsibility for themselves and their condition. They are passive in terms of acting responsibly and tend towards aggression (aimed at themselves and/or others). In order to get them to listen and ultimately change behavior, one must first get their acceptance through demonstration of superior competence leading to an acceptance of external authority.

- Quadrant III: Resigned—The resigned patients, characterized by sadness, often exhibits low assertiveness levels and shows little responsibility. Seligman (1975) described this as learned helplessness. They feel powerless and helpless. They are not longer capable of acting in a purposeful manner. They victimize themselves. The best way to overcome their attitude is to show positive encouragement and direct lead, and maybe later some empathy, all pursued in tiny *baby steps*.

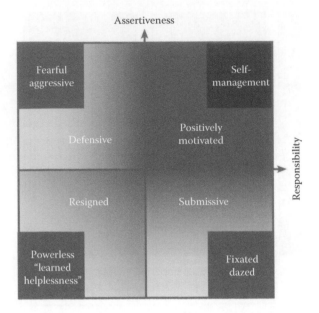

FIGURE 10.12
Core emotional states in relation to assertiveness and responsibility.

- Quadrant IV: Submissive—The submissive, serving patients, characterized by a lack of self-confidence, demonstrate a high degree of responsibility and compliance but little self-determination. They are the classic followers. It takes an empathetic tone, proof of competence, and meaningful stimuli to bring about a change from obedience to personal responsibility.

10.3 Early Results

10.3.1 Genetic Fingerprint of the Customer Personality

The ViQ helps to determine the individual characteristics of a person or target group. Its result can be read like a genetic code. The derived psychological profile of a person is as unique as a fingerprint. We have performed more than 500,000 ViQ tests worldwide until now.

Sections 10.3.2 through 10.3.6 describe results of five exemplary studies using our profiling method. The presented case studies demonstrate how the distinct *fingerprint* of the investigated population determines their particular behavior pattern. Understanding this *fingerprint* and related communication preferences provides a base for conceptualization and design of our group-specific interventions.

We measure our results in a type of *fingerprint* diagram, which displays the type of personality analysis we gain from our studies. The diagrams display z-standardized means of the six measured psychological dimensions. The z-values themselves are retrieved from a comparison of the measured profile against a representative norm, which we determine based on our collected reference profiles. Positive values indicate a higher than normal activation of a personality system, while negative values mark a reduced activation level.

10.3.2 Case Study: Smoker/Nonsmoker Study

10.3.2.1 Project

In this explorative study, we determined differences of psychographic characteristics of smokers and nonsmokers. The *fingerprint* (Figure 10.13) in Section 10.3.2.3 shows the results.

10.3.2.2 Result

In a representative sample of more than 1000 people, smokers exhibited lower scores in the thinking (T) dimension. This demonstrates that smokers tend to be less governed by their deliberate acts and more by their impulses. While the z-standardized values of the other five psychographic characteristics were around the mean of a representative norm (±0.10), thinking values of smokers were on average −0.67, which is 2/3 of a standard deviation below the representative norm. The strong connection between low thinking and smoking was demonstrated further by the fact that this trait also correlated strongly with the number of cigarettes smoked per day: People smoking less than 10 cigarettes per day had an average value in thinking of z = −0.55, people smoking 10–20 cigarettes of z = −0.66, and people smoking more than 20 cigarettes a day of only z = −0.74. These results are statistically significant ($p < 0.05$).

10.3.2.3 Relevance for Population Management

Smoking cessation efforts and other health behavior change initiatives can address group segments on the basis of their psychographic characteristics. It is possible to achieve initial acceptance followed by efficiency and long-lasting results. Our system can deliver

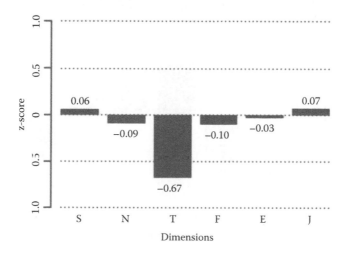

FIGURE 10.13
The *genetic fingerprint* of the personality of smokers.

specifically designed outreach campaigns to dramatically increase the efficiency of established smoking cessation initiatives, particularly those with an mHealth emphasis.

10.3.3 Case Study: Leading Global Software Supplier

10.3.3.1 Project

We developed the communication guidelines for individual customer segments of a multinational software company. The segmentation was achieved by classifying several occupational groups analyzed in individual studies. Participants in the various studies were invited
to take a test on a secured area of the company's website or via other platforms (Figure 10.14).

10.3.3.2 Result

This customer segment is defined by decision makers of communal organizations in rural
Germany, like hospitals, public water supply, and public administration; it is characterized
by a very strong average level of the thinking dimension (z = 0.56). The thinking value is
higher than ½ standard deviation compared to the German norm. It predicts for customers of this segment to be responsive to rational, systematic, and analytic arguments. The
very low level of intuition suggests that these arguments should not be abstract but must
be very concrete and specific.

Elevated levels of extraversion and feeling predict that customers of this segment are
quite prone to social events and close relationship to sales and counseling persons. Quite
astonishing was the very low value in judging, which means that these people are on average perceiving types who like to decide and behave unconventional and do not like being
closely guided.

Based on these results, we designed group-specific communication also considering the
company's corporate identity umbrella. The personality-specific communication proved
to be extremely effective, resulting in an average increase in sales of roughly 20%. The
insights are also of strategic importance to the company in terms of optimizing the companies approach to other target groups.

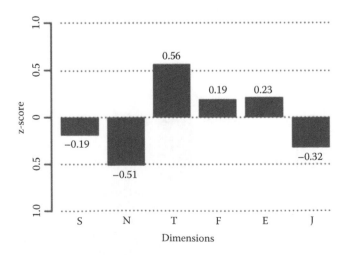

FIGURE 10.14
Description of specific psychographic customer profiles for the various segments.

10.3.3.3 Relevance for Population Management

Personality profiling and population segmentation provide the opportunity for significant improvements in counseling, sales, and marketing and, consequently, increased intervention acceptance and compliance. This case example relates well to the unique challenges of population management.

10.3.4 Case Study: Major Retailer

10.3.4.1 Project

We demonstrated the significance of seemingly small differences between a customer segment and the general population norm.

10.3.4.2 Result

As depicted in Figure 10.15, the deviation of customers of a major retailer from the population norm was only $z = 0.13$ for feeling, $z = -0.08$ for judging (i.e., customers were on average perceiving types), and $z = 0.05$ for intuition. This psychological characteristic of intuition, feeling, and perceiving was replicated for men and women, different age categories, customers, prospects, and even fans on Facebook, partly with much larger effect sizes. Altogether 50,000 persons participated in the study.

The analysis demonstrated that the customer base exhibited a clear psychographic alignment that matched the company's brand model. A comparison to the country's norm showed that the company was reaching around 50% of the population with its identity and positioning.

By aligning themselves with this segment, the company was able to achieve clear communication and an unambiguous market position, helping to differentiate from its competitors. With our support, the company is still using these insights to inform and tune its communication across all channels.

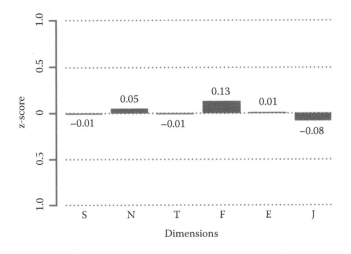

FIGURE 10.15
The *genetic fingerprint* of customers of a major retailer.

10.3.4.3 Relevance for Population Management

This large-scale population assessment and segmentation enabled in-depth insights into the retailer's customer base. It demonstrates that even small effect sizes can translate to major differences in the composition of target groups. The same sort of effect can be exhibited in the population management area using NeuroIPS-based positioning and corresponding interventions for large groups of patients.

10.3.5 Case Study: Leading Health Plan

10.3.5.1 Project

Approach to increase registration rates for the company's disease management program. We applied both our NeuroIPS method and our intervention expertise to design communication and interventions for patients with chronic diseases.

Approximately 4000 patients (asthma, type 2 diabetes, chronic obstructive pulmonary disease (COPD), coronary heart disease, multimorbidity) took part in the study. The majority of participants were over 50 years of age with equal ratio of men to women.

10.3.5.2 Result

We were able to demonstrate clear differences between the psychographic profile of the study cohort and the general population. These differences could be clearly linked to both chronic illnesses in general and additionally to disease specific effects.

For many years, our customer has been using these insights to inform its communication with insured parties and patients, leading to significant increases in the enrollment in their disease management programs.

10.3.5.3 Relevance for Population Management

Segmentation and the downstream matching of communication to target groups boost significant potential when it comes to enhancing the use and efficiency of established measures and programs. This direct experience with a specific cohort of diagnosed individuals has significant relevance to one of the higher-cost patient groups to be addressed in delivering successful outcomes for population management.

10.3.6 Case Study: Leading Diagnostics Company

10.3.6.1 Project

The objective was to develop global guidelines for communicating with diabetes patients. To this end, we studied around 10,000 people with type 2 diabetes in Europe, the United States, South America, and China (Figure 10.16).

10.3.6.2 Result

Over time, people with type 2 diabetes (but not those with type 1 diabetes) develop a personality that is increasingly characterized by greater discipline and less spontaneity and zest for life. This effect is independent of sociodemographic variables and culture.

The study showed that values for dimensions of sensing and judging are significantly higher and values for dimension feeling are lower compared to the norm. This psychological pattern of high sensing–judging with low feeling is known in segments of the

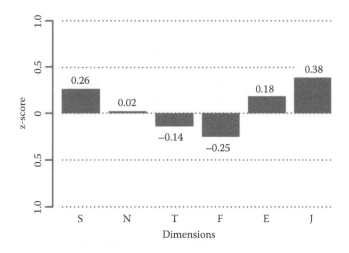

FIGURE 10.16
The *genetic fingerprint* of studied people with type 2 diabetes.

population that have to show very high levels of self-control and conscientiousness, for example, army officers, accountants, and technicians in safety relevant branches.

What differentiates people with diabetes from these occupational segments, however, is the low thinking. We suppose that chronic diseases influence both the decision-making systems thinking and feeling. While the psychological pattern of high sensing–judging with low feeling presumably is adaptive to a more compliant lifestyle, low thinking is not adaptive (as seen in the demonstrated case study with smokers).

In a different sample of 1000 persons with diabetes, we were able to show that communication designed using NeuroIPS was about twice as effective as communication designs based on other approaches.

10.3.6.3 Relevance for Population Management

These insights can be used to design, optimize, and establish successful interventions for people with diabetes and other chronic diseases—a significant objective for population management.

10.4 Conclusion

The field of mHealth is an area in which technical, medical, psychological, and educational discipline must join together to form striking applications. Its future success will primarily be influenced by the imperative of profitability, overshadowing the known expertise in this discipline. Even the most powerful intervention and the most sophisticated technology will fail to be adopted if it is not possible to deliver them cost-effectively.

From our point of view, the combination of two key success factors has the potential to achieve both cost-effectiveness and successful impact on the addressed population. These factors are completely dependent and affect each other in their effectiveness.

The first success factor is automating connectivity. To be best positioned to succeed, applications expected to reach and support larger groups must be fully automated.

The requirements are challenging: future programs should provide solutions that integrate in-depth analytics, personalization, personal support (teaching, coaching, motivating), as well as moderated micro groups and actively managed escalation systems. The support (e.g., content, frequency) has to be adaptable to the current situation of the participant. At the same time, the systems should be completely scalable.

In order to deliver volume and value with a genuine patient-centered approach, it will be mandatory to develop individualized interventions. This is the second key success factor.

Personalization can be considered as a *megatrend*, which is strongly reinforced by demographic and technological changes. Therefore, we are convinced that the time has now come for an individualized configuration of interventions to the specific profile of each participant. Determining the psychographic and personal characteristics of individual patients or representative population segments is crucial for this to be achieved. The identification of the implicit, that is, nonconscious personality traits is especially important for creating a tailored communication profile. Congruence between the personality of a participant and the program interventions ensures sustainable stimulation of the patient's intrinsic motivation and results in the achievement of successful outcomes related to enrollment, engagement, and ongoing behavior change.

We have taken a crucial step in direction towards a new generation of powerful and cost-effective mHealth applications by combining both automation and individualized interventions in our system. Programs and services that fail to take this step will be unable to achieve meaningful behavior change in their target populations that yield the necessary long-term profitability to remain viable.

Questions

1. What is next in disease management?
2. What are the pressing problems of health plans?
3. Why are big pharma and diagnostic companies required to enhance their customer relations?
4. What can ACOs and physicians do to enhance patient engagement?
5. What is driving the future of employer based health and wellness programs?
6. Why is the sophistication of patient engagement lacking behind general progress in health care?

10.A Appendix

10.A.1 Interventions

Table 10.A.1 provides a list of commonly used interventions, which can be integrated in our outreach activities.

TABLE 10.A.1

Examples for Interventions of Optional MassineBoecker's Outreach Activities

Area	Topic
Content	Website content
	MMS retrieval by smartphones
	Online videos
	Printed materials, brochures, flyers, mail items
	Online training units
	Smartphone applications
Doctor–patient relationship	Doctor intervention coordination for medical home and ACO projects
	Preparation for doctor's appointments
	Follow-up after doctor's appointments
	Therapy-monitoring systems with doctor intervention
	Support for home visits
Devices	Home monitoring for a wide range of biometrics, such as weight, glucose, activity, blood pressure
	Device maintenance
	Online oxygen therapy support
	Transfer of glucose data and interpretation
	Medical alarm systems
	Therapy monitoring for nonchronic illnesses
Training	Interactive and individual at the patient's request
	Online self-management training
	Illness-specific focus groups (online)
	Local training groups
	Campaigns for increased health awareness
	Automated telephone messages
Patient-to-patient support	Local self-management patient groups
	General online forums
	Monitored and moderated general online forums
	Illness-specific online forums (patient/patient/group)
	Moderated illness-specific forums (patient/group/moderator)
Health management	Active membership
	Phone-based and/or automated coaching
	Event-triggered automatic intervention
	Escalation system
	Dynamic intervention adjustment
	Goal, activity, and stats tracking
	Reminders (goals, medication, appointments)
	Hospital discharge management
	Health programs with games and challenges
	General text messaging

References

Balcetics, E. and Dunning, D. (2006). See what you want to see: Motivational influences on visual perception. *Journal of Personality and Social Psychology, 91*, 612–625.

Bischof, N. (1985). *Das Rätsel Ödipus: Die biologischen Wurzeln des Urkonfliktes von Bindung und Autonomie*. Munich, Germany; Zurich, Switzerland: Piper.

Camerer, C., Loewenstein, G., and Prelec, D. (2005). Neuroeconomics: How neuroscience can inform economics. *Journal of Economic Literature, XLIII*, 9–64.

Henderson, J.M. and Hollingworth, A. (1999). High-level scene perception. *Annual Review of Psychology, 50*, 243–271.

Jung, C.G. (1923/1971). *Theory of the Types*. New York: Penguin Books.

Jung, C.G. (1994). *Psychologische Typen*. Olten, Switzerland: Walter-Verlag

Kuhl, J. (2000). A functional-design approach to motivation and volition: The dynamics of personality systems interactions. In M. Boekaerts, P.R. Pintrich, and M. Zeidner (Eds.), *Self-Regulation: Directions and Challenges for Future Research* (pp. 111–169). New York: Academic Press.

Kuhl, J. (2001). *Motivation und Persönlichkeit: Interaktion psychischer Systeme*. [*Motivation and Personality: Interactions of Mental Systems*]. Göttingen, Germany: Hogrefe.

Kuhl, J., Scheffer, D., Mikoleit, B., and Strehlau, A. (2010). *Persönlichkeit und Motivation in Unternehmen*. Stuttgart, Germany: Kohlhammer.

Murray, H.A. (1938). *Explorations in Personality*. New York: Oxford University Press.

Murray, H.A. (1943). *Thematic Apperception Test Manual*. Cambridge, U.K.: Harvard University Press.

Myers, I. (1987). *Introduction to Type*. Palo Alto, CA: Consulting Psychologists Press.

Scheffer, D. and Heckhausen, H. (2008). Trait theories of motivation. In J. Heckhausen and H. Heckhausen (Eds.), *Motivation and Action*, 2nd ed. (pp. 42–86). Cambridge, U.K.: Cambridge University Press.

Scheffer, D. and Kuhl, J. (2005). *Erfolgreich Motivieren* [*Motivating Successfully*]. Göttingen, Germany: Hogrefe.

Scheffer, D. and Kuhl, J. (2008). Volitionale Prozesse der Zielverfolgung [Volitional basis of goal pursuit]. In U. Kleinbeck and Schmidt (Eds.), *Enzyklopädie der Psychologie: Wirtschafts-, Organisations- und Arbeitspsychologie*. Göttingen, Germany: Hogrefe.

Scheffer, D. and Loerwald, D. (2008). Messung von Persönlichkeitseigenschaften mit dem Visual Questionnaire (ViQ)—Attraktivität als Nebengütekriterium. In W. Sarges and D. Scheffer (Hrsg.), *Innovative Ansätze für die Eignungsdiagnostik* [*Innovative Approaches in Personnel Selection*]. (pp. 51–63). Göttingen, Germany: Hogrefe.

Scheffer, D., Loerwald, D., and Mikoleit, B. (2006a). Persönlichkeit und individuelle Motivation. Vom Umgang mit der Krankheit Diabetes. *Diabetes News, 3*, S. 9.

Scheffer, D., Loerwald, D., and Mikoleit, B. (2006b). Persönlichkeit und individuelle motivation. Der Effekt der Wahrnehmung. *Diabetes News, 6*, S16.

Scheffer, D. and Mikoleit, B. (2013). Personal goals. In W. Sarges (Hrsg.), *Management Diagnostik*. 4th ed. (pp. 301–308). Göttingen, Germany: Hogrefe.

Seligman, M.E.P. (1975). *Helplessness. On Depression, Development and Death*. San Francisco, CA: Freeman and Company.

Zaltman, G. (2003). *How Customers Think*. Boston, MA: Harvard Business School Press.

11

mHealth Translation Framework: Behavioral Design and Development to Improve Self-Management in Chronic Health Conditions

John Dinsmore and Ronan McDonnell

CONTENTS

11.1 Introduction

It is a startling reality but today, we face a world in which the onus for managing our own long-term health will increasingly fall on our shoulders, whether this is via self-management practices and/or formal or informal care management assistance. Overstretched health services need patients to become empowered, particularly in primary care settings, to develop self-management strategies to offset care costs to which assistive mHealth platforms will play a pivotal role. To empower individuals requires the provision of techniques and strategies to evolve new lifestyle changes, combined with the ability to adopt new practices and modify existing behaviors (behavioral change) to those best suited to managing one's health, particularly as we age and develop potentially debilitating chronic conditions such as diabetes, stroke, and chronic obstructive pulmonary disease (COPD).

Studies have shown that adopting healthy behaviors, improving treatment and self-management practices, can offset the progression of many chronic conditions (e.g., Al-Mallah et al., 2006; Hamman et al., 2006; Effing et al., 2009). However, the amount of information that can be traditionally conveyed within primary health-care services to improve self-management practices (e.g., consultations and information sheets) is unfortunately limited. Modern approaches such as randomly searching the web via engines such as *Google* are also laced with potential difficulties as they can present individuals with nonvalidated, inaccurate, and unstructured health content that may indirectly have a negative impact on rehabilitation and disease management. Developing condition-specific mHealth technologies can help avoid these issues and also possibly be an excellent catalyst to behavioral change by providing individuals with a tailored and personalized approach to health-care management.

Modern mHealth technology has two key features that endear it as a behavioral change catalyzing agent: firstly, its diversity or versatility and, secondly, the ability to deliver nonstatic service and treatment approaches. To capture these value points, we firstly have to explain the concept of self-management. Defined broadly across literature as a systematic, multifaceted, educational approach to patient care, self-management aims to teach skills necessary for individuals to carry out condition-specific routines via healthy behavior change in order to control and improve their health and well-being. As an umbrella concept, self-management is composed of multiple sections, which include improving psychological and physical health, adhering to treatment plans, monitoring ones condition, implementing relevant changes based on health fluctuations, improving quality of life, and maintaining social function (Dinsmore, in press). The key challenge in mHealth self-management platform development is the ability of software designers and developers to effectively translate human factors, which engage, motivate, and sustain healthy lifestyle changes and outcomes into a digital technology platform that provides a seamless everyday experience to living and managing with a chronic condition.

This is a daunting task to deliver with self-management generally a complex operation. However, the versatility of modern mHealth technologies (and accessories) with numerous interactive and feedback features such as text, video, GPS, pedometers and the ability to read vital signs combined with the ability to deliver nonstatic and ever-changing content, user experiences, user interfaces, and services means that not only can we incorporate face-to-face knowledge and learning experiences into portable and remote-based learning pathways but we can also define a potentially new effective model for primary health care, assisted via patient technology empowerment.

This chapter will explore critical translational synergies necessary to develop a holistic user-centered design (UCD) of mHealth platforms that address the physical, psychological, and social needs necessary for increasing patient self-efficacy, empowerment, and ultimately behavioral change in chronic illness self-management practices. Outlining a framework to collect and understand data, shape the behavioral change experience, and improve the complexity of interactions between stakeholders (e.g., patients, designers, developers, health professionals, health-care services) in the mHealth chronic self-management process and primary health-care ecosystem is key to the successful delivery and growth of the mHealth self-management field. Within this framework, commonalities in design for chronic illness related mHealth technologies that can assist in the translation of methodologies and models across differing conditions. Finally we shall discuss the need for health service redesign to accommodate, sustain, and scale mHealth technologies.

11.2 mHealth, Behavioral Change, and Self-Management

Behavioral change has been a field flourishing steadily since the 1970s, yet mHealth is in its infancy and thus it is difficult to gauge presently the effectiveness of behavioral change models and theory when incorporated into mHealth. The academic interpretation of behavior change in health identifies that action, thoughts, feelings, and physical function are tightly interlinked and therefore affect the way a patient manages his or her illness. There are multiple behavioral change techniques (BCTs) that address lifestyle changes; these include providing information and feedback, goal setting, tailoring the user experience, prompting, problem solving, plans for social change, contract for change, relapse planning, preventative coping strategies, giving data an emotional hook, and developing social interactions with peer support. Many techniques have been derived from models that include the transtheoretical model (TTM) (Prochaska and DiClemente, 1984; Prochaska et al., 2002); the theory of planned behavior; the theory of reasoned action (TRA); the health belief model; and the attitude, social influence, and self-efficacy model (ASE model) (Rosenstock, 1966; Fishbein and Ajzen, 1975; Ajzen and Fishbein, 1980; De Vries et al., 1988, 1995; Ajzen, 1991). Behavioral change models primarily focus on building patient confidence by shaping perceptions to behavior necessary to reach an achievable goal. Changing one's perceptions is core to behavioral change and to do so involves techniques that include building a positive internal locus of control and delivering correct knowledge to influence the perception of recovery and the ability to weigh pros and cons to new behaviors.

Common theories in mHealth have centered primarily on the social–cognitive theory, in COPD, and in obesity; for example, the use of peer-based exercise videos may be used for individuals to support the adoption of new behaviors. To date, only one systematic review of mHealth trials presently exists that has comprehensively reviewed studies concentrating on technology-based interventions delivered to consumers (Free et al., 2013). This key review sought to quantify the effectiveness of mobile-technology-based interventions delivered to health-care consumers for health behavioral change and management of chronic conditions. From 26,221 studies reviewed for inclusion, only 75 met the mHealth inclusion criteria. Of the 75, 26 behavior change studies (all in high-income countries), totaling 10,706 participants (sample sizes 17–5,800), addressed interventions aimed to increase healthy behaviors (e.g., smoking cessation, calorie intake, sexual behavior), increase activity (e.g., exercise), and prevent disease progression or onset (health behavior change interventions). The remaining 49 studies focused on disease management trials with a total of 6832 participants (sample sizes 16–273 participants). Interventions studied were designed to improve the self-management of conditions (such as diabetes and cardiovascular conditions) and therapeutic interventions (including those providing psychological support). Technologies currently used in the behavioral change research studies included mobile phones, personal digital assistants (PDAs), and handheld computers. Primary functions used to induce change included SMS, MP3, MP4/video, and telephone use. Due to the pace of technology change, however, many of these studies are unfortunately now outdated, as the use of PDAs, for example, and many of the mobile phone models have been replaced by tablets and smartphones in the modern-day context.

Of interest in this review is that only 7 of the 26 located studies used behavioral change theories to specifically construct their intervention, where the maximum number of techniques used 18 (median was 6) (e.g., Free et al., 2009, 2011). Providing feedback to the intervention used was the main BCT used, followed by goal setting, providing information on the consequences of the behavior, personalizing the intervention, using prompts to help self-manage, and identifying and implementing problem-solving interventions. The overall findings within this review toward health-based behavioral change technologies are mixed. However, an important finding was that short-term evidence for the use of psychological support interventions could be clinically important if proven beneficial in a long-term study.

The question is, therefore, why has there been a poor adoption rate of BCTs throughout the whole design and development process from baseline concept to full product development? The answer may lie in the complexity to both combine and deliver both self-management and behavioral change into a personalized platform that will have a widespread effect across various individual users.

Presently, the most successful research combining these two areas has been via face-to-face interventions; in COPD, for example, this has included programs in smoking cessation, use of breathing apparatus, medicinal self-management, and exercise. These programs have also fundamentally understood and addressed the processes of care and daily interactions between patients, family, carers, friends, and health professionals as part of a self-management ecosystem. Today, in mHealth, our knowledge of behavioral change is focused primarily on the translation of these successful studies to technology.

Overall, the premise remains that BCTs are primarily used to assist in the development of *structured educational programs* for patient self-management. These programs are defined as "comprehensive in scope, flexible in content, responsive to an individual's clinical and psychological needs and adaptable to his or her educational and cultural background" (NICE, 2003: 14). Their aim is to help modify behavior by primarily improving perceived internal control of one's condition (Worth and Dhein, 2004). This in effect improves patient empowerment and readiness to treat one's own condition. Unfortunately, literature to date does not fully specify the exact process to knowledge acquisition from these programs. Yet there is a broad consensus that providing strong, easily digestible feedback during the action and implementation phases is important to improving behavioral change programs, for example, using smartphone applications to accurately track weight loss and calorie intake as part of an obesity weight-loss program.

Changing an individual's behavior is also a time-consuming process. It has been suggested that a minimum of 3 months should be allowed before assessing effects of a supported behavioral change program (e.g., Wempe and Wijkstra, 2004). Long-term educational initiatives that introduce new learning stimuli (via user-friendly platforms) to avoid drop-off may improve the behavioral change process; however, it is still unclear from behavioral change models what is the best practice methodology for patients maintaining effective self-management practices after exiting a behavioural change programme.

11.3 Framework to Facilitate Effective Behavioral Change in mHealth

The *iceberg effect* (Dinsmore, in press) is when a research, design, and/or development team focuses too heavily on the visual usability of the technology (tip of the iceberg) and subsequentially fails to fully address the (potentially complex) core underlying social and psychological factors crucial to developing an efficient and effective relationship between

TABLE 11.1

Six-Stage Framework to Overcoming the *Iceberg Effect*

Step	Description
1.	The end user is the primary target—needs to be understood by the design and development team (user centered, sensitive, inclusive design).
2.	Communication and education is key to avoiding a translational dilemma.
3.	Combining self-management and behavioral change—via knowledge delivery and learning programs.
4.	Sustaining change—providing a continuous platform to sustain new healthy learned behaviors.
5.	Iterate and innovate—continuously develop the user experience.
6.	Technology adoption and health-care service redesign.

the health-care user and technology. To avoid the *iceberg effect* a six stage framework is suggested by Dinsmore to assist the behavioural change process.

Key to most components in this framework and highlighted in step 2 of Table 11.1 is that effective communication and education between patients, carers, health-care professionals, and technology development teams is crucial to producing an effective behavioral change model and thus avoiding a *translational dilemma*. By this we mean that on one side, mHealth design and development teams need to be aware of the backdrop to respective chronic conditions and that management requires considerable self-care effort from patients. On the other side, health professionals need to be aware of the limitations to the technology. Ensuring this translation is effectively maintained is a crucial piece of the framework.

Sustaining change and avoiding drop-off or compliance issues are also major challenges for mHealth. Developing programs constantly available for patient participation as part of a *real-time*, interactive, ongoing service may increase an individual's motivation to sustain engagement. Issues detrimental to sustaining change include limited knowledge and skill in technology use, poor personal motivation to make changes, and lack of specific competencies to engage in any lifestyle modification processes aside from any technology use.

Another key issue in the framework mentioned earlier to help induce behavioral change is the ability for a strong design and development innovation process and the preempting of technology changes in society. Spotting technology trends should be balanced by potential patient affordability, which could also be a predictor for societal adoption. For example, the rise in *smart* televisions could be central to a future home-based self-management learning and knowledge dissemination pathway. Indeed, there is an argument, that to assist health service redesign bespoke costly health-care devices should be reduced in favor of currently existing, user-friendly, mass-market technologies such as the iPad and smartphones.

11.4 Key Challenges

11.4.1 Issues for Designers and Developers

Presently, there is no generally agreed framework for approaching the meeting of health and technology to induce behavioral change in chronic conditions. As stated previously, this is not surprising due to the complexity of merging these fields in existing health services. As a best practice, it is agreed that a user centred design (UCD) approach with reference to the health condition and cultural context is important. UCD is a complex process in mHealth, which involves a multidisciplinary team of interface designers, software developers, and usability specialists (including psychologists and ethnographers). The aim is to create a strong model of learning and/or feedback to the mental model of what the patient expects to learn and achieve from engaging in particular activities (Weinschenk, 2011). This may also involve a training element if we are to induce a new mental learning model. Educating designers and developers in the health-care field is a timely and patient process but a necessary step forward to avoid the *iceberg effect* (Dinsmore, in press).

To help quantify potential BCTs, Abraham and Michie (2008) introduced a taxonomy based on 26 BCTs. However, problems with variability in the reporting of techniques in studies coupled with the fact that the typologies do not address mHealth specifically mean to a certain extent that designers and developers are in the dark as to what BCTs are needed to tackle condition-specific problems. Focusing on various design strategies as part of a development and research process will help better define and validate BCTs for mHealth development.

Indeed, the importance of the involvement of users in the design of technology has been known for many years. Techniques such as UCD and participatory design mandate that users (or in the case of participatory design, the stakeholders) are significantly involved in the design process from as early as possible (Mao et al., 2005). This is to ensure that the product designed as a result of this process is usable by that group, as well as incorporating the needs of the group into the design of the end product. However, a large proportion of devices and systems are often designed by those who may not have an appreciation for the difficulties that certain users face when using technology, such as physical impairment and a lack of experience with technology (Sayago and Blat, 2010). Furthermore, many people are excluded from using products as a result of designers not taking account of the end user's functional capabilities (Keates and Clarkson, 2003). As such, it is important to gain a good understanding of the problems faced by users when designing for them.

Instead of describing the design and development of each of the myriad of different user groups, health conditions, and cultural contexts that need to be considered when designing and developing mHealth technology, we will focus on one relatively large group that is particularly affected (overly so, in fact, compared to other age groups) by chronic disease as well as physical and mental impairments, that being older adults. According to the Centre for Disease Control in the United States, older people suffer particularly from chronic disease, with 80% having at least one chronic condition and 50% having at least two (National Center for Chronic Disease Prevention and Promotion, 2011). However, we believe that many of the issues that affect older adults when it comes to mHealth technology design and development can be applied across user groups, conditions, and contexts.

There are a variety of methods that have been employed to design for older people or those with disabilities, such as universal design, inclusive design, and accessible design. However, in addition to the needs of chronically ill people being catered for in the technology itself, the methods used to assess these needs must also be tailored for this user group. Issues such as the length of interviews, attendance, question design, and terminology used during interviews and requirements gathering need to be well planned when chronically ill and particularly older adults are involved (Barrett and Kirk, 2000).

11.5 User-Sensitive Inclusive Design: Case Studies

The importance of involving users throughout the design process, whether that be through UCD, participatory design, or other design processes that recommend involving users has been widely reported (e.g., Abras et al., 2004; Sharp et al., 2007).

UCD also encompasses processes where users influence how a design takes shape. It places the user not as part of the team but as a voice communicated to the designer through the researcher (Sanders, 2002). Abras and colleagues in 2004 stated that there may be a number of ways that the users can be involved, but the important point is that users are involved, such as consulting them about their needs via interviews or making them partners in the design process.

There are many reported benefits to the use of UCD. The *ultimate test* of a product's usability depends on measuring users' experience with it (Dix et al., 2004). The approach can ensure a system "will make the most of human skills and judgment, will be directly relevant to the work in hand or other activity, and will support rather than constrain the user" (Sharp et al., 2007). However, while UCD can help ensure that a product is suitable for its intended purpose, the main disadvantage of it is cost, in both financial and time terms, as it takes a significant period of time to research, gather, and analyze the data that inform the design. Furthermore, UCD teams can benefit from a multidisciplinary group including psychologists and sociologists who can understand user requirements and explain these to the technical part of the team (Abras et al., 2004); however, this can incur extra financial costs.

Participatory design is another type of design process that not just involves the user but focuses on involving all stakeholders, such as workers and management in the research process as there may be different goals for each group, even if some of the groups are not end users of the product. Participatory design first appeared in the 1970s in Scandinavia in reaction to the neglect of workers' interests in the introduction of systems into the workplace. In participatory design, the user "becomes a critical component of the process" (Sanders, 2002). Workshops, lectures, and courses are used alongside techniques from the ethnographic world including interviews and observation (Kensing and Blomberg, 1998) in the participatory design process.

However, as stated previously, care needs to be taken when using these processes to design technology particularly for older and chronically ill people, as they present much greater challenges than traditional user groups (Newell et al., 2007) such as a greater occurrence of impairments. For example, 65% of those with visual impairments are over the age of 50, despite this group only accounting for 20% of the population (United Nations, 2007). Furthermore, older people, such as those with chronic conditions, may have different needs and wants and may operate in a different environment to other age groups, and as such, a standard piece of technology may not be suitable for them (Gregor and Newell, 2001). In an effort to encourage more understanding of the needs of these different user groups, many of whom live with one or more chronic conditions, there are studies and techniques that have tried to mimic the impairments and constraints of people with a variety of impairments. These include simulating visual impairments in software so designers can experience the impairments for themselves (Goodman-Deane et al., 2007). Other examples include the *Third Age Suit*, developed by the Ford Motor Company to simulate reduced joint mobility so designers understand what it feels like to be older (Ford Motors, 2011), while Hitchcock and Taylor (2003) designed wearable devices to address such limitations. However, it is important that these custom systems for older and disabled people can also interact with more mainstream technology. Shneiderman (2000) suggested that this universal usability (where more than 90% of all households successfully use information and communication technology [ICT] services at least once a week) contains three important issues: technology variety (wide range of hardware and software are supported), user diversity (wide range of skills, abilities, and impairments), and bridging gaps in user knowledge, which can help older and chronically ill people use technology more easily and successfully.

However, it is also very important that the needs of older and chronically ill people are considered and incorporated from the earliest stage possible. Indeed, while UCD (defined as involving users in the design of a given system they will use) is considered the most appropriate design process in most projects or studies (Mao et al., 2005), it is not necessarily always suitable for older people or those with impairments or chronic illnesses.

Newell and Gregor (2002) have described many of the problems with using techniques for designing with older and disabled people such as UCD, design for all, and accessible design. They describe design for all as "a very difficult, if not impossible task," with the danger that by making products accessible to people with certain disabilities, it may make the product more difficult to use for those without any or with different types of impairments. They proposed a further development on UCD, which they call *user-sensitive inclusive design* (USID).

USID differs from UCD in a number of ways, with inclusivity replacing the word universal, as inclusive design is more achievable than universal design due to the potential to alienate one group in attempting to design for another, while *sensitive* is used instead of *centered* as it can be difficult to obtain a proper sample user group due to recruitment difficulties (Newell and Gregor, 2002).

As an example of where USID is more appropriate, Gregor and Newell (2001) reported the use of a standard UCD methodology in the design of a web browser for visually impaired users, reporting that a more inclusive design (they recommended USID) would have been more appropriate. In the designers' attempts to ensure a homogenous user group, they did not account for the diversity that exists in older population groups. As a result, the researchers stated that the designers were *genuinely surprised* that older users could not use the browser, even though "five minutes spent observing a visually impaired older adult trying to get going with the software would have made this clear."

This design methodology seems more suitable for older and impaired users, such as those with chronic conditions, but there are also specific techniques that must be adapted to allow technology to be designed using USID. These include recruitment methods, including the choice of location so that it is accessible for people with mobility problems (Newell and Gregor, 2002) or keeping information-eliciting techniques such as focus groups and interviews shorter as some users can have limited attention spans or may become tired more quickly (Barrett and Kirk, 2000). Furthermore, carer involvement or individuals such as medical experts who may not be users of the system but could have a valuable contribution and input into the research should also be included in the research process, if deemed relevant.

However, we believe that there is scope to extend USID, especially in the health and self-management of chronic disease, by incorporating psychological and behavioral change techniques into modern technology and software. By grounding our technology not just in the way that humans interact with technology but also examining habit formation/retention and learning techniques, technology could be a powerful tool in helping people to self-manage their diseases and ultimately improve health outcomes.

11.5.1 Iterative Development in Software Design

Once the information has been gathered, requirements specified, and development of the technology started, the iterative process of showing it to users for feedback is crucial. Dickinson and colleagues (2005) state that *think-aloud interviews*, where a user describes in detail their thought process as they use the technology, can highlight inconsistencies in the interface and can give "rich and interesting information, but will increase the cognitive complexity of the task." They recommend starting any user testing protocol on the second run of using the system, especially for those with any cognitive impairment.

Other techniques that have been used include logging the usage of a developed system for errors and iteratively changing the interface based on feedback and where users are encountering errors (Wherton and Prendergast, 2009; Doyle et al., 2010).

11.6 Development of a Framework

The theory of innovation diffusion (Rogers, 2003) states that individuals are more likely to consider technology use only when it is relevant to them. Thus, developing a framework to better capture data, understand self-management needs and requirements, and ultimately inform behavioral change is paramount in the redesign of an mHealth primary care ecosystem. Validating mHealth technologies within such an ecosystem will be judged not only by review of system analytics to vital signs or functional measurements but also by research conducted to capture the emotional engagement and motivation of the user with the technology, application, or service provided.

Developing a framework to collect and analyze data in real time will allow for a greater understanding of differing BCTs and related platform design and development features tailored to various condition groups and self-management needs. mHealth has the capability to monitor many diverse issues in real time and if so may increase emotional and motivational engagement as part of holistic treatment rather than tackling one issue in isolation. For example, in COPD, the use of mHealth to provide exercise video intervention will need design features to ensure that those on oxygen are advised to its appropriate use during exercise with levels monitored as with increased exercise more oxygen will be used. Finally advice with breathing techniques should be provided to prevent exacerbations brought on by increased activity. A study by Rose Biomedical (2004) with COPD patients showed that if correctly administered, a distinct correlation existed between improved health benefits, motivation, and behavioral change with video-based interventions that provide a medium to disseminate knowledge as part of a self-management and rehabilitation plan. However, several recent randomized controlled trials (RCTs) using video tele–health care have also shown that even though those that use the technology report high levels of satisfaction, up to 80% did not use the technology long term fearing it may replace visitations by nursing staff (Mair et al., 2006).

In obesity, a similar problem exists with research showing that mHealth-related treatments that focus solely on improving functional outputs appear to have little to no effect (Shaw et al., 2006). There is a need to use mHealth in a more holistic capacity outside the traditional biomedical model to one that also envelops psychosocial, medicinal, and carer support factors.

11.6.1 Health Service Redesign and Framework

The greatest challenge facing mHealth is societal adoption of technology and the need for health services to redesign current practice to incorporate this adoption. Educating stakeholders in technology use is crucial to this process. Presently in some high-income countries, the level of subscription to mobile services is greater than the population number (Ofcom, 2009), and in developing countries, the mobile technology and telecommunications sectors are among the fastest growing with high geographical network coverage (Donner, 2008); in 2011 alone, there were 491.4 million smartphone units sold worldwide (IDC, 2012). A need exists to leverage this rise in mHealth to redesign the primary health-care ecosystem as part of a patient-empowered, multifaceted user experience.

Retrofitting or disruption of the home ecosystem *consumer market* (smartphones, TVs, tablets) may be the best option to facilitate the adoption of technology to catalyze behavioral changes via an easily accommodated user experience. However, issues still exist when repurposing market products for health care including data security, dealing with maintenance issues, and ensuring feedback to patients is provided within best clinical practice guidelines and within the cultural context.

Vital to a health system redesign will be the extraction of data for use by clinicians. The merging of data recorded at clinical encounters with key validated day-to-day data from patients would seem like a likely future direction for mHealth. While some clinical illnesses, such as diabetes, already have established methods of saving user data for analysis with a health-care professional, there are not many existing health records that allow for the incorporation of patient-gathered data into the records.

Part of the issue arises out of clinically validating these data—for example, if an individual is due to take a medication for a month and supplied with a bottle that records when it is opened, that is not proof that the medicine was taken, merely that the bottle was opened. Bottles that intend to increase the adherence levels are already in active development (Adheretech, 2013) by tracking not only if the bottle was opened but whether any medication was taken out, communicating this over mobile phone networks to the patient via text or to their friends, family, or doctor (in a recognized health data format) if the patient so wishes. However, this brings a focus onto privacy issues regarding patient-recorded data—who owns it, who is allowed access to it (and who grants that access), where and how is it stored, and can (should?) it be anonymized and used in broader research?

Overall, though accurate, regular data for clinicians regarding how individuals adhered to treatment regimens over a fixed period could be invaluable (and cost effective) for health systems. In fact, improved patient empowerment with increased clinical involvement may increase the self-management process via mHealth.

Improving social interaction among users such as developing peer support networks may also be core to adoption of an mHealth ecosystem. Impaired, ill, and older people can be at increased risk of social isolation and loneliness. There have been many investigations into increasing social interaction among older people, as well as several reviews into these interventions in order to identify the common features of the most successful ones. Cattan and colleagues (2005) performed a review of 21 different health promotion interventions for their effectiveness, which showed that successful interventions typically

- Are group interventions providing social support
- Target-specific groups that have an educational or training input
- Should be within an existing service or evaluate that service

- Use validated measuring tools to evaluate the intervention
- Allow participants some level of control

Findlay (2003) also conducted an examination of the evidence for 17 interventions to reduce social isolation among older people. It was found that involving older people in the planning, implementation, and evaluation of interventions makes them more likely to succeed.

Infrastructural improvements are also crucial to inclusive redesign. Health systems incorporating new technologies also need to strategically design services to facilitate the rise of mHealth as part of new models of treatment and care. For example, this may include better training methods for community-based health professionals or better broadband connectivity in rural areas. Effective government and industry cooperation to improve infrastructure and produce affordable commercial mHealth facilitator technologies is important to induce change. However, it is recognized that building a relevant and effective health-based support system into the home environment in line with clinical guidelines, recommendations, and treatments is a major challenge for developers aiming to deliver validated, subjective, secure, and accessible technologies (Bayliss et al., 2007).

An effective mHealth delivery framework requires inclusion of constantly updated multimedia and support content to provide knowledge, rehabilitation, medical device use, and peer perspectives all within cultural context. The *apps* market is where many designers and developers are now focusing their attention. Healthcare *apps* will play an increasingly important role in mHealth self-management practices, tracking behaviors and personal feedback and informing health-care professionals of patient progress.

Yet tighter regulations for validating stand-alone software and applications are critical to avoid *apps* that not only provide poor content but also lack strong design to target the condition they are developed for. On diagnosis of a chronic condition, individuals can feel vulnerable; thus, it is crucial to provide relevant information in an accessible, tailored, and digestible format to improve self-management and coping practices at this point.

The future of mHealth is also tending toward using modern technology to merge data across care settings so that data can be (appropriately) accessed anywhere. With mHealth, it could be argued that the home is becoming one of these care settings. Furthermore, this merging of medical technology with consumer technology is already underway, with add-ons for the iPad that can measure your heart rate, blood glucose levels, and many other data points.

The key issue in the merging of medical data generated inside and outside of a clinical encounter/setting is ensuring that the data, especially generated via consumer technology, are medically sound and valid. Extreme care must be taken with this technology so that it is properly validated, tested, and approved. In the United States, software can now be classified as a medical device and requires rigorous, standardized testing if it is to achieve this classification. However, the medical and health informatics communities need to be proactive in ensuring that it is clear to consumers when technology has been developed and approved as a medical device and when it has not.

11.6.2 Are Electronic Health Records/Patient Health Records the Foundation Stone to Service Redesign?

In recent years, there has been an increase in the use of electronic personal health records (PHRs), which allow a patient "to access their medical records, self-management tools, and new avenues of communication with their health care providers" (Tenforde et al., 2011).

While evidence to support the clinical value of PHRs remains somewhat limited due to a paucity of studies, they can be used in conjunction with BCTs. One particular study (Ralston et al., 2009), which conducted an RCT using a PHR and care manager providing individual patient advice for the intervention group versus usual care for the control, saw improved glycemic control levels for the intervention versus the control group. However, it was suggested that this was due to the care manager rather than the availability of the PHR. Furthermore, it is suggested that the resources for such a care manager would be beyond the ability of most practices, which is exactly where a customized, behavioral-change-based software system could bring benefits. Finally, the intervention group in this study was granted complete access to the electronic medical record (EMR) used by their primary care provider, with the study suggesting future work is needed to identify the best way to share electronic records with those who have chronic illnesses, given the benefits of EMR access for patients.

11.6.3 Research Framework

To finalize this chapter, it is important to suggest a research framework moving forward to improve best practice for mHealth design, development, validity, adoption, engagement, and motivational behavioral change. Presently, we have an existing knowledge base that informs clinical guidelines, clinicians, and best practice in self-management. This is used to also inform our development of current mHealth. As we move forward, it is crucial that we build a framework that allows for the collection and feedback of data both separately and together from patients, carers (informal and formal), and health professionals to better understand the dynamics of the self-management field and behaviors that exist within it. Empowering the patient and providing better guidance to health professionals is critical to improved self-management. Data from these stakeholders need to be fed into three key research fields collaboratively moving forward: (1) *health informatics*, (2) *behavioral change*, and (3) *self-management program development*. Using health informatics to better understand the massive amount of data collected in mHealth will be crucial to discovering BCTs, shaping self-management programs, and designing and developing mHealth to facilitate these programs. Mastering this area in real time will also allow us to truly discover how best to tailor new programs to various condition and refine which combination of condition specific BCT's will work to best support scalable adoption of mHealth.

11.7 Future Directions

One of the core focuses of mHealth in chronic illness is behavioral change and adoption of healthy lifestyle self-management practices for patients. A second focus is on developing a better understanding and relationship between all players in the primary care ecosystem (patients, carers, health professionals, policy makers, family) to facilitate and support behavioral changes in disease-specific groups. As both behavioral change and self-management are complex fields, research is needed to better understand the specific behavioral change and self-management techniques in relation to cultural and disease-specific contexts. Reaching this goal will require improved translation of information between all stakeholders in the primary care ecosystem, to understand and overcome not only the barriers to mHealth adoption but also how best to deliver psychoeducation learning programs to patients. Improved research in the fields of self-management, behavioral change, and health informatics will be as essential as the design and development of future mHealth technologies.

Redesign and development of the home-based primary care ecosystem is another core component to the future progression of self-management mHealth-supported behavioral change. Improving social engagement may occur in mHealth if we perceive mHealth platforms as personalized *companions* to our daily routine rather than as clinical devices. Incorporating personalized mHealth applications into existing consumer technology rather than developing increasingly bespoke and expensive new technologies may be key to improving societal adoption of mHealth. This has already begun, with accessories that measure blood pressure and heart rate already available for consumer systems like the iPad. As such, it is important that the software underlying these systems is medically sound and interoperable with modern health systems and electronic health records. We envisage that this trend toward the home-based ecosystem will continue and expand, with accessories moving beyond simply recording information but also offering medical feedback and advice to patients, as well as (with the patient's consent) communicating some of these data back to the clinician as and when necessary. Keeping patients motivated and engaged to use mHealth is critical, so it is important that the advice mHealth-based applications provide to patients is tailored and changes as the needs and wants of the patients change.

Central to the new tailored approach will be the merging of a plethora of design and development techniques such as UCD with BCTs. This is necessary so that mHealth applications and devices are designed not just to be usable by as wide a group as possible but that they also engage with the psychological needs of those users, so that behavioral change and, ultimately, improved health outcomes are possible.

Finally, it is important that new regulations that classify software as a stand-alone medical device do not stifle progress to adopting mHealth technologies into the home environment.

11.8 Conclusion

Using mHealth to facilitate behavioral change is in its infancy, with more large design and development research studies needed to discover the *best practice* structure and content to self-management/behavioral change programs in each disease-specific condition. Existing research however suggests that mHealth can play a potentially important role in improving self-management practices and improving health and well-being, in line with reducing the burden on overstretched health services. However, it must be realized that the way technologies and their users will behave is hard to predict and that effective design strategies must be employed to validate and produce effective mHealth uses in chronic illness self-management. Tackling the subjective, daily needs of individuals from various cultural and geographic contexts is a core design challenge to producing effective long-term models of behavioral change.

Building design support frameworks to motivate, sustain, and scale mHealth use in community-based environments such as the home as part of a new self-management paradigm will be a difficult task. While mHealth is now beginning to realize the importance of social, cognitive, and behavioral change theory, few studies actually incorporate this holistically into the design and development process. Realization of the need for addressing issues such as the *iceberg effect* and *translation dilemma* is important to progress forward. Health service redesign to adopt new mHealth technologies is an important step to facilitating societal adoption of health-care technologies (particularly mobile), of which the patient health record may be a critical cornerstone. Development in the research fields of *health informatics* and *behavioral change/self-management* may see positive disruption of the current health-care ecosystem, as we better understand mHealth users' needs in relation to their condition. This may ultimately inform new and improved clinical best practice guidelines for future mHealth introduction and use.

Questions

1. What is behavioral change with reference to chronic illness?

2. Why has it been difficult to define one behavioral approach to the design and development of mHealth technologies?

3. Name three key BCTs that can be used as part of the mHealth design and development process. Give an example of how they can be incorporated into a mobile application to address a chronic health condition.

4. Outline some of the techniques designers can use when designing for users with disabilities or impairments. Discuss techniques both for the design of the product itself and for gathering requirements (e.g., focus group design). Give some real-world examples of both.

5. What are some of the difficulties that may arise with combining patient and clinician data?

References

Abraham C. and Michie S. (2008). A taxonomy of behavior change techniques used in interventions. *Health Psychology* 27: 379–387.

Abras C., Maloney-krichmar D., and Preece, J. (2004). User-centered design. *Encyclopaedia of Human-Computer Interaction*. Thousand Oaks, CA: Sage Publications.

Adheretech ed. (2013). Adheretech. Available at http://www.adheretech.com (Accessed September 4, 2013).

Ajzen I. (1991). The theory of planned behavior. *Organizational Behavior and Human Decision Processes* 50: 179–211.

Ajzen I. and Fishbein M. (1980). *Understanding Attitudes and Predicting Social Behavior*. Englewood Cliffs, NJ: Prentice Hall.

Al-Mallah M.H., Tleyjeh I.M., Abdel-Latif A.A., and Weaver W.D. (2006). Angiotensin converting enzyme inhibitors in coronary artery disease and preserved left ventricular systolic function: A systematic review and meta-analysis of randomized controlled trials. *Journal of the American College of Cardiology* 47: 1576–1583.

Barrett J. and Kirk S. (2000). Running focus groups with elderly and disabled elderly participants. *Applied Ergonomics* 31(6): 621–629.

Bayliss E., Bosworth H.B., Noel P.H., Wolff J.L., Damush T.M., and Mciver L. (2007). Supporting self-management for patients with complex medical needs: Recommendations of a working group. *Chronic Illness* 3: 167–175.

Cattan M., White M., Bond J., and Learmouth A. (2005). Preventing social isolation and loneliness among older people: A systematic review of health promotion interventions. *Ageing and Society* 25(1): 41–67.

De Vries H., Blackbier E., Kok G., and Dijkstra M. (1995). The impact of social influences in the context of attitude, self-efficacy, intention and previous behavior as predictors of smoking onset. *Journal of Applied Social Psychology* 25: 237–257.

De Vries H., Dijkstra M., and Kuhlman P. (1988). Self-efficacy: The third factor besides attitude and subjective norm as a predictor of behavioral intentions. *Health Education Research* 3: 273–282.

Dickinson A., Eisma R., Gregor P., Syme A., and Milne S. (2005). Strategies for teaching older people to use the world wide web. *Universal Access in the Information Society* 4: 3–15.

Dinsmore J. (in press). Avoiding the "Iceberg Effect"–The need to incorporate a behavioral change approach to technology design in chronic illness. In Eds. D. Prendergast and C. Garattini. *Ageing and the Digital Lifecourse*. Berghahn Books.

Dix A., Finlay J., Abowd G., and Beale R. (2004). *Human-Computer Interaction*. Harlow, U.K.: Pearson.

Donner J. (2008). Research approaches to mobile use in the developing world: A review of the literature. *The Information Society* 24: 140–159.

Doyle J., Skrba Z., McDonnell R., and Arent B. (2010). Designing a touch screen communication device to support social interaction amongst older adults. *HCI 2010, the 24th BCS Conference on Human Computer Interaction*, ACM, Dundee, Scotland.

Effing T., Monninkhof E., van der Valk P., Zielhuis G., Walters H., van der Palen J., van Herwaarden C.L., and Partidge M.R. (2009). Self-management education for patients with chronic obstructive pulmonary disease. *Cochrane Database of Systematic Reviews* 17(4): CD002990.

Findlay R. (2003). Interventions to reduce social isolation amongst older people: Where is the evidence? *Ageing and Society* 23(05): 647–658.

Fishbein M. and Ajzen J. (1975). *Belief, Attitude, Intention and Behavior: An Introduction to Theory and Research*. Reading, MA: Addison-Wesley.

Ford Motors. (2011). Third age suit. Available at http://corporate.ford.com/newscenter/press-releases-detail/pr-ford26rsquos-26lsquothird-age-34447 (Accessed July 3, 2014).

Free C., Knight R., Robertson S., Whittaker R., Edwards P. et al. (2011) Smoking cessation support delivered via mobile phone text messaging (txt2stop): A single-blind, randomised trial. *Lancet* 378: 49–55.

Free C., Phillips G., Galli L., Watson L., Felix L., Edward P., Patel V., and Haines A. (2013). The effectiveness of mobile-health technology-based health behavior change or disease management interventions for health care consumers: A systematic review. *PLoS Medicine* 10(1): e1001362. doi:10.1371/journal.pmed.1001362.

Free C., Whittaker R., Knight R., Abramsky T., Rodgers A., and Roberts I.G. (2009). Txt2stop: A pilot randomised controlled trial of mobile phone-based smoking cessation support. *Tobacco Control* 18: 88–91.

Goodman-Deane J., Langdon P.M., Clarkson P.J., Caldwell N.H.M., and Sarhan A.M. (2007). Equipping designers by simulating the effects of visual and hearing impairments. *Proceedings of the Ninth International ACM SIGACCESS Conference on Computers and Accessibility*. ACM, Tempe, AZ.

Gregor P. and Newell A.F. (2001). Designing for dynamic diversity: Making accessible interfaces for older people. *Proceedings of the 2001 EC/NSF Workshop on Universal Accessibility of Ubiquitous Computing: Providing for the Elderly*, pp. 90–92. Alcr do Sal, Portugal.

Hamman R.F., Wing R.R., Edelstein S.L., Lachin J.M., Bray G.A. et al. (2006). Effect of weight loss with lifestyle intervention on risk of diabetes. *Diabetes Care* 29: 2102–2107.

Hitchcock D. and Taylor A. (2003). Simulation for inclusion—True user-centred design. *Include 2003 Conference Proceedings*. London, U.K.

International Data Corporation (IDC). (2012). Smartphone market hits all-time quarterly high due to seasonal strength and wider variety of offerings, according to IDC. IDC website. Available at http://www.idc.com/getdoc.jsp?containerId=prUS23299912 (Accessed November 16, 2012).

Kasikci M.K. (2011). Using self-efficacy theory to educate a patient with chronic obstructive pulmonary disease: A case study of 1 year follow up. *International Journal of Nursing Practice* 17: 1–8.

Keates S. and Clarkson P.J. (2003). Countering design exclusion: Bridging the gap between usability and accessibility. *Universal Access in the Information Society* 2: 215–225. Available at http://dx.doi.org/10.1007/s10209-003-0059-5.

Kensing F. and Blomberg J. (1998). Participatory design: Issues and concerns. *CSCW Conference* 7: 167–185. Available at http://dx.doi.org/10.1023/A:1008689307411.

Mair F.S., Goldstein P., Shiels C., Roberts C., Angus R., O'Connor J., Haycox A., and Capewell S. (2006). Recruitment difficulties in a home telecare trial. *Journal of Telemedicine and Telecare* 12(Suppl 1): S26–S28.

Mao J., Vredenburg K., Smith P.W., and Carey T. (2005). The state of user-centered design practice. *Communications of the ACM* 48(3): 105–109.

National Center for Chronic Disease Prevention and Promotion, Centre for Disease Control. (2011). Healthy aging. http://www.cdc.gov/. Available at http://www.cdc.gov/chronicdisease/resources/publications/aag/pdf/2011/Healthy_Aging_AAG_508.pdf (Accessed September 2, 2013).

National Institute for the Health and Clinical Excellence NICE. (2003). *Guidance on the Use of Patient-Education Models for Diabetes*. London, U.K.: National Institute for Clinical Excellence.

Newell A., Arnott J., Carmichael A., and Morgan M. (2007). Methodologies for involving older adults in the design process. Eds. Newell A., Arnott J., Carmichael A., and Moran M. *Universal Access in Human Computer Interaction. Coping with Diversity*. Berlin, Germany: Springer. pp. 982–989.

Newell A.F. and Gregor P. (2002). Design for older and disabled people—Where do we go from here? *Universal Access in the Information Society* 2: 3–7. Available at http://dx.doi.org/10.1007/s10209-002-0031-9.

Ofcom. (2009). The consumer experience: Telecoms, internet and digital broadcasting 2009. Evaluation Report.

Prochaska J. and DiClemente C. (1984). *The Transtheoretical Approach: Crossing Traditional Boundaries for Therapy*. Homewood, IL: Dow-Jones-Irwin.

Prochaska J.O., Reeding C.A., and Evers K.E. (2002). The transtheoretical model and stages of change. In Eds. K. Glanz and B.K. Rimer, *Health Behavior and Health Education: Theory, Research and Practice*, 3rd edn., pp. 99–120. San Francisco, CA: Jossey-Bass.

Ralston J., Hirsch B., Hoath J., Mullen M., Cheadle A., and Goldberg H. (February 2009). Web-based collaborative care for type 2 diabetes. *Diabetes Care* 32(2): 234–239.

Rogers E.M. (2003). *Diffusion of Innovations*, 5th edn. New York: The Free Press.

Rose Biomedical. (2004). Chronic obstructive pulmonary disease; NIH funded study identifies new tool to help individuals with COPD. *Obesity, Fitness and Wellness Week*. Atlanta, GA: NewsRX.

Rosenstock I.M. (1966). Why people use health services. *Milbank Memorial Fund Quarterly* 83(4): 1–32.

Sanders E.B. (2002). From user-centered to participatory design approaches. *Design and the Social Sciences: Making Connections*, pp. 1–8. London, U.K.: Taylor & Francis.

Sayago S. and Blat J. 2010. Telling the story of older people emailing: An ethnographical study. *International Journal of Human-Computer Studies* 68: 105–120.

Sharp H., Rogers, Y., and Preece, J., 2007. *Interaction Design; Beyond Human Computer Interaction*. New York: Wiley.

Shaw K., Gennat H., O'Rourke P., and Del Mar C. (2006). Exercise for overweight or obesity. *Cochrane Database of Systematic Reviews* 18(4): CD003817.

Shneiderman, B. (2000). Universal usability. *Communications of the ACM* 43: 84–91.

Tenforde M., Jain A., and Hickner J. (2011). The value of personal health records for chronic disease management: What do we know? *Family Medicine-Kansas City* 43(5): 351

United Nations Department of Economic and Social Affairs. (2007). World population ageing 2007. Available at http://www.un.org/en/development/desa/population/publications/pdf/ageing/WorldPopulationAgeingReport2007.pdf.

Weinschenk S.M. (2011). *100 Things Every Designer Needs to Know About People*. Berkeley, CA: New Riders. Pearson Education.

Wempe J.B. and Wijkstra P.J. (2004). The influence of rehabilitation on behavior modification in COPD. *Patient Education and Counselling* 52: 237–241.

Wherton J. and Prendergast D. (2009). The building bridges project: Involving older adults in the design of a communication technology to support peer-to-peer social engagement. Eds. Wherton J., and Prendergast D. *HCI and Usability for E-Inclusion*, pp. 111–134. Heidelberg, Germany: Springer.

Worth H. and Dhein Y. (2004). Does patient education modify behavior in the management of COPD? *Patient Education and Counselling* 52: 267–270.

12

Neuropsychiatric mHealth:
Design Strategies from Emotion Research

Thomas D. Hull

CONTENTS

12.1 Introduction

Neuropsychiatric disorders or psychopathology (these terms will be used interchangeably) offers a significant and largely untapped opportunity for mHealth. Unlike other domains of medicine, the treatment for neuropsychiatric disability can be delivered directly through technology itself through utilizing carefully constructed instructional media designed to formulate a life perspective, build skills, and ultimately lead to altered neurodynamics and structural changes in the brain (Karlsson 2011; Linden 2006; Stahl 2012). Rather than simply being a way to track adherence, supplement visits to health-care facilities, or measure certain biomarkers in assisting clinicians, mHealth can directly deliver the psychoeducational material, social support, and therapeutic activities that clinical science suggests are ameliorative for psychiatric and psychological disorders.

On the other hand, neuropsychiatric disorders are like other domains of medicine in that they carry a high level of societal burden (Hyman 2008; Jason and Ferrari 2010; Kessler et al. 2009); have therapies that are evidence based through both effectiveness and efficacy research (Dickerson and Lehman 2006; Seligman 1995); are as or more efficacious than certain well-established medical practices in cardiology, geriatrics, vaccination procedures, and ophthalmological surgeries (Minami et al. 2008; Wampold 2007); and have a disproportionate amount of patient need relative to qualified practitioners in rural areas (Hinrichsen 2010). Technology offers a significant advantage in resolving cost and logistical problems compared to merely training more professionals. It was noted in a national study commissioned by President Kennedy in 1963 that mental health professionals were sadly lacking and may never be able to meet the demand of society-wide neuropsychiatric needs (Kennedy 1963), and in spite of a nearly 50-year effort to increase trained personnel per capita, this has continued to be the case (Hoge et al. 2007).

Nevertheless, it is not the purpose of this chapter to argue for the many economic benefits and advantages of neuropsychiatric mHealth as ample evidence has been presented elsewhere (Kazdin and Blase 2011). It is also not the intent to provide an overview of quantified self, tailored health messaging, personality trait identification, or behavioral reinforcement applications as this research has existed in mHealth for some time (e.g., Bauer et al. 2003; Boschen 2009; Klasnja and Pratt 2012; Morris et al. 2010). Therefore, in addition to avoiding redundancy, overlooking these types of applications results from the belief that mHealth can go beyond the already encouraging effects these applications offer for those struggling with neuropsychiatric disorders. To wit, the evidence derived from research on emotion and self-regulation presented in this chapter provides several compelling reasons to think that mHealth can be designed to treat neuropsychiatric disturbances directly, as opposed to merely measuring symptoms or providing information to oneself or one's health-care provider.

By addressing what is unique about mHealth in this domain, namely, that the technology itself can function as a direct intervention, we hope to break ground and provide a framework for continued research on the effectiveness and treatment outcomes of neuropsychiatric mHealth. This focus and approach inevitably avoids the important but ancillary activities of planning the delivery of intervention, appropriateness of this treatment for particular populations, the specific technology medium (i.e., handheld vs. personal computer, technology-assisted therapy streaming vs. self-paced automated programs), or validated techniques of measurement and assessment. Instead, we seek to draw attention to avenues for the more effective design of the mHealth intervention. In other words, it is the *medicine* itself and not the packaging or delivery of it that is important to us here.

The design characteristics of psychotherapeutic treatment are largely overlooked in the extant literature. A review of the literature on technology-based neuropsychiatric interventions reveals that each study chooses one particular orientation to therapy and tests its effectiveness (i.e., through a naturalistic design) or efficacy (i.e., through a randomized, controlled design). This leads to the reporting of results for a small handful of therapies (usually cognitive behavioral therapy [CBT] and acceptance and commitment therapies). There are three problems with this. First, it ignores the common mechanisms of change that underlie all therapies and the finding that all forms of validated therapies have a similar magnitude of effect (Lambert and Ogles 2004; Leibert 2011). Secondly, it fails to consider the advantages of drawing on multiple forms of therapy in conjunction with each other that utilizing only one style of therapy alone cannot match. Third, it obstructs the progress of translational research. Comparative studies of efficacy are important, but randomization can often get in the way of the central pursuit in mHealth, which is designing robust and effective interventions for natural populations of consumers. Obtaining a broadly applicable set of design principles for neuropsychiatric disorders that derive from recent findings in psychopathology and affective science may provide more practical utility.

One final thought by way of introduction. *Therapeutic design* is not a commonly employed concept among clinicians. The reason for this is that clinicians are typically exposed to a few popular systems of therapy, each of which assumes one view of human nature to which you either subscribe or you do not. Whichever view of human nature feels intuitively right to the individual clinician (and that also has some empirical backing) generally becomes that clinician's orientation, and it is assumed that they endeavor to master it in order to derive the full benefits of the manualized form that was vetted through research. In other words, if one intends to apply a standardized and validated form of treatment, there is no room for a design space. Nevertheless, at one point, nearly two-thirds of therapists have thought of themselves as having an *eclectic* orientation drawing on several models (Slife and Reber 2001), and attempts at integrating models of therapy are currently being made (Castonguay 2011). For some, this works against efficacy studies of therapy or at least complicates matters (Garland et al. 2010).

However, since neuropsychiatric mHealth will be designing for a technology medium that has no personal preferences and that does not need to enhance its skill through practice, it is possible to more effectively standardize the treatment (Carroll et al. 2008) while simultaneously drawing on the strengths of multiple forms of therapy. The advantages of this approach will become clearer in the following.

12.2 Overview

The structure of this chapter is as follows. An overview and evidence for the common factors approach to psychotherapy is provided. Common factors is the term for research that seeks to understand why different therapies get the same results. The goal in this section is to provide a rationale for diffusing the *horse racing* that commonly happens in psychotherapy research and to resist the trend to engage in it yet again as neuropsychiatric treatment moves on to technology-based interventions and mHealth. This is followed by a look at the transdiagnostic approach to psychopathology. *Transdiagnosticism* is motivated by the observation that nearly all forms of neuropsychiatric disturbance involve criteria specifying emotional problems and dysregulation. The transdiagnostic model enables the identification of a common mechanism of change, namely, emotion regulation, which highlights the utility of emotion research. Lastly, a design framework is presented that draws on the burgeoning findings of affective science for enhancing neuropsychiatric mHealth.

12.3 Common Factors in Psychotherapy

Psychotherapy works (Lambert and Ogles 2004; Seligman 1995) and various approaches to therapy work more or less equally well (see Leibert 2011 for a review). Psychotherapy is as, and often more (Shedler 2010a), effective than drug treatments for emotional problems and is more enduring in its effects while simultaneously avoiding the potentiality of drug resistance (Imel et al. 2008). This understanding comes after decades of research and leaves off with a variety of questions. What is it about psychotherapy that makes it so effective as a healing agent? And why, in spite of the theoretical disagreements about the fundamental nature of what heals people, does each model return roughly the same magnitude of effects?

The popularly named *common factors* approach to psychotherapy is an attempt at an answer. It is an empirical critique of certain assumptions that remain in psychotherapy following its historical development out of the medical and neurological disciplines. The central assumption it challenges is that a particular therapeutic model or intervention should have its own unique *ingredients* that assert a causal role in recovery. It is beyond the scope of this writing to provide an overview of the historical genesis of these conditions (but see Wampold 2010 for a review). Nevertheless, as a result of this critique, a set of models have been proposed that seek to unify the effects of therapy under a set of processes and factors that are shared among the various theoretical orientations (e.g., Frank and Frank 1991; Garfield 1995; Marmor 1962). When the similarities of these models are viewed together, they center on the idea that the specific technique or set of techniques employed by a given therapeutic school are much less important than the relationship developed between the therapist and client, their agreement about the change goals to be pursued, and addressing the client's individual characteristics. Thus, emphasis is placed on the social and relational structure of the healing setting common to all therapeutic treatments rather than on the details of the treatment itself, and this relationship between the therapist and his or her client is referred to as the therapeutic alliance (Wampold 2010).

When the therapeutic alliance is strong, the outcomes of therapy are strong, regardless of the therapy approach. If the alliance is weak, the outcomes are weak (Horvath et al. 2011). We conclude from this that psychotherapies work to employ relationships along with instruction to ameliorate neuropsychiatric disturbance. These relationships also appear to operate from a distance through media in which the therapist is absent (Norcross et al. 2003). While this is an insight into the design process, it fails to explain how a social relationship and authoritative recommendations from the therapist reduce neuropsychiatric symptoms. To explore potential mechanisms of change, the transdiagnostic approach to psychopathology is presented.

12.4 Transdiagnostic Approaches to Neuropsychiatric Disorder and Psychopathology

The transdiagnostic view of psychopathology derives from the ubiquity of emotion dysregulation seen across disorders. A cursory glance at the *Diagnostic and Statistical Manual of Mental Disorders*, fourth edition, text revision (or DSM-IV-TR; American Psychiatric Association 2000; Kring 2008), and at the recently released DSM-V (American Psychiatric Association 2013), is sufficient to identify the widespread prevalence of emotion dysregulation diagnostic criteria. Historically, disorders have been viewed as discrete categories, but recent reviews of the cognitive and behavioral processes of neuropsychiatric disturbances provide evidence that these categories may have far more overlap than previously thought (Harvey et al. 2004; see also Barlow et al. 2004; Borsboom and Cramer 2013; Patrick and Bernat 2006; Thayer and Brosschot 2005).

First, the rates of comorbidity (i.e., meeting the diagnostic criteria for more than one disorder) are exceptionally high. In the National Comorbidity Survey, Kessler and colleagues (1994) identified comorbidity rates at nearly 80% when measuring across the life span.

Additionally, even when looking just at the past 12 months of an individual's life, 45% of people who met criteria for one disorder also met criteria for another (Kessler et al. 2005). It has also been found that treatments and techniques that work for one disorder also have effective applicability in a range of other disorders (Barlow et al. 2004; Hayes et al. 1999; Persons et al. 2003; Tsao et al. 2002). From the transdiagnostic perspective, neuropsychiatric disorder just is a difficulty in regulating one's emotions. To see why this is not just a superficial concern but is closer to the heart of psychopathology, it is necessary to discuss what emotions are.

Emotions are not just vestigial reflexes from our evolutionary past; they are functional and adaptive for survival (Frijda 1986; Keltner and Kring 1998; Lang et al. 1990; Levenson 1994; Tooby and Cosmides 2008). They coordinate our physiology (i.e., heart rate, metabolism, blood pressure, gustatory activity) and our subjective states (i.e., the love, anger, and fear that we experience), direct our attention to important aspects of our environment, and motivate our behavior, including not just individual approach–avoid responses (Davidson 2004) but also social cues and expressions through the face and gestures (Ekman 1992; Matsumoto et al. 2008). All of this activity is designed to help us act efficaciously in a given situation; therefore, a dissociation between the various components or a general breakdown in the emotional process can be very incapacitating.

In order for the functionality of emotions to be adaptive, they must be triggered by and employed within the appropriate context to the right amount and in the right way. Ongoing and recent research is indicating that it is a breakdown or mismatch between emotional response and context that leads to the burden of neuropsychiatric disorder. Given the global nature of the emotional response, with all of the physiological, subjective, and behavioral activity that it entails, it is easy to see how it is that psychopathology can be so disruptive of one's life and of the lives of loved ones.

12.5 Context (In)Sensitivity and the Contextual Model of Emotion

The contextual view of emotion is derived from evidence of the dynamic interaction between emotion regulation and a range of contextual factors. In a recent review, Aldao (2013) identified some of the factors that determine the efficacy of an emotional response such as the aspects of the person (i.e., personality, demography), features of emotionally evocative stimuli, how the individual selects among regulation strategies, and how the outcome of the emotional response is measured. Sensitivity to context is made more complex by the additional factors of one's ability to monitor goals (Carver and Sheier 1982), one's current mood and affect (Russell and Barret 1999), motivation styles (Ryan and Deci 2000), and the social milieu (Taylor et al. 1996). Maladaptive emotional responses occur when aspects of this rich contextual tapestry are misperceived, thus either failing to trigger an emotion or triggering an inappropriate emotion. If this insensitivity to context is persistent, then neuropsychiatric disorder can result.

An example may be of use here. In one study, Rottenberg et al. (2002) found that individuals struggling with depression felt just as sad watching a film with strong loss themes as when watching a happy or comedic film. What is especially interesting is that the sadness of the depressed individuals was no greater than those not struggling with depression that watched the same sad film. In other words, depression is not characterized by feeling *more* sadness than other people, but rather by feeling sadness in contexts in which sadness is not typically called for.

Fortunately, this is an alterable condition. Rottenberg and colleagues (2005) report that previously depressed individuals who were no longer experiencing depressive symptoms also no longer exhibited context insensitivity. In addition, Coifman and Bonanno (2010) used a prospective design to show that recently bereaved individuals who were context sensitive had marked reductions in depressive symptoms 18 months later in comparison with those who were context insensitive. From these studies and others (e.g., Gehricke and Shapiro 2000; Larson et al. 2007; Rottenberg and Gotlib 2004), it is apparent that a mismatch between context and emotional responses contributes to the development and maintenance of neuropsychiatric disorder. We have also seen that recovery from these disorders is a corollary with a return to context sensitivity (Rottenberg et al. 2005).

These effects highlight an interesting obstacle into recovery from neuropsychiatric disturbance. Successful emotion regulation requires that one learn (or relearn) how to create a match between one's emotional response and the appropriate context. The challenge is that there are so many contexts that could possibly be out of sync that it is unlikely for there to be a silver bullet or single technique that will succeed in returning sufferers to health. This view should disabuse us of a notion that is easy to take away from the section on emotion dysregulation as a transdiagnostic factor for neuropsychiatric disorder. Namely, that if pathology in general has emotion dysregulation as a common feature, then perhaps there is a single, optimal way to learn to regulate emotion that will apply across all disorders. This is often assumed in the methods and results of efficacy research on therapeutic models. As we have already seen, the common factors model and the similar effect sizes among psychotherapy orientations already argue against this. Finer-grained research within the affective sciences provides additional evidence against the search for a single and uniformly effective emotion regulation strategy as well.

12.6 Regulatory Flexibility

Although a tradition of theorizing about coping and emotion regulation that stressed a dynamic and contextual perspective had influence (e.g., Folkman and Lazarus 1985), there has been a more recent trend of models that assume that certain regulation strategies are universally more effective and adaptive than others. Baker and Barenbaum (2007) propose that problem-focused strategies are more adaptive than emotion-focused strategies. Similarly, John and Gross (2004) have sought to identify unhealthy (i.e., expressive suppression) as seen against healthy (i.e., reappraisal) strategies. Both models have been troubled by a series of findings. The relation between specific strategies and overall mental health is highly variable (Folkman and Moskowitz 2004), the within-person consistency of strategy use is far from uniform (Cheng 2001), and adaptive consequences for the assumedly unhealthy strategies have been found (Aldao 2013; Austenfield and Stanton 2004; Carver and Connor-Smith 2010; Webb et al. 2012). This has led researchers to investigate the alternative construct of regulatory flexibility (e.g., Gruber et al. 2011; Sheppes et al. 2012; Westphal et al. 2010).

Regulatory flexibility is a measure of the difference between one's ability to deploy a variety of strategies for increasing (enhancing) and decreasing (suppressing) emotional responsiveness. An individual who expresses a greater range of positive and negative emotions is said to be more flexible than someone who can only suppress negative emotion or only enhance positive emotion. While potentially counterintuitive at first, the logic behind this finding is fairly straightforward. If, as discussed, emotions are coordinated and adaptive responses to the environment, then consistently regulating them in one way should tend toward decreased well-being. On the other hand, gaining and employing the ability to express the right emotion at the right time or to regulate away the wrong emotion in the right way at the right time will tend toward better overall adjustment.

Unfortunately, many individual psychotherapeutic orientations assume the uniform efficacy of their techniques given their particular theories of change (e.g., reappraisal in CBT, reinforcement schedules and tiny habits in behaviorism, relationships in interpersonal therapy). Given its dominance in psychotherapy research and mHealth applications, the CBT model will be further explored.

Reappraisal in CBT is a technique that seeks to shift the way one thinks about a certain stimulus and thus how one will respond to it (Beck 1967). For example, when shown an image of an angry dog, one could reappraise the scene as a dog who is smiling or that is trying to show you how clean its teeth are. Gross and Thompson (2007) have amassed a significant literature demonstrating the utility of this strategy. But in spite of its beneficial outcomes, people are less likely to employ it in high-intensity contexts (Sheppes et al. 2012), it is only associated with improved well-being among individuals with high life stress (Troy et al. 2010), and it can even be maladaptive when deployed in situations involving controllable stress events (Troy et al. 2013). The latter finding is of particular note because it illustrates how, when in a malleable situation, it does not serve the individual well to regulate away the desire and ability to change it.

In fairness, the overly simple presentation of the preceding CBT is a caricature. CBT employs a variety of strategies, the complete elucidation of which goes beyond the scope of this chapter. Nevertheless, there is a tendency to deemphasize flexibility in the model, and each of the preferred CBT strategies share a close family resemblance (see Papa et al. 2012, for a more thorough discussion of CBT). An additional source of evidence for the inflexibility of relying solely on a CBT model is that in some cases its ameliorative effects on neuropsychiatric disorder begin to wane after about 6 months (Shedler 2010b). The flexibility model predicts that by buttressing one therapeutic model with another or several others, patients will gain a broader range of skills, techniques, and regulation strategies for more flexible and longer-lasting effects.

12.7 General Design Principles

It has been argued earlier that the greatest opportunity for neuropsychiatric mHealth is in the theoretically rich design of the intervention itself. By failing to pay close attention to the programming and the selection of therapeutic models, neuropsychiatric mHealth could find itself relegated to the curiosities of the past as the ameliorative effects of the applications dissipate with time or are unable to address the wide variety of recovery goals individuals set for themselves. Affective science offers insights for why designers must incorporate multiple models into their applications and even provides ideas for how to do so effectively. Before turning to case studies of neuropsychiatric mHealth applications, it is important to mention a few remaining design principles.

First, while it might seem to follow from the material presented, it would be unwise to present the variety of regulatory techniques outside of the respective theoretical and rhetorical contexts of each therapy (Frank and Frank 1991). As Frank and Frank note, each therapeutic system has its own internal logic, worldview, and process. What might seem like superficial accouterments can in actuality be part of the glue that holds the regulation techniques together within their appropriate frames of reference. In addressing context insensitivity, it would be a mistake to decontextualize the individual therapeutic systems by stripping out only the techniques themselves. Instead, each system should be allowed to stand in an interdependent relation to the others selected and designed for without a destruction of their histories, identities, and language.

In traditional brick-and-mortar approaches to therapy, this would be difficult and costly to accomplish. But mHealth provides the opportunity to expose patients to a wider range of therapeutic systems affordably, with greater standardization, and in geographic areas where finding a single practitioner can be a challenge, let alone a cadre of specialists who could offer the comprehensiveness required for successfully addressing neuropsychiatric disturbance through increasing flexibility. The result is a tighter, more consistent, and broad-based intervention for underserved populations and for the market in general.

Second, in addition to designing for multiple approaches to therapy, one should also give thought to the sequencing of the therapeutic approaches based on the strengths and weaknesses of each so that they compliment and build on each other. CBT is relatively easy to learn and apply, and it can contribute to rapid symptom reduction (Leibert 2011). Given these characteristics, it could be an excellent system for initiating a neuropsychiatric intervention. However, it is not very exploratory and may leave areas for potential treatment unaddressed and thus longer-lasting health benefits untouched. In this case, a system designed for flexibility would begin to introduce individuals to treatment modalities based on other models, such as those that are psychodynamic, spiritual, or more relational in nature.

Third, although neuropsychiatric disorders have been stressed here, the aforementioned design model also applies to health problems in which psychological technique is supportive of other goals, such as weight loss (see Kessler 2009 for an example). In the few interventions that explicitly and deliberately incorporate psychological change principles, CBT is assumed to be the model of choice. This is a disservice to the patient because it transgresses evidence-based practice and fails to take into account the findings of affective science. While requiring a little more effort, the solution is to simply incorporate additional models. The vast majority of therapeutic models have several manuals and guidance materials that are just as readily available as CBT models and will serve to enhance outcomes for the end user.

Finally, it should be noted that it is not necessary for a person to be experiencing full-blown neuropsychiatric symptoms to generate benefit from this model. None of us perfectly engage with each context and we all have areas of our lives that we would like to improve, even if it has not progressed to the point of pathology. Working to increase the number of regulation strategies at our disposal and identifying new contexts to deploy them in can offer benefits to anyone.

12.8 Applications and Cases

Neuropsychiatric mHealth tends to fall into three broad classes of applications: the psychoeducational, the interpersonal, and the facilitative, or some combination. Each provide virtual environments in which flexible therapeutic design can be incorporated for improved outcomes. The nuances and practices will differ but the core rationale remains the same. The cases discussed will necessarily be brief as many are in the very early stages. The following information is intended more as avenues for exploration than examples of what *works* or has empirical evidence demonstrating its effectiveness.

12.8.1 Psychoeducational

Psychoeducation is the delivery of content designed for therapeutic purposes and behavior change. In substance, it is no different than the didactic features of face-to-face treatment in which skills and techniques are taught to the patient by the clinician. This content can be designed, complete with exercises and feedback, for delivery through technology.

Candeo (http://candeobehaviorchange.com/) is an example of a depth methodology. Depth methodologies focus on a single class of behaviors and address as many aspects of the behavior as possible. It operates in the problematic sexual behavior domain and collects and publishes outcomes data on its program (Hardy et al. 2010). It utilizes a CBT model in the initial phases of treatment that is focused on reappraising the triggering stimuli of addictive behavior. Users are moved through the CBT training modules complete with interactive exercises and the receipt of feedback. At the conclusion of this regimen, principles and practices derived from positive psychology, psychodynamic, and relational models of psychotherapy are presented with their respective exercises and feedback. The program is notable for employing live *coaches* who act as facilitators and sources of support in addition to the interactive forums and journaling tools of the application.

Candeo's depth methodology is apparent from its focus on a single behavior and its utilization of multiple therapeutic systems within that behavior. The outcomes reported so far are promising, though stricter experimental procedures are certainly needed in future studies to increase confidence in the effectiveness of this approach.

Health Media (http://www.wellnessandpreventioninc.com/) by contrast uses a breadth methodology. It is a large company that offers a wide range of psychoeducational courses for a great many disorders. It too uses a coaching model, but the coaching in this case is largely algorithmic and automated. As a proprietary and business-to-business-driven company, it is difficult to acquire much in the way of information about its methods. Nevertheless, it serves as an example of a different style of design in which a large number of *flatter* courses are offered to the market.

There are a small handful of psychoeducational programs available that fall somewhere between the depth and breadth methodologies in each behavior market. As research increases on these programs, a key question to ask is to what extent depth programs deliver improved outcomes on a wider variety of measures than breadth programs, if they do.

12.8.2 Interpersonal

Interpersonal applications vary significantly in their dynamics but share the same common core of connecting people with each other who have a similar background and goals for change. This category can also be divided into two subtypes, namely, communication based vs. project based.

Communication-based interpersonal applications have been around for some time now. These include virtual groups, chat rooms, and support e-mail lists. In The Rooms (http://intherooms.com/) is an example. Signing up is free and they offer forums, chatting services, directories of local 12-step group meetings, and video meetings related to a variety of addictions, especially substance use and abuse behaviors. Many people struggling with psychopathology or otherwise disturbed emotional functioning tend to feel very isolated from those around them. Since one of the purposes of emotion is to send and receive social signals, when this is dysregulated, it makes engaging in satisfying social interactions a challenge. Online groups provide a safe and less intense form of social interaction when organized topically and by a behavior common to each user. In this way, they provide a more understanding and sympathetic communicative environment, thus providing a kind of training area where it is possible to renew one's feel for affirming and rewarding social interaction.

Project-based interpersonal applications on the other hand are nearly nonexistent. Their model however is based on the alternative reality gaming paradigm as put forward by McGonigal (2011). Alternative reality gaming is a scenario in which people attempt to solve game quests or goals in the real world with each other. Mobile devices often act as a way for the real-world interaction to be organized and for players to find each other and to use check-ins or other methods of enhancing, tracking, and reporting gameplay actions.

While this project-based architecture can be used for *gamifying* the treatment of neuro-psychiatric disorders, one significant alteration is possible. Rather than completing tasks in an alternate fantasy or game world, project-based reality gaming could focus on motivating players to complete tasks related to the demands of their actual living. The goal is to enhance the reinforcement and motivation for completing self-development tasks or projects in the real world. Many therapies recognize the importance of repeating new skills and regulation strategies until individuals are capable of effectively applying them. Neuropsychiatric reality gaming is simply a way of reducing the habituation to practice, while enhancing learning through greater and more compelling rewards and reinforcement structures.

12.8.3 Facilitative

The final class of neuropsychiatric application is the facilitative. This medium is less automated than the previous two and relies on a synchronous connection between a clinician or other professional resource and the individual seeking help. Examples of this kind of application are Skype therapy, online coaching (often through instant messaging or a virtual bulletin board), and mobile-led groups.

One of the primary challenges is that the professional guidelines for many of these applications have not been firmly established yet. Whether a therapist can Skype with someone out of the state in which they are licensed or out of the country in which they reside is a complex issue because it is not immediately clear which agency should be the governing authority. On the other hand, there are a number of benefits. Many patients move away from a therapist that they like and continued interaction can be very desirable and beneficial. Skype (or any other kind of video conferencing software) also holds out the possibility of treating people in more remote, rural areas where qualified help is otherwise unavailable. As long as both participants are comfortable with the medium and can see each other on video, thereby enabling the clinician to gauge body language and other subtle cues, the potential for a good outcome should remain high given the literature on the therapeutic alliance discussed previously. One way to build flexibility into this paradigm is to create a virtual clinic that houses multiple therapists, each of whom are expert in particular therapeutic models. Another way is to supplement the video therapy with asynchronous online training that provides the theory and techniques from other approaches to therapy.

Mobile-led groups are another solution for achieving a one-to-many match between a qualified professional and a group of patients. Typically a remote area identifies a set of individuals who are in need of assistance or treatment. They can then select someone from among themselves or another person in the local area or village to facilitate the session by managing the mobile technology. Once together and ready to begin, a professional can call in and hold a group session. It is important for this method that the emphasis is on group cohesiveness given the distance away from the clinician. Facilitating a level of comfort between participants enables them to support each other in the clinician's absence.

There is not a lot of explicit research on mobile-led groups. Indeed, there may be none. This stems in part from the largely international utility of the paradigm, which is then compounded by ambiguities of professionalism surrounding the exporting of one's licensure. However, the Substance Abuse and Mental Health Services Administration (SAMHSA) is beginning to release guidelines regarding the provision of e-therapy (SAMHSA 2009). E-therapy by this definition is any kind of therapy that takes place in a live (i.e., synchronous) but virtual setting. As organizations continue to experiment with this modality and as guidelines continue to be released, a more mature and open industry should begin to take shape.

12.9 Conclusion

The neuropsychiatric domain will be a large growth area for mHealth. Anywhere from 15% to 45% of people worldwide are suffering from a diagnosable neuropsychiatric disorder (Kazdin and Blase 2011). The reach and maturity of both the development and research of neuropsychiatric mHealth applications is very young. This immaturity is reflected in the somewhat cursory and ill-defined set of principles contained in this chapter. As interest in this domain increases, so will our understanding and thereby our design strategies. Nevertheless, this chapter has endeavored to open up a new set of resources from the affective sciences to help bootstrap neuropsychiatric mHealth into a more mature perspective derived from the long history of psychotherapy and emotion research. This includes the following:

- Breaking down the barriers between treatments and disorders through the common factors and transdiagnostic approaches to psychotherapy.
- This allows for greater freedom and depth in the selection of treatments to employ within a given population of interest.
- The greater the variety in design and content, the more likely individuals are to benefit from the flexible coping strategies learned from each school of thought.
- This will enhance our ability to ask new questions of our research activities in the nascent medium of mHealth instead of rediscovering old answers from previous research on psychotherapy.
- Lastly, as more and more individuals begin to take advantage of the logistical, financial, and motivational benefits of neuropsychiatric mHealth, we can have more confidence that our interventions will have a positive and lasting impact on those we are striving to serve.

In addition to the many aforementioned possibilities, there is also good reason to think that the therapy and behavior change industries are approaching the maturity needed for business models that support large-scale applications (see Christensen et al. 2009 for discussion). Once health-care innovators catch this vision, there will be exponential growth in the market and in research on how to elaborate the present models for improved and more accessible care. In order for these promises to be realized, it is predicted that research will need to cease taking place in *behavior and disorder silos* and instead look to more general principles of emotion research for investigating the common underlying processes of change into and out of neuropsychiatric disturbance. These findings can then be tailored to specific contexts of application and individual circumstances.

12.10 Future Directions

Neuropsychiatric mHealth has tended to lag behind other domains, including its sister disciplines of behavior modification and the quantified self. The next 5–10 years will, however, see a significant amount of growth in research on stand-alone (i.e., without a clinic element) neuropsychiatric mHealth applications for a variety of behaviors. This growth will be prompted by technological advances, shifts in the culture of business innovation for health care, and the increasingly unmet need and resulting social burden of neuropsychiatric disorder. It will also grow in the wake of advances in psychotherapy research. The number of efficacious emotion regulation techniques and our understanding of the causes of emotion dysfunction will open the way for more standardized and targeted approaches.

Once a well-documented research base is established, the models will be elaborated through findings in clinical practice and emotion science research. These findings will tend toward the incorporation of new frameworks of psychopathology that address comorbidity, flexible emotion regulation, and multiple models of psychotherapy. Applications will become increasingly mobile, increasingly dynamic, and increasingly mainstream. The exemplary growth of popular interest in psychology and neuroscience over the last 20 years will continue and provide an enhanced focus on neuropsychiatric topics. The importance of this development cannot be overstated. Many people live with untreated disturbances because psychology and psychotherapy are simply unknown to the general public. But there are signs that this is changing and will continue to change in the years ahead.

Applications that fail to isolate significant aspects of their design for measurement will fall behind. The potential to use the fine-grained nature of the research in affective science and emotion regulation in predictive and iterative frameworks is promising. As applications on mobile devices increase in interactivity, real-time data can be generated that measure and track flexible regulation (or the lack thereof) and *training prompts* could be developed that encourage individuals to draw on sources of regulation that they do not commonly employ. It may even be possible to predict when an individual will struggle the most and proactively engage them in coping exercises or opportunities to plan ahead for improved regulation.

Another growth area will be the incorporation of *brain training* exercises targeted to known neurocircuits underlying pathology. There is some skepticism of this approach, but it will either be born out through more research or else researchers and designers will move past the reliance on cognitive fMRI tasks as training activities and instead develop more exercises similar to cognitive bias modification (e.g., Enock and Mcnally 2013) that tap broader and deeper cognitive mechanisms for change.

If we look at the 10-year mark and beyond, it may be possible to obtain adapters that connect mobile devices to mobile EEG equipment (e.g., http://emotiv.com/ or http://advancedbrainmonitoring.com/) for rough but usable analyses of neurodynamics. Once research has established correlations between the readings of consumer EEG or similar devices and well-being, it may be possible to do mental health *check-ins* as are done with electronic measurements of blood sugar for diabetics currently. The potential to incorporate various kinds of neurofeedback methods on the go will increase as well.

The translational science identified earlier will be a central project, but so will the basic science of emotion research. The growing emphasis on context in the study of emotion will identify sets of context features for which certain regulation strategies are best suited, as well as context features that predict poor utility of certain strategies. In addition, connections will likely be made between a variety of cognitive processes (e.g., attention, memory, visual scanning, social referencing) and emotion dysregulation. By understanding the dynamic feedback between disrupted processes and disrupted affect, it may be possible to interrupt the cycle and return the individual to healthy functioning.

Acknowledgments

The author would like to thank the helpful and insightful comments from two anonymous reviewers, as well as the oversight of the editors in the preparation of this chapter.

Correspondence concerning this chapter should be addressed to Thomas D. Hull (tdh2120@tc.columbia.edu).

Questions

1. What sets neuropsychiatric mHealth apart from other health domains?
2. What is the central insight from research on the transdiagnostic model of psychopathology?
3. How is current research characterizing emotion?
4. How does context insensitivity lead to psychopathology?
5. Describe regulatory flexibility. How is it related to well-being?

References

Aldao, A. The future of emotion regulation research: Capturing context. *Perspectives on Psychological Science*, 8(2) (2013): 155–172.

American Psychiatric Association. *Diagnostic and Statistical Manual of Mental Disorders*, 4th edn. *text rev.* (Washington, DC: American Psychiatric Association), 2000.

American Psychiatric Association. *Diagnostic and Statistical Manual of Mental Disorders*, 5th edn. (Washington, DC: American Psychiatric Association), 2013.

Austenfield, J.L. and A.L. Stanton. Coping through emotional approach: A new look at emotion, coping, and health-related outcomes. *Journal of Personality*, 72(6) (2004): 1335–1364.

Baker, J.P. and H. Berenbaum. Emotional approach and problem-focused coping: A comparison of potentially adaptive strategies. *Cognition & Emotion*, 21(1) (2007): 95–118.

Barlow, D., L.B. Allen, and M.L. Choate. Toward a unified treatment for emotional disorders. *Behavior Therapy*, 35 (2004): 205–230.

Bauer, A., R. Percevic, and E. Okon et al. Use of text messaging in the aftercare of patients with bulimia nervosa. *European Eating Disorders Review*, 11 (2003): 279–290.

Beck, A.T. *Depression: Clinical, Experimental, and Theoretical Aspects.* (New York: Harper & Row), 1967.

Borsboom, D. and A.O.J. Cramer. Network analysis: An integrative approach to the structure of psychopathology. *Annual Review of Clinical Psychology*, 9 (2013): 91–121.

Boschen, M.J. Mobile telephones and psychotherapy: II. A review of empirical research. *Behavior Therapist*, 32 (2009): 175–181.

Carroll, K., S. Ball, and S. Martino et al. Computer-assisted delivery of cognitive-behavioral therapy for addiction: A randomized trial of CBT4CBT. *The American Journal of Psychiatry*, 165 (2008): 881–888.

Carver, C.S. and J. Connor-Smith. Personality coping. *Annual Review of Psychology*, 61(1) (2010): 679–704.

Carver, C.S. and M.F. Scheier. Control theory: A useful conceptual framework for personality—Social, clinical and health psychology. *Psychological Bulletin*, 92(1) (1982): 111–135.

Castonguay, L.G. Psychotherapy, psychopathology, research and practice: Pathways of connections and integration. *Psychotherapy Research*, 21(2) (2011): 125–140.

Cheng, C. Assessing coping flexibility in real-life and laboratory settings: A multimethod approach. *Journal of Personality and Social Psychology*, 80(5) (2001): 814–833.

Christensen, C.M., J.H. Grossman, and J. Hwang. *The Innovator's Prescription: A Disruptive Solution for Health Care.* (New York: McGraw-Hill), 2009.

Coifman, K.G. and G. Bonanno. When distress does not become depression: Emotion context sensitivity and adjustment to bereavement. *Journal of Abnormal Psychology*, 119(3) (2010): 479–490.

Davidson, R. Well-being and affective style: Neural substrates and biobehavioral correlates. *Philosophical Transactions of the Royal Society of London Series B*, 359 (2004): 1395–1411.

Dickerson, F.B. and A.F. Lehman. Evidence-based psychotherapy for schizophrenia. *Journal of Nervous and Mental Disease*, 194 (2006): 3–9.

Ekman, P. Facial expression of emotion: New findings and questions. *Psychological Science*, 3(1) (1992): 34–38.

Enock, P.M. and R.J. Mcnally. How mobile apps and other web-based interventions can transform psychological treatment and the treatment development cycle. *The Behavior Therapist*, 36 (2013): 56, 58, 60, 62–66.

Folkman, S. and R.S. Lazarus. If it changes it must be a process: Study of emotion and coping during three stages of a college examination. *Journal of Personality and Social Psychology*, 48(1) (1985): 150–170.

Folkman, S. and J.T. Moskowitz. Positive affect and the other side of coping. *American Psychologist*, 55(6) (2004): 745–774.

Frank, J.D. and J.B. Frank. *Persuasion and Healing: A Comparative Study of Psychotherapy.* (Baltimore, MD: Johns Hopkins University Press), 1991.

Frijda, N. *The Emotions.* (Cambridge, U.K.: Cambridge University Press), 1986.

Garfield, S.L. *Psychotherapy: An Eclectic-Integrative Approach.* (New York: Wiley), 1995.

Garland, A.F., M.S. Hurlburt, L. Brookman-Frazee et al. Methodological challenges of characterizing usual care psychotherapeutic practice. *Administration and Policy in Mental Health*, 37 (2010): 208–220.

Gehricke, J.G. and D. Shapiro. Reduced facial expression and social context in major depression: Discrepancies between facial muscle activity and self-reported emotion. *Psychiatry Research*, 95(2) (2000): 157–167.

Gross, J.J. and R.A. Thompson. Emotion regulation: Conceptual foundations. In *Handbook of Emotion Regulation* (New York: Guilford Press), 2007, pp. 3–24.

Gruber, J., I. Mauss, and M. Tamir. A dark side of happiness? How, when, and why happiness is not always good. *Perspectives on Psychological Science*, 6(3) (2011): 222–233.

Hardy, S., J. Ruchty, T.D. Hull et al. A preliminary study of an online psychoeducational program for hypersexuality. *Sexual Addiction and Compulsivity*, 17 (2010): 247–269.

Harvey, A., E. Watkins, and W. Mansell et al. *Cognitive Behavioral Processes across Psychological Disorders: A Transdiagnostic Approach to Research and Treatment* (New York: Oxford University Press), 2004.

Hayes, S.C., K.D. Strosahl, and K.G. Wilson. *Acceptance and Commitment Therapy: An Experiential Approach to Behavior Change* (New York: Guilford Press), 1999.

Hinrichsen, G.A. Public policy and the provision of psychological services to older adults. *Professional Psychology: Research and Practice*, 41 (2010): 97–103.

Hoge, M.A., J.A. Morris, and A.S. Daniels et al. *An Action Plan for Behavioral Health Workforce Development*. (Washington, DC: Department of Health and Human Services), 2007.

Horvath, A.O., A.C. Del Re, C. Fluckiger, and D. Symonds. Alliance in individual psychotherapy. *Psychotherapy*, 48(1) (2011): 9–16.

Hyman, S.E. A glimmer of light for neuropsychiatric disorders. *Nature*, 455(16) (2008): 890–893.

Imel, Z.E., M.B. Malterer, and K.M. McKay et al. A meta-analysis of psychotherapy and medication in unipolar depression and dysthymia. *Journal of Affective Disorders*, 110 (2008): 197–206.

Jason, L.A. and J.R. Ferrari. Oxford house recovery homes: Characteristics and effectiveness. *Psychological Services*, 7 (2010): 92–102.

John, O.P. and J. Gross. Healthy and unhealthy emotion regulation: Personality processes, individual differences, and life span development. *Journal of Personality*, 72(6) (2004): 1301–1333.

Karlsson, H. How psychotherapy changes the brain. *Psychiatric Times*, 28(8) (2011): 1–5.

Kazdin, A. and S.L. Blase. Rebooting psychotherapy research and practice to reduce the burden of mental illness. *Perspective on Psychological Science*, 6(1) (2011): 21–37.

Keltner, D. and A. Kring. Emotion, social function and psychopathology. *Review of General Psychology*, 2 (1998): 320–342.

Kennedy, J.F. Mental illness and mental retardation. *American Psychologist*, 18(6) (1963): 280–289.

Kessler, D. *The End of Overeating*. (New York: Rodale), 2009.

Kessler, D., S. Aguilar-Gaxiola, and J. Alonso et al. The global burden of mental disorders: An update from the WHO World Mental Health surveys. *Epidemilogia e Psichiatria Sociale*, 18 (2009): 23–33.

Kessler, R.C., W.T. Chiu, O. Demler et al. Prevalence, severity, and comorbidity of 12 month DSM-IV disorders in the national comorbidity survey replication. *Archives of General Psychiatry*, 62 (2005): 617–627.

Kessler, R.C., K.A. McGonagle, S. Zhao et al. Lifetime and 12-month prevalence of *DSM-III- TR* psychiatric disorders in the United States: Results from the National Comorbidity survey. *Archives of General Psychiatry*, 51 (1994): 8–19.

Klasnja, P. and W. Pratt. Healthcare in the pocket: Mapping the space of mobile- phone health interventions. *Journal of Biomedical Informatics*, 45(1) (2012): 184–198.

Kring, A. Emotion disturbances as transdiagnostic processes. In *The Handbook of Emotions* (New York: Guilford Press, 2008), pp. 691–705.

Lambert, M.J. and B.M. Ogles. The efficacy and effectiveness of psychotherapy. In *Bergin and Garfield's Handbook of Psychotherapy and Behavior Change* (New York: Wiley), 2004, pp. 139–193.

Lang, P.J., M.M. Bradley, and B.N. Cuthbert. Emotion, attention, and the startle reflex. *Psychological Review*, 97 (1990): 377–395.

Larson, C.L., J.B. Nitschke, and R. Davidson. Common and distinct patterns of affective response in dimensions of anxiety and depression. *Emotion*, 7(1) (2007): 182–191.

Leibert, T.W. The dimensions of common factors in counseling. *International Journal for the Advancement of Counselling* 33 (2011): 127–138.

Levenson, R. Human emotion: A functional view. In *The Nature of Emotion* (New York: Oxford University Press), 1994, pp. 123–126.

Linden, D.E.J. How psychotherapy changes the brain—The contribution of functional neuroimaging. *Molecular Psychiatry*, 11 (2006): 528–538.

Marmor, J. Psychoanalytic therapy as an educational process. *Science and Psychoanalysis*, 5 (1962): 286–299.

Matsumoto, D., D. Keltner, S. Michelle et al. Facial expressions of emotion. In *The Handbook of Emotions* (New York: Guilford Press), 2008, pp. 211–234.

McGonigal, J. *Reality is Broken: Why Games Make Us Better and How They Can Change the World*. (New York: Penguin Books), 2011.

Minami, T., B. Wampold, and R.C. Serlin et al. Benchmarking the effectiveness of psychotherapy treatment for adult depression in a managed care environment: A preliminary study. *Journal of Consulting and Clinical Psychology*, 76 (2008): 116–124.

Morris, M.E., Q. Kathawala, and T.K. Leen et al. Mobile therapy: Case study evaluations of a cell phone application for emotional self-awareness. *Journal of Medical Internet Research*, 12 (2010): e10.

Norcross, J., J. Santrock, and L. Campbell et al. *Authoritative Guide to Self-Help Resources in Mental Health*. (New York: Guilford Press), 2003.

Papa, A., M. Boland, and M.T. Sewell. Emotion regulation and CBT. In *Cognitive Behavior Therapy: Core Principles for Practice* (New York: Wiley), 2012, pp. 273–323.

Patrick, C.J. and E.M. Bernat. The construct of emotion as bridge between personality and psychopathology. In *Personality and Psychopathology* (New York: Guilford Press), 2006, pp. 174–209.

Persons, J.B., N.A. Roberts, and C. Zalecki. Anxiety and depression change together during treatment. *Behavior Therapy*, 34 (2003): 149–163.

Rottenberg, J. and I.H. Gotlib. Socioemotional functioning in depression. *Mood Disorders*, (2004): 61–77.

Rottenberg, J., J. Gross, and I.H. Gotlib. Emotion context insensitivity in major depressive disorder. *Journal of Abnormal Psychology*, 114(4) (2005): 627–639.

Rottenberg, J., K.L. Kasch, J. Gross et al. Sadness and amusement reactivity differentially predict concurrent and prospective functioning in major depressive disorder. *Emotion*, 2(2) (2002): 135–146.

Russel, J.A. and L.F. Barrett. Core affect, prototypical emotional episodes, and other things called 'Emotion': Dissecting the elephant. *Journal of Personality and Social Psychology*, 76(5) (1999): 805–819.

Ryan, R. and E. Deci. Self-determination theory and facilitation of intrinsic motivation, social development, and well-being. *American Psychologist*, 55(1) (2000): 68–78.

SAMHSA—Substance Abuse and Mental Health Services Administration. *Considerations for the Provision of E-Therapy*. (Rockville, MD: Author), 2009.

Seligman, M.E.P. The effectiveness of psychotherapy: The consumer reports study. *American Psychologist*, 50(12) (1995): 965–974.

Shedler, J. Getting to know me. *Scientific American*, 21 (November/December 2010a): 53–57.

Shedler, J. The efficacy of psychodynamic psychotherapy. *American Psychologist*, 65(2) (2010b): 98–109.

Sheppes, G., S. Scheibe, G. Suri et al. Emotion regulation choice: A conceptual framework and supporting evidence. *Journal of Experimental Psychology: General*, 143(1), (2014): 163–181.

Slife, B.D. and J.S. Reber. Eclecticism in psychotherapy: Is it really the best substitute for traditional theories? In *Critical Issues in Psychotherapy: Translating New Ideas Into Practice* (New York: SAGE Publications, Inc.), 2001, pp. 213–234.

Stahl, S.M. Psychotherapy as an epigenetic 'Drug': Psychiatric therapeutics target symptoms linked to malfunctioning brain circuits with psychotherapy as well as with drugs. *Journal of Clinical Pharmacy and Therapeutics*, 37 (2012): 249–253.

Taylor, S.E., H.A. Wayment, and M. Carrillo. Social comparison, self-regulation and motivation. In *Handbook of Motivation and Cognition, Vol. 3: Interpersonal Context* (New York: Guilford Press), 1996, pp. 3–27.

Thayer, J. and J.F. Brosschot. Psychosomatics and psychopathology: Looking up and down from the brain. *Psychoneuroendocrinology*, 30 (2005): 1050–1058.

Tooby, J. and L. Cosmides. The evolutionary psychology of the emotions and their relationship to internal regulatory variables. In *The Handbook of Emotions* (New York: Guilford Press), 2008, pp. 114–137.

Troy, A.S., A.J. Shallcross, and I. Mauss. A person-by-situation approach to emotion regulation: Cognitive reappraisal can either help or hurt, depending on the context. *Psychological Science*, 24(12) (2013): 2505–2514.

Troy, A.S., F.H. Wilhelm, A.J. Shallcross et al. Seeing the silver lining: Cognitive reappraisal ability moderates the relationship between stress and depressive symptoms. *Emotion*, 10(6) (2010): 783–795.

Tsao, J.C.I., J.L. Mystkowski, B.G. Zucker et al. Effects of cognitive behavioral therapy for panic disorder on comorbid conditions: Replication and extension. *Behavior Therapy*, 33 (2002): 493–509.

Wampold, B. Psychotherapy: The humanistic (and effective) treatment. American Psychologist, 62, (2007): 857–873.

Wampold, B. Research evidence for the common factor models: A historically situated perspective. In *The Heart and Soul of Change*, 2nd edn. (Washington, DC: APA), 2010, pp. 49–81.

Webb, T.L., E. Miles, and P. Sheeran. Dealing with feeling: A meta-analysis of the effectiveness of strategies derived from the process model of emotion regulation. *Psychological Bulletin*, 138(4) (2012): 775–808.

Westphal, M., N.H. Seivert, and G. Bonanno. Expressive flexibility. *Emotion*, 10(1) (2010): 92–100.

Section IV

mHealth Social Perspectives

13

Social Perspectives on Health: mHealth Implications

Hossein Adibi

CONTENTS

13.1 Introduction

Although commercial interest is the main driver of technological inventions, unintended social outcomes may overtake the commercial interest in the long term. For example, telephone engineers explained that the invention of the fixed telephone in the late nineteenth century in the United States was made for the business world and not for social conversation (Flinchy, 1997). This has been paralleled in the early twenty-first century by the advent of the mobile phone. The mobile phone was originally created for adults for business use (Aoki and Downes, 2003). However, today, the mobile phone has become the most popular form of electronic communication globally. This is not surprising due to the fact that human beings as *social beings* constantly explore, intentionally or unintentionally, social dimensions of new inventions and use them to enhance their social interactions. Today, adaption of the mobile phone has become an integral part of billions of peoples' daily lives, amazingly enhancing their communication on global context. In fact, the mobile phone has turned from a technological tool to a social tool. It has been used in many contexts and areas including health. It is one of the most growing trends changing the dissemination of information on health and well-being as well as delivery of health-care services. Some aspects of health-care delivery have been discussed in Chapter 1, and the focus of this section is on the use of mHealth in social sciences.

Traditional design analysis typically focuses on the system evaluation from the technological point of view, whereas social analysis looks at the way technology is incorporated into the users' activities. Based on social viewpoints, social and technical issues are inter-dependent and therefore cannot be separated. Social factors, such as aging, gender, and culture, can affect user's behavior toward the use of mHealth; on the other hand, charac-teristics of individual personal attitude and personal philosophy may override these fac-tors. Recognizing effective social factors in mHealth allows mobile health-care providers to understand users' requirements and improve the design of the new mobile health-care sys-tems. There are many social-related applications of mHealth, including gender, social class, and aged care, which this section will explore. Discussion on social perspective of health and illness provides us clear ground of complexity in this area but also shed light to progress.

13.2 Social Perspective and mHealth

The social perspective of health focuses on the social patterns of health and illness. Health, health care, and illness cannot be understood in isolation. Patterns of disease, illness, treat-ment, and provision of health services are crucially influenced by social class, ethnicity, gender, age, and disability. Thus, social and economic conditions and their effects on peo-ple's lives determine their health status and their risk of illness. For example, the lower an individual's socioeconomic position, the worse their health.

Social perspective also seeks social explanations for the illness rather than just focusing on biological and psychological explanations. It has become a common knowledge now that it is the living and working conditions that fundamentally shape why some groups of people get sicker and die sooner than others. Studies of morbidity (illness) and mor-tality (death) have consistently shown that the poor have the highest rates of illness. For example, Turrell (1999) and his colleagues by analyzing the data on health inequality con-clude that "socioeconomic differences in health are evident for both females and males at every stage of the life-course (birth, infancy, childhood and adolescence, and adulthood)." Similarly, Waitzkin (2000) by undertaking social determinant approach explains how poor living and working conditions such as poverty, discrimination, lack of educational and employment opportunities, and inadequate nutrition and housing directly influence the state of health and illness.

Another obvious factor supporting social perspective is life expectancy. The average life expectancy at birth of people living in least developed countries of the world is around 20 years less than that for developed countries such as Australia, which has an average life expectancy of 81.4 years (WHO, 2008). It should be emphasized here that the high life expec-tancy of Australian people "is not due to any biological advantage in the Australian gene pool, but is rather reflection of our distinctive living and working conditions" (Germov, 2009:4).

However, the fundamental cause for the existence of such high discrepancy is found to be within social determinant of health. "Social determinants of health refers to the complex, integrated and overlapping social structures and economic systems that include social and physical environments and health services" (WHO, 2010).

With further explanation, the WHO's White Paper provides detailed descriptions of social determinants by stating that "five determinants of population health are gener-ally recognized in the scientific literature biology and genetics (e.g., sex), individual behavior (e.g., alcohol or injection drug-use, unprotected sex, smoking), social environ-ment (e.g., place of residence, crowding conditions), built environment (i.e., buildings, spaces, transportation systems, and products that are created or modified by people), and health services (e.g., access and quality of care, insurance status)." However, it is interesting to point out that historically many public health efforts have focused on individual behaviors, whereas factors such as social environment, physical environment, and health services are not controllable by individual but affect the individual environment.

In order to further enhance our understanding of social perspective of health and illness, it is necessary to consider a comprehensive definition of health here. The definition of health has changed several times during the course of history as focus has shifted from perceiving health as a purely biological or physiological matter to understanding that health is affected not just by biological factors but also by psychological and sociological circumstances. One of the most comprehensive definitions of health is presented by the World Health Organization. Preamble to the Constitution of the World Health Organization (WHO, 1946) defines health as "a state of complete physical, mental and social well-being and not merely the absence of disease or infirmity." This definition has not been amended since 1946 but has been enhanced by a number of declarations and charters. For example, Ottawa Charter identifies three basic strategies for health promotion. These are *advocate*, *enable*, and *mediate*. In describing *advocate*, it states that "good health is a major resource for social, economic and personal development, and an important dimension of quality of life. Political, economic, social, cultural, environmental, behavioural and biological factors can all favour or harm health. Health promotion aims to make these conditions favourable, through advocacy for health." The *enable* strategy relates to achieving equity in health. The Ottawa Charter on *mediate* strategy emphasizes that "the prerequisites and prospects for health cannot be ensured by the health sector alone. Health promotion demands coordinated action by all concerned, including governments, health and other social and economic sectors, non-government and voluntary organisations, local authorities, industry and the media" (WHO, Ottawa 1986).

It is on this basis that in social sciences the emphasis of mHealth is on the contribution of the provision of social and personal resources of individuals, groups, and communities. However, some of the most relevant areas in social science in relation to mHealth will be discussed in this chapter. The first area of discussion is culture and analysis of the impact of mobile technology on culture as well as the accommodation of mHealth in wider social contexts.

13.3 Culture and mHealth

Culture is a way of life. As our way of life is changing, culture is changing too. One of the most obvious drivers of change is technology. This is not a new phenomenon, rather it is as old as when human beings used tools to challenge nature and attempted to overcome the environmental barriers they were facing. In fact, it is true that humans, by changing nature, have changed themselves: the way we live. Technology has had different effects in different eras on human existence. However, during the last two centuries, the impact of technology has been enormous. In fact, the pace of change has been so fast that other institutions in society (such as education, economy, commerce, law institutions, and political structures) have not been able to respond properly and on time. This phenomenon has been investigated and researched by sociologists and it is called *cultural lag* by Ogburn (1922).

Cultural lag theory suggests that a period of maladjustment occurs when the nonmaterial culture is struggling to adopt to new material conditions. Ogburn in explaining his theory distinguishes four stages. These include innovation, accumulation, diffusion, and adjustment. Adjustment of nonmaterial culture takes a very slow process due to its resistance to the use and impact of new technology. Any serious retardation of this process may cause cultural lag.

For example, the use of mobile phone in health encountered resistance from physicians, patients as well as academia, and others. However, as the technology has progressed and also responded to some extent to the demands in the area, the gradual adjustment process became smoother. By applying cultural lag theory to mHealth, it is possible to state that mobile phone technology has been partly successful in the process of accumulation and diffusion in the area of health. Yet it is progressing through full adjustment process in order to be used safely and efficiently in many areas of health now and in the future. Its enormous potential is progressively acknowledged and mHealth's use is attracting more support. However, at present, mHealth is still facing serious challenges from part of the culture resisting change and this will continue for some time to come. By acknowledging this process, it is fair also to state that the pace of change is so great that the adaptation and adjustment do not require centuries or several decades to overcome major cultural or institutional barriers. Rather, there are signs that the time zone will be limited due to advancement of smartphone technology to be sufficiently used in mHealth mostly in developed countries as well as in developing countries. One of the drivers of this change would be dissemination of information and knowledge in the areas of cultural and social aspects of health and illness.

In this context, sociology considers health and illness as social constructions. Different cultures have different understandings of health and illness. "People also within societies are differentially located with respect to access to 'expert' knowledge in this area. Cultural understandings and structure of health care play important parts in determining health and illness" (Germov, 2009:177). Thus, mHealth will have much wider applications and bring new changes in various areas in society.

In the context of culture and media, there are growing awareness and use of mobile phones with little resistance in a number of areas including the rise of mobile learning, mobile commerce, mobiles for information and entertainment, mobiles as a games platform as well as getting shaped new culture around devices such as Blackberries. Mobile phones have been used progressively in every area of our existence including health and illness situations.

For example, mobile phones will ease patients' communications not only with nurses and doctors but also with people close to patients and fellow patients who are suffering from the same disease. This is a visible reflection of the changing characteristics of our cultural norms and cultural worlds. Indeed, mobile phones create a new form of social and cultural diversity. It is highly observable that diversity has become the defining characteristic of our social and cultural worlds. The global intercultural contacts are increasing with enormous speed as a result of global immigration, speed of travel, and the use of mobile phones. In the area of health care, evidence indicates that the use of mobile phones will enhance doctor–patient interactions leading to improvement of patients' health status. Tirado (2011:2) in his research indicates that "the health of Hispanics and other minority populations can be improved by accessing mobile devices to receive vital health messages, monitor their conditions, and receive other health-related wireless intervention."

Our social world in which traditional, social, cultural, and geographical boundaries have given way to increasingly complex representations of identity, creates new challenges and new demands for social scientists. It is within such a world that the use of smartphones in health and illness attracts growing attention in various parts/components of our culture and society. Some of these areas that are closely relevant to mHealth are chosen to be discussed in this chapter.

13.4 mHealth and Interactions between Doctors and Patients

The ability to communicate is a basic human skill. The interaction between doctors and patients involves the forming of a relationship and the gathering and giving of information. Its purpose is to promote the physical, emotional, and social well-being of patients and their families. Communication skills are fundamental to the interaction and involve verbal, nonverbal, and written methods of communications.

A critical part of all doctor–patient interactions involves eliciting information from the patient. The core skills that are needed to facilitate the process of information gathering are skills that help to facilitate patient involvement in the medical interview in a way that enables the doctors to arrive at an accurate diagnosis of a patient's problem or symptoms.

The crucial importance of this communication has been realized decades ago, and academic programs have been established to prepare medical students to greatly enhance their communication skills. For example, a national conference on the importance of communication was held in March 1992 in Canada, and all medical schools participated in various programs of the conference (WHO MNH, 1993).

One of the outcomes of the conference was to recommend training workshops and other methods to enhance communication skills of medical students and prepare them to have a healthy and meaningful interaction with patients. The training on communication skills becomes one of the most emphasized and rapidly growing sector of medical education in recent years. Students have shown that there are significant associations among many aspects of the clinicians' interpersonal skills and aspects of the patient's motivation and satisfaction (WHO MNH, 1993).

mHealth can enhance the communication between doctors and patients through the flow of information. Increasingly, literature in this area indicates the growing direct use of mobile phones as an alternative to face-to-face patient/doctor visits (Caffery and Smith, 2010; Heaney et al., 2010). Thus, smartphones can add a distinctive opportunity for the exchange of further information leading to better communication between doctors and patients with much improved outcomes. In addition, patient communications with family and friends become an essential element of support and comfort. Furthermore, since smartphones have more functionality such as camera and video capability and music players, patients can benefit from their use enormously. With regard to privacy and safety issues of patients in health-care facilities, appropriate and sensible use of mobile phones should help ensure that patients and clients remain safe from intrusion and they are treated with dignity.

Finally, mHealth has the capacities to improve patients' engagement and empower them to better express and explain their opinion and views on their health conditions. mHealth has the capacity to make health care safer by giving patients tools to manage their own health. The use of mHealth also has interesting implications in the area of gender.

13.5 Gender and Health

The far-reaching differences between men and women have inspired curiosity, poetry, romance, and polemics for centuries, but it was only in recent history that feminism introduced the concept of gender several decades ago to distinguish between biological sex and the social construction of masculinity and femininity. This distinction has enabled health researchers to move beyond comparative analyses of male and female morbidity and mortality rates. The sexes differ in the types of lives that they typically live, and these differences can influence health.

Researchers use the concept of gender to study the impact of women's status in society on their heath and health care. For example, women have higher rates of illness, but men die younger. Women have more frequent illness and disability, but the problems are usually not life threatening. In contrast, men suffer more from life-threatening diseases, and these cause more permanent disability and earlier death for them.

Sex differences in health are the outcome of differential risks acquired from roles, lifestyles, and preventative health practices. However, we need to know the economic, social, and political mechanisms of prejudice and discrimination that lead to ill health and poor health care for women and men. For example, certain occupations are significant sources of injury and disease. The sexual division of labor has tended to concentrate men in the occupations in which such hazards are greatest. These include construction, mining, waterside work, and farming, which are mainly done by men.

Since the health hazards of women's work such as office job, nursing, or household work are often hard to detect, thus they may be underestimated (Strazdins and Broom, 2004). However, males, especially young men and adolescents, are more likely to engage in a range of dangerous activities such as risky driving, contact sports, and physical aggression. Consequently, males suffer higher rates of nonintentional and intentional injury (Broom, 2009).

However, the implication of sexism in medicine has been that doctors were ignoring women's health concerns in medical research and interventions. In addition, evidence points to the fact that women were not informed of the availability of other treatments or "the side effects of drugs/therapies, and labelling of women's problems as 'psychosomatic' rather than 'real'" (Broom, 2009:139).

In the area of men's health, it has also became known that elements of masculinity can be hazardous to men's health. But recent changes in social, political, economic, and technological areas affecting men and women's health altered the context in which health is created and managed. One of the distinguished characteristics of this change is that attention is placed on gender and not just on women or men.

The World Health Organization by acknowledging the existence of these gender biases and warning undesirable health consequences states: "Gender biases in power, resources, entitlements, norms and values, and the way in which organisations are structured and programmes are run damage the health of millions of girls and women. The position of women in society is also associated with child health and survival – of boys and girls" (WHO, 2008). While this has wider implications in the use of health resources, it also creates great potential for mHealth applications in gendering health. In fact, mHealth can be seen to have a potential to bridge the gender health discrepancies.

Discussions and research on gendered health have also opened up the fields of gay and lesbian health research and led to the study of the transgendered. However, the underlying change in this context relates to the power relations and accessibilities to health resources and the view of how mHealth revolutionizes this context in distant future. All this requires keeping research in this area active and ongoing.

It should be emphasized here that the distinct gender roles, norms, values, and behaviors give rise to gender inequalities. But within our societal change, gender norms and values are not fixed. They evolve over time, vary substantially from place to place, and are subject to change. Thus, the poor health consequences resulting from gender differences and gender inequalities are not static either. They can be changed. One of the drivers of the change is technology, and in this context, mobile technology and mHealth have enormous potential and can play a positive distinctive role to bring change in a number of ways:

- It can store relevant health information for both genders on various distinctive issues. This is particularly important for countries where there is gender segregation and less health knowledge is publicly available in some sensitive areas. By mHealth application, easy access to this information is possible by just pushing a button, as the technology could proceed in this direction.
- Now that people and particularly young people are *addicted* to mobile phones and they are using them everywhere and anywhere, it is possible to feed this *captive audience* health information for preventative purpose and for enhancing the quality of health. This is a delicate task particularly for women and girls and particularly in developing countries. In fact, this may become a health tool/health search tool. The friendly and attractive design of the programs and their applications attracts users' attention when they confront particular issues.
- Another particular area also would be sex education particularly for girls and women in developing countries.

Of course all these need to be researched, explored, and technically advanced in order to be implemented in a more efficient and productive manner. Another aspect of gender inequality relates to power and control. One of the prevalent unhealthy examples is the occurrence of domestic violence and how mHealth can be developed and used to save people lives that are in such dangerous situations.

13.6 Domestic Violence and mHealth

Our home is considered to be a comfort zone. Yet for many women, a home is a place of pain and humiliation (Adibi, 1998). Domestic violence is common and widespread. Domestic/family/partner violence happens in many forms including physical, sexual, emotional, spiritual, social, economic, and psychological abuse. It is a form of violence that can occur within any relationship and is about the power and control. The National Council to Reduce Violence against Women and Children (NCRVWC) found that "a central element of domestic violence is that of an ongoing pattern of behaviour aimed at controlling one's partner through fear (for example, by using violent or threatening behaviour)." It goes on to state that "the violent behaviour is part of a range of tactics used by the perpetrator to exercise power and control and can be both criminal and non-criminal in nature."

Domestic violence is a complex issue and there is no single cause that leads to abuse. However, there are a number of risk factors associated with perpetrators and victims of violence. For example, alcohol is a significant risk factor causing domestic violence particularly in indigenous communities (Mitchell, 2011:2). In the United States, "domestic violence is the leading cause of injury among women of reproductive age" (World Bank, 1993:50). Also, women appear to be at particularly high risk when they are pregnant (Stark and Flincraft, 1991). Similar situation exists in Australia. The Women's Safety Survey reported that 338,700 women were victims of physical violence, and 180,400 of these assaults were committed by the women's partners or former partners, nearly three times as many as those where the attackers were strangers (67,300) (ABS 1996:19). While any injury can leave emotional scars, "the psychological impact of being attached by a parent or loved and trusted partner is particularly devastating and personally debilitating" (Germov, 2009:136).

The legal recognition of domestic violence as crime is a recent phenomenon in many countries. For example, *The Guardian* reports that "earlier this year [2013], a powerful image was circulated in Saudi Arabia. A woman with one black eye stared bleakly out. Underneath her niqabed face was written the simple phrase in Arabic: 'And what is hidden is greater,' a saying that highlight and tackle domestic violence in the country. However, this incident led Saudi Arabia's cabinet passed a legal ban on domestic violence and other forms of abuse against women for the first time in the kingdom's history" (*The Guardian*, 2013).

Domestic violence is a serious health problem and it affects people in all stages of life. Governments and community centers by providing support aim to prevent domestic violence. For example, *the make the call campaign* aims to prevent domestic violence by prompting friends, family members, neighbors, and colleagues who support someone they know is being abused to make the call for support, advise, and referrals to prevent serious harm from occurring (Queensland Government: Department of Community, Child Safety and Disability Services).

The Brisbane Domestic Violence Service (BDVS) is a free service and aims to support women and children to ensure they are safe and free from fear of domestic violence. However, in providing advice, the most comprehensive list of advice and support is produced by Laura Crites, Department of the Attorney General, Hawaii, and presented by the Center for Prevention of Abuse. These include the following:

- Call the police if you see or hear evidence of domestic violence.
- Speak out publicly against domestic violence.
- Take action personally against domestic violence when a neighbor, a coworker, a friend, or a family member is involved or being abused.
- Encourage your neighborhood to be concerned and watch out for domestic violence as with burglaries and other crimes.
- Reach out to support someone who you believe is a victim of domestic violence and/or talk to a person you believe is being abusive.
- Help others become informed, by inviting speakers to your church, professional organization, civic group, or workplace.
- Support domestic violence counseling programs and shelters.

While these services are very valuable and can assist victims greatly, there are other cases where victims are in a very dangerous situation and only one phone call in haste can save their lives. This is the situation that only mHealth can assist. The following two cases clearly illustrate the significance of mHealth in these situations:

1. Case One: A simple mobile phone with a hotline to the police saved the life of Ruth. She was experiencing domestic violence and abuse at the hands of her former husband. For years, she was shoved, kicked, and bruised. She tolerated this situation by telling herself that she was to blame for his aggression. Fearing for her life after one particularly terrifying episode, Ruth finally went to the police. She was given a mobile phone that with just one push of a button would connect straight to the police. Ruth kept the phone with her when she felt in danger to use it. It was not long before Ruth in a dangerous situation pressed the button and the call alerted the police to her location. Within minutes, officers were on the scene and the years of abuse finally drew to close (*Telegraph*, 2013).

2. Case Two: This case is about how smartphone apps can help victims of domestic violence. The smartphone is called *Aurora* and developed by Komosion for Women NSW, part of the NSW Government of Family and Community Services. It was launched by the NSW minister of Family and Community Services. Minister Goward in launching this phone said: "This app is the first of its kind in the world to combine information with access to help services, the ability to create a trusted network of friends who can be easily contacted with an agreed message and a GPS system to 'call-for-help' alert" (ABC News, 2013).

These cases clearly show how smartphone apps can develop in relation and in response to health needs in various areas. However, this requires the cooperation and close working of technologists with social scientists to complement both technological innovations with sociological aims. In exploring further dimensions of the use of smartphones in the area of well-being, social capital has become a growing resource of individuals, and its role in the context of mHealth will be discussed here.

13.7 Social Capital and mHealth

One of the distinguished aspects of mobile phones is not only increasing the volume of one's communication but also one could argue that it enhances one's social capital as well. It is the purpose of this section to explore the interconnection of social capital and mHealth. In order to explore this relationship, it is necessary to create a relevant context, since social capital has been discussed in various contexts and sometimes with controversies.

However, our focus is to demonstrate how mHealth and social capital of individuals are closely connected. Research in the area of social settings and health clearly indicates that hygiene, warm houses, and the capacity to grow and buy good food have implications for health (Edmondson, 2003). Another researcher also suggests that *sense of community* and *community empowerment* contribute to public health (Shiell and Hawe, 1996). In the same context, Wilkinson (1996) points to the significance of psychosocial factors on health as well. The question is, "How do these factors that are caused by social inequalities relate to social capital?"

In this context, Baum (1999:177); that social capital as part of a broader project is "to reduce inequality in society." Thus, by enhancing the status of social capital, it becomes possible to reduce inequalities and for our purpose inequalities in health area by adapting mHealth strategy. Putnam (1993) concludes that social capital is about social networks, norms of reciprocity, and trust. This in turn creates opportunity for individuals to enhance their social capital in order to enhance their health status.

Other researchers see social capital in a much broader context. For example, Coleman (1994:300) considers social capital as a set of resources in family, and community per se, about which generalizations can be made: It is useful for the development of young people because it establishes trust, mutual obligation, information, norms and sanctions, and so on. In this context, mHealth can be considered attracting more attention. Winter et al. (2000) further their emphasis on health and suggest that in these definitions, the assumption is that access to social capital will lead to improved health outcomes by lowering stress and providing outlets for social interaction and opportunities for enhancing control over one's life through democratic participation in community life.

By analyzing these definitions in the context of health, it becomes clear that social capital can be a very useful resource in assisting individuals and families in the time of illness and distress. This is particularly important in developing countries where the dominant form of the family is still an extended one. In this context, mHealth can be used not only to provide ongoing support and constant contacts with patients but also assist patients in early recovery. An example will clarify the significance of mHealth in this context. It has become a norm in Iran that when a person is hospitalized, it is a requirement that another person in the family or a friend needs to stay with the patient almost all the time in the hospital. In such a situation, a mobile phone not only plays a crucial role for family members and friends to keep their contacts with the patient but also assists patient to communicate with a physician who is not available in the hospital all the time. So patients can receive ongoing suggestions such as changing the doses of drugs, information on new drugs, and preparation for a particular procedure in advance. Therefore, family or social network and the trust in physician are obvious characteristics of social capital exemplified here. Furthermore, in this context, mHealth has the potential to provide social support for the patient when they are discharged from the hospital by keeping constant contacts with doctors, family members, and friends. This in turn will assist patients with speedy recovery.

13.8 Equity and mHealth

Equity means social justice or fairness or the state of being just. In other words, "equity is the justice applied in circumstances covered by law yet influenced by principles of ethics and fairness" (The Free Dictionary, n.d.). However, health equity is the absence of unfair differences in health services and outcomes among groups of people. "Choosing to eat healthy food, being physically active, limiting alcohol consumption and not smoking requires people to be empowered to make these choices. It means that the healthy choice must be physically, financially and socially the easier and more desirable choice relative to the less healthy option. This is not always the case, particularly with decreasing social position" (Friel, 2009:1).

By analyzing the existing data on geographical and socioeconomic factors, it is possible to identify vulnerable populations who could become equity-oriented target for health intervention. Findings of research indicate that a significant proportion of the world's population have unhealthy diet including energy-dense nutrient-poor foods and harmful quantity of alcohol. "The harmful health consequences of these behaviours, and the inequity in their social distribution, are, however, indicative of both market failure and failure by government to protect the health of all its citizens" (Friel, 2009:22).

Health equity depends on and is influenced by the following:

1. Government policies in political, economic, health, social, and cultural sectors are the major cause of the existing health inequity in society. In fact, every aspect of government policies affects health and health equity directly or indirectly. These include finance, education, economy, housing, transport, and health, to name a few. For example, while health may not be the main aim of finance policies of the government, it has a strong bearing on health and health equity.

2. Another major factor causing health inequity is the inconsistency of government policies for health. For example, the government's trade policy encourages production and consumption of food high in fat and sugar. This is in contradiction with the government official health policy. In fact, the government health policy must encourage the production of fruits and vegetables if it is serious to promote the health of population. Therefore, policy coherence is crucial if the intention is to reduce health inequity. Furthermore, reinstatement of a strong public sector is needed and requires policies, legislation, and regulations that tackle the underlying social causes of health inequity.

3. The second major factor leading to inequity in health relates to the fact that there is a visible lack of an individuals' empowerment to challenge and change the unfair health distribution resources in society. Therefore, empowering vulnerable people to challenge the existing health inequity and demand organized and systematic improvement of health status of population is a vital strategy. In this context, mHealth can play a very important role by organizing local groups on health equity. However, mHealth can play a significant role by promoting the health of individuals and families through the provision of relevant and on-time health-related information. One example would be to encourage people to purchase fruits and vegetables from *farmers markets*, which are much cheaper than in corporate supermarkets. This can be done through supplying relevant and timely messages and establishing a network to promote the status of health of vulnerable people via mobile/cyber local communities.

These strategies also can be used in developing countries. Due to high illiteracy among the poor and people living in remote areas, the emphasis should be placed on information in plain language with culturally appropriate strategies in delivering relevant and on-time health-related information.

For example, in this context, the greatest forgeable benefit of mHealth in developing countries will be through the development of low-cost audible and readable message delivery. This information can be used through the utility of mHealth to improve patient follow-up, and compliance with treatment, as a starting point.

Furthermore, mHealth has the potential and can make significant contributions in achieving equitable care particularly to the poor and those residing in rural and remote areas and regions, those who have no voices through media.

Another implication of this would be opening a gateway to broader issues of human rights. Not only can mHealth play a significant role in equity health context, but it can also pave the way for addressing health as a human right issue and moving to address other human right issues in larger contexts.

This is in line with the WHO's advocacy standing and support. The WHO in White Paper (2010) on the determinants of health concludes: "We will actively promote awareness, engagement, and action on striving toward fairness in policies, services, access, and environmental conditions" (p. 7). Thus, there are growing demands for health equity expressed in various forms and contents, and mHealth has the potential to play a decisive and ongoing role in our constant changing world.

13.9 Future Directions

Social perspective on health provides a comprehensive context through which mHealth can be viewed holistically. The global world in the twenty-first century is confronted with a number of very serious challenges. In discussing these challenges, the Millennium Project asserts that "the world is improving better than most pessimists know, but future dangers are worse than most optimists indicate." With this interesting statement, the report highlights the challenges ahead by stating that "the world is getting richer, healthier, better educated, more peaceful, and better connected and that people are living longer, yet half the world is potentially unstable. Protesters around the world show a growing unwillingness to tolerate unethical decision making by power elites. An increasingly educated and Internet-connected generation is rising up against the abuse of power. Food prices are rising, water tables are falling, corruption and organized crimes are increasing, environmental viability for our life support is diminishing, debt and economic insecurities are increasing, climate change continues, and the gap between the rich and poor continues to widen dangerously" (UN Millennium Project, 2012). All these, of course, have serious immediate and long-term effects on the health status of people.

Among the 15 major challenges (see UN Millennium Project, 2012), health-care delivery is a pervasive global challenge with huge ramifications for costs and human well-being. Emerging population dynamics including longer life expectancy and lower birth rates are also challenging traditional approaches. It is within this context that mobile technology can certainly play an important role in providing health services to millions of people.

In fact, the growing innovation of mobile sophistication and its applications in the area of health has evolved into a new mHealth field. Mobile health adds modern technology advances to health-care systems to let users exploit heath information and health-care services anytime and anywhere. This will lead to a number of significant changes in health-care delivery including information flow between doctor and patient. "All aspects of communications, interaction, and information flow will become mediated (and monitored) by electronic tools" (Weiner et al. 2012:8).

However, future technological advances and applications of mHealth will be fully realized if full considerations are given to the social perspective of health. It is within this context that we need to properly and objectively address and overcome a number of serious challenges and issues in order to be on the right track in achieving the WHO's objectives of *health for all*. Some of the most significant challenges for full adaption and effective use of mHealth include the following:

1. Approach to mHealth should be interdisciplinary and include involvements from a number of professional disciplines: "how best should we pull together the required expertise which will include specialists with backgrounds in: interpersonal and

mass communications, human factors, clinical sciences, health informatics and IT, computer science and engineering, public health/population sciences, and health management and policy?" (Weiner, 2012:6).

2. The second focus should be on the efficient use of mHealth within the patient-centered needs of each individual, as well as the needs of the community.

3. The third challenge relates to digital divide or access to the Internet through smartphones between social groups and within the countries. "Evidence in developed countries indicates that this divide has narrowed over the past decades. For example, overall penetration of the Internet within American homes has increased from 40% in 2000 to nearly 75% in 2009; mobile handset ownership in the United States is higher among Latinos and African-Americans than among whites (87% vs. 80%), and access, while lower in lower SES groups, is increasing across the socioeconomic spectrum" (Tirado, 2011:3). However, despite the promise of shrinking the digital divide along racial/ethnic and socioeconomic lines, evidence points to a new digital divide between younger and older consumers. Whether this disparity is related to age or generational issues has yet to be definitely determined.

4. The fourth challenge relates to the fact that the aging population is increasing rapidly. Providing continuous medical care for elderly people living in their homes represents a major challenge. Therefore, the demographic change in the twenty-first century demands new strategies in health care for the elderly. Allowing people to stay in their home environment as long as possible is an important goal for health for all. "Innovative information and communication technologies are generally supposed to play an important role in retaining elderly patients in their home environment" (Schulke, op.cit, 2010).

5. In the context of *health for all*, the other great challenge of mHealth at present and in the future relates to rural population of the world. According to the UN, "by the middle of 2009, the number of people living in urban area (3.42 billion) had surpassed the number living in rural areas (3.41 billion)" (UN, 2012). Although the percentage of the world's rural population is decreasing, the process is very slow and the numbers are still in billions. It is obvious that the majority are living in developing countries, but also some are living in developed countries including Latino, African-American, and native American populations in the United States and indigenous populations in Australia.

These people often must travel great distances to see a physician. "While tele-heath has done much to link, those general practitioners with specialists in distant medical centers, it has done little to narrow the gap between patients and doctors" (Tirado, 2011:2). However, mHealth has the great potential to support communication between health-care providers and their patients to increase effective monitoring and management of chronic disease.

6. Another huge challenge for using mHealth relates to social class and increasing income disparities. The cost of access to mHealth services for lower-income populations, particularly in developing countries, is extremely high and out of reach of hundreds of millions of people. Although situations are different in developed countries, there are serious affordability issues. "Mobile technology has an inherent dynamic that drives the market toward ever more costly hardware and software applications. With the rise of iPhones and Android-powered smart phones

available only with post-paid cell phones contracts, many low-income users are relegated to using cheaper basic prepaid phones that lack the ability to support health-related mobile applications available to the more affluent health-conscious user" (Tirado, 2011:3). Therefore, there is a need for a government intervention to discuss and negotiate this issue with mobile technology industries and mobile phone carriers in order to find a realistic solution within the existing economic framework.

7. Finally, there are issues related to patient privacy and cultural, linguistic, and appropriateness of health-related applications of mHealth. For example, Michalski (2007) expresses concern that "the people building these technologies have little understanding of the implications of what they're creating, and little recourse to call a halt when they do spot something amiss." In stressing the pervasive nature of these new technologies on the health care of patients, Upkar (2007) argues these "devices must be designed to offer intuitive interfaces that can learn with and from individuals." Resolving these issues requires adaption and reinforcing the interdisciplinary approach to mHealth again. This requires that experts and professionals in all related fields work together in order to ensure all concerns have been addressed and reflected in the use of mobile technology, particularly in the area of health.

13.10 Conclusions

In our world of dangerously widening gap between rich and poor and affluent and vulnerable populations, we are facing discrimination on the basis of gender, age, ethnicity, social class, and disability on a daily basis. In the area of health, our global community is facing exponential increase in health-care costs. Present governments/nations of the world are not ready or able to bear the costs. This requires structural change in a number of institutions including the economy itself. However, within this undesirable reality of today, technological advancement could provide a window of opportunity. In particular, mHealth appears to be one of the solutions through which health-care information and services could be provided to millions of people in order to improve the quality of their lives and reduce the cost and affordability of dominant health-care delivery systems.

In this context, however, particular attention needs to be paid to empower the venerable populations in a global context to demand health equity as their human right and have access to required health services to improve the quality of their lives. Such a public awareness and wakening may provide real opportunities for institutional change, at least in health-care delivery systems.

In the context of interdisciplinary approach, mobile phone technology has been turned into a social and encyclopedia information/research tool. Smartphones have great potential to enrich the use of mHealth, which is attracting growing attention from academia and equally from the media. If the trends continue in a progressive manner to overcome the existing challenges and to respond to the needs of people including vulnerable people, then mHealth may revolutionize the health research and service delivery in years to come.

Questions

1. How do you define health?
2. What is cultural lag theory in relation to mHealth?
3. How do mobile phones enhance the interaction between doctors and patients?
4. What are the effects of occupations on the health of men and women?
5. How can smartphone apps help victims of domestic violence?
6. What factors influence health equity?
7. Briefly discuss three challenges that mHealth needs to deal with in order to become more effective now and in the future.

References

ABC News. *Phone app to help victims of domestic violence*. http://www.abc.net.au/news/2013-05-06/smartphone-techology. Accessed on August 31, 2013.

Adibi, H. (1999). Health homes, in Honari, M. and Boleyn, T. (eds.), *Health Ecology: Health, Culture and Human-Environmental Interactions*. Rutledge, London, U.K., pp. 193–206.

Aoki, K. and Downes, E.J. (2003). An analysis of young people's use of and attitudes toward cell phones. *Telematics and Informatics*, 20, 349–364.

Australian Bureau of Statistics (1996). *Women's Safety Australia*, Cat.no. 4128.0, ABS, Canberra, Australian Capital Territory, Australia.

Australian Institute of Health and Welfare (AIHW) (2008). *Australia's Health 2008*, Cat. No. AUS 99. AIHW, Canberra, Australian Capital Territory, Australia.

Baum, F. (1999). The role of social capital in health promotion. Australian perspectives. *Health Promotion Journal of Australia*, 9(3), 171–178.

Brisbane Domestic Violence Service (BDVS), http://www.bdvs.org.au/. Accessed on August 31, 2013.

Broom, D. (2009). Gender and health, in Germov, J., (ed.) *Second Opinion: An Introduction to Health Sociology*, 4th edn. Oxford University Press, South Melbourne, Victoria, Australia.

Caffery, L. and Smith, A.C. (2010). A literature review of email-based telemedicine. *Studies in Health Technology and Informatics*, 161, 20–23.

Coleman, J.S. (1994). *Foundations of Social Theory*. Belknap Press, Cambridge, MA.

Crites, L. (1992). Preventing domestic violence, Prevention Communique, March 1992, Crime Prevention Division, Department of the Attorney General, Honolulu, HI.

Edmondson, R. (2003). Social capital: A strategy for enhancing health? *Social Science & Medicine*, 57, 1723–1733.

Flinchy, P. (1997). Perspectives for a sociology of the telephone. *The French Journal of Communication*, 5(2), 149–160.

Forgas, J.P. (2012). Affective influences on interpersonal behaviour: Towards understanding the role of affect in everyday interactions, in Forgas, J.P. (ed.), *Affect in Social Thinking and Behaviour*, Taylor & Francis Group, Hove, U.K.

The Free Dictionary (n.d.). www.thefreedictionary.com/equity.

Friel, S. (2009). *Health Equity in Australia: A Policy Framework Based on Action on the Social Determinants of Obesity, Alcohol and Tobacco*. National Centre for Epidemiology and Population Health, the Australian National University, Canberra, Australian Capital Territory, Australia.

Germov, J. (2009). *Second Opinion: An Introduction to Health Sociology*, 4th edn. Oxford University Press, South Melbourne, Victoria, Australia.

The Guardian, Saudi Arabia's domestic violence law is a first step to changing attitudes, http://www.theguardian.com/commentisfree/2013/aug/30/saudi-arabia, Accessed on August 31, 2013.

Heaney, D., Elwyn, G., and Sheikh, A. (2010). The quality, safety and content of telephone and face-to-face consultations: A comparative study. *Quality and Safety in Health Care*, 19, 298–303.

Michalski, J. (2007). I am worried that people building these technologies have little understanding of their implications. *Mobile Persuasion: 20 Perspectives of the Future of Behaviour Change*, in Fogg, B.J. and Eckles, D. (eds.). Captology Media, Stanford, CA.

Michell, L. (2011). Domestic violence in Australia—An overview of the issue, Parliament of Australia. http://www.aph.gov.au/About_Parliament/Parliamentary_Departments/Parliamentary_Library/pubs/BN/2011-2012/DVAustralia. Accessed on August 31, 2013.

National Council to Reduce Violence against Women and Children (NCRWC). (2009). Background paper *to Time for Action: The National's Council's Plan to Reduce Violence against Women and Children*, 2009–2021, Department of Families, Housing, Community Services and Indigenous Affairs (FaHCSIA), Canberra, Australia, viewed 31 August 2013, http://www.dss.gov.au/our-responsibilities/women/programs-services/reducing-violence/the-national-plan-to-reduce-violence-against-women-and-their-children-2010-2022.

Nguyen, Angela-Minh Tu D. and Benet-Martinez, V. (2010). *Culture*, Bicultural Identities in a Diverse World, Crisp Richard, J. (ed.). Blackwell Publishing Limited, Malden, MA.

Norman, C. (2009). Health promotion as a systems science and practice. *Journal of Evaluation in Clinical Practice*, 15, 868–872.

Ogburn, F.W. (1922). *Social Change with Respect to Culture and Original Nature*. Digitalized by Microsoft Corporation in 2007.

Pew. (2009). Pew Internet & American Life Project, http://www.rewinternt.org/. Accessed on August 31, 2013.

Putnam, R. (1993). *Making Democracy Work: Civic Traditions in Northern Italy*. Princeton University Press, Princeton, NJ.

Queensland Government: Department of Community, Child Safety and Disability Services, http://www.communities.qld.gov.au/communityservices/violence-prevention/about-domestic-and-family-violence-prevention.

Schulke, A.M., Plischke, H., and Kohls, N.B. (2010). Ambient Assistive Technologies (AAT): Sociotechnology as a powerful tool for facing the inevitable sociodemographic challenges? *Philosophy, Ethics, and Humanities in Medicine*, 5(1), 8.

Schweitzer J. and Synowiec C. (2012). The economics of eHealth and mHealth. *Journal of Health Communication*, 17(Suppl 1): 73–81.

Shiell, A. and Hawe, P. (1996). Health promotion, community development and the tyranny of individualism. *Health Economics*, 5(3), 241–247.

Stark, E. and Flincraft, A. (1991). Spouse abuse, in Rosenberg, M. and Finley, M. (eds.), *Violence in America: A Public Health Approach*. Oxford University Press, New York, pp. 161–181.

Strazdins, L. and Broom, D. (2004). Acts of love (and work): Gender imbalance in emotional work of women's psychological distress. *Journal of Family Issue*, 25, 356.

Telegraph, http://www.telegraph.co.uk/women/womens-life/9913548/International-Womens-Day-2013-It-saved-my-life-The-mobile-phone-giving-domestic-violence-victims.

Tirado, M. (2011). Role of mobile health in the care of culturally and linguistically diverse US populations. *Prospectives in Health Information Management*, 8, Winter, le. PMCID: PMC 3035829.

Turrell, G., Oldbenburg, B., McGuffog, I., and Dent, R. (1999). *Socioeconomic Determinants of Health: Towards a National Research Program and a Policy and Intervention Agenda, Centre for Public Health Research*, School of Public Health, Queensland University of Technology, Brisbane, Queensland.

Tushes, M. (2007). Commentary: Gender and health. *Journal of Public Health Policy*, 28, 319–321, Palgrave Macmillan Ltd, http://www.palgrave-journals.com/jphp/journal/v28/n3/full/3200141a.html.

UN Millennium Project. (2012), Gender, Women and Health. http://www.google.com.au/search?q=UN+Millennium+Project+(2012)+Gebder%2C+Wmen+and+Health&ie=utf-8&oe=utf-8&aq=t&rls=org.mozilla:en. Accessed on February 02, 2013.

Upkar, V. (2007). Pervasive healthcare and wireless healthcare monitoring. *Mobile Networks and Applications*, 12(2–3), 124.

Verbrugge, L.M. (September 1985). Gender and health: An update on hypotheses and evidence. *Journal of Health and Social Behaviour*, 26, 156–182.

Waitzkin, H. (2000). *The Second Sickness: Conditions of Capitalist Health Care*, 2nd edn. Rowman & Littlefield, Lanham, MD.

Weiner, J.P. (2012). Doctor-patient communication in the e-health era. *Israel Journal of Health Policy Research*, 1, 33. www.ijhpr.org/content/1/1/33. Accessed on September 2, 2013.

Weiner, J.P., Fowles, J.B., Chan, K.S. (2012). New paradigms for measuring clinical performance using electronic health records. *International Journal of Quality in Health Care*, 24(3), 200–205.

WHO. (1946). Preamble to the constitution of the World Health Organisation as adapted by *the International Health Conference*, New York, June 19–22, 1946 by the representatives of 61 States and entered into force on April 7, 1946. http://www.who.int/about/definition/en/print.html.

WHO. (1986). Ottawa Conference, http://www.who.int/healthpromotion/conference/previous/ottawa.html. Accessed on August 31, 2013.

WHO. (1993). Doctor-patient interaction and communication, http://whqlibdoc.who.int/hq/1993/WHO_MNH_PSF_93.11.pdf. Accessed on August 31, 2013.

WHO. (2008). World Health Report, http://www.who.int/social_determinaants/. Accessed on August 31, 2013.

WHO. (2010). NCHHSTP White Paper on Social Determinants of Health. Establishing a holistic framework to reduce inequalities in HIV, Viral hepatitis, STDs, and tuberculosis in the United States, http://www.cdc.gov/socialdeterminants/docs/sdh-white-paper-2010.pdf. Accessed on August 15, 2013.

Wilkinson, R. (1996). *Unhealthy Societies: The Affliction of Inequality*. Rutledge, London, U.K..

Winter, I., Cox, E., Latham, M., and Baum, F. (2000). *Social Capital and Public Policy in Australia*. Australian Institute of Family Studies, Melbourne, Victoria, Australia.

World Bank. (1993). *World Development Report 1993: Investing in Health*. Oxford University Press, New York.

14

Study of the Effective Factors in Mobile Health-Care Success: Sociotechnical Perspective

Pantea Keikhosrokiani, Nasriah Zakaria, Norlia Mustaffa, and Ibrahim Venkat

CONTENTS

14.1 Introduction

Applying advanced telecommunication technologies in health care becomes a controversial issue nowadays. Mobile health-care systems (MHSs) involve the use of superior and reliable communication techniques to deliver biomedical signals over long distances. As MHSs are playing a critical role in human life, assessment is required for these kinds of systems in order to prove their success and failure. Traditional design analysis typically focuses on the system assessment from the technology point of view, whereas sociotechnical analysis looks at how the technology will incorporate into the user's activities. Based on sociotechnical viewpoint, social and technical issues are complementary and they cannot be separated. This study aims to provide a critical review of the research in this domain and to offer our perspective on possible directions for future mobile health. The focus of this study is on sociotechnical factors that will affect mobile health-care performance. We review sociotechnical factors of different MHSs in order to recognize their effective factors.

Subsequently, we categorize sociotechnical factors with an objective to find the most important factors in mobile health-care performance. After studying sociotechnical factors and classifying them, we propose a sociotechnical model for MHSs. Based on the concluding remarks, this study allows mobile health-care providers to understand user needs and improve the design of the new MHSs.

14.2 Background

In this study, we utilize sociotechnical factors, assessment of the systems, and mobile computing factors in order to find the important sociotechnical factors in the success of MHS. By synthesizing some important sociotechnical factors gathered from previous studies, we propose a new sociotechnical model for MHS assessment in order to have a successful system. This study can assist health-care providers to design new MHS by considering important sociotechnical factors that will affect MHS performance. For this purpose, a review on sociotechnical models in health care, assessment of MHS, and mobile computing in health care is presented in this section.

14.2.1 Sociotechnical Models in Health Care

Sittig and Singh (2010) introduced an 8D model as illustrated in Figure 14.1, which is specifically designed to address the sociotechnical challenges involved in the design, development, implementation, use, and evaluation of health-care systems. The 8D model consists of hardware and software; clinical content; human–computer interface; people; workflow and communication; organizational policies and procedures; external rules, regulations, and pressures; and system measurement and monitoring. These eight dimensions are not independent, sequential, or hierarchical, but rather they are interdependent and interrelated concepts similar to compositions of other complex adaptive systems (Sittig and Singh, 2010).

In order to recognize the factors that shape human and system performance in health care, LeBlanc et al. (2011) tried to gain a deeper understanding of performance shaping factors (PSFs) that can enhance or degrade performance. The researchers depicted different ways on how simulations have been utilized to study PSFs. Furthermore, the researchers also discussed about directions of the future researches as well as methodological approaches in order to study PSFs. They classified the researches to study PSFs using simulation, which

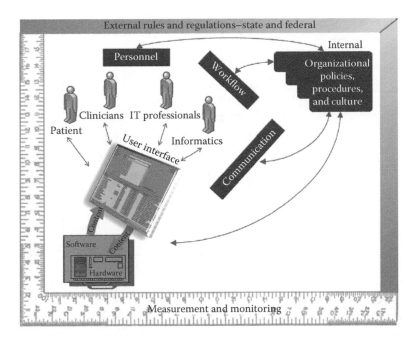

FIGURE 14.1
Illustration of the complex interrelationships between the eight dimensions of the new sociotechnical model. (From Sittig, D.F. and Singh, H., *Qual. Safety Health Care*, 19, i68, 2010.)

is stratified into five levels of analysis: *individual, teams, technology/task, work environment*, and *systems* (Sittig and Singh, 2010). This study is similar to Vincent's human factors framework (Vincent et al., 1998) for the study of risk and safety in medicine, with the omission of the *patient* level of analysis as simulation generally excludes the patient from the experience.

14.2.2 Assessment of Mobile Health-Care Systems

Assessment is a systematic determination for identifying the feasibility, value, and quality of the technology, policy, and program using different methods, strategies, rules, and regulation. In the health-care area, assessment can be a comparison between particular clinical issues such as "psychotherapy versus drug treatment for mental disorders" or it can be an assessment of a program or system for providing health-care service. On the other hand, assessment can focus on both sides of the matters (Telemedicine and Medicine, 1996). The main purpose of system assessment is to measure the particular issue of the system that has an impact on the number of successful events generated by that system. In the telemedicine area, *assessment criterion* is used as a measurement, standard, and indicator to judge or depict the outcome of the health-care system. Patient satisfaction, quality of care, quality of service, accessibility, cost, etc., are some examples of evaluation criteria. Although many health-care assessments concentrate on individual patient care, assessment of the entire population or a specific vulnerable group needs to be more accurate (Cronholm and Goldkuhl, 2003).

Telemedicine uses information and communication technologies in order to provide health care to the beneficiaries over a long distance. Such systems use advanced communication technologies to support various applications. Although telemedicine provides unique services, it is still lacking in providing evidence pertaining to its effectiveness. Thus, assessment of telemedicine practicality, value, and affordability is needed in order to improve the performance of telemedicine. Researchers are interested in the assessment

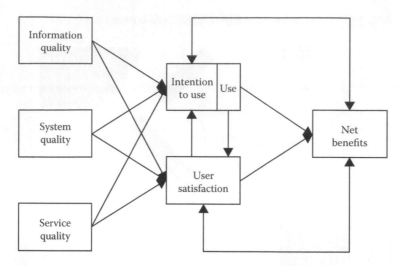

FIGURE 14.2
Updated D&M IS success model. (From DeLone, W.H. and McLean, E.R., *J. Manage. Inform. Syst.*, 19, 9, 2003.)

of information and communication technologies in telemedicine with the rapid technological growth in telecommunication technologies. Telemedicine assessment is needed to identify the effects, values, feasibility, or other qualities of technologies, programs, and principles in this area (Field, 1996). Assessment of health-care systems could be useful not only for studying systems or clinical performance but also for effectively answering the question: "To what extent will such systems be used or how efficiently will they be used?"

Many researchers have been conducted to identify parameters inherent on the success of information system (IS). As a result, IS success has been classified into system quality, information quality, intention to use/usage, user satisfaction, individual impact, and organizational impact (DeLone and McLean, 1992). Shannon and Weaver (1948) allocated efficiency and accuracy of the system that provide information to the technical level, success of the information in carrying the meaning to the semantic level, and information effect on the receiver to effectiveness level (Shannon and Weaver, 1948). Mason (1978) defined influence as occurred events that take place during the end of IS (Mason, 1978). These influences consist of information evaluation, information application, and information receipt that might change recipient behaviors and system performance. Because of changes in the management and role of ISs, DeLone and McLean (2003) had updated the success model as shown in Figure 14.2.

In the updated model, quality has three major elements, which include *information quality*, *service quality*, and *system quality*. These elements must be controlled separately as they have impact on *use* and user *satisfaction* elements. They allocated *intention to use* element as an attitude, whereas *use* as a behavior. Some researchers used *intention to use* element but the majority of the researchers prefer to utilize *use* element as a link between attitude and behavior that is difficult to measure. *Use* and *user satisfaction* dimensions are interrelated, but greater *user satisfaction* will be defined with a positive experience of *use* dimension; likewise, *user satisfaction* element has the same impact on *use* and *intention to use* (DeLone and McLean, 2003). In addition, these kinds of *use* and *user satisfaction* elements will make a certain *net profit*.

14.2.3 Mobile Computing in Health Care

Modern technologies promote the progress in health care and reform the current healthcare systems. Utilizing mobile applications in health care can be considered as emerging and enabling technologies that have been used in many countries for emergency, general care, monitoring, and so on (Ammenwerth et al., 2003; Haux, 2006). For instance, various wireless technologies such as mobile computing, wireless network, and global positioning systems (GPSs) have been utilized in Sweden for ambulance care (Geier, 2001). Another example of using these technologies is in emergency trauma care in the Netherlands (Jan ten Duis and van der Werken, 2003).

There are several important sociotechnical factors in order to determine the success of modern technologies in health care. For instance, "how modern technology is compatible with healthcare practitioners' current working conditions; what kinds of training programs, resources and support were provided; what incentives were used to get health-care professionals to use the system" (Ammenwerth et al., 2003). Moreover, some studies indicated that "mobile device size, access procedures, ease of use, mobile interface, and training and support" are the most important factors for new application usage in health care (Gebauer and Shaw, 2004; Lee and Benbasat, 2004). Later, Wu et al. (2007) presented a revised technology acceptance model to examine what determines the acceptance of MHS by health-care professionals. Identification of the factors that are likely to be significant antecedents of planning and implementing mobile health care to enhance professionals' MHS acceptance was conducted by Wu et al. (2007). The results indicated that perceived usefulness, perceived ease of use, compatibility, and MHS self-efficacy are important determinants to users' behavioral intent (Wu et al., 2007).

14.3 Needs and Demands for Improving Mobile Health-Care Performance

With advancement of communication technologies, health-care systems have been dramatically improved in the past few years. Delivery of health care depends on the variety of people that interact with each other as well as using different technologies and services. Advancement of technologies has in turn increased the complexity of health-care systems that plays a significant role in the multiple vulnerabilities, failures, errors, and finally the performance of those health-care systems (Kohn et al., 2000). As complexity is inherent in health care, it is important to identify the multiple system elements, their interaction, and their impact on quality of care, performance of the health-care system, as well as understanding the key adaptive role of people in the system.

14.4 Classification of Sociotechnical Factors Affecting on Mobile Health-Care Performance

Although mobile-based health-care systems exist for many years, methods to assess the output and outcome of those systems are still challenges for decision makers or those who want to measure the performance of such systems (Ammenwerth et al., 2004; Kaplan, 2001). Many sociotechnical factors have a strong effect on the performance and success of MHS. Understanding of those factors can assist health-care providers to recognize user need and to improve the design of MHS. For this purpose, we have identified and classified the effective sociotechnical factors in MHS based on the previous studies. Figure 14.3 depicts an overview of the classification of sociotechnical factors in MHS.

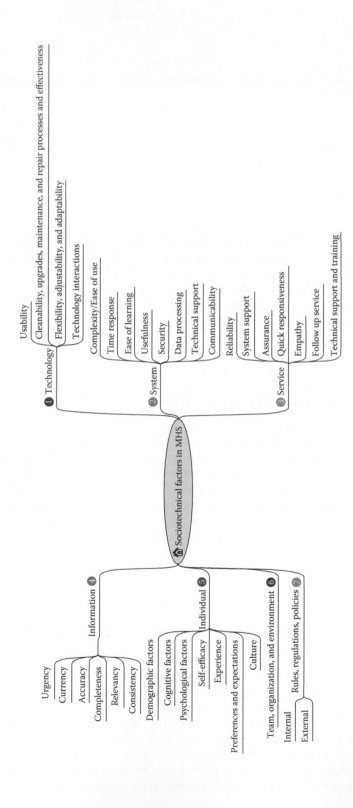

FIGURE 14.3

Classification of sociotechnical factors in MHS.

14.4.1 Technology

The introduction of information technology (IT) in health care has a deep effect on the performance of the health-care systems. Many researchers have studied IT adoption and diffusion recently. Adoption and diffusion is defined as the decision to accept or invest in a new technology. The key objective of those IT researches is to access the value of the IT to the organization or project. Thus, organizations can manage their IT resources better and enhance overall effectiveness. Another approach is to examine the determination of IT adoption and usage by individual users (Masrom and Hussein, 2008). With the advancement of new technologies especially in health care, health-care services such as monitoring patients anytime from anywhere, rescuing patients in emergency, tracking the patient's current location, and many other services become possible. MHS involve the use of superior and reliable communication techniques to deliver biomedical signals over long distances. As MHS is based on the new technologies, technology itself is one of the core factors in the success of MHS (Wu et al., 2007). The main sociotechnical factors associated with technology are as follows: (1) usability; (2) clean ability, upgrades, maintenance, and repair processes and effectiveness; (3) flexibility, adjustability, and adaptability; (4) technology interactions; and (5) accessibility as depicted in Table 14.1 (LeBlanc et al., 2011).

14.4.2 System

As mentioned by DeLone and McLean, *system quality* is one of the most important factors in the success of an IS. With higher system quality comes higher user satisfaction and increased use of the system, which in turn can be the concluding factors of a successful IS (DeLone and McLean, 2003; Pérez-Mira, 2010). System quality here refers to the collaboration of software and hardware to measure how well they work together. Some of the significant factors, which may affect system quality for MHSs, are (1) complexity/ease of use, (2) time response, (3) ease of learning, (4) usefulness, (5) security, (6) data processing, (7) technical support, and (8) communicability as shown in Table 14.2.

Some parameters have an important role in system quality. *Ease of use* is one of the effective factors in system quality defined as user satisfaction, convenience, and pleasure from using the health-care system. *Ease of learning* is another important factor in system quality that depicts whether the system is easy to learn by users or not. Another factor that will influence on the quality of the system is usefulness of the system. The studies on system quality are often associated with system performance. Examples of system quality measures are ease of use, ease of learning, response time, *usefulness*, availability, reliability, completeness, system flexibility, and easy access to help (Anderson and Aydin, 2005; Lippeveld, 2001; Seddon et al., 1994; Smith, 1999).

TABLE 14.1

Important Sociotechnical Factors Associated with Technology

Assessment Factor	References
Usability	LeBlanc et al. (2011), Subasi et al. (2009), Leitner et al. (2007), Islam et al. (2007), Chien et al. (2012), Dahl et al. (2009)
Clean ability, upgrades, maintenance, and repair processes and effectiveness	LeBlanc et al. (2011)
Flexibility, adjustability, and adaptability	LeBlanc et al. (2011)
Technology interactions	LeBlanc et al. (2011)
Accessibility	Subasi et al. (2009), Waldorf and Che (2010), Tanser et al. (2006), Paez et al. (2010)

TABLE 14.2

Important Sociotechnical Factors Associated with System

Assessment Factor	References
Complexity/ease of use	Wu et al. (2007), Anderson and Aydin (2005), Lippeveld (2001), Smith (1999), Seddon et al. (1994)
Time response	Varshney and Vetter (2002), Pérez-Mira (2010), Schewe (1976), Miller (1968), Conklin et al. (1982), Srinivasan (1985), Nickerson (1969), Au-Yeung (2007)
Ease of learning	Anderson and Aydin (2005), Lippeveld (2001), Smith (1999), Seddon et al. (1994)
Usefulness	Wu et al. (2007), Anderson and Aydin (2005), Lippeveld (2001), Smith (1999), Seddon et al. (1994)
Security	Yusof et al. (2006)
Data processing	Chatterjee et al. (2009), Lu et al. (2005), Lapinsky et al. (2001)
Technical support	Yusof et al. (2006)
Communicability	Nilakanta and Scamell (1990), Minnick et al. (1994), O'Connor et al. (2011), Gebauer and Shaw (2004), Chatterjee et al. (2009), Riemer and Frobler (2007)

There is a positive relationship between the number of *communication* channels in an IS and user satisfaction (Nilakanta and Scamell, 1990). Wireless devices were shown to improve nurses' satisfaction in a Chicago hospital by providing multiple channels of communication (Minnick et al., 1994). O'Connor et al. (2011) stated that network availability and network stability affect system quality (O'Connor et al., 2011). There is evidence that mobile devices are suitable for supporting synchronous and asynchronous communication, especially in a health-care setting (Chatterjee et al., 2009; Gebauer and Shaw, 2004). Riemer and Frobler (2007) argued that users prefer to use technologies that offer multiple channels of communication.

Data processing capability will affect the use of the system as well as satisfaction among users (Chatterjee et al., 2009). The data processing capability of a mobile device has been alleged to affect use of the technology (Lapinsky et al., 2001; Lu et al., 2005). Additionally, higher data processing capability of a PDA has a positive impact on health care as it offers stronger decision-making support for experts in this area (Baumgart, 2005). Decision support capability of a mobile device will increase user satisfaction (Chatterjee et al., 2009).

Response time, which is typical networking statistic (Varshney and Vetter, 2002), will influence system quality (Pérez-Mira, 2010; Schewe, 1976). Other researchers argued that response time delays will change user behavior and decrease user satisfaction (Conklin et al., 1982; Miller, 1968; Srinivasan, 1985). It is a vital fact that the response time is essential for any interactive system. The faster the response time, the better the impression it will make on users and thus affect the user's behavior, as delay is considered a *psychological factor* in the user's mind (Nickerson, 1969). In health-care systems, especially for patients who need vital care, treatment must be given as early as possible (Au-Yeung, 2007).

Security and technical support are two more important factors that will affect health-care system quality. The health-care environment requires unique security and management in order to protect patient information, patient privacy monitoring, and secure access to network resources and ensures network availability for everyone. Technical support is needed for reducing response time, improving communication services, reducing unplanned downtime, and enhancing usability throughout. Generally, technical support can improve health-care system performance. Both of these factors will positively affect the use of the health-care systems as well as user satisfaction (Yusof et al., 2006). Based on DeLone and McLean (2003), the use of the system and user satisfaction have a positive effect on the success of the health-care systems (DeLone and McLean, 2003). Thus, security and technical support are considered as two important factors that could significantly contribute to the success of health-care systems.

14.4.3 Service

Service quality is another factor that has a positive impact on the use of an IS and user satisfaction (Pérez-Mira, 2010). ICT and especially mobile applications facilitate health-care system delivery around the world. Health-care systems became more accessible and affordable by using electronic services. The quality of such services is a main concern for health-care systems (Akter and Ray, 2010; Akter et al., 2010a,b). The quality of mobile health services has been defined as the user's opinion about the overall excellence or superiority of the system (Akter et al., 2010a, Zeithaml, 1987). Some of the effective factors in health-care service quality are illustrated in Table 14.3. These factors consist of (1) reliability, (2) system support, (3) assurance, (4) quick responsiveness, (5) empathy, (6) follow-up service, and (7) technical support and training.

User satisfaction has been shown to be positively affected by the *reliability* of the system and the availability of the particular service to the users (DeLone and McLean, 2003). Lack of reliability and efficiency negatively affect the quality of the services (Akter and Ray, 2010; Akter et al., 2010b). Most traditional health-care systems focus on service quality, and to assess it they often use the SERVQUAL model, which consists of five dimensions: reliability, responsiveness, assurance, empathy, and tangibles (Parasuraman et al., 1986). Akter et al. (2010) discovered the four main themes (system reliability, system availability, system efficiency, and system privacy) in their qualitative results that can determine a customer's assessment of the quality of health-care system (Akter et al., 2010a). Some of the important aspects that need to be taken care of are reliability, critical performance, and time-sensitive nature pertaining to the medical information associated with the health-care industry during their deployment phase (Istepanian et al., 2004).

Another component of user satisfaction and use of the system is *system support*. Poor user support may result in losing of customers (DeLone and McLean, 1992, 2003). Full commercial support of PDA software has been claimed as a key reason for widespread usage in the mobile technology environment (Gillingham et al., 2002). PDAs have become highly functional and useful in health care. Lu et al. (2005) argued that technical and organizational support of PDAs will improve their usability in enhancing clinical practice (Lu et al., 2005).

TABLE 14.3

Important Sociotechnical Factors Associated with Service

Assessment Factor	References
Reliability	DeLone and McLean (2003), Akter and Ray (2010), Akter et al. (2010 a,b), Parasuraman et al. (1986), Istepanian et al. (2004)
System support	DeLone and McLean (1992, 2003), Gillingham et al. (2002), Lu et al. (2005)
Assurance	Parasuraman et al. (1986), Lin and Hsieh (2011), Büyüközkan and Çifçi (2012)
Quick responsiveness	Yusof et al. (2006)
Empathy	Yusof et al. (2006)
Follow-up service	Yusof et al. (2006)
Technical support and training	Wu et al. (2007)

As mentioned earlier, *assurance*, which is related to the reputation and competence of the health-care system, is one of the five dimensions of the SERVQUAL model (Parasuraman et al., 1986). In their qualitative and quantitative study, Lin and Hsieh (2011) have developed a 20-item 7D SSTQUAL scale to assess self-service technology (SST) services. This SSTQUAL scale also includes assurance, along with functionality, enjoyment, security, design, convenience, and customization. Assurance has been defined as the *reputation and competence* of the SST provider (Lin and Hsieh, 2011). In the past, the service quality assurance dimension has been defined as the "knowledge and courtesy of employees and their ability to inspire trust and confidence." In order to evaluate health-care service quality, Büyüközkan and Çifçi (2012) utilized the six dimensions of tangibles, responsiveness, reliability, information quality, assurance, and empathy (Büyüközkan and Çifçi, 2012).

Some other factors such as *quick responsiveness, empathy, follow-up service, technical support, and training* can be considered as service quality dimensions. These factors have a positive effect on the use of the systems as well as user satisfaction and finally will affect on the system success (Wu et al., 2007; Yusof et al., 2006).

14.4.4 Information

Information quality is another important factor in the IS success model. DeLone and McLean define information quality as the quality of the information that has been produced by the specific IS. The quality of an IS can be proved if the system output is accurate, current, timely, reliable, concise, and relevant (Bailey and Pearson, 1983; Huh et al., 1990; Nelson and Todd, 2005). The important factors in information quality are as follows: (1) urgency, (2) currency, (3) accuracy, (4) completeness, (5) relevancy, and (6) consistency as listed in Table 14.4.

Immediacy of feedback, which is the ability of a system to support rapid bidirectional communication, contributes as a vital parameter for any typical communication system. Once a user is engaged with a specific technology, immediacy of feedback or interacting *on-demand* is necessary to satisfy the user (Daft and Lengel, 1986; Dennis and Valacich, 1999; Gebauer and Shaw, 2004). *Urgency* of the task is crucial especially in the health-care field wherein workers need to deal with emergency contingencies and work under time pressure (Berg, 1999; Chatterjee et al., 2009). Some researchers claim that mobile devices are highly suitable for the performance of tasks, which require immediate access to information (Cousins and

TABLE 14.4
Important Sociotechnical Factors Associated with Information

Assessment Factor	References
Urgency	Dennis and Valacich (1999), Gebauer and Shaw (2004), Daft and Lengel (1986), Berg (1999), Chatterjee et al. (2009), Cousins and Robey (2005), Kakihara and Sorensen (2001), Perry et al. (2001)
Currency	Bailey and Pearson (1983), King and Epstein (1983), Molla and Licker (2001), Wang and Liu (2007), Yun et al. (2006)
Accuracy	Bailey and Pearson (1983), Mahmood (1987), Miller and Doyle (1987), Gorla et al. (2010), Huh et al. (1990), Nelson and Todd (2005), Jan et al. (2004), Liao and Kao (2008), Tibballs and Weeranatna (2010), Moule (2000)
Completeness	Yusof et al. (2006)
Relevancy	Yusof et al. (2006)
Consistency	Yusof et al. (2006), Southon et al. (1997)

Robey, 2005; Kakihara and Sorensen, 2001). Furthermore, mobile device have the ability to support task performance anytime and anywhere (Perry et al., 2001).

According to Bailey and Pearson (1983) and King and Epstein (1983), *currency* of information, which is due to real-time services, influences the results of IS assessment (Bailey and Pearson, 1983; King and Epstein, 1983). An e-commerce system needs to reduce customer uncertainty in terms of information quality. In order to do so, information must be relevant, understandable, and complete as well as current (Molla and Licker, 2001). Wang and Liu (2007) utilized currency as one of their initial technical quality criteria and measurable indicators by which to evaluate the quality of health information on the Internet. They defined currency for health-care information as *date of creation* and *date of last update* (Wang and Liu, 2007). To manage an efficient location-based service, a real-time GIS platform is needed to deal with the dynamic status of moving objects (i.e., the patient) (Yun et al., 2006). Especially in the case of emergency, ambulances must reach a patient's location rapidly (Schmid and Doerner, 2010).

Accuracy is a significant dimension of information quality (Bailey and Pearson, 1983; Gorla et al., 2010; Mahmood, 1987; Miller and Doyle, 1987). Accuracy can be defined as "an agreement with an attribute about a real world entity, a value stored in another database, or the result of an arithmetic computation" (Gorla et al., 2010). According to Huh et al. (1990), accuracy, completeness, consistency, and currency are the four dimensions of an IS (Huh et al., 1990). Nelson and Todd (2005) used the combination of accuracy, completeness, currency, and format as measures of information quality (Nelson and Todd, 2005). Since accuracy is important in a positioning system, Jan et al. (2004) proposed a cell-based positioning scheme to improve the accuracy of a positioning system (Jan et al., 2004), and Liao and Kao (2008) designed a novel method of utilizing mobile user orientation information to improve prediction accuracy (Liao and Kao, 2008). Time and accuracy are two important factors of health care, especially in cardiac arrest diagnosis (Moule, 2000; Tibballs and Weeranatna, 2010).

Two other criteria used for information quality are *completeness* of the information and their *relevancy*. The information transferred by a patient's mobile device to the hospital must be complete and relevant. On the other hand, the information sent by hospitals to the doctors, physicians, and patients need to be complete and relevant as well (Yusof et al., 2006). Another important factor in information quality is *consistency* of the information. Health-care providers must ensure the consistency of information in direction and interactive between all levels of the health-care systems (Southon et al., 1997; Yusof et al., 2006).

14.4.5 Individual

Another important factor that will affect the performance of MHS is individual factors such as (1) demographic factors (age, gender, ethnicity, socioeconomic, and cultural), (2) cognitive factors (memory, attention, vigilance, reaction time, information processing, problem solving, error propensity, recognition, and recovery), (3) psychological factors, (3) self-efficacy, (4) experience, and (5) preferences and expectations (LeBlanc et al., 2011). Table 14.5 depicts the list of sociotechnical factors associated with the individual.

Yu et al. (2009) focused on the antecedent variables, including social influence factors such as subjective norm and image, that were examined together with demographic variables including age, job level, long-term care work experience, and computer skills with regard to their impact on caregivers' acceptance of health. Based on their results, all of the factors had effect on the acceptance of a health IT application except age and long-term care work experience (Yu et al., 2009).

TABLE 14.5

Important Sociotechnical Factors Associated with Individuals

Assessment Factor	References
Demographic factors	Yu et al. (2009)
Cognitive factors	LeBlanc et al. (2011)
Psychological factors	Rubin et al. (2008), Powell-Cope et al. (2008)
Self-efficacy	Wu et al. (2007)
Experience	LeBlanc et al. (2011)
Preferences and expectations	LeBlanc et al. (2011)
Culture	Zakour (2004), Ebrahimi et al. (2010)

14.4.6 Teams, Organization, and Environment

Another effective factor in the performance of mobile health care is *team and organizational factors* such as interpersonal communication, teamwork skills and capabilities (leadership/followership, resource allocation, situation awareness, coordination), teamwork competences/processes (conflict and conflict resolution, team workload and cognition), and team composition (culture, gender, expertise, experience, professions, hierarchy) (LeBlanc et al., 2011).

Each staff member in health-care community can be considered as part of the health-care team within their unit, organization, hospital, etc. The impact of each member on the patient and finally health-care system performance is influenced by other members of the team and their communication, support, supervision, and so on. Management actions and decisions have a strong impact on the team. In addition, the team's environment is another impressive factor in the performance of the health-care systems as it has an effect on the team. In turn, an organization could be affected by external environmental factors such as commercial and financial issues (Vincent et al., 1998). Consequently, teams, organization, and environment are complementary factors in the performance of health-care systems especially in mobile-based health-care systems as communication between team members, teamwork, and the environment are necessary. In mobile-based health-care systems, professionals need to access and input medical or patient information from anywhere, at any time. They need to communicate with each other in a proper environment in order to save patients' lives. Thus, with a good teamwork and in a healthy environment, MHS can facilitate efficient and effective patient care information input and access at the point of patient care (Wu et al., 2007).

14.4.7 Rules, Regulations, and Policies

Rules, regulations, and policies incorporate with (1) internal policies, procedures, and culture and (2) external rules, regulations, and pressures (Sittig and Singh, 2010). *Internal organizational policies, procedures, and culture* will affect other sociotechnical factors. For instance, based on the internal organizational policies, procedures, and culture, when the budget will be allocated for each organization, the leader will decide how the budget must be divided to purchase hardware, software, etc. The organizational leaders are responsible for this task and they must define rules and regulation for implementation, use, monitoring, and assessment of the health-care systems. It is compulsory to ensure that the workflow engaged with operating those health-care systems is consistent with policies and procedures. Internal policies, procedures, and culture usually are created in response to the external rules and regulations (Sittig and Singh, 2010). *External rules, regulations, and pressures* are those factors that will be assigned from outside of the organization mostly from the government. These kinds of rules and regulations have an effect on the internal organizational policies, procedures, and culture. For instance, there are few national developments that will affect the delivery of health-care systems including (1) "the initiative to develop the data and information exchange capacity to create a national health information network" (Office of the National Coordinator for Health Information Technology, Department of Health and Human Services, 2009); (2) "the initiative to enable patients to access copies of the clinical data via personal health records" (Sittig, 2002); and (3) "clinical and IT workforce shortages" (Detmer et al., 2010).

14.5 Sociotechnical Model for Mobile Health-Care System

As MHSs are playing a critical role in human life, assessment is required for this kind of systems in order to prove their success and failure. Traditional design analysis typically focuses on the system assessment from the technology point of view, whereas sociotechnical analysis looks at how technology will incorporate into the user's activities. From the perspective of sociotechnical aspect, social and technical issues are complementary and they cannot be separated. Hence, based on the previous studies, we classify the sociotechnical factors to seven groups including (1) technology; (2) system; (3) service; (4) information; (5) individual; (6) teams, organization, and environment; and (7) rules, regulations, and policies. Based on these components, we design a model for sociotechnical factors in MHS as illustrated in Figure 14.4.

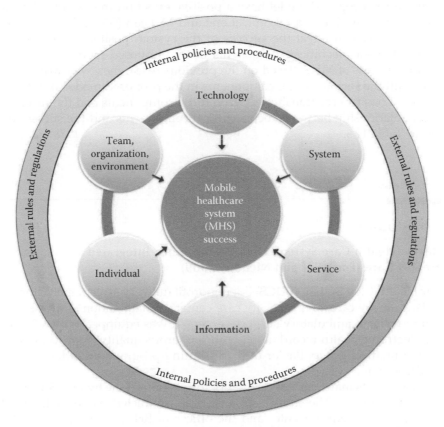

FIGURE 14.4
Proposed sociotechnical model for MHS.

TABLE 14.6

Positive Relationships of the Sociotechnical Factors and MHS Success

Positive Relationship	References
Technology → MHS success	LeBlanc et al. (2011), Wu et al. (2007), Holden and Karsh (2010)
System → MHS success	LeBlanc et al. (2011), DeLone and McLean (2003), Chatterjee et al. (2009), Jen and Chao (2008)
Service → MHS success	DeLone and McLean (2003), Chatterjee et al. (2009)
Information → MHS success	DeLone and McLean (2003), Chatterjee et al. (2009), Jen and Chao (2008)
Individual → MHS success	LeBlanc et al. (2011), Yusof et al. (2006), DeLone and McLean (2003), Jen and Chao (2008)
Teams, organization, and environment → MHS success	LeBlanc et al. (2011), Yusof et al. (2006), DeLone and McLean (2003), Jen and Chao (2008)
Rules, regulations, and policies → MHS success	Sittig and Singh (2010)

These key sociotechnical factors influence the performance and success of MHS. The factors must be studied in relation to each other and they cannot be viewed independently. In the proposed model, some components are more tightly coupled than others are. For instance, technology and system, service, and information are dependent on one another and other social components mostly depend on these components. Most of the components in the proposed model have a positive effect on the use of MHS as well as user satisfaction from MHS. Based on DeLone and McLean (2003), the *use* of the system and user *satisfaction* will affect the success of the system positively. Therefore, we can conclude that the sociotechnical components in the proposed model influence the performance and success of MHS. Based on the previous researches, we clarify the relationship between all the sociotechnical components in the proposed model and the success of MHS. We find a positive relationship between those components and the success of MHS. The positive relationship between sociotechnical components and the success of MHS is illustrated in Table 14.6.

14.6 Case Studies

14.6.1 Baylor Health Care System: High-Performance Integrated Health Care (Emswiler and Nichols, 2009)

The Baylor Health Care System (BHCS) is a nonprofit delivery system including a network of hospitals, primary care, and specialty care centers, rehabilitation clinics, senior health centers, and affiliated ambulatory surgery centers. It was equipped with a 3000-bed system using electronic health record along with numerous quality improvement tools and 450-physician medical group. Baylor utilizes training programs in order to improve the quality of the system.

This study presents many successful quality innovations done by BHCS as well as organizational achievements. The study's site visit was conducted on April 10 and 11, 2008, at Baylor University Medical Center and the offices of BHCS's Institute for Health Care Research and Improvement in Dallas, Texas. Firstly, the researchers start the observation of BHCS from historical and organizational perspective. Then they explore the BHCS's quality infrastructures. Additionally, they examine training programs as well as physicians and doctors. They tried to assess BHCS's quality innovations and utilized the lesson learned to improve the delivery of care. BHCS starts to measure the actual delivery of clinical preventive services (CPSs).

BHCS was recognized and appreciated by many patients in 2008. It was awarded the annual National Quality Healthcare Award from the National Quality Forum and Modern Healthcare. Moreover, it received the Leapfrog Patient-Centered Care Award for the engagement and leadership of the BHCS Board in patient-centered care. A few weeks later, Steve Hines and Maulik Joshi published an article, *The Variation in Quality of Care Within Health Systems*, naming BHCS the third highest-performing system in the United States in quality performance, out of 73 systems ranked (Hines and Joshi, 2008). HealthTexas Provider Network won the 2008 Top Leadership Teams in Healthcare Award in July for the medical group practices category because of their superb leadership teamwork. The quality of care has always been an integral part of the BHCS vision of and commitment to best care. Ultimately, health-care quality improvement at BHCS is about leadership in the drive to provide safe, timely, effective, efficient, equitable, and patient-centered medical care.

14.6.2 Trauma Care Systems in the Netherlands (Jan ten Duis and van der Werken, 2003)

Dutch trauma surgeons (Dutch Trauma Society) were concerned about the quality of care to (multi)trauma patients, in the prehospital as well as the in-hospital setting as they saw a lack of standardization in delivery systems for trauma care, especially for patients with multiple injuries. Thus, public health inspectorate and the government decided to enhance teaching and training, reorganization, regionalization, and implementation where all partners in trauma care were involved. They discovered the important factors in quality improvement including the regionalization of ambulance care, the introduction of mobile medical teams, the availability of trauma helicopters, the categorization of hospitals, the designation of trauma centers, and the given responsibility of these centers in the regionalization of trauma care. All of these factors will affect not only the individual organizations but also all of the entire chain of trauma care.

Moreover, nationwide trauma registration database has been designed to include all in-hospital trauma patients. Although the quality of care was improved, shortage of intensive care beds, the impossibility to use the helicopter service at night, the shortage in the number of mobile medical teams at night, and the slowness in executions of agreements between contracting parties remain as a serious concern. This shows that not only medical partners and hospitals but all government and insurance companies should take their responsibility in this.

14.7 Future Directions

All dimensions of the proposed sociotechnical model for MHS establish a new paradigm for the study of MHS. Assessment of the system can be done before the final stage of implementation by considering sociotechnical factors in order to improve the quality and safety in MHS. We anticipate that additional study of the proposed sociotechnical model and their complex interactions will yield further refinements to this model and, ultimately, improvements in the quality and safety of the MHS that translate to better health and welfare for the patients. Moreover, identification of sociotechnical factors has significant potential to generate important knowledge that will inform efforts to improve the education and training of health-care providers, enhance new technological design, and optimize work environments and health-care systems. The proposed sociotechnical model can be expanded in the future. Some factors such as trust, risk, safety, and anxiety can be added to the model to enhance the proposed model. Moreover, national culture dimensions can be assessed to understand cultural differences effect in MHS. Finally, sociotechnical factors can be classified based on the model of MHS in the future.

14.8 Concluding Remarks

All dimensions of the proposed sociotechnical model for MHS establish a new paradigm for the study of MHS. Assessment of the system can be done before the final stage of implementation by considering sociotechnical factors in order to improve the quality and safety in MHS. We anticipate that additional study of the proposed sociotechnical model and their complex interactions will yield further refinements to this model and, ultimately, improvements in the quality and safety of the MHS that translate to better health and welfare for the patients. Moreover, identification of sociotechnical factors has significant potential to generate important knowledge that will inform efforts to improve the education and training of health-care providers, enhance new technological design, and optimize work environments and health-care systems. The proposed sociotechnical model can be expanded in the future. Some factors such as trust, risk, safety, and anxiety can be added to enhance the proposed model. Moreover, national culture dimensions can be assessed to understand cultural differences effect in MHS. In conclusion, sociotechnical factors can be classified based on the model of MHS in the future.

Questions

1. What are the key sociotechnical factors in MHS?
2. What are the important elements of technology that are effective in the success of MHS?
3. What are the important elements of a system that are effective in the success of MHS?
4. What are the important elements of service that are effective in the success of MHS?
5. What are the important elements of information that are effective in the success of MHS?
6. What are the key individual factors that are effective in the success of MHS?
7. How can sociotechnical factors affect the success of MHS?

References

Akter, S., D'Ambra, J., and Ray, P. 2010a. Service quality of mHealth platforms: Development and validation of a hierarchical model using PLS. *Electronic Markets*, 20, 209–227.

Akter, S., D'Ambra, J., and Ray, P. 2010b. User perceived service quality of m-health services in developing countries. *18th European Conference on Information Systems*. Pretoria, South Africa: University of Pretoria. http://ro.uow.edu.au/cgi/viewcontent.cgi?article=4188&context=commpapers

Akter, S. and Ray, P. 2010. mHealth—An ultimate platform to serve the unserved reviews. *IMIA Yearbook*, pp. 94–100, Stuttgart, Germany: Schattauer.

Ammenwerth, E., Brender, J., NykäNen, P., Prokosch, H.-U., Rigby, M., and Talmon, J. 2004. Visions and strategies to improve evaluation of health information systems: Reflections and lessons based on the HIS-EVAL workshop in Innsbruck. *International Journal of Medical Informatics*, 73, 479–491.

Ammenwerth, E., Gräber, S., Herrmann, G., Bürkle, T., and König, J. 2003. Evaluation of health information systems—Problems and challenges. *International Journal of Medical Informatics*, 71, 125–135.

Anderson, J. and Aydin, C. 2005. Overview: Theoretical perspectives and methodologies for the evaluation of healthcare information systems. *Evaluating the Organizational Impact of Healthcare Information Systems*. New York: Springer.

Au-Yeung, S.W.M. 2007. *Response Times in Healthcare Systems*. London, U.K.: Imperial College.

Bailey, J. and Pearson, S. 1983. Development of a tool for measuring and analyzing computer user satisfaction. *Management Science*, 29, 530–545.

Baumgart, D.C. 2005. Personal digital assistants in health care: Experienced clinicians in the palm of your hand? *The Lancet*, 366, 1210–1222.

Berg, M. 1999. Patient care information systems and health care work: A sociotechnical approach. *International Journal of Medical Informatics*, 55, 87–101.

Büyüközkan, G. and Çifçi, G. 2012. A combined fuzzy AHP and fuzzy TOPSIS based strategic analysis of electronic service quality in healthcare industry. *Expert Systems with Applications*, 39, 2341–2354.

Chatterjee, S., Chakraborty, S., Sarker, S., Sarker, S., and Lau, F.Y. 2009. Examining the success factors for mobile work in healthcare: A deductive study. *Decision Support Systems*, 46, 620–633.

Chien, T.-N., Hsieh, S.-H., Cheng, P.-H., Chen, Y.-P., Chen, S.-J., Luh, J.-J., Chen, H.-S., and Lai, J.-S. 2012. Usability evaluation of mobile medical treatment carts: Another explanation by information engineers. *Journal of Medical Systems*, 36, 1327–1334.

Conklin, J.H., Gotterer, M.H., and Rickman, J. 1982. On-line terminal response time: The effects of background activity. *Information and Management*, 5, 169–173.

Cousins, K.C. and Robey, D. 2005. Human agency in a wireless world: Patterns of technology use in nomadic computing environments. *Information and Organization*, 15, 151–180.

Cronholm, S. and Goldkuhl, G. 2003. Strategies for information systems evaluation—Six generic types. *Electronic Journal of Information Systems Evaluation*, 6, 65–74.

Daft, R.L. and Lengel, R.H. 1986. Organizational information requirements, media richness and structural design. *Management Science*, 32, 554–571.

Dahl, Y., Alsos, O., and Svanæs, D. 2009. Evaluating mobile usability: The role of fidelity in full-scale laboratory simulations with mobile ict for hospitals. In: Jacko, J. (ed.) *Human-Computer Interaction. New Trends*. Berlin, Germany: Springer.

DeLone, W.H. and McLean, E.R. 1992. Information system success: The quest for dependent variable. *Information System Research*, 3, 60–95.

DeLone, W.H. and McLean, E.R. 2003. The DeLone and McLean model of information systems success: A ten-year update. *Journal of Management Information Systems*, 19, 9–30.

Dennis, A.R. and Valacich, J.S. 1999. Rethinking media richness: Towards a theory of media synchronicity. In *Proceedings of the 32nd Annual Hawaii International Conference on System Sciences, 1999. HICSS-32*. 10 pp.

Detmer, D.E., Munger, B.S., and Lehmann, C.U. 2010. Medical informatics board certification: History, current status, and predicted impact on the medical informatics workforce. *Applied Clinical Informatics*, 1, 11–18.

Ebrahimi, N., Singh, S.K.G., and Tabrizi, R.S. 2010. Cultural effect on using new technologies. *World Academy of Science, Engineering and Technology*, 46, 901–905.

Emswiler, T. and Nichols, L.M. 2009. *Baylor Health Care System: High-Performance Integrated Health Care*. New York: Commonwealth Fund.

Field, M.J. 1996. *Telemedicine A Guide to Assessing Telecommunications in Health Care*. Washington, DC: National Academy Press.

Gebauer, J. and Shaw, M.J. 2004. Success factors and impacts of mobile business applications: Results from a mobile e-procurement study. *International Journal of Electronic Commerce*, 8, 19–41.

Geier, J. 2001. Saving lives with roving LANs-Swedish ambulance becomes a true mobile platform. Network World, (Retrieved march 27, 2006 from http://www.networkworld.com/reviews/2001/0205bgside.html).

Gillingham, W., Holt, A., and Gillies, J. 2002. Hand-held computers in health care: What software programs are available? *The New Zealand Medical Journal*, 115(1162): U185.

Gorla, N., Somers, T.M., and Wong, B. 2010. Organizational impact of system quality, information quality, and service quality. *Journal of Strategic Information Systems*, 19, 207–228.

Haux, R. 2006. Health information systems—Past, present, future. *International Journal of Medical Informatics*, 75, 268–281.

Hines, S. and Joshi, M.S. 2008. Variation in quality of care within health systems. *Joint Commission Journal on Quality and Patient Safety/Joint Commission Resources*, 34, 326–332.

Holden, R.J. and Karsh, B.-T. 2010. The technology acceptance model: Its past and its future in health care. *Journal of Biomedical Informatics*, 43, 159–172.

Huh, Y.U., Keller, F.R., Redman, T.C., and Watkins, A.R. 1990. Data quality. *Information and Software Technology*, 32, 559–565.

Islam, R., Ahamed, S., Talukder, N., and Obermiller, I. 2007. Usability of mobile computing technologies to assist cancer patients. In: Holzinger, A. (ed.) *HCI and Usability for Medicine and Health Care*. Berlin, Heidelberg, Germany: Springer.

Istepanian, R.S.H., Jovanov, E., and Zhang, Y.T. 2004. Guest editorial introduction to the special section on m-Health: Beyond seamless mobility and global wireless health-care connectivity. *IEEE Transactions on Information Technology in Biomedicine*, 8, 405–414.

Jan, R.-H., Chu, H.-C., and Lee, Y.-F. 2004. Improving the accuracy of cell-based positioning for wireless networks. *Computer Networks*, 46, 817–827.

Jan ten Duis, H. and van der Werken, C. 2003. Trauma care systems in the Netherlands. *Injury*, 34, 722–727.

Jen, W.-Y. and Chao, C.-C. 2008. Measuring mobile patient safety information system success: An empirical study. *International Journal of Medical Informatics*, 77, 689–697.

Kakihara, M. and Sorensen, C. 2001. Expanding the "mobility" concept. *ACM SIGGroup Bulletin*, 22, 33–37.

Kaplan, B. 2001. Evaluating informatics applications-some alternative approaches: Theory, social interactionism, and call for methodological pluralism. *International Journal of Medical Informatics*, 64, 39–56.

King, W.R. and Epstein, B.J. 1983. Assessing information system value: An experimental study. *Decision Sciences*, 14, 34–45.

Kohn, L.T., Corrigan, J.M., and Donaldson, M.S. 2000. *To Err is Human: Building a Safer Health System*. Washington, DC: National Academies Press.

Lapinsky, S.E., Weshler, J., Mehta, S., Varkul, M., Hallett, D., and Stewart, T.E. 2001. Handheld computers in critical care. *Critical Care*, 5, 227–231.

LeBlanc, V.R., Manser, T., Weinger, M.B., Musson, D., Kutzin, J., and Howard, S.K. 2011. The study of factors affecting human and systems performance in healthcare using simulation. *Simulation in Healthcare*, 6, S24–S29.

Lee, Y.E. and Benbasat, I. 2004. A framework for the study of customer interface design for mobile commerce. *International Journal of Electronic Commerce*, 8, 79–102.

Leitner, G., Ahlström, D., and Hitz, M. 2007. Usability of mobile computing in emergency response systems—Lessons learned and future directions. In: Holzinger, A. (ed.) *HCI and Usability for Medicine and Health Care*. Berlin, Heidelberg, Germany: Springer.

Liao, I.E. and Kao, K.-F. 2008. Enhancing the accuracy of WLAN-based location determination systems using predicted orientation information. *Information Sciences*, 178, 1049–1068.

Lin, J.-S.C. and Hsieh, P.-L. 2011. Assessing the self-service technology encounters: Development and validation of SSTQUAL scale. *Journal of Retailing*, 87, 194–206.

Lippeveld, T. 2001. Routine health information systems: The glue of a unified health system. In Keynote address at the *Workshop on Issues and Innovation in Routine Health Information in Developing Countries*, March, 2001. Potomac, MD, pp. 14–16.

Lu, Y.-C., Xiao, Y., Sears, A., and Jacko, J.A. 2005. A review and a framework of handheld computer adoption in healthcare. *International Journal of Medical Informatics*, 74, 409–422.

Mahmood, M.A. 1987. System development methods—A comparative investigation. *MIS Quarterly*, 11, 293–311.

Mason, R.O. 1978. Measuring information output: A communication systems approach. *Information & Management*, 1, 219–234.

Masrom, M. and Hussein, R. 2008. *User Acceptance of Information Technology: Understanding Theories and Models*. Selangor Darul Ehsan, Malaysia: Venton Publishing (M) Sdn. Bhd.

Miller, J. and Doyle, B.A. 1987. Measuring the effectiveness of computer-based information systems in the financial services sector. *MIS Quarterly*, 11, 107–124.

Miller, R.B. 1968. Response time in man-computer conversational transactions. *Proceedings of the December 9–11, 1968, Fall Joint Computer Conference, Part I*. San Francisco, CA: ACM.

Minnick, A., Pischke-Winn, K., and Sterk, M. B. 1994. Introducing a two-way wireless communication system. *Nursing Management*, 25, 42–47.

Molla, A. and Licker, P.S. 2001. E-commerce systems success: An attempt to extend and respecify the Delone and Maclean model of IS success. *Journal of Electronic Commerce Research*, 2, 131–141.

Moule, P. 2000. Checking the carotid pulse: Diagnostic accuracy in students of the healthcare professions. *Resuscitation*, 44, 195–201.

Nelson, R.R. and Todd, P. A. 2005. Antecedents of information and system quality: An empirical examination within the context of data warehousing. *Journal of Management Information Systems*, 21, 199–235.

Nickerson, R.S. 1969. Man-computer interaction: A challenge for human factors research. *IEEE Transactions on Man-Machine Systems*, 10, 164–180.

Nilakanta, S. and Scamell, R. W. 1990. The effect of information sources and communication channels on the diffusion of innovation in a data base development environment. *Institute for Operations Research and the Management Sciences (INFORMS)*, Linthicum, MD, 36, 24–40.

O'Connor, Y., O'donoghue, J., and O'reilly, P. Year. 2011. Understanding mobile technology post-adoption behaviour: Impact upon knowledge creation and individual performance. In *Proceedings—2011 10th International Conference on Mobile Business, ICMB 2011*, Como, Italy, pp. 275–282.

Office of the National Coordinator for Health Information Technology, Department of Health and Human Services. 2009. *American Recovery and Reinvestment Act of 2009, Title XIII—Health Information Technology, Subtitle B—Incentives for the Use of Health Information Technology*, Section 3013, State Grants to Promote Health Information Technology Planning and Implementation Projects.

Paez, A., Mercado, R., Farber, S., Morency, C., and Roorda, M. 2010. Accessibility to health care facilities in Montreal Island: an application of relative accessibility indicators from the perspective of senior and non-senior residents. *International Journal of Health Geographics*, 9, 1–15.

Parasuraman, A., Zeithaml, V.A., Berry, L.L., and Marketing Science, I. 1986. *Servqual, a Multiple-Item Scale for Measuring Customer Perceptions of Service Quality*. Cambridge, MA: Marketing Science Institute.

Pérez-Mira, B. 2010. *Validity of Delone and Mclean's Model of Information Systems Success at the Web Site Level of Analysis.* Doctor of Philosophy. Baton Rouge, LA: Louisiana State University

Perry, M., O'hara, K., Sellen, A., Brown, B., and Harper, R. 2001. Dealing with mobility: Understanding access anytime, anywhere. *ACM Transactions on Computer-Human Interaction,* 8, 323–347.

Powell-Cope, G., Nelson, A. L., and Patterson, E. S. 2008. Patient care technology and safety. *Patient Safety and Quality: An Evidence-Based Handbook for Nurses.* Rockville, MD: Agency for Healthcare Research and Quality.

Riemer, K. and Frobler, F. 2007. Introducing real-time collaboration systems: Development of a conceptual scheme and research directions. *Communications of the Association for Information Systems (CAIS),* 20, 204–225.

Rubin, G. J., Cleare, A. J., and Wessely, S. 2008. Psychological factors associated with self-reported sensitivity to mobile phones. *Journal of Psychosomatic Research,* 64, 1–9.

Schewe, C. 1976. The management information system user: An exploratory behavioral analysis. *Academy of Management Journal,* 19, 577–590.

Schmid, V. and Doerner, K.F. 2010. Ambulance location and relocation problems with time-dependent travel times. *European Journal of Operational Research,* 207, 1293–1303.

Seddon, P., Kiew, M.-Y., and Patry, M. 1994. A partial test and development of the DeLone and McLean model of IS success. The International Conference on Information Systems (ICIS) at AIS Electronic Library (AISeL), ICIS 1994 Proceedings, *Australian Journal of Information Systems,* 4, 90–109.

Shannon, C.E. and Weaver, W. 1948. A mathematical theory of communication. *Bell System Technical Journal,* 27, 379–423, 623–656.

Sittig, D.F. 2002. Personal health records on the internet: A snapshot of the pioneers at the end of the 20th century. *International Journal of Medical Informatics,* 65, 1–6.

Sittig, D.F. and Singh, H. 2010. A new sociotechnical model for studying health information technology in complex adaptive healthcare systems. *Quality and Safety in Health Care,* 19, i68–i74.

Smith, J. 1999. *Health Management Information Systems: A Handbook for Decision Makers.* Buckingham, U.K.: Open University Press.

Southon, F.C.G., Sauer, C., and Dampney, C.N.G. 1997. Information technology in complex health services: Organizational impediments to successful technology transfer and diffusion. *Journal of the American Medical Informatics Association,* 4, 112–124.

Srinivasan, A. 1985. Alternative measures of systems effectiveness: Associations and implications. *MIS Quarterly,* 9, 243–253.

Subasi, Ö., Leitner, M., and Tscheligi, M. 2009. A usability and accessibility design and evaluation framework for ict services. In: Holzinger, A., and Miesenberger, K. (eds.) *HCI and Usability for e-Inclusion.* Berlin Heidelberg, Germany: Springer.

Tanser, F., Gijsbertsen, B., and Herbst, K. 2006. Modelling and understanding primary health care accessibility and utilization in rural South Africa: An exploration using a geographical information system. *Social Science & Medicine,* 63, 691–705.

Telemedicine, Committee on Evaluating Clinical Applications of Telemedicine, and Institute of Medicine. 1996. *Telemedicine: A Guide to Assessing Telecommunications for Health Care.* Washington, DC: The National Academies Press.

Tibballs, J. and Weeranatna, C. 2010. The influence of time on the accuracy of healthcare personnel to diagnose paediatric cardiac arrest by pulse palpation. *Resuscitation,* 81, 671–675.

Varshney, U. and Vetter, R. 2002. Mobile commerce: Framework, applications and networking support. *Mobile Networks and Applications,* 7, 185–198.

Vincent, C., Taylor-Adams, S., and Stanhope, N. 1998. Framework for analysing risk and safety in clinical medicine. *British Medical Journal,* 316, 1154.

Waldorf, B. and Che, S. 2010. Spatial models of health outcomes and health behaviors: THE role of health care accessibility and availability. In: Páez, A., Gallo, J., Buliung, R.N., and Dall'erba, S. (eds.) *Progress in Spatial Analysis.* Berlin, Germany: Springer.

Wang, Y. and Liu, Z. 2007. Automatic detecting indicators for quality of health information on the web. *International Journal of Medical Informatics*, 76, 575–582.

Wu, J.-H., Wang, S.-C., and Lin, L.-M. 2007. Mobile computing acceptance factors in the healthcare industry: A structural equation model. *International Journal of Medical Informatics*, 76, 66–77.

Yu, P., Li, H., and Gagnon, M.-P. 2009. Health IT acceptance factors in long-term care facilities: A cross-sectional survey. *International Journal of Medical Informatics*, 78, 219–229.

Yun, J.-K., Kim, D.-O., Hong, D.-S., Kim, M. H., and Han, K.-J. 2006. A real-time mobile GIS based on the HBR-tree for location based services. *Computers & Industrial Engineering*, 51, 58–71.

Yusof, M.M., Paul, R.J., and Stergioulas, L.K. 2006. Towards a framework for health information systems evaluation. In *Proceedings of the 39th Annual Hawaii International Conference on System Sciences, 2006*. Hawaii, *HICSS'06*, IEEE, pp. 95a. http://ieeexplore.ieee.org/xpls/abs_all.jsp?arnumber=1579480&tag=1

Zakour, A.B. 2004. Cultural differences and information technology acceptance. In *Proceedings of the 7th Annual Conference of the Southern Association for Information Systems*, Georgia, pp. 156–161.

Zeithaml, V.A. 1987. *Defining and Relating Price, Perceived Quality, and Perceived Value*. Cambridge, U.K.: Marketing Science Institute.

15

Implementing mHealth in Low- and Middle-Income Countries: What Should Program Implementers Consider?

P.J. Wall, Frédérique Vallières, Eilish McAuliffe, Dave Lewis, and Lucy Hederman

CONTENTS

15.1 Background and Introduction

Growing interest within the health-care sector to capitalize on the widespread uptake of mobile communication technologies combined with rapid improvements in telecommunications infrastructure in low- and middle-income countries (LMICs) has resulted in a dramatic increase of mHealth initiatives in recent years (Collins 2012, Purkayastha et al. 2013). The International Telecommunications Union estimates that in 2014 there are almost 7 billion mobile phone subscriptions worldwide, with LMICs accounting for over three quarters of these subscriptions. This represents a mobile penetration rate of 90% across all LMICs, with Africa expected to have a penetration rate of almost 70% by the end of 2014 (International Telecommunications Union, 2014). Carving their niche as one of the world's most ubiquitous modern technologies, mobile phones are now more accessible than to a bank account, electricity, a toilet, or clean water (World Bank 2012).

When introducing mHealth in any setting, it is important to keep in mind the interrelatedness of the technology and the social, political, environmental, and cultural contexts within which it is embedded. The term *sociotechnical* is frequently used to refer to the interaction between people and society (social) and the machine and technology (technical) (Walker 2008). Though complex regardless of where the implementation takes place, paying attention to these factors is particularly important in LMICs, where program implementers may not be familiar with the prevailing social, political, and human contexts.

15.2 Chapter Objectives

The primary objective of this chapter is to discuss the more practical considerations of implementing mHealth in an LMIC, focusing specifically on technical, environmental, social, and human considerations. We believe that when taken into account by program implementers, these considerations will improve the chances of an mHealth initiative being successfully implemented, scaled, and sustained. Unless otherwise specified, we use the term *end user* broadly to refer to anyone within the health system who might utilize an mHealth device.

In addition to a review of key examples from the existing mHealth literature, the authors draw primarily from their experiences implementing mHealth as part of an ongoing longitudinal, action research program in Sierra Leone. This is supplemented by data from five in-depth key informant interviews conducted with mHealth program managers across Malawi, Mozambique, Zimbabwe, India, and Sierra Leone. A series of questions were developed to help guide the key informants through describing their specific mHealth program across a technical, environmental, or social and human capacity considerations. Transcripts were then analyzed using a thematic framework approach whereby responses were categorized under the identified themes. Subthemes were subsequently identified within each of these larger areas.

15.3 Scramble for mHealth in LMICs

High penetration of mobile phones has created a critical mass of infrastructure with the possibility of leveraging solutions across a variety of health-care challenges commonly found in LMICs. In contexts where poor road infrastructure and a lack of access to transport are common, mHealth has the potential to improve timely health data collection and transfer (Tomlinson et al. 2009). Moreover, mHealth could facilitate disease surveillance, improving the ability to identify and manage disease outbreaks and epidemics through more time-efficient health information systems (Robertson et al. 2010), and allow for more efficient diagnosis and greater adherence to medicines, medical procedures, and protocols (DeRenzi et al. 2008, Routen et al. 2010). Similarly, mHealth could enhance case management and offer a greater continuity of care for patients through better access to patient histories and electronic medical records (Rotheram-Borus et al. 2011), improve patient communication (Siedner et al. 2012), and provide medical adherence support (Pop-Eleches et al. 2011, Zurovac et al. 2011).

mHealth also has the potential to improve working conditions for the largely overburdened and underpaid health workers in LMICs (Iyengar and Florez-Arango 2013). Through better time efficiency and improved health worker support, mHealth could help address the performance and retention issues present across all cadres of health workers (Willis-Shattuck et al. 2008, Kallander et al. 2013). In the case of LMICs, where many governments rely on lower cadres of health workers such as community health workers (CHWs), mHealth is seen as a promising tool to help support community-level health programs (Strachan et al. 2012). Given that CHWs vary considerably in terms of their training, experience, and literacy levels, the availability of timely clinical information, learning materials, and regular household visit reminders are seen as important mHealth tools that can be used to empower CHWs in their role bridging communities and formal health systems (Braun et al. 2013, Kallander et al. 2013).

Despite its potential, the design, implementation, and adoption of mHealth are not without a wide range of challenges and risks (Kahn et al. 2010, Manda and Msosa 2012, Chigona et al. 2013). In LMICs, the implementation of mHealth programs is predominantly led and funded by international and government donors and largely takes the form of short-term pilot projects (McNamara 2003, Curioso and Mechael 2010, WHO 2011). This externally driven proliferation of pilot programs has led to many paralleled and largely uncoordinated efforts in LMICs. Over the last few years, it has become increasingly evident just how difficult it is to sustain mHealth projects beyond pilot implementation (Anderson and Perin 2009, Curioso and Mechael 2010). Even in the case where mHealth initiatives manage to progress beyond pilot phase, many fail to achieve success at scale (Anderson and Perin 2009, Curioso and Mechael 2010). This has resulted in claims that the mHealth sector suffers from a particularly debilitating case of *pilotitis* (Kuipers et al. 2008). Therefore, and although there is great potential to improve access to the quality of health care in remote settings, mHealth presents its own set of challenges that must be taken into consideration if one is to implement an initiative in a LMIC.

15.4 Technological Considerations

Recent years have seen an increase in the use of mHealth applications, many of which are primarily designed for use in LMICs (Macleod et al. 2012, Chatfield et al. 2013). The following section presents some of technological considerations for implementing mHealth in LMICs.

15.4.1 Hardware

Many of our interviewees reported the use of lower-end feature phones as part of their mHealth initiatives. The most reportedly used phones included the Nokia 2700, the Nokia C2-01, and a variety of other low-end Android phones (such as those manufactured by Micromax, Spice, and Karbonn). In comparison to higher-end smartphones, mHealth initiatives that make use of low-end mobile phones are arguably better suited to LMICs (Sanner et al. 2012). Factors such as greater affordability, increased local access for repairs and servicing of the phones, lowered susceptibility to theft (Tomlinson et al. 2009), more efficient battery power, and a greater user familiarity with low-end mobile phones resulting in higher mobile phone literacy have all been identified as characteristics that make low-end phones particularly adaptable to LMICs (Sanner et al. 2012).

Much of the literature suggests that solar chargers are becoming increasingly common in areas of nonexistent or sparse power supply (e.g., Hoefman 2000, Boyce 2012, Kallander et al. 2013, McCord et al. 2013). However, in both our own and our informants' experience, chargers are problematic for a variety of reasons including their susceptibility to breakage, their inability to charge a fully drained battery, and their ineffectiveness in cloudy or rainy conditions, which in some LMICs can last an entire season. Taking these factors into account, low-end phones are more appropriate for use in LMICs at the time of writing. However, this may change as the price of smartphones continues to decrease and the availability of smartphones specifically designed to cater to local markets increases. As such, program managers should constantly reassess which type of phone is most appropriate for use in their specific context taking into account the aforementioned factors.

15.4.2 Software and the Importance of Localization

The most common mHealth software in LMICs includes Internet browser solutions, interactive voice response (IVR), plaintext short message service (SMS), and locally installed handset and SIM card applications (Sanner et al. 2012). MOTECH, ChildCount, CommCare, Text to Change, eMOCHA®, and DHIS Mobile are only a few of the many examples of mHealth software solutions specifically designed for use in an LMIC. Today, with the increasing popularity of open-source code, rarely are mHealth programs built from scratch and global efforts are being made to reduce duplication and encourage innovation (Klungsöyr et al. 2008). An example of this is the Open mHealth, which seeks to build an open-source software architecture to enable custom integrations for specific environments.

When implemented in a different context, the mHealth software should still be easily understood to the end user. To enhance the understanding of the end user, many of our informants suggested the integration of audiovisual content. One interviewee recounted that when viewed in groups, the audiovisual content provided a platform to discuss the cultural taboos of diabetes in India, ranging from the marriage prospects of females with diabetes to the perceptions of diabetes as a lifestyle disease. Whether visual or audio, it is necessary to translate any existing content into local languages and to adapt any visual content to ensure it is culturally appropriate and aligned to local visual literacy. Audiovisual content should be relevant and engaging, preferably reflecting a common social scenario. Careful consideration should also be given to the use of conversational language in multimedia files as user acceptability may vary in terms of gender, perceived age, and socioeconomic status associated with particular voices, intonations, and accents (GrameenFoundation 2011).

The temptation exists for many program implementers to say that a software program *works* once the software is seen to carry out the functions for which it was designed. No matter how well it is designed, however, software localization must also be met with user acceptability. The iteration process is a continuous one and one that must be carried throughout the entire project cycle, integrating feedback from end users into future versions of the software. Unfortunately, a user-centric design appears to be the exception rather than the rule in mHealth programs (Braun et al. 2013).

15.4.3 End-User Flexibility

The localization process may be inhibited where certain functions of the mHealth hardware or software are blocked. The most common reason given by program implementers as to why certain features are blocked is to prevent misuse or *abuse*. This might include curtailing the capability to make personal calls, limiting camera functions, or placing restrictions on the secure digital (SD) card. In contrast, an unrestricted phone gives the end user free reign over the mobile's full functions.

Where there are no limits placed on the mobile phones, end users may load personal content (music, pictures etc.) on the SD cards, which has in some cases, according to informants, led to the accidental deletion of the mHealth software. This can be problematic, as in many instances the end user will require assistance to reload it (Pascoe et al. 2012). This can be avoided by burning the application into the internal memory of the mobile handset (Purkayastha et al. 2010).

It is the authors' experience that unrestricted use can also lead to innovative and unforeseen uses of the technology. One of our informants recounted how the health worker played the audio file of the mHealth software on loop in the health center. This was likened to patients listening to a *conversation* and was reported to be an effective way of delivering health messages. In another example, the informant reported that health workers used the camera on the phone to take pictures of ailing children to send to the health center. The health center worker was then able to assist the health worker to make a remote diagnosis and to refer the patient to the appropriate health facility. Similar innovative use of the camera has also been reported in the literature (Purkayastha et al. 2010).

In the cases where the use of an unrestricted phone was monitored, rarely did informants report any abuse of the phone. This supports the idea that end users should have the autonomy to use the full feature set of the phone and the associated software if innovations in mHealth are to be facilitated. This is particularly important in LMICs, where, when it comes to innovative uses for the mobile phone, there are many shared lessons to be learned (Svensson and Wamala 2012).

15.5 Environmental Considerations

When implementing mHealth solutions, it is important to remember that the technology itself is not independent of the environment within which the mHealth project is taking place (Manda and Herstad 2010). This next section discusses some of the environmental considerations that implementers should factor in conjunction with the technological considerations discussed earlier.

15.5.1 Infrastructure and Power Supply

For many health center workers in LMICs, it is not uncommon for health data to be recorded in several paper registers before being manually aggregated and physically transferred from the health center to district-level management teams prior to being digitized and aggregated for analysis (Purkayastha et al. 2010). Poor road infrastructure compounded by a lack of access to transport further delays this process. Sending health information via mobile phones has the potential to address some of the structural barriers prevalent across all levels of a health system (Tomlinson et al. 2013). In the case of medical emergencies, mHealth can strengthen referral systems, reducing the delays faced by patient travel (Tamrat and Kachnowski 2012). One informant spoke of a program that delivered patient HIV test results directly to a health worker's mobile. They reported a much faster turnaround time for test results as well as time savings for patients with negative results, who did not need to return to the health centre for test results. mHealth can also improve supply chain management of essential drugs and diagnostic tools, helping to prevent stock-outs in already poorly resourced health centers (Barrington et al. 2010).

Despite increasing network coverage, mobile network connectivity in rural areas of many LMICs remains unreliable (Chigona et al. 2013). The lack of electricity in LMICs is also seen as a constraint (Chetley 2007). While there does need to be some level of mobile phone signal and availability of electricity within reasonable reach, both our own and our interviewees' experiences suggest that these factors do not seem as fundamental as one might expect. As one informant described, "people have learned to cope" and most individuals manage to overcome limited mobile network coverage and an inconsistent power supply. Informants described how individuals adapt to patchy mobile network signals by traveling to specific spots or placing their phones up certain trees. Similarly, our informants recalled how inconsistent power supply is often overcome through a variety of local innovations and improvisations. One informant spoke of how users manage to *find electricity* through such local innovations and improvisation as charging mobile phones on car batteries, swapping batteries with those traveling to areas with power, the use of solar chargers, or charging phones in local kiosks equipped with a generator.

To maximize its potential at scale, an mHealth solution must be reliable across a large variability of network coverage and across inconsistent and irregular power supplies. To this end, the end user should have the ability to work offline, with data stored and retrieved directly onto the mobile phone and sent only when the network connection is sufficient. With the rapidly changing infrastructure in many LMICs however, such technological considerations will likely be the least salient rate-limiting factor for the implementation of mHealth in the future.

15.5.2 Ethical Implications

Beyond the use of firewalls, data encryption, and server security, there exist additional privacy concerns that program implementers must consider around the access, storage, utilization, and sharing of individual patient data in LMICs. Program implementers should consider *who* in a household has access to the phone, which may vary from one context to the next. In Sierra Leone, for example, it is not uncommon for a household or neighbors to share a single phone (Magbity et al. 2011). In the case where the phone is shared, access may be limited for those family members who must first seek permission before they can use the phone (Rotheram-Borus et al. 2012). Access and utilization of the phone have further implications for the transfer of what may be considered culturally sensitive information, such as receiving information on family planning or, as one informant described, receiving the results of an HIV test.

It is important for implementers to keep in mind that the use of a phone may not be appropriate in all cases. One informant questioned their use of a mobile phone to record data and to aid in the discussions of sensitive issues such as suspected child abuse or gender-based violence in Malawi. Similarly, the use of mHealth had an unexpected adverse effect when women receiving text messages were physically assaulted by their husbands, who suspected the messages were a sign of infidelity (Rotheram-Borus et al. 2012). In addition, the large number of stakeholders often involved in the implementation of mHealth in LMICs implies that implementers need to set clear data-sharing policies. Among the various stakeholder groups discussed in the following section, implementers should consider who is most appropriate to host the data and what level of access the various groups involved should have to these data.

15.5.3 Creation of Strong Stakeholder Networks

Building powerful networks to establish early buy-in and ownership from key stakeholders is imperative. The authors' experience suggests that strong stakeholder networks can be more easily formed and sustained when the various participants and organizations are transparent about their motives, objectives, and outputs for an mHealth initiative. It follows that there may be implicit inconsistencies between the motives of various stakeholders for implementing mHealth in LMICs.

In the creation of strong networks, key agents from the following sectors, each with their own strengths and expertise, should be included in all phases of the implementation cycle. Firstly, the ministry of health can provide guidance on relevant national policies and ensure that all health information systems are interoperable and aligned to their needs within the ministry. Failure to secure government involvement in the Gambia, for example, led to Vodafone having to suspend the scale-up of their mHealth project (Vodafone 2013). Contrastingly, the strong ownership of a RapidSMS system for maternal and child health in Rwanda contributed to the successful scale-up of an mHealth tool for CHWs (Ngabo et al. 2012). However, technical and other infrastructural weaknesses of local ministries of health to support new technology solutions may mean that alliances with other partners and stakeholders may also be necessary (Manda and Sanner 2012).

Our experience indicates that communication regulatory bodies and NGOs can also play an important role in negotiations with mobile network operators (MNOs). Given that regulatory bodies are those responsible for issuing the licenses to MNOs, they are likely better positioned to negotiate affordable rates for the wide-scale implementation of mHealth, whether that be out of pocket for end users or available free of charge as a type of public good. Likewise, the presence of NGOs in some of the more remote places of LMICs, combined with their strong experience in project management, means NGOs are well positioned to coordinate implementation efforts and inform national mHealth policies. Industry research institutions and academics alike can also offer important support and expertise to develop the evidence base for mHealth in each context.

Lastly, and perhaps most importantly, is the importance of involving end users and beneficiary groups. One of the most valuable roles of implementers is to engage end users and encourage them to participate in the process of developing and implementing mHealth. User groups are the foundation of any mHealth program, and their involvement is key throughout every step of the design, development, iteration, implementation, and scaling processes (mHealthAlliance 2011). Engaging a large number of stakeholders, however, requires an increased level of coordination among stakeholders, many of which have different, and often conflicting, motivations for the scaling of mHealth solutions.

15.5.4 Cost Considerations

Cost remains one of the most significant barriers to health-care access in LMICs. Features such as remote diagnosis and digital health data entry can significantly reduce travel time and opportunity costs and imply potential savings for patients and health workers alike. If an mHealth solution is going to be widely adopted by health service providers across different cadres, it must be seen to provide tangible benefits (Archer 2009), offering savings in terms of both time and out-of-pocket expenditures (Purkayastha et al. 2010).

Cost also has important implications for the implementation and scale-up of mHealth programs (Sanner et al. 2012). Increasing mobile phones in circulation as part of an mHealth program has important implications for the number of training sessions, training manuals, as well as both hardware and software updates. There is some evidence that technological interventions are heavily reliant on external financial support and thus prone to collapse as soon as external funding ceases to exist (Manda and Sanner 2012).

Availability of adequate funding is essential if technology-driven interventions are to be maintained (Lucas 2008). This is not always easy to achieve, and costs may not always be readily discernible from the results of any pilot implementation (Lucas 2008). There are many examples of mobile solutions where costs remain unknown because the mHealth project has never been brought to scale (Manda and Msosa 2012).

15.6 Social and Human Capacity Considerations

Thus far, this chapter has discussed important technological and environmental considerations for the implementation of mobile phone solutions in LMICs and elsewhere. Given that a sociotechnical approach calls for mHealth implementers to consider the interrelatedness of the socio (people and society) and the technical (software and technology), program implementers would be remiss if they also did not factor in the human component of mHealth.

15.6.1 Human Motivation and Human Capacity

The effect of technology on improving health outcomes hinges predominantly on the capacity and motivation of those using the technology (Toyama 2010). Therefore, in order for the technology to achieve what was intended, even the most well-designed mHealth solution must be accompanied by a competent and motivated user (Purkayastha et al. 2010).

It is generally quite easy to get people excited about an mHealth initiative. The innovative component and unique possibilities offered by mHealth are powerful incentives for generating early interest in an mHealth program. Unfortunately, initial positive responses to an mHealth initiative do not necessarily translate into an increase in the uptake of the technology (Chaiyachati et al. 2013). For this reason, evaluating the motivations of end users remains an important monitoring component of implementing mHealth programs (Vallières et al. 2013).

The assumption that once an mHealth project is up and running the technology will simply fit into the environment and be easily adopted by the user has been described by

Shozi et al. (2012) as *fallacy*. Key to mediating against this is the adoption of a participatory approach, allowing end users to participate and feedback into health information technology (Hostgaard et al. 2011). mHealth cannot be seen as a substitute for the human intellect or motivation of the person using the mobile, and ultimately, an mHealth solution should empower and enhance the competence of the end user, allowing them to gain greater recognition as part of the larger health system (Purkayastha et al. 2010).

15.6.2 Human Technical Capacity

Individuals should be given the technology they need to enhance their work, within the limitations of their knowledge and familiarity with the technology. Given that the potential benefits of technology are dependent on the absorptive capacity of its user (Toyama 2010), a solution should cater to its specific users, many of whom will vary in terms of literacy rates, preferred means of communication (SMS, voice call), age, and familiarity with mobile phones. In India, for example, an informant described how adult users were most comfortable making voice calls and receiving texts, whereas younger users were more inclined to send text messages as well. Likewise, it is also important for medical professionals to have an appropriate level of technical competence if they are to be able to use the mHealth applications to their full potential (Orlikowski 2000). In LMICs, this level of technical competence may not always exist at the community level, reinforcing the importance for mHealth to be simple and easy to use (Purkayastha et al. 2013).

Local technical expertise is also important to support the mHealth solution in-country. In addition to the aforementioned overreliance on external sources of funding to drive the technology forward, a high dependence on an external agency for technological support further undermines the sustainability of an mHealth solution over time. Engaging with local developers has revealed important implications for program sustainability in that it not only builds local capacity, but it also generates more contextually appropriate, transferable, and effective solutions (Curioso and Mechael 2010).

15.6.3 Literacy Rates

Beyond technological capacity, the use of certain applications requires that users possess a sufficient level of literacy skills (Manda and Msosa 2012). In many LMICs, we are faced with low and often mixed literacy rates. Some mobile phone features, such as SMS, heavily rely on users' writing competencies (Manda and Herstad 2010). This is also true when an mHealth application requires that the end user input, transfer, collect, and process data as well as follow instruction from the written text (Chigona et al. 2013). In a context with high rates of illiteracy, the use of IVR is strongly recommended since it is compatible with a variety of different handsets (Sherwani et al. 2009).

Surprisingly, when supported by strong audiovisual cues such as images and IVR, illiterate users are often able to navigate solutions. Over time, and with sufficient training and practice, it seems illiterate users come to recognize certain prompts and words and are able to associate this with the content of the written text. As one of our informant interviewees reported, some illiterate CHWs even began to enter simple text. Illiterate users do, however, require additional and often separate training from literate users. One of our informant interviewees suggested that in some cases, it may be appropriate to invite family members to participate in the training, so as to assist their family member in becoming more familiar with the phone.

15.6.4 Human Resources for Health

The global human resource for health crisis has resulted in an overburdened and heavily under-resourced health staff (WHO 2006). Health worker acceptance of a new mHealth initiative can be inhibited if new workflows and procedures are considered additional work at any level of the health system when compared to the status quo ante. In this sense, the implementation of mHealth must be perceived as a means to alleviate health worker workload (Iyengar and Florez-Arango 2013). As one informant described, alleviating health worker workload cannot be synonymous with passing on work to a lower cadre of health worker, who may become overburdened themselves. The adoptability is reduced when the end user, regardless of the cadre of health worker, cannot see the tangible benefits of the technology. Transitioning into a computerized system may require a period whereby both systems are run in parallel. Careful consideration must be given to how this transition phase is communicated to end users. In the case of health workers, it should be clear as to how the incoming system will ultimately improve working conditions and reduce workload (Ganesan et al. 2012).

mHealth solutions are commonly introduced using a top-down approach (Braun et al. 2013). Changes to protocol are typically enforced by a person of authority with the aim of ensuring health workers adhere to policies, protocols, and workflows as a requirement. As such, many programs miss an opportunity to enhance both health worker and patient participation in the design and implementation processes (Braun et al. 2013). Though important to obtain buy-in from key decision makers and managers in order to introduce new technology into an existing workflow, adoptability is increased when end users are made part of the design and implementation process (Hostgaard et al. 2011, Darby et al. 2012).

15.6.5 Human Capacity for mHealth Monitoring and Evaluation

For improved effectiveness of mHealth, there is a need for changes in both the practice and the research of mHealth. Most of the mHealth programs that have been implemented thus far lack substantial and vigorous monitoring and evaluation, resulting in a largely underwhelming evidence base demonstrating the efficiency and effectiveness of mHealth to ultimately improve health outcomes (Curioso and Mechael 2010, Tomlinson et al. 2013). Consequently, many of the barriers to and gaps in mHealth scale and sustainability result from limited knowledge of what works, how it works, and how much it will cost (Mechael et al. 2010). Too often the focus of mHealth programs is less on whether health outcomes are improved rather than does the technology *work* (Labrique et al. 2013). Moreover, there is a dearth of evidence demonstrating how mHealth impacts on patient service utilization, access, health worker productivity, supportive supervision, and quality of care.

Given that health ministries in LMICs are already working under heavily constrained budgets, as well as under the considerable pressure of international donors and agencies to prioritize certain diseases, evaluations should demonstrate the cost-effectiveness of mHealth innovations. Programs should bear in mind the type of evidence that is of interest to governments, data that will garner support and encourage governments and/ or investors to expand mHealth programs beyond proof of concept. Strong evidence is thus required if mHealth initiatives are to be included in overall health strategies and interventions (WHO 2011). In addition, program implementers must be careful not to create parallel health monitoring systems. Health monitoring data collected as part of an mHealth initiative should be aligned to the existing monitoring system, complementing rather than duplicating the efforts of the ministry of health. Steps toward improving monitoring and evaluation need to address the lack of human capacity, training, and resources for data to be collected, analyzed, and interpreted within a country's own health centers and health ministries.

15.7 Case Studies

Mobile technology is evolving rapidly. Over the coming years, the mobile phone is likely to become significantly more powerful, and mobile apps are likely to become progressively more sophisticated. The software and hardware is also expected to decrease in cost, and may increasingly begin to be designed, developed, and manufactured in LMICs. Indeed, this is already happening with the production of the Elikia smartphone in Congo. Advances are also being made in data transmission technologies, and mobile operators are scrambling to rollout 3G and 4G phone services across Africa and Asia. Such rapid evolution in this field will present both huge opportunities and challenges for global health, and specifically for mHealth in LMICs.

While these advances are exciting from a technological perspective, there are also many non-technological factors to consider. The potential for mobile technologies to become ubiquitous and to weave themselves into the fabric of our daily lives will provide unique possibilities to significantly reconfigure the way we work and live. This is likely to have a dramatic effect on healthcare provision, particularly in LMICs, where increasingly faster 4G data transfer speeds will facilitate the transmission of greater quantities of data and use of the cloud. This, combined with the increasing availability and sophistication of low-cost hardware that can be added-on to mobile devices (e.g. ultrasound scanners, biometric sensors, microscopes, etc.), is likely to allow more advanced diagnostic procedures to be administered by local health workers with medical diagnosis being provided in real time remotely by more qualified professionals. The potential may even exist for the transmission of various types of data between the patient and the remote health professional, and for patients to self diagnose. This type of remote analysis and diagnosis may mitigate to some extent the chronic lack of qualified health professions and doctors in many LMICs. However, the technology is not a replacement for the human capacity, which still requires huge investment in LMICs.

Such advances may also facilitate the training and education of both health workers and patients. The opportunity to virtually attend lectures and conferences in many of the world's top universities is likely to become increasingly available, and many educational institutions are providing much of this online content free of charge. This will make formal health education and training available to many of the worlds most underprivileged and remote health providers. Such education and training will present an opportunity to reconfigure their individual work practices, and perhaps even the overall health care structures of the country or region.

It is difficult to state with any degree of certainty where such a rapid evolution of mobile technology will take us. What is certain however is that it will undoubtedly present new opportunities and challenges to the mHealth program implementer in LMICs.

Many of these opportunities and challenges will be technological in nature, but there will also be various social, political, and cultural issues to consider. The introduction of advanced technology to an LMIC always has the potential to clash with pre-existing societal structures and norms, and the more advanced the technology the bigger the potential clash. Existing socio-technical frameworks and approaches will have to evolve with the technology if future mHealth initiatives are to have the potential to scale beyond pilot and sustain.

15.8 Conclusions

A sociotechnical approach views any information system as a social system that cannot be transferred physically in the same way as software applications or a piece of hardware. In the same way that giving someone a hammer will not build a home, simply introducing a mobile phone and its associated software is unlikely to improve health outcomes on its own. Implementers are likely to have to negotiate a multiplicity of interacting sociotechnical factors that are both within and outside their control (Manda and Sanner 2012).

Technocentric approaches have largely dominated the field of development informatics to date, and a more interdisciplinary approach is necessary if we are to address the challenges currently facing the implementation of mHealth (Walsham 2012). The mobile phone should be seen as a component of a larger health information infrastructure including paper forms, the networks, and the preexisting data flow processes (Purkayastha et al. 2010), social factors, market-based incentives, regulatory frameworks, and local cultures (mHealth Alliance 2012). Consideration of such sociotechnical process may also empower individuals and communities, leading to greater social change and an improved quality of life, and would substantiate the widely held belief that mHealth is a tool for human development (Chigona et al. 2013).

Our objective in this chapter was to highlight some of the more common technological, environmental, and human capacity considerations of implementing mHealth in LMICs. Though we recognize that program implementers may not be able to influence and control all of the sociotechnical factors covered in this chapter, many are within the control of mHealth program implementers. What our informant interviews and review of the literature have highlighted is that despite its lure, mHealth cannot be viewed as a panacea to address the health problems of LMICs. Like any other global health initiatives, mHealth programs face a variety of diverse challenges and problems that must be examined within the context in which they are being implemented.

The primary objective of an mHealth initiative should be to provide solutions to specific problems that prevent the health system from delivering health interventions that are already known to be effective (Labrique et al. 2013). Amidst the mounting pressure to scale interventions beyond the pilot phase, the importance of strong evidence, proven cost-effectiveness, and a sound public health agenda must not be neglected in the interests of corporate growth or the desire to increase a market share of mobile subscriptions. Where the impetus and funding for mHealth projects in these contexts originate from a top-down approach, the adoptability and therefore the sustainability of mHealth programs are implicitly threatened. As such, mHealth projects must be deeply rooted within the technological, environmental, and social contexts within which they are deployed. The best way to ensure this happens is through the regular and consistent involvement of the very people the technology is meant to serve.

Questions

1. What are some of the specific health challenges that mHealth can help address in LMICs?

2. What are some of the ethical questions implementers should consider before implementing mHealth in an LMIC?

3. Identify some of the key stakeholder groups that could be included in the implementation of an mHealth program in LMICs. What strengths and weakness does each stakeholder group bring to the table?

4. Discuss some of the human implications of implementing mHealth.

5. What are the key challenges that program implementers should consider when implementing mHealth programs in LMICs?

6. What is a socio-technical approach, and what needs to be considered when adopting such an approach to implementing an mHealth initiative in an LMIC?

References

Anderson, R. and N. Perin. Case studies from the Vital Wave mHealth Report 2009. Available from http://homes.cs.washington.edu/~anderson/docs/2009/mHealthAnalysis_v1.pdf. Accessed 1st July 2014.

Archer, N. 2009. Mobile E-health: Making the case. In *Medical Informatics: Concepts, Methodologies, Tools, and Applications*, pp. 95–106. Hershey, PA: IGI Global.

Barrington, J., O. Wereko-Brobby, P. Ward, W. Mwafongo, and S. Kungulwe. 2010. SMS for Life: A pilot project to improve anti-malarial drug supply management in rural Tanzania using standard technology. *Malar. J.* 9:298. doi: 10.1186/1475-2875-9-298.

Boyce, N. 2012. The Lancet Technology: March, 2012. *The Lancet* 379 (9822):1187.

Braun, R., C. Catalani, J. Wimbush, and D. Israelski. 2013. Community health workers and mobile technology: A systematic review of the literature. *PLoS One* 8 (6):e65772. doi: 10.1371/journal. pone.0065772.

Chaiyachati, K.H., M. Loveday, S. Lorenz, N. Lesh, L.M. Larkan, S. Cinti, G.H. Friedland, and J.E. Haberer. 2013. A pilot study of an mHealth application for healthcare workers: Poor uptake despite high reported acceptability at a rural South African community-based MDR-TB treatment program. *PLoS One* 8 (5):e64662. doi: 10.1371/journal.pone.0064662.

Chatfield, A., G. Javetski, and N. Lesh. 2013. *CommCare Evidence Base*. New York: Dimagi.

Chetley, A. 2007. Improving health, connecting people: The role of ICTs in the health sector of developing countries. A framework paper, InfoDev.

Chigona, W., M. Nyemba, and A. Metfula. 2013. A review on mHealth research in developing countries. *J. Community Inform.* 9 (2).

Collins, F. 2012. How to fulfill the true promise of "mHealth": Mobile devices have the potential to become powerful medical tools. *Sci. Am.* 307 (1):16.

Curioso, W.H. and P.N. Mechael. 2010. Enhancing 'M-health' with south-to-south collaborations. *Health Aff. (Millwood)* 29 (2):264–267. doi: 10.1377/hlthaff.2009.1057.

Darby, J., J. Black, D. Morrison, and K. Buising. 2012. An information management system for patients with tuberculosis: Usability assessment with end-users. *Stud. Health Technol. Inform.* 178:26–32.

Ganesan, M., S. Prashant, and A. Jhunjhunwala. 2012. A review on challenges in implementing mobile phone based data collection in developing countries. *J. Health Inform. Dev. Ctries* 6 (1):366–374.

GrameenFoundation. 2011. *Mobile Technology For Community Health in Ghana: What It Is and What Grameen Foundation Has Learned So Far*. Washington, DC: Grameen Foundation.

Hoefman, B. and H. van Beijma. 2000. mHealth in Practice: Mobile technology for health promotion in the developing world. *Future* 1990:1999.

Hostgaard, A.M., P. Bertelsen, and C. Nohr. 2011. Methods to identify, study and understand end-user participation in HIT development. *BMC Med. Inform. Decis. Mak.* 11:57. doi: 10.1186/1472-6947-11-57.

International Telecommunications Union. (2014). The world in 2014: ICT Facts and Figures. Geneva, Switzerland: International Telecommunications Union.

Iyengar, M.S. and J.F. Florez-Arango. 2013. Decreasing workload among community health workers using interactive, structured, rich-media guidelines on smartphones. *Technol. Health Care* 21 (2):113–123. doi: 10.3233/thc-130713.

Kahn, J.G., J.S. Yang, and J.S. Kahn. 2010. 'Mobile' health needs and opportunities in developing countries. *Health Aff.* 29 (2):252–258.

Kallander, K., J.K. Tibenderana, O.J. Akpogheneta, D.L. Strachan, Z. Hill, A.H. Ten Asbroek, L. Conteh, B.R. Kirkwood, and S.R. Meek. 2013. Mobile health (mHealth) approaches and lessons for increased performance and retention of community health workers in low- and middle-income countries: A review. *J. Med. Internet Res.* 15 (1):e17. doi: 10.2196/jmir.2130.

Klungsöyr, J., P. Wakholi, B. Macleod, A. Escudero-Pascual, and N. Lesh. 2008. OpenROSA, JavaROSA, GloballyMobile—Collaborations around open standards for mobile applications. Paper read at *1st International Conference on M4D Mobile Communication Technology for Development*, Karlstad, Sweden.

Kuipers, P., J.S. Humphreys, J. Wakerman, R. Wells, J. Jones, and P. Entwistle. 2008. Collaborative review of pilot projects to inform policy: A methodological remedy for pilotitis? *Aust. New Zealand Health Policy* 5:17. doi: 10.1186/1743-8462-5-17.

Labrique, A.B., L. Vasudevan, E. Kochi, R. Fabricant, and G. Mehl. (2013). "mHealth innovations as health system strengthening tools: 12 common applications and a visual framework." *Global Health: Science and Practice* no. 1 (2): 160–171.

Lucas, H. 2008. Information and communications technology for future health systems in developing countries. *Soc. Sci. Med.* 66 (10):2122–2132. doi: http://dx.doi.org/10.1016/j.socscimed.2008.01.033.

Macleod, B., J. Phillips, A.E. Stone, A. Walji, and J.K. Awoonor-Williams. 2012. The architecture of a software system for supporting community-based primary health care with mobile technology: The Mobile Technology for Community Health (MoTeCH) initiative in Ghana. *Online J. Public Health Inform.* 4 (1). doi: 10.5210/ojphi.v4i1.3910.

Magbity, E., H. Ormel, H. Jalloh-Vos, K. de Koning, E. Mbalu Sam, and H. van Beijma. 2011. mhealth for maternal and newborn health in resource-poor and health system settings, Sierra Leone Feasibility Study Report. http://r4d.dfid.gov.uk/PDF/Outputs/Misc_MaternalHealth/technicalbrief-mhealth-SierraLeone.pdf. Accessed July 1, 2014.

Manda, T.D. and J. Herstad. 2010. Implementing mobile phone solutions for health in resource constrained areas: Understanding the opportunities and challenges. In *E-Infrastructures and E-Services on Developing Countries*, A. Villafiorita, R. Saint-Paul, and A. Zorer, eds., pp. 95–104. Berlin, Germany: Springer.

Manda, T.D. and Y. Msosa. 2012. Socio-technical arrangements for mHealth: Extending the mobile device use and adoption framework. In *e-Infrastructure and e-Services for Developing Countries*, R. Popescu-Zeletin, K. Jonas, I.A. Rai, R. Glitho, and A. Villafiorita, eds., pp. 208–217. Berlin, Germany: Springer.

Manda, T.D. and T.A. Sanner. 2012. Bootstrapping information technology innovations across organisational and geographical boundaries: Lessons from an mHealth implementation in Malawi. Paper read at *IRIS*. No. 35. Tapir Akademisk Forlag.

McCord, G.C., A. Liu, and P. Singh. 2013. Deployment of community health workers across rural sub-Saharan Africa: Financial considerations and operational assumptions. *Bull. World Health Org.* 91 (4):244–253b.

McNamara, K.S. 2003. *Information and Communication Technologies, Poverty and Development: Learning from Experience*. Washington, DC: World Bank.

Mechael, P., H. Batavia, N. Kaonga, S. Searle, A. Kwan, A. Goldberger, L. Fu, and J. Ossman. 2010. Barriers and gaps affecting mHealth in low and middle income countries: Policy white paper: Columbia university. Earth institute. Center for global health and economic development (CGHED): with mHealth alliance.

mHealthAlliance. 2011. *Participatory Design for mHealth*. Washington, DC: mHealth alliance.

mHealth Alliance. 2012. *Leveraging Mobile Technologies to Promote Maternal & Newborn Health: The Current Landscape & Opportunities for Advancement in Low-Resource Settings*. Oakland, CA: The Center for Innovation & Technology in Public Health Public Health Institute.

Ngabo, F., J. Nguimfack, F. Nwaigwe, C. Mugeni, D. Muhoza, D.R. Wilson, J. Kalach, R. Gakuba, C. Karema, and A. Binagwaho. 2012. Designing and implementing an innovative SMS-based alert system (RapidSMS-MCH) to monitor pregnancy and reduce maternal and child deaths in Rwanda. *Pan Afr. Med. J.* 13:31.

Orlikowski, W.J.. 2000. Using technology and constituting structures: A practice lens for studying technology in organizations. *Org. Sci.* 11:404–428.

Pascoe, L., J. Lungo, J. Kaasbøll, and I. Koleleni. 2012. Collecting integrated disease surveillance and response data through mobile phones. Paper read at *IST-Africa*, May 9–11, 2012.

Pop-Eleches, C., H. Thirumurthy, J.P. Habyarimana, J.G. Zivin, M.P. Goldstein, D. de Walque, L. MacKeen et al. 2011. Mobile phone technologies improve adherence to antiretroviral treatment in a resource-limited setting: A randomized controlled trial of text message reminders. *AIDS* 25 (6):825–834. doi: 10.1097/QAD.0b013e32834380c1.

Purkayastha, S., T.D. Manda, and T.A. Sanner. 2013. A post-development perspective on mHealth—An implementation initiative in Malawi. Paper read at 2013 *46th Hawaii International Conference on System Sciences (HICSS)*, January 7–10, 2013, Wailea, HI.

Purkayastha, S., S. Sahay, and A. Mukherjee. 2010. Exploring the potential and challenges of using mobile based technology in strengthening health information systems: Experiences from a pilot study. Paper read at *Sixteenth Americas Conference on Information Systems*, August 12–15, 2010, Lima, Peru.

Robertson, C., K. Sawford, S.L. Daniel, T.A. Nelson, and C. Stephen. 2010. Mobile phone-based infectious disease surveillance system, Sri Lanka. *Emerg. Infect. Dis.* 16 (10):1524–1531. doi: 10.3201/eid1610.100249.

Rotheram-Borus, M.J., L. Richter, H. Van Rooyen, A. van Heerden, M. Tomlinson, A. Stein, T. Rochat et al. 2011. Project Masihambisane: A cluster randomised controlled trial with peer mentors to improve outcomes for pregnant mothers living with HIV. *Trials* 12:2. doi: 10.1186/1745-6215-12-2.

Rotheram-Borus, M.J., M. Tomlinson, D. Swendeman, A. Lee, and E. Jones. 2012. Standardized functions for smartphone applications: Examples from maternal and child health. *Int. J. Telemed. Appl.* 2012:973237. doi: 10.1155/2012/973237.

Routen, T., L.F. Silas, M. Mitchell, J. van Esch, N. Lesh, J. Lyons, and R. Badiani. (2010). Using Mobile Technology to Support Family Planning Counseling in the Community. *Paper read at International Conference on ICT for Africa*, at Yaoundé, Cameroon.

Sanner, T.A., L.K. Roland, and K. Braa. 2012. From pilot to scale: Towards an mHealth typology for low-resource contexts. *Health Policy Technol.* 1:155–164.

Sherwani, J., S. Palijo, S. Mirza, T. Ahmed, N. Ali, and R. Rosenfeld. 2009. Speech vs. touch-tone: Telephony interfaces for information access by low literate users. *Proceedings of the 3rd International Conference on Information and Communication Technologies and Development (ICTD'09)*, IEEE Press, Piscataway, NJ, pp. 447–457.

Shozi, N.A., D. Pottas, and N. Mostert-Phipps. 2012. A socio-technical perspective on the use of mobile phones for remote data collection in home community based care in developing countries. In *e-Infrastructure and e-Services for Developing Countries*, R. Popescu-Zeletin, K. Jonas, I. Rai, R. Glitho, and A. Villafiorita, eds., pp. 135–145. Berlin, Germany: Springer.

Siedner, M.J., J.E. Haberer, M.B. Bwana, N.C. Ware, and D.R. Bangsberg. 2012. High acceptability for cell phone text messages to improve communication of laboratory results with HIV-infected patients in rural Uganda: A cross-sectional survey study. *BMC Med. Inform. Decis. Mak.* 12:56. doi: 10.1186/1472-6947-12-56.

Strachan, D.L., K. Kallander, A.H. Ten Asbroek, B. Kirkwood, S.R. Meek, L. Benton, L. Conteh, J. Tibenderana, and Z. Hill. 2012. Interventions to improve motivation and retention of community health workers delivering integrated community case management (iCCM): Stakeholder perceptions and priorities. *Am. J. Trop. Med. Hyg.* 87 (5 Suppl):111–119. doi: 10.4269/ajtmh.2012.12–0030.

Svensson, J. and C. Wamala. 2012. M4D: Mobile communication for development. In *SANORD Symposium*, Aarhus, Denmark.

Tamrat, T. and S. Kachnowski. 2012. Special delivery: An analysis of mHealth in maternal and newborn health programs and their outcomes around the world. *Matern. Child Health J.* 16 (5):1092–1101. doi: 10.1007/s10995-011-0836-3.

Tomlinson, M., M.J. Rotheram-Borus, L. Swartz, and A.C. Tsai. 2013. Scaling up mHealth: Where is the evidence? *PLoS Med.* 10 (2):e1001382. doi: 10.1371/journal.pmed.1001382.

Tomlinson, M., W. Solomon, Y. Singh, T. Doherty, M. Chopra, P. Ijumba, A.C. Tsai, and D. Jackson. 2009. The use of mobile phones as a data collection tool: A report from a household survey in South Africa. *BMC Med. Inform. Decis. Mak.* 9:51. doi: 10.1186/1472-6947-9-51.

Toyama, K. 2010. Can technology end poverty? *Boston Rev.*

Vallières, F., E. McAuliffe, I. Palmer, E. Magbity, and A.S. Bangura. 2013. Supporting & strengthening maternal, neonatal, and child health services using mobile phones in Sierra Leone: A research protocol. *Harvard Afr. Policy J*, 8, 46–51.

Vodafone. 2013. Sustainability report. Vodafone Group.

Walker, G.H. 2008. A review of sociotechnical systems theory: A classic concept for new command and control paradigms. *Theor. Issues Ergon. Sci.* 9 (6):479–499.

Walsham, G. 2012. Are we making a better world with ICTs? Reflections on a future agenda for the IS field. *J. Inform. Technol.* 27:87–93.

WHO. 2006. *World Health Report 2006*. Geneva, Switzerland: World Health Organization.

WHO. 2011. mHealth: New horizons for health through mobile technologies: Second global survey on eHealth. In *Global Observatory for eHealth Series*. Geneva, Switzerland: World Health Organization. http://www.who.int/goe/publications/goe_mhealth_web.pdf. Accessed July 1, 2014.

Willis-Shattuck, M., P. Bidwell, S. Thomas, L. Wyness, D. Blaauw, and P. Ditlopo. 2008. Motivation and retention of health workers in developing countries: A systematic review. *BMC Health Serv. Res.* 8 (1):247.

World Bank. 2012. *Information and Communications for Development: Maximizing Mobile*. Washington, DC: World Bank.

Zurovac, D., R.K. Sudoi, W.S. Akhwale, M. Ndiritu, D.H. Hamer, A.K. Rowe, and R.W. Snow. 2011. The effect of mobile phone text-message reminders on Kenyan health workers' adherence to malaria treatment guidelines: A cluster randomised trial. *Lancet*. doi: 10.1016/s0140-6736(11)60783-6.

Further Reading

CommCare for Anteatal Services in Nigeria. USAID mHealth Compendium, Volume 2, Technical Report 2012. http://www.jhsph.edu/departments/international-health/_documents/USAIDmHealthCompendiumVol2FINAL.pdf. Accessed July 1, 2014.

Mobile Technology for Community Health in Ghana: What it is and What Grameen Foundation Has Learned So Far (2012). Washington DC: Grameen Foundation. http://www.grameenfoundation.org/sites/grameenfoundation.org/files/MOTECH-Lessons-Learned-Sept-2012.pdf. Accessed July 1, 2014.

mHealth in Ethiopia - *Strategies for a New Framework (2011)*. Vital Wave Consulting. http://www.vitalwaveconsulting.com/pdf/2011/mhealth%20Framework%20for%20ethiopia%202011.pdf. Accessed July 1, 2014.

Section V

mHealth Education, Adoption, and Acceptance

16

Role of mHealth in Nursing: A Conceptual Framework

Phillip Olla, Christie Butrico, and Chaz Pichette

CONTENTS

16.1 Introduction

Over the last decade, the mobile health (mHealth) phenomenon has developed into one of the most significant technological advances in digital health. The mHealth trend is amalgamating health-care professionals, academic researchers, and technology industry disciplines worldwide to achieve innovative solutions in the areas of health-care delivery and biomedical sectors. The impact of mHealth on the nursing domain is not fully clear; there are tremendous opportunities and challenges ahead for nurses aiming to adopt mHealth into their field. Around the turn of the century, electronic health (eHealth) emerged and was utilized by health-care providers as the world of information technology expanded. As medical records transitioned from paper charting to electronic charting, the next logical step in this progression was mobile charting. Mobile health, also called mHealth, is the term used to encompass retrieving and manipulating eHealth information on mobile devices such as personal digital assistants (PDAs), smartphones, or tablet personal computers (PCs).

PDAs were originally devices that stored an individual's personal information such as their appointment calendar and address book. The first smartphones were those that combined PDA and paging capabilities to a mobile telephone and eventually added Internet access. Today, a smartphone is a much more complex device with features that mimic PCs. Smartphones are built on a mobile interface and include functions as phones, PDAs, pagers, and computers. They typically have Internet access and run third-party, add-on applications (apps) all on one compact handheld mobile device. They have the aptitude to be more beneficial than traditional PDAs because they have the capacity and versatility to access and view medical data such as orders, medications, adverse effects, and lab values and retrieve videos of obscure procedures that need to be reviewed before performing patient care.

Nurses today are more technologically inclined, as most in the younger generation were introduced to computers at a young age (Hendrich et al., 2008). "According to a recent survey conducted by Wolters Kluwer Health's Lippincott Williams & Wilkins (LWW), 71 percent of nurses are already using smartphones for their job. The survey included responses from 3,900 nurses and nursing students in Massachusetts" (Dolan, 2012, para. 1). Mobile electronic devices improve nursing care by providing access to evidence-based resources, acting as an educational tool for nurses and/or patients, and enabling users to create a library of apps that enhance nursing practice.

Nurses in particular have the ability to use smartphones during their practice in a variety of ways. Smartphones can become medical devices, store data, and contain nursing apps of the users' choice. Apps can be stand-alone, which means they will run on the device independently, or bundled, which often requires an Internet connection to operate.

This chapter is structured as follows. Section 16.2 provides a review of the literature, discussing point of care, reference, educational, and communication. Section 16.3 describes a conceptual framework that was created from the literature describing the following components: mobile device, features and functionality, medical apparatus, data storage, and nursing meaningful use. Prior to the conclusion, Section 16.4 presents a discussion that illuminates the challenges of integrating mHealth into the nursing practice.

16.2 Literature Review

This section provides an overview of the literature that focuses on the deployment of smartphones in the nursing domain. After researching a vast quantity of articles, four main areas where nurses use mobile devices in clinical practice emerged. The areas are point of care, reference, educational, and communication.

16.2.1 Point of Care

Dr. Doran provided a report of three different studies in Canada on the use of PDAs and clinical decision-making in 2009. PDAs and smartphones have the ability to bring evidence-based resources to nurses at the point of care, which is when nurses and patients interact. As a resource to obtain evidence-based guidelines, nurses gave mobile devices a moderately high rating (Doran, 2010; Hawkins, 2012). Research shows having access to reliable information for nursing interventions helps minimize variation in clinical practices and improves the quality of care nurses provide. Smartphones allow immediate and effective communication between colleagues via e-mail, text, or voice, allowing for a decrease in the amount of adverse events in health care. These devices allow the user to store a library at their fingertips. This aids in providing accurate information on drug-to-drug interaction thereby decreasing medication errors (Doran, 2009).

A researcher in 2011, at a 135-bed medical center in Oklahoma, reveals that nurses use iPod Touch and an app to help prevent medication errors and adverse drug events. Nurses scanned the bar code of the drug with the iPod Touch and the patient's medical bracelet and then wait for the app to verify the medication, dose, time, and patient from the electronic health record (EHR). "In 2011, the hospital pharmacy dispensed about 750,000 drugs" and "the hospital's medication error rate was just 1.84 per 10,000 doses" (Williams, 2012, p. 99). The evidence clearly shows a decrease in medication-related mistakes in the hospital setting.

Nurses find mobile devices useful in the field, such as during home care visits. A recent study in Quebec, involving 139 nurses, implemented mobile computing devices for home care visits. Results show the mobile computing devices change the nurse's intervention approach from a narrowly focused prescribed task to seeing the person as a whole (Pare et al., 2011). The nurses also recognize improvement on the quality of their notes to the patient records and more importantly an "increase in the quality of care they provided to patients" (Pare et al., 2011, p. 316). This modern approach to home health care allows the nurse better management of the patient's clinical data, scheduling, and care planning and enhances the communication with team members. Because benefits of holistic care are more evident today, the use of patient-specific technology enables team members to better communicate with one another. This increased communication will enable each member of the care team to more readily recognize they are part of a much larger whole and more easily access and share information.

Physicians and nurses are able to send prescription orders to pharmacies, look up drug references and drug-to-drug interactions, and review medical conditions from mobile apps. Nurses are able to use an app and the iPod Touch in direct patient care for medication administration, care interventions, specimen tracking, and team communication (Williams, 2012). This quick, easy access to information often gives the nurse more time to spend face-to-face with the patient.

16.2.2 Reference

Frequently used medical applications are drug references, which provide information on thousands of medications, doses, safety information, adverse drug–drug interactions, and more (Williams, 2012). Popular drug reference apps include ePocrates®, Skyscape©, and Lexi-Drugs®. These apps receive high ratings among health-care providers (Wyatt and Krauskopf, 2012, p.14). The U.S. National Library of Medicine's application, PubMed on Tap, allows access to current research and clinical information at the users' fingertips.

16.2.3 Education

Mobile devices are now being used in direct patient care in unexpected, yet often over-looked and obvious ways. Recently, a hospital nurse provided her patient, a middle-aged man admitted for a severe forearm injury that required surgical reconstruction, with written discharge instructions for wound care. Soon, the patient and family members "realized that changing the dressing looked easier on paper than in real life. Family members were reluctant to even look at the arm, much less to dress the wound" (Holt et al., 2011, p. 48). As the patient reached for his mobile phone to call a friend for help, the nurse was struck with inspiration. "She noticed that the patient's phone had both camera and voice memo capabilities, discovered that the patient was familiar with these functions, and realized that the mobile phone" would be a great way to provide detailed, step-by-step wound management instruction for the patient to refer to while at home (Holt et al., 2011, p. 48). Because the nurse used the patient's own phone to take the pictures and they were not copied to another device, the Health Insurance Portability and Accountability Act (HIPPA), which protects personal health information, was not violated as the patient had full control of the photos. This simple, yet ingenious intervention empowered the patient to confidently care for himself and allowed his wound to heal without infection or a return visit to the hospital (Holt et al., 2011).

16.2.4 Communication

Many health-care professionals value the ease and efficiency of communications with colleagues offered via smartphones or tablets. Some indicate the integrated camera functionality as a useful tool, which enables them to take a photo of a wound and send it to a colleague to collaborate on treatment options (Garrett and Klein, 2008). Another article notes the increase of smartphones in medical practice as "hospitals are providing clinical staff with smartphones rather than pagers because smartphones are more efficient. The ability to send longer text messages eliminates the phone tag that often occurs with pagers" (Moore, 2011, p. 393). Communications via voice, text, video, and e-mail are all tools clinicians use in practice. The fact that all this can be achieved on one device prevents health-care workers from having multiple pagers hanging from their waistband or in their pockets. "Nurses may need to go through a complex process to find the responsible physician for a patient using traditional numeric paging. In Toronto General Hospital, the use of smartphones simplifies reaching the responsible physician" (Mosa et al. 2012, p. 11).

16.3 Conceptual Framework

Smartphones and tablets enable nurses to provide direct patient care based on the most current, evidence-based research available using the advanced technology our society is embracing. The innovative devices and apps existing today, and those planned for the

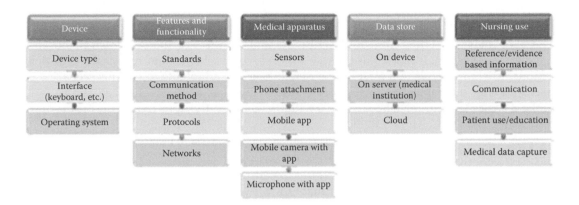

FIGURE 16.1
Conceptual nursing framework for mHealth apps.

near future, provide immediate care, even from a distance, in ways many never imagined. Nurses now have the ability to chart patient information, retrieve lab results, access medical reference materials, and educate their patients from mobile devices, all of which contribute in revolutionizing the health-care industry. Teaching opportunities exist for nurses to provide care for these populations and improve their quality of life by obtaining, identifying, and using mobile devices and apps that can assist them.

Key factors impacting the adoption of mHealth for the nursing practice are compiled into a framework (Figure 16.1) that encompasses the following components: mobile device, features and functionality, medical apparatus, data storage, and nursing meaningful use. Each one of components will be discussed in the following subsections.

16.3.1 Mobile Devices

It all starts with the device. What device will enable the nurse to more easily and quickly perform their job with handheld technological support? In addition to smartphones, some nurses prefer using tablet devices such as Apple's iPad or Samsung's Galaxy, as the features between the smartphones and tablet devices are very similar. They often use the same app (some apps are designed with variation to fit the larger screen of the tablets) and have long battery life, so charging is rarely needed throughout the day. A recent article in the *American Journal of Nursing* (*AJN*) by an emergency department registered nurse (RN) Megen Duffy mentions the dissimilarities between smartphones and tablets saying "it's tempting to view the iPad as a giant iPhone that doesn't make phone calls" but goes on to point out key differences (Duffy, 2012, p. 59). The size is an obvious deviation as tablets are larger and therefore have larger screens than smartphones. Both are lightweight devices, much more so than standard laptop computers, but the smartphone can fit in most nurses scrub pockets, which might also be the case for the upcoming iPad mini. Many smartphones and tablets have Bluetooth capabilities, which allow user to attach apparatuses such as a wireless microphone, keyboard, or medical attachments such as glucose meters. Wireless keyboards are handy if the user is doing a lot of typing. There are many Bluetooth keyboards on the market in varying sizes, some of which can fold into a smaller size for easier portability and storage.

The size of the device and functionality is now a key differentiator for device manufactures. The key items to consider are the interface (virtual keyboard vs. physical), organizational policy on mobile devices, and support for operating systems (OSs).

16.3.1.1 User Interface

The user interface (UI) is used to describe the look and feel of the on-screen menu and navigational system. It encompasses how it works, color scheme, and responses to button actions. It is imperative that a UI that is easy to use and adaptable to various scenarios is used. Having the capability to support gestures, widgets, and personalized setups will make it easier to customize how a nurse will interact with the apps and OSs.

16.3.1.2 Operating System

An OS is a system of software that programs or controls a computer, tablet, or smartphone. The leading mobile OSs today are arguably iOS by Apple Inc., Windows by Microsoft, Android by Google Inc., BlackBerry OS by Research In Motion (RIM), and Symbian OS by Nokia and Accenture. Mobile apps work with OSs the same way cellular networks work with mobile phones; a smartphone is initially designed to work on a specific network and is run by a specific OS. More and more apps are designed to be compatible with various OSs, or the app is written for multiple OS platforms allowing more consumers to have access to the app, regardless of which OS their mobile device utilizes. There are two considerations when selecting a device. What support is available in the institution for each mobile OS? Do the main apps run on the OS that is being considered as a standard for the health-care organization? As of January 2014, the iOS by Apple is the dominant OS in health care. Research by Manhattan Research in 2011 revealed that 75% of physicians surveyed possessed at least one Apple device. (Comestock, 2013) Additionally, a Vitera survey conducted in 2012 of health-care professionals backed up this high number, revealing that 60% of respondents used an iPhone and 45% owned an iPad (Morris, 2012).

16.3.2 Features and Functionality

Standard smartphones include communication abilities via audio, video, text, and e-mail, have a camera with video recording features, and store enormous amounts of data. Smartphones have advanced technology incorporating digital storage, sensors, microphones, and cameras that rival those sold individually. The camera function on Apple's latest iPhone camera is an 8-megapixel camera with high-definition video recording, autofocus, LED flash, and more. This quality rivals many stand-alone digital cameras on the market. The vast quantity of storage space is the key feature to today's smartphone and users can often choose the amount of memory they prefer. Battery life, which varies depending on the specific device and usage, is often comparable, and most will operate all day only needing recharging nightly. This is an important feature when compared to laptop computers that often require recharging after a few hours of continuous use. One other feature that is now standard is the accelerometer. An accelerometer is a sensor embedded within the smartphone that measures the tilting motion and orientation of a mobile phone. All these features are creating opportunities for developers to harness the power of the devices to create innovative apps to address specific health-care problems.

16.3.3 Medical Apparatus

There are now a variety of medical devices that require the use of a smartphone to operate them. Devices can be categorized into five groups (Figure 16.2). The first group contains apps such as mobile microscope that requires a lens attachment to support an app

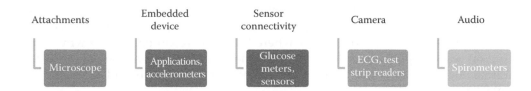

FIGURE 16.2
Medical apps.

to diagnose diseases of the blood such as malaria (Breslauer et al., 2009). The next group is the most popular and practical. Persuasive apps are loaded to the devices that support the nurse or patient. Examples include disease diaries, drug calculators, and medication adherence solutions (Ragnemalm and Bång, 2012).

The next group of apps relies on connectivity software to interact with an external medical device such as a glucose meter to capture medical parameters into a software app (Baron et al., 2012). There are connectivity protocols such as Bluetooth low energy, ZigBee, near-field communication (NFC), and even cables. One of the most innovative uses of the mobile phone camera is for medical diagnostics; the camera has been used to decode test strips and capture ECGs and imaging for dermatology applications. There have been some recent innovations in using the audio jack to capture sound. An example of these types of apps are mobile devices that were used to capture peak flow (Larson et al., 2012b).

16.3.4 Data Storage

Memory, or digital storage capability, enables users to store information, contacts, calendars, pictures, music, movies, mobile apps, and application data. Many smartphones and tablets are available with different storage capacity and are priced according to the amount of memory the user wants.

16.3.4.1 On-Device Storage

Apple's newest iPhone and iPad are available with 16, 32, or 64 GB storage capacities. Larger, more complex files require more storage space. A single picture will take up less memory than a movie. Larger apps, such as drug reference manuals, will take up more memory than a drip calculation. The reference manual stores an entire book on the device, where the drip calculation is a mathematical program that requires the user to input specific numbers. Individual calculations for all drugs are not stored on a drip app, just the calculated equation needed to solve the math, and thus require far less on-device storage.

16.3.4.2 On Site Servers

A server is a central computer that houses enormous amounts of data others can access via a network. Urban hospitals and corporations that use private, closed servers require login access. The information and data hospitals create and store, such as lab and test results, radiology imaging, and health information, which is usually encrypted for security, are stored at a central location. Corporations provide extensive security measures, depending on the sensitivity of data stored, at this central location. Other computers access the server information through the private network, often hardline (as opposed to wireless), for viewing and data entry. Individual computers, or workstations, located throughout the hospital, do not store all the data on each computer, but rather each computer is able to access the central server where the data are stored. This is why a nurse working on one computer workstation is able to log off, move down the hall or to a different floor, log on to the server from another computer workstation, and access the same information.

16.3.4.3 Cloud Computing

The latest addition to smartphone and tablet storage is wireless third-party servers, now commonly called cloud computing. Cloud computing is based on sharing of resources and applications to achieve coherence and economies of scale. At the underpinning of cloud computing is the broader concept of converged infrastructure and shared services. Because both technology and security measures are progressing for wireless computing, virtual storage sources are now growing in popularity as well as capacity. Smartphones and tablets now have the ability to quickly, and wirelessly, access secure data, which users store in advance, with a third-party server, usually for a fee.

16.3.5 Nursing Use

Recent reports suggest that both Google's Play app store for Android devices and Apple's app store have around 700,000 applications available, with the reported number of medical apps available reaching 40,000 over the summer (Gold, 2012; Oliver, 2012). With this staggering number of apps, nurses need to be cognizant of where the apps are designed. As mentioned earlier, there are a number of references, calculation, educational, journal, and database apps available for nurses and health-care providers designed by accredited, respected sources. However, there are also apps developed by relatively unknown sources. Many apps have disclaimers when initially opened, which require user compliance to access the app (WHO, 2011). This is a safeguard developers' implement that prevents legal responsibility if the app used in medical practice provides unfavorable results. Nurses may need to do some research regarding the source before using these apps in their medical practice. They have been taught to use credible information, and they need to apply that knowledge not only to paper reference materials but to digital ones as well. While the Federal Drug Administration (FDA) is attempting to regulate the growing number of apps, it is a daunting task, which will be discussed later.

16.3.5.1 Reference and Evidence-Based Information

As discussed previously, reference apps are the most commonly used among medical professionals. Also available are IV compatibility, drug interaction, and pill identifier apps. Pharmacology apps are accessible in subspecialty categories including, but not limited to, pediatric, neurology, cardiovascular, antibiotic, and anesthesiology. A wealth of evidence-based information is obtainable in the form of free apps that require a subscription, similar to magazines, newspapers, or journals consumers receive via paper or digital mail. The American Nurses Association (ANA), *Journal of the American Heart Association* (*JAHA*), PubMed on Tap, *AJN*, and the *New England Journal of Medicine* (*NEJM*) name a few publications in the countless number available for mobile devices. Many employers or universities will provide their employees or students access to these articles, so it is beneficial to ask the employer or institution before purchasing.

16.3.5.2 Communication

There are now businesses dedicated to health-care communications via smartphones. One such company, Voalte, Inc., developed an app for nurses that other health-care workers, such as physicians, are requesting. Voalte One, an app designed for the iPhone, is now used in 22 hospitals in the United States, including Cedars-Sinai in Los Angeles and Massachusetts General. The app allows nurses to communicate with each other at point of care by text or voice, alerts the nurse to patient alarms, and provides instant response to the patients through a room callback element. Due to the versatility, nurses "send out communications as they [think] of them, freeing them to continue with their tasks. For recipients, the texts [allow] them to respond when they are able" (Pollick, 2012, "The Apple Advantage," para. 11). This communication at point of care allows nurses to be more efficient, safe, and productive while improving patient and nurse satisfaction.

16.3.5.3 Patient Use and Educational Apps

As mentioned earlier, smartphones are assisting nurses with patient education for care and treatment compliance. Other apps can help patients track physical activity, diet, sleep cycles, infant activity, and much more. Mobile medical apps prove to be a great source of information for health-care providers as well as for patients. The industry is expanding at a startling rate creating an exciting environment for those in the field. Some apps provide wait times for nearby emergency departments or urgent care clinics and give directions to each (Williams, 2012). Other apps assist with virtual physician visits and enable patients to communicate with physicians without actually making a trip to the doctor's office (Maslowsky et al., 2012).

16.3.5.4 Medical Data Capture

More technical apps are emerging on the market, like the iBGStar, the first FDA-approved blood glucose meter and the corresponding app designed for a smartphone. The apparatus, which directly connects to any Apple iOS device (iPhone, iPod Touch, or iPad) and works in sync with the iBGStar diabetes manager app, helps patients with their self-monitoring blood glucose (SMBG) (Tran et al., 2012). The app immediately displays and stores the patient's blood sugar reading and has multiple features, including the ability to share the results via e-mail to another person, such as a parent or health-care worker. "Studies have shown the device to be 99.5% accurate and to meet the International Organization for Standardization's standards for glucose monitors" (Tran et al., 2012, p. 77). Tran goes on to mention other devices still in development that may provide an alternative means for checking glucose readings without using a lancet. If successful, the lancet-free device would prevent millions of patients from pricking their fingers multiple times a day to check their sugar level.

16.4 Adoption Challenges

As beneficial as smartphones and tablets are to nursing practice, challenges impede the progression of this technology. Integration of the latest technology to practical use takes time, especially if a corporation is large. Smaller hospitals are able to integrate new, evidence-based technologies faster than larger hospitals. Bar Code Medication Administration (BCMA) was devised by a nurse in Kansas in 1995 and was introduced to 161 Veterans Affairs hospitals from 1999 to 2001 (Putzer & Park, 2010). However, it has yet to be incorporated into many hospitals today. With nearly two decades since conception and integration, BCMA has proven time and again to increase patient safety and reduce medication administration errors. Despite these sterling results, BCMA has yet to be implemented in some of the nation's top hospitals. This is an example of the assimilation of technology that can lag behind its development in our society. A possible solution could include the incorporation of the newer technology to specific patient care units of bigger hospitals as opposed to waiting for the entire institution to adopt the change at once. Beginning with units that have a higher portion of medication administration would benefit to the earlier use of the BCMA technology such as intensive care units (ICUs), operating rooms (ORs), or cardiac units.

16.4.1 FDA Regulation

In July of 2011, the FDA responded to the growing number of medical applications and issued a draft guidance for the regulation of mobile medical apps. Not all apps, such as reference guides, will be regulated. Those that enable users to input patient-specific information and transform the mobile device into a medical device, like a blood glucose monitor or stethoscope, and those that are an extension of a currently used medical device, such as a remote display of bedside monitors, will be regulated (Barton, 2012).

16.4.1.1 Regulated Apps

When a drug reference guide is used as a source of information in the medical practice to check medication dosage, some feel that it becomes a medical device. Numerous mHealth apps range from tips and tool for consumers to medical information for health-care workers. Many believe regulation of health apps for portable electronic devices is needed in order to ensure patient safety when medical apps are used as medical tools. The reality is that many nurses use apps that come with a legal disclaimer when the app is originally opened. The disclaimer is a legal statement so the app's creator is not legally responsible if a medical error occurs when the app is used in medical practice. Nurses, as all health professionals, need to know who is creating the apps they download and use their educated judgment on whether the source is credible.

16.4.1.2 FDA Cost and Delay

With this relatively new FDA regulation comes a hefty monetary cost and time delay in releasing apps to the public, of which many consumers are unaware. The FDA uses a "premarket approval (PMA) pathway to evaluate and approve technologies that are truly novel and pose a high potential risk to patients using them" and a different evaluation, called 501(k), for low- to medium-risk devices (Makower et al., 2010, p. 12). According to a survey of over 200 medical technology (medtech) companies during the summer of 2010, the release time of devices using PMA, compared with similar, if not the same devices to the European Union (EU) and their regulation authorities, took an average of 24 months longer in the United States, from first communication with FDA to approval (Howard, 2011). For higher-risk devices, the FDA's approval took an average of 43 months longer than EU's. That's two to three and a half years of device lag for U.S. patients.

Perhaps the FDA is not giving itself enough credit. Devices approved by the FDA between 2004 and 2009 had a 99.6% success rate without a Class I recall. In addition to the time suspension, the average monthly cost of the 510(k) evaluation process is $520,000, while the PMA process is $740,000 monthly (Makower et al., 2010). This is a tremendous expense each company is required to pay, especially when they have fewer than 50 employees, which includes 80% of medtech companies (Makower et al., 2010). While this particular survey focuses on medical devices, the same, if not more, suspensions and costs are expected to occur with mobile medical apps, as they will also be evaluated using the 501(k) process. These FDA regulations are meant to protect U.S. patients, but many believe the lengthy postponements are proving to be a disservice and a hindrance to the U.S. medical community as the technology is available to patients around the world before it is available to the country that is the world's leading producer of such advancements.

16.4.2 Monetary Concerns

16.4.2.1 Device Value

The financial cost of the device itself is a deterrent to consumers. The price of the iPad mini starts at $329, the iPad (third generation) at $499, the iPod Touch (fifth generation) at $299, and the iPhone 5 at $199 with a mobile service plan that includes a 2-year contract. Older-generation models are available for reduced prices depending if consumers buy them used or refurbished. Google's original Nexus 7 tablet starts at $199 and the newer Nexus 10 starts at $399. Verizon wireless Droid phones start at $99 including a 2-year contract, with the price increasing depending on the smartphone preferred. While the newest technology is expensive, the slightly older models are available for reduced prices if the consumer takes time to look for sales or secondhand stores. Technological device turnover is relatively quick, with newer model or generation smartphones and tablets releasing every 12–16 months, such as iPhone since its original release in 2007. The older models are often able to receive software updates from the OS companies for a few years, allowing the platform to update despite being the previous device generation.

16.4.2.2 Price of Apps

Another challenge is the monetary price of apps for consumers. While many apps are available free of charge, some carry a hefty price tag. Some apps provide a free *lite* version of their app with limited information and offer another app for a price with significantly more information, often ranging from $0.99 to $5.99. Other more complex apps can cost $50, $100, and beyond. The Davis Drug Guide, a drug reference book many nursing schools refer to while educating their students, is available as an app for the same price as the paper version. The app on a smartphone will take considerably less space in the nurse's pocket than a thick book, which is a key benefit for nurses and nursing students alike.

16.4.3 Security

Security may be a concern for mHealth on smartphones and tablets if patient health information (PHI) or EHR is accessed, stored, or inputted to the device. Before nurses are able to view their work e-mail on smartphones or tablets, most institutions of employment will walk them through the setup process to ensure PHI is secure. This often includes using only secure networks, user authentication, and individual passwords. Even after this is done, many times nurses cannot send e-mail from their mobile device because of the security measures in place that prevent smartphones and tablets from storing information. This is a safeguard institutions implement before access to secure information, such as employee e-mail, is allowed on personal devices.

16.4.3.1 Patient Health Information

In many situations, viewing secure PHI on a mobile device is similar to viewing the information on the computers at large hospitals. The data are not stored on each device or computer itself but accessed through the device or computer. This is how nurses can access their patient's information for various computers in a hospital and do not need to travel back to one specific computer. The PHI, like an x-ray image or a patient chart, is often stored on the hospital server, which is then accessed through a secure network that meets HIPPA regulations. This is frequently achieved through an Advanced Encryption Standard (AES) encryption or a virtual private network (VPN) that usually involves a unique individual identification name and password for each employee (Luxton et al., 2012).

Institutions and health-care facilities often have security measures prepared and primed before a device is linked to the facility and its specific hospital programs that store PHI. If a nurse is careless and leaves a paper with PHI in a public area, that nurse is in violation of HIPPA and can be held accountable. This also applies to digital PHI. The nurse should take the same precautions they have done throughout their career and keep patient information private and secure. Just as all computers in hospitals are password protected, nurses should password protect their personal smartphones or tablets, which is included on all OSs, if they use them at work to store or view patient information.

16.4.3.2 App Security

Many apps used in nursing today do not actually input PHI that can lead back to an individual patient. For example, an app that helps set an IV drip rate or configure a medicine dose involves inputting numbers, and possibly a specific weight, into the app, but such data would not lead to a specific patient if an unknown person stumbled across the device. If the nurse is entering PHI, such as on admission, to a tablet, information is not saved or stored on the device itself but sent over a secure connection for storage on the hospital server (Yadalla & Shankar, 2012). The tablet is only a relay device, used to key or view information.

16.5 Conclusion

This chapter has provided a framework to understand the mHealth innovation that is being harnessed by nursing to improve care. As nurses find new ways to improve their own clinical practice with mobile devices, their patients will benefit from the efficiency gained by receiving more individual time with the nurse, safer care, more time to provide individual and family teaching, enhanced collaboration with the health-care team, and empowering patient-centered care. Our society has changed nursing practice in many ways, particularly by the amount of documentation necessary to prevent litigation. The use of mobile devices during clinical practice helps nurses quickly access evidence-based information, as well as allow access to the complete patient record at the point of care.

As the practice of nursing continues to evolve, nurses must continue to stay current with technology, while not losing focus of why technology can and should be used.

___

Case Studies

1. Amy is a student nurse and has just begun her first clinical rotation in the hospital. As part of the orientation to the unit, she is introduced to the tablet devices the nurses carry with them. She is encouraged to explore the device and become familiar with it. Later that day she is asked by one of the patients about a medication that she is unfamiliar with. While in the room she uses a drug reference application on the tablet to assist her in answering all of the questions pertaining to the unfamiliar medication. This is an example of which domain of clinical practice where nurses use mobile devices? Select all that apply.

   a. Point-of-care
   b. Educational
   c. Communication
   d. Reference

2. Steve works in the purchasing department of a small rural hospital. To compete with the larger urban hospitals, he is considering investing in mobile devices. In his search for the right product, he considers the weight of the device and feels a light but durable device is best suited for the job. Steve finds that these devices come in all sizes and feels a smaller device would work best. He also wants something that is easy to use and chooses an operating system that most of the employees are familiar with. In his decision for the right mobile device to purchase, Steve's considerations represent which component of mHealth for nursing?

   a. Component 1—Mobile device
   b. Component 2—Features and functionality
   c. Component 3—Medical apparatus
   d. Component 4—Nursing use
   e. All of the above

3. Monique is a manager on a nursing unit and is shopping for a new mobile device. As part of her job, she regularly sends and receives emails containing patient-specific information. One of the many features that convinced her to upgrade to this particular model was the ability receive email to her device on the go. After purchasing her new device, Monique contacted the IT representative where she works to help assist her in setting up her work email. The representative explained that only a secure network could be used to access the email and the steps she must follow to set up her account password. What are the multiple security precautions in place to protect ?

   a. Patient health information
   b. Reference materials
   c. Personal photos
   d. Personal documents

Questions

Four questions on the four main areas where healthcare personnel use mobile devices:

1. Utilization of a mobile device in situations where the nurse and patient interact describes which area where nurses frequently use mobile devices?
   (a) Reference
   (b) Point-of-care
   (c) Educational
   (d) Communication

2. Dosage calculation apps, medical abbreviation apps, and field-specific apps are examples of what types of mobile applications?
   (a) Reference
   (b) Point-of-care
   (c) Educational
   (d) Communication

3. What is the key differentiator between smartphones and tablets for device manufacturers?
   (a) Functionality
   (b) Weight
   (c) Size
   (d) Battery life
   (e) a and c

4. In 2011, what was the dominant operating system in healthcare?
   (a) Windows by Microsoft
   (b) Android by Google
   (c) iOS by Apple
   (d) BlackBerry OS by Research In Motion

5. Which group of mobile health applications relies on connectivity software to interact with an external medical device and communicate that information into a software application?
   (a) Embedded device
   (b) Camera
   (c) Audio
   (d) Sensor connectivity

6. Data storage in the form of wireless third-party servers is more commonly referred to as _____.
   (a) On-device storage
   (b) Cloud computing
   (c) On-site servers
   (d) Private network

7. How many of the 700,000 combined Google Play app store and Apple App store applications are medical apps?

   (a) 20,000

   (b) 50,000

   (c) 40,000

   (d) 100,000

8. The utilization by the SpiroSmart app of a smart phone's microphone to measure lung function is an example of what facet of Nursing Use of applications?

   (a) Reference apps

   (b) Evidence-based practice apps

   (c) Data collection apps

   (d) Communication apps

9. The FDA's Premarket Approval (PMA) evaluation process carries a monthly fee of how much?

   (a) $720,000

   (b) $520,000

   (c) $640,000

   (d) $500,000

10. How do health-care organizations ensure security of patient health information and electronic health records accessed by mobile devices?

   (a) Use of secure networks

   (b) User authentication

   (c) Password protection

   (d) Limiting access to secure information

   (e) All of the above

References

Baron, J., McBain, H., and Newman, S. (2012). The impact of mobile monitoring technologies on glycosylated hemoglobin in diabetes: A systematic review. *Journal of Diabetes Science and Technology*, 6, 1185–1196.

Barton, A. (2012). The regulation of mobile health applications. *BMC Medicine*, 10(1), 46. doi:10.1186/1741-7015-10-46.

Breslauer, D. N., Maamari, R. N., Switz, N. A., Lam, W. A., and Fletcher, D. A. (2009). Mobile phone based clinical microscopy for global health applications. *PLoS One*, 4, 7. doi:10.1371/journal.pone.0006320.

Comestock, J., (2013). Manhattan: 72 Percent of Physicians Have Tablets. MobiHealth News. Retrieved November 6, 2012 from http://mobihealthnews.com/21733/manhattan-72-percent-of-physicians-have-tablets/

Dolan, B. (2012, May 1). Survey: 71 percent of US nurses use smartphones. Mobihealthnews.com. Retrieved November 7, 2012, from http://mobihealthnews.com/17172/survey-71-percent-of-us-nurses-use-smartphones/.

Doran, D. (2009). The emerging role of PDAs in information use and clinical decision making. *Evidence-Based Nursing, 12*(2), 35–38. doi:10.1136/ebn.12.2.35.

Doran, D. (2010). Supporting evidence-based practice for nurses through information technologies. *Worldviews on Evidence-Based Nursing, 126(2)* 1–13. doi:10.1111/j.1741–6787.2009.00179.x.

Duffy, M. (2012). Tablet technology for nurses. *The American Journal of Nursing, 112*(9), 59–64. doi:10.1097/01.NAJ.0000418927.60847.44.

Garrett, B. and Klein, G. (2008). Value of wireless personal digital assistants for practice: Perceptions of advanced practice nurses. *Journal of Clinical Nursing, 17*(16), 2146–2154. doi:10.1111/j.1365-2702.2008.02351.x.

Gold, J. (2012, June 27). FDA regulators face daunting task as health apps multiply. Usatoday30. Usatoday.com. Retrieved November 19, 2012, from http://usatoday30.usatoday.com/news/health/story/2012-06-22/health-apps-regulation/55766260/1.

Hawkins, M. (2012, August 15). Vitera Healthcare Solutions: Mobile HER solution. Retrieved November 15, 2012 from http://viterahealthcare.worldpress.com.

Hendrich, A., Chow, M. P., Skierczynski, B. A., and Lu, Z. (2008). A 36-hospital time and motion study: How do medical-surgical nurses spend their time? *The Permanente Journal, 12*(3), 25–34.

Holt, J. E., Flint, E. P., and Bowers, M. T. (2011). Got the picture? Using mobile phone technology to reinforce discharge instructions. *The American Journal of Nursing, 111*(8), 47–51. doi:10.1097/01.NAJ.0000403363.66929.41.

Howard, P. (2011, January 26). Three areas where Obama can help America innovate. *The Examiner 126,* 16–17.

Larson, E. C., Goel, M., Borello, G., Heltshe, S., Rosenfeld, M., and Patel, S. N. (2012). *SpiroSmart: Using a microphone to measure lung function on a mobile phone.* Paper presented at the 2012 ACM Conference on Ubiquitous Computing, Pittsburg, PA. Abstract retrieved from http://dl.acm.org/citation.cfm?id=2370261

Larson, E. C., Goel, M., Boriello, G., Heltshe, S., Rosenfeld, M., and Patel, S. N. (2012b). SpiroSmart. In *Proceedings of the 2012 ACM Conference on Ubiquitous Computing — UbiComp'12* (p. 280). New York: ACM Press. doi:10.1145/2370216.2370261.

Luxton, D. D., Kayl, R. A., and Mishkind, M. C. (2012). mHealth data security: The need for HIPAA-compliant standardization. *Telemedicine and E-Health, 18*(4), 284–288.

Makower, J., Meer, A., and Denend, L. (2010). FDA impact on US medical technology innovation: A survey of over 200 medical technology companies, Arlington (Virginia). Retrieved November 14, 2012, from http://www.inhealth.org/doc/Page.asp?PageID = DOC000188.

Maslowsky, J., Valsangkar, B., Chung, J., Rasanathan, J., Cruz, F. T., Ochoa, M. et al. (2012). Engaging patients via mobile phone technology to assist follow-up after hospitalization in Quito, Ecuador. *Telemedicine and E-Health, 18*(4), 277–283. doi:10.1089/tmj.2011.0156.

Moore, S. (2011). Texts, tweets, and patient portals. *Oncology Nursing Forum, 38*(4), 393. doi:10.1188/11.ONF.393.

Morris, M., (2012) Vitera Healthcare. Retrieved November 19, 2012 from http://www.viterahealthcare.com/company/Pages/pr_ViteraHealthcareSolutionsStudyIndicatesThattheMajorityofHealthcareProfessionalsAreInterestedinaMobileEHR Solution.aspx.

Mosa, A. S. M., Yoo, I., and Sheets, L. (2012). A systematic review of healthcare applications for smart-phones. *BMC Medical Informatics and Decision Making, 12*(1), 67. doi:10.1186/1472-6947-12-67.

Oliver, S. (2012, September 26). Google android store reaches 25 billion downloads, 675,000 apps. Appleinsider.com. Retrieved November 19, 2012, from http://appleinsider.com/articles/12/09/26/google-android-reaches-25-billion-downloads-675000-apps.

Pare, G., Sicotte, C., Moreault, M. P., Poba-Nzaou, P., Nahas, G., and Templier, M. (2011). Mobile computing and the quality of home care nursing practice. *Journal of Telemedicine and Telecare, 17*(6), 313–317.

Pollick, M. (2012, November 12). Catching two waves: Digital healthcare and smartphones. Heraldtribune.com. Retrieved November 19, 2012, from http://www.heraldtribune.com/article/20121112/ARTICLE/121119936/2107/BUSINESS?p = all&tc = pgall.

Putzer, G.J. and Park, Y. (2010). The effects of innovation factors on smartphone adoption among nurses in community hospitals. *Perspectives in Health Information Management/AHIMA, American Health Information Management Association, 7(Winter),* 21–27.

Ragnemalm, E. L. and Bång, M. (2012, June 6–8). Persuasive technology and mobile health: A systematic review. In *The 7th International Conference on Persuasive Technology, Linkoping, Sweeden,* 45–48.

Tran, J., Tran, R., and White, J. R. (2012). Smartphone-based glucose monitors and applications in the management of diabetes: An overview of 10 salient "apps" and a novel smartphone-connected blood glucose monitor. *Clinical Diabetes, 30,* 173–178.

WHO. (2011). *mHealth: New Horizons for Health through Mobile Technologies,* Global Observatory for eHealth series — Volume 3, WHO, Geneva, Switzerland, viewed November 13, 2011.

Williams, J. (2012). The value of mobile apps in health care. *Healthcare Financial Management, 66*(6), 96–101.

Wyatt, T. and Krauskopf, P. (2012). E-health and nursing: Using smartphones to enhance nursing practice. *Online Journal of Nursing Informatics, 16*(2), 12–23.

Yadalla, H. K. and Shankar, V. (2012). Professional usage of smart phone applications in medical practice. *International Journal of Health and Allied Science, 1*(2), 44–46. doi:10.4103/2278-344X.101656.

17

mHealth Contents and Services Delivery and Adaptation Challenges for Smart Environments

Tayeb Lemlouma, Sébastien Laborie, Philippe Roose, Abderrezak Rachedi, and Kenza Abdelaziz

CONTENTS

17.1 Introduction

According to the United Nations projections [1], in 2050, the old-age dependency ratio of the population aged 65 years or over will approximate 51.70% of the population aged 20–64. For instance, this ratio will approximate 48.2% in France, 62% in Germany, and 44.10% in the United Kingdom (Figure 17.1). In the health-care domain, the evaluation of the autonomy or dependency of a person is of high importance. Indeed, such evaluation is used by the professionals of health to identify the persons' needs for assistance, services, and allowance. It also enables to take the right decision about keeping the person in the health institution, nursing home, or independently at home with or without any health-care monitoring. In a smart home environment, and more globally in a smart city environment, particular attention and care should be made for dependent people, in particular for elderly since they are left on their own at home or in the city. Mobile health (mHealth) sensors and applications provide the ability to continuously monitor persons with different sources of information and in different places. A continuous dependency evaluation is of high concern since it can detect the changes regarding the people's abilities to achieve elementary daily tasks. In this chapter, we discuss the proposition of a framework for automatic and flexible dependency evaluation that can notify any changes of the elderly dependency and hence allow providing him, over time, with the required help. We address the main issues related to such a framework and show how to take benefits from

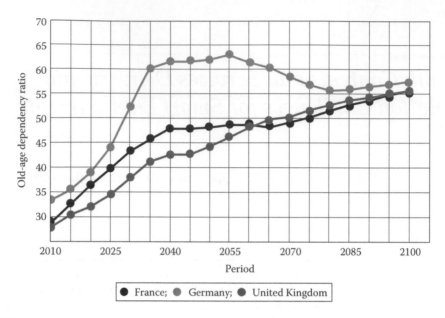

FIGURE 17.1
Old-age dependency in France, Germany, and United Kingdom.

modeling the context for mHealth services and applications. We consider the heterogeneity of elderly profiles and service sources that can come from anywhere from home or the city. The dependency evaluation takes benefits from the description model of the elderly and the dynamic composition of services. We focus on the activities of daily living (ADLs) as defined in the French Autonomy Gerontology Iso-Resources Group (AGGIR) model [2]. The automatic way in observing the different elderly activities allows avoiding human errors and patient ignorance that face the current use of the AGGIR tool and similar evaluation methods used in geriatrics [3–5].

Health smart homes represent a complex and interesting application domain. It involves various end users and persons (health professionals, patients, elderly, dependent persons), dedicated norms, various vendors' devices, and many types of applications. The complexity of existing architectures makes offered health-care services rigid and not always adapted to the resident profile and its evolution over time. The evolution of the resident life, in terms of needs and preferences, usually involves applying significant modifications on current systems on the hardware/device architecture, activation/deactivation of sensors, and deployment/disabling of services.

In order to validate our approach, we target the French dependent population. Especially, we focus our study on the AGGIR dependency model used in the French medical field. Our objective is to provide an open and flexible architecture as well as an extensible model linked to opened data that can refer to a wide variety of services (mHealth hardware sensors and software; simple or composed services).

One of the most challenging issues is to design an adaptable architecture for different homes and people profiles. Moreover, the architecture should be optimized not only to provide multiservices to assist people (like indoor localization, collection of health-relevant data) but also to detect and to rapidly tackle security issues while ensuring QoS. These features must be implemented while minimizing the overhead in terms of processing and communications.

Parameters such as clinician requirements, the nature of the data to be collected, the information reporting, the end-user (patient or subject) modeling, and the physical environment (e.g., includes both indoor and outdoor environments) must be taken into account for the proposed protocols and mechanisms. Moreover, it is important to define different types of logical entities used in the architecture to perform the following tasks: (a) technology selection and coexistence, (b) designing of multiservices interaction model, (c) deploying strategies to localize and target patient or subject, (d) designing threats model, (e) detecting security attacks, (f) making decisions on the level of security and QoS, and (g) defining the interaction between the entities to maintain or to adapt the QoS and security.

For each end-user and home profiles, the targeted architecture should be able to propose the appropriate network technologies and handle their coexistence. It should also consider the existing network topology, the design of the communication protocols, and the control mechanisms of the distributed smart objects for achieving an optimal behavior in terms of bandwidth and power consumption.

The main objective is finding how to design and build a smart object network solution that addresses a particular clinical research need. The network used in the mHealth field includes a combination of technologies that collect data about patient activity and analyze these data to extract clinical knowledge and/or validate a research hypothesis. In this context, the most important technology criteria are mainly the cost, scalability, fault tolerance, localization, routing, wireless regulatory issues, energy efficiency, medical regulatory issues, long-term usability, and privacy and security aspects. The objective is to select the best technology combination to collect and analyze health-relevant data and to provide an appropriate feedback to the patient, as well as to deliver relevant information to the clinician.

If we consider the case of wireless technologies' coexistence, mHealth devices work alongside other devices both within and beyond the wireless sensor network (WSN), without interfering with them or being disrupted. The amount of available spectrum is limited and the competition for particular channels may be problematic. Systems outside WSNs represent a potential problem here; possible competitors for spectrum include wireless LANs (Wi-Fi), mobile smartphones, personal digital assistants (PDAs), Bluetooth networks, ZigBee networks, wireless modems, and medical systems such as wireless medical telemetry service (WMTS) and wireless body area network (WBAN).

17.2 Case Studies

We present here two case studies based on the French system:

1. Enriching homes and cities by advanced and optimized mHealth services
2. Providing monitoring services that automatically evaluate the person's dependency

The consideration of these situations helps to develop smart environments with mHealth applications that provide adapted and context-aware services and contents. In our study, we investigate useful cases and situations related to mHealth services for the daily life of an elderly or a dependent person. These situations are based on our previous experiences related to mHealth services for elderly and dependent persons in heterogeneous environments [6–9].

17.2.1 Enriching the Home Environment with Advanced and Optimized mHealth Services

The scope of this case study involves health smart homes and digital homes. Health smart homes are those supporting health and health-care-based services, while digital homes are those exploiting multimedia devices such as home media servers and renderers to provide a universal media access. We consider the context of smart and digital homes in order to provide mHealth services with various devices that simplify the acceptance of the mHealth solutions and foster the user-active mode. The digital dimension of the case study aims at making current digital architectures tuned toward health needs of assistance of the resident. The flexibility is increased by taking into account existing architectures that vary from the simplest ones (e.g., including only a simple operator box) to more complex architectures (such as DLNA-like architectures [10]) that could integrate home-to-home communication.

The objective here is to provide the elderly, patients, and dependent persons with services and appropriate supports in order to tackle the difficulties of their everyday lives. Services could include video or audio notifications to take medicine, alerting appropriate persons (a caregiver, a health-care assistant, a referring doctor, or the closest family member around the house) if the resident fell, tracking the way the resident is eating in order to advise him some receipts, providing the resident with regular video or audio calls from family or medical assistance, keeping a social contact (display multimedia contents about family events), and maintaining a sense of participation in society and fighting feelings of loneliness.

Notification to take medicines (drug reminder): the drug reminder is an example of a local mHealth service that is generally very useful for some person profiles, such as those suffering from certain diseases, like Alzheimer. Thus, when provided, this service will be in charge of reminding the resident of the drugs that he should take at specific times. In order to follow a user-centric approach, such a service should be based on contextual data stored locally at the home box/gateway (where data can be easily updated) and/or remotely on the cloud.

Contact with the outside: when the health smart home suspects an urgent event (like a heart attack or fall for some profiles), some actions must be triggered rapidly as a response to the event, for example, contacting a qualified person to confirm the transmitted alert and then calling the emergency if needed. This type of complex service may require the consultation of some data stored in several databases or to perform some mappings between a set of data, which are obtained from different sources, such as a repository of services, contact lists related to the resident, their availabilities, their addresses, or their positions. Hence, several contextual parameters can be considered to refine the establishment of a contact with the outside as a response to an alert. The provided service has to avoid human systematic calls to emergency when it is not necessary. It will also enhance the user satisfaction who expects a rapid care, and a real-time answer that depends on the context of the alert. For instance, the system should be able to identify the person to contact, depending on the type of incident and the availability of persons registered in some databases and having a relation with the resident. The system could identify categories of people who are able to intervene (neighbor, relative, doctor) with an order of preference. Tables 17.1 and 17.2 are given to show an example of a simple association between possible detected events and contact lists. Moreover, each type of incident can be assigned to a set of contacts and theirs associated priorities, which depend on the nature of the incident (Table 17.1).

TABLE 17.1

Association between Detected Events and Contacts

Event	Contact List Identifier
Fall (time, location, etc.)	L1
Heart attack (time, location, etc.)	L2

TABLE 17.2

Contact List *L*1

Contact	Type	Real-Time Availability	Contact Action
Marie Doeuf	Family	False	Phone/SMS
Julien Doeuf	Family	True	Phone
Philippe Dupond	Neighbor	True	SMS/e-mail
Louise Le Sage	Relative	False	Phone

According to these priorities and contact availabilities (at the moment of the event), one or more contacts will be contacted (using phone calls, SMS, e-mail, etc.) with the available and appropriate technology. Indeed, a contact with a higher priority is contacted first. In case of unavailability, the system will try to contact the next one and so forth (Table 17.2).

Mastering the resident context can rely on the set of mHealth sensors existing in the home as well as the existing devices that can be considered as nonnative sensors, such as smartphones, home boxes, desktops, and webcams. This can also be refined depending on the home hardware capabilities (home profile).

Home-to-home mHealth services: to ensure a best quality of experience (QoE) for the user, services are discovered and provided within his home. However, the heterogeneity of home profiles can imply different software and hardware capabilities that vary from one home to another. Furthermore, the mobility of the user can make some of the residential services unavailable. For these two identified situations (i.e., limitations in the home capabilities and the user's mobility), the home-to-home mHealth communication can provide the user with new additional services. We identify the following cases:

- The user is outside the home and so he could count on the nearest smart home to use a service (e.g., notification to take medicines). Of course such services have to be authenticated.

- The user is outside the home, and the nearest smart home could provide him with localization services based on the user's new position.

- The user is at home and he can interact with a new service available in a neighbor smart home that has more software/hardware capabilities.

Notice that those three items raise some security and thrust problems. We will not address them in this chapter.

Energy and resources saving: mHealth services play a major role to ensure an optimal energy saving of the home. Elderly and dependent persons are particularly affected by this aspect due to their fragile health state and their dependency that makes them unaware of the energy aspects. This objective can be reached by the use of an interconnected set of presence sensors, distributed energy meters (using ZigBee and Z-Wave), and software agents connected to the home box/gateway. The energy-saving service can implement different strategies that ensure a good distribution of the energy inside the home, avoid unnecessary consumptions, and make the energy use tailored to the context of the resident. Only useful energy is provided when it is needed, for example, turning off the light when no presence is detected, turning off the TV set when the resident is sleeping or no user's action is detected within a given time [7], and closing the tap to prevent water flood and reduce its consumption.

Optimizing the mHealth services in complex architectures of the digital home: the objective of a digital home is to offer an interoperable network that ensures using and sharing multimedia-based services in a seamless environment regardless of existing devices and sources (formats, location, delivery methods and protocols, etc.). Unfortunately, most of the existing home networks do not meet this objective yet [11]. Indeed, users are still unaware of the existence of some technologies such as the Digital Living Network Alliance (DLNA) [10] already integrated in about 74% of existing consumer electronics (CEs) but only with 6% of real users [10]. Many reasons explain this situation. We can particularly quote the complexity of the proposed technologies inside the home network, the intercommunication and cooperation between different industries and norms, the heterogeneity of devices and media formats used in different services, and finally, the lack of any intelligent components or services that help terminals and users to find, configure, and connect their terminals in order to use the available services in the best possible and automatic way.

The key motivation of this study is to enable the access to the media content of mHealth services with heterogeneous mobile terminals and without any user's preliminary configurations and settings. In addition, we aim at optimizing the network traffic generated by the components of complex architectures (such as DLNA) by allowing only the necessary traffic when needed. For the home network optimization, our proposition is to move toward a request/answer model. Consequently, no traffic will be generated if the user or the application does not ask the network for something. Therefore, any transmitted data will be an answer to the users' needs or requests. Modern CEs and mobile terminals used in mHealth are an appealing platform for advanced functionalities and applications. However, many devices are still not compatible with some protocols needed in complex digital home architecture such as the multicast protocol [12] used for discovering and announcing services and contents. To improve the accessibility to different mHealth services and consider the existing heterogeneity of devices and formats, we adopt the use of open standards and protocols. This approach provides an easy integration of heterogeneous services and sources.

In complex digital home architectures, the use of the hypertext transfer protocol (HTTP) allows to solve many compatibility issues between the different architecture components and to optimize the network traffic. Indeed, HTTP only transports content that answers the users' (or applications') request. Moreover, HTTP is widely compatible with the majority of limited terminals; it does not require any advanced settings. The protocol's transport does not make any restriction on the format of transmitted data, so it is easier to add media encoding and adaptation methods for limited devices. Capabilities and preferences can be described in specific profiles, defined using Semantic Web technologies, and *linked data* principles can be coupled with the available HTTP TCN negotiation [13]. Furthermore, the use of HTTP only requires a web browser, currently embedded in the majority of mobiles and terminals with limited capabilities.

Enabling the mHealth services in homes with limited capabilities: Robles and Kim [14] concluded in their review that a main challenge of installing a smart home system is balancing the complexity of the system and its usability. Based on the work of Anderston and Ayclin [15], half of all computer-based information systems related to health-care fail due to user resistance and staff interference. The main reason is that users were asked to significantly alter traditional workflow patterns to accommodate the system, rather than the system accommodating the users. According to a study carried out within 300 people aged 65–85 years old [16], the elderly are constantly facing to the ever-evolving technology and need an appropriate support in order to satisfactorily tackle the difficulties of their everyday lives. The study shows that 99.4% of the elderly are able to handle the TV sets, while only 67.7% of them are able to use a simple wireless phone.

Providing smart media services through existing home devices, like TV sets, allows the elderly and disabled to have a personalized content adapted to their needs without leaving their homes and without any complex home architectures. Furthermore, enabling smart services with the user's familiar devices simplifies the use of advanced technologies and makes their integration at home easily accepted by the elderly and patients. The trends in smart homes research indicate an increasing popularity of using middleware. The use of middleware is efficient to integrate heterogeneous multivendor devices that coexist in the same mHealth system [17].

Considering the change and evolution over time of the user's preferences makes provided services and content dynamically adapted to current preferences. Taking into account the context of the mHealth ecosystem makes it tuned toward the preferences and needs of the elderly. Considering the profile of the user and the capabilities of his home (home profile) increases the acceptance of advanced technologies and services by the elderly. The users' acceptance of the proposed system represents a key criterion in our approach. Unfortunately, this aspect was generally ignored in previous context-aware approaches.

17.2.2 Automatic and Context-Aware Evaluation of People Dependency in Homes and Cities

The dependency of persons is defined as the ability of a person to achieve elementary tasks of daily living without the help or stimulation of a third party. In the geriatrics domain, different methods and tools have been defined to evaluate this dependency using different metrics and factors, such as age, medical status, mobility, and fall risk. The main reason of the existence of such different methods is the determination of what is a basic activity of daily living (ADL, called also BADL). Methods are mostly based on the ADL definitions of Katz and Akpom [18] and Maheney and Barthel [19] and the instrumental daily activities (IADL) of Lawton and Brody [20]. Among these very numerous scales, we quote, for instance, Resident Assessment Instrument (RAI) [3], AGGIR, Functional Autonomy Measurement System (SMAF) [4], and Medical Outcomes Study Short-Form (MOS-SF)-36 [5]. Existing evaluation methods are achieved using questionnaires and manual scales. Consequently, they are usually subject to human errors or patient ignorance while answering the questionnaires.

mHealth capabilities provide the ability to include any kind of services and observation sources. Indeed, mobile devices (with some cheap extensions) are today being used as sensors in several health applications such as mHealth information access and delivery. However, mHealth faces challenges related to handling and integrating these heterogeneous sources of information. This situation is particularly true when the eHealth ecosystem aims at providing complex composed services. Our study focuses on making the evaluation of the person's dependency automatic and more flexible by integrating different sources and considering the different person's activities. Despite the specification of services per daily activity that we aim at identifying in order to validate the proposed framework, we do not make any restriction neither on the kind of used services (hardware sensors, software or manual inputs) nor in data sources used in handling the elderly dependency. Our framework benefits from the collaboration of these different sources and opened technologies that facilitate the processing of heterogeneous services and sources. This use case considers the main ADL as a whole and does not focus on a particular activity like some previous works. This choice is done because of the heterogeneity of profiles regarding the monitored persons; also, we believe that when a person's dependency degree changes, this is the consequence of not only one activity of the person but also several of them. Consequently, such changes should be quickly notified to health-care providers.

Most context-aware approaches refer to the Dey's context definition [21] as *any information that can be used to characterize the situation of an entity*. An entity includes a person, a place, or an object. This case study focuses on the evaluation of the dependency in order to help health-care professionals to provide services and monitoring without keeping persons in health institutions. Hence, the context considered in this study is represented by a set of the different variables that influence the elderly dependency evaluation. In order to validate our approach for this case study, we target the French dependent population and in particular elderly dependent persons. In France, in order to evaluate the needs of an elderly person in terms of medical assistance and allowance, the AGGIR model is widely used [2]. In order to understand the approach, we discuss in the following the application of the AGGIR evaluation.

The AGGIR evaluation is manually achieved by medical reviewer, such as the referring doctor or a medical assistant. Each daily activity of a person is qualified by the medical reviewer using four possible adverbs: spontaneously (S), completely/totally (T), usually (U), and correctly (C). The S adverb means that the activity can be done without any stimulation. The T adverb means that the person can do all of the actions involved in the activity. The U adverb means that the activity is done regularly; the frequency of the execution depends on the nature of the activity. The C adverb involves the quality of the person's realization, safety, and conformance with the recommended usage of the activity. According to a logical condition involving these adverbs, the activity is evaluated with the following modalities: A, B, or C. A given activity is evaluated with the modality A, if the person can achieve the activity with the following condition: $S \wedge T \wedge C \wedge U$; the condition of the C evaluation is $\neg S \wedge \neg T \wedge \neg C \wedge \neg U$; the B evaluation is given if the two previous conditions are not met. The A evaluation means that the person is completely autonomous in achieving the given activity, B means that the person is partially dependent, and C means that the person is dependent and cannot achieve the activity alone. The AGGIR model considers 17 variables describing the ADL, such as the mental coherence, hygiene, orientation, and mobility. Each variable evaluates the dependency degree of achieving a given activity.

Based on a multivariate analysis involving more than 5000 persons, the model has identified 10 discriminated variables, 8 of them are used in the classification of the dependent persons into 13 profile ranks and 6 groups. The 13 profile ranks concern the losses of autonomy of dependent persons. The 6 groups, called *iso-resource* groups or GIR, reduce the number of profile ranks and concern the needs of assistance and allowances. The GIR algorithm computes the iso-resource group number (1–6) based on predefined association between profile ranks and groups. For instance, rank 1 is associated to GIR 1 and ranks 2–7 are associated to GIR 2 (Table 17.3). The first group (GIR 1) represents the persons that are completely dependent, while the last group (GIR 6) represents autonomous persons.

To identify the profile rank of a person, the model uses eight classification functions that compute the classification scores. The person is classified as belonging to the profile rank for which he or she has the highest classification score (Table 17.3).

TABLE 17.3

Association between Profile Ranks, Classification Scores, and GIR Groups

Profile Ranks	Score Condition	GIR
1	$S1 \geq 4380$	1
2	$4140 \leq S1 < 4380$	2
3	$3390 \leq S1 < 4140$	
4	$S2 \geq 2016$	
5	$S3 \geq 1700$	
6	$1432 \leq S3 < 1700$	
7	$S4 \geq 2400$	
8	$S5 \geq 1200$	3
9	$S6 \geq 800$	
10	$S7 \geq 650$	4
11	$S8 \geq 4000$	
12	$2000 \leq S8 < 4000$	5
13	$S8 < 2000$	6

Scores conditions are tested in a sequential order of the classification functions from S_1 to S_8. Classification functions are defined as

$$S_i = \sum_{k=1}^{8} W_{ik}$$

where

S_i is the score of the ith function

W_{ij} is the weight for the jth variable modality (i.e., A, B, or C)

Table 17.4 presents the different weights of the different variable modalities regarding S_1 and S_2 functions. Let us consider the example of a person with the following evaluation:

TABLE 17.4

Weights of the Classification Functions $S1$ and $S2$

Activity		W_{1i}	W_{2i}	Activity		W_{1i}	W_{2i}
Coherence	C	2000	1500	Eating	C	60	60
	B	0	320		B	20	0
	A	0	0		A	0	0
Orientation	C	1200	1200	Elimination	C	100	100
	B	0	120		B	16	16
	A	0	0		A	0	0
Hygiene	C	40	40	Transfers	C	800	800
	B	16	16		B	120	120
	A	0	0		A	0	0
Dressing	C	40	40	Int. moving	C	200	−80
	B	16	16		B	32	−40
	A	0	0		A	0	0

coherence = C, orientation = C, hygiene = C, dressing = A, eating = C, elimination = A, transfers = A, and interior moving = B.

We have $S_1 = \sum_{k=1}^{8} W_{1k} = 3332$ and $S_2 = \sum_{k=1}^{8} W_{2k} = 2660$ (Table 17.4)

To identify the profile rank, the score conditions are tested first with the S_1 score; if there is no satisfied condition, the score is then tested with S_2 and so forth until the last condition of S_8 ($S_8 < 2000$). Here, the score of S_1 (i.e., 3332) does not satisfy the S_1 score conditions (Table 17.3). However, the score of S_2 (i.e., 2760) satisfies the $S_2 \geq 2016$ condition (Table 17.3). Hence, the person's profile rank is 4 and his iso-resource group is 2.

17.3 Modeling and Using Context Information for mHealth Applications

Context awareness began to be investigated at the beginning of the 1990s with the emergence of mobile computing. Two pioneer investigations were the *Active Badge* system [22], developed at the Olivetti Research Lab, and the active map [23] at Xerox PARC. Since then, a wide range of context-aware systems has been reported. As *location* is considered as a common piece of context used in application development, most of these systems focus on it while neglecting other context information. Location-aware systems targeted different applications, such as tour guide* applications [24] and advertising systems in smart spaces [25]. These systems use different technologies to acquire location information, such as global positioning system (GPS), underlying communication infrastructures, cameras, and card reader.

Because context is more significant than only *location*, some works combine different context information, such as time, user location, activity, and interest to build context-aware systems.

Examples of these systems are the GUIDE system [26] and the conference assistant [27]. These systems are typically proprietary and depend on the applications for which they are built. A survey of mobile location and context-aware systems is presented in [28]. Some other works, such as the context toolkit, Service-Oriented Context-Aware Middleware (SOCAM) [29], and Context Broker Architecture (CoBrA) [30], focus on providing frameworks for context awareness that enable an easy and a rapid prototyping of context-aware applications. The study carried out in [31] introduces and compares different context-aware frameworks based on various design criteria, such as architecture, context model, resource discovery, and sensing.

Context awareness is also a key technology in the vision of health care. The work presented in Rodriguez et al. [32,33] reports studies conducted in a hospital. These papers present an ambient intelligence environment for hospitals, called Simple Agent Library for Sope Applications (SALSA), based on autonomous agents. Three scenarios for information management and interaction in such environments were developed. In Kofod-Patersen and Aamodt [34], another work addresses context awareness in hospital environments. Authors use case-based reasoning to provide context classification. A context in this chapter includes location, identity and role of present people, the time of the day, and any artifacts that are present, such as patient lists. The work in Corchado et al. [35] describes the Geriatric Ambient Intelligence (GerAmi) system that helps health-care facilities with the challenges of caring for Alzheimer's patients.

In the literature, and especially in the works cited earlier, we have noticed that profiles, in particular, user profiles, in the mHealth domain are poor semantically, not generic,

---

* http://www.cs.berkeley.edu/_dey/context.html

and not dynamic enough to take into account the evolving situation of a dependent person [2,36,37]. Even for standard mobile device profiles (like UaProf [38], CC/PP [39], and WURFL*), they contain mainly static descriptive values and do not express rich semantic constraints related to the user or the surrounding context. Furthermore, mixing all heterogeneous existing profiles inside an mHealth application may lead to an explosion of profiling information. Consequently, the management, the storage, and the access of all these information could lead to a real bottleneck, thus decreasing the efficiency of systems. This is an even more important challenge in the mHealth field because relevant and optimized profile modeling are more than necessary since in some critical situations, the system has to retrieve the relevant information at a given time and make the relevant decisions as soon as possible (i.e., providing the most appropriate services).

In smart environments, we propose to consider three kinds of profiles: (1) data related to the user, (2) characteristics of devices that are available inside the smart environment, and (3) multimedia services that could provide or process any kind of information:

- *The user profile*: This profile contains information about a person. A wide variety of sources could provide such kind of information. Of course, the data sources could be stored locally or remotely (e.g., medical records, relatives or friends, visited places, agendas); thus, the user profile integrates and/or refers to such data. Of course, some parts of the user profile may be public, while other parts may be private (i.e., restricted access).

- *The home profile*: The home profile describes all the capabilities of a smart environment in terms of hardware and software. For instance, it details which kind of equipment are available inside a specific home, such as sensors and fixed or handheld devices.

- *The service profile*: Each service is described by a service profile. More precisely, it details the action it supports, its types of inputs and outputs, its availability, its current efficiency, its execution constraints, its location, etc.

In order to illustrate the complementarity of such profiles (Figure 17.2), the medication reminder is a classic example related to the mHealth domain as presented in Section 17.2.1. The implementation of such a service can be based on data stored locally on the set-top box (where data can be easily updated) and/or remotely on some remote servers. This service can provide reminders using several media formats (e.g., text, HTML, audio or video) and its execution can be adaptable relying on the following:

- Data stored in the *user profile*: disabilities and their degrees (e.g., hearing level), location, preferences, etc.

- Capabilities of the smart environment for displaying the reminders, in terms of hardware and software (*home profile*).

- According to the resident location, the nearest mHealth equipment or sensor could be used. In fact, we can use a smartphone, a tablet, a TV, or an FM radio frequency for the drug reminder. Note that the resident location could be easily obtained thanks to mHealth presence sensors and indoor techniques (*service profile*).

- Hardware and software capabilities of another smart environment: to ensure services continuity when the user profile indicates a location different from its own smart environment and close to another one, like a neighbor.

---

* http://wurfl.sourceforge.net

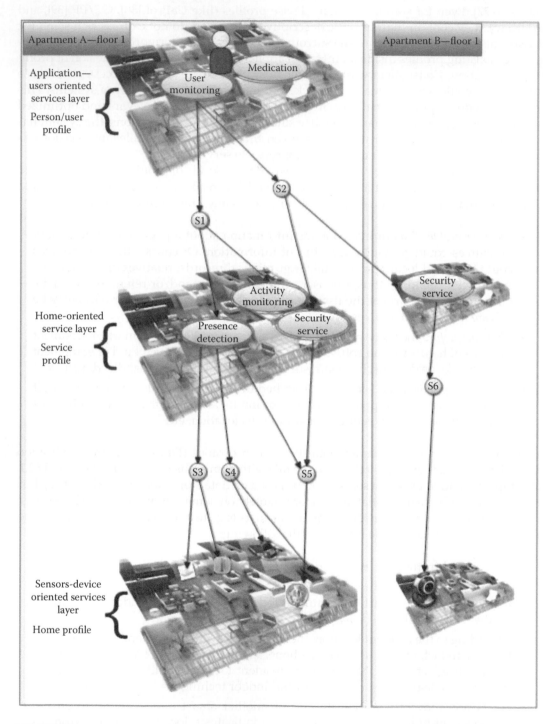

FIGURE 17.2
The complementarity of the user/home/service profiles.

In order to propose an optimal model that complies with the previous requirements and constraints, Semantic Web technologies and linked data principles are interesting ways to specify such profiles for the following reasons:

- *Profile expressiveness*: Since many data have to be connected together (internal data and external data, which may be available in the linked open data, cloud), we use the RDF W3C standard in order to reach this objective in a compatible way. This standard supports ontologies (like RDFS and OWL) that can specify flexible semantic vocabularies (such as general concepts and subconcepts) as well as complex inference rules (such as the *subsumption*, the *disjunction*, and other combinations).

- *Profile storage optimization*: Thanks to the inherent property of RDF, we can refer to URLs in descriptions. In our context, the benefits of using URLs are twofold. Firstly, it allows reducing the file size of a given profile by distributing some of its facets and/or default information in several data centers. Secondly, in the smart environment amount of profiles, each profile may refer to a common set of data; hence, these data sets are stored once and are not duplicated. The former allows optimizing the profile access, while the latter reduces the all storage space.

- *Profile handling optimization*: A wide variety of services may be invoked to provide up-to-date data, such as the person social contacts and their availabilities, the current weather, the electronic health record, GPS localizations, and available connected sensors. These services may be annotated with rich semantic descriptions (like SAWSDL [40], WSMO*) in order to better retrieve and compose them. Usually, profiles contain static and dynamic values that are updated regularly. In profiles, for particular properties, instead of hard-coded values and periodic updates, we propose to refer to sets of potential services that might provide on demand the relevant information at a given time.

Taking into account the use case presented in Section 17.2 (i.e., determining automatically AGGIR values and the GIR score of a person), we have illustrated in Figure 17.3 how user/home/service profiles may be structured and linked.

On Level 1, the home profile can be instantiated thanks to devices' and captors' descriptions. In our experiments, we have used the *iCasa* [41] platform in order to get all the capabilities of a smart environment, for example, get the locations of the presence sensors and the provided values. Indeed, the *iCasa* platform provides support to build digital home applications. In particular, its simulator allows developers to test visually and interactively their own digital home applications. Consequently, we can exploit the data provided by this platform in order to encode, using RDF, all the required descriptions and values.

On Level 2, the service profile corresponds to a repository of ambient services. Depending on the mHealth application, these high-level services may correspond to presence detection (this ambient service may refer to several presence captors) and hygiene monitoring (this ambient service may refer to the elements in the fridge, the use of the shower, etc.).

On Level 3, the user profile is instantiated, updated, and refined. Concerning the use case presented in Section 17.2, the user profile may contain—initially—basic information

---

* http://www.wsmo.org

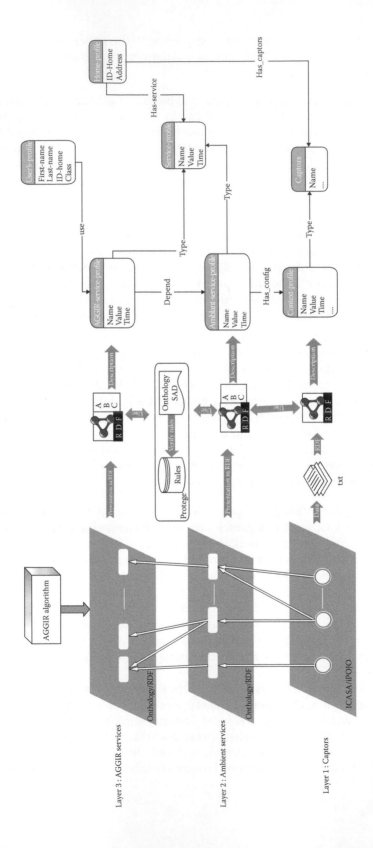

FIGURE 17.3
User/home/service models and dependencies.

about the patient (e.g., name, address, and phone number). Then, for determining the GIR score of the patient, AGGIR values need to be specified. The service profiles can be used to select all ambient services that can determine the AGGIR values. When all AGGIR values are determined, one can compute the GIR score. All the data about the evaluation process of the AGGIR values with ambient services and the GIR score can be stored in the user profile. Hence, it is easy to detect if the patient's degree of dependency changes from one GIR category to another one. It is important to mention that the monitoring of a human activity was studied, separately, in some previous works such as in Le et al. [42] for the human mobile/immobile state detection and in Kokkinos et al. [43] for the fall detection. Hence, the automatic evaluation was proposed, each time, for a separate activity but, unfortunately, without the reasoning on the whole profile of the person, without the consideration of the other possible activities, and without the identification of required actions to perform as an answer to a detected situation. Indeed, a detected situation of fall can be very risky for some profiles of patients but it can be considered as *a normal situation* for other profiles. We notice also that when elderly are left on their own at home with an initial given degree of dependency, one of the important aspects of the mHealth ecosystem is to detect the dependency change that can be achieved only by considering the whole profile. We believe that detecting the dependency change is more important than considering a given daily activity independently from its context. If the person is still in the same degree of dependency, he or she will continue to receive the usual help and allowance; however, if this degree changes, the automatic evaluation must trigger appropriate alarms in time.

Figure 17.4 presents an example of a person profile that corresponds to a potential instance of an evaluation process encoded in RDF/XML [44]. At the beginning of the profile, you will find details about the patient. In the illustrated example, the profile corresponds to a person named *John Doeuf, 55 years old*. Thereafter, for a specific date, some evaluation variable values are detailed.

For each AGGIR variable, we propose to refer in our user profile to the ambient service descriptions that have been used with their confidence values. For instance, in Figure 17.4, the coherence variable value (i.e., C) has been computed thanks to a specific ambient service with a high confidence value, that is, 0.9 (confidence values are set between 0 and 1; the higher the value is, the higher is the confidence). Of course, thanks to the service URL, here http://127.0.0.1/sensor01/rawdata#SensorOutput, it is possible to access other detailed descriptions in its service profile, for example, the service characteristics, the service location, and the service owner.

17.4 Future Directions

This chapter discussed the objective of providing mHealth services in smart environments and making automatic the dependency evaluation process, which is usually done manually in the medical field. We focused on providing mHealth services based on the context and considering the dependency change within the medical AGGIR model. Our proposition tries to move the current situation toward a flexible architecture and an extensible model linked to opened data referring to a wide variety of services.

```
<rdf:RDF xmlns:rdf="../22-rdf-syntax-ns#" xmlns:sad="http://www.irisa.fr/sad#"
 xmlns:rdfs="../rdf-schema#" xml:base="../sad#">
<rdf:Description rdf:nodeID="AGGIRprofile">
 <rdf:type rdf:resource="../sad#Profile"/>
 <sad:patientInformation>
  <rdf:Description rdf:nodeID="PatientInformation">
   <sad:name>John Doeuf</sad:name>
   <sad:email>John.Doeuf@example.org</sad:email>
   <sad:address>Place des Champs Elysée</sad:address>
   <sad:phoneNumber>0123456789</sad:phoneNumber>
   <sad:age>55</sad:age>
   <sad:ss>158016403506712</sad:ss>
  </rdf:Description>
 </sad:patientInformation>
 <sad:evaluations>
  <rdf:Description rdf:nodeID="Evaluation1">
   <rdf:type rdf:resource="../22-rdf-syntax-ns#Bag"/>
   <sad:creationDate>
        2013-01-10 15:30:00
   </sad:creationDate>
   <sad:lastModification>
        2013-02-18 09:00:00
   </sad:lastModification>
   <rdf:li>
    <rdf:Description
         rdf:about="../sad#coherence_Communication">
         <rdf:type rdf:resource="../sad#Variable"/>
         <sad:value>C</sad:value>
         <sad:sensorDataSource
  rdf:resource="http://127.0.0.1/sensor01/rawdata#SensorOutput"/>
         <sad:sensorConfidence>0.9</sad:sensorConfidence>
    </rdf:Description>
   </rdf:li>
   <rdf:li>
    <rdf:Description rdf:about="../sad#orientation_In_time">
     <rdf:type rdf:resource="../sad#Variable"/>
     <sad:value>A</sad:value>
     <sad:sensorDataSource
  rdf:resource="http://192.168.1.3/sensor06/rawdata#SensorOutput"/>
     <sad:sensorConfidence>0.75</sad:sensorConfidence>
    </rdf:Description>
   </rdf:li>
   ...
  </rdf:Description>
 </sad:evaluations>
</rdf:Description>
</rdf:RDF>
```

FIGURE 17.4
An RDF/XML example of a term declared using our metamodel, which corresponds to our multilevel approach.

17.5 Conclusion

We have identified priority variables for the dependency changes according to the context of the elderly (GIR groups). We described our vision of a flexible architecture that considers the ability to integrate various sources (services, sensors, and data) and notify the dependency changes. We also propose RDF model and inference engine. The RDF model includes several facets like home profile and user profile. The objective is to manage heterogeneous sensors and provide services and a dynamic composition of services. This goal is achieved based on both software component and web technology composition in order to best select suitable ones. Such dynamic composition allows to provide new rich services that consider the user's quality of services and acceptance and to apply required adaptation to meet the user's satisfaction.

In the context of smart environments, the delivery and adaptation of mHealth contents and services raise many challenges to be solved within a short delay of a few years. An interesting solution should be to propose a user-centric approach that aims at guaranteeing secured and context-aware health-care services that satisfy the QoE of the targeted users: elderly and dependent persons. Nonetheless, achieving this user-centric approach requires considering the heterogeneous aspect of the existing communication platforms and services as well as the need for efficiently exploiting the contextual environment (e.g., patients' medical information and needs, patients' physical location). In such an approach, the main objective will be to develop a secure context-aware communication and service platform over a heterogeneous networking environment. This platform aims at facilitating the elderly and dependent persons' experience in their daily life. To achieve this objective, a particular interest will be given to the context awareness and interoperability problems, and dynamic and adaptive approaches will be explored at both the networking and service layers. Furthermore, given the critical nature of the information exchanged in health-care scenarios, the security issue is yet another important aspect that has to be addressed.

One key novelty should be that the user-centric approach is used in the sense that the security mechanisms, networking protocols, and services are triggered and managed according to user's (patient's) preferences, experience, and level of dependency while taking into account the environmental context.

Such ambition raises several technical problems as how to provide an autonomic adaptive IP-based end-to-end communication architecture over heterogeneous networks encompassing WSN segments and WAN. The communication architecture should include adapted QoS routing and reliable transport functionalities and needs to elaborate on the necessary interfaces between the communication architecture on the one hand and both context-aware services and security on the other hand. Service deployment and execution should be highly dynamic and continuously adapted (both services composition and deployment). This is the step toward long-life applications well known as *eternal* applications.

Questions

1. Are current mHealth applications and devices adapted to elderly and dependent persons and why?
2. Why does this chapter focus on the AGGIR model?
3. Can the proposed approach be applied in other dependency evaluation tools?

4. Why does the approach consider different profiles such as the home profile?

5. Why do we need to compose new services?

References

1. United Nations, Department of Economic and Social Affairs, Population Division. World Population Prospects: The 2012 Revision, Volume I: Comprehensive Tables ST/ESA/SER.A/336, 2013.

2. Ministry for Labour Social Relations Family and Solidarity, AGGIR (Autonomy Gerontology Iso-Resources Group) model: The national standardized instrument determining the attribution of the specific dependence allowance in France, *Official Journal of the French Government of 21 August 2008*, 2008–2821, 2008.

3. Achterberg, W., Pot, A., van Campen, C., & Ribbe, M. Het Resident Assessment Instrument (RAI): een overzicht van internationaal onderzoek naar de psychometrische kwaliteiten en effecten van implementatie in verpleeghuizen. Tijdschr Gerontol Geriatr, 30, 2, 1999.

4. R. Herbert et al., The Functional Autonomy Measurement System (SMAF): Description and validation of an instrument for the measurement of handicaps, *Journal of Age and Aging*, 17, 293–302, 1988.

5. M. Horney, Measuring and monitoring general health status in elderly persons: Practical and methodological issues in using the sf-36 health survey, *Journal of Gerontologist*, 36, 571–583, 1996.

6. T. Lemlouma, S. Laborie, and P. Roose, Toward a context-aware and automatic evaluation of elderly dependency in smart homes and cities, in *Proceeding of the IEEE 14th International Symposium and Workshops on a World of Wireless, Mobile and Multimedia Networks (WoWMoM)*, pp. 1–6, Madrid, Spain, 2013.

7. T. Lemlouma and M. A. Chalouf, Smart media services through TV sets for elderly and dependent persons, in *Wireless Mobile Communication and Healthcare*, Vol. 61 of *Lecture Notes of the Institute for Computer Sciences*, Social Informatics and Telecommunications Engineering, pp. 30–40, Springer, 2013.

8. T. Lemlouma, A. Rachedi, and M. A. Chalouf, A new model for NGN pervasive eHealth services, in *Proceedings of the International Symposium on Future Information and Communication Technologies for Ubiquitous HealthCare* (Ubi-HealthTech), pp. 1–5, Jinhua, China, 2013.

9. T. Lemlouma, UNIVERSALLY: A context-aware architecture for multimedia access in digital homes, in *Advanced Infocomm Technology*, Vol. 7593 of *Lecture Notes in Computer Science*, pp. 128–137, Springer, 2013.

10. In-Stat (website), UPnP and DLNA–Standardizing the Networked Home, Ref. #IN1004647RC, http://www.in-stat.com/.

11. L. Socher, The digital home: Highly promising, highly complex, *Annual Review of Communications*, 61(1), 237–244, 2008.

12. S. Deering, Host Extensions for IP Multicasting, IETF, RFC 1112 - https://tools.ietf.org/rfc/rfc1112.txt

13. K. Holtman and A. Mutz, Transparent Content Negotiation in HTTP, IETF, RFC 2295 - http://www.ietf.org/rfc/rfc2295.txt

14. R. J. Robles and T.-H. Kim, Review: Context aware tools for smart home development, *International Journal of Smart Home*, 4(1), 1–12, 2010.

15. J. Anderston and C. Aydin, Evaluating the impact of health care in-formation systems, *International Journal of Technology Assessment in Health Care*, 13(2), 380–393, 1997.

16. Z. Roupa, M. Nikas, E. Gerasimou, V. Zafeiri, L. Giasyrani, E. Kazitori, and P. Sotiropoulou, The use of technology by the elderly, *Health Science Journal*, 4(2), 118–126, 2010.

17. M. Alam, M. Reaz, and M. Ali, A review of smart homes-past, present, and future, *IEEE Transactions on Systems, Man, and Cybernetics, Part C: Applications and Reviews*, 42(6), 1190–1203, 2012.

18. S. Katz and C. Akpom, A measure of primary sociobiological functions, *International Journal of Health Services*, 6(3), 492–508, 1976.

19. F. Mahoney and D. Barthel, Functional evaluation: The Barthel index, *Maryland State Medical Journal*, 14(2), 61–65, 1965.

20. M. Lawton and E. E. Brody, Assessment of older people: Self-maintaining and instrumental activities of daily living, *Journal of Gerontologist*, 9, 179–186, 1969.

21. A. Dey, Understanding and using context, *Personal and Ubiquitous Computing Journal*, 5(1), 4–7, 2001.

22. R. Want, A. Hopper, V. Falcao, and J. Gibbons, The active badge location system, *ACM Transactions on Information Systems*, 10(1), 91–102, 1992.

23. R. Want, B. N. Schilit, N. I. Adams, R. Gold, K. Petersen, D. Gold-berg, J. R. Ellis, and M. Weiser, The ParcTab ubiquitous computing experiment, *IEEE Personal Communications*, 2, 28–43, 1995.

24. S. Long, R. Kooper, G. D. Abowd, and C. G. Atkeson, Rapid prototyping of mobile context-aware applications: The cyberguide case study, in *Proceedings of the Second Annual International Conference on Mobile Computing and Networking, MobiCom'96*, (New York), pp. 97–107, ACM, 1996.

25. A. Asthana, M. Cravatts, and P. Krzyzanowski, An indoor wireless sys-tem for personalized shopping assistance, in *Mobile Computing Systems and Applications*, pp. 69–74, DOI:10.1109/MCSA.1994.512737 - IEEE Press - Santa Cruz, USA, 1994.

26. N. Davies, K. Cheverst, K. Mitchell, and A. Friday, 'caches in the air': Disseminating tourist information in the guide system, in *Proceedings of the Second IEEE Workshop on Mobile Computer Systems and Applications, WMCSA'99*, (Washington, DC), p. 11, IEEE Computer Society, 1999.

27. A. K. Dey, D. Salber, G. D. Abowd, and M. Futakawa, The conference assistant: Combining context-awareness with wearable computing, in *Proceedings of the Third IEEE International Symposium on Wearable Computers, ISWC'99*, (Washington, DC), p. 21, IEEE Computer Society, 1999.

28. G. Chen and D. Kotz, A survey of context-aware mobile computing research, Technical Report, Hanover, NH, 2000.

29. T. Gu, H. K. Pung, and D. Q. Zhang, A service-oriented middleware for building context-aware services, *Journal of Network and Computer Application*, 28, 1–18, January 2005.

30. H. Chen, T. Finin, and A. Joshi, An ontology for context-aware pervasive computing environments, *Special Issue on Ontologies for Distributed Systems, Knowledge Engineering Review*, 18, 197–207, 2003.

31. M. Baldauf, S. Dustdar, and F. Rosenberg, A survey on context aware systems, *International Journal of Ad Hoc Ubiquitous Computing*, 2, 263–277, June 2007.

32. M. D. Rodríguez, J. Favela, E. A. Martinez, and M. A. Munoz, Location-aware access to hospital information and services, *IEEE Transactions on Information Technology Biomedicine*, 8, 448–455, December 2004.

33. M. D. Rodríguez, J. Favela, A. Preciado, and A. Vizcaıno, Agent-based ambient intelligence for healthcare, *AI Communications*, 18, 201–216, August 2005.

34. A. Kofod-Petersen and A. Aamodt, Contextualised ambient intelligence through case-based reasoning, in *Proceedings of the Eighth European Conference on Advances in Case-Based Reasoning, ECCBR'06*, (Berlin, Germany), pp. 211–225, Springer-Verlag, 2006.

35. M. Corchado, J. Bajo, and A. Abraham, Gerami: Improving health-care delivery in geriatric residences, *IEEE Intelligent Systems*, 23, 19–25, March 2008.

36. A. H. Maslow, A theory of human motivation, *Psychological Review*, 50, 370–396, 1943.

37. V. Henderson, *The Nature of Nursing: A Definition and Its Implications for Practice, Research and Education*, New York: Macmillan, 1966.

38. WAP Forum, User agent profile, specifications, Open Mobile Alliance, WAP-248-UAPROF-20011020-a, Version 20, 86 pages, October 2001.

39. G. Klyne, F. Reynolds, C. Woodrow, H. Ohto, J. Hjelm, M. H. Butler, and L. Tran, Composite Capability/Preference Profiles (CC/PP): Structure and vocabularies 1.0, recommendation, W3C, January 2004.

40. J. Farrell and H. Lausen, Semantic annotations for WSDL and XML schema, W3C recommendation, http://www.w3.org/TR/sawsdl/, August 2007.

41. C. Escoffier, S. Chollet, and P. Lalanda, Lessons learned in building pervasive platforms, *The 11th Annual IEEE Consumer Communications and Networking Conference*, Las Vegas, CA: États-Unis (2014).

42. X. H. B. Le et al., Health smart home—Towards an assistant tool for automatic assessment of the dependence of elders, *IEEE Engineering in Medicine and Biology Society (EMBS 2007)*, August 2007, 3806–3809, doi:10.1109/IEMBS.2007.4353161.

43. M. Kokkinos, N. Doulamis, and A. Doulamis, Local geometrically enriched mixtures for stable and robust human tracking in detecting falls, *International Journal of Advanced Robotic Systems*, doi: 10.5772/54049, 2013.

44. D. Beckett and B. McBride, RDF/XML syntax specification (Revised), recommendation, W3C, February 2004.

18

Investigating mHealth Education and Training for the Health-Care Sector

Mahmoud Numan Bakkar and Elspeth McKay

CONTENTS

18.1 Introduction

This chapter commences with a discussion on mobile health (mHealth) education/training and use, then on the notion of distributed instructional training strategies, followed by a section on instructional preferences, namely, problem-based learning (PBL). Next, there are sections on the key advantages for instructional mHealth programs for medical training environments, instructional theories, and key mHealth education/training difficulties. Three case studies are provided to show how the rural areas of the developing world are already adopting mHealth training programs. We end this chapter with mHealth future directions.

18.1.1 Background and Definition

A world statistics report published by the International Telecommunication Union (ITU) in 2011 showed that the number of mobile phone users reached 5.3 billion global users; 2 billion out of the 5.3 billion are using the Internet, which is about one-third of the world population (Iheed Institute, 2011). Therefore, because of their popularity, mobile devices are now being used in every aspect of people's lives around the world, playing a crucial role in global health-care education, for example, increasing the awareness of diseases, like diabetic prevention programs, remote patient monitoring (Figure 18.1), data collection, and health-care training programs (Davies et al., 2012).

This chapter discusses the deployment of mobile usage in health-care education through different technological modes. For example, many health-care practitioners use them to retrieve the latest medical knowledge through regular learning management systems (LMSs), for things like accessing disease simulations. mHealth education/training is also distributed through specially devised instructional strategies that enable health-care educators to conveniently facilitate their training programs online. Thereby, adopting a systematic approach to the design of these strategies, based on the learners' needs, is considered a key educational/training advantage in medical education, for example, problem-based instruction.

Yet the actual deployment of mHealth education/training faces many key difficulties and practical barriers, such as how to apply the theoretical background with best medical practice during the training delivery that suits various individual instructional preferences. Moreover, successful facilitation of mHealth education/training programs may also be affected by other barriers such as budget restrictions and competing priorities within the hospitals (Instructional Design Fusions, 2010).

However, there are benefits of mHealth education/training that clearly arise in developing countries and their rural areas, for example, low infrastructure cost, high-security communication protocols and mobile user-friendly devices, and the use of a short message service (SMS)–based Human immunodeficiency virus infection/acquired

FIGURE 18.1
Diabetes testing—Rowville Health. (From Rowville Health 2014, *Diabetes testing—Rowville Health*, digital image, Rowville Health, viewed January 10, 2014, http://rowvillehealth.com/portfolio/diabetes-testing-and-education/.)

immunodeficiency syndrome (HIV/AIDS) education system for the prevention of HIV disease through increased awareness of that disease (Ebad, 2013).

18.1.2 Education/Training and Use

Mobile devices in the health-care industry are used for many educational/training purposes by medical staff, patients, and health-care students, for example, SMS for appointment reminders, telemedicine, accessing patient records, treatment, and health-care awareness or retrieving health-care knowledge on the latest disease treatments. In Kenya, mHealth tools are being used to enable the government to reduce the administrative tasks of managing the HIV disease awareness programs (Iheed Institute, 2011). Moreover (Figure 18.2), health-care students can derive the benefit of using their mobile devices to access their instructional programs in different ways, for example, downloading information using their handheld devices during their university lectures (Luanrattana et al., 2012).

Nowadays, there are different usage modes for mobile devices in education/training. As mentioned before, an LMS is one of the most basic access modes for mobile devices in mHealth education/training. Since the knowledge and information in the health-care industry is developing every day, so too is the use of the mobile applications that the medical staff deploy to easily retrieve any information they need to then share with their patients (Ducut and Fontelo, 2008). These patients can also locate their own health-care knowledge easily, by using their own mobile devices. However, on the one hand, to avoid medical errors, the health-care staff should have the initial ability to access, evaluate, and analyze the health-care knowledge, while on the other hand, patients should have access to increase their awareness of certain health-care issues (Ducut and Fontelo, 2008). Nevertheless, it is preferable that the patients refer back to their health-care specialist to obtain their prescription and diagnosis. LMS are also helpful in creating different medical simulations for medical students (Ducut and Fontelo, 2008).

FIGURE 18.2
Nursing education. (From Microsoft Office Clip Art, 2014, *Nursing education*, digital image, Clip Art used with permission from Microsoft, Redmond, WA.)

In health-care studies, it has been noted that there is a difficulty for students to interview patients for their study needs and to gain an actual medical/health-care experience (Ducut and Fontelo, 2008). Some hospitals do not operate as educational institutes and therefore may not have clear procedures and rules that permit students to approach patients. However, some hospitals are dedicated to work as educational/training hospitals, and as such, students can attend regular workshops with a health-care mentor or medical field specialist who works with them and the patients. Using mHealth simulations reduces the ethical dilemma between educational technology adoption and ensuring patients' rights, by helping the students to conjure up their medical patient cases. For example, anatomic case models simulation is used in teaching task-based clinical situations. Further, using simulators in health-care training makes it easy and convenient for the students to attend their training any time and without any limitations. Repeating their simulations many times over enables particular medical skill acquisition and reduces errors, thereby improving patient health and safety (Ducut and Fontelo, 2008).

Another use of the mobile devices in health-care education involves the knowledge acquisition assessment process. This testing regime includes user-friendly mobile devices that effectively manage a large number of assessments. These digital devices are more secure and accurate with real-time results obtained with less effort and time and at a low cost. Students in health-care studies use their mobile device tools such as MedCalc for clinical computations. There are some applications that provide access to a variety of health-care processes, such as clinical decisions, clinical notes, and prescription writing. Moreover, they can be useful for synchronizing with the hospital's health-care information systems (HISs) and the hospital's electronic medical records (EMRs), as well as for documenting administrative work, health-care studies, and medical care (Ducut and Fontelo, 2008).

Finally, medical staff also benefit from the virtual training environments by way of social media tools such as social networks, blogs, and podcasts (Figure 18.3). These Web 2.0 technologies enable increased collaboration between trainees and their facilitator, thereby helping everyone to share their updated knowledge easily, wherever they are in the world by accessing their smartphones and receiving podcasts (Ducut and

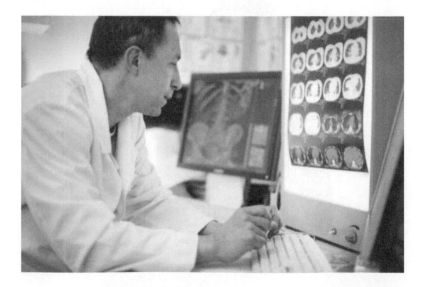

FIGURE 18.3
Medical staff. (From Microsoft Office Clip Art, 2014, *Medical staff*, digital image, Clip Art used with permission from Microsoft, Redmond, WA.)

FIGURE 18.4
Mobile devices. (From Microsoft Office Clip Art, 2014, *Mobile devices*, digital image, Clip Art used with permission from Microsoft, Redmond, WA.)

Fontelo, 2008). Using such smart devices motivates medical students to participate in real-time autonomous learning/training experiences (Figure 18.4).

18.2 Instructional Training Strategies

mHealth educational programs are most often intended to be used by either patients or medical trainees (students and health-care workers) to relay clear instructions. According to andragogy theories, which are used to design, develop, and deliver web-mediated instructional materials as effective adult knowledge acquisition, to promote that a trainee/learner needs to know (prior to the instruction): why they are undergoing a certain instructional program; that they are responsible for their own learning/training decisions; and that they prefer to be viewed as self-directed (Hewitt-Taylor, 2003). Life and work experience provides a rich learning resource that has the potential to effectively motivate people to learn what they need to achieve. Andragogy theories depict that adults are task-centered learners where internal instructional motivators are the main factors that ignite an adult's responsiveness. In contrast, theories of pedagogy stipulate that younger learners (children) are subject-centered learners.

18.2.1 Distributed Instructional Strategies

mHealth education/training programs give rise to the convenience of distributed instructional strategies, according to Willems et al. (2012); and as such, they enhance the instructional performance outcome levels. To achieve this end, the instructional content designer needs to apply a systematic approach to the preparation of their health-care training content. The instructional design approach should be based on identifying the learner needs, application of mHealth standards, and implementing sound instructional formats (Willems et al., 2012). Furthermore, the similarity in smartphone software platforms, which usually employ common mobile applications, makes it much easier for the distribution of their training programs among the various mobile devices.

Drawing on Gagné's nine instructional events when using a mobile device for educational/training purposes, his instructional tenets say it is necessary to rely upon clear instructional strategies that include gaining the learner's attention by informing the expected outcomes of the training, stimulating recall of prerequisite learning, presenting new materials, providing learning guidelines, eliciting performance, providing feedback about corrections, assessing performance, and enhancing retention and recall (Tough, 2012). Consequently, the mHealth training application developer should give design consideration toward instructional strategies such as attracting the attention of their user, by using the best color with a consistent design and functionality; presenting a clear course objective; enabling the learner to navigate into the new materials easily and follow a set of clear instructional guidelines, provided either visually or by using audio instructions; and then measuring the learner/trainee's performance efficiently, by providing a feedback report that simulates the weakness and the strength in the trainee's knowledge, to enhance the retention rate of the mHealth education/training.

Finally, mHealth education/training represents an up-to-date example in distributed learning environments; the old mobile devices could not enable the face-to-face instructional mode and used only the traditional eLearning tools. However, the newer smartphones can more easily enable the distributed delivery mode of eLearning, which combines both the time and distance with instructional delivery, with a face-to-face facility. This duality mode presents a blended learning/training experience combining both the distance and face-to-face instructional environments. This distinction is one of the advantages for implementing mHealth education/training in rural and distant community places (Willems et al., 2012).

18.2.2 Problem-Based Learning

By adopting a problem-based instructional stance, it affords the trainees to practice in a real-life scenario that is based on a constructivist (andragogy) cognition model. The process starts by identifying a problem in a case study, which may have different scenarios, aligned with the instructional objectives. Next, there is an examination of the learner's performance using an assessment instrument such as questioning, which is based on the expected training outcomes.

Typically there are different instructional designs and procedures for implementing a PBL course that may point out the following:

1. Problem identification
2. A problem discussion, with the learning facilitator
3. Creating and defining a set of hypotheses, related to the problem with potential solutions
4. Identifying the learning issues in the problem
5. Opportunities to research the instructional/learning issues
6. Reporting the findings and the opportunity to reexamine the learning problem
7. Presenting a formal solution for the learner's feedback (Elrod et al., 2005)

Mobile technology can facilitate the previous steps to implement PBL in health-care education/training. Research findings reveal that there are many factors that impact the performance of PBL in health-care education/training, indicating that there is no difference in the performance between the traditional classroom and the PBL class. However, a learner's attitude may be affected using the latter approach (McLinden et al., 2006). For example, McLinden et al. (2006) revealed that motivation and satisfaction increased and the learner's intrinsic interest was higher than in the traditional classroom. In addition, they also showed increased effectiveness of interaction among the learner/trainee.

It is proposed here that when PBL strategies are implemented using mHealth education/training, they will assist to center the learner, focus on specific tasks, and enhance the collaborative learning research knowledge base. They will also present a clear example of the efficiency and effectiveness of combining the ICT tools and the PBL delivery modes (McLinden et al., 2006).

The traditional use of PBL was in professional development for students in the higher educational settings, using ICT tools. This approach (especially using mobile technology) enabled health-care students to be independent. In addition, they became more autonomous learners using the research tools to search for their medical knowledge and to select, analyze, evaluate, and present the information (Elrod et al., 2005).

Finally, we say that there will be an enhanced learning experience in medical education/training, when it employs mobile ICT tools that embed a video processing capability, such as virtual simulation with 3D technology, which can easily be delivered using the mobile communications network technologies such as 3G and 4G mobile networks. Moreover, this approach to the instructional design will foster an increased awareness for the benefits of interactive learning/instructional opportunities, for example, attendance at a remote surgery and remote diagnosis for people living in rural areas.

18.3 Key Learning Advantages

Employing mHealth education/training programs has many advantages in health-care studies, for example, expanding the size of the cohort, since the traditional classroom is limited to a certain number of participants; instead, mHealth training environments enable more participants from different regions to attend the training, thereby gaining from the shared knowledge acquisition and training experiences (Hewitt-Taylor, 2003).

18.3.1 Enhanced Classroom Environments

mHealth expands the geographic borders of the traditional classroom while under the control of a central instructional management system, thereby reducing the extra effort and facility required to arrange the more traditional classroom training environment.

Location and time flexibility is also a major element of change in using the mHealth training; the participant attends their training based on their own schedule without any work duty pressures and overlapping priorities. Also, the decentralized location of the training room presents major advantages. Since attending some training locations requires such logistical issues as arranging to buy travel tickets, organizing hotel bookings, and being in a remote place, with long days away from the family. Instead, mHealth training enables the medical staff to attend to their education while staying at home with their family, without the extra cost of tickets and hotel bookings. Moreover, coping with road traffic in major cities is also a major issue, whereby using the mHealth training reduces the need to use the municipal transport systems and henceforth reduces the traffic on the road.

18.3.2 Convenient to Manage Work Scheduling

Since health-care work is often pressured, the hospital management cannot allow large numbers of staff to leave their rostered workstations to attend training classes; instead, they are required to maintain the continuity of service delivery to their patients (Figure 18.5). Hence, using an mHealth training program assists management to resolve this issue. By implementing such time flexibility allows the health-care trainees to accommodate their work responsibilities with their professional development training needs, thereby creating *on-demand* opportunities for the staff to focus on their actual training needs, by selecting the relevant parts of the course as they need them and skipping particular instructional modules they may already know about (Hewitt-Taylor, 2003), (Figure 18.6).

In the future, mHealth training tools can provide an ideal theoretical background for most medical issues; accommodate efficient and effective health-care training within the workplace environment, by reducing unnecessary travel to and from the workplace; and searching time for acquiring new knowledge. In addition, mHealth training tools can be used to source current medical discussions, thereby providing sound explanations for trainees.

18.3.3 Instructional Design and Learning

The most successful mHealth training tools deliver the best programs when they are underpinned by robust instructional design and appropriate learning theories. According

FIGURE 18.5
Enhanced classroom environments. (From Microsoft Office Clip Art, 2014, *Convenient to Manage Work Scheduling*, digital image, Clip Art used with permission from Microsoft, Redmond, WA.)

FIGURE 18.6
Convenient to manage work scheduling. (From Microsoft Office Clip Art, 2014, *Convenient to Manage Work Scheduling*, digital image, Clip Art used with permission from Microsoft, Redmond, WA.)

to Mat-Jizat (2012), *behaviorism, cognitivism,* and *constructivism* ideologies shape the instructional design characteristics for effective online instruction. It should be noted here that there are differences between instructional theories and learning theories (Reigeluth, 1983). The latter helps us to understand "how people learn and understand the learning process," while the principles of instructional design concentrate on the "instructional delivery modes that focus on the task analysis processes" to follow the course objectives that develop the sound instructional activities and instructional content materials that will enable the trainee to reach the expected instructional/learning outcomes (Reigeluth, 1983).

Similarly in mHealth education/training, the participant's behavior should be considered while developing the learning content (Figure 18.7). For instance, predetermined participant behavior will affect the instructional outcome. As such, the principles of instructional design are used to identify learning subskills and the associated tasks that assist the trainee to acquire the expected instructional outcomes (Mat-Jizat, 2012).

The second set of theories to be considered for implementing mHealth education is the cognitive theories, which focus on people's internal mental processes during their learning (McKay, 2000a) and as such look beyond the behavior as described by Mat-Jizat (2012). The instructional design adopted for mHealth applications therefore should consider cognitive theory either in the delivery phase of the instruction/training or in the assessment phase (McKay, 2000b). The third set of theories that are important for mHealth education/training design is from constructivism as they define the construction of individual learning strategies, when the learner creates new ideas and constructs new concepts (Mat-Jizat, 2012).

FIGURE 18.7
mHealth education—key difficulties. (From Microsoft Office Clip Art, 2014, *mHealth education key difficulties,* digital image, Clip Art used with permission from Microsoft, Redmond, WA.)

18.4 Key Difficulties

In this section, we list six key difficulties with mHealth education/training; the first two deal with time and cost, while the next three deal with the individual's personal difficulties, with the last arising as unrelated technical barriers to the adoption of mHealth education.

The first aspect regarding difficulties for mHealth deployment is about knowing how to apply the correct theoretical background to the health-care practices during the training sessions. This means there are two different views of learning/instruction, which raise concerns for certain training facilitation skills that are needed to direct the instructional facilitation (McKay and Henschke, 2009). As such, they are not easily delivered using the currently available (generic) mHealth training tools. For example, some online medical training laboratories should be assessed by a medical laboratory specialist (Hewitt-Taylor, 2003).

The second aspect relates to the time flexibility of using mHealth educational tools and the cost to the individual trainee. In other words, it is assumed by the employer that the trainee accommodates their own time for their training, without being paid or funded. Some courses require extra time and effort to complete, which can lead the trainee to request for study leave. In this case, the hospital management needs to negotiate with the trainee to either extend their time off or, in some cases, provide funding for the training. For instance, study leave for the trainee can be distributed during the week to avoid allowing the whole departmental staff leaving their workstations at the same time. Flexibility is also an issue that is related to course content delivery mode. Said a little differently here, this means that in a face-to-face delivery mode, some participants may be aware of a particular topic covered during the class. In this case, using mHealth training tools can afford them to skip some of the more familiar topics and focus on the acquisition of new knowledge.

The third aspect relates to the likelihood of the following personal barriers to adoption of mHealth education/training programs, such as individual learning preferences, whether they can afford attending the training programs; the individual's (cognitive) approach to organize and process the information (McKay, 2000a,b); external constraints, which may affect them, such as the family commitment and full-time work; the ability to resolve and manage their participation needs for the mHealth education class as required; and negotiation with the management to resolve any outcome issues.

The fourth aspect stems from the trainee/student's reflection and interaction with the course content, which includes their willingness to interact with their actual health-care practice and experience, with their new instructional/training topics. If this is the case, the mHealth tools should assist them to achieve that end. In helping the trainee to develop new knowledge, mHealth tools should enable the trainee to express their critical reflections, thereby improving their interactive participation through an evaluation process of their learning/instructional experiences.

The fifth aspect defines that some trainee/students may not study enough or complete the mHealth training program. This nonperformance can be resolved by using different assessment processes to measure their final outcome achievement levels (McKay, 2000b). This performance measurement aspect arises as a result of the need to cater for flexibility of the learning preferences (McKay and Izard, 2012; McKay and Martin, 2010) and the declaration of the assessment process to the knowledge navigation/instructional schedule such that it assists the learner to organize their instructional/training topics (McKay and Henschke, 2009). Put more simply, this means to afford them the opportunity to easily follow the instructional/training assessment, without being overloaded with extra work and their everyday life commitments (McKay, 2007).

The sixth aspect relates to a number of interrelated issues, for instance, the adoption of ICT tools and the level of the participant's information technology (IT) skills (McKay, 2003), the availability of the mobile network coverage (McKay, 2006), the ability to acquire the mobile device (McKay, 2005) that is required to run the training application and the using of this device, and the technical issues and the support that help to resolve any practical aspects while learning using the mHealth educational tools (McKay and Henschke, 2009).

18.5 Case Studies

18.5.1 MAMA Bangladesh and South Africa

The Mobile Alliance for Maternal Action (MAMA) program (Figure 18.8) uses mobile phones to help mothers to promote their antenatal care, assist HIV-positive mothers, and provide them with needed health-care information to prevent transmission of the disease to their babies. MAMA Bangladesh was launched in December 2012 to help the poor women in Bangladesh. In this instance, the application provided statistics of maternal, neonatal, and child health (MNCH), showing that every hour, one woman dies due to pregnancy-related complications in Bangladesh. Therefore, it can be assumed that using such mHealth applications helps in educating pregnant women, thereby reducing the risk of them losing their life or their baby (Iheed-Institute, 2011).

MAMA South Africa was officially launched in May 2013, in two South African cities (Hillbrow and Johannesburg). MAMA program aims to increase the mother's awareness and educate them to adopt healthy behaviors; sending a free SMS text message twice a week to the pregnant mother starts from the fifth week of pregnancy until her baby is 1 year old. The mother can customize the service to use any of the five languages provided or receive more details from the mHealth program provider. She can also visit the online web-based portal http://askmama.mobi/and share her story online or read more articles about pregnancy and babies (Iheed-Institute, 2011).

The instructional training strategies adopted for MAMA Bangladesh incorporate sending a voice message that involves a role-play to present real-life scenarios of characters that includes the pregnant women's doctor, mother-in-law, and husband, to help with his wife's decision-making process (Iheed-Institute, 2011).

FIGURE 18.8
MAMA Mobile Community/SA's week-by-week guide to pregnancy and infants. (MAMA SA 2014, MAMA Mobile Community/SA's week-by-week guide to pregnancy and infants, digital image, MAMA is a partnership between the U.S. Agency for International Development (USAID), Johnson & Johnson, the United Nations Foundation, and BabyCenter and operates through a secretariat hosted by the mHealth Alliance 2014, viewed January 10, 2014, http://askmama.mobi.)

The evaluation of MAMA South Africa reveals that as of April 2013, over 17,500 South African women had used this service, and 80% of the mothers gained new knowledge about their child's milestone development, vaccination dates, and children's foods. Information available to pregnant mothers includes illness signs, how to avoid the transmission of HIV to their children, and improved nutrition. The program also provides the mothers with the ability to share information and advice to each other. By late 2013, more than 52,000 mothers and guardians had subscribed to the service; and 17% of the subscribers were under the poverty line and receiving the service free (Iheed-Institute, 2011).

18.5.2 Health Education and Training in Africa Program Pilot in Ethiopia

The Health Education and Training in Africa (HEAT) program (Figure 18.9) was launched in 2011 in partnership between the Ethiopian Federal Ministry of Health (FMOH), the Regional Bureaus, the U.K.'s Open University (OU), UNICEF, the World Health Organization (WHO), and the African Medical and Research Foundation (AMREF). This program was funded by a grant of £250,000 given by the Ferguson Trust and $4 million from UNICEF (Iheed-Institute, 2011).

This health-care program came to change the health-care education delivery across Africa; it targets health-care workers to support their training needs among the different communities in Africa and works with FMOH to train 31,000 health extension workers. The HEAT program helps to standardize the training curriculum, delivering it directly to the regional communities and reducing the possibility for variations in the training content among the different colleges (Iheed-Institute, 2011).

It should be noted here that HEAT also uses a blended approach to its instructional strategies to ensure the consistency of the online service delivery quality, which does not disturb the existing health-care service delivery in the villages. HEAT integrates the instructional theory and the professional health-care practice by training and working at the same time (Iheed-Institute, 2011).

The distance learning initiative Teacher Education in Sub-Saharan Africa (TESSA) deployed the HEAT program through the web portal http://www.open.ac.uk/africa/heat/our-work/heat-ethiopia, which was accessed by 400,000 primary school teachers in 12 African countries. This online health-care initiative is facilitated by Ethiopian educators and the U.K.'s OU. It delivers rich educational/training resources, having freely accessed instructional content for primary school level education (Iheed-Institute, 2011).

The evaluation of the HEAT program reveals that there is an increased interest among the government and other interested agencies to adopt the HEAT program. The online numbers show that in the first quarter of 2011, it commenced with 1000 students, 100 distance learning tutors, and 100 practical skills tutors (Iheed-Institute, 2011).

FIGURE 18.9
HEAT program. (From The Open University, 2014, *HEAT program*, digital image, United Kingdom, viewed January 10, 2014, http://www.open.ac.uk/africa/heat/our-work/heat-ethiopia.)

Finally, from the aforementioned evaluation, HEAT is proving to be a valuable educational/training health-care resource, which develops the knowledge acquisition and the skill development of the African health worker. As such, it is supporting the further development of new distance learning experts in Africa. Furthermore, after securing more funding, the program developers expect to expand the HEAT training program to be deployed in six more African countries: Zambia, Tanzania, Rwanda, Ghana, Kenya, and Nigeria (Iheed-Institute, 2011).

18.5.3 African Medical and Research Foundation in Kenya

AMREF provides distance learning/training programs (Figure 18.10) for nursing registration in Kenya. The number of registered nurses in Kenya is very low. The online statistics reveal that for every 27,000 people, there is only one registered nurse; 85% of the nurses identified in 2005 were only trained to a certificate level, as Africa did not have a registered nurse diploma at the time. What is more, the available resources to train newly registered nurses is limited, and only 100 nurses each year could qualify for the registered diploma out of the 20,000 total number of nurses in Kenya (Iheed-Institute, 2011).

This distance learning project is aimed to shift nurses' training from the traditional classroom to a distance learning/training mode. At this point, this project involves a system of paper-based instructional/teaching materials, with partnership between AMREF, the Nursing Council of Kenya, the Kenyan Ministry of Health, Accenture (a management consulting, technology services and outsourcing company helping clients in South Africa become high-performance businesses; see http://www.accenture.com/za-en/Pages/index.aspx), and the Kenya medical training colleges. A team from this partnership has developed four computer-based training modules that are also delivered by a mobile phone. The program was funded by Accenture (eLearning Nurse Upgrading Programme), the Flying Doctors Society of Africa, and the Fresenius Foundation (AMREF Virtual Nursing School) (Iheed-Institute, 2011).

Finally, the evaluation of this program shows that 7000 students were enrolled between 2005 and 2011. Moreover, the same program has now been replicated in Uganda and Tanzania supported by AMREF. The program announces its successes saying,

> … we are proud of the programme, as nurse managers are reporting an improvement in the quality of nursing care. With improved nursing care, we are confident of our contribution in steering our country towards meeting the health-related MDGs. (AMREF, 2011)

FIGURE 18.10
AMREF Health Africa program. (AMREF, 2014, *AMREF Health Africa programme*, digital image, Kenya, viewed January 10, 2014, http://amref.org/.)

18.6 Future Directions

mHealth education and health sector training programs share the same motivation for growth and development. The continually increasing trend toward ICT tool adoption reflects the integral part technology is playing in the health-care industry. These trends indicate the applicability of mHealth education/training applications as

- Patients' mobile device solutions for their health tracking and medications' compliance
- Mobile applications that allow patients to check their own medical records or retrieve desired health-care information
- Remote monitoring of the mobile application for the patient's health care
- Continuing medical educational/training applications
- Communication applications between the patients and their physicians
- Health awareness applications (Fusions, 2010)

However, despite the future trends towards increasing availability of mHealth education/training tools, there are barriers of adoption toward this technology that include the following:

- Budget and competing priorities in the hospital governance.
- Lack of implementation efficiencies.
- Different hardware standards, platform compatibility, and lack of standard hardware configurations, giving rise to possible compatibility issues between health-care devices.
- Patient privacy and information security.
- Rules and regulations that manage the use of mHealth educational tools.
- A slowdown in the growth rate of smartphone users means that most people wishing to access mHealth education is still low. Hence, many users may not be able to access mHealth educational tools because they do not have a smartphone device (Fusions, 2010).

Finally, it is expected that the smartphone usage however will continue to rise more than tablets and cell phones. The two most popular mobile operating systems iOS and Android are the two dominant mobile operating systems on the market. Yet there is concern about the emerging and fancy mobile applications that are unreliable. Furthermore, there is doubt about the general use and security of cloud computing (Iheed-Institute, 2011). While another view says that mHealth education is just one part of the wider learning environment (Iheed-Institute, 2011), a blended approach that includes face-to-face and virtual learning is better (Fusions, 2010). Therefore, mHealth education/training is just one element of the whole online training debate that should be integrated with other instructional strategies to benefit trainee/students and their health-care studies.

18.7 Conclusion

In recent decades, the mobile communications industry has developed rapidly. There has been a trend toward high-level capacity performance and scope of the subsequent availability of ICT tools. Current mobile users constantly change their devices to keep abreast of the latest technology. In the health-care sector, for instance, the sophisticated features of the mobile services attract keen interest from a diverse range of medical practitioners worldwide (Iheed-Institute, 2011). Of late, the mHealth industry has emerged in an intuitive manner as medical staff and clinicians adopt these digital devices in their personal life and in their work environment. From our observations, the health industry in general requires secure mobile-enhanced educational and training strategies to ensure the quality of their service provision is maintained in a safe and secure environment. The three case studies outlined in this chapter illustrate the potential benefits of mHealth applications in achieving better health outcomes.

Questions

1. Define the mHealth education.
2. Mention three usages of mobile devices in health education/training programs.
3. Explain how PBL assists in mHealth education/training.
4. Mention four key difficulties in mHealth education/training.

References

AMREF. 2011. AMREF training [Online]. http://amref.ichameleon.com/where-we-work/our-work-in-kenya/training-7000-nurses-in-kenya/ (accessed January 6, 2014).

Davies, B. S., Rafique, J., Vincent, T. R., Fairclough, J., Packer, M. H., Vincent, R., and Haq, I. 2012. Mobile Medical Education (MoMEd)—How mobile information resources contribute to learning for undergraduate clinical students—A mixed methods study. *BMC Medical Education*, 12, 1.

Ducut, E. and Fontelo, P. 2008. Mobile devices in health education: Current use and practice. *Journal of computing in Higher Education*, 20, 59–68.

Ebad, R. 2013. Impact of mobile telephony innovations in mHealth education and awareness programs. *VSRD International Journal of Computer Science & Information Technology*, 3, 2231–2471.

Elrod, G. F., Coleman, A. M., Shumpert, K. D., and Medley, M. B. 2005. The use of problem-based learning in rural special education preservice training programs. *Rural Special Education Quarterly*, 24, 28–32.

Fusions, I. D. 2010. The future of mhealth learning | Instructional Design Fusions [Online]. http://instructionaldesignfusions.wordpress.com/2010/12/23/the-future-of-mhealth-learning/ (accessed September 18, 2013).

Hewitt-Taylor, J. 2003. Facilitating distance learning in nurse education. *Nurse Education in Practice*, 3, 23–29.

Iheed-Institute. 2011. mHealth education: Harnessing the mobile revolution to bridge the health education and training gap in developing countries, Ireland.

Luanrattana, R., Win, K. T., Fulcher, J., and Iverson, D. 2012. Mobile technology use in medical education. *Journal of Medical Systems*, 36, 113–122.

Mat-Jizat, J. E. M. 2012. Investigating ICT-literacy assessment tools: Developing and validating a new assessment instrument for trainee teachers in Malaysia. RMIT University, Melbourne, Victoria, Australia.

McKay, E. 2000a. Instructional strategies integrating the cognitive style construct: A meta-knowledge processing model (Contextual components that facilitate spatial/logical task performance). Unpublished Doctoral Dissertation (Computer Science and Information Systems). Total fulfilment: Available online from http://tux.lib.deakin.edu.au/adt-VDU/public/adt-VDU20061011.122556/, Deakin University, Melbourne, Victoria, Australia.

McKay, E. 2000b. Measurement of cognitive performance in computer programming concept acquisition: Interactive effects of visual metaphors and the cognitive style construct. *Journal of Applied Measurement*, 1, 257–291.

McKay, E. 2003. Managing the interactivity of instructional format and cognitive style construct in Web-mediated learning environments. Zhou, W., Nicolson, P., Corbitt, B. and Fong, F. (Eds.) *Advances in Web-Based Learning-ICWL 2003*. Springer, Berlin, Germany.

McKay, E. 2005. Human-computer interaction closes the digital divide: A multicultural, intergenerational ICT case study. *Proceedings of the 2005 South East Asia Regional Computer Science Confederation (SEARCC) Conference*, Vol. 46, Sydney, New South Wales, Australia. Australian Computer Society, Inc., Darlinghurst, New South Wales, Australia, pp. 29–33.

McKay, E. 2006. The effectiveness of rich internet application for education and training. *International Journal for Continuing Engineering Education and Life-Long Learning*, 16, 151–155.

McKay, E. 2007. Planning effective HCI to enhance access to educational applications. *Universal Access in the Information Society*, 6, 77–85.

McKay, E. and Henschke, K. 2009. A new perspective on competency management: Implemented through effective human-computer interaction. Tatnall, A., Visscher, A., Finegan, A. and O'Mahony, C. (Eds.) *Evolution of Information Technology in Educational Management*. Springer, New York.

McKay, E. and Izard, J. 2012. Investigating online training in government agencies: Designing adaptive web-based instructional programmes to reskill the workforce. *International Journal of Business Research*, 12, 69–82.

McKay, E. and Martin, J. 2010. Mental health and wellbeing: Converging HCI with human informatics in higher education. *Information in Motion: The Journal Issues in Informing Science and Information Technology*, 7, 339.

McLinden, M., McCall, S., Hinton, D., and Weston, A. 2006. Participation in online problem-based learning: Insights from postgraduate teachers studying through open and distance education. *Distance Education*, 27, 331–353.

Reigeluth, C. M. 1983. Meaningfulness and instruction: Relating what is being learned to what a student knows. *Instructional Science*, 12, 197–218.

Tough, D. 2012. A focus on Robert Gagné's instructional theories: Application to teaching audio engineering. *Music and Entertainment Industry Educators Association*, 12, 209–220.

Willems, J., Haigh, B., Haigh, C., D'amore, A., and Johnston, D. 2012. Towards a three-dimensional definition of distributed learning in the context of clinical training in rural medicine. http://www.odlaasummit.org.au/admin/_files/programme/1359415305_willems-haigh-haigh-et-al.pdf, pp. 143–150 (accessed June 29, 2014).

19

Importance of National Culture on Using Mobile Health-Care Systems (MHS)

Pantea Keikhosrokiani, Banafsheh Farahani, Nasriah Zakaria, Norlia Mustaffa,
Hassan Sadeghi Amini, Gilnaz Purhossein, and Sayedmehran Mirsafaie Rizi

CONTENTS

19.1 Introduction

Mobile health-care system (MHS) utilizes emerging mobile communication and network technologies used specifically for the delivery of biomedical data anytime from anywhere. One of the main concerns of MHS is adoption with new technologies. Since there are various cultures along with different distinguished boundaries, MHS users will behave inconsistently in order to accept the new technologies. In line with this, it is tried to examine cultural effects on the acceptance of MHS and compare it with some other user's characters such as age and gender. In this regard, uncertainty avoidance (UA), which is one of the dimensions of national culture that was provided by Hofstede, has been used to detect the influence of culture on the acceptance of MHS. For this purpose, a survey questionnaire was distributed among smartphone users in three countries including Iran, Malaysia, and America among developing and developed countries to determine users' behavior with different national culture backgrounds in interaction with MHS. Trust and anxiety are used as user acceptance factors in order to check whether UA has a significant relationship with user acceptance or not. In addition to descriptive analysis, one-way analysis of variance (ANOVA) is applied to verify the influence of national culture on trust and anxiety. If culture plays a significant role in the process of the acceptation, MHS providers can use this information to evade culturally related problems for future projects.

19.2 Background

19.2.1 Mobile Health-Care System

Information and communication technology can be applied in health-care services. Utilizing wireless network, mobile communication, and pervasive devices in health care leads to the emergence of new health-care systems. Advancement of the new technologies makes mobile health-care systems more realistic and feasible. For instance, portable and wearable devices can continuously monitor a patient with chronic disease anytime from any place. In addition, wireless sensors can exchange medical data via wireless networks. Thus, mobile computing not only provides a new opportunity for patients but also allows physicians to access patients' data anytime and anywhere (Wu et al., 2007; Boukerche and Yonglin, 2009).

19.2.2 User Acceptance of New Technology

Introduction of information technology (IT) in health care has a deep effect on the performance of the health-care systems. Many researchers have studied IT adoption and diffusion recently. Adaption is defined as the decision to accept or invest in a technology. The key objective of those IT researches is to access the value of the IT to the organization or project. Thus, organizations can manage their IT resources better and enhance overall effectiveness. Another approach is to examine the determination of IT adaption and usage by individual users (Masrom and Hussein, 2008). Moreover, investigating the factors associated with user acceptance of the new technology has been an important research in information system (IS) area. User acceptance is defined as the willingness within a user group to employ IT for the task it is designed to support (Dillon and Morris, 1996). Several models are developed to investigate the factors affecting the acceptance of new technologies in organizations. The theoretical models used to study user acceptance, adoption, and IT usage behavior comprise the theory of reasoned action (TRA), theory of planned behavior (TPB), technology acceptance model (TAM), unified theory of acceptance and use of technology (UTAUT), and innovation diffusion theory (IDT) (Masrom and Hussein, 2008).

19.2.2.1 User Acceptance of Mobile Health-Care System

With the advancement of new technologies in health care, adaption and acceptance of new technology turn into important factors in MHS. Increasing interest in end users' reactions to MHS has elevated the importance of theories that predict and explain health IT acceptance and use. The end users can be patients, doctors, physicians, health-care providers, and so on. Many studies investigate user adaption and acceptance of MHS such as Wu et al. (2007), Zhengchuan et al. (2008), Wan Ismail et al. (2012), Wu et al. (2011), Cocosila (2013), Hung and Jen (2012), and Holden and Karsh (2010).

19.2.2.2 Important Factors in User Acceptance of Mobile Health-Care System

Modern technologies promote the progress in health care and reform the current health-care systems. Utilizing mobile applications in health care can be considered as emerging and enabling technologies that have been used in many countries for emergency, general care, monitoring, and so on (Ammenwerth et al., 2003; Haux, 2006). For instance, various wireless technologies such as mobile computing, wireless network, and global positioning systems (GPSs) were utilized in Sweden for ambulance care (Geier, 2001). Another example is emergency trauma care in the Netherlands (Jan ten Duis and van der Werken, 2003).

There are many important factors to examine what determines MHS acceptance by health-care professionals and health-care users. Wu et al. (2007) consider some variables (such as technical support and training, compatibility, and MHS self-efficacy, perceived usefulness, and perceived ease of use) to model health-care professionals' MHS acceptance in the health-care environment. These factors will affect behavioral intention of users to use mobile computing in health care. In addition, Jen and Chao (2008) used information quality, system quality, user satisfaction, and *mobile health-care anxiety* as factors that are significantly connected with mobile patient safety service system success. Gilson (2005), Abelson et al. (2009), and Schee et al. (2006) mentioned that *trust* measurement turns out to be an important indicator of support for a health-care system.

19.2.2.3 Cultural Differences and User Acceptance

People in developed and developing countries faced many differences socially and economically. The differences are not only based on social and economical but also on the cultures of each country. However, many researchers considered the existence of advanced technological infrastructure as well as popularization of new technologies in society as the root of significant differences. Therefore, many studies were conducted in order to evaluate the cultural differences in the acceptance of new technologies. Such studies regarding the cultural differences in the acceptance of new technologies include those of Zakour (2004), Thaker (2008), Ebrahimi et al. (2010), and Teo et al. (2008).

19.2.3 National Culture

The nature of culture is pervasive and broad, which requires a detailed study of language, knowledge, laws, religions, food, customs, music, art, technology, work patterns, products, and other artifacts that give the society its distinctive flavor. A great number of definitions are provided, and in this regard, Merchant and Merchant (2011) affirm that: "as early as 1952, researchers identified more than 160 definitions of culture, and in 1994, it was estimated that culture has been defined in approximately 400 ways." By having diverse definitions, culture affects people's deeds in the society due to their ideas, values, attitudes, and normative or expected patterns of behavior. Hall (1976) states that: "Culture is not genetically inherited, and cannot exist on its own, but is always shared by members of a society." Moreover, Hofstede (1991) gives another definition of culture as "the collective programming of the mind which distinguishes the members of one group from another" (p. 1), which is passed from generation to generation. Culture is the society personality and it helps to distinguish one society from another (Schiffman and Kanuk, 2000). De Mooij (2010) refers to culture as the glue that binds groups together. He mentioned that culture includes whatever has worked in the past such as shared beliefs, attitudes, norms, roles, and values that were usually transferred from generation to generation within the groups. According to the objective of consumer behavior studies (the influence of culture on consumer behavior), Schiffman and Kanuk (2000) defined culture as "the sum total of learned beliefs, values, and customs that serve to direct the consumer behavior of member of a particular society" (p. 322).

Cultural dimensions are often known as an incisive effect on the success or failure of transferring new technology inside firms. In recent and previous years, worldwide studies showed the importance of studying how culture effects on the adoption of new technologies in organizations (Deal and Kennedy, 1982). Culture, as a system of shared values and assumptions, is very fundamental for any organizational activity. It includes how organizations function, how employees interact, and how decisions are made. Culture indicates a core set of values, which describe the attitudes that employees use to attune themselves toward change and their approaches to face the introduction of something new (Kanter, 1989). Hofstede (1997) argued that culture is the only most important factor accounting for success or failure in organizations.

Though there are diverse definitions of culture in the literature, and according to the importance of values among different culture, there is general agreement about using (Hofstede, 1980) cultural model (dimensions of value) in social science study. Hofstede (1980) developed a model of five dimensions of national culture that helps to understand basic value differences. In his model, he distinguishes cultures according to power distance, individualism/collectivism, masculinity/femininity, UA, and long-term orientation. A great number of studies have highlighted the mentioned model although there are other cultural models used in different research.

According to Brebbia and Zubir (2011), "In 1980, the Dutch management researcher Greet Hofstede first published the result of his study of more than 100,000 employees multinational IBM in 40 countries" between 1967 and 1973. He was challenging to find value dimensions toward cultural differences (Jandt, 2010). Hofstede classified four dimensions as power distance, individualism/collectivism, masculinity/femininity, and UA for national culture values. He identified the fifth dimension as long-/short-term orientation by the help of Bond in 1984. The second survey, which helps to develop the fifth dimension, was called *Chinese value survey*, which was used to identify Asian typical culture.

19.3 Important Factors in Acceptance of Mobile Health-Care System

There are many important factors associated with the acceptance and use of MHS; however, we only focus on mobile *health-care anxiety, trust,* and *UA* as a cultural element in this chapter.

19.3.1 Mobile Health-Care Anxiety

Mobile health-care anxiety is "the main reason for prevention of computer technology adoption is defined as trepidation, fear, concern and hesitation of damaging the device, being embarrassed" (Heinssen et al., 1987). Jen and Chao (2008) defined mobile health-care anxiety as "a high anxious response towards interaction with the mobile patient safety information system" that has a negative effect on the quality of care that physicians provide for the patient. As the results show, information quality, system use, user satisfaction, mobile health-care anxiety, impact on the individual, and impact on the organization are all factors that are significantly connected with mobile patient safety service system success. This concluded that mobile patient safety ISs play a role in improving patient services and reducing patient risk, as well as showing that the use of these communication systems may unintentionally be a factor in physician anxiety (Jen and Chao, 2008). Abelson et al. (2009) defined anxiety as the antonym of trust (Abelson et al., 2009).

19.3.2 Trust

Trust has been defined as "a set of beliefs that other people would fulfill their expected commitments under conditions of vulnerability and interdependence" (Rousseau et al., 1998). Alali and Salim (2011) used trust as one of the factors in their *virtual communities of practice* success model, noting that trust influences user satisfaction as well as intention to use (Alali and Salim). Trust measurement turns out to be an important indicator of support for a health-care system (Gilson, 2005; Schee et al., 2006; Abelson et al., 2009). Because of the importance of trust in the health-care area, Lord et al. (2010) evaluated "the effect of patient physician trust on how British South Asian (BSA) and British White (BW) patients cope when diagnosed with cancer." Wright et al. (2004) identified trust as the most significant factor in doctor/patient communication (Wright et al., 2004).

19.3.3 Uncertainty Avoidance

The range of avoidance of uncertainty and ambiguity among people is related to this part of national culture value. People with high UA culture care about rules and they do not like to take risk, while people with low UA culture more willingly accept risk. They feel that there should be as few rules as possible and they are open to have different behavior when needed. Hofstede (2001) states that: "Uncertainty avoidance reflects the degree of comfort members of a culture feel in unfamiliar or unstructured situations and the extent to which a society tries to control the uncontrollable."

Individuals with high UA are more concerned with security in life, feel a greater need for consensus and written rules, and are intolerant of deviations from the norm. In contrast, individuals with low UA are less concerned with security, rely less on written rules, and are more risk tolerant (Hofstede, 1984). Individuals with high UA are less likely to take risks and are more intolerant of deviations from an established code of ethics. Luthar and Luthar (2002) state: "Indeed, cultures scoring high in the UA index use rules and codes as means of minimizing or avoiding risk. Uncertainty accepting cultures are more accepting of opinions other than their own."

Sigala and Sakellaridis (2004) argue that: "People vary in the extent that they feel anxiety about uncertain or unknown matters, as opposed to the more universal feeling of fear caused by known or understood threats. So, people differ in their avoidance of uncertainty, creating different rituals and having different values regarding formality, punctuality, legal-social requirements and tolerance for ambiguity." Yoo and Donthu (2005) found that high UA customers expect to have higher service quality, while individuals with low UA have lower expectation in this matter. Moreover, UA is positively correlated with *reliability, responsiveness, assurance, and empathy* but negatively with *tangibles*. People in low UA "do not need precise and explicit details such as job descriptions, product descriptions, and product use instructions. In contrast, people in a higher uncertainty avoidance society need to control the environment, events, and situations" (Yoo and Donthu, 2005).

Hofstede (1984) asserts that in countries with high UA culture, risk should be controlled by conservatism, law, and order, while in countries with low UA, uncertainty and risk taking is not that much serious. Therefore, the second group is more open to change and novelty, more willing to take risks, and more tolerant for diversity and ambiguity. Regarding the influence of reference group in UA discussion, Bathaee and Pechtl (2011) believe that, "since individuals in high uncertainty avoidance cultures may feel strong need for consensus and believe more on experts and their knowledge (Hofstede, 1988), the informational influence of reference-groups would be more needed, and is therefore stronger, among individuals with high level of uncertainty avoidance." They quote from Hofstede (1980), "those with higher avoidance of uncertainty find deviant persons or ideas dangerous and try to avoid them. In other words, the utilitarian influence of reference-groups could be higher in these cultures and people try to conform to the group because of the rejection risk."

Hofstede (2001) indicated that there is an implicit paradox in linking culture with change. In the first glance, necessarily, culture has traditional and stable quality, so how can you have a *culture of change*? Yet this is exactly that thing that innovative organization needs. The most widely cited theory in culture and organizational culture is from Geert Hofstede's work (Ogden and Cheng). In his official website, he describes all steps were done by him and his group or others to enhance the theory later. Zakour (2004) based on Mintsberg's classification of organizational structure classified organizations in four different categories by connecting different cultures to correlate the power distance and UA dimensions. According to his findings, there is a high correlation between the Muslim religion and the Hofstede dimensions of power distance (PD) and UA scores.

As shown in his website in Malaysia's page, the combination of two high scores, UA and PD, can make societies that are highly rule oriented and pay attention to laws, rules, regulations, and controls to decrease the amount of uncertainty, and in contrast, inequalities of power and wealth have been allowed to grow within the society. These cultures somehow like to follow a system that contains different categories of clan that does not allow significant upward mobility of their citizens. When these two dimensions (power distance [PD] and uncertainty avoidance [UA]) are mixed, it creates a situation where leaders really have ultimate power and authority, and the rules, laws, and regulations will be developed by those who are in power and who can reinforce their own leadership and control (Ebrahimi et al., 2010).

The study has been done by Ebrahimi et al. (2010) who tried to explore the influence of culture on the intention to use new technology.

> With doing Pearson's correlation analysis, researchers found that there is a positive but not strong relationship between culture's dimensions and intention to use new technologies. Strength of relationships may be affected by medium sample size and can be increased by choosing larger sample sizes for future studies. The three Culture dimensions included Power Distance, Uncertainty Avoidance, and Collectivism/ Individualism have positive significant correlation with Behavioral Intention to use new technologies. (Ebrahimi et al., 2010, p. 1032)

Consequently, individuals from high UA culture will have fewer tendencies to use IT as compared to individuals from low UA culture because of the lack of social presence in IT, which can accentuate uncertainty. Data analysis revealed that the TAM does not hold for certain cultural orientations. Most significantly, low *UA* seems to nullify the effects of perceived ease of use and perceived usefulness that are important factors in the acceptance of new technology (McCoy et al., 2007). The social influence exerted by important persons will be much more important in determining IT use in cultures seeking to avoid uncertainty than in cultures comfortable with uncertainty. Actually, in order to deal with uncertainty and ambiguity, individuals in strong UA cultures are very concerned with the establishment and respect of rules; therefore, the subjective norms will be more important as guidance to behavior than for individuals in weak UA cultures, which rely more on their proper competence to evaluate a situation (Zakour, 2004).

19.4 Research Design and Methodology

19.4.1 Research Instrument

We used an online questionnaire as a tool to gather data from Iran, Malaysia, and America. The questionnaire consists of four main parts including demographics, smartphone users, chronic disease experience, and assessment part. Focus of assessment part was on *trust* and *anxiety* as these two are the most effective factors of *UA* from the national culture dimension. We can observe the cultural differences in trust and anxiety more than any other factors in technology acceptance theories. Thus, the main concentration of this study is to check the effect of national culture (UA dimension only) on trust and anxiety. The type of questions in the first three parts was optional, while in the assessment part, data were collected using a five-point Likert-type scale from one, being *strongly agree*, to five, being *strongly disagree*.

19.4.2 Subject

The goal of this study is to highlight the role of national culture in the use of MHS. Therefore, the subject of this study was smartphone users from three countries including Iran, Malaysia, and America. A survey questionnaire was distributed to gather data from three countries and the convenience sampling was used for data collection. A total number of 326 responds were collected from three countries (Iran, Malaysia, and America). Among the total respondents, 100 were Iranians, 122 Malaysians, and 104 Americans.

19.4.3 Hypothesis

Based on the literature in part 3 of this study, we can conclude that trust and anxiety are two effective factors in the acceptance and use of MHS. Moreover, UA culture that is of the dimensions of culture significantly affects acceptance and use of new technologies. Therefore, this study evaluates the effect of culture (UA) on trust and anxiety of MHS users. The result will assist up prove the importance of culture in acceptance of MHS.

H_T: Culture (UA dimension) has a significant relationship with trust in MHS.
H_A: Culture (UA dimension) has a significant relationship with anxiety in MHS.

19.4.4 Analysis Method

Descriptive analysis as well as ANOVA was used using SPSS software. Descriptive analysis assists us to describe and compare the main features of the collected data quantitatively, whereas ANOVA is used to analyze the differences between group means and their associated procedures (such as *variation* among and between groups). Hence, ANOVA helps us to measure the effect of nationality on trust and anxiety and finally on the acceptance of MHS.

19.5 Data Analysis and Results

19.5.1 Descriptive Analysis

This section presents descriptive analyses of demographic data providing a profile of respondents including frequency values. All tables in this section present a profile of the surveyed respondents with regard to age, gender, and education, smartphone users, and their experience about chronic disease. In this research, basic descriptive analysis, using frequency table, was used to provide a demographic profile of respondents. The second part of data analysis is to understand smartphone users and their opinion about the usefulness of smartphones for health care. The third part discusses about chronic disease, whether the respondents experience it on their own or someone in their family. All questions were required to be answered, so there is no missed question. Respondents were from Iran, Malaysia, and America with the total number of 326 (Iranians, 100; Malaysians, 122; Americans, 104).

The information shown in Table 19.1 reveals that for Iranians, just over half the sample (64%) were female and 36% were male. In America, about 61.5% of respondents were female, while 38.5% were male, and in Malaysia, 55.7% were female and 44.3% were male. The age group from 18 to 25 was 36% for Iranians, and the majority of respondents between ages 18 and 25 in America and Malaysia were 46.2% and 27.9%, respectively. Another age group of 26–33 for Iranians was 38%, Americans 32.7%, and Malaysians 52.5%. There were only 7.7% respondents younger than 18 years old from America. The age rate of 34–41 was 20% for Iranians, 3.8% for Americans, and 14.8% for Malaysians, and only 6% for Iranians were in the age of 42 and above followed by 9.6% for Americans and 4.9% for Malaysians.

In education, the data in Table 19.1 illustrate that majority of Iranians (58%) were postgraduates, 36% were undergraduates, and only 6% had diplomas. There were 44.2% undergraduate respondents from America, followed by 19.2% postgraduates, 15.4% with diplomas and also high school, and 5.8% with professional degrees. The majority of the respondents from Malaysia were in undergraduate level (60.7%), followed by 31.1% postgraduates, 6.6% with professional degrees, and only 1.6% high school.

TABLE 19.1

Demographic Data

		Iranian (%)	American (%)	Malaysian (%)
Age	Under 18	—	7.7	—
	18–25	36	46.2	27.9
	26–33	38	32.7	52.5
	34–41	20	3.8	14.8
	42 and above	6	9.6	4.9
Gender	Male	36	38.5	44.3
	Female	64	61.5	55.7
Education	High school	0	15.4	1.6
	Diploma	6	15.4	0
	Undergraduate	36	44.2	60.7
	Postgraduate	58	19.2	31.1
	Professional degree	0	5.8	6.6

In the second section as shown in Table 19.2, which was about smartphone users, 100% of the respondents from Iran own smartphones. The majority of the smartphone models in Iran was HTC brand (30%), followed by Samsung (24%), Nokia (20%), Sony (16%), and a small number of Blackberry, iPhone, and LG. In America, the percentage of smartphone users was 86.5% in which 48.1% of respondents owned iPhone surprisingly and 17.3%

TABLE 19.2

Data Associated with Smartphone

		Iranian (%)	American (%)	Malaysian (%)
Smartphone/tablet owners	Yes	100	86.5	91.8
	No	0	13.5	8.2
Smartphone/tablet model	Samsung	24	17.3	54.9
	iPhone	4	48.1	21.3
	Motorola	0	3.8	0
	Blackberry	4	1.9	3.3
	HTC	30	5.8	0
	Nokia	20	5.8	3.3
	iPad	0	0	1.6
	Sony	16	0	6.6
	LG	2	3.8	0
	Other	0	13.5	9
Duration of using smartphone	Less than 1 year	24	30.8	26.2
	1–3 years	56	51.9	57.4
	4–7 years	20	15.4	11.5
	More than 8 years	—	1.9	4.9
Experience of using smartphone	Poor	10	13.5	11.5
	Fair	22	17.3	14.8
	Good	50	28.8	55.7
	Very good	18	40.4	18
Usefulness of smartphone for health care	Yes	90	92.3	90.2
	No	10	7.7	9.8

Samsung, followed by HTC (5.8%), Nokia (5.8%), Motorola (3.8%), Blackberry (1.9%), and others (13.5%). On the other hand, in Malaysia, 91.8% of respondents owned smartphones, in which most owned Samsung (54.9%), followed by iPhone (21.3%), and other models such as Sony (6.6%), Nokia (3.3%), and Blackberry (3.3%), and about 9% did not mention their mobile phones.

Most of the respondents from Iran, Malaysia, and America have been using smartphones for 1–3 years. About 56% of Iranians have been using smartphones for 1–3 years, 24% less than 1 year, and 20% for 4–7 years in which 50% have mentioned their good experience of using smartphones, 22% fair, 18% very good, and 10% poor. Accordingly, 90% of Iranians think that smartphones are useful for health care. In America, 51.9% used their smartphones between 1 and 3 years, followed by 30.8% less than 1 year, 15.4% for 4–7 years, and 1.9% more than 8 years. 40.4% of the respondents have very good experiences of using smartphones, 28.8% good, 17.3% fair, and 13.5% poor. 92.3% of Americans' opinions were that smartphones are useful for health care. As it can be seen, 57.4% of respondents from Malaysia were using smartphones for 1–3 years, followed by 26.2% for less than 1 year, 11.5% for 4–7 years, and only 4.9% more than 8 years. In this regard, 55.7% of Malaysians have good experience of using smartphones, 18% very good, 14.8% fair, and 11.5% poor. Almost 90.2% of Malaysians believe that smartphones are useful for health care.

The next part of the questionnaire was about the experience of respondents about chronic disease. Respondents were asked if they have any experience of chronic disease or if anyone in their family has experienced. The number of Iranians who experienced chronic disease was 20%. On the other hand, about 78% of Iranians have experienced chronic disease in their family and 94% think that health-care applications for monitoring chronic disease patients are useful as illustrated in Table 19.3.

Based on Table 19.3, in America, about 28.8% experienced chronic disease. In addition, 57.7% of respondents from America had someone in their family that experienced chronic disease. 88.5% of Americans believe that health-care applications for monitoring chronic disease patients are useful. In this regard, in Malaysia, the number of respondents who experienced chronic disease was only 9.8%. Respondents who have experienced chronic disease in their family were 63.9%, and 90.2% have agreed with the usefulness of health-care applications for monitoring chronic disease patients.

19.5.2 Analysis of Variance Result

As mentioned before, the focus of this study is to understand the importance of national culture in the acceptance and use of MHS. Trust and anxiety are two effective factors on the acceptance and use of new technologies. Hence, we decided to verify the role of

TABLE 19.3

Data Associated with Chronic Disease

		Iranian (%)	American (%)	Malaysian (%)
Experience of chronic disease	Yes	20	28.8	9.8
	No	80	71.2	90.2
Anyone with chronic disease in their family	Yes	78.0	57.7	63.9
	No	22.0	42.3	36.1
Usefulness of health-care applications for monitoring chronic disease patients	Yes	94.0	88.5	90.2
	No	6.0	11.5	9.8

TABLE 19.4

Trust and Anxiety Questions

Factor	Sign	Questions
Trust	T1	I would like to use MHS, if it is trustworthy.
	T2	Personal information protection will be important in using MHS.
	T3	MHS must have people that I can rely on.
	T4	I will allow MHS to use my health data for my treatment.
Anxiety	A1	I feel fearful about using MHS for chronic disease.
	A2	It scares me to think that I could lose a lot of information using MHS.
	A3	I hesitate to use MHS for fear of making mistakes I cannot correct.

national culture on trust and anxiety of the users toward using MHS. Based on our questionnaire, trust is divided into four parts (T1, T2, T3, T4), whereas anxiety is separated to three parts (A1, A2, A3). These parts indicate the number of questions in trust and anxiety sections. Table 19.4 illustrates details of trust and anxiety dimensions based on our questionnaire. This part of the questionnaire was collected using a five-point Likert-type scale from one, being *strongly agree*, to five, being *strongly disagree*.

Table 19.5 shows the one-way ANOVA of culture as an influential factor, which affects the trust and anxiety of users. As mentioned before, only UA from other dimensions of national culture was considered in this study and the analysis as well. Table 19.5 shows that UA culture partially influences MHS users' trust. The last column of Table 19.5 depicts

TABLE 19.5

Significance Analysis of UA on MHS Users' Trust and Anxiety

Factor			Sum of Squares	df	Mean Square	F	Sig.[a]
Trust	T1	Between groups	11.257	2	5.629	4.828	0.009
		Within groups	376.571	323	1.166	—	—
		Total	387.828	325			
	T2	Between groups	1.252	2	0.626	0.460	0.632
		Within groups	439.631	323	1.361		
		Total	440.883	325			
	T3	Between groups	3.512	2	1.756	1.654	0.193
		Within groups	342.844	323	1.061		
		Total	346.356	325			
	T4	Between groups	8.112	2	4.056	3.265	0.039
		Within groups	401.286	323	1.242	4.828	
		Total	409.399	325			
Anxiety	A1	Between groups	0.554	2	0.277	0.240	0.787
		Within groups	372.281	323	1.153		
		Total	372.834	325			
	A2	Between groups	25.089	2	12.544	10.970	0.000
		Within groups	369.365	323	1.144		
		Total	394.454	325			
	A3	Between groups	7.720	2	3.860	3.605	0.028
		Within groups	345.850	323	1.071		
		Total	353.571	325			

[a] Values less than 0.05 are significant.

the significance of UA on T1, preference of using trustworthy MHS services (0.009 < 0.05), and T4, allowance to use health data for treatments (0.039 < 0.05), between three groups of nationality in this study.

Based on the results illustrated on Table 19.5, UA culture significantly influences anxiety between Iranians, Malaysians, and Americans as three groups of nations in this study. As it is shown Table 19.5, UA culture does not have a significantly influential role on MHS, feeling fearful about using MHS for chronic disease, but it significantly affects A2, scared to lose a lot of information using MHS (0.000 < 0.05), and A3, hesitant to use MHS for fear of making mistakes that cannot be corrected (0.028 < 0.05).

Although all of the results shown in Table 19.5 are not significant, the significant results indicate that culture has a significant effect on trust and anxiety that prove the validity of the proposed hypotheses, HT and HA, respectively.

19.6 Case Studies

19.6.1 Cultural Competence in Health-Care Systems: A Case Study (Perez et al., 2006)

This is a study in California about cultural competence training need in a health service system. Many studies have been done about how important culturally competent health-care systems are and how they can eliminate long-standing disparities in health status of people with different race, ethnic, and cultural background. The study's major aim was to identify the training needs of workers in a public health agency. The survey instrument was based on the cultural competence framework developed by the Arizona Department of Health Services (1995).

The questionnaire included six questions related to demographic information and the other seven sections dealing with knowledge levels related to training on cultural competence available at the workplace. For conducting the survey, through stratified random sampling from a population of 3217 workers at a health services system in California, a total of 900 employees including administrators and staff were selected to participate. Participants were from five departments in the agency: (1) children and family services, (2) community health, (3) administration, (4) employment and temporary assistance, and (5) adult services.

The number of returned surveys that were usable was 360, an overall response rate of 40% with 55% of administrators and 35% staff. The data were analyzed through SPSS. T-test was used to analyze continuous data for independent samples and nominal data were analyzed using the Pearson chi-square and Yates correlation factor.

The study has found that administrators are more likely participating in activities related to cultural awareness twice than staff. The findings also indicated that more administrators than staff perceived that contract or in-house translators were available. This disconnects with frontline personnel raising the questions about quality of the services being delivered. In addition, both administrators and staff indicated that consumers could not have access to a direct employee that can communicate in their own native language.

In regard to training needs, more administrators than staff indicated that training about cultural awareness and competence was needed. The findings reflect the need of additional training in specific areas like cross-cultural communication, ways to involve clients in their prevention and treatment process, and health-related beliefs across cultures. Providing health service to multiethnic population is very sensitive and local health service agencies must respond to the needs for cultural training through their employees.

19.6.2 Cultural Aspects of Primary Health Care in India: A Case-Based Analysis (Worthington and Gogne, 2011)

Belief systems and moral values belong to humans by their very nature, and for many of them, culture and religion play a significant role in their life especially in Eastern countries like India. Although humans' cultural and religion considerations can have a positive effect on their lives, the negative impact of norms bounded by culture and belief on their well-being is inevitable.

While health care in India is trying to serve all population, sometimes, due to some reasons like lack of literacy, culture practice, and religion belief that affect directly the health-seeking behavior, people cannot obtain proper and on-time medical help. Besides, people, no matter if they are educated or not, prefer traditional and local religious healers rather than Western forms of treatment.

S was a woman, who was brought to a hospital with symptoms of hard breathing, palpitation, and numbness. The symptom was accompanied by anxiety, aggression, voice changes, and delirium. Her family referenced her condition to *ancestral spirits*, which was a reflection of their traditional and cultural belief. They took her to a temple in order to take consultation from a religious healer. She improved for a few days only, and eventually, she was taken to a local psychiatrist. However, she refused psychiatric help due to the stigma of such a therapy.

After a period of psychotherapy, it is identified that she was under pressure of an unwilling marriage arranged by her family. On one hand, she had an ambition to continue her study, and on the other hand, she was shy to disaccord with her family, and this inner conflict caused the aforementioned symptoms. Gradually, during some psychoeducative sessions for both the patient and her family, the misconceptions of traditional and religious healing were clarified without any insult to their religious beliefs and it was suggested not to force her in to getting married until she was completely ready.

As it is obvious, the tremendous effect of culture and religion considerations on people's life is common in India and other Eastern countries. Consequently, it is compulsory to consider culture and the religious practice of people in order to achieve satisfactory patient outcomes. As is mentioned before, the issue neither is about education of people nor about the priority of a system of therapy over the other (traditional system and modern system), but it is about the coexistence of different systems as it common in some countries like China.

19.7 Future Directions

This study can be expanded in the future by focusing on other dimensions of culture such as power distance and masculinity/femininity. Rather than the culture dimensions mentioned in this study, monochronic/polychronic time can be evaluated as one of the culture dimensions in the acceptance of new technologies in health care (Zakour, 2004). Hall and Hall (1969) divided societies into two categories based on the time use: monochronic or polychronic. According to them, individuals from monochronic time-use societies tend not to focus on more than one activity at the same time. They prefer to have control on their time and stick to a schedule. In contrast, individuals from polychronic time-use societies have more tendencies to focus on more activities in a block of time and are willing to be free in terms of switching between activities. Thereupon, the first group will accept IT more in order to control their time and schedule, while the second group will accept it less as it might confine their freedom of using time comfortably (Lindquist and Kaufman-Scarborough, 2007). Moreover, cultural differences can be tested on other factors of technology acceptance in health care such as subjective norm, self-efficacy, performance efficacy, attitude, and satisfaction.

19.8 Concluding Remarks

With the advancement of new technologies in health care, the adaption and acceptance of new technology turns into important factors in MHS. Increasing interest in end users' reactions to MHS has elevated the importance of theories that predict and explain health IT acceptance and use. People in developed and developing countries faced many differences from social and economical points of view. The differences are not only social and economical but also based on cultures of each country. However, many researchers considered the existence of advanced technological infrastructure as well as popularization of new technologies in society as the root of significant differences. Therefore, many studies were conducted in order to evaluate the cultural differences in the acceptance of new technologies.

Consequently, this book chapter attempts to investigate the importance of culture (UA dimension) in the trust and anxiety of users in accepting MHS. Trust and anxiety are two effective factors of acceptance of new technologies in health care. Culture has five dimensions including power distance, individualism/collectivism, masculinity/femininity, UA, and long-term orientation. This study's focus was on the UA dimension only. The cultural comparison was among three different developing and developed countries. The total number of 326 data was collected from Iran, Malaysia, and America using survey questionnaire. Among 325 total responses, 100 were from Iran, 122 from Malaysia, and 104 from America. Descriptive analysis and ANOVA were conducted to verify the hypothesis and show the significance of culture on trust and anxiety as two effective factors of acceptance of MHS.

There are few points from descriptive analysis that need to be highlighted. Based on the results, the majority of the respondents were educated with at least undergraduate degrees in all three countries. This shows educated people had more interest to fill up the questionnaire as we aimed to ask many people to fill up the questionnaire but some of them rejected. The surprising part is the smartphone model. Among Iranian respondents, the majority used HTC device, while in Malaysia and America, Samsung and iPhone, respectively. This can be because of the advancement of technology and infrastructure in each country, price, market, and so on. There is no respondent in Iran who is using a Smartphone for more than 8 years, and this can be again because of the growth of new technologies in each country as well as the importance and interest of people in using smartphones in their daily life. Americans are the most experienced users of smartphones among the three countries (Iran, Malaysia, and America). Most of the respondents from all three countries agree on the usefulness of smartphones in health care. This means they accepted smartphones to be used in MHS. For the experience of the chronic disease between family members, Iranians experience it the most. Finally, most of the respondents agree that smartphones are useful in monitoring chronic disease patients.

Based on the significant result of ANOVA, it can be concluded that culture differences affect trust and anxiety significantly and therefore the acceptance of MHS. It also confirmed the proposed hypotheses. Table 19.5 presents that UA culture partially influences MHS users' trust. The last column of Table 19.5 depicts the significance of UA on T1, preference of using trustworthy MHS services ($0.009 < 0.05$), and T4, allowance to use health data for treatments ($0.039 < 0.05$), between three groups of nationality in this study. Based on the results illustrated on Table 19.5, UA culture significantly influences anxiety between Iranians, Malaysians, and Americans as three groups of nations in this study as depicted in Table 19.5, UA culture does not have a significantly influential role on MHS, feeling fearful about using MHS for chronic disease, but it significantly affects A2, scared to lose a lot of information using MHS ($0.000 < 0.05$), and A3, hesitant to use MHS for fear of making mistakes that cannot be corrected ($0.028 < 0.05$).

The results of this study can be useful for health-care providers in order to evade culturally related problems for future projects. It goes on to suggest that health-care providers need to compare their organizational culture with successful organizations.

Questions

1. What is national culture?
2. What are the five dimensions of culture based on Hofstede?
3. What is uncertainty avoidance?
4. What is trust and how will it affect health care?
5. What is anxiety in health care?
6. What are the new dimensions of cultures that can be effective in the acceptance of new technologies in health care?

References

Abelson, J., Miller, F. A., and Giacomini, M. 2009. What does it mean to trust a health system? A qualitative study of Canadian health care values. *Health Policy*, 91, 63–70.

Alali, H. and Salim, J. 2011. Information system success and acceptance theories: Towards developing a "virtual communities of practice" success model. In: *Retrieval (STAIR), 2011 International Conference on Semantic Technology and Information*, IEEE, http://ieeexplore.ieee.org/xpl/articleDetails.jsp?arnumber=5995807, June 28–29, 2011. pp. 306–312.

Ammenwerth, E., Gräber, S., Herrmann, G., Bürkle, T., and König, J. 2003. Evaluation of health information systems—Problems and challenges. *International Journal of Medical Informatics*, 71, 125–135.

Bathaee, A. and Pechtl, H. 2011. Culture affects consumer behavior-theoretical reflections and an illustrative example with Germany and Iran, Univ., Rechts-und Staatswiss. Fak: Greifswald, Germany.

Boukerche, A. and Yonglin, R. 2009. A secure mobile healthcare system using trust-based multicast scheme. *IEEE Journal on Selected Areas in Communications*, 27, 387–399.

Brebbia, C. A. and Zubir, S. S. 2011. Management of Natural Resources, Sustainable Development and Ecological Hazards, *Transactions on Ecology and The Environment*, 148(3), WIT Press. http:// www.witpress.com/, ISSNL 1743–3541.

Cocosila, M. 2013. Role of user a priori attitude in the acceptance of mobile health: an empirical investigation. *Electronic Markets*, 23, 15–27.

Deal, T. and Kennedy, A. 1982. *Corporate Cultures: The Rites and Rituals of Corporate Life*, Addison-Wesley Co.: London, U.K.

De Mooij, M. and Hofstede, G. 2010. The Hofstede model: Applications to global branding and advertising strategy and research. *International Journal of Advertising*, 29, 85–110.

Dillon, A. and Morris, M. G. 1996. User acceptance of new information technology: Theories and models. In M. Williams (ed.) *Annual Review of Information Science and Technology*, Vol. 31, Medford, NJ: Information Today, pp. 3–32.

Ebrahimi, N., Singh, S. K. G., and Tabrizi, R. S. 2010. Cultural effect on using new technologies. *World Academy of Science, Engineering and Technology*, 46, 1029–1033.

Ferraro, G. P. 1994. *The Cultural Dimension of International Business*, Prentice Hall: Englewood Cliffs, NJ.

Geier, Jim. 2001. Saving lives with roving LANs-Swedish ambulance becomes a true mobile platform. Network World, (Retrieved march 27, 2006 from http://www.networkworld.com/reviews/2001/0205bgside.html).

Gilson, L. 2005. Editorial: Building trust and value in health systems in low- and middle-income countries. *Social Science & Medicine*, 61, 1381–1384.

Hall, E. T. 1976. *Beyond Culture*, Anchor: Garden City, NY.

Hall, E. T. and Hall, E. T. 1969. *The Hidden Dimension*, Anchor Books: New York.

Haux, R. 2006. Health information systems—Past, present, future. *International Journal of Medical Informatics*, 75, 268–281.

Heinssen, R. K., Glass, C. R., and Knight, L. A. 1987. Assessing computer anxiety: Development and validation of the computer anxiety rating scale. *Computers in Human Behavior*, 3, 49–59.

Hofstede, G. 1980. *Culture's Consequences: International Differences in Work-Related Values*, SAGE Publications: Beverly Hills, CA.

Hofstede, G. 1984. *Culture's Consequences: International Differences in Work-Related Values*, Sage Publications: Beverly Hills, CA.

Hofstede, G. 1988. The Confucius connection: From cultural roots to economic growth. *Organizational Dynamics*, 16, 4–21.

Hofstede, G. 1991. *Cultures and Organizations: Software of the Mind*, McGraw-Hill: London, U.K.

Hofstede, G. 1997. *Cultures and Organizations: Software of the Mind*, rev. ed., McGraw-Hill: New York.

Hofstede, G. 2001. Culture's recent consequences: Using dimension scores in theory and research. *International Journal of Cross Cultural Management*, 1, 11–30.

Holden, R. J. and Karsh, B.-T. 2010. The technology acceptance model: Its past and its future in health care. *Journal of Biomedical Informatics*, 43, 159–172.

Hung, M.-C. and Jen, W.-Y. 2012. The adoption of mobile health management services: An empirical study. *Journal of Medical Systems*, 36, 1381–1388.

Jan ten Duis, H. and van der Werken, C. 2003. Trauma care systems in the Netherlands. *Injury*, 34, 722–727.

Jandt, F. E. 2010. *An Introduction to Intercultural Communication: Identities in a Global Community*, Sage Publications: Los Angeles, CA.

Jen, W.-Y. and Chao, C.-C. 2008. Measuring mobile patient safety information system success: An empirical study. *International Journal of Medical Informatics*, 77, 689–697.

Kanter, R. M. 1989. The new managerial work. *Harvard Business Review*, 67, 85–92.

Lindquist, J. D. and Kaufman-Scarborough, C. 2007. The Polychronic—Monochronic Tendency Model PMTS scale development and validation. *Time & Society*, 16, 253–285.

Lord, K., Ibrahim, K., Kumar, S., Rudd, N., Mitchell, A. J., and Symonds, P. 2010. Measuring trust in Healthcare professionals: A study of ethnically diverse UK cancer patients. *Clinical Oncology*, 24, 13–21.

Luthar, V. K. and Luthar, H. K. 2002. Using Hofstede's cultural dimensions to explain sexually harassing behaviours in an international context. *International Journal of Human Resource Management*, 13, 268–284.

Masrom, M. and Hussein, R. 2008. *User Acceptance of Information Technology: Understanding Theories and Models*, Venton Publishing (M): Selangor, Darul Ehsan Malaysia.

Mccoy, S., Galletta, D. F., and King, W. R. 2007. Applying TAM across cultures: The need for caution. *European Journal of Information Systems*, 16, 81–90.

Merchant, J. E. and Merchant, S. 2011. Management styles and cultural differences: Bridging the productivity gap. In: *The Implementation of Information Technology*. International Business & Economics Research Journal (IBER), 3.

Ogden, H. J. and Cheng, S. 2005 Age, gender and country effects on cultural dimensions in Canada and China, in *proceeding of 11th cross-cultural research conference*, Puerto Rico, USA.

Perez, M. A., Gonzalez, A., and Pinzon-Perez, H. 2006. Cultural competence in health care systems: A case study. *California Journal of Health Promotion*, 4, 102–108.

Rousseau, D. M., Sitkin, S. B., and Burt, R. S. 1998. Not so different after all: A cross-discipline. *View of Trust Academy of Management Review*, 23, 393.

Schee, E. V. D., Groenewegen, P. P., and Friele, R. D. 2006. Public trust in health care: a performance indicator? *Journal of Health Organization and Management*, 20, 468–476.

Schiffman, L. and Kanuk, L. 2000. *Consumer Behaviour*, Prentice Hall: Melbourne, Victoria, Australia.

Sigala, M. and Sakellaridis, O. 2004. Web users' cultural profiles and E-service quality: Internationalization implications for tourism web sites. *Information Technology & Tourism*, 7, 13–22.

Teo, T., Wong, S. L., and Chai, C. S. 2008. A cross-cultural examination of the intention to use technology between Singaporean and Malaysian pre-service teachers: An application of the Technology Acceptance Model (TAM). *Educational Technology & Society*, 11, 265–280.

Thaker, S. N. 2008. Understanding the role of culture in the health-related behaviors of older Asian Indian immigrants, a dissertation submitted to the graduate faculty of the university of Georgia, available in Google Scholar.

Wan Ismail, W. K., Chan, P. H. K., Buhari, N., and Muzaini, A. 2012. Acceptance of Smartphone in Enhancing Patient-Caregivers Relationship, Journal of Technology Management and Innovation, 7(3), 71–79.

Wright, E. B., Holcombe, C., and Salmon, P. 2004. Doctors' communication of trust, care, and respect in breast cancer: Qualitative study. *British Medical Journal*, 328, 864.

Wu, I.-L., Li, J.-Y., and Fu, C.-Y. 2011. The adoption of mobile healthcare by hospital's professionals: An integrative perspective. *Decision Support Systems*, 51, 587–596.

Wu, J.-H., Wang, S.-C., and Lin, L.-M. 2007. Mobile computing acceptance factors in the healthcare industry: A structural equation model. *International Journal of Medical Informatics*, 76, 66–77.

Yoo, B. and Donthu, N. 2005. The effect of personal cultural orientation on consumer ethnocentrism: Evaluations and behaviors of US consumers toward Japanese products. *Journal of International Consumer Marketing*, 18, 7–44.

Zakour, A. B. 2004. Cultural differences and information technology acceptance. In: *Proceedings of the Seventh Annual Conference of the Southern Association for Information Systems*, IGI Publishing, pp. 156–161.

Zhengchuan, X., Chenghong, Z., and Ling, H. 2008. Examining user acceptance of mobile services. In: *International Conference on Wireless Communications, Networking and Mobile Computing, 2008. WiCOM '08. Fourth October 12–14, 2008.* pp. 1–4.

Section VI

mHealth Aged Care and Aging Population

20

Opportunities for mHealth in an Aging Population Context

Brenda M.B. Reginatto

CONTENTS

20.1 Introduction

Virtually all countries in the world will experience a profound demographic transformation in the coming decades. The proportion of people aged 60 and above, which currently represents 11% of the world population, is expected to double by 2050 [1]. Such demographic changes are expected to cause a profound impact on the prevalence of chronic conditions common in old age, resulting in a significant hike in the costs of health-care provision.

This chapter will explore the potential for mHealth solutions to positively disrupt health-care provision for older adults. The terms *older adults* and *older people* are used in this chapter as recommended by international gerontology communities in order to tackle ageist perceptions often associated with other terms such as *the elderly* or *elderly people* [2,3].

mHealth solutions in this context not only have the potential to enable better management of chronic conditions but also support older adults to live safer and more connected lives, in their place of choice. Additionally, mHealth solutions in this arena have the potential to support family caregivers, professional caregivers, and other health-care professionals to better care for the older person.

This chapter is structured as follows: Section 20.2 reviews demographic, health, and economic trends, making the case for the growth of the mHealth industry in an aging society. Section 20.3 explores the key stakeholders in this industry. Section 20.4 offers an overview of the areas where mHealth is currently supporting older adults and the different mHealth solutions currently available. Section 20.5 discusses the challenges to achieving wider mHealth adoption among older adults. Section 20.6 presents two case studies to illustrate how mHealth solutions are addressing population aging issues in practice. Section 20.7 provides an overview of future trends in mHealth solutions for older people and is followed by concluding remarks.

20.2 Setting the Scene: Why Population Aging Represents an Opportunity for mHealth

20.2.1 Population Aging and Its Impact on Health Care

Population aging is one of the most important global demographic trends observed in history. It can be defined as the process through which older individuals become a proportionally larger share of the total population, as a result of declining fertility rates and increasing life expectancy. Greater longevity has been achieved in many countries due to a combination of improved nutrition, health-care services, education, and economic well-being and is considered by many a *triumph of development* [1,4].

This unprecedented phenomenon is taking place in nearly all countries in the world at a fast pace. Worldwide, the population aged 60 or over is growing at a faster rate than the total population and is expected to more than triple by 2050, from 600 million to 2 billion people [5]. Remarkably, the number of people aged 80 years or over is growing faster than any younger age group within the older population [1,6].

At the same time population is aging, the prevalence of chronic diseases common in old age is expected to increase significantly in both higher- and lower-income countries, having an important impact in health-care expenditure [7]. In the United States, the prevalence of cardiovascular diseases, including hypertension, coronary heart disease, and stroke, is projected to increase by 10% over the next 20 years, and its direct cost is set to almost triple, from $272.5 billion in 2010 to $818.1 billion in 2030 [8]. The global health expenditure for diabetes is projected to increase between 30% and 34% over the same period [9].

Other health conditions common in later life are also expected to follow similar trends. The total number of people with dementia worldwide is set to nearly double in the next 20 years, reaching 65.7 million people by 2030. The global costs of the disease, estimated around US$604 billion in 2010, are projected to increase even more quickly than its prevalence [10].

Also of concern is the high incidence of falls and related injuries among older age groups. Approximately 28%–35% of people over the age of 65 fall each year and injure themselves, increasing to 32%–42% of those aged 70 years or more. The number of injuries caused by falls is projected to double by 2030, potentially resulting in a significant increase in hospital and long-term care costs [11].

The impact of such trends on the demand and financing of health-care services has become central to international agendas [1]. Shifting the focus of health-care services from curative to preventative strategies and empowering older people to take active control over their health are being recognized as key approaches to achieve more cost-effective and sustainable health-care services [12,13].

Another essential element is the recognition that, in order to live longer and healthier lives, older people will require much more than medical services. Research shows that other aspects such as the ability to remain living independently in one's home of choice and staying connected with family and friends have an important impact in older people's health outcomes and should be equally supported through a more holistic approach to health care [14,15].

20.2.2 Older Adults and the Use of Technology

The use of technology among older adults, although lower when compared to younger age groups, has been growing significantly in recent years. In the United States, the percentage of adults aged 65+ online increased from 38% in 2008 to 53% in 2012. In the United Kingdom, 56% of adults aged 65–74 had Internet access at home in 2012, compared to 35% in 2007. While smartphone ownership is still low among this age group—around 13% of Americans aged 65+ and 12% of adults 65–74 in the United Kingdom owned a smartphone in 2012—this is also rapidly increasing [16,17].

The use of technology among the *baby boomer* generation—those born between 1946 and 1964—is significantly higher. Both in the United States and in the United Kingdom, nearly 80% of adults in this age group are online and a third of them already own a smartphone

[16,17]. This group is also the fastest-growing age segment to adopt social networking technologies. These trends deserve attention for a few reasons. Since 2011, approximately 10,000 baby boomers have turned 65 every day. The sheer size of this generation combined with their higher level of education, wealth, and technology literacy compared with previous cohorts is leading companies to rethink product and service design to cater for the demands of this *new older population* [4,6]. Moreover, great part of baby boomers currently provides some sort of support to an older relative. Many of them do so while coping with their own family and professional responsibilities [1], potentially requiring an entirely different range of products and services to support them in this role.

20.2.3 Role of mHealth in Reshaping Health Care for Older Adults

The potential for mHealth technologies to address some of the challenges posed by population aging is quickly capturing the attention of business and technology developers around the world [18]. The combination of unobtrusive sensing, wireless communication, ubiquitous computing, and social networking offered by mHealth applications opens endless opportunities for the development of solutions that can help older people live healthier and more independent lives [6,19].

As explored elsewhere in this book, mHealth technologies have the potential to enable health-care professionals provide timely, cost-effective, and accessible services to the wider population. This is particularly relevant to the management of chronic conditions common in old age. Solutions in this scenario could allow health-care professionals monitor their older patients remotely in order to detect, and thus prevent, complications and crises that can lead to acute episodes. Moreover, mHealth technologies have the potential to provide older adults—and their caregivers—tools that can assist them in taking more active control over their health and self-manage their conditions more effectively.

mHealth solutions that support safety and independence are also becoming increasingly popular among older adults and their caregivers. Sensors designed to detect emergencies, such as falls, can play an important role in enabling older adults to remain living at home as opposed to moving into residential care. Additionally, mHealth solutions have the potential to help older people stay connected with their communities [15]. This is particularly significant given the recently established link between loneliness, functional decline, and death among older populations. Studies have shown that older people who feel lonely are more likely to experience decline in mobility and the ability to complete activities of daily living and are also at increased risk of death. Therefore, solutions that allow older people to remain connected with their communities also have the potential to benefit their health [14].

20.3 Who Are the Key Stakeholders in mHealth for Older Adults?

Different stakeholders may be involved and benefit from the use of mHealth solutions for older adults. Consider, for example, technologies designed to support the management of chronic conditions (Figure 20.1). In this context, mHealth solutions include vital signs monitors that connect wirelessly to smartphone applications, allowing older adults to track, analyze, and share their health outcomes. In this scenario, health-care professionals, both in primary care and in the hospital sector, may use the data shared to remotely monitor the

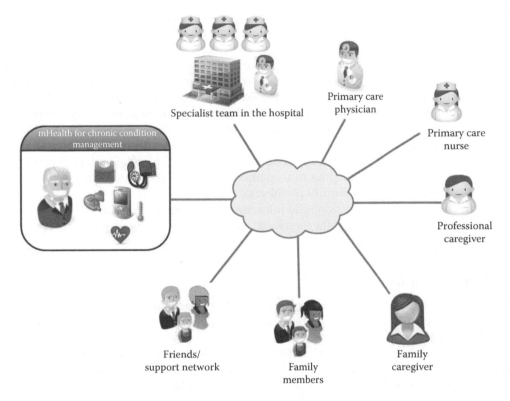

FIGURE 20.1
mHealth for chronic condition management stakeholders. (Adapted from Reginatto, B.M.B., *Int. J. Adv. Life Sci.*, 4(3 and 4), 63, 2012.)

older person's health and to provide timely intervention when necessary (the exact professionals involved may vary depending on the country's health-care system configuration) [18]. Family and professional caregivers can make use of the solutions' range of reminders, alerts, and decision-making tools to better manage the care of the older person. Distant family members can enjoy greater peace of mind and become more engaged in their relative's care if receiving updates on the older person's health status. Finally, many mHealth solutions for chronic condition management are also integrated to social networking platforms, allowing older adults to engage and share their experiences with friends and other people in similar circumstances.

In this chapter, we will focus on three main groups of stakeholders, namely, older adults, family caregivers, and providers involved in the care of older adults. This classification is based on the "aging in place triangle of relationships" suggested by Orlov (2013) and has been adapted to focus specifically on the context of mHealth solutions.

Since the challenges faced by older adults have been already discussed earlier, this section will further explore the characteristics, roles, and challenges faced by family caregivers and providers involved in the care of older adults.

20.3.1 Family Caregivers

A family caregiver is here defined as an unpaid individual (a spouse, partner, family member, friend, or neighbor) involved in assisting the older adult with activities of daily living (e.g., bathing, dressing), instrumental activities of daily living (e.g., paying bills, arranging doctor's appointments, hiring paid services), and/or medical tasks (e.g., helping the person to take medications correctly and to adhere to treatment recommendations) [20,21].

Even in countries with well-developed formal care systems, most of the care of older people is provided by their families. In Sweden and Denmark, around 40% of the population reported providing some sort of unpaid assistance [22]. In the United States, it is estimated that at least 43.5 million people provide unpaid care to a relative age 50 or older, representing 19% of the total adult population. Many family caregivers take on this role while also balancing their own work and household duties, which can be very stressful and overwhelming [1]. Some of the challenges identified by this stakeholder group include care coordination issues (e.g., time management, effective communication with healthcare professionals and other family members), care provision issues (e.g., keeping track of medical records and medication schedules, monitoring relative's vital signs, preventing and detecting falls), as well as other personal issues (e.g., social isolation, impact on own physical and emotional health) [20,23].

20.3.2 Long-Term Care Providers

Since the role of mHealth in primary and secondary care has been explored elsewhere in this book, we will here focus on how mHealth technology can support professionals providing long-term care to older adults, both in homecare and residential care settings. Professional caregivers in homecare settings include nurse aides and homecare aides, who usually complement the care provided by family caregivers in the older person's home [20]. There is increased demand for these services as both older adults and their family caregivers express a clear preference for services that allow the older person to remain living at home and retard the need for residential care [24]. Professional caregivers' responsibilities include assisting the older person with activities of daily living and medical tasks, as well as liaising with family members and other health-care professionals regarding the older person's progress [20,25].

As the older person's support needs evolve, they may have to move into a residential care setting, such as assisted living facilities or nursing homes. Assisted living facilities normally offer support with everyday activities such as meals, medication management, or bathing/dressing, while nursing homes usually provide additional skilled care to cater for older adults with more complex health-care needs.

Employees and managers in residential care organizations often face challenges due to certain characteristics of the services provided. For example, residents in these facilities frequently have multiple chronic conditions requiring complex geriatric care. This consists not only of clinical care but also assessments of individuals' functional abilities and cognitive and mental health, all of which must be taken into consideration when making patient-centered care decisions. Another challenge lays on the fact that care decisions are shared across multiple disciplines and are often made by physicians who are not at the facility. Nurses are primarily responsible for administering and managing care plans, while caregivers and other allied health professionals are directly delivering patient care. Coordinating care provision and decision making among all professionals and communicating these effectively to all parties, including family members, can therefore be a difficult goal to achieve [25,26].

mHealth solutions that aim to address the needs of older adults and assist family caregivers and long-term care professionals to provide more efficient and coordinated care are further explored in the next section of this chapter.

20.4 Exploring the Landscape of mHealth Solutions for Older Adults

This section offers an overview of existing mHealth technologies designed to support older adults. Solutions in this space can be divided into five broad categories, depending on the overall purpose of the technology: (1) health and wellness solutions, (2) safety solutions, (3) social connectedness solutions, (4) family caregiving solutions, and (5) professional caregiving solutions. Within these categories, technologies may be further divided into subclasses, based on the type of data collected and the specific needs addressed. A summary of these categories and subclasses can be found in Table 20.1.

Different models to categorize technology solutions that support older adults can be found in the literature. The framework below combines elements of two of these models and has been adapted to focus specifically on mHealth solutions [15,27].

20.4.1 Health and Wellness Solutions

This category includes solutions that primarily support the health and wellness of older adults by monitoring physiological and mental health status. Devices that fall in this group include vital signs monitors and body-worn sensors that can measure, for example, activity levels and heart rate. The information gathered by these devices is sent wirelessly to smartphone applications, allowing data tracking and analysis. Several applications allow the user to share their information with family members and health-care professionals [15].

This category also includes solutions specifically designed to monitor and manage chronic conditions common in old age (e.g., diabetes and heart failure). The ability to set automatic alerts that raise attention to abnormal values is an important feature of these technologies. Some of these tools also provide automated virtual coaching designed to educate and empower users to manage their health and chronic conditions more effectively.

TABLE 20.1

mHealth for Older Adult Categories and Subclasses

Health and wellness solutions	Vital signs monitors
	Subjective health risk assessment
	Activity monitors
	Medication adherence systems
	Cognitive training
Safety solutions	Environment sensors/motion sensors
	Mobile PERS and GPS tracking devices
	Video monitoring
Social connectedness solutions	Online social networking platforms (general, specific for older adults, or specific for people suffering from specific chronic conditions)
	Mobile applications offering user-friendly interfaces to mobile devices
Family caregiving solutions	Task management and tracking tools
	Communication and care coordination tools
	Educational tools
Professional caregiving solutions	Tools for data collection, record keeping, and communication
	Care coordination tools for residential care settings

20.4.1.1 Vital Signs Monitors

Particularly relevant for chronic condition management, these monitors measure vital signs such as weight, blood pressure, blood glucose, heart rate, temperature, ECG, galvanic skin response, gait, and hydration. While many vital signs monitors consist of traditional devices (e.g., weight scales, blood pressure cuffs, and blood glucose meters) that connect wirelessly to the Internet, the market for wearable patches and wrist-worn devices that continuously measure multiple vital signs is growing rapidly [28].

20.4.1.2 Subjective Health Risk Assessment

While body-worn sensors can collect numerous physiological data, health-care providers still need to assess specific patient health risks that cannot be objectively measured. This is especially useful in the remote management of chronic conditions such as chronic pulmonary disease. Existing solutions in this group include mobile applications through which older people can answer subjective questions and score their symptoms [15].

20.4.1.3 Activity Monitors

Activity monitors are being increasingly used by older adults who want to enhance their physical activity levels, either as an adjunct to chronic condition management or simply to pursue a healthier life style. These monitors can measure steps taken, speed, activity levels, calories spent, amount of time spent in rest or without getting up, and sleep patterns. Most activity monitors are wrist-worn devices, while others can be carried in the pocket. Smartphones also contain accelerometers that enable their use as activity monitors. Some activity-monitoring mobile apps also allow meal logging and offer personalized health coaching based on activity, sleep, and diet data gathered [15].

20.4.1.4 Medication Adherence Systems

As the name says, the ultimate goal of these systems is to enhance adherence to prescribed medication, a task that can be particularly challenging for older adults who are on complex medication regimens. This group includes devices, such as multiday container trays, smart pill bottles, or smart lids on pill containers, give light, and sound notification at the time the medication is supposed to be taken. If the person does not open a compartment of a pillbox or the lid of a pill container after receiving this first alert, the device sends a reminder alert via automated phone call, text messaging, or e-mail. Automatic alerts can also be sent to a remote caregiver if a dose is missed or if medication doses need refilling. Medication compliance history is stored in the cloud and can be shared with health-care professionals or family members as required [28].

Medication adherence systems also include smartphone mobile applications that work independently from smart pill containers. These applications allow the older person or a caregiver to enter data about the prescribed medication and set reminder notifications at scheduled times. Certain applications use gamification elements to increase medication adherence, awarding points when the person takes his/her medication on time. In some cases, these points can be later exchanged for real prizes [29].

20.4.1.5 Cognitive Training

Games designed to stimulate brain function are widely available both online and through mobile applications and have shown good results in improving cognitive ability of older adults [15,27]. Certain game applications have the ability to collect information about the older person's cognitive function over time, which can be helpful to health-care professionals looking after the older person's mental health.

20.4.2 Safety Solutions

This category includes mHealth solutions designed to support older people's safety and independence. Although these products may not collect physiological data or facilitate the delivery of medical services as the previous category, safety solutions can have a profound positive impact on the lives of older adults who are beginning to develop functional and cognitive limitations but wish to remain in their homes. Motion sensors are being used as an unobtrusive way of monitoring older adults' activities of daily living, such as getting out of bed, bathing, and toileting, and can send alerts to family members' or caregivers' mobile devices if activities are out of sync with usual behavior patterns.

Other products in this group include environment sensors that can identify other potential problems, ranging from falls to gas leaks in the home, and notify a family member or caregiver in case of emergency. Certain solutions even enable caregivers to remotely control appliances and utilities in the home of the older person, using their own mobile devices.

The new generation of personal emergency response systems (PERS) and global positioning service (GPS) tracking devices also falls in this category. These solutions allow the older person to call for help and be located at any time, from anywhere. Caregivers can receive updates with the older person's location in their mobile devices, which is particularly valued by those caring for people with cognitive impairment, who can more easily get lost.

20.4.2.1 Environment Sensors/Motion Sensors

Environment and passive sensor technologies used for home safety have developed greatly in recent years and are now the most advanced and well-diffused solutions in the *aging-in-place* market. More traditional products that fall in this category include sensors that can detect situations that are potentially dangerous to the older person's health: whether doors are properly locked, stove on/off status, and presence of smoke, humidity, and fire. Alarms can be set up so that whenever a potentially urgent situation is detected, an alert is automatically sent to a caregiver's mobile device. Some of these solutions also allow caregivers to use their mobile devices to control certain functions remotely, such as dim lights and lock unlocked doors [15,27].

More sophisticated solutions include wireless motion sensors that can learn the daily activity patterns of the older person and send alerts to the caregiver's smartphone when movement aberrations are detected. These sensors can be placed anywhere in the house (doors, bed, chair, toilet) and can detect unusual patterns including nighttime wakefulness or excessive daytime sleeping, falls, difficulty getting up from a seated position, and inactivity in the kitchen area (e.g., refrigerator and kitchen cabinets not opened as expected). Certain solutions in this category also allow caregivers to check the older person's status at any time through a mobile or web-based application [6,15,27,28].

20.4.2.2 Mobile PERS and GPS Tracking Devices

Products for personal safety, including PERS and GPS tracking devices, have evolved so that they can be activated while outside of the home, providing an important element of protection for older adults. Traditionally offered in the form of pendant alarms worn around the neck, PERS are now offered through a range of more discreet wearable devices, cell phones, or smartphone applications. Some of these products require the touch of a button to alert an emergency response service, while others can detect emergency events, such as falls, automatically.

Also gaining increasing attention are GPS tracking systems specifically designed for older people suffering from early stage Alzheimer's disease or mild cognitive impairment. These solutions aim to enhance the safety of older people who tend to wander off home or are at risk of getting lost, as well as to provide peace of mind to their caregivers [15,27]. Wearable devices and smartphone applications in this category allow the older person to easily call for help in case they get lost and can also be set to send regular location updates to a caregiver's smartphone [30]. In a similar manner, shoes with built-in GPS tracking systems can instantly alert a caregiver via short message service (SMS) or e-mail if the wearer of the shoes wanders outside of a preset location determined by the caregiver [15].

20.4.2.3 Video Monitoring

Video monitoring systems are another example of solutions that aim to support older people leaving on their own while providing family members with peace of mind. Through these applications, caregivers can access a direct video feed on their smartphone or tablet, which allow them to monitor the older person's status while still being mobile [28].

20.4.3 Social Connectedness Solutions

This category includes technologies that support adults to remain connected with their families and communities. These may range from general or older adult–specific online social networking platforms to mobile applications that offer more user-friendly interfaces to smartphones and tablets to allow older adults to communicate more easily with their family and friends [15,28].

Several mHealth applications designed to support older adults to manage chronic conditions are integrated with social networking platforms that allow them (and/or their caregivers) to connect, share experiences, and learn from other people in similar circumstances [15,28,31].

20.4.4 Family Caregiving Solutions

While a number of the aforementioned mHealth technologies can also help family caregivers provide more effective care or enjoy greater peace of mind, this category includes solutions exclusively designed to support family caregivers in their role. Caring for an older relative is a big undertaking and can often be overwhelming [32]. For many family caregivers, this is a role that comes unexpectedly after a crisis and several of them don't feel prepared to deal with the different responsibilities involved [33].

Until recently, most of the technology support in this space was coming from online portals that offer caregiving tips and best practice advice and that match service providers with family caregivers who are searching for specific services. These platforms are now evolving to include extensive user reviews to assist caregivers in making informed decisions [15].

This category, however, has developed quickly and now includes mobile tools that allow caregivers to track and manage daily tasks, medication schedules, doctor appointments, and billable work. Certain applications also allow uploading of medical documents, helping caregivers to keep all medical information centralized and easy to access.

Improving communication and care coordination is another aim of many mobile solutions in this group. Through shared calendars, caregivers can allocate tasks to other family members, friends, or professional caregivers, who can then input information about how they got on with their tasks. An activity feed displays tasks completed and relevant comments, keeping everyone informed about the current status of the care of the older person [34–36].

Solutions that aim to improve family caregiver well-being and decision-making capacity include applications designed to provide relevant educational content that can support them by offering practical caregiving advice as well as coping strategies (see Section 20.6.1).

20.4.5 Professional Caregiving Solutions

This category includes mHealth solutions designed to support professional caregivers providing homecare services or residential care for older adults. mHealth platforms in this group can be used to make care workflow more efficient, by enabling professional caregivers to collect health data, record daily activities, and communicate findings with other health-care professionals and family members in a quicker and easier manner [15,37].

Another example that falls in this category includes solutions for care coordination in residential care settings (see Section 20.6.2). These mobile applications are designed to facilitate data collection by all health-care professionals involved in the care of the residents, such as doctors, nurses, care assistants, and allied health-care professionals. All data entered are analyzed and will trigger instant and automated notifications to key staff members, should abnormalities be detected. These solutions also allow nurse managers to schedule, track, and notify different professionals about assessment outcomes and daily caregiving tasks, enabling them more control over the care of each resident. Moreover, solutions in this group can facilitate communication with family members. Through their mobile devices, family members can access a dashboard that keeps them informed of how their relative is progressing. Family members can also receive automated notifications about their relative's activities as they take place [38].

20.5 Barriers to Wider Adoption of mHealth among Older Adults: Facts and Myths

Despite its potential benefits, none of the aforementioned mHealth solutions has reached mass adoption among older adults and/or their caregivers yet. Barriers to health technology adoption among older adults have been extensively discussed in the literature. This analysis offers an overview of such obstacles and assesses their relevance in the specific context of mHealth and current adoption trends.

20.5.1 Acceptance Barriers

Earlier studies have suggested that lack of previous experience with computers and poor information and communications technology (ICT) skills may cause anxiety among older users and lead to technology rejection [39]. While this may be the case for many older adults, it is important to highlight that the use of and the interest in technology are rapidly increasing in this age group.

Even though technology uptake has been slower among adults aged 65+ when compared to their younger peers, studies have shown that the number of older adults online and their use of mobile phones, laptop, or tablet computers has grown significantly in recent years [3,16]. Additionally, the use of the Internet for health purposes has also increased among the older population [40]. A recent survey of older adults aged 50+ indicates that while the use of mobile technologies to support health remains relatively low in this group (only 11% currently use mobile technology to track their health over time and 4% use the same to send information about their health status to a health-care professional), around half of the respondents indicated interest in using mobile technology to support their health. When family caregivers were asked the same question, around 17% reported currently using mobile technology to help track the health of the person they care for. While the use of mobile technologies that allow caregivers to be informed of the older person's health indicators or if they need help remains low (3%), nearly 4 in 10 expressed interest in using these types of solutions [41]. Considering these trends, it is possible to suggest that poor ICT literacy among older people is a barrier that will gradually become insignificant and should not hinder wider adoption in the future [18].

Other acceptance barriers documented in the literature include confidentiality and privacy concerns, which may limit mHealth uptake by both older people and health-care professionals [28,40,42,43]. A recent survey showed that 81% of older adults aged 65+ would be willing to give up some of their privacy in order to remain living at home for as long as possible [23]. This finding may suggest that privacy concerns are not a significant barrier to adoption, should the perceived benefits of the technology outweigh its impact on user privacy. Successful initiatives at health-care system level have demonstrated that confidentiality concerns can also be overcome once adequate privacy controls are in place. In Denmark, for example, health data are securely shared through an official eHealth portal. In this context, patients have access to this website and can easily control privacy functions, including monitoring who has accessed or modified their personal medical records [44]. Considering that technical controls (e.g., encryption, electronic identification) are available and widely used in other electronic services to ensure user privacy (e.g., online banking), the challenge here remains how to communicate this message clearly to users—be them older adults, caregivers, or health-care professionals—in order to gain their trust.

Lack of awareness of the available technology and its potential benefits by older adults is another existing barrier to mass adoption of mHealth solutions by older adults [18,28,45]. While awareness of traditional PERS devices is considerably high among older adults, most people in this age group remain unaware of mHealth technologies, including medication management and falls detection solutions [23]. Once more, in order to achieve wider adoption, it is vital that mHealth companies find effective means to communicate with this population, such as trusted referral channels [45].

20.5.2 Technology Design Barriers

Technology design issues have been highlighted in several studies [3,28,39,45–48] and remain a challenge to overcome. Evidence shows that technology solutions that are difficult to use, nonintuitive, disruptive to user's lifestyle, stigmatizing, or are not perceived as addressing a real need tend to be abandoned by users over time [28,31]. Poor market segmentation strategies and lack of sufficient involvement of end users during the development stages of mHealth solutions may account for many of these issues.

Targeting consumers simply by age group, such as 65+ or 80+, and assuming homogeneity of needs and abilities across these large age ranges have been considered a poor strategy to engage with older consumers [3]. In order to avoid the pitfalls of cohort thinking, mHealth developers should work closely with older adults throughout the design process to fully understand not only their needs, skill level, and physical/cognitive abilities but also their preferences, lifestyles, and aspirations [3,47,49].

Designing for "us as we age," rather than for "them who are already old," has been suggested as a better approach to designing technology that is relevant, usable, and desirable for older adults. This approach is believed to encourage designers to think more about the huge variations among older consumers and possibly pay more attention to the look, feel, and desirability of the technologies produced. It may also encourage designers to think about aging as a continuous process and develop solutions that can adapt over time as people's needs change [3].

20.5.3 Financial Barriers

The fact that older people are often at greater risk of lower incomes may limit their capacity to purchase mHealth products and services, particularly if these are not covered by health insurance schemes. Similarly, physicians and long-term care providers will be hesitant to invest in solutions that impact existing organizational workflows and operational processes unless they have financial or other incentives to do so. Such financial challenges have been acknowledged by different studies and public–private partnerships have been suggested as ways of overcoming them [28,39,42,44,45].

20.5.4 Barriers to Health-Care Professional Adoption

As seen so far, challenges to mass mHealth adoption by older adults exist and deserve attention if these solutions are to realize their full potential. It is important to highlight, however, that some of the most pressing barriers to wider mHealth adoption are shown to be connected with the reluctance of health-care professionals, and not older adults, to embrace this new form of health-care provision [18].

Studies have pointed to different reasons for limited technology uptake by health-care professionals, including organizational barriers (e.g., lack of organizational willingness to change, absence of champions in the health-care system), policy and legislation barriers (e.g., lack of medicolegal frameworks applicable to mHealth, unclear data protection legislation), and evidence-based barriers (e.g., lack of robust evidence for mHealth effectiveness and return on investment) [18].

The main obstacle to wider mHealth penetration in the health-care system, however, remains the lack of clear incentives for health-care professionals to embrace the technology [18,43,47]. It has been acknowledged that the absence of reimbursement arrangements significantly discourages health-care professionals to incorporate mHealth solutions to their services [39,46–49].

In order to address these barriers, several countries have used government financial incentives as effective policy tools to incentivize technology adoption among health-care professionals [18]. Pay for performance schemes are also being recognized as a valuable approach to incentivize health-care professionals to shift away from procedure-driven models. In the United Kingdom, for example, around 15% of general practitioners' (GPs') salaries are based on their performance against a set of quality measures, which is incentivizing health-care professionals to embrace mHealth solutions that help them better coordinate patient care [50].

20.6 Case Studies

20.6.1 The Caregiver Support Tool*

20.6.1.1 Background

Progressive decline in functional status of persons with dementia is emotionally and financially exhausting for family caregivers [52]. Caring for someone with dementia can lead to depression, sleep disturbance, anxiety, and loss of companionship. Difficulties in dealing with behavioral problems, aggression, and incontinence can also contribute to the stress [52–54]. These factors are also found to be strong predictors for patient referral into residential care settings [55]. Therefore, addressing caregiver consequences not only affects their perceived ability to cope but also plays an important role in helping people with dementia remain at home for longer and reducing cost to health-care systems.

In reality, however, the well-being of family caregivers is often not taken into account, advised upon, or monitored. The time and resources required to routinely conduct self-report assessments in an outpatient setting are challenging, and for this reason, this information is not routinely collected.

In 2011, researchers at Technology Research for Independent Living (TRIL) Centre† developed the caregiver support tool (CST), a computerized system for assessing and supporting the psychosocial well-being of family caregivers for persons with dementia. The following case study gives an overview of the solution's concept and summarizes the initial results of the exploratory home deployment.

20.6.1.2 Solution Overview

The overall objective of the study was to develop a computerized instrument for tracking the psychosocial well-being of spousal caregivers living with a person with dementia using self-administered scales. It was intended that the support tool would use the quantitative information gathered in order to determine the needs of the caregiver in real time and provide support by way of informative documents and videos to address these issues.

The CST was developed over the course of 9 months through a participatory design process involving TRIL team members and individuals involved in the care of persons with dementia. This was done to ensure that the design was grounded in an understanding both of user needs and practices. The subsequent 2-week home deployment of the tool was an exploratory study to investigate how the system was used in the context of real homes outside the laboratory and how it might be improved to encourage social support and questionnaire completion.

The tool took the form of an app, called The Care Companion, and was loaded onto the Motorola Xoom 10 in. tablet. The assessments were based on validated scales, and the questions asked were designed to encourage the caregiver to think about the rewards of caregiving, their own well-being, and any difficulties they may have faced. Using the assessment scores, the tool was able to make new relevant content available to the caregiver including articles, videos, guides, and tips on a wide range of topics like Alzheimer's disease, looking after health and well-being and coping strategies.

Additionally, the supportive content integrated multimedia presentations sharing personal accounts from other caregivers and health-care professionals. The goal was to emulate

---

* Case study adapted from Intel, 2011 [51].
† http://www.trilcentre.org/.

an empathetic support network in place of actual social contact with these individuals as the tool did not support networked communication.

In order to reduce potential anxiety around technology introduction for the caregiver alone, the tool also provided the opportunity for shared experiences between the caregiver and their spouse. This included the ability to review personal photos, mementos, or movies together as well as the inclusion of simple fun and enjoyable one- and two-player games.

20.6.1.3 Implementation Outcomes

After the 2-week home deployment, exit interviews were conducted with all participants to gather feedback on the users' experience. Upon analysis of the data collected, researchers found that the relevance of the content depended largely on participants' level of caregiving experience, with most experienced caregivers finding the information presented too general to be of use to their particular case. Additionally, all participants felt that the professional content was aimed at a *generic caregiver* and did not consider specific caregiving typology. There was a general preference toward caregiver-generated content, which participants recognized as valuable in providing empathetic support to caregivers.

In general, participants found that they adapted the use of the tool as and when they needed it, rather than sticking to a prescribed routine of filling in the questionnaires every 2 days. This finding suggests that, in order to be successfully adopted, the tool must be allowed to build itself into the lives and routines of participants organically, rather than being imposed externally. The ability to use the tool to share photographs with their spouse was also seen as valuable, as it allowed participants to integrate the tool into other areas of life. Participants pointed that, in order to be viable in the long term, support tools must provide not only information but also engagement, entertainment, and a real sense of relationship bridging.

The usability of the device was one of the strong points to come through in the data, with most participants expressing the ease of use of the tablet and intuitive feel to it. In general, participants found the questionnaires the least engaging aspect of the deployment, with all stating in one way or another that they only completed the questionnaires because they were told to by an external party. It was suggested that a longer period between questionnaire completion would have been more desirable to motivate users to continue completing the questionnaires full time.

20.6.1.4 Concluding Remarks

This case study described the initial results of the exploratory home deployment of a computerized system for assessing and supporting the psychosocial well-being of family caregivers for persons with dementia. The data generated provided insights into how the system is used in a home context and provided further support in understanding participants' perceptions of the system and how it could be effectively implemented.

Participants generally found the tool a positive addition to their lives, providing support and useful information on the management of dementia, particularly with regard to providing empathetic support on the experience of other caregivers via user experience videos. The challenge for such a system was designing it in a way that users would go back to it regularly to complete the online questionnaires. Feedback from users suggests that the system must be integrated into the full spectrum of caregivers' lives if questionnaire uptake is to be completed long term. Additionally, interval periods between questionnaire completion must be extended if users are not to become apathetic.

In conclusion, even though this particular tool would require further refinement and adjustment for long-term care suitability, this study has demonstrated the viability of the CST as a welcome addition to future home health-care plans. As the cost of such technologies decreases, and their distribution becomes more widespread, it is expected that such tools will become more commonly used to support the monitoring and management of family caregiver well-being.

20.6.2 Communication and Care Coordination Platform for Residential Care Facilities*

20.6.2.1 Background

As discussed in Section 20.3.2, achieving efficient care coordination is a challenge frequently faced by residential care organizations. Barriers to effective and timely communication among all professionals involved in care planning and delivery include the mix of in-house staff and off-site providers, the fact that many staff members work different shifts, and the use of one-to-one communication methods (e.g., phone calls, paper notes, letter, e-mail, and fax). Maintaining family members informed and engaged in the care of their relatives has been also identified as a significant challenge in this context.

The following case study examines how mHealth tools can support residential care facilities to improve care coordination, decision making, and communication among clients, professionals, and family members. For the single purpose of illustration, Caremerge, a communication and care coordination platform, was chosen to demonstrate how these issues are being tackled in practice.

This case study is structured as follows: the first section presents an overview of the solution's main features. Enablers and barriers to implementation are discussed in the second section, followed by an analysis of outcomes obtained to date and concluding remarks.

20.6.2.2 Solution Overview

Caremerge is an American health-care technology company founded in 2012 with the goal of revolutionizing communication and care coordination in residential care facilities. In April 2013, the company was selected to be part of the GE Health Growth Acceleration Program in partnership with Startup Health. Since then, Caremerge has been widely recognized by its contributions to innovation in the long-term care industry [38].

Caremerge is a communication and care coordination platform, consisting of mobile- and web-based solutions. In brief, the mobile application allows professionals to access patient information, as well as to capture and share relevant data with all relevant parties, in a simple and timely manner. The solution also allows family members to remain up to date with regard to their relative's progress. Specific features include the following:

- Ability to allocate and monitor the status of care tasks. In this scenario, personal care tasks, for example, can be allocated to a specific staff member, who can easily tick it off using their mobile device when the task is completed.

---

* Data presented in this case study were gently provided by Asif Khan, CEO at Caremerge (private communication).

- Ability to set e-mail and/or SMS notifications and alerts. This can be done to ensure timely follow-up of outstanding tasks or to automatically notify all relevant professionals of important notes and events.

- Ability to set up group chats to share relevant information with multiple parties at the same time. For example, a family member can start a group chat with the nurse, physician, and activities manager to voice a concern regarding their family member's low mood and discuss strategies to encourage greater participation in appropriate activities.

- Family members can access real-time information about their relative's progress. In this scenario, family members can, for example, check on whether their relative had their meal or attended activities as planned on any given day.

- Ability to send personalized voice reminders to residents regarding upcoming activities. Residents can opt for receiving a phone call to remind them of the starting time and location of activities of interest.

- Ability to track and measure performance against a range of quality metrics. Through this feature, care managers can, for example, monitor frequency and reason for hospital readmission and the facility's level of compliance with recommended preventive measures (e.g., periodic medical reviews and immunization programs). The tool also enables care managers to monitor whether residents are receiving appropriate support in the different dimensions of wellness. In this scenario, a wellness chart is used to illustrate residents' participation levels in the different activities offered, giving an insight on whether residents' holistic needs and preferences are being fulfilled.

20.6.2.3 Solution Implementation

According to Khan, to this date, implementation of the Caremerge platform has been largely successful. The CEO explains that the short time to get the system up and running (the basic version of the system can be implemented in less than 4 h in a 100-bed facility) and the fact that minimal training is required due to its intuitive design have played a significant role to successful implementation.

Additionally, the perceived benefits to employees are believed to be crucial to its fast adoption. Khan explains that the high staff turnover observed in residential care settings is often associated to lack of tools to organize daily tasks, poorly structured routines, and frustration due to excessive paper work and miscommunication. Postimplementation feedback revealed that employees found that these issues were addressed by Caremerge tools, allowing them to become more productive in their workplace.

One of the only challenges to speedy implementation of Caremerge systems is the need for integration with existing solutions already in use in some facilities. However, Khan believes that system integration issues should not be a significant problem going forward, as systems are becoming more interoperable.

Another aspect that is likely to change in the near future is how much residents interact with the system. Khan observed that the majority of older people living in residential care facilities are not yet using mobile devices. For this reason, Caremerge has incorporated the use of personalized phone messages to remind residents of activities that are about to start. Feedback obtained from residents indicates that this feature has been much appreciated, as it allows them to keep track of their own schedule. In certain facilities, however, younger residents are already using the solution themselves to self-check-in when attending activities. This level of interaction with the system is expected to increase in coming years as younger generations of older adults, who are more familiar with mobile technology, start to move into residential care.

20.6.2.4 Implementation Outcomes

Since its launch in the summer of 2012, the number of facilities using Caremerge systems has grown fast. The company was expecting to have their solutions implemented in over 170 facilities across the United States by the end of 2013, with plans to expand to the European market in the near future.

According to Khan, the overall feedback from residential care managers to this date has been very positive. They believe that the care coordination platform greatly supports the provision of patient-centered care, reducing the time spent with administrative tasks and allowing all professionals involved to focus on the *person in the middle*. The CEO highlights, however, that the system itself is only a tool to achieve better quality care and that ultimately it is up to care managers and providers to make meaningful use of the intelligence provided by the system and turn this into improved care outcomes.

20.6.2.5 Concluding Remarks

This case study aimed to illustrate how mHealth solutions can address some of the challenges faced by residential care facilities. As the number of older people requiring complex geriatric care increases with population aging, solutions that support care coordination and stakeholder communication will be ever more needed. Caremerge's experience has shown that understanding the needs of all stakeholders involved and delivering technology that is easy to use are the first steps to truly revolutionize residential care and improve the lives of older people who rely on their services.

20.7 Future Trends in mHealth Solutions for Older People

A number of mHealth solutions are currently under development and will be available to older adults, family, and professional caregivers in the near future. Some examples of these technologies are briefly described in the following:

- *Falls prevention solutions*: Different mHealth solutions that aim to detect risk of falls and support on falls prevention interventions (rather than simply falls detection) are being developed. These include algorithms to predict falls, body-worn sensors as well as sensor-embedded insoles, and walking sticks that can detect balance impairment and send data wirelessly to the user's or clinician's mobile device [15,28,56].

- *Chronic condition management solutions*: A range of both invasive and noninvasive technologies that aim to support chronic condition management are currently under research. Examples include implantable glucose and blood pressure monitors and smart contact lenses capable of monitoring glucose, cholesterol, temperature, inflammation, infection, and fatigue. These technologies are capable of transmitting the data collected to the user's mobile devices, potentially increasing capacity for self-management [28]. Other solutions that can support chronic condition management currently under development include smartphone cases that can be used to detect heart rhythm problems and prevent strokes, as well as sensor mats capable of detecting the development of foot ulcers in diabetic patients [28,57].

- *Medication management solutions*: In the near future, medication management solutions are likely to include devices capable of readjusting the dosing schedule when users are late in taking their medications, as well as context-aware reminders tailored to identify the best time and place for the user to take their medication [28]. Smart pills containing ingestible sensors have also been developed in order to objectively determine whether a person has taken their medication and should reach the wider market soon [15].

- *Environmental sensor solutions*: This category includes smart bed systems that aim to support residential care providers to prevent falls, reduce pressure ulcers, and improve health management. A combination of sensors placed under the mattress and a mobile app will enable providers to monitor their client's motion, heart rate, and respiratory rate while they are in bed [58].

- *Continence management solutions*: Smart continence systems are being developed to support caregivers, mainly those caring for older people with dementia. They consist of sensor-embedded disposable pads that can detect and alert caregivers when the older person becomes incontinent. These solutions not only allow caregivers to perform timely pad changes (reducing the risk of dermatitis/rash and urinary infection) but also allow them to learn the person's continence patterns and identify the optimum times to help them in using the toilet [59].

20.8 Conclusion

This chapter explored the potential for mHealth solutions to positively disrupt health-care provision for older adults. mHealth technologies in this context are being designed not only to enable better management of chronic conditions but also to support older adults to live safer and more connected lives, in their place of choice. mHealth solutions in this arena are also supporting family caregivers, professional caregivers, and other health-care professionals to better care for the older person, which is seen as crucial to increase quality and reduce health-care costs.

As the costs of technology continue to drop and older adults, family caregivers, and care providers become increasingly more familiar with the use of mobile devices to manage health issues, it is possible to expect that these technologies will significantly change the way we grow old and how we care for our older population.

Questions

1. What is population aging and how can mHealth help in addressing its challenges?
2. In what areas can mHealth solutions directly support older adults?

3. How can mHealth solutions support family members who are caring for an older relative?

4. How can mHealth solutions support professional caregivers working with older adults?

5. What are the main challenges to wider mHealth adoption associated with technology design and how can designers address these issues?

References

1. UNFPA and HelpAge International. 2012. 2012 Ageing in the twenty-first century: A celebration and a challenge. http://www.unfpa.org/webdav/site/global/shared/documents/publications/2012/UNFPA-Exec-Summary.pdf (accessed September 15, 2013).
2. Dahmen, N.S. and R. Cozma. 2009. Media takes: On aging. International Longevity Center, USA and Aging Services of California. http://www.mailman.columbia.edu/sites/default/files/Media_Takes_On_Aging.pdf (accessed September 15, 2013).
3. Roberts, S. 2010. The fictions, facts and future of older people and technology. International Longevity Centre, London, U.K. http://www.ilcuk.org.uk/images/uploads/publication-pdfs/pdf_pdf_118-2.pdf (accessed September 15, 2013).
4. Kalache, A. 2013. The longevity revolution: Creating a society for all ages. Government of South Australia. http://www.thinkers.sa.gov.au/KalacheReport/files/inc/f0c2e5815c.pdf (accessed September 15, 2013).
5. World Health Organization and US National Institute of Aging. 2011. Global health and aging. http://www.who.int/ageing/publications/global_health/en/ (accessed September 15, 2013).
6. Orlov, L.M. 2013. Technology for aging in place: 2013 market overview. Aging in place technology watch. http://www.ageinplacetech.com/files/aip/Market%20Overview%207-5-2013.pdf (accessed September 15, 2013).
7. World Health Organization. 2011. Global status report on noncommunicable diseases 2010. http://www.who.int/nmh/publications/ncd_report_full_en.pdf (accessed September 15, 2013).
8. Heidenreich, P.A., J.G. Trogdon, O.A. Khavjou et al. 2011. Forecasting the future of cardiovascular disease in the United States: A policy statement from the American Heart Association. *American Heart Association. Circulation.* 123(8): 933–944.
9. Zhang, P., X. Zhanga, J. Brownb et al. 2010. Global healthcare expenditure on diabetes for 2010 and 2030. *Diabetes Research and Clinical Practice.* 87(3): 293–301.
10. World Health Organization. 2012. Dementia: A public health priority. http://whqlibdoc.who.int/publications/2012/9789241564458_eng.pdf (accessed September 15, 2013).
11. World Health Organization. 2007. WHO global report on falls prevention in older age. http://www.who.int/ageing/publications/Falls_prevention7March.pdf (accessed September 15, 2013).
12. World Health Organization. 2008. 2008–2013 action plan for the global strategy for the prevention and control of noncommunicable diseases. http://whqlibdoc.who.int/publications/2009/9789241597418_eng.pdf (accessed September 15, 2013).
13. World Economic Forum. 2013. Global risks 2013. 8th edn. http://www3.weforum.org/docs/WEF_GlobalRisks_Report_2013.pdf (accessed September 15, 2013).
14. Routasalo, P. and K.H. Pitkala. 2003. Loneliness among older people. *Reviews in Clinical Gerontology.* 13(4): 303–311.
15. Lindeman, D., R. Ghosh, S. Ratan, and V. Steinmetz. 2013. The new era of connected aging: A framework for understanding technologies that support older adults in aging in place. Center for Technology and Aging, CITRIS/Public Health Institute. http://www.techandaging.org/ConnectedAgingFramework.pdf (accessed September 15, 2013).

16. Smith, A. 2012. 46% of American adults are smartphone owners. Pew Research Center's Internet & American Life Project. http://pewinternet.org/Reports/2012/Smartphone-Update-2012.aspx (accessed September 15, 2013).

17. Ofcom. 2013. Adults' media use and attitudes report. http://stakeholders.ofcom.org.uk/binaries/research/media-literacy/adult-media-lit-13/2013_Adult_ML_Tracker.pdf (accessed September 15, 2013).

18. Reginatto, B.M.B. 2012. Understanding barriers to wider telehealth adoption in the home environment of older people: An exploratory study in the Irish context. *International Journal on Advances in Life Sciences.* 4(3 and 4): 63–76.

19. Mehregany, M. and E. Saldivar. 2012. Opportunities and obstacles in the adoption of mHealth, in *MHealth: From Smartphones to Smart Systems*, eds. R. Krohn and D. Metcalf. Chicago, IL: Healthcare Information Management and Systems Society (HIMSS).

20. Alzheimer's Association. 2013. Alzheimer's disease facts and figures. http://www.alz.org/downloads/facts_figures_2013.pdf (accessed September 15, 2013).

21. National Alliance for Caregiving and AARP. 2009. Caregiving in the U.S.: A focused look at those caring for the 50+. http://assets.aarp.org/rgcenter/il/caregiving_09.pdf (accessed September 15, 2013).

22. OECD. 2011. Help wanted? Providing and paying for long-term care. http://www.oecd.org/health/longtermcare/helpwanted (accessed September 15, 2013).

23. Barrett, L.L. 2011. Healthy@Home 2.0. AARP. http://assets.aarp.org/rgcenter/health/healthy-home-11.pdf (accessed September 15, 2013).

24. Resnick, H.E. and M. Alwan. 2010. Use of health information technology in home health and hospice agencies: United States, 2007. *Journal of the American Medical Informatics Association.* 17(4): 389–395.

25. Mohamoud, S., C. Byrne, and A. Samarth. 2009. Implementation of health information technology in long-term care settings: Findings from the AHRQ health IT portfolio. AHRQ. http://www.leadingage.org/uploadedFiles/Content/About/CAST/Resources/AHRQ_HIT_Portfolio.pdf (accessed September 15, 2013).

26. Regional Dementia Network and The Community Gateway CIC. 2013. Maximising the potential for the use of assistive technology: An information toolkit to support people with Dementia, their Carers and Dementia services. Improvement and Efficiency West Midlands. http://www.iewm.net/at-toolkit/ (accessed September 15, 2013).

27. Alwan, M., D. Wiley, and J. Nobel. 2007. State of technologies in aging services. Center for Aging Services Technologies (CAST). http://www.leadingage.org/uploadedFiles/Content/About/CAST/Resources/State_of_Technology_Report.pdf (accessed September 15, 2013).

28. Office of Disability, Aging, and Long-Term Care Policy. 2012. Report to congress: Aging services technology study. U.S. Department of Health and Human Services (DHHS). http://aspe.hhs.gov/daltcp/reports/2012/ASTSRptCong.pdf (accessed September 15, 2013).

29. PillJogger. 2013. Welcome. http://www.pilljogger.com/ (accessed September 15, 2013).

30. Saltzman, M. 2011. Smartphone apps for 'Aging in Place'. http://www.aarp.org/technology/innovations/info-06-2011/smartphone-apps-aging-in-place.html (accessed September 15, 2013).

31. Orlov, L.M. 2011. Connected living for social aging: Designing technology for all. AARP. http://www.aarp.org/content/dam/aarp/technology/innovations/2011_04/Connected-Living-for-Social-Change.pdf (accessed September 15, 2013).

32. Age UK. 2013. Advice for carers: A practical guide. http://www.ageuk.org.uk/Documents/EN-GB/Information-guides/AgeUKIG13_Advice_for_carers_inf.pdf (accessed September 15, 2013).

33. Department of Health. 2010. Recognised, valued and supported: Next steps for the carers strategy. https://www.gov.uk/government/uploads/system/uploads/attachment_data/file/213804/dh_122393.pdf (accessed September 15, 2013).

34. CareTree. 2013. Why CareTree? http://www.caretree.me/why-choose-home-care-software-caretree/ (accessed September 15, 2013).

35. Unfrazzled Care. 2013. Unfrazzle App-Overview. http://unfrazzledcare.com/overview/ (accessed September 15, 2013).
36. Home Touch. 2013. How It Works. http://www.myhometouch.com/how-it-works (accessed September 15, 2013).
37. Healthcare Information and Management Systems Society (HIMSS). 2014. Case Study: Decreasing Costs and Improving Outcomes Through Community-Based Care Transitions and Care Coordination Technology. https://s3.amazonaws.com/cdn-cah/HIMSS-case-study.pdf (accessed June 24, 2014).
38. Caremerge. 2013. Solutions-Family Engagement. http://www.caremerge.com/web/solutions/#Contextual (accessed June 24, 2014).
39. Broens, T.H.F., R.M.H.A. Huis in't Veld, M.M.R. Vollenbroek-Hutten, H.J. Hermens, A.T. van Halteren, and L.J.M. Nieuwenhuis. 2007. Determinants of successful telemedicine implementations: A literature study. *Journal of Telemedicine and Telecare*. 13(6): 303–309.
40. Anderson, J.G. 2007. Social, ethical and legal barriers to e-health. *International Journal of Medical Informatics*. 76(5–6): 480–483.
41. Barrett, L.L. 2011. Health and caregiving among the 50+: Ownership, use and interest in mobile technology. AARP. http://assets.aarp.org/rgcenter/general/health-caregiving-mobile-technology.pdf (accessed September 15, 2013).
42. Goodwin, N. 2010. The state of telehealth and telecare in the UK: Prospects for integrated care. *Journal of Integrated Care*. 18(6): 3–10.
43. Kubitschke, L. and K. Cullen. 2010. ICT & ageing–European study on users, market and technologies. Final report. Brussels, Belgium: Commission of the European Communities.
44. Castro, D. 2009. Explaining international IT application leadership: Health IT. The Information Technology and Innovation Foundation. http://www.itif.org/files/2009-leadership-healthit.pdf (accessed September 15, 2013).
45. Alwan, M. and J. Nobel. 2011. State of technology executive summary. Center for Aging Services Technologies (CAST). http://www.leadingage.org/State_of_Technology_Executive_Summary.aspx (accessed September 15, 2013).
46. Runge, A. and F. Feliciani. 2010. Telemedicine: From technology demonstrations to sustainable services, in *The European Files: The Telemedicine Challenge in Europe*. European Commission. Brussels. http://ec.europa.eu/information_society/activities/health/docs/publications/2010/2010europ-files-telemedicine_en.pdf (accessed September 15, 2013).
47. Alwan, M. and J. Nobel. 2008. State of technology in aging services according to field experts and thought leaders. American Association of Homes and Services for the Aging (AAHSA). www.agingtech.org (accessed September 15, 2013).
48. Pique, J.-M. 2010. Impact on the restructuring of healthcare, in *The European Files: The Telemedicine Challenge in Europe*. European Commission. Belgium. http://ec.europa.eu/information_society/activities/health/docs/publications/2010/2010europ-files-telemedicine_en.pdf (accessed September 15, 2013).
49. COCIR. 2010. COCIR Telemedicine toolkit. For a better deployment and use of Telehealth. European Coordination Committee of the Radiological, Electromedical and Healthcare IT Industry. http://www.cocir.org/uploads/documents/-903-cocir_telemedicine_toolkit_march_2010.pdf (accessed September 15, 2013).
50. NHS. 2010. GP payments calculation service Memorandum of Information (MOI). http://www.connectingforhealth.nhs.uk/systemsandservices/gpsupport/gppcs/gppcsmoi.pdf (accessed September 15, 2013).
51. Intel. 2011. White paper on the caregiver support tool. Internal report. Dublin (unpublished).
52. Burns, A. and P. Rabins. 2000. Carer burden in dementia. *International Journal of Geriatric Psychiatry*. 15(S1): S9–S13.

53. McCurry, S.M., L.E. Gibbons, R.G. Logsdon, M. Vitiello, and L. Teri. 2003. Training caregivers to change the sleep hygiene practices of patients with dementia: The NITE—AD project. *Journal of the American Geriatrics Society*. 51(10): 1455–1460.
54. Chenoweth, B. and B. Spencer. 1986. Dementia: The experience of family caregivers. *The Gerontologist*. 26(3): 267–272.
55. Hope, T., J. Keene, K. Gedling, C.G. Fairburn, and R. Jacoby. 1998. Predictors of institutionalization for people with dementia living at home with a carer. *International Journal of Geriatric Psychiatry*. 13(10): 682–690.
56. Redd, C.B. and S.J.M. Bamberg. 2012. A wireless sensory feedback device for real-time gait feedback and training. *Mechatronics, IEEE/ASME Transactions*. 17(3): 425–433.
57. Lau, J.K., N. Lowres, L. Neubeck et al. 2013. iPhone ECG application for community screening to detect silent atrial fibrillation: A novel technology to prevent stroke. *International Journal of Cardiology*. http://dx.doi.org/10.1016/j.ijcard.2013.01.220 (accessed December 20, 2013).
58. BAM Labs, Inc. 2013. BAM labs Touch-free Life Care (TLC) system. http://www.bamlabs.com/ (accessed September 15, 2013).
59. Fish, P. and V. Traynor. 2013. Sensor technology: A smart way to manage continence. *Australian Journal of Dementia Care*. 2(1): 35–37.

21

mHealth for Elder Care:
Survey of Existing Solutions and Future Road Map

Marilia Manfrinato, Kami Shalfrooshan, and Rozita Dara

CONTENTS

21.1 Introduction

Nowadays, the world is experiencing a continued growth in the elderly population [1–3]. According to recent studies [4], the population of people aged 65 years and over will almost double between 2000 and 2050. Due to advancements in medical care and a priority on care for the elderly, average life expectancy has increased and medical issues related to aging are gaining prominence. This will increase the demand for services that can facilitate this modern trend [2–5]. Ultimately, the outcome of this growth is the cost that the aging population represents to the health-care system [6,7]. On the one hand, this growth can increase the pressure on health-care facilities, time of health-care professionals in home visits that are more common in the elderly population, and the cost of treatment themselves. On the other hand, the quality of health-care service provided to the elderly cannot be traded off to reduce the cost.

Elderlies often have the desire to remain at home, maintain their independence and quality of life [2,3]. However, the aging process brings with it health deterioration [8] and disabilities beyond chronic conditions [2,8,9]. Memory loss, cognitive impairment, incapacity to accomplish daily activities on their own [9], tendency to falls [10,11], and diminished mobility [2,11] are some of the causes of the need for institutionalization or the loss of autonomy of the elderly. In the search for solutions to the current situation of the elder care, informatics applications, mobile health (mHealth), and smart home technologies seem to be a promising approach to solving or improving some of these problems [1,3,5] as well as reducing the costs of the health-care system [6,7]. With the new mobile technology, health care could be more convenient than ever for the patient and for the professionals. Although there are always limitations when it comes to the adoption of technology [12] (i.e., information privacy, reliability of the technology, and usability), attempts are underway to overcome these challenges and improve the quality of life for the aging population.

In this chapter, we intend to provide an overview of the elder care health issues and mHealth solutions. Contributions of this study are as follows: (1) we first highlight the most common types of the elderly health issues; (2) we then discuss mHealth solutions in the context of six predefined themes; (3) following to the introduction of the themes, we provide a survey of the existing mHealth solutions for geriatric care; and lastly; (4) we highlight some of the challenges and future road map of the mHealth solutions.

21.2 Elder Care

There are a few key areas that are important in the treatment of the elderly patients:

- A quick treatment can be the decisive factor in the survival of an elderly patient, making monitoring programs increasingly necessary [13,14]. Telemedicine appears to be a good way to improve the monitoring and care of the patients with multiple chronic illnesses, allowing information exchange to occur at any time or place [7,13]. This has led to the use of telemonitoring solutions for observing vital parameters of the elderly, mainly those related to cardiovascular and respiratory systems like apnea and heart diseases [15]. The platform for monitoring these patients needs to be able to automatically collect biometrics from the patient, process them, and then send alerts to a remote health-care provider if the parameters, including heart rate, blood oxygenation level, and blood pressure, are abnormal [7].

- Asthma, hypertension, and heart failure are some of the chronic diseases that can benefit from the use of telemonitoring [16]. Researchers reported that the use of such technologies can enable more efficient follow-up procedures which allows the caregivers to notice early changes in the health state of the patients and also providing a better control of the diseases [16].

- Diabetes is another chronic disease that the use of mobile monitoring has received a lot of attention. Elder people with diabetes have more comorbidities than the others [17], and glycemic control is an important parameter that can be reached with the use of mHealth [16].

- This is also a real problem for aging adults and can transpire without perception of the patients. It comes down to the caregiver to evaluate this parameter objectively [18]. Even in a moderate intensity and amount, physical activity can improve the health condition and prevent diseases like hypertension, cancer, and stroke [19]. Falls are a key issue for the elderly population as they can worsen sedentary behavior simply because of the fear of falling notwithstanding the risk of injury and fractures [10].

- Obesity is another problem that is exacerbated by the reduction of physical activity. It also aggravates the decay of the physical condition and the fragility of the elderly, being a risk factor for mobility in such population [20,21]. The existing platform for monitoring these patients also includes collection of biometrics, images, and videos.

- With the prevalence of long-term prescriptions and chronic illnesses, medication control is becoming an important part of older adults' lives. With mHealth, it is possible to monitor and evaluate the medication intake by the patients [1,22,23].

- Dementia is a chronic and slowly progressing neurodegenerative disorder that is rarely reversible [24] and causes cognitive impairment [25]. One example is Alzheimer's disease, for which incidence, associated with other dementias, is growing rapidly [26]. Mobile devices have shown a potential for monitoring the cognitive health and have been extensively used in the reminiscent therapy. This consists of a nondrug treatment, utilizing prompts that encourage the patients to talk about their past experiences [26].

- A new approach in telemedicine is the teledermatology, helping with diagnostic, monitoring, and therapy of aging adults, mainly those who do not have easy or direct access to dermatologists [27]. This approach has shown to be successful in the geriatric care and can overcome the existing challenges that these patients are dealing with on a daily basis. Teledermatology can significantly improve patients' quality of life since it does not require patients' presence in the hospital and can be performed by general practitioners [27].

With the increasing number of mobile phone users and the development of sensors and apps, elder care has entered into a new era of managed care. In the following sections, we provide an overview of some of the existing solutions.

21.3 Mobile and Health Care

To facilitate and standardize our research concerning geriatric care, we utilized a predefined categorization for mHealth [21,22]. The categories are defined in the following section:

21.3.1 Education and Awareness

This category refers to the patients' control of information about their options and health status in a broader context. Education and awareness are important to patients as they help patients make informed decisions about their health. Currently, awareness campaigns are mainly presented via television and radio. Mobile offers the ability to send information directly to a patient's phone, such as various treatment methods, nearby health services, and even disease management. This information can be tailored to suit the individual patient's needs. Since mobile tools are discreet, some successful projects in developing countries have enabled patients to get information on their phones that would be considered taboo in their country.

Education and awareness have been enabled through the use of games and applications to transmit knowledge on health topics. They have also been enabled through provisional access to health science publications or databases.

21.3.2 Remote Data Collection

Data collection is an important factor when dealing with any new technology as it allows for the evaluation of the efficiency of existing programs and the design of the new ones. Another crucial use of remote data gathering is the benefit to public health as the databases collect information in real time. For populations that do not have easy access to hospitals, the information to make decisions during treatments can, therefore, be made more reliable. In addition, mobile devices that can be used for data collection are both abundant and accessible in most countries around the world.

21.3.3 Remote Monitoring

Remote monitoring allows observing multiple conditions as well as detecting medication and measurement of vital signs. This solution allows for easy access to relevant information about patients for the health professionals. This is a significant advancement in health care mainly for those who do not have access to hospitals. Furthermore, monitoring chronic conditions has proven to be an important factor in improving survival rates.

21.3.4 Communication and Training for Health-Care Workers

Mobile technology can easily be a source of information. The use of these devices can provide support to health-care workers, allowing them to be more capable and self-sufficient. In addition to improving health-care practices, mHealth can also close the distance between health-care centers, facilitating communication between health-care professionals.

21.3.5 Disease and Epidemic Outbreak Tracking

Real-time detection of potential disease outbreaks is paramount in preventing devastating epidemics. Currently, public health officials rely on radio reports, satellite, and written letters to gain information about potential outbreaks and new diseases. These methods are often expensive and slow, providing information at a slower rate than the spread of the disease. Mobile devices offer professionals instant access to up-to-date data regarding diseases and are much cheaper than the current systems.

21.3.6 Diagnostics and Treatment Support

Diagnostics are an extremely important aspect of health care. The inability to diagnose a condition can have severe consequences. Health-care workers, and even patients, must rely on only their knowledge of specific diseases to come up with treatment options. mHealth applications can help health-care workers make more informed diagnosis even in very remote areas, reducing the need for patients to visit a hospital or clinic. Some examples of this technology include *appointment reminders* that pertain to sending messages by short message service (SMS), or voicemail. Appointment reminders could be aimed at achieving treatment compliance, avoiding the aggravation of diseases and other challenges such as drug resistance and reminding patients of schedules and appointments like consultations/immunizations. Decision support systems, which include the development of software algorithms to provide counseling information to health-care workers, are useful for supplementing patient records.

21.4 mHealth Solutions for Elder Care

We aim to provide a review of the state-of-the-art mHealth solutions for elder care. In this section, we review the existing solutions in the context of the categories discussed earlier.

21.4.1 Education and Awareness

Most of the existing applications of education and awareness in the elder care are through mobile phone apps [1,17], games [23], and telemonitoring systems [24].

In the study by King et al. [17], three apps were created and tested from the behavioral science-informed user experience design (BSUED). These applications were used to guide the strategies, comprehending analytics, and affective and social motivational frames to stimulate the elderly and help them avoid sedentary behavior. The data collected via smartphones were transmitted to the project's local servers and monitored by the researchers [17]. The findings of this study found that apps were well accepted by the participants that ranged from 45 to 81 years old who had no previous experience with smartphones. They also found that the apps were able to significantly increase the frequency of exercise per week for patients and to reduce the number of hours that the patients remained seated [17].

A casual educational game was designed for mobile phones called *OrderUP!* to encourage people to have a healthier nutrition. The benefit of using games as a form of education is that it helps people think about their own health while keeping them engaged. The game had positive outcomes such as raising consciousness, with self-free evaluation, and helping relationships and counterconditioning.

The program was implemented for the Nokia N95 cell phone platform. The game consisted of the player assuming the role of a server in a restaurant, and the goal was to make the healthier meal recommendations to the customers. Doing this, the person can keep the game running as long as possible. The dishes presented in the game were similar to the typical diet of the population analyzed in the study. All the participants revealed that playing *OrderUP!* initiated a change in their nutritional behavior and a self-evaluation process [23].

Some other studies have evaluated the use of telemedicine in older patients with diabetes. IDEATel [24], for example, was created by the Centers for Medicare and Medicaid services. This system consisted of a web-enabled computer that could upload blood glucose and blood pressure to be shared in real time with the dietician, nurse, or the case manager through videoconferencing [24]. This study showed significant improvement for the levels of glycated hemoglobin, blood pressure, and lipids indicating that the diabetes self-efficacy is improved with the use of telemedicine.

All the studies in this category have shown the potential benefits of using mHealth technologies to improve education and raise awareness. However, the issue is for the elderly patients, the medium of delivering these applications, especially in the form of games, could be unfamiliar or simply unappealing limiting the success of these technologies in this targeted population.

21.4.2 Remote Data Collection and Remote Monitoring

Data collection and remote monitoring go hand in hand when dealing with the elderly. Most of the devices are not just for gathering data. They also work on the monitoring of welfare and chronic conditions, considering the natural deterioration of the health of the population [8]. Wearable and nonwearable sensors, telemonitoring using videos, mobile phones, and devices capable of collecting vital signs are some of the technologies used in these categories.

There are different types of sensors for monitoring systems. For example, in [2], the authors discuss residences specialized in monitoring and promoting independence of the elderly. The paper provides an overview of different types of monitoring systems such as physiological monitoring, functional monitoring/emergency detection and response, safety monitoring and assistance, security monitoring and assistance, social interaction monitoring and assistance, and cognitive and sensory assistance. Moreover, wearable sensors such as accelerometers, temperature sensors, electrodermal activity sensors, heart rate sensors, and oximetry sensors are discussed in [25]. The data collected by these sensors are transmitted to an external computer or mobile phone via Bluetooth, *Zigbee*, or RFID stands for Radio-Frequency IDentification (RFID) [25].

Demiris et al. [9] designed a new approach to evaluate the health of the aging adults in order to intervene and prevent (or reduce health) damages due to the age. In this study, the authors evaluated physiological/functional (vital signs, mobility), social, spiritual, or cognitive/mental parameters using a screening platform for wellness.

The platform consisted of three parts: GAITRite, a telehealth kiosk, and CogniFit. All three collected data in secure web servers and transferred data to the Excel files. The telehealth kiosk stored the patients' profiles with nutritional and educational content. The wireless devices also captured blood pressure, heart rate, glycemia, oxygen saturation, and patient weight. The CogniFit is a brain fitness web-based software solution that enables the assessment of cognitive abilities including awareness, eye–hand coordination, speed of processing, response time, visual scanning, and working memory. The GAITRite includes an electronic walkway with sensor pads and a software system that provides a footprint analysis, spatial parameters, temporal definitions, and switching levels of gait [9].

In another study, health information from elderly was collected by an mHealth monitoring system, which consisted of a wireless wearable body area network (WWBAN), an intelligent central node (ICN), and an intelligent central server [6].

The WWBAN consisted of sensors that collected the biometrics of the patient and sent this information via Bluetooth to the ICN. ICN, in a smartphone, collects and processes the data that it receives from WWBAN and communicates that with the ICS via general packet radio service/universal mobile telecommunications system (GPRS/UMTS). The ICS stores the information in a database. These data can later be accessed and analyzed by the family and the doctor through a web application. When anything outside normal limits is detected, the system sends alerts to family and doctor, and in case of emergency, it also provides the patient's current location [6].

These technologies are very useful as they provide health-care professionals with a way to measure important medical parameters on a long-term basis. This also, in turn, provides a better data set to make treatment options more accurate and reliable. In addition, this adds a level of convenience for both professionals and patients as monitoring does not take place in a hospital environment but can be implemented from one's home.

Another interesting development is the decentralized telemonitoring system that can assess the physical activity of the patients [16]. The sensor is a wearable motion detector attached to the belt at waist level. In these systems, sensors around the home capture data and send them to a *distribute data server* (DDS). The data are then processed and stored in a multimedia card (MMC). The authors have proposed an algorithm for real-time activity identification encompassing still postures, postural changes, dynamic movements, and detection of falls. In the case of detecting an urgent situation or emergency, caregivers can be notified by mobile phone through text messages or e-mails. This study has some advantages over other existing systems due to its decentralized architecture. This results in the lower cost, more privacy of health data, and better preservation of data.

Furthermore, there have been many attempts for solutions that can remotely gather patients' data. The aim of such solutions is to bring down the cost, provide patients with more control over their treatment options, improve long-term monitoring, and alert doctors of emergency situations. A summary of the uses and potential benefits can be found in the following:

- Matsui et al. developed a heart rate variability (HRV) monitoring system to measure vital signs such as cardiac and respiratory and autonomic nerve activity by a noncontact mechanism. The system allows elderly people to be evaluated without physical and mental stress. The system can substitute the use of ECG for monitoring the HRV and can alert caregivers in cases of autonomic dysfunction [4]. As the HRV alteration can be a symptom of severe diseases, the noncontact system might be more appropriate for long-term monitoring of HRV than the ECG. It can also provide baseline to preventative care of monitored patients.

- Medical sensors networks (MSNs) were developed to monitor changes in the vital signs of participants. The MSNs could potentially decrease the number of doctor's visits and the necessity of collecting vital signs manually. They can also capture medical emergency events. The sensors can detect, collect, and process one or more physiological signals continuously [7].

- iCare offers a solution for personal remote monitoring health, life assistance, information system, and medical guidance. The system consists of four parts: the devices (sensors, medical devices), a smartphone that receives the data from the sensors, a server, and an emergency call center [14].

- A smart home based on a cognitive sensor network was developed in order to monitor daily activities and warn the appropriate professionals or family members in the case of any abnormal activity through the use of a phone call, text message, or e-mail. The system consisted of integrating various sensors that communicate via radio frequency and a controller that received and processed the data received from the sensors [5]. Some of the sensors collected data about the use of electrical appliances such as toasters and microwave ovens to evaluate the daily activities of the patient. The water use and sleeping pattern of the elderly people were also monitored. In this system, a panic button was integrated for the patient in case of an emergency or for urgent assistance [5]. The project was tested and was shown to be successful in various conditions.

- A hospital, in the north of Singapore, developed a pilot program to provide care for residents in nursing homes using telemedicine. Monitors, cameras, and microphones were used to enable teleconsultations, multidisciplinary team meetings, mortality audits, family conferences, and nurse training sessions. The authors believed that this kind of technology can bring patients, family, nurses, and physicians closer together. Although the teleconsultation could never replace a physical consultation, this technology is capable of connecting patients, physicians, and nurses [28].

In the study by Shah et al. [13], data including patient demographics, medication lists, and medical histories were analyzed and evaluated for the eligibility of telemedicine by an emergency physician and two geriatricians. During the acute care, technicians obtained the historical information of the patient, vital signs, images, and videos to analyze if that treatment needed the real presence of the physician or could be performed via telemedicine [13].

The authors concluded that almost half of all the consultations, due to acute illness, could have been handled by telemedicine. They affirm that an efficient telemedicine program can reduce time and effort needed for each meeting between physicians and patients. They also concluded that this approach can reduce cost, for example, transportation fee, and a large number of the elderly patients prefer the comfort of using these solutions over traveling [13].

The paper by Wang and Liu [26] presented some of the new approaches that are being created for health monitoring including methods and apparatus using the tongue image, mobile diagnostic system using audible sound, systems modifying the battery pack of a mobile phone, multifunctional mobile phones performing hearing and vision tests, a mobile phone detecting egg, a mobile phone with a vibration module for health care, a mobile phone with an ear temperature sensor, and a diagnosis system by analyzing a facial image and voice characters.

The survey by Naeemabadi et al. [15] developed a monitoring system composed of

- A mobile station, responsible for recording vital parameters and creating a database with the information collected
- A short-range communicator that will transmit data collected by the mobile station wirelessly to a home station
- A home station that processes the vital signs received, correcting error frames
- A long-range communicator that transfers the processed data to a medical center enabling the control of vital signs and health conditions in the medical care station
- A medical care station that cares of the patient remotely [15]

This system was created based on the necessity of the elderly to have an independent life. In fact, the confidence ratio is at the level at which care of this kind could be deemed appropriate, making it possible to apply this technology [15].

Researchers developed a telepresence robot for interpersonal communication (TRIC). This solution was applied in a home environment that allowed the elderly to remain in their residence of choice, enabled monitoring of their health status by their caregivers, and maintained contact with their families. Through controlling the camera, the actions, and the real-voice communication, the caregiver could monitor and interact with the elderly [27].

For its second application, a DDS processed and stored the monitored data transmitted by the sensors remotely. This enabled patients to request the information to be stored via an Internet web browser. This system enabled e-mail and text communication in the case of an emergency [27]. The advantage of the use of TRIC in remote care of the elderly is mobility of the devices and the capability to connect the patients to their family or caregiver [27].

The study by Takahashi et al. [8] aimed to assess the effectiveness of home telemonitoring for reducing the number of hospitalizations and emergency department visits and for improving the functional status of patients when compared with the typical care. The results of a survey indicated that telemonitoring provided an effective tool for monitoring the changes in health status of the elderly [8].

21.4.3 Communication and Training for Health-Care Workers

In our study of the literature, we were not able to find many papers that have focused on this particular category in the context of geriatric care. Most of the existing work have focused on the communication between patients and health-care workers, but not between professionals. Mobile technology can assist in communications over long distances. Therefore, relationships between health professionals across countries or even continents can be established in order to share ideas, provide training about state-of-the-art solutions and approaches, and, finally, train professionals in remote locations.

A case was presented by Low et al. [28] in which Ms C, a 92-year-old woman of frail health, was living in a nursing home. She chose not to be transferred to an acute care hospital in the event of severe illness due to her weakened state. The researchers decided to have a teleconsultation in the last moments of her life since she was living in a remote area and the researchers could not visit her frequently, but they wanted to monitor her health status in real time. As she was in the dying process, researchers were able to talk to her and observe the process. The physicians were pleased to have chosen this approach because the physical visits could have caused them to lose the last moments of the patient's life since they could not travel frequently to the remote area. The telemedicine also provided assistance to the nurse that was with the patient during her last moments of life. This gave the nurse confidence because she had someone guiding and supervising her in the task of preparing for the patient's death [28].

21.4.4 Disease and Epidemic Outbreak Tracking

For this category, similar to the previous one, we could not find specific studies related to disease and epidemic outbreak tracking occurring in the elderly population. However, mobile technologies have the capability to increase awareness of outbreaks for the elderly as they tend to be more vulnerable.

21.4.5 Diagnostics and Treatment Support

A wide variety of devices can be found, in the literature, for diagnostic and treatment support of older adults, ranging from telemonitoring to accelerometers in mobile phones. The monitoring of biometrics falls under this category as well.

Shah et al. [13] claims that almost half of the consultations for acute illnesses could be handled by telemedicine. They also argue that an efficient telemedicine program can reduce time and effort needed for each meeting between physicians and patients. Furthermore, they suggest that this approach can reduce costs, that is, transportation, and improve quality of life as a large number of the elderly patients prefer to stay home rather than traveling back and forth to a physician's office.

Treatments could sometime be provided by the patients themselves allowing more freedom and control. As an example, in the study discussed in Section 21.4.1, that is, IDEATel [24], the authors were able to show significant improvement regarding the levels of glycated hemoglobin, blood pressure, and lipids by enabling the patient to control and use telemedicine solutions.

In the study by Boman et al. [18], a portable system to measure body weight, blood pressure, and pulse rate, called CheckUp, was developed to improve the monitoring process of international normalized ratio (INR) values, an important parameter of the control of warfarin treatment. The INR was transmitted to CheckUp bag via Bluetooth, and then the data were transmitted to the hospital via mobile phone. This enabled follow-ups and dosage control of the anticoagulant treatment by the professionals that could determine the correct and safe doses of warfarin for each patient [18]. This study examined the efficacy of the CheckUp system in reducing the processing time of patient monitoring in comparison to usual procedures. Another benefit was the reduction of the transport time, especially to the primary health center (PHC) located far from the analyzing laboratory, for example, PHCs located in rural areas [18].

One barrier for the use of β-blockers is the possible adverse effects of this drug, including severe bradycardia. In order to reduce this side effect and to ensure the right dose of the drug, researchers developed a telemonitoring program to observe the heart rate of CHF patients from their homes [19]. The results showed that the patients monitored by telemedicine achieved the right dosage of carvedilol faster than the control group. This accomplishment was important for the treatment since the efficiency of carvedilol is dose dependent. The study presented a relevant clinical advantage of telemonitoring and, also, reduced costs of the health-care system [19].

A study by Lee and Carlisle [10] developed and examined a computer algorithm for fall detection using mobile phone technology. The researchers then compared motion signals obtained by an accelerometer built in a cell phone with the signals obtained by a regular accelerometer. The results of the study showed significant specificity and sensibility of detection by the mobile phone, supporting the usage of such mobile solutions for detecting falls [10].

Lastly, in the mobile assistant for the elders (MASEL) project, an alarm system is designed to remind patients which drugs and what doses should be taken. The decision system is personalized and the prescription is made by physicians remotely. MASEL provides an interface between medical professionals, patients, and family when it comes to drug usage time [1]. The program has three different profiles: (1) one enables the management of the patient's data for the time and doses as well as the kind of medication that are being taken by the patient, (2) the family profile report provides family or friends information about the treatment, and (3) the patient profile is used to alert and confirm when they have taken each medicine [1].

21.5 Successful Case Studies

mHealth solutions have been successfully used to improve the quality of life of the elderly across the globe. In Denmark, for example, Skaevinge project [29] became a national phenomenon for empowering elderly patients to become more independent and to enable them to stay longer at home instead of moving to the nursing homes. This project was later adopted in Japan, Australia, Germany, and several other countries. In one of the Skaevinge projects, Elda, from Germany, reported how the use of social media on mobile helped her family avoid hospitalization of their 94-year-old father. A social media app connected several family members together as well as a local health-care provider to remotely monitor their father's health and provide care whenever needed at home.

Another example was a pilot project by Roanoke Chowan Community Health Center (RCCHC) [30] in which remote monitoring was established for elderly with cardiovascular disease, diabetes, and hypertension. The case study participants have indicated that despite their limited use and familiarity with computer technology, they were happy with the outcome of the project and were willing to adopt the program.

21.6 Future Road Map

There are numerous opportunities to improve the quality of life for older patients using mHealth technologies. Some of the opportunities have been discussed in this chapter and many other innovative research ideas exist in the literature. These technologies are poised to transform the future of health care and our society as a whole.

The solutions that have been developed can manage everything from tracking patients' medication to reminders about upcoming appointments, preventing falls, reminding patients about their medicine intake, remote monitoring, or simply as a way of remotely informing patients of test results [1,13–24].

From a technological perspective, the future of mHealth solutions involves making sure that information can be transmitted securely and that systems are robust enough to deal with large quantities of patients if they are all to be managed at the same time. Furthermore, more attention can be focused on sending information via mobile devices in situations that might need the patients to have received information quickly or in countries that lack efficient mail delivery systems. However, both the quality of information provided [25] and the assured delivery of the information need to be fully investigated on to make the technology more reliable [7].

Another aspect of the technology that can be made available to the elderly is the use of mobile technologies as a way to allow older patients keep their independence while still being provided viable health-care avenues. There are a great number of varying systems that allow remote monitoring of patients; however, further research is necessary to certify the efficiency and applicability of all of the devices that can be implemented [13,27]. In some studies, bigger samples are required [9]; in others, a longer study period is necessary [4]. With the focus of mHealth being emergency care, a more efficient system is also something to be concerned about [14].

Another motivation for the development of mHealth technologies is decreasing cost of health care to deal with increased pressure on services and to provide wider access to a growing elderly population. Elder care services can be expensive for poor families and thus are often used by upper–middle class families [12]. Future work should be focused on technologies that can decrease the frequency of expensive hospital visits with the use of smartphone apps along with peripherals that can measure vital signs in small clinics or even at home instead. Also, devices that can help improve everyday activities such as robots that assist in cleaning and bathing or even taps that prevent arthritic pain mean that assistance can be provided without the need of health-care professionals. Yet, this technology is in its infancy and further advancements could mean a better allocation of resources.

For elderly patients who are lonely, any kind of interaction can be beneficial, although some people may argue that telemedicine is not the same as face-to-face interaction for some patients. The telemedicine approach could be even more beneficial due to geographical constraints making it harder or even impossible to have face-to-face meetings [28]. The issue to focus on in future research is to develop technology that will be able to increase patients' social interactions, which is very beneficial for people's morale especially if they are usually alone.

One of the main issues in adoption of new technology especially among older adults is the usability of the technology in question. It has been noted that mHealth technologies such as *Google Health* failed because the developers did not consider the industry itself and there was a compromise on user experience [31]. Future work should address user usability at all points as technology is useless unless someone is willing to use it. Therefore, there should also be consideration put into making a product, which is engaging or with a small amount of education becomes relatively simple to use.

Lastly, future research could evaluate the efficacy of social media as a platform for education and promotion of health-care initiatives that could lead to increased awareness of important issues.

21.7 Conclusion

The application of mHealth for the elderly shows many advantages and also challenges. In this chapter, we reviewed the existing mHealth solutions and their applications. We further explored some of the challenges and potential future directions of mHealth technology for geriatric care.

Questions

1. What are some of the factors affecting the adoption of mobile health applications by the elderly?

2. What treatments in elderly care have mobile health applications shown to be most effective in?

3. Elderly isolation is one of the major concerns of our society. What mobile solutions have been developed to address this issue?

References

1. Sánchez, M.A., Beato, E., Salvador, D., and Martín, A. Mobile Assistant for the Elder (MASEL): A practical application of smart mobility, highlights in PAAMS, *Advances in Intelligent and Soft Computing*, 89, 325–331. Springer, Heidelberg, Germany, 2011, doi:10.1007/978-3-642-19917-2_39.
2. Demiris, G. and Hensel, B.K. Technologies for an aging society: A systematic review of "smart home" applications, 33–40, 2008.

3. Zhongna, Z., Wenqing, D., Eggert, J., Giger, J.T., Keller, J., Rantz, M., and Zhihai, H. A real-time system for in-home activity monitoring of elders, *Proceedings of 31st Annual International Conference in IEEE Engineering in Medicine and Biology Society*, Minneapolis, MN, pp. 6115–6118, 2009.

4. Matsui, T., Yoshida, Y., Kagawa, M., Kubota, M., and Kurita, A. Development of a practicable non-contact bedside autonomic activation monitoring system using microwave radars and its clinical application in elderly people, *Journal of Clinical Monitoring and Computing*, 27(3), 351–356, 2013. doi: 10.1007/s10877-013-9448-3.

5. Gaddam, A., Mukhopadhyay, S.C., and Gupta, G.S. Elder care based on cognitive sensor network, *IEEE Sensors Journal*, 11, 574–581, 2011.

6. Bourouis, A., Feham, M., and Bouchachia, A. Ubiquitous mHealth Monitoring System for Elderly (UMHMSE), *International Journal of Computer Science and Information Technology*, 3(3), 74–82, 2011.

7. Hu, F., Xiao, Y., and Hao, Q. Congestion-aware, loss-resilient bio-monitoring sensor networking for mHealth applications, *IEEE JSAC*, 27(4), 450–465, 2009.

8. Takahashi, P.Y., Hanson, G.J., Pecina, J.L., Stroebel, R.J., Chaudhry, R., Shah, N.D., and Naessens, J.M. A randomized controlled trial of telemonitoring in older adults with multiple chronic conditions: The Tele-ERA study, *BMC Health Services Research*, 10(255). doi: 10.1186/1472-6963-10-255.

9. Demiris, G., Thompson, H.J., Reeder, B., Wilamowska, K., Zaslavsky, O. Using informatics to capture older adults' wellness, *International Journal of Medical Informatics*, 82(11), 232–241, 2011.

10. Lee, R.Y.W. and Carlisle, A.J. Detection of falls using accelerometers and mobile phone technology, *Age Ageing*, 40(1–7). doi: 10.1093/ageing/afr050.

11. Albert, M., Kording, K., Herrmann, M., and Jayaraman, A. Fall classification by machine learning using mobile phones, *PLoS One*, 7(5), e36556, 2012.

12. Bookman, A. and Kimbrel, D. Families and elder care in the twenty-first century, *Future Child*, 21(2), 117–140, 2011.

13. Shah, M.N., McDermott, R., Gillespie, S.M., Philbrick, E.B., and Nelson, D. Potential of telemedicine to provide acute medical care for adults in senior living communities, *Academic Emergency Medicine*, 20(2), 162–168. doi: 10.1111/acem.12075.

14. Lv, Z., Xia, F., Wu, G., Yao, L., and Chen, Z. iCare: A Mobile Health Monitoring System for the Elderly, IEEE/ACM Int. Conf. on Green Computing and Communications, Hangzhou, China, 699–705, 2010.

15. Naeemabadi, M., Zabihi, M., Ordoubadi, B.S., Saleh, M.A., Khalilzadeh, M.A., and Ordoubadi, M.S. Tele-homecare system design for elderly, *Proceedings of Fifth International Conference on AICT*, Baku, Azarbaijan, pp. 1–5, 2011.

16. Yang, C.C. and Hsu, Y.L. Development of a wearable motion detector for telemonitoring and real-time identification of physical activity, *Telemedicine Journal and e-Health*, 15, 62–72, 2009.

17. King, A.C., Hekler, E.B., Grieco, L.A., Winter, S.J., Sheats, J.L., Buman, M.P., Banerjee, B., Robinson, T.N., and Cirimele, J. Harnessing different motivational frames via mobile phones to promote daily physical activity and reduce sedentary behavior in aging adults, *PLoS One*, 8(4), e62613, 2013. doi: 10.1371/journal.pone.0062613.

18. Boman, K., Davidson, T., Gustavsson, M., Olofsson, M., Renström, G.B., and Johansson, L. Telemedicine improves the monitoring process in anticoagulant treatment, *Journal of Telemedicine and Telecare*, 18(6), 312–316, 2012. doi: 10.1258/jtt.2012.120319.

19. Antonicelli, R., Mazzanti, I., Abbatecola, A.M., and Parati, G. Impact of home patient telemonitoring on use of beta-blockers in congestive heart failure, *Drugs and Aging*, 27, 801–805, 2010.

20. Guo, X., Sun, Y., Wang, N., ZeyuPeng, J., and Yan, Z. The dark side of elderly acceptance of preventive mHealth services in China, *Electronic Markets*, 23(1), 49–61, 2013.

21. UN Foundation-Vodafone Foundation Partnership. *mHealth for Development: The Opportunity of Mobile Technology for Healthcare in the Developing World*. Washington, DC: Berkshire, 2009.

22. World Health Organization. mHealth: *New Horizons for Health through Mobile Technologies*. Global Observatory for eHealth series, Vol. 3, 2011, http://whqlibdoc.who.int/publications/ 2011/9789241564250_eng.pdf?ua=1

23. Grimes, A., Kantroo, V., and Grinter, R.E. Let's play!: mHealth games for adults, *Proceedings of the UbiComp*, September 26–29, 2010, Copenhagen, Denmark.

24. Trief, P., Teresi, J.A., Eimicke, J.P., Shea, S., and Weinstock, R.S. Improvement in diabetes self-efficacy and glycaemic control using telemedicine in a sample of older ethnically diverse individuals who have diabetes: The IDEATel project, *Age Ageing*, 38, 219–225, 2009.

25. Fletcher, R.R., Poh, M.Z., and Eydgahi, H. Wearable sensors: Opportunities and challenges for low-cost healthcare, *Proceedings of International Conference on IEEE Engineering in Medicine and Biology Society*, Buenos Aires, Argentina, 1763–1766, 2010.

26. Wang, H. and Liu, J. Mobile phone based healthcare technology, Recent Patents on Biomedical Engineering, 2009.

27. Tsai, T., Hsu, Y., Ma, A., King, T., and Wu, C. Developing a telepresence robot for interpersonal communication with the elderly in a home environment, *Telemedicine and e-Health*, 13(4), 407–424, 2007.

28. Low, J.A., Beins, G., Lee, K.K., Koh, E. Last moments of life: Can telemedicine play a role? *Palliative and Supportive Care*, 11, 1–3, 2013.

29. World Health Organization. A compendium of primary care case studies, last visited January 31, 2014, http://www.who.int/hrh/nursing_midwifery/compendium_hrh_studies.pdf.

30. NORC. patient provider telehealth network—Using telehealth to improve chronic disease management, last visited January 31, 2014, http://www.healthit.gov/sites/default/files/pdf/ RCCHCandPHS_CaseStudy.pdf

31. Neil, V. Sculley: Health tech needs usability, not flash, InformationWeek Healthcare, January 13, 2012. Last visited August 28, 2013. http://www.informationweek.com/healthcare/ leadership/sculley-health-tech-needs-usability-not/232400350.

22

mHealth for Aging Populations: Community, Participation, Connectivity, and Mobile Technologies for Older People

Elspeth McKay and Jennifer Martin

CONTENTS

22.1 Introduction

We first identify the older population to set the context for the discussion on their community participation, connectivity, and mobile device usage. We then provide the theoretical background to frame our chapter and orient the reader to understand the complexity of the relationship between older persons and their desires and needs for connectivity through specialized information and communications technology (ICT) tools. Central to this discussion are the free mobile device applications that are currently available at no cost to the consumer, like the free kiosks placed in some Australian public/municipal libraries. There are some mobile applications that are used to promote a safe environment for social networking, while others afford language translation to provide added comfort for older people who may wish to revert to their linguistic origins later in life. We use personas and scenarios to identify some mobile applications that are designed for older people that include medical, health and fitness, maintaining a quality lifestyle, well-being and sourcing aged care, and general products and day-to-day necessities.

22.2 Background

Worldwide, people are living longer due to the decline in infant mortality, the control of infectious diseases, and improvements in nutrition and living standards. Aging reflects trends in mortality, frequency of chronic disease, and maintenance of autonomy. Attention to these independent, though related, variables will increase the proportion of the population surviving disease free to an advanced age. The expected life span is likely to increase alongside the number of years a person is disease free. However, projections indicate that the number of years that an older person is expected to live with loss of independence and loss of autonomy is also likely to increase, particularly due to the increase in dementia (Barry, 2003).

The literature describes older adults as anyone over the age of 40 at the lower end of the scale to being *over 75* at the other (Wagner et al., 2010); we have therefore limited our focus to concentrate on the upper end, where our definitions have been accessed through a social cognitive approach to draw out the human dimensions of human–computer interaction (HCI).

Australians have one of the highest life expectancy rates in the world ranking sixth among organisation for economic co-operation development (OECD) countries (AIHW, 2012). The life expectancy for Australian men is 79.5 years, only slightly lower than the 79.9 years recorded for males from the highest ranking country Switzerland. Australian women have a life expectancy of 84 years, likewise only marginally lower than 84.6 for Japan, the highest ranking country for females. Life expectancy projections for both Australian men and women have increased by 25 years over the past century. However, the life expectancy for aboriginal and Torres Strait Island peoples is approximately 12 years lower than the general population (AIHW, 2012). Approximately one-quarter of Australia's population aged over 65 speaks a language other than English. Migrant elderly are a significant and growing proportion of the population but continue to be underrepresented within aged care services (Abraham and Martin, 2014). Alongside this population growth is increasingly complex health needs. Health is difficult to define with wide-ranging definitions from a narrow physical focus on the absence of disease of infirmity to a social determinants model. For the past 60 years, the World Health Organization (WHO, 2012) has taken a holistic view of health incorporating the biological, psychological, and social aspects. It is this broader view of health that has guided policy and service development in Australia. This includes "both physical and mental dimensions, within a context that includes genetic, cultural, socioeconomic and environmental determinants" (AIHW, 2012, p. 2). Migrant elderly are a significant and growing proportion of the Australian population. Unlike many other countries around the world, the Australian-born population is fluent in only one language, English. Approximately 25% of Australia's population aged over 65 years speaks a language other than English, with the majority of these people born overseas. They have significant heath needs yet their usage of aged care services is low (Martin, 2013).

The age profile of the Australian population is changing as the population is growing, aging, and living longer. Federal government estimates suggest that by 2050, more than

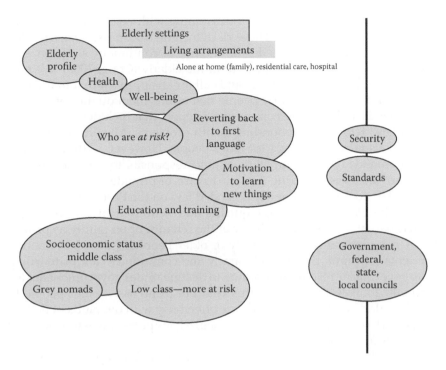

FIGURE 22.1
Aging population context.

1 in 20 workers will be employed in aged care (Butler, 2012). The changing population profile and increased aged care workforce have seen a rapid increase on health expenditure as a percentage of gross domestic product. This is within large-scale reforms of the health sector with a rethinking of traditional roles of health professionals and their educational and training requirements. Many of the new roles will be in direct care as well as prevention to identify health and well-being issues early. A focus is on better alignment and coordination of services that are cognizant of new and emerging technologies such as e-health, telehealth, and the use of avatars as well as face-to-face practices (see Figure 22.1). In addition to direct clinical work, this gives rise to new opportunities for system designers and managers, consultants, and application developers (Cormack, 2012).

22.2.1 What Are Their Needs?

Of specific importance to older people are disorders that affect hearing and vision, dental problems, incontinence of urine and feces, and intellectual failure, particularly dementia. Cancer is a major health problem in Australia followed by cardiovascular disease and mental disorders with a high correlation between physical health problems and mental disorders (AIHW, 2012).

One-third of those aged 65 years and over and three-quarters of those aged 65 years and over are taking medication on a regular basis. It is not unusual for an older person to be discharged from the hospital on ten or more medications (Martin, 2012). Concerns have been raised about an unexplained increase in the use of sedatives, tranquillizers, and anti-psychotic medications when older people enter residential care (White, 2013). This is particularly concerning as the older population is a more diverse physiological group and therefore much more prone to adverse reactions of drug therapy. Of particular concern is the extensive use of sedatives for insomnia with research findings suggesting that the risks outweigh the benefits. These risks include dizziness, disorientation, loss of balance and falls, and in some cases death. The use of continuous sedation in palliative care, referred to as *deep sedation*, can be fatal in the elderly (Craig, 2008). We too are concerned about the medicalization of *old age* and call for an approach that maintains the dignity, well-being, and integrity of older people that afford them quality of life with access to family and friends and ongoing community support that includes the latest ICT tools. Loneliness is a key factor in compounding health problems (Martin, 2012). The family is the greatest single source of support for older people. However, with increased population mobility, demands of paid employment, and responsibility for their own children, adult children are often not spending the time to care, or provide companionship, for their elderly parents. The controversial newly amended law on the Protection of the Rights and the Interests of the Elderly in China, requiring adult children to visit or keep in touch with their aged parents, is a stark reminder of the reliance upon family as primary carers (Min, 2013). A comprehensive British study of older people in disaster situations found that governments rely upon this *invisible* family and community care primarily provided by women as government resources continue to decline under government austerity measures (Dominelli, 2013). ICT cannot fill this void but it can assist older people to remain connected, particularly with children who are often busy with demands of their own families as well as work commitments. This is particularly important when children are living and working in different cities.

Eric Erikson (1980) developed a concept of personality development linked with biological development over three decades ago that still remains relevant to social work practice today across all cultures. He argued that psychological and physical development were linked and identified stages of physical development and associated psychological tasks at each stage. The ages specified by Erikson for each developmental stage were an approximate guide allowing for wide variation according to individual differences with these stages overlapping. Erikson's final developmental stage was late adulthood, for those aged 60–65 years and over (Erikson, 1980). He identified *integrity* as the main developmental task and *living with dignity*. This required having a sense of meaning and purpose in life as well as order. Integrity resulted in joy for living in contrast to despair that often developed in older people, particularly in Western industrialized countries, due to unresolved issues of loss and grief and loneliness. Issues of loss and grief are a particular issue for refugees and asylum seekers as well as older new immigrants who do not speak the English language and who are reliant upon younger family members for communication.

A postmodern view of aging focuses on living well in the community with an emphasis on choice, control, coordination, and empowerment (Martin, 2012). This requires consideration of hopes and dreams and strengths and resources in addition to needs and risk factors. Positive choices are highlighted in relation to the type and quality of care required. Older people who are also carers need to be seen as people with their own needs and desires not purely as resources. An understanding of the dynamic and fluid nature of culture, which respects the diversity of practices within and between different cultural groups, is required. Increasingly however, culture is ignored in the design and development of generic aged care services driven by cost reductions (Dominelli, 2013). Cost-saving measures are also incurred in the training and development of the aged care workforce. The complexity of needs of older people is not matched by staffing capabilities, training, and award structures. Again, ICT tools can be of assistance in the provision of easily accessible translation services as well as education and continuing professional development for the aged care workforce as discussed at the end of this chapter. Those with dementia have particularly complex needs.

22.2.2 Dementia

Dementia is the third main cause of death in Australia (AIHW, 2012). While generally associated with the elderly, dementia is not a normal part of the aging process. Many older people well advanced in years retain full intellectual functioning. Its occurrence is determined by genetic predisposition, family history, and general health and well-being. Dementia is common in older people but quite rare in middle aged or young people. Dementia of the Alzheimer's type affects 6% of people over 65 years in the Australian population. This percentage rises to 11% for those aged over 75 years. It is estimated that half of all permanent residents in Australian aged care facilities are diagnosed with dementia and have high care needs (AIHW, 2011). Dementia is a condition for which there is no known cure or medical treatment and which results in the gradual deterioration of memory, intellect, and the ability for self-care. The onset of dementia is slow and insidious and leads to major personality alteration, markedly affecting the individual's ability to continue to relate to the world around them. In the early stages of dementia, the person has failing memory and changes in personality and behavior with depression not uncommon. In later stages, the person is increasingly confused and may become upset and irritable and may also become incontinent (DOHA, 2011). In advanced stages, the person affected will not recognize loved ones or even her or his own face in the mirror. It is a condition that presents a social problem of considerable magnitude in countries, like Australia and China, with an aging demography. Dementia is difficult to diagnose and is a major challenge in terms of its management over often quite long periods of time.

22.3 Aged Care Services

Aged care legislation provides the regulatory framework for policy, principles, and standards of care for older people. This includes all levels of government and private providers covering both residential and community services. It is widely acknowledged that the multifaceted Australian health system is difficult to navigate. "This web of public and private providers, settings, participants and supporting mechanisms is nothing short of complex" (AIHW, 2012, p. 17). Multiple service providers are located in a range of private and public settings. This is supported by various legislative, funding, and regulatory arrangements with responsibilities distributed across all three levels of federal, state/territory, and local government, nongovernment organizations, and individuals. For many older people from culturally and linguistically diverse (CALD) backgrounds, particularly new and emerging communities, navigating this service system on their own can be daunting (ACA, 2010). An Australian study of the health and welfare needs of the Somli elderly highlighted gaps in services and where services did exist, some people were frightened to initiate service access due to language difficulties and/or past negative experiences with authorities. An elderly Somali man is quoted as saying, "No-one told us anything about services in Australia. We had no help. We were waiting for our children to help us" (Abraham and Martin, 2014, p. 83). The study found that there was heavy reliance upon assistance provided by family or trusted members of the community. Extended or acute aged care services were only sought as a last resort, and it was usually for crisis management when there is a health emergency (Abraham, 2012). ICT tools can assist; see Section 22.5 further on in this chapter.

There is a growing burden of chronic disease, workforce pressures, and "unacceptable inequities in health outcomes and access to services" (DOHA, 2011, p. 9). Australia's aging population means that a significant strain will be placed on the health budget as currently fiscal spending on people aged 65 years and over is four times greater than the rest of the population (HWA, 2011). Health needs tend to become more complex as people age. A 2009 study of older Australians living in the community found that approximately half (49%) of those aged 65–74 years had five or more chronic conditions. For those aged 85 years and over, this increased to 70% (AIHW, 2011). A priority for strategic planning for sustainable health care into the future is on wellness, prevention, and primary health care. These are also priority areas identified by the WHO as populations around the world are managing and planning for growing and aging populations. The WHO slogan of "good health adds life to years" sends a message of older people having active and productive lives. A focus is on the resourcefulness and roles of older people in their families and communities as opposed to a disease and deficits model of aging (WHO, 2012). Health-care workers however have been criticized for their controlling role in access to services and for fostering a dependency view of older people with policy and practices dominated by risk management (Martin, 2013).

The current adoption of ICT by older people supports this notion (Lee, 2004), showing they are capable users of computerized information systems (ISs) according to Czaja and Lee (2008). However, usability assistance may be required with applications designed for people with dementia and other health-care needs. The emphasis is on a strengths approach rather than a deficits model (Martin, 2012).

Attention is increasingly focused on the well-being of older people. However, inconsistent research findings show a simultaneous high life satisfaction score alongside high suicide rates (Chong, 2007). A recent study of older Australians found that the majority of respondents, 84%, felt positive about their quality of life. However, the researchers lamented the lack of data available on cause of death and hence no analysis of mortality patterns, including suicide (AIHW, 2012). A substantial proportion of older people are diagnosed with mental illness. In 2007, 8% of older people living in the community were found to have a mental illness with this figure increasing for those in residential settings (ABS, 2009). The scarcity of reliable research on the health and well-being needs of older people from all backgrounds is concerning. The social work profession is well placed to make a significant contribution by including the *voices* of older people in the design, delivery, and evaluation of services (Powell, 2007). In Britain, Lena Dominelli (2013), professor of social work, argues that health care should be viewed within the context of social development with strategies developed to assist minority ethnic elderly to age with care (p. 10). For Dominelli, social development is about *putting people first* advocating for community-oriented models of practice. These include mainstream and ethno-specific services working together and developing strong linkages.

For older people without health insurance, the high costs of treatment relative to income can mean lengthy delays in treatment on public hospital waiting lists and reduced medical options. The established link between poor health outcomes and poverty requires policies targeted specifically at addressing the health and economic needs of low-income earners. A study of older people in China found that high health expenditure was a major cause of poverty for those living in rural areas (Liu, 1998). Efforts are required to improve health-care system infrastructure, accessibility and affordability, and suitability, such as provision of mobile devices for loan.

The Australian health workforce faces both challenges and opportunities as major changes are occurring in the structure and financing of services as well as new systems for performance reporting and accountability including an expansion of e-health (AIHW, 2012). Heath Workforce Australia (HWA) was established in 2010 as an initiative of the Council of Australian Governments (COAG) to address the challenges of providing a skilled health workforce that meets the increased projected demands into the future. The National Health Workforce Innovation and Reform Strategic Framework for Action 2011–2015 is a new approach to workforce development across the health and education sectors (HWA, 2011). A social determinants model of population health and well-being views health broadly to include nutrition, housing, education, employment, family and community stability, and safety. This approach necessarily broadens the definition of health to include a range of organizations and services not previously considered to be under the umbrella of health. A focus is on wellness, prevention, and primary care and balancing the priorities of consumer and community needs with cost efficiency. It is about reconfiguring the health workforce and education in Australia to prepare and support this workforce.

Models of education traditionally provided by universities are not considered appropriate to meeting future health workforce demands in Australia. This claim is based on research conducted by Frank et al. (2010) of 20 international educational and professional leaders that identified numerous areas where professional education failed to meet health challenges. Brownie et al. (2011a) argue that there is a disjuncture between professional education and practice models with competencies not considered to be well matched to changing population needs. This research finding is attributed to the tendency of various professions to act in isolation and in competition with each other. The authors argue that shared competencies across health systems are crucial for meeting future challenges. Interprofessional education is considered essential.

The path to the successful implementation of collaborative care involves interprofessional education, which is based on the clear articulation of the competencies that are essential to effective teamwork and the delivery of health care (Brownie et al., 2011b, p. 22). We argue this includes collaboration between the social and applied sciences and business IS.

22.4 ICT for Older People

The mobile device applications on offer these days provide a generic approach to their usability. The main criticism here is that the older user cannot adjust their delivery and/or input mode (Taylor and Rose, 2005); one exception would be the ability to alter text and graphical image size; however, doing this is often clumsy and more complicated than it should be. It is a case of "what you see is what you get" otherwise known by technology developers as the WYSIWYG (pronounced wiz-ee-wig) syndrome. Perhaps the reason for this oversight is that until recently, there has been a strong commitment by IS designers to build ubiquitous applications. Personas and scenarios assist in tailoring the design and application process to the needs of the user population.

22.4.1 Personas and Scenarios

A persona is a hypothetical construct that embodies the main features of the user population. The persona assists in understanding information communication technology needs of older people to informing design and accessibility and ultimately suitability. However, it must not become a replacement for active user involvement. For the purpose of this chapter, we have focused on goal-directed design by creating fictional personas (Cooper, 2004; Martin and McKay, 2010). The personas form the basis for creating activity scenarios, short stories, or narratives to provide greater detail about older people and their particular needs, interests, and concerns.

An analysis of the social contextual factors identified in Table 22.1 assists in scenario development. Both Nick and Brenda live in the community, Nick with his wife and Brenda on her own. Even though Brenda lives alone, she is in better health than Nick and has a more extensive social support network. As Nick's dementia progresses, he will need increased assistance in the home with support for his wife also as primary carer. He may revert back to his first language and require service providers who are familiar with his language and cultural needs as discussed earlier. Nick can benefit from mobile computing devices to assist him to remain mentally and socially active as well reminders for medication in his preferred language. Adjustable font size, voice activation, and talking applications all assist older people with vision and hearing loss (Czaja and Lee, 2008). Education and training in use of social networking sites and the Internet can assist Nick to remain mentally and socially active; the earlier this is provided, the better, given his potential cognitive decline and decreased motivation and ability to learn new things. Cost is an issue for Nick as a pensioner without accumulated funds to support him and his wife. Brenda could potentially be invited to run education and training sessions for older people such

TABLE 22.1

Personas and Scenarios of Older People

Persona: Nick	Persona: Brenda
76 years old	93 years old
Male	Female
English speaking: fourth language	English speaking
Retired laborer	Retired school principal
Married	Single
Migrated to Australia as a refugee from Bosnia 9 years ago	Born in Australia
No close family in Australia apart from his wife and a few friends	Oldest of three of eight surviving siblings, with numerous nieces and nephews and offspring
Lives with his wife in a flat	Lives alone in an apartment
Aged pension	Superannuation
Diagnosed with dementia 1 year ago	In good health, high blood pressure, successful back surgery 10 years ago
Takes medication daily	Takes medication daily
Attends day program at local community center	Plays bridge and mahjong and attends church each week, attends concerts, visits family regularly, family historian
Comfortable using basic functions on a computer such as e-mail and word processing	Advanced computer skills
Internet use for basic searches via computers at local library; computer at home without Internet	Regular user of social networking sites mostly Facebook and Internet—home computer with Internet

as Nick as to the benefits of social networking and Internet use. Brenda may benefit from mobile devices if her health declines but is not in need of them currently. She is able to afford services now yet this may change as she ages and her savings dwindle. Security is an issue for both Nick and Brenda who will both potentially require affordable personal and home mobile electronic security devices. Further possibilities relevant to both Nick and Brenda are discussed in the remainder of this chapter.

Given the aging dilemma and the increased dependence the general community has for receiving most of their information via the Internet, older people will also increasingly rely upon ICT tools to enhance their quality of life and well-being (Melander-Wilkman, 2008). Researchers have been trying to understand the general impact of ICT for an aging population (Steyaert et al., 2006). Since the proliferation of digital technologies around the world, awareness is growing of the possibility of shaping of the technology, to the needs of specific user populations rather than the IS design being the sole purview of the computer scientists. Yet in terms of social theory, technology has been seen as an *off-stage phenomenon* and not taken too seriously (Howcroft et al., 2004). Because more and more people in the community participate in various online activities using a variety of mobile technologies, the IS designers need to engage with the social scientists and older people themselves to ensure the IS needs of older people such as Nick and Brenda are met.

Figure 22.2 explains the relationships between the special need context of the older people and the HCI framework involved in facilitating mobile health (mHealth) programs. Over the past decade, there have been considerable changes and developments in the use of technology. Within the IS design community, there is an emerging awareness for addressing the health and well-being of older people. For instance, the Usable Technology for Older People: Inclusive and Appropriate (UTOPIA) project was funded by the Scottish

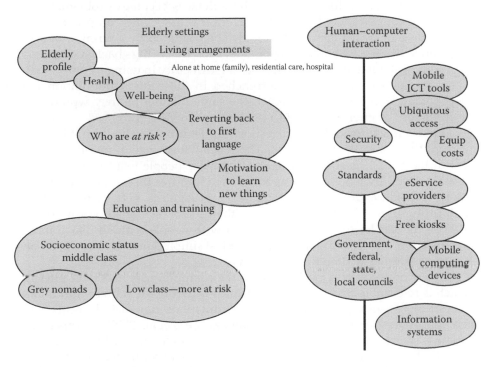

FIGURE 22.2
Well-being and ICT nexus for older people.

Higher Education in collaboration with the Universities of Glasgow, Napier, and Abertay Dundee. UTOPIA was focused on the helping industry tackle the challenges of an increasingly elderly user population. However, the project's main aim was to educate industry IS developers on how to gather the needs and wants of older users (Newell, 2002).

Consider the pace at which mobile technologies are entering the HCI arena. We know that all around the world, people who are wealthy enough and have interests in mobile information and communications technology (mICT) tools are adopting applications at a rapid rate. These mICT tools include phones, laptops, iPads, and tablet computers, which now drive a new wave of connectivity not seen before in human history. Yet this is where the whole technology-driven vista becomes problematic! Over two decades ago, warning bells rung out (Dreyfus and Dreyfus, 1986) that still reverberate today. The worry was then that the next generation would defer many of their usual life skills to this type of computational machinery. This warning reverberates today (McKay, 2008); we must ensure older people are included in the emerging social connectivity afforded by mICT tools (Czaja and Lee, 2008).

These days, many people simply rely on intuition to make a computer work (McKay, 2008); they access mobile devices with relative ease. For others however, it can be an uncomfortable ride, with unseen difficulties lurking everywhere. For those of us who are more familiar with computers, all we really need to do is switch it on and get going. Nevertheless for some, using a computer or mobile device remains a difficult aspiration, especially for the elderly, as there may be other things that affect their HCI, like not being able to buy one or like knowing how to connect a computer to the Internet. Generally speaking, these days, computers are easy to use. It is not necessary to know why a computer follows a particular booting protocol before it can work for us. All the same, the reluctance for an inexperienced user toward adopting mICT tools is understandable. One explanation for why using these devices becomes problematic for older people may be to suggest there has been an event that associates computer use as being unpleasant. These awkward memories can also be triggered by a number of seemingly unrelated events, like seeing something on television that brings back such memories or even just simply watching another person having difficulty in using a computer (McKay, 2008). Computer phobia can become entrenched in the general population requiring expert psychological intervention in much the same manner as they would be treated for other anxieties (Wagner et al., 2010).

The HCI environment for older people shown in Figure 22.3 should not be very much different from the rest of the population. There is a wide range of mobile ICT tools that is too vast to list here. It is sufficient to say that older people want to tap into them all (Steyaert et al., 2006). They do see ICT as a means to improve their quality of life (Steyaert et al., 2006), for instance, providing faster access for diagnosis, better treatment, monitoring, and alarming from a distance; supporting their social relationships with friends who may not be as mobile; leisure, connecting them with people with similar interests; and telework, education, and learning new things (or just simply a means to brush up on something they were familiar with in times gone by).

In the past, a connection was made between the computer user and their physical well-being, which is relevant to this discussion when considering the human limitations experienced in our more senior years. Some of these limitations may include levels of hearing, arm reach, muscular strength, and visual distance (Te'eni et al., 2007). To this end, mobile devices such as smartphones and tablet computers offer older people lightweight usability alternatives to the cumbersome fixed-line phones, personal computer (pc), and laptops.

Consideration for decreasing functional ability is relevant to IS design, in so far as the changing nature of our sensory/perceptual processes, motor abilities, response speed, and

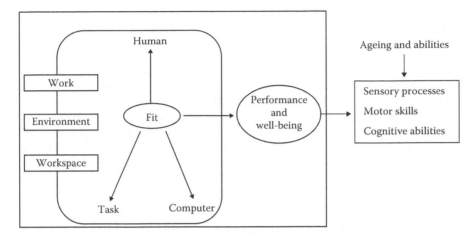

FIGURE 22.3
Inclusive HCI for participation of older people. (Adapted from Te'eni, D. et al., Physical engineering, In *Human Computer Interaction*, Wiley, Hoboken, NJ, pp. 67–84, 2007.)

cognitive processes (Czaja and Lee, 2008). Growing older, our eyesight changes often making it more difficult to perceive and comprehend visual information (consider the differences between our personas Nick and Brenda mentioned earlier). Mobile devices offering synthetic speech may be characterized by a particular digital distortion, making the comprehension more difficult for older people. IS designers therefore need to allow for people with declining eyesight and hearing faculties.

Ubiquitous access has grown to mean efficient and effective HCI between people and their computing devices (Preece, 2005; Salber et al., 1998). To ensure that older people are included means that attention is given to equipment design and cost and security and standards of eService provision. Care and attention to accommodate age-related changes in people's functions simply mean that consideration be given to display screens, placement, size, shape, and labeling of controls (Czaja and Lee, 2008), with special consideration also given to design and layout of the instructional manuals.

Federal and state government entities, including local councils, have the ability to provide older people free kiosks and mobile computing devices for loan through community programs that are usually based in public/municipal libraries (http://www.library.sunshinecoast.qld.gov.au/).

22.4.2 Impact on ICT Tool Development through Increased Complexity of Health Needs

As mentioned in Section 22.2.1, the complexity of older peoples' health needs means that eventually, there will be growing numbers with a greater incidence of chronic and degenerative diseases (Frenk et al., 1989) in Kinsella et al. (2001). The adoption of ICT tools for elder health care is slowly taking place (Newell, 2002). However, most of the online programs or software applications currently address a narrow range of mHealth categories. They provide health and welfare information/advice without providing the full range of ordinary daily activities that would connect older people with their broader needs to support social contact with family and friends (Eggermont and Vandebosch, 2009).

22.4.3 Impact on ICT Tool Development for Older Australians from Culturally Diverse Backgrounds

As people grow older, they often express a preference for services that are provided in their own language and that are culturally appropriate (Czaja and Lee, 2008). The media links socioeconomic status and technology more often referring to the general population as a whole and more specifically related to disparity in technology use and its impact on academic/industrial sector performances (Sun and Benton, 2008). Instead, information technology (IT) should be treated as an infrastructure commodity. For instance, we take no notice of the community utilities provided by our government/local councils like streetlights and water/sewer services and broadband/wireless networks (Nelson, 2006). It makes sense to consider the social and psychological aspects of computerization (Bradley, 2006), showing there is a convergence between human needs and technology applications (Table 22.2).

In the kitchen, there are many mobile devices that enable older people to maintain their connectivity with others while they engage in their daily activity. For instance, there are online ordering companies where shopping lists can be sent directly to the store, with same day delivery.

Household management and maintenance activities that lend themselves to mobile technologies involve washing machines and garden watering devices; older people who prefer to remain in their own homes appreciate the opportunity to maintain garden and service utilities (water, gas, and electricity).

Personal caring activities are perhaps the best examples of present-day mHealth facilitation. We are now seeing mobile technologies that improve the quality of older people's lives to educate themselves and the people who look after them (Heywood-Smith, 2013). People wish to speak directly with their doctors in person (Eggermont and Vandebosch, 2009);

TABLE 22.2

ICT Activities in the Home

In the Kitchen	Entertainment	Personal Care
Cooking	Watching TV and movies	Sleeping/resting
Dining	Playing	Showering
Shopping	Internet browsing	Beauty treatment
	Reading	
Maintenance	**Communication and Information**	**Working Hobbies**
Cleaning	Learning	Intellectual
Laundering	Socializing	Artistic
Recycling/managing	Information acquiring (news)	Music
Waste	Phoning	
Storing	eCommunication	
Supplying water and power		
Caring	**Household Management**	**Love**
Treating sick persons	Paying bills	
Care of the elderly	Handling mail	
Childcare	Security	
Caring for pets		

Source: Adapted from Bradley, G., *Social and Community Informatics: Humans on the Net*, Routledge, Taylor & Francis Group, New York, 2006.

only these technologies must be easy to use and keep the possibility of their physical limitations in mind, for instance, enhanced sound for mobile phone devices and visual displays that are easy to read and comprehend.

Along with health and well-being, entertainment serves as an important feature of mHealth for older people. One of the current challenges facing us is to know how best to develop input devices and interface design, such as speech recognition and help menus that accommodate older people's needs (Czaja and Lee, 2008).

Searching for health information involves knowing how to get the best out of the mobile communication and information devices. Staying in touch is vital for older people. Research has shown that older adults are receptive to using e-mail to connect with friends and family, reducing the isolation that is often experienced by the elderly (Eggermont and Vandebosch, 2009).

Since our adoption of mobile devices around the home, the division between working and hobbies has been blurred. The plethora of handheld devices means that older people can learn how to keep doing the things they like, for instance, reading the news online instead of walking down to the store to buy the newspaper. Similarly, the *grey nomads* keep track of job advertisements as they roam from place to place. Often, these people take on itinerate positions that others in the community pass up.

Finally, the Bradley (2006) model identifies *love* as a necessary aspect of life; we too choose to identify this with elder lifestyle. Love in terms of mHealth for us means keeping the zest for wellness and lifestyle, with mobile devices that afford our elders with high levels of community connectivity and participation.

22.5 Mobile Applications for Older People

It is commonly known by IS designers that the *user experience* is foremost in their website planning, while purchase and payment are regarded as secondary importance (Nielsen and Norman, 2000). The underlying premise has been to make people happy about their online experiences and they will return as happy customers and keep paying for their online products (Dancrai, 2013). Unfortunately, this popular marketing dogma does not apply to the older customers who may not have the ability to pay.

The mHealth ICT tool selection framework for this chapter involved locating available iPad applications, which is specifically related to older people, then conducting a usability features evaluation. Categories identified in the first instance included promoting a safe environment for social networking, medical, health and fitness, and lifestyle and sourcing aged care products. The usability features were generally disappointing, with very little interaction possible between the iPad application and the user. These usability features are discussed next.

22.5.1 mICT Tools That Promote a Safe Environment for Social Networking

The trouble with the mobile computing environment is that it has grown exponentially (Sears and Jacko, 2008). Never before has humankind experienced such rapid change. We have not really had enough time to step back and reflect on the ripple effect of the digital medium per se, let alone understand the deleterious effects upon our social interaction. Patricia Wallace (1999) published the first recognition that there were serious problems with people's online behavior. Unfortunately, while there are some wonderful connections people can make with other people through the Internet, the converse is also true. Patricia discussed the full range of bad behavior; humans act differently online than when they are face to face (Wallace, 1999).

However, it is too easy to make friends online, trapping an unwary person into their revealing private information to an unknown source. Understandably, some older people may turn to their mobile devices because they are lonely. Social networking sites encourage such immediate membership. Yet research reveals that at times, too much information given to a mobile device is not a good idea. For instance, the relationship between people's social anxiety and the interactional fidelity of virtual humans (Kang and Gratch, 2012) is pertinent to an unwary older person wishing to connect with other people online. Kang and Gratch (2010) focused on the issue of nonverbal feedback (the kind we are accustomed to receiving in face-to-face interaction). They investigated across three experimental conditions where participants interacted with real human videos and virtual humans in computer-mediated interview interactions. Their results revealed that socially anxious people revealed more information and greater intimate information about themselves when interacting with a virtual human when compared with real human video interaction, whereas less socially anxious people did not show this difference (Kang and Gratch, 2010).

To safeguard our older people in such circumstances, IS design must include ethical rules and regulations that ensure the social networking environment is safe. This protection emanates from the heuristics for user interface design, the *rules of thumb* IS designers use to plan the digital products. The problem is that this set of usability guidelines assumes a generic population profile. Sadly, while there is an emerging awareness among the web design community for addressing health and well-being needs of our older citizens, there is no account for the changing nature of aged-related human cognition.

There is one free iPad application that falls within the category of aged care standards and accreditation. It is called the *Results and processes guide*; it presents the Aged Care Act 1997 standards: management systems, staffing, and organizational development; health and personal care (residents' physical and mental health); residential lifestyle (to achieve active control of our own lives within the residential care service and in the community); physical environment and safe systems (for safe and comfortable environs to ensure quality of life and welfare of residents, staff, and visitors for Sydney, Melbourne, Adelaide, Brisbane, Perth, and Divisional offices).

According to Czaja and Lee (2008), our sensory processes alter as we age. The most obvious is our visual ability that may affect older people's mobile computing activities. With increased age, some older people's visual perception may become troublesome; we lose contrast (and color) sensitivity with a decreased awareness for the blue region. Another feature of aging and computer usage relates to our heightened sensitivity to glare. Moreover, Kline and Schieber (1985) were able to show that older people have problems with their searching skills and the ability to detect visually embedded objects (Czaja and Lee, 2008; Kline and Schieber, 1985). The problem is that mobile devices often rely upon textual messages. Given the aversion older people may have toward dealing with glary screens, the management of the mobile's visual resources should be kept simple.

The social networking application for the aged called *Keep-in-Touch* (Fares, 2013) asks for personal location. Yet, the security of such an application seems a little suspect as there is no way of telling how secure the information an aged user (or their family members) provides. There are two free aged care iPad medical category applications. The *Aged Care Locator* (Ask.com, 2013) appears to provide the following: locate suburb, find me, and show distance (10–100 km). The *BEAT-D (Build Environment)* (UoW, 2013) offers dementia training study centers in most Australian states. Initiated by the University of Wollongong and Guardian Software Development partnership, this iPad application is designed to help assess the suitability of a build environment for the purpose of accommodating people with dementia. It has been developed by the NSW/ACT Dementia Training Study Centre, a center supported by the Australian Department of Health and Aging. There is no fee for this application or for the services associated with it, while the *Geriatric Glossary* application costs $0.99.

22.5.2 Availability of mICT Applications for Language Translation

On the one hand, at this point, apart from the Google translation service, there are no translation services that run on mobile computing devices. Google does, however, offer a free statistical multilingual translation facility that converts textual documents from one language into another. On the other hand, there are companies that provide translation at a cost (http://www.acrosstranslations.com/blog). This service provides many country to country translations into and from places like traditional Chinese and simplified Chinese as well as the countries listed in the following table.

Arabic	Bosnian	Croatian
Danish	Dutch	English
Farsi	French	German
Greek	Indonesian	Italian
Japanese	Korean	Macedonian
Portuguese	Russian	Serbian
Spanish	Thai	Vietnamese

22.5.3 mICT Applications for the Aged That Include Medical, Health and Fitness, and Lifestyle

There are a number of applications that are available to mobile devices. The Australian government offers access to a comprehensive set of community services that include Centrelink, Child Support, and Medicare (DHS, 2013). While these services are impressive, there are no official government websites designated toward older people.

Are You Ok (Miura, 2013) is a free iPad utilities category application for the aged, which provides voice response and a simple message sending facility (I'm out; I'm busy; and I need help); the first impression of this iPad utility is that it may be designed for the Japanese culture. The *Retirement Fund Calculator* (ACMA, 2013) offers an iPad application utility category costing $0.99.

There are two free health and fitness iPad applications designed for the aged community. The first is the *Wellness & Lifestyles Services* (Heywood-Smith, 2013). It says it is bringing health care to you and lists 12 items shown as ADLs, BEHs, and CHC categories: with choice that form 'current daily rate with a $ value; new daily rate; and increase per annum. An initial impression of this application is that it seems to be related to health-related training programs for aged care. The second health and fitness iPad application is called *Smart Ageing*; it is offered by the BH Institute Mobile Learning Portal and appears to be available for members only. There are four lifestyle free iPad applications that require Wi-Fi technology to download, each one involving bible verses on one main element of the human life cycle: *Raise me up HD*, KenMac Technology; *LifeCycle: Born_HD*; *LifeCycle: Sickness_HD*; and *LifeCycle: Rest_HD*.

22.5.4 mICT Tools That Source Aged Care Products

There is one free iPad application in the reference category for the aged; *Evadale Healthcare* (AppsUnloaded, 2013) provides a comprehensive range of home improvement devices: fall prevention (doormats, touch lamps, night-lights, retractable hose reel, clothesline, visual and hearing impaired phones), bathroom (toilet aids, shower and bath seats, etc.), home improvements (wireless/portable doorbell), mobility aids (wheelchairs, ramps, shopping and walking frames, oxygen bottle carrier, etc.), aids for daily living (long shoe horn), beds and pressure care (community bed packages, bed support rails, overbed table, safety crash mat, etc.).

22.6 Discussion and Conclusion

In this chapter, we have examined the literature concerning an aging population to identify the impact of ICT on their health and welfare needs for quality mHealth services. There are two distinct HCI streams that are involved: the medical dimension, where the full range of health disorders may be addressed by ICT tools for information provision, and the recognition for the practice of mHealth service provision through public access programs that provide educative applications for managing health and welfare of the general community. More specialized ICT tools remain within the relative security of the hospital/medical environment, largely to provide patients' record keeping. These include telemedicine, electronic transmission of x-rays and test results, and file sharing between medical practitioners, hospital nursing staff, and administration. We have also identified a range of ICT services available to assist people as they age, including the use of online translating services. Because of the multidisciplinary context involved in such online health-care provision, where medical, psychology, social welfare, and IS development combine, we have used a social cognitive lens when exploring the HCI needs of older people.

Personas and scenarios have been used to explore the mHealth needs of older people and to assist in profiling user needs and the potential of ICT service provision, to meet these needs. It is inevitable that older people will have very different usage patterns, for instance, when compared with younger people, differences will occur between older people. These differences will come about because of peoples' changing physiology and cognitive performance issues, even when they remain healthy. The most common ICT usage for older adults includes communication and social support, leisure and entertainment, information seeking health, information seeking, education and training, and productivity (Wagner et al., 2010, p. 873).

System developers are required to be mindful that older people will want to engage with ICT tools that are appropriate for their specific needs. IS designers must comprehend the interactive relationships between the special need context of the older people and the HCI framework that is involved in facilitating the mHealth systems. To understand this context, it is essential that accurate personas and scenarios are developed to ensure quality and relevance of these services.

Questions

1. What are the future challenges to IS design strategies for the development of ubiquitous mHealth services for older people?

2. What strategies would be useful to engage older people in the development of appropriate mHealth service development?

3. Should there be an mHealth standards commission led by government agencies' initiatives or the industry sector?

4. How can we maintain quality control and relevance for mHealth services for older people?

5. How can issues of access and equity be addressed so that mHealth technologies are available and affordable for all older Australians?

6. What ethical issues and dilemmas may arise with increased use of mHealth services for older people?

References

Abraham, N. (2012). *What Are the Barriers for Somali and Assyrian Chaldean Communities in Accessing Mainstream Health and Welfare Services?* RMIT University, Melbourne, Victoria, Australia.

Abraham, N. and Martin, J. (2014). Cultural safety with new and emerging communities. In H. K. Ling, J. Martin, and R. Ow (eds.), *Cross Cultural Social Work.* pp. 151–165, Palgrave Macmillan: Melbourne, Victoria, Australia.

ABS. (2009). Australian Bureau of Statistics. *National Survey of Mental Health and Wellbeing: Summary of Results, 2007, ABS cat. no. 4326.0.*, ABS: Canberra, Australian Capital Territory, Australia.

ACA. (2010). Aged Care Australia. *Help Staying at Home: Services for Culturally and Linguistically Diverse People*, Department of Health and Ageing, Commonwealth of Australia: Canberra, Australian Capital Territory, Commonwealth of Australia.

ACMA. (2013). ACFI calculator. (Electronic version), from http://acma.net.au/acfi-calculator/. Accessed on June 28, 2014.

AIHW. (2011). Australian Institute of Health and Welfare. *Australia's welfare 2011. Australia's welfare series no. 10. Cat. no. AUS 142.* ACT: AIHW.

AIHW. (2012). Australian Institute of Health and Welfare, (2012). *Australia's health series, No:13, Cat. No: AUS 156.* ACT: AIHW.

AppsUnloaded. (2013). Evadale healthcare. [Electronic version], from http://itunes.apple.com/vg/app/evadale-healthcare/id612275524?mt=8.

Ask.com. (2013). Aged care locator app. [Electronic version], from http://au.ask.com/web?q=aged+care+home+blackburn&qsrc=178&o=7101&l=sem.

Barry, P. (2003). *Mental Health and Mental Illness*, Lippincott: Philadelphia, PA.

Bradley, G. (2006). *Social and Community Informatics: Humans on the Net.* Routledge, Taylor & Francis Group: New York.

Brownie, S., Bahnisch, M., and Thomas, J. (2011a). *Competency Based Education and Competency Based Career Frameworks: Informing Australia Health Workforce Development*, University of Queensland & HWA: Adelaide, South Australia, Australia.

Brownie, S., Bahnisch, M., and Thomas, J. (2011b). *Exploring the Literature: Competency Based Education and Training and Competency Based Career Frameworks*, University of Queensland & HWA: Adelaide, South Australia, Australia.

Butler, M. (2012). *Health Workforce Insights*, Issue No. 4, Adelaide: Health Workforce Australia.

Chong, A. M. (2007). Promoting the psychosocial health of the elderly—The role of social workers. *Social Work in Health Care*, 44(1–2), 91–109.

Cooper, A. (2004). *The Inmates Are Running the Asylum: Why High Tech Products Drive Us Crazy and How to Restore the Sanity*, 2nd edn., Pearson Higher Education: Indianapolis, U.K.

Cormack, M. (2012). *Health Workforce Insights*, Issue No. 4, Adelaide: Health Workforce Australia.

Craig, G. (2008). Palliative care in overdrive: Patients in danger. *American Journal of Hospice and Palliative Care*, 25(2), 139–145.

Czaja, S. J. and Lee, C. C. (2008). Information technology and older adults. In A. Sears and J. A. Jacko (eds.), *The Human-Computer Interaction Handbook: Fundamentals, Evolving Technologies and Emerging Applications*, 2nd edn., Vol. 1, pp. 777–792. Taylor & Francis Group: New York.

Dancrai. (2013). 15 Years of providing IT support and service to SME, Corporate and Government clients. [Electronic version], from http://www.dancrai.com.au/. Accessed on July 25, 2014.

DHS. (2013). Express plus mobile apps. [Electronic version], from http://www.huamservices.gov.au/customer/services/express-plus-mobile-apps. Accessed on July 25, 2014.

DOHA, Department of Health and Ageing. (2011). *General Practice Programs to Improve Access and Outcomes in Mental Health*. ACT: Commonwealth of Australia. Canberra.

Dominelli, L. (2013). Mind the gap: Built infrastructures, sustainable caring relations, and resilient communities in extreme weather events. *Australian Social Work*, 66(2), 204–217.

Dreyfus, H. L., and Dreyfus, S. E. (1986). *Mind over Machine: The Power of Human Intuition and Expertise in the Era of the Computer*. Free Press: New York.

Eggermont, S. and Vandebosch, H. (2009). Towards the desired future of Elderly and ICT. *Annual Meeting of the International Communication Association*. Sheraton, New York.

Erikson, E. H. (1980). *Identity and Life Cycle*, W.W. Norton: New York.

Fares, J. (2013). Keep-in-touch. [Electronic version], from http://www.keepitouch.net.au/main.html. Accessed on July 25, 2014.

Frank, J., Chen, Z., Bhutta, J., Cohen, N., and Crisp, T. (2010). Health professionals for a new century: Transforming education to strengthen health systems in an interdependent world. *Lancet*, 376, 1923–1958.

Frenk, J., Frejka, T., Bobadilla, J. L., Stern, C., Sepulveda, J., and Jose, M. (1989). The epidemiological transition in Latin America. Paper presented at the *International Population Conference*, New Delhi, India.

Heywood-Smith, N. (2013). About wellness & lifestyles. [Electronic version], from http://www.wellnesslifestyles.com.au/why-us/about-us/. Accessed on July 25, 2014.

Howcroft, D., Mitev, N., and Wilson, A. (2004). What we may learn from the social shaping of technology approach. In J. Mingers and L. Willcocks (eds.), *Social Theory and Philosophy for Information Systems*, pp. 329–371. John Wiley & Sons Ltd: West Sussex, U.K.

HWA. (2011). Health Workforce Australia. *National Health Workforce Innovation and Reform Strategic Framework for Action 2011–2015*. Commonwealth of Australia: Adelaide, South Australia, Australia.

Kang, S.-H. and Gratch, J. (2010). Virtual humans elicit socially anxious interactants' verbal self-disclosure. *Computer Animation and Virtual Worlds*, 21, 473–482.

Kang, S.-H. and Gratch, J. (2012). Socially anxious people reveal more personal information with virtual counselors that talk about themselves using intimate human back stories. [Electronic version], from http://www.sinhwakang.net/CYBERTHERAPY2012_sKang.pdf

Kinsella, K., Velkoff, K., and Velkoff, V. A. (2001). *An Aging World: 2001*. US Government Printing Office: Washington, DC.

Kline, D. W. and Schieber, F. J. (1985). Vision and aging. In J. E. Birren and K. W. Schaie (eds.), *Handbook of the Psychology and Aging*, pp. 296–331. New York: Van Nostrand Reinhold.

Lee, A. S. (2004). Thinking about social theory and philosophy for information systems. In J. Mingers and L. Willcocks (eds.), *Social Theory and Philosophy for Information Systems*, pp. 1–26. Wiley: West Sussex, U.K.

Liu, Y., Hu, S., Fu, W., and Hsaio, W.C. (1998). Is community financing necessary and feasible for rural China? In M.L Barer, T.E. Getzen, and G.L. Stoddart (eds.), *Health, Health Care and Health Economics: Perspectives on Distribution*. New York: Wiley.

Martin, J. (2012). *Mental Health Social Work*. Ginninderra Press: Adelaide, South Australia, Australia.

Martin, J. (2013). Building a culturally diverse and response aged care workforce. In H.K. Ling, J. Martin and R. Ow (eds.), *Cross Cultural Social Work: Local and Global*. Palgrave Macmillan: Melbourne, Victoria, Australia.

Martin, J. and McKay, E. (2010). Developing Information Communication Technologies for the Human Services: Mental Health and Employment. In Martin, J. & Hawkins, L. *Information Communication Technologies for Human Services Education and Delivery*. New York: IGI Global.

McKay, E. (2008). *The Human-Dimensions of Human-Computer Interaction: Balancing the HCI Equation* 1st edn. Vol. 3. IOS Press: Amsterdam, Netherlands.

Melander-Wilkman, A. (2008). *Ageing well: mobile ICT as a tool for empowerment of elderly people in home heal care and rehabilitation*. Luleå University of Technology Luleå.

Min, H. (2013), Law for elderly creates a legal conundrum. *Shanghai Daily*, Monday, July 8, p. 4.

Miura, E. (2013). Are you ok? [Electronic version], from http://itunes.apple.com/au/app/are-you-ok/id569772749?mt=8.

Nelson, M. (2006). Technology and socio-economic status. [Electronic version], from http://electmikenelson.blogspot.com.au/2006/03/technology-and-socio-economic-status.html

Newell, A. F. (2002). HCI and older people. [Electronic version], from http://www.dcs.gla.ac.uk/utopia/workshop/newell.pdf.

Nielsen, J. and Norman, D. A. (2000). Web-site usability: Usability on the Web isn't a luxury. [Electronic version], from http://www.informationweek.com/773/web.htm.

Powell, J. (2007). Promoting older people's voices—The contribution of social work to inter-disciplinary research. *Social Work in Health Care*, 44(1–2), 111–126.

Preece, J. (2005). Online communities: Design, theory, and practice. *Journal of Computer-Mediated Communication*, 10(4), Article 1.

Salber, D., Dey, A. K., and Abowd, G. D. (1998). *Ubiquitous Computing: Defining an HCI Research Agenda for an Emerging Interaction Paradigm*. GVU Center & College of Computing, Georgia Institute of Technology: Atlanta, GA https://smartech.gatech.edu/bitstream/handle/1853/3438/98–01.pdf.

Sears, A. and Jacko, J. A. (2008). Future trends in human-computer interaction. In A. Sears and J. A. Jacko (Eds.), *The Human-Computer Interaction Handbook: Fundamentals, Evolving Technologies and Emerging Applications* (2nd ed., Vol. 1, pp. 1281–1290). New York: Taylor & Francis Group.

Steyaert, S., Eggermont, S., and Vandebosch, H. (2006). Towards the desired future of the elderly and ICT: Policy recommendations based on a dialogue with senior citizens, *Second International Seveill Seminar on Future-Oriented Technology Analysis: Impact of FTA Approaches on Policy and Decision-Making*. Seville, Spain.

Sun, C. Y. and Benton, D. (2008). The socioeconomic disparity in technology use and its impact on academic performance. Paper presented at the *Society for Information Technology & Teacher Education International Conference*. Chesapeake, VA.

Taylor, T. and Rose, J. (2005) Bridging the divide: Older learners and new technologies, AVTEC conference, available at http://pdf.aminer.org/000/370/053/women_and_technology_the_next_step_in_bridging_the_divide.pdf.

Te'eni, D., Carey, J., and Zhang, P. (2007). Physical engineering. In *Human Computer Interaction*: Developing effective organization Systems, pp. 67–84. Wiley: Hoboken, NJ.

UoW. (2013). BEAT-D (Built environment assessment tool: Dementia). [Electronic version], from http://itunes.apple.com/au/app/beat-d-build-environment-assessment/id527112326?mt=8.

Wagner, N., Hassanein, K., and Head, M. (2010). Computer use by older adults: A multi-disciplinary review. [Electronic version], 26, from http://www.business.mcmaster.ca/IS/head/Articles/Computer%20use%20by%20older%20adults.pdf.

Wallace, P. (1999). *The Psychology of the Internet*. Cambridge University Press: Cambridge, U.K.

White, C. (2013). Sedative prescribing among older people doubles with move into residential care. *British Medical Journal*, 3(346), f1272.

WHO. (2012). *Ageing and Life Course*. WHO Press: Geneva, Switzerland.

Section VII

mHealth Regional, Geographical, and Public Health Perspectives

23

Current Situation and Challenges for mHealth in the Latin America Region

Ana Lilia González, Alejandro Galaviz-Mosqueda, Salvador Villarreal-Reyes, Roberto Magana, Raúl Rivera, Manuel Casillas, and Luis Villasenor

CONTENTS

Over the last 60 years, the epidemiological profile (EP) in Latin America and the Caribbean has shifted from communicable diseases toward chronic and degenerative diseases. This shift, altogether with the unbalanced distribution of physicians within Latin American countries, has posed new challenges to the health-care systems (HCSs) in the region. Several of these challenges could be addressed by properly implementing mHealth services such as teleconsulting, telemonitoring, and raising awareness. Considering the coverage and penetration of cellular networks within Latin America and the Caribbean, the deployment of these types of mHealth services is technically feasible. Consequently, mHealth has the potential to offer new and novel paradigms in health-care delivery, which in turn can help Latin American HCS to cope with the new challenges they are facing. However, in order to fully integrate the mHealth services within Latin American HCS, several challenges related to legislation, technology, and users/physicians acceptance must be solved. In this chapter, an overview of the current situation and challenges of mHealth projects in the Latin America and the Caribbean region is presented. Furthermore, the potential that mHealth has to offer for the countries within this region is exemplified by means of the Mexican case analysis.

23.1 Introduction

The region from Río Bravo (México) to Tierra del Fuego (Argentina and Chile) including the Caribbean Islands is called Latin America. This region comprises three zones: *Central America* from Mexico to Panama; *South America*, beginning in Colombia and covering the southern region of the American continent; and the *Caribbean*, which comprises the set of islands located southeast of the Gulf of Mexico and U.S. mainland, east of Central America, and north of South America. The countries in this region share similar historical and cultural contexts, and in general, the area is characterized by a set of emerging economies where economic and social inequities prevail (World Bank 2013).

An important trend in the region is the migration of rural population to urban centers, which has given rise to the tendency of *megacities* (ONU-Habitat 2012). Rural population migrates to urban centers looking for better access to employment, transportation, health care, and other services. However, an unplanned urbanization can exacerbate inequalities among the population due to scarceness and uneven distribution of basic services (ONU-Habitat 2012). This is reflected in the access to health-care services, since typically a higher proportion of health personnel is located within wealthy urban areas, leading to an unbalanced geographical distribution of physicians (Dussault and Franceschini 2006).

Another important trend in the region is that life expectancy (LE) has been increasing over the last years. This is a consequence of local and international efforts focused on providing access to a minimum health-care level, safer water supplies, and sanitation facilities, among others (WHO/UNICEF Joint Water Supply and Sanitation Monitoring Programme 2005). Although LE has been increasing in Latin America, this does not necessarily imply that the healthy life expectancy (HALE) follows the same trend (Salomon et al. 2012), as the prevalence of health problems such as obesity and diabetes has been rising in the region (Webber et al. 2012). Consequently, the Latin American EP already shows a shift toward a major incidence of chronic degenerative diseases among the population. Therefore, treatment and prevention of these diseases are now the major concern for the HCSs in Latin America.

The unbalanced physician distribution and the EP changes toward chronic degenerative diseases pose the major challenges to the HCS in Latin America. While the unbalanced physician distribution causes an increase on the total cost of the HCS, the shift toward chronic degenerative diseases has caused an increase in the demand of health-care services for long-term treatments. Additionally, the need for more and better health-care services in rural areas is still a concern in Latin America (ONU-Habitat 2012). Therefore, there is a necessity in the region for the provision of health-care services by means of different and novel paradigms. In this sense, several of the problems faced by the HCS could be addressed by implementing effective public and private telemedicine solutions (Webb et al. 2013). Furthermore, as it is explained in this chapter, because of the growth and penetration of cellular networks, the implementation of mHealth solutions has the potential to play a major role in HCS improvement (PricewaterhouseCoopers México 2012).

It should be noted that in Latin America, there are already several mHealth projects covering the private, social, and public sectors (Iwaya et al. 2013). However, there are still several issues that must be addressed to fulfill the potential offered by mHealth. For example, the lack of standards and norms is still an open issue in several countries. Additionally, several of the currently available hardware and software for mHealth have been designed for developed countries where there is a wide coverage of high-data-rate wireless networks. Therefore, these issues must be addressed to fully exploit the potential offered by the eventual adoption of mHealth solutions within the HCS in Latin America.

23.2 Demographic and Technological Prospects in Latin America

The EP reflects the morbidity and mortality patterns as well as the risk factors in a specific region. Thus, the EP must be considered in the design and implementation of effective health-care initiatives that improve the population health. An adequate use of information and communication technologies (ICTs) can increase the effectiveness of traditional HCSs, as evidenced by the proliferation of electronic health records, hospital information systems, and telemedicine solutions implemented across the world. In this context, mHealth has the potential of opening the landscape for a whole new spectrum of health-related services, whose aim should be addressing the challenges posed by specific EPs. Therefore, in order to adequately analyze the use of mHealth in the Latin American region, it is important to contextualize its EP, the density and distribution of physicians in the region, and the available ICT infrastructure. As such, this section presents a brief overview of these aspects.

23.2.1 Latin America Epidemiological Profile

An EP portrays morbidity and mortality tendencies in specific geographical regions. Therefore, the EP represents an overview of social, economic, demographic, cultural, and health-care determinants that are involved in the incidence of specific diseases (Omram 2001). From the government point of view, decision making in the management of health plans and programs should be consistent with the most common diseases afflicting the population, their causes, and impact. Therefore, information from EPs should be used as an indicator for present and future needs of infrastructure, material, and human resources that are (or will be) needed to cope with the diseases that inflict a major social and economic impact.

As previously mentioned, an important trend in Latin America and the Caribbean is the LE index increase. In fact, LE in the region has grown from 29 years in 1900 to 74 in 2010 (Pan American Health Organization 2012). Thus, the population in Latin America is progressively ageing, with an increase in the proportion of people with 60 years or more from about 5.5% in 1950 to about 9.7% in 2010. Furthermore, it is expected that in 2050, the elderly population will represent above 24% of the total population in the region (Saad et al. 2009). It is noteworthy that the HALE index has not followed the same growth rate as the LE index. For instance, for female population in Mexico, the LE in 2010 was 78.4 years, while the HALE was 69.1 years (Salomon et al. 2012). Therefore, it is natural that noncommunicable diseases related to the ageing process are now more prevalent among the population and are as well the main cause of deaths in the elderly (see Figure 23.1).

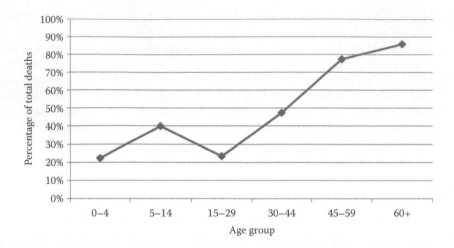

FIGURE 23.1
Deaths caused by noncommunicable diseases as a percentage of the total number of deaths in the corresponding age group. (From World Health Organization, *Causes of Death 2008 Summary Tables*, Health statistics and informatics Department, World Health Organization, Geneva, Switzerland, 2011.)

The EP of Latin America and the Caribbean shows that currently noncommunicable diseases cause most of the reported deaths (see Figure 23.2), prevailing, among others, type 2 diabetes and ischemic heart diseases (World Health Organization 2011). This kind of diseases is more prevalent in elderly and overweight/obese people. Thus, it can be concluded that part of the shift in the Latin American EP has been caused by the ageing population trend altogether with unhealthy lifestyles adopted by the population.

The EP shift toward chronic degenerative diseases is facing the HCS to new challenges in the private and public sectors, as this kind of diseases requires long-term treatments representing an important percentage of the patient and HCS expenditure. For instance, in Mexico, nearly US$4524 million were required in 2012 for the treatment of diagnosed diabetic patients. This is 15% more than the amount required in 2011 for the same purpose (Hernández-Ávila et al. 2013) and almost 19% more than the expenditure on *Seguro Popular*, which is a public HCS that aims to serve nearly 46% of the population (Hernández-Ávila et al. 2013).

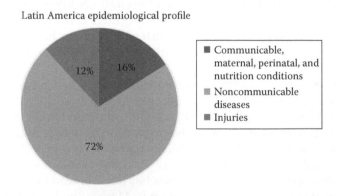

FIGURE 23.2
Most common causes of death in Latin America and the Caribbean. (From World Health Organization, *Causes of Death 2008 Summary Tables*, Health statistics and informatics Department, World Health Organization, Geneva, Switzerland, 2011.)

23.2.2 Density and Distribution of Physicians in Latin America

Although some progress toward alleviating the burden of chronic diseases has been made in Latin America, there are other challenges that must be addressed by the region HCS. For example, as previously mentioned, physicians tend to be established in clusters within the most urbanized regions (Dussault and Franceschini 2006). Thus, health-care services can be less accessible for rural and certain sectors of urban population because of geographical (e.g., rural zones) and/or economical (e.g., a lack of resources for traveling toward the health-care center) reasons. Although there are some countries in Latin America whose physician density is far below the World Health Organization (WHO) recommendation of 25 physicians per 10,000 inhabitants (e.g., 9.3 in Guatemala, 9.2 in Peru, and 3.7 in Honduras) (Organización Mundial de la Salud 2012), several countries of the region are not that far from this recommendation (e.g., 19 in Mexico, 17.8 in Chile, and 19.5 in Brazil) (Guillou et al. 2011; Scheffer et al. 2011; Sistema Nacional de Salud 2011). However, as the trend of unbalanced geographical distribution of physicians is followed in Latin America, within the countries of the region, there are significant areas that are underserved. For example, in Brazil, the Federal District has 40.2 physicians per 10,000 inhabitants, while Maranhao only has 6 (Scheffer et al. 2011). In Chile, there are 21.2 physicians per 10,000 inhabitants in the central region versus 11.7 in the south region (Guillou et al. 2011). This behavior is observed as well in Mexico, where there are 39 physicians per 10,000 inhabitants in Mexico City, while in Chiapas, there are only 12 (Sistema Nacional de Salud 2011). In this context, it is important to note that the concentration of physicians in wealthier regions can lead to underutilization of qualified personnel, while the total cost of the HCS is increased (Dussault and Franceschini 2006).

23.2.3 mHealth Potential in Latin America

The use of ICT can help Latin American HCS to address several of the challenges posed by the changes in the EP and the unbalanced distribution of physicians. This can be achieved by providing novel paradigms for health-care provision. For example, through mHealth-based teleconsulting, the HCS can expand its coverage, delivering basic or specialized health-care services to populations located far from the health centers. Additionally, mHealth-based telemonitoring can increase the capacity of HCS by reducing the hospital bed occupancy rate. Furthermore, the implementation of raising awareness campaigns and wellness services through mHealth can help to decrease the incidence of chronic degenerative diseases prevalent in the region.

It is important to note that for a successful implementation of mHealth services, the availability and accessibility of wireless communications infrastructure must be considered. This is a consequence of the different data rates needed for each particular mHealth service. Therefore, in order to properly design and implement any mHealth solution, it is crucial to take into account the capacity of the ICT infrastructure available in the deployment region. Thus, in the next subsection, an overview of the available ICT infrastructure in Latin America is presented.

23.2.4 Communications Infrastructure in Latin America

In Latin America, the penetration of fixed-telephone subscriptions has remained below 20% over the last 10 years (ITU 2012). This means that a large part of the population does not have access to this service and as a consequence may not have direct access

TABLE 23.1

Cellular Generations and Their Data Rates

Generation	Technology	Data Rates
2	GSM	50–60 kbps
2 transitional (2.5, 2.75)	EDGE/GPRS, CDMA2000	70–130 kbps
3	WCDMA, 1xEVDO,	200 kbps–1.4 Mbps
3 transitional (3.5, 3.75, 3.9)	HDSPA, HDSPA+, EVDO (Rev. A and B), LTE	14.4–172 Mbps
4	LTE advanced	Potentially >100 Mbps

to health-care services provided by this means. On the other hand, mobile services were initially considered as an exclusive prerogative for high-income users. However, prepaid services opened the market for a wider range of users. In fact, mobile cellular subscriptions in Latin America have reached a penetration rate above 100% in the last years. Although the penetration of mobile data services is still in a value near 19%, there is a raising trend in the number of high-speed mobile data subscriptions. Furthermore, according to a forecast made by Ericsson in 2012, by 2017, 3G/4G (e.g., WCDMA/HSPA, LTE) mobile subscriptions in Latin America will be dominant (Ericsson 2012). It is noteworthy that HSPA is able to provide peak data rates of up to 14 Mbps, while with HSPA+ and LTE, higher peak data rates can be achieved (see Table 23.1). Thus, considering the current and future data rates offered by the mobile cellular networks in the region, there exists the possibility of implementing a significant variety of mHealth services, ranging from low data rate (e.g., short message service (SMS) appointment remainders) to high data rate (e.g., teleconsulting) applications. For example, in (Kim et al. 2009), it is shown that an acute stroke can be diagnosed by means of mHealth telediagnosis with data rates above 400 kbps, which is within the capabilities of current cellular technologies such as HSPA, HSPA+, and LTE. Thus, cellular networks in Latin America represent a prominent connectivity option because of its large penetration rate and the increasing trend of high-speed data subscriptions. Therefore, the implementation of mHealth services over cellular networks has the potential of addressing several of the challenges faced by Latin American HCS.

23.3 mHealth in Latin America

As mentioned in Section 23.2.3, several of the challenges faced by public and private health-care providers can be addressed through the proper implementation of mHealth services. With this in mind, several Latin America countries are already exploring the potential of mHealth through the implementation of diverse projects focused on strengthening different areas of the HCSs.

When implementing a particular mHealth program, it is necessary to consider several key issues related to specific constraints of the deployment scenario like the mHealth service purpose (e.g., data collection and reporting), deployment circumstances (e.g., emergency or disaster situation), target users (e.g., health physicians, medical specialists, or general population), and addressed topics (i.e., chronic diseases and lifestyle changes).

Thus, specific deployment scenarios as defined by the WHO will be used to introduce several representative mHealth projects implemented in Latin America (Kay 2011):

- *Health call centers/health-care telephone helpline*: In this scenario, health-care professionals offer triage services through a voice call. For public health-care centers, this service can help in reducing unnecessary visits, with the consequent resources savings. On the other hand, the private sector can exploit an existing market for this kind of services. Examples in Latin America are Telemedic in Colombia and Mexico, Telemed in Dominican Republic, and MedStar Telehealth Services in Trinidad and Tobago, all of which are private services that the users pay (Ivatury et al. 2011).

- *Emergency toll-free telephone services*: Their aim is to provide a quick response to emergency events with professional health support. An example of a service working within this mHealth scenario is the toll-free 112 number that mobile users can dial in Colombia, Costa Rica, Ecuador, and Mexico to request an emergency service.

- *Treatment compliance*: Reminders are sent through SMS or voice calls to patients in order to achieve treatment adherence. Important issues such as drug resistance can be addressed by means of this service (Curioso and Kurth 2007). The *X out TB* project implemented in Nicaragua is an example within this mHealth scenario. This project focused on adherence to tuberculosis (TB) treatments, offering an award (free mobile minutes) when patients showed high compliance rates (Trafton 2008).

- *Appointment reminders*: Reminders about appointments, immunization, and treatment results (among others) are sent to patients by using voice messages or SMS. An example of a project addressing this mHealth scenario was implemented in Brazil (Da Costa et al. 2010), where SMS reminders were sent to patients to improve the attendance rate. The authors reported an attendance increase of 6.5% when reminders were sent.

- *Community mobilization and health promotion*: It is performed by sending SMS for health promotion and alerting large target groups about health campaigns. Although it was not performed by means of SMS, a project focused on attending this scenario was a pilot program implemented in Honduras for chronic disease health promotion (Piette et al. 2011). In this project, patients received interactive voice-over IP calls from one server in the United States. These messages were aimed to promote a proper management of diabetes.

- *Raising awareness*: This scenario is focused on addressing one of the stronger challenges in the region: motivating the population to become actively involved in taking care of themselves. Thus, projects within this category are focused on raising the population awareness about the risks and consequences of different diseases. Raising awareness is commonly implemented using SMS. However, it can be addressed by means of games and other kinds of friendly applications to increase the public acceptance. As an example, the *Get the Message Campaign* implemented in the Caribbean was aimed to inform the population about noncommunicable diseases by means of SMS (Healthy Caribbean Coalition 2010).

- *Mobile telemedicine*: According to the WHO Global Observatory for eHealth (World Health Organization 2011), this category is defined as communication and consultation between health-care professionals aimed at improving diagnosis and treatment. For example, the *Penn Librarians & Doctors Teaming Up* project allows physicians in Guatemala to get a second opinion about their patients from specialized physicians located in the United States (Penn Medicine 2011).

- *Public health emergencies*: This scenario is addressed through several mobile services like SMS, voice calls, and even e-mails. These mHealth services are mainly focused on informing and supporting the population in emergency cases like natural disasters and infectious outbreaks. Through these services, rescue and assistance support can be better coordinated and consequently more effective. For example, in the project implemented in Haiti after the 2010 earthquake, information about population displacements was estimated by means of the data sent from mobile phones to the base stations (Bengtsson et al. 2011). This information can be useful when planning HCS responses, such as tracking and attending possible epidemiological outbreaks.

- *Health surveys and surveillance*: The projects in this scenario are focused on health-related data collection and reporting (surveys) and disease tracking (surveillance) by means of mobile devices. Health surveys can provide relevant information for the execution of well-defined health-care actions, while surveillance can be used to support preventive health actions and to keep track of epidemiological outbreaks. Two relevant examples within this category are eMOCHA® in Colombia (Johns Hopkins Center for Clinical Global Health Education 2013) and *SEAMUS* in Haiti, Dominican Republic, and Colombia (Organización Panamericana de la Salud/Organización Mundial de la Salud 2010). The eMOCHA® project is aimed at eradicating dengue by means of smartphones. In this project, field workers collect water samples, surveys, and geographic and biological information and send it through the cellular network for its analysis and review. In *SEAMUS*, field workers analyze the water quality and send the results to a storage central by means of SMS. Then, this information is analyzed in order to trigger alerts if needed.

- *Patient monitoring*: This mHealth scenario considers the management, tracking, and treatment of patients based on mobile devices. Thus, projects within this scenario could be especially useful in the treatment and management of chronic degenerative diseases. An example is the *AToMS* project, whose aim is to remotely monitor the heart condition in emergency situations by transmitting electrocardiogram (ECG) information through the cellular network (Correa et al. 2011). The ECG information is reviewed by an available cardiologist before the patient reaches the hospital, thus reducing the intervention delay.

- *Information initiatives*: They are focused on providing access to scientific publications and databases by means of mobile devices. As an example, consider the *Penn Librarians & Doctors Teaming Up* project, where physicians in Guatemala were provided with access to the University of Pennsylvania medical database (Penn Medicine 2011).

- *Decision support systems*: They are implemented with algorithms that, based on medical information about patients, can suggest a diagnosis. In mHealth scenarios, this kind of services can be accessed through mobile devices. Decision support systems can be extremely useful if availability or access to physicians is limited. An example of this kind of projects is *BabyCare* (Costa et al. 2010), which is a Brazilian project implemented in PDAs aimed at supporting the diagnosis of children from underprivileged communities. Another project within this scenario is *InteliMed* (Menezes Júnior et al. 2011), whose aim is to support asthma diagnosis using computational intelligence.

- *Patient records*: In the context of mHealth, electronic patient records are stored at clinical databases to be accessed through mobile devices. The anywhere access to these electronic records can facilitate a more precise and quick support of treatments and diagnosis. A relevant project in this area is presented in (Murakami et al., 2004), which involves the use of PDAs to periodically access the records of patients with cardiac problems located in the intensive care units.

From the previously presented examples, it can be seen that in Latin America, there are and there have been several efforts to implement mHealth solutions. It is noteworthy that several of these projects intend to strengthen the public HCS in areas such as capacity (e.g., mobile telemedicine, decision support systems, and patient monitoring), effectiveness (e.g., treatment compliance, appointment remainders, patient records, and information initiatives), planning (e.g., health surveys and surveillance), prevention (e.g., health promotion and raising awareness), and emergency response (public health emergencies, emergency toll-free telephone services, and decision support systems). Additionally, from the private point of view, the implementation of mHealth projects represents a business opportunity that has not been fully exploited (PricewaterhouseCoopers México 2012). However, in order to fully exploit the potential that mHealth has to offer to Latin American HCS, several regulatory, political, social, and technological aspects must be addressed in the region. In Section 23.5, the challenges for the deployment of mHealth solutions in Latin America will be treated with more detail.

23.4 mHealth Potential in Latin America: The Mexico Case Example

In this section, the potential that mHealth offers to address several of the problems faced by Latin American HCS is analyzed through the Mexico case example. Mexico is one of the largest and most populated countries in Latin America, and it was chosen because several of the most important health-related trends commonly found in Latin America can be clearly observed in Mexico, particularly the following:

- The EP shift toward chronic degenerative diseases. As evidenced by the fact that heart diseases and diabetes mellitus are the first and second causes of death in the country.

- The increase in the prevalence of overweight and obesity among the population. In fact, Mexico tops the list of countries with overweight/obesity problems, as above 69% of its adult population is overweight or obese, which represents an 8.7% increase compared with that in 2000 (Gutiérrez et al. 2012).

- The unbalanced geographical distribution of physicians. As evidenced by the fact that while large urban centers like Mexico City have 39 physicians per 10,000 inhabitants, in other regions like Chiapas, there are just 12 physicians per 10,000 inhabitants.

The Mexican government has been making important efforts to control the epidemic nature of chronic diseases, particularly diabetes and heart diseases. These efforts have included prevention, education, and early detection campaigns aimed to alleviate the economic burden that represents the long-term treatment of these diseases (e.g., nearly US$4524 million was required in 2012 to treat diabetes alone). This presents an opportunity for the development of mHealth solutions designed to support raising awareness and health promotion campaigns. Furthermore, the provision of mHealth services like mobile telemedicine, treatment compliance, and remote patient monitoring can help to increase the capacity of the Mexican HCS and reduce its expenditure.

Besides the EP shift, another important issue observed in Mexico is that its heterogeneous demographic and geographic conditions affect the way health care is delivered to the population. For example, Mexico has big urban centers like Mexico City with a vehicular density of 41 vehicles per 100 inhabitants. Therefore, traffic jams are a common problem in the city with the consequent travel time increments and even the obstruction of emergency services. Additionally, there are several rural regions such as Tulijá Tseltal-chol in Chiapas, where sometimes people in life-threatening conditions (such as obstetric emergencies) have to travel up to 4 h in order to reach the emergency room (Nazar Beutelspacher et al. 2011). Hence, traffic jams and difficult geographical conditions in Mexico can impose important delays for the provision of health services. In this context, the deployment of mHealth services like mobile telemedicine and remote patient monitoring would enable the provision of several health-care services without the need of transporting the patient, or even while the patient is being transported toward a hospital. Additionally, because of Mexico geographical characteristics, management of natural disasters such as earthquakes or hurricanes, altogether with epidemic outbreaks, is a big challenge for the HCS capacity. Thus, programs focused on quantifying health needs and providing information within the emergency situation could have a great impact in helping the HCS effectiveness.

23.4.1 mHealth Application Scenarios in Mexico

In Mexico, several efforts aimed to introduce the ICTs in the health-care practice have already been made. First, in 1985, the national program called CEMESATEL was implemented with the aim of providing remote education to physicians through satellite links. During the 1990s, several telemedicine services over satellite links were implemented within the Institute for Social Security and Services for State Workers (ISSSTE for its acronym in Spanish), which is one of the major social health-care providers in the country (Gertrudiz 2010). Later, in 2004, the National Center for Technological Excellence in Health (CENETEC) was created to support emerging initiatives related to the inclusion of ICT within the health system.

In the context of the Mexican case, different mHealth projects could help the HCS to cope with several of the main challenges that it is currently facing. However, in order to define specific mHealth scenarios in Mexico, it is necessary to first describe the hospital and mobile communications infrastructure available in the country.

23.4.1.1 Hospital and Mobile Communications Infrastructure in Mexico

As previously mentioned, Mexico is facing an unbalanced geographical distribution of physicians. Additionally, the same phenomenon is observed in the distribution of hospital infrastructure. For example, states like Colima have more than 50 primary health centers (PHCs) per 100,000 potential users, while the Estado de Mexico has just 13.6. This does not necessarily mean that states or cities with low PHC densities will have relatively less doctors, as Mexico City has less than 10 PHC per 100,000 potential users (even lower than Estado de Mexico) and 39 physicians per 10,000 inhabitants (the largest in the country). Furthermore, when considering that physicians tend to be established in clusters within the most urbanized regions (Dussault and Franceschini 2006), it can be concluded that access to health-care services might be limited by geographical and/or transportation barriers. An important consequence of this issue is the productivity loss and its socioeconomic impact, caused by the lack of access to health services and the waste of time that implies traveling during long periods to have access to health care.

Regarding the wireless communications infrastructure available in Mexico, the mobile cellular network has reached a penetration of 86.7 subscriptions per 100 inhabitants. Furthermore, during the period from April to June 2013, mobile broadband subscriptions increased almost 20%, reaching 14.45 million (Secretaría de Comunicaciones y Transportes 2013). This means that the cellular network offers a viable connectivity alternative for the delivery of health-care services. In this sense, the implementation of mHealth services could help to mitigate some of the consequences of having an unbalanced distribution of physicians and hospital infrastructure, by providing novel ways to deliver health-care services to the population.

23.4.1.2 Application Scenarios and Some Examples of mHealth Projects in Mexico

Based on the previous discussion and considering aspects such as the EP, the use of mHealth services to address the following scenarios can help to strengthen and improve the Mexican HCS:

1. *Public health*: The implementation of health campaigns to encourage a healthy lifestyle can be made through mobile devices. Additionally, public health campaigns to monitor the current state of the population health can be performed by this mean. For instance, body mass index, blood glucose, and weight information can be gathered by means of digital monitors, and then this information can be sent over the cellular network toward a central database for its analysis. *a-Prevenir*© and *CardioNET* are two important projects in this sense. *a-Prevenir*© was developed in mid-2013 by the government-sponsored CICESE Research Center (Villarreal-Reyes et al. 2013). *CardioNET* was motivated by the high prevalence of cardiac conditions in Mexico (Déglise et al. 2012).

2. *Primary health-care delivery*: Considering the cellular network penetration and its development toward enabling wireless broadband access, it is feasible to implement several telemedicine solutions aimed to provide primary health-care services for population segments with little or no access to health care. An example is *AmaneceNET*, which is a pilot program operating in the state of Chiapas aimed at pregnant woman living in rural zones (Instituto Carlos Slim de la Salud 2009).

3. *Patient monitoring*: In chronic degenerative diseases, treatment compliance is a key issue to achieve a good quality of life. An example of a project with this aim is the *Dulce Wireless Project*, which focuses in diabetes prevention and treatment support (International Community Foundation 2012). Additionally, continuous mHealth monitoring can be used for better diagnosis or treatment adjustment. An example is *Remote Fetal Monitor*, which is a project developed by the government-sponsored

CIDESI Engineering Center. The system focuses on monitoring high-risk pregnancies in order to prevent, control, or make interventions in case of abnormalities (Arciniega Montiel 2012).

4. *Emergency response*: The prehospital trauma care system extends health services out of the hospitals and it is particularly useful in life-threatening scenarios. For example, onboard patient monitoring during emergency transportation services can help in diagnosis support and/or in adequately preparing the emergency room to receive the patient. An important effort in this case is the *Telecare System for Moving Vehicles* (*TSMOV*). It was developed in 2007 by CICESE Research Center and Medica Sur, which is a private hospital located in Mexico City. In this project, an ambulance was equipped with medical devices to monitor different biophysiological variables like ECG, oxygen saturation, heart and breathing rate, blood pressure, and temperature. The devices were mounted in an ambulance and the gathered data were sent toward the health-care facilities through the cellular network (Rivera-Rodríguez et al. 2012).

5. *Public emergencies*: During public emergencies such as epidemic outbreaks and natural disasters, the emergency response systems can use mobile data communications devices to provide a more coordinated response. Additionally, emergency care services can be deployed in the areas where they are most needed. For example, during the 2009 H1N1 outbreak in Mexico, an assisted triage was implemented via SMSs (Lajous et al. 2010). The triage protocol used a survey to evaluate the current health status of individuals. Another example is *Dengue Monitoring with Cell Phones*, which is a project aimed at the control and surveillance of dengue and malaria diseases by means of mobile devices (Lozano-Fuentes et al. 2012).

Several of the presented projects focus in the prevention, tracking, and treatment compliance of chronic degenerative diseases. This is expected because of the EP shift toward this kind of diseases. Nevertheless, it is noteworthy that attention to vulnerable groups is important as well.

23.4.2 Examples of mHealth Traffic Profiles

When designing an mHealth service for a particular application scenario, it is necessary to characterize the traffic profile of the service based on the scenario requirements. Once this has been done, proper network planning can be performed with the aim of fulfilling the traffic requirements in terms of bandwidth, delay and packet loss tolerance, latency, jitter, and reliability.

Different mHealth scenarios will lead to different traffic profiles and hence to different network requirements. For example, a particular mobile telemedicine service might require establishing a videoconference session, and hence the use of mobile networks providing medium to high data rates will be needed. In contrast, a raising awareness service may only need the sporadic transmission of SMSs, and thus, the use of medium/high data rates will not be required.

As previously mentioned, there are different mHealth scenarios that can be implemented with the available technological infrastructure in Mexico. Particularly mobile patient monitoring, teleconsulting, and raising awareness are services feasible to implement. Therefore, a generic traffic profile for these three services is provided in this section.

Teleconsulting can be performed through audio or videoconference with different quality requirements. Generally, video codecs are classified in terms of the provided data rates and the supported resolutions and frame rates. Thus, the videoconference data rate and image quality are closely related to the desired video size and the particular codec used. Similarly, audio codecs provide different bitrates depending on the desired quality. Therefore, depending on the kind of service and the network capacity, different codecs can be used to implement the teleconsulting service. Table 23.2 shows typical bit rates and resolutions for different video codecs. A basic teleconsulting service can be established using the lowest-bit-rate video codec, as long as there is no need to visually verify a particular health condition.

Mobile patient monitoring is focused on acquiring different biophysiological data such as blood pressure, blood glucose, temperature, weight, ECG, and oxygen saturation in blood, among others. Thus, the traffic profile for mobile patient monitoring will be determined by the characteristics of the variables to monitor (e.g., sampling frequency, bits per sample) and the required periodicity of the information. The biophysiological variables most commonly found in mobile patient monitoring scenarios are shown in Table 23.3 (Gallego et al. 2005; Vouyioukas et al. 2007; Skorin-Kapov and Matijasevic 2010; Adibi 2012).

Raising awareness looks to inform the population about the risks and consequences of different diseases. As previously mentioned, raising awareness is commonly implemented using SMSs. This kind of messages can be encoded in three different ways: 7-bit characters, 8-bit characters, and 16-bit characters. However, regardless of the encoding used, the size of an SMS in GSM networks is 1120 bits (Wilde and Vaha-Sipil 2010). Although the frequency of the SMSs sent will depend on the specific campaign, the generated traffic for this kind of services can be regarded to be very low.

TABLE 23.2

Different Video Codecs and Their Provided Data Rate

Codec	Bit Rate	Resolution
H.261	40 kbps–2 Mbps	176 × 144, 352 × 288
H.263	64 kbps–1.2 Mbps	128 × 96, 704 × 576, 1408 × 1152
MPEG-2	5–25 Mbps	352 × 420 to 1920 × 1080
MPEG-4/AVC	40 kbps–10 Mbps	176 × 144 to 1920 × 1080
DV	30 Mbps	720 × 480
HDV	25 Mbps	1080i

TABLE 23.3

Different Biophysiological Variables and Their Required Data Rate

Biophysiological Variable	Sampling Frequency (Hz)	Bits per Sample	Delay Sensitivity	Data Rate (bps)
Electroencephalography (EEG)	256 (×24 Ch.)	16	~1 s	98,304
Electrocardiography (ECG)	200 (×3 Ch.)	12–16	~1 s	7,200–9,600
Glucose monitoring	40–200	16	NO	640–3,200
Blood pressure	120	16	NO	1,920
Blood oxygen (SpO$_2$)	60 (×2 Ch.)	16	NO	1,920
Cardiac output	40	16	NO	640
Respiration	50	6	NO	300
Body temperature	0.2	12	NO	2.4

23.4.3 mHealth Model Example: Diabetes Prevention in Mexico

This section introduces an mHealth model for diabetes prevention. This disease was chosen because of the high prevalence of chronic diseases in Mexico. The proposed model was developed based on the WHO recommendations for diabetes prevention. Preventive and educative actions are key in reducing the incidence rate of this disease, as overweight and obesity can significantly increase the risk of suffering type 2 diabetes (WHO Media Centre 2013).

According to the WHO, there are three preventive stages related to diabetes and its treatment (WHO Study Group on Diabetes Mellitus 1985). During the first stage, healthier lifestyles and better dietary patterns are encouraged. In the second stage, periodic health screenings for population at risk or people above 45 years old are recommended. Finally, the third stage is aimed at avoiding complications (e.g., glaucoma) in diagnosed patients.

In the context of mHealth diabetes prevention in Latin America and Mexico, the first stage has been commonly addressed through SMS. In this stage, sending one encouragement message of 140 bytes per day can be used for primary diabetes prevention purposes. With the introduction of data connectivity services over the cellular data networks, the use of alternative mobile messaging applications providing more functionalities than SMS may be explored.

Because of the large territorial extension, a nomadic kiosk to perform public health campaigns in rural and marginalized areas can be considered for the second stage of diabetes prevention. The kiosk will measure biophysiological variables like blood pressure, weight, temperature, blood glucose, and heart rate. General information of the patient and the collected data will then be transmitted toward a central or local database through the cellular network. This information can then be stored in an electronic patient record such that a physician can later analyze it. This will enable the early detection of diseases like diabetes or hypertension among the population. Considering the increasing coverage and data rates of mobile broadband services, an mHealth teleconsulting service can be established between a patient and a physician to receive health advice. A traffic profile to implement this service is presented in Table 23.4. From this table, it can be seen that with the inclusion of teleconsulting services, the kiosk would require a data link of about 304 kbps, which is feasible to obtain from 3G cellular networks.

For the third stage, treatment adherence is vital. Thus, raising awareness campaigns can be useful to inform the patient about the possible consequences of uncontrolled diabetes. Additionally, self-monitoring is also very important. Therefore, the use of SMS-based applications to report the blood glucose value is proposed. This information can be used to update the electronic patient records for treatment compliance monitoring. For both applications, sending three SMSs per day (equivalent to 3360 bits per day) is considered to

TABLE 23.4

Diabetes Prevention Traffic Profile at Stage 2

Variable	Data
Patient profile including	≈500 bytes (once during the handshake)
Name, e-mail, age, sex, height, temperature, glucose, weight, heart rate, blood pressure	
Audio	64 kbps
Video	240 kbps
Traffic profile	≈ 304 kbps

be adequate. As in the case of first-stage diabetes prevention, the use of alternative mobile messaging applications providing more functionalities than SMS may be explored.

As it can be seen from the provided traffic profiles for each prevention stage, it is feasible to implement the proposed mHealth model over the existent cellular networks available in Mexico.

23.5 mHealth Challenges and Future Trends in Latin America

As previously mentioned, mHealth services have the potential to strengthen different areas of the private and public HCSs in Latin America. In this context, by using mHealth services, public HCS can improve its coverage and capacity without significantly increasing its economic burden (PricewaterhouseCoopers México 2012). In fact, different individual projects have been implemented in the region showing that mHealth services are a feasible solution for the main issues faced by Latin American HCS (Iwaya et al. 2013). Nevertheless, there are different challenges that have to be addressed in order to fully integrate the mHealth solutions within the HCSs. Particularly regulatory, technological, and socioeconomic conditions are relevant issues to be addressed. In this section, an overview of these challenges is presented.

The first challenge to overcome for the deployment of mHealth services in Latin America is not necessarily associated with its technical feasibility. In fact, although implementing a particular mHealth solution can be technically possible, from the public HCS point of view, it is necessary to guarantee its affordability, especially for low-income populations. Thus, economical support can overweight technical feasibility. For this reason, when proposing the implementation of specific mHealth services, an economic study considering possible savings must be included.

In Latin America, the digital literacy rate is low; thus, although the penetration rate of mobile telephony is high, actions focused on improving the digital literacy rate are needed in order to achieve a greater impact of any proposed mHealth solution. Furthermore, efforts aimed at incentivizing the use of mobile technologies in the everyday medical practice is a challenge that has yet to be addressed (Mechael et al. 2010).

Another issue is that most of the mHealth proposals implemented in Latin America are deployed as isolated small-scale pilot projects, designed for very specific applications. Thus, interoperability among the different projects and HCS is not assured. Consequently, sharing information between different mHealth initiatives even within the same HCS becomes difficult.

Regarding the wireless infrastructure, 2G cellular networks are still the predominant technology in Latin America (Ericsson 2012). Thus, several mHealth projects previously implemented in the region focused on low-data-rate services (e.g., SMS raising awareness). However, a significant growth of mobile broadband subscriptions and coverage is expected within the next years. Thus, the deployment of mHealth services like teleconsulting over wider areas will be feasible.

Although implementing mHealth teleconsulting is currently feasible in certain countries, a major challenge arising in this scenario is the use of electronic signatures for prescriptions and patient record updates. For example, the electronic signature is accepted in Mexico by the internal revenue service for taxpaying purposes. Nevertheless, it is not accepted by the Mexican HCS. Thus, this is a barrier to overcome for the implementation of advanced teleconsulting services.

Privacy of health-related data is a major legal concern in several Latin American countries. However, most of the reviewed projects did not focus their efforts in securing the patients' privacy. Thus, including mechanisms that guarantee the privacy of the transmitted data is still an open issue.

23.6 Conclusions

In order to achieve the adoption of mHealth as a means of delivering health services to the population, the challenges previously described must be addressed by joint efforts and initiatives brought forward by key players such as policy makers, users, health-care providers, technology providers, and academia. For example, Latin American governments could invest in the public founding of mHealth projects in order to attract the interest of developers and the academic community. Additionally, health-care providers can actively encourage the use of mobile technology by physicians.

In conclusion, mHealth services in Latin America are a promising option to cope with the existing shortcomings of its HCS. Thus, the key players in Latin America must perform actions that promote and encourage the adoption of mHealth within their health-care ecosystem.

Questions

1. After this chapter lecture, which is the biggest challenge remaining for the mHealth deployment in the Latin America region?
2. Considering traditional public health-care services such as consultation, emergency response, and epidemiological surveillance, how can the mHealth paradigm improve patient attention?
3. The use of SMSs to perform preventive actions is a common practice in Latin America. Which are the advantages and disadvantage of this technological connectivity option?
4. Which factors should be considered when designing an mHealth service to be deployed within a specific scenario?
5. Several mHealth projects in Latin have been focused on communicable diseases. Provide a feasible explanation for this fact.

References

Adibi, S. 2012. Link technologies and BlackBerry Mobile Health (mHealth) solutions: A review. *IEEE Transactions on Information Technology in Biomedicine* 16 (4): 586–97.

Arciniega Montiel, Sadot. Monitor Fetal Remoto, Tecnología Para El Cuidado de La Salud [Remote Fetal Monitor. Healthcare Technology]. News. *Con-Ciencia*, June 19, 2012. Accessed July 21, 2014. http://blogs.eluniversal.com.mx/weblogs_detalle16537.html.

Bengtsson, L., Xin, L., Thorson, A., Garfield, R., and Schreeb, J.v. 2011. Improved response to disasters and outbreaks by tracking population movements with mobile phone network data: A post-earthquake geospatial study in Haiti. Ed. Peter W. Gething. *PLoS Medicine* 8 (8): e1001083. doi:10.1371/journal.pmed.1001083.

Correa, B.S.P.M., Gonçalves, B., Teixeira, I.M., Gomes, A.T.A., and Ziviani, A. 2011. AToMS: A ubiquitous teleconsultation system for supporting AMI patients with prehospital thrombolysis. *International Journal of Telemedicine and Applications* 2011: 1–12.

Costa, C. L. B., Pinto, V.C., Cardoso, O.L., Baba, M.M., Pisa, I.T., Palma, D., and Sigulem, D. 2010. BabyCare: Apoio À Decisão Na Atenção Primária Materno-Infantil Com Computadores de Mão [BabyCare: Decision support to maternal and child primary care with handheld computers]. *Ciência & Saúde Coletiva* 15: 3191–3198.

Curioso, W H, and Kurth, A E. 2007. "Access, Use and Perceptions Regarding Internet, Cell Phones and PDAs as a Means for Health Promotion for People Living with HIV in Peru." *BMC Medical Informatics and Decision Making* 7, 1: 24.

Da Costa, T.M., Salomão, P.L., Martha, A.S., Pisa, I.T., and Sigulem, D. 2010. The impact of short message service text messages sent as appointment reminders to patients' cell phones at outpatient clinics in São Paulo, Brazil. *International Journal of Medical Informatics* 79 (1): 65–70.

Déglise, C., Suzanne Suggs, L., and Odermatt, P. 2012. Short Message Service (SMS) applications for disease prevention in developing countries. *Journal of Medical Internet Research* 14 (1): e3. doi:10.2196/jmir.1823.

Dussault, G. and Franceschini, M.C. 2006. Not enough there, too many here: Understanding geographical imbalances in the distribution of the health workforce. *Human Resources for Health* 4 (12).

Ericsson, AB. 2012. *Traffic and Market Report On The Pulse of the Networked Society.* White Paper. Ericsson, June 2012.

Gallego, J.R., Hernandez-Solana, A., Canales, M., Lafuente, J., Valdovinos, A., and Fernandez-Navajas, J. 2005. Performance analysis of multiplexed medical data transmission for mobile emergency care over the UMTS channel. *IEEE Transactions on Information Technology in Biomedicine* 9 (1): 13–22.

Gertrudiz, N. 2010. E-Health: The case of Mexico. *Latin American Journal of Telehealth* 2 (2): 127–167.

Guillou, M., Jorge Carabantes, C., and Verónica Bustos, F. 2011. Disponibilidad de Médicos Y Especialistas En Chile [Availability of doctors and specialists in Chile]. *Revista Médica de Chile* 139 (5): 559–570.

Gutiérrez, J.P., Rivera-Dommarco, J., Shamah, T., Oropeza, C., and Hernández-Avila, M. 2012. *Encuesta nacional de salud y nutrición 2012: Resultados nacionales [National Survey of Health and Nutrition 2012: National Results].* Cuernavaca, México: Instituto de Salud Pública (Mex).

Healthy Caribbean Coalition. 2010. Get the Message: An Advocacy Campaign in the Caribbean. Experiencia de eSalud: Campaña "Get The Message" En El Caribe. Accessed June 28, 2014. http://www.paho.org/ict4health/index.php?option=com_bpform&view=show&fid=13&lang=en.

Hernández-Ávila, M., Gutiérrez, J.P., and Reynoso-Noverón, N. 2013. Diabetes Mellitus En México. El Estado de La Epidemia [Diabetes Mellitus in Mexico. Epidemic Status]. *Salud Pública Méx* 55 (2): 129–136.

Instituto Carlos Slim de la Salud. 2009. Que Más Personas Vivan Más Y Mejor [More People Live Longer and Better]. Accessed June 27, 2014. http://www.carlosslim.com/pdf/icss_informe2009.pdf.

International Community Foundation. 2012. Annual Report 2012. California. Accessed June 27, 2014. http://www.icfdn.org/publications/ar2012/ar2012.php.

ITU. 2012. Latin America and the caribbean key statistical highlights. Accessed June 27, 2014. http://www.itu.int/net/newsroom/Connect/americas/2012/docs/americas-stats.pdf.

Ivatury, G., Moore, J., and Bloch, A. 2011. A doctor in your pocket: Health hotlines in developing Countries. *Innovations: Technology, Governance, Globalization* 6 (1): 119–153.

Iwaya, L.H., Gomes, M.A.L., Simplício, M.A., Carvalho, T.C.M.B., Dominicini, C.K., Sakuragui, R.R.M., Rebelo, M.S., Gutierrez, M.A., Näslund, M., and Håkansson. P. 2013. Mobile health in emerging Countries: A survey of research initiatives in Brazil. *International Journal of Medical Informatics* 82 (5): 283–298.

Johns Hopkins Center for Clinical Global Health Education. 2013. *eMocha*. eMocha Center for Clinical Global Health Center, Accessed July 21, 2014. http://main.ccghe.net/content/emocha

Kay, M. 2011. *mHealth New Horizons for Health through Mobile Technologies*. Geneva, Switzerland: World Health Organization. http://www.who.int/goe/publications/goe_mhealth_web.pdf.

Kim, D.-K., Yoo, S.K., Park, I.-C., Choa, M., Bae, K.Y., Kim, Y.-D., and Heo, J.-H. 2009. A mobile telemedicine system for remote consultation in cases of acute stroke. *Journal of Telemedicine and Telecare* 15 (2): 102–107. doi:10.1258/jtt.2008.080713.

Lajous, Martín, Leon Danon, Ruy López-Ridaura, Christina M. Astley, Joel C. Miller, Scott F. Dowell, Justin J. O'Hagan, Edward Goldstein, and Marc Lipsitch. September 2010. Mobile Messaging as Surveillance Tool during Pandemic (H1N1) 2009, Mexico. *Emerging Infectious Diseases* 16, 9: 1488–89.

Lozano-Fuentes, S., Ghosh, S., Bieman, J.M., Sadhu, D., Eisen, L., Wedyan, F., Hernandez-Garcia, E., Garcia-Rejon, J., and Tep-Chel, D. 2012. *Using Cell Phones for Mosquito Vector Surveillance and Control*. In 'SEKE', Knowledge Systems Institute Graduate School, San Francisco Bay, CA, pp. 763–767.

Mechael, P., Batavia, H., Kaonga, N., Searle, S., Kwan, A., Goldberger, A., Li, F., and Ossman, J. 2010. *Barriers and Gaps Affecting mHealth in Low and Middle Income Countries: Policy White Paper*. New York: Columbia University. Earth institute. Center for global health and economic development (CGHED): with mHealth alliance.

Menezes Júnior, Júlio Venâncio de, Raphael José D'Castro, Francisco Marinho Moreira Rodrigues, Cristine Martins Gomes de Gusmão, Nilza Rejane Sellaro Lyra, and Silvia Wanick Sarinho. 2011. InteliMed: Uma Experiência de Desenvolvimento de Sistema Móvel de Suporte Ao Diagnóstico Médico [InteliMed: An Experience of Development of Mobile Support System In Medical Diagnosis]" *Revista Brasileira de Computação Aplicada* 3 (1): 30–42. doi:10.5335/rbca.2011.004.

Murakami A. et al. 2004. Acesso a Informações Médicas através do Uso de Sistemas de Computação Móvel [Access to Medical Information Systems through the Use of Mobile Computing]. In IX Congresso Brasileiro de Informática em Saúde—CBIS2004, Brazil.

Nazar Beutelspacher, Austreberta, Benito Salvatierra Izaba, Magdalena del Carmen Morales Domínguez, Annette Hartman, and Mehujael Rodríguez Mazariegos. 2011. Estudio Cualitativo de Barreras de Demanda Y Oferta Con Enfoque a Nivel Local Y Comunitario Y Cambio de Comportamientos En Municipios Prioritarios de Chiapas, México [Qualitative Study of Barriers Demand And Offer With Focus on Community and Local Level and Behavior Change in Priority Municipalities of Chiapas, Mexico]. Iniciativa Salud Mesoamérica 2015 (ISM 2015). Banco Interamericano de Desarrollo (BID). Accessed July 21, 2014. http://www.iadb.org/WMSfiles/products/SM2015/Documents/website/SM2015%20M%C3%A9xico%20-%20Estudio%20de%20barreras.pdf.

Omram, A.R. 2001. The epidemiologic transition: A theory of the epidemiology of population change. *Bulletin of the World Health Organization* 79 (2): 161–170.

ONU-Habitat. 2012. *Estado de Las Ciudades de América Latina Y El Caribe [State of the Latin America and the Caribbean Cities Report]*. UN-HABITAT. Accessed July 21, 2014. http://www.onuhabitat.org/index.php?option=com_docman&task=cat_view&gid=362&Itemid=18.

Organización Mundial de la Salud. 2012. *Estadísticas Sanitarias Mundiales [World Health Statistics]*. Organización Mundial de la Salud.

Organización Panamericana de la Salud/Organización Mundial de la Salud. 2010. SEAMUS: Sistema de Transmisión de Información Basado En SMS [SEAMUS: System for Information Transmission Based on SMS]. Accessed June 28, 2014. http://www.paho.org/ict4health/index.php?option=com_bpform&view=show&fid=16&lang=en.

Pan American Health Organization. "Health Determinants and Inequalities." In *Health in the Americas: Regional Outlook and Country Profiles*, 12–56. Washington, DC: PAHO, 2012. Accessed June 27, 2014. http://www2.paho.org/saludenlasamericas/dmdocuments/hia-2012-chapter-2.pdf.

Penn Medicine. 2011. Penn Librarians and Doctors Team up to Help Guatemalan Patients. *Penn Medicine News: Penn Librarians and Doctors Team up to Help Guatemalan Patients.* Accessed July 21, 2014. http://www.uphs.upenn.edu/news/News_Releases/2011/03/penn-guatemala-technology-training/.

Piette, J.D., Mendoza-Avelares, M.O., Ganser, M., Mohamed, M., Marinec, N., and Krishnan, S. 2011. A preliminary study of a cloud-computing model for chronic illness self-care support in an underdeveloped country. *American Journal of Preventive Medicine* 40 (6): 629–632. doi:10.1016/j.amepre.2011.02.014.

PricewaterhouseCoopers México. 2012. Convergence and opportunities in the healthcare sector doing business in Mexico. A guide for smart investments. Accessed September 14, 2014. http://www.pwc.com/es_MX/mx/publicaciones/archivo/2012-09-doing-business-health.pdf.

Rivera-Rodríguez, R., Serrano-Santoyo, A., Tamayo-Fernández, R., and Armenta-Ramade., A. 2012. Sistema Móvil de Teleasistencia Médica Para La Atención En Tiempo Real de Casos de Urgencia [Mobile Telecare System for Real-Time Medical Emergency Care]. *Ingeniería, Investigación Y Tecnología* 13 (1): 1–8.

Saad, P.M., Miller, T., and Martínez, C. 2009. Impacto de Los Cambios Demográficos En Las Demandas Sectoriales En América Latina [Impact of Demographic Change on Sectorial Demands in Latin America]. *Revista Brasileira de Estudos de População* 26 (2): 237–261. doi:10.1590/S0102-30982009000200006.

Salomon, J.A., Wang, H., Freeman, K., Vos, T., Flaxman, A.D., Lopez, A.D., and Murray, C.J.L. 2012. Healthy life expectancy for 187 Countries, 1990–2010: A systematic analysis for the Global Burden Disease Study 2010. *The Lancet* 380 (9859): 2144–2162. doi:10.1016/S0140-6736(12)61690-0.

Scheffer, M., Biancarelli, A., and Cassenote, A. 2011. *Demografia Médica No Brasil: Dados Gerais E Descrições de Desigualdades [Medical Demography in Brazil: General Information and Descriptions of Inequities].* São Paulo, Brazil: Conselho Regional de Medicina do Estado de São Paulo.

Secretaría de Comunicaciones y Transportes. *Diagnostico E Indices de Producción Del Sector Telecomunicaciones [Telecommunications Sector Diagnosis and Indexes].* Segundo Trimestre 2013. México: Secretaría de Comunicaciones y Transportes, August 26, 2013. Accessed July 21, 2014. http://www.ift.org.mx/iftweb/wp-content/uploads/2014/01/ITEL_Segundo_Trimestre_2013.pdf.

Sistema Nacional de Salud. 2011. Boletín de Información Estadística. Recursos Humanos, Físicos Y Materiales [Bulletin of Statistical Information. Human Resources, Physical and Material]. 30. México: Secretaría de Salud.

Skorin-Kapov, L. and Matijasevic, M. 2010. Analysis of QoS requirements for E-health services and mapping to evolved packet system QoS classes. *International Journal of Telemedicine and Applications* 2010: 1–18. doi:10.1155/2010/628086.

Trafton, Ann. 2008. "Eradicating TB with ... Cell Phone Minutes?" *Massachusetts Institute of Technology News.* Accessed July 21, 2014. http://web.mit.edu/newsoffice/2008/tb-cellphone-tt0604.html.

Villarreal-Reyes, S., Rivera-Rodríguez, R., Gómez-Velazco, L.E., Castañeda-Segura, R., Tamayo-Fernández, R., Valenzuela-Godoy, P., Barrera-Juárez, O., Martínez-Aragón, E., and Soto-Olivares, C.Y. 2013. *A-Prevenir* (version 1.0). México: CICESE.

Vouyioukas, D., Maglogiannis, I., and Komnakos, D. 2007. Emergency M-health services through high-speed 3G systems: Simulation and performance evaluation. *SIMULATION* 83 (4): 329–345. doi:10.1177/0037549707083113.

Webb, C.L., Waugh, C.L., Grigsby, J., Busenbark, D., Berdusis, K., Sahn, D.J., and Sable, C.A. 2013. Impact of telemedicine on hospital transport, length of stay, and medical outcomes in infants with suspected heart disease: A multicenter study. *Journal of the American Society of Echocardiography* 26 (9): 1090–1098. doi:10.1016/j.echo.2013.05.018.

Webber, L., Kilpi, F., Marsh, T., Rtveladze, K., Brown, M., and McPherson, K. 2012. High rates of obesity and non-communicable diseases predicted across Latin America. Ed. Noel Christopher Barengo. *PLoS One* 7 (8): e39589. doi:10.1371/journal.pone.0039589.

WHO Global Observatory for eHealth, and World Health Organization. 2011. mHealth new horizons for health through mobile technologies. Accessed June 27, 2014. http://www.who.int/goe/publications/goe_mhealth_web.pdf.

WHO Media Centre. 2013. Obesity and overweight. 311. Accessed June 27, 2014. http://www.who.int/mediacentre/factsheets/fs311/en/.

WHO Study Group on Diabetes Mellitus, ed. 1985. *Diabetes Mellitus: Report of a WHO Study Group on Diabetes Mellitus.* Geneva, Switzerland: WHO.

WHO/UNICEF Joint Water Supply and Sanitation Monitoring Programme. 2005. *Water for Life: Making It Happen.* Geneva, Switzerland: World Health Organization: UNICEF.

Wilde, E. and Vaha-Sipil, A. 2010. *URI Scheme for Global System for Mobile Communications (GSM) Short Message Service (SMS). RFC.* Accessed July 21, 2014. http://tools.ietf.org/html/rfc5724.

World Bank. 2013. GINI index. GINI Index Data Map. Accessed June 27, 2014. http://data.worldbank.org/indicator/SI.POV.GINI/countries?order=wbapi_data_value_2011%20wbapi_data_value%20wbapi_data_value-last&sort=asc&display=map.

World Health Organization. 2011. *Causes of Death 2008 Summary Tables.* Geneva, Switzerland: Health statistics and informatics Department, World Health Organization. Accessed June 27, 2014. http://www.who.int/gho/...disease/global_burden_disease_DTH6_2008.xls.

24

mHealth Issues, Challenges, and Implications: Ethics of Mobile Health Care—The Case of Sweden

Elin Palm

CONTENTS

24.1 Introduction

Smartphones and mobile applications have the potential to reshape and reform health-care provision, enabling personalized health monitoring outside traditional care units (Istepanian, 2004; van Halteren et al., 2004). Aging populations are typically seen as a serious challenge to prevailing health-care systems since extended life length (in most developed countries) is taken to imply that individuals will spend a longer part of their lives with complications and chronic diseases* (UN 2009). Concomitantly, the need for nursing, care, and prevention of complications associated with chronic conditions is taken to increase (Steg et al., 2006; OECD report, 2010). Information and communication

---

* Age-related chronic diseases such as diabetes; lung, locomotor, and sensory disorders; cardiovascular disease; and various forms of cancer are estimated to rise. With advancing age, elderly people are said to suffer from cognitive, visual, and hearing impairments and frailty due to disease, which may hinder independent living (Comyn et al., 2006).

technology (ICT)-based devices and systems are developed to meet the alleged economic and organizational challenges stemming from aged populations, and future health care will most likely be facilitated and distributed by means of, for example, sensors, actors, cameras (Palm, 2012), smartphones, and mobile applications (Coeckelbergh, 2010; Palm, 2013). An increasing availability of high (mobile)-bandwidth and mobile telecommunications system makes mHealth possible (van Halteren et al., 2004). Combined with a miniaturization of computers and sensors, ICT-based applications and services become mobile, wearable in clothing and in gadgets (Palm, 2010, 2012; Mittelstadt et al., 2011)— even possible to insert under the skin (Collste, 2011). Notably, mobile devices enable remote care provision. By means of mobile devices, individuals in need of support can receive medical advice from a distance, record their own health data, and forward these electronically to physicians or specialists irrespective of their location. Such devices can monitor a person's health condition continuously, identify deteriorating health conditions in real time, and alert care recipient and care provider in case of an occurrence. In this way, (routine) health care can be provided outside traditional care units such as in the homes of patients. Importantly, the possibility of monitoring temporarily and chronically ill patients without having to accommodate them in hospitals or wards has been considered a major economic and organizational advantage for the national health-care services in Europe (Palm, 2012). To a greater extent than traditional care, mobile-based care is involving the care recipient in his or her care process. A potential benefit is that such care can empower care recipients and promote their comfort. At the same time, this novel form of care may give rise to ethical considerations. mHealth provision implies health-related and person-specific data collection that may infringe on care recipients' privacy.

In this chapter, the focus is on the role of mobile devices in the future Swedish health care and on the potential ethical implications of a shift to mHealth. By means of a model for ethical technology assessment (eTA), analyzing the effects of novel technology on certain key values at an early stage of technology development and implementation, benefits and drawbacks of mHealth care will be identified. eTA also encircles relevant stakeholders and investigates how the different stakeholders' rights and interests are affected by certain technologies and technological systems (Palm and Hansson, 2006; Palm, 2013). By applying this assessment model on mHealth technology, ways in which positive aspects can be safeguarded and negative features avoided will be suggested.

24.2 Ambitions of the Future Swedish Health Care

Mobile devices are taken to influence health care significantly (van Halteren et al., 2004). Ten years ago, the following vision was outlined: "The increased availability, miniaturization, performance, enhanced data rates, and the expected convergence of future wire-less communication and network technologies around mobile health systems will accelerate the deployment of mHealth systems and services within the next decade" (Istepanian, 2004). Key findings of a global survey of mHealth initiatives conducted by the World Health Organization in 2011 are that even if mHealth applications have entered the market and been tested in several health-care settings, strategic implementation and integration in existing health-care systems are still lacking. "Policy-makers and administrators need to have the necessary knowledge to make the transition from pilot programmes to strategic large-scale deployments" (WHO report on mHealth, 2011). Whereas call centers, mobile applications, and SMS-based appointment reminders are commonplace, more advanced functions like health surveys (use of mobile devices for health-related data collection), surveillance of health-care recipients (e.g., measuring treatment compliance), and decision support are rare. That is, as of yet, mobile devices are not integrated in the health-care systems and utilized for chronic disease management as envisioned by care planners and technology developers. Sweden is no exception. Whereas a large number of test beds and living labs have been set up and visions for future mHealth have been presented, (European Commission Information Society and Media Directorate-General, 2011) a substantial shift to mHealth has not materialized yet. Encompassing and well-integrated mHealth systems are still discussed as something to be realized in the future. In a 2010 report, the Swedish Association of Local Authorities and Regions (SALAR) (responsible for strategic national health-care planning) expressed a rather bold vision for the future Swedish health care (Larsson, 2010; SALAR report, 2010). During the next decade, it was declared that the number of care places within traditional care institutions such as hospitals should be reduced by 50%. As far as possible, individuals in need of health care should be offered assistance and support in their homes rather than in wards. Home-based care provision and health-care services should be delivered by means of different types of ICT-based services and via mobile communication devices. This care should be of comparable standard to the traditional care and in addition be more flexible and easy to access. By mobile care, individuals in need of care should be able to get support more quickly and frequently (SALAR report, 2010). Briefly put, in Sweden, technology-assisted home care is considered a low-cost care alternative of equal standard to traditional care (SALAR report, 2010), which promotes security and independence of individuals in need of care and support (Szebehely and Trydegård, 2011), empowering outpatients.

Even if a complete shift to an integrated mHealth system has not been realized yet, technology developers offer a broad range of mobile solutions today and mobile devices and some systems are tested in Swedish health-care provision. Mobile devices enable monitoring and sharing of health information and feed into electronic patient records. More specifically, mobile applications have been developed to treat chronic illnesses such as diabetes and heart diseases. Today, smartphone-based glucose meters such as the *GlucoLog* record glucose levels instantaneously, transmit data to care provider electronically, and remind patients when they need to undertake glucose tests (A. Menarini Diagnostics, 2014). Cardiac diagnostics can be conducted by mobile devices, and Ericsson mHealth offers a solution for remote patient monitoring, supporting measurement of vital signs such as blood pressure, pulse rate, and oxygen saturation. Collected health data are transferred over mobile networks from patients to health-care providers (Ericsson, 2012). Importantly, mobile devices are employed not only within the somatic but also within the psychological care. One of the most popular medical applications in Sweden today (bought via App Store) is *iSelfhelp* (Kubicek and Boye, iSelfhelp—Depression, 2011), developed within the field of cognitive behavioral therapy (CBT). This application provides the user with a self-assessment tool in the form of an electronic dairy where he or she can register how he or she feels over time. The self-assessment tool is intended as a complement to traditional therapy in the treatment of individuals suffering from insomnia, anxiety, and depression.* This information will later serve as a basis for evaluation according to a manual—the same tool as used in traditional cognitive-based therapy.†

Technology developers have introduced m-based remote care as a care form that not only improves efficiency but also promotes the quality of care: "Not only can it help improve health services for employees of enterprises, it also enables effective and more efficient service delivery for healthcare providers and new revenue streams for operators" (Ericsson, 2014).

Technology developers and care planners typically emphasize the many positive aspects of mHealth. Certainly, these technologies may not only carry important benefits, but they may also raise ethical concerns that ought to be reflected on, preferably at an early stage before the technology is entrenched in society. For instance, a shift in care locus from the hospital to the domestic sphere may imply a profound change in the nature of care and have a strong impact on the situation of and the relation between formal and informal care providers and care recipients (Palm, 2012). In the following sections, implications of mobile devices and an mHealth system will be investigated by means of a brief technology assessment focusing on ethically relevant aspects.

24.3 Ethical Technology Assessment

Novel technologies often give rise to social and ethical problems—health-care technologies being an exception. In most cases, the ethical problems that appear are not of a radically new kind, but rather novel twists on age-old ethical issues such as individual liberty and equality. Completely new ethical problems seldom, if ever, appear (Johnson, 1985). Values that are prevalent in most ethical debates surrounding information and communication technology can be encircled (Brey, 2000). Drawing on these insights, we should learn from previous lessons and investigate whether novel technology affects values and interests that typically come in conflict. The model for eTA (Palm and Hansson, 2006) to be used in this chapter proposes a checklist consisting of nine values that should be used to assess novel technologies or novel uses of already existing technologies: (1) dissemination and use of information; (2) control, influence, and power; (3) impact on social contact patterns; (4) privacy; (5) sustainability; (6) human reproduction; (7) gender, minorities, and justice; (8) international relations; and (9) impact on human values. These values have been crucial in relation to previous technological innovations and are likely to be so in the future as well (Palm and Hansson, 2006). An assumption underlying this approach is that artifacts never can be neutral. On the contrary, they are influenced by human values and shape and form the context in which they are used and those using them. For instance, health-care technologies and medical technologies may alter our understanding of the issues at stake in health care, the role of care providers and patients, and the way in which care is conducted (Coeckelbergh, 2013).

The aims of eTA are to (1) evaluate emerging technologies and (2) influence technology development—preferably before the artifacts have reached the market. Once technologies

---

* Psychology Lab CBT is a therapy tool support, 2014.
† Research has shown that ICT-based self-treatment devices can have a certain supportive function in the treatment of mild anxiety (Andersson et al., 2012a, Andersson, 2012) and for individuals with eating disorders (Andersson et al. 2012b) but that they cannot replace traditional therapy. Thus, applications like *iSelfhelp* should not be used as stand-alone devices but serve as complement to therapy.

have been introduced at the market and implemented in care services, amendments are significantly more difficult to make than during the design and testing stage. Based on ethical assessment results, technology developers and those responsible for the introduction of certain technical solutions can adjust the technology, promote positive aspects of the technology, and avoid negative implications such as far as possible.

An ethical assessment can encircle concerned parties' needs, interests, and opinions (Palm, 2013). Moreover, it can show who is affected by a certain technology, in what way, and to what extent and may contribute to a fair distribution of benefits and drawbacks. In the case of mobile technology developed for health-care use, the following parties should be concerned:

24.3.1 Care Recipients (Patients)

Health-care professionals include general practitioners, special practitioners, nurses, district nurses, and home-help service personnel:

- Caring kin persons/kin persons
- Technology developers
- System providers
- System owners
- System users
- County council politicians
- Health technology buyers

A full-fledged eTA would take into consideration the rights and interests of all these stakeholders (Palm, 2013).

Importantly, not only users benefit from eTA but not only technology developers and care providers. Negative impact may be costly in many respects, not only from an economical perspective. Users' well-being is central to any technology developer—and perhaps particularly so regarding health-care technology. By carrying out an ethical assessment at an early stage, certain problems can be avoided before the technology enters the market. Thus, technology assessment can be in the interests of those developing new technical solutions. If aspects in need of modification are pointed out at the prototype stage, technology developers are more likely to alter the design than what they would be later on in the developmental process when a change is more inconvenient and costly (Palm and Hansson, 2006).

Ideally, eTA should be undertaken in dialogue with technology developers throughout the whole chain of research and development and have the form of a continuous dialogue rather than a single evaluation at a specific point in time. In practice, it is rather difficult to conduct eTA since it requires a close cooperation between technology developers and ethicists. Although it can be possible in multidisciplinary research and development projects,* it is seldom the case that single companies involve ethical expertise. Yet important attempts are made in Sweden, for example, by New Tools for Health, to integrate ethicists in projects where health-care technologies are developed

---

* Examples of how eTA has been used in relation to novel health-care technology can be found in the book *Interdisciplinary Assessment of Personal Health Monitoring* summing up the research conducted within the Framework 7 project Personalized Health Monitoring Ethics (PHM) (edited by Schmidt and Reinhoff, 2013).

(see New Tools for Health, http://www.newtoolsforhealth.com/). In most practical cases, however, a more feasible option is to analyze the technology reactively, once the technology already is on the market but before it has matured and permeated. At this stage, the ways in which technology is implemented may still be modified. Attempts may be made by care planners and care providers to avoid negative implications where different choices of usage are possible. Importantly, the examples of mobile devices chosen here are all of recent date and allow for such modifications. Findings presented in this chapter are communicated to New Tools for Health project leaders who can influence the direction of current and future projects on mobile technology within the health-care sector.

Among the many stakeholders listed earlier, this chapter focuses on the impact of mHealth on care providers, care recipients, and informal care providers and the (possibly altered) relations between them. Within this rather limited assessment, the main attention will be on the three of the aspects listed: *control, influence, and power, impact on social contact patterns*, and *privacy*. Furthermore, mobile technologies discussed here are not primarily devices that serve to inform or remind patients about health-care appointments, that is, those devices most frequently used today (WHO report, 2011), but rather mHealth solutions that are under development, aiming at providing health monitoring and more qualified support. Mobile-based monitoring system for home care includes at least one diagnostic device in a patient's home, a communication device for transmitting health information collected by the diagnostic device to an information retrieval device, which processes and presents the information to a professional care provider, who may start a health-relevant action for the patient. The technology may be used (1) for monitoring patients with specific chronic conditions, for example, heart disease or diabetes, or (2) for having a more general monitoring function aiming at helping elderly, weak, and perhaps somewhat demented people.

24.3.2 Control, Influence, and Power

Typically, mHealth involves the participation and active contribution of a (remote) patient, thus an important element of self-care. Instrumented self-care is increasingly promoted by care planners and technology developers (cf. Fex, 2010). Wisely designed, it is said that mobile-based care regimes can maximize care recipients' ability to manage on their own. Moreover, it is argued that this care form can empower care recipients and strengthen their autonomy (cf. Fex, 2010). Self-care instruments come in the form of wearable devices enabling physiological monitoring, on-site analysis, and automatic alerts (in case of anomalies), for example, wrist- and armbands, smart clothes, and smartphones (Giannetsos et al., 2011). Smartphone applications can track the user's movement, evaluate his or her exercise output, and share such data with health-care professionals. *Body area networks*, or sensors worn in or beneath clothing, can be combined with smartphone applications to provide longitudinal health tracking (Boulos et al., 2011). Data generated from such devices allow the bearer to learn about his or her health status and parameters that increase or decrease his or her condition (Mittelstadt et al., 2011). Better informed, he or she can make more conscious lifestyle choices. Thus, individuals in need of care can become more active in health matters and take responsibility for their health and well-being (Coeckelbergh, 2013). This potential, it is argued, should be utilized to promote an increased health awareness among actual and presumptive patients. Ideally, citizens should be more concerned with and actively taking responsibility for their health. "Individuals must play a role in taking care of their own health, and therefore citizens' and patients' participation and empowerment need to be regarded as core values in all health-related work at the EC level" (Commission of the European Communities).

Certainly, unrestricted access to real-time high-quality data is an aspect that generally is taken to promote awareness and well-informed decision making. However, collection, processing, and transfer of health-related data may also imply an external control labelled *somatic surveillance* that can affect the care recipient negatively. Somatic surveillance has been described as a technological monitoring of and intervention into body functions (Monahan and Wall, 2007). Somatic surveillance systems depend upon three interrelated processes: first, the translation of corporeal information (e.g., heart rate, hormone levels, and temperature) into data, usually by means of sensors applied directly to or embedded within the body; second, the communication of those data across networks, effectively situating bodies as *nodes* on larger information networks; and third, the intervention into body functions through various sociotechnical feedback mechanisms. In the final part of the process, bodies are not only informatized but controlled in various ways. The systems, in other words, do not merely generate data; they also produce social norms, technical constraints, and network commands (Monahan and Wall, 2007).

At a first glance, a constant measuring of health parameters influencing the care recipient's behavior and making his or her more health concerned may seem an exclusively positive aspect. However, continuous health checks may also induce stress and, in a sense, curb the care recipient's autonomy. Subject to somatic surveillance, he or she may feel forced to internalize and act according to standards that are not his or her own but set (and enforced) by the health-care provider. Moreover, to be under continuous monitoring may increase the care recipient's feeling of being a patient. Under such conditions, individuals receiving care and support may come to think of themselves, foremost in terms of health/ill health, transforming their identities to that of patients first and foremost (Collste, 2011). That is, an informatization of the body may have a strong impact on the care recipient's self-understanding.

Moreover, mobile devices enable flexible care provision. With access to a broad range of self-care technologies, care recipients' chances to influence and design a care model that suits their preferences, needs, and overall situation are significantly strengthened. The opportunity to be involved in and influence the care process is said to empower care recipients and to promote their autonomy and self-determination. A greater involvement of the care recipient in the care process may be a means to rectify a traditional power asymmetry between care provider and care recipient. However, whether the technology empowers or hampers the care recipient is likely to depend on the extent to which the latter feels capable of taking on the responsibility that self-care requires, that is, the decision making and the involvement necessary (e.g., interpretation of and response to health data) under this new care regime. To an ordinary, healthy person, thinking of a future situation in which he or she would need assistance and care, self-care may seem an ideal, but once chronically ill, the situation may look completely different to the person with a factual care need. Then, many aspects of self-care may seem overly demanding. The extent to which self-care is considered attractive may also vary with digital literacy (Golau and Fombona, 2012; Ng, 2012), that is, whether the user is familiar with digital technologies (digital natives) or unskilled in that regard (digital immigrants). Individuals with prior knowledge of and experiences from digital technologies are more likely to feel comfortable with and empowered by self-care devices. Certainly, self-care is typically promoted as an optional complement to ordinary care. And, as long as a voluntary option, self-care may be positively received. At the same time, the cost-saving features of self-care are stressed by policy makers, care planners, and technology developers alike. Given the need to cut costs within health care, what are optional care forms today may, if proven to be as cost-effective as believed today, become mandatory in the future. And if not freely chosen, it may be perceived as burdensome by care recipients and as something that infringes on their autonomy. Already today, for persons living in remote areas like in the northern parts of Sweden, remote care may be the only means of getting continuous health check-ups and feedback. Hence, the chances that individuals have to consent or reject should be considered before mobile care is promoted as an individual choice.

24.3.3 Impact on Social Contact Patterns

Mobile care devices enable remote attendance, consultation, support, and health checkups and reduce the need for face-to-face meetings. This means that caregivers and care receivers are no longer in the same location but communicate and exchange information via mobile devices (Coeckelbergh, 2013). In that way, chronically ill and frail elderly people can stay at home while their medical condition is monitored by nurses and other care workers. That is, the technology may both reduce the number of hospital visits and hospitalizations and even delay the need for a move to residential care. Whereas care recipients may benefit in terms of convenience and comfort (remaining at home), this care form may hamper the relations not only between care provider and care recipient but also between care providers. Under an mHealth system, physical meetings can, to a large extent, be replaced by *virtual meetings*. By reducing direct intervention between the agents, remote care may alienate the care provider from the care recipient and from other care workers, rendering the care practice more impersonal and less virtuous. It has been noted that direct, practical, and physical engagement seems to be discouraged under an mHealth system (Coeckelbergh, 2013). Limited physical encounters may be to the detriment of both care recipients and care providers. In order to develop the know-how (tacit knowledge) and skills characteristic of their health-care profession, it has been stressed that nurses and doctors need to interact with patients. Reduced unmediated meetings render their training more difficult (Coeckelbergh, 2013). Care recipients may increasingly seek health information via applications and the Internet. In that sense, they miss skilled physicians' examination relying on visual and tactile, embodied understanding (Coeckelbergh, 2013). However, it is important to consider that ICT-mediated care provision can, but must not *necessarily*, reduce the conditions under which skilled engagement and craftsmanship can develop. Care provision has always involved technology and wisely used mobile technology can enhance the quality of care. Mobile devices may in many ways facilitate communication between care provider and care recipient, and traditional ways of communicating may have to take a new form. mHealth may be advisable in some but not all situations and contexts. For instance, remote care may function well in cases where doctor and patient and therapist and client have already established an alliance but may be more challenging if the relation must be established in the *virtual* world. For such reasons, it has been suggested that care provision based on mobile devices should be used in combination with *old* forms of interaction and care, rather than implemented as an alternative that operates on its own (Coeckelbergh, 2013). Technology-based support may function well if part of a trusted structure and backed up with support from the care provider. Empirical studies conducted in Sweden on the expectations and experiences of elderly people living with different types of remote attendance at home indicate that technology can convey security and comfort (Essén, 2008). Elderly Swedish couples setup with ICT-based support systems in their homes have expressed great content with their technological assistance. Some of the reasons for this content were that they felt more safe, more cared for, and less isolated with than without these systems. However, the study subjects stressed that the key to their positive attitudes and the success of (this form of) remote care was the underlying relation to the care provider. They characterized this relation by words like *respect, mutual recognition*, and *trust* (Essén, 2008). This indicates that mobile care solutions may be of great value if integrated in a well-functioning care relation. Certainly distance and alienation are typical problems within modern health-care provision, but importantly, in many cases, alienation stems from bureaucratic rules and requirements (based on economic rationality and cost-efficiency) rather than on the technology mediating care (Yakhlef and Essén, 2013).

Care planners, care providers, and technology developers alike often stress that "despite a care need, individuals should be able to, as far as possible, function unaided in a location of their choice" (SALAR, 2010). Being *unaided* tends to be equated with not having to rely on the immediate involvement of care professionals. Independence is assumed to be a desirable thing and something that can be promoted by offering individuals in need of support: alternatives to chose from and the ability to remain at home and instrumented care, reducing the need for direct involvement of care providers (Palm, 2012). Certainly, self-care may enhance the quality of care and promote independence. However, it may also cause stress and insecurity and replace interpersonal contacts, hence be to the detriment of social contacts. Importantly, social contact networks have been established as key to health and well-being. Conversely, (perceived) social isolation contributes to dissatisfaction and has become an increasing problem in relation to home care (Cacioppo et al., 2002; Burns and Haslinger-Baumann, 2008). Whereas inpatients benefit from daily interaction with health-care professionals, recipients of self-care technology may become socially isolated during their convalescence (Palm, 2013). Rather than making care recipients independently proper, that is, able to manage on their own, they will remain dependent on assistance, whether it comes from flesh and blood persons or from technology. Set up with advanced monitoring devices at home, the user may need consultancy and support, and if this cannot be had from the care provider, he or she may have to turn to family and friends. Remote care may thus alter existing relations of dependence: decreasing care recipients' dependence on formal care providers while increasing their need for support from informal care providers, that is, nonprofessional and nonremunerated care agents (Palm, 2013), and reliance on technological systems. In the long run, remote care and concomitant reduced contact with care providers may lead to weakened social contacts—even to social isolation. And, as of yet, the effects of being dependent on machines are not properly investigated. Social isolation is a risk faced by all patient types but in particular by elderly people with chronic diseases and limited social networks who, to an increasing extent, receive care provision.

24.3.4 Privacy

A common concern regarding all surveillance-capable technology is that of potential privacy intrusions. Constant monitoring is almost by default said to hamper personal autonomy. However, it should be recognized that in the case of mobile care, certain aspects of privacy may be enhanced, whereas other dimensions may be negatively affected. In order to demonstrate this, Beate Rössler's triptych model of privacy will be used. Her model distinguishes between (1) informational privacy, (2) local privacy, and (3) decisional privacy. Beate Rössler argues that privacy is necessary for individuals to be able to form and enact their own goals and express their identity and values. The acts of deciding what to reveal of oneself to others, compartmentalizing information, creating personal space, and influencing relationships and surroundings are all expressions of self-government.

Thus, Rössler anchors privacy in personal autonomy (Rössler, 2005). Privacy is described as the spatial and intellectual sphere that enables individuals to decide over matters that concern them, to control who has access over information about themselves (both direct sensory access and stored data), and to establish and develop different types of relations. Conversely, privacy intrusions are unwarranted intrusions into the private realm, personal decisions, and personal data, respectively (Rössler, 2005:44). On this account, privacy is not everything that is private or secret but that which is related to autonomy.

What then is the scope of privacy? Arguably, *private* can apply to actions, relations, and decisions irrespective of where these take place. Importantly, what we consider private and rightfully private is based on social constructions and the conventional view of privacy as belonging only to the traditionally private sphere and not to public and semipublic areas. However, rather than understanding particular realms or spaces as private, dimensions of an individual's actions and interests can be private (Rössler in:. Rössler (Ed) 2004:7). Likewise, rather than understanding certain types of information as sensitive, we should recognize that information can be neutral in some contexts and sensitive in other contexts.

In relation to mHealth, it is important to consider all dimensions of privacy and not only focus informational privacy as most assessments do. Certainly, mHealth devices deal with person-specific and potentially privacy-sensitive data that require adequate respect and information security. Although care recipients may get an increased access to their health data, they are not necessarily aware of how the information gathered about them is processed and used by health-care providers. Over time, a large amount of data (person specific and health related) have been collected—information that may be compiled, analyzed, and transferred in ways beyond the data subject's awareness. Hence, care recipients facing mobile-based remote care must get necessary and sufficient information regarding all stages of data usage in order to be able to consent the use of the systems in a meaningful way, to decide under what conditions they wish to use them, etc. They also need to be able to influence how data about them should be used, who should have access (decisional privacy), etc.

Furthermore, instrumented home-based care implies that care provision takes place in the most private of domains, that is, in the care recipients' homes. In many ways, this implies great benefits in terms of convenience. Frail persons must not spend time and energy traveling to and from health-care institutions in order to receive the support they need. But there are also potential drawbacks that merit attention. A significant aspect of local privacy is a person's ability to control and restrict others' access to his or her private domain (home or dwelling place). Under an mHealth system, care providers may get virtual (remote) and physical (direct) access to the patients' home. Information on care recipients' everyday life may become visible to a larger extent than if care is received outside the home. Some of this information is relevant to care provision, but aspects unnecessary for the care process may also be disclosed by means of, for example, smart health-care gadgets. Hence, care recipients' ability to control their private domain may be significantly circumscribed with surveillance-capable care equipment at home such as movement sensors registering and reporting information about activity and inactivity, health status, and his or her doings to a care unit. Home-care technology can continuously register, store, and transfer data from the private realm of the care recipient's home to the care provider. For instance, sensors integrated in wearable monitoring device can feed information about him or her (Palm, 2012). Remote health data collection and monitoring may imply the gathering of surplus—and potentially privacy invasive—information such as user location and behavior connected with health parameters. Wearable monitors enable longitudinal tracking and health and behavior parameters can be linked and evaluated. Spatial information, or how a user moves, enables tracking of health-related behaviors in private spaces, such as sleeping patterns and falls (Mittelstadt et al., 2011).

Privacy in relation to home-care technology is a highly complex matter. Some dimensions of privacy may be protected, while other aspects are intruded upon. Traditionally, the home or dwelling place has been a secluded place, where we can restrict others' access and introspection why monitoring at home may be perceived of as privacy sensitive. However, by means of remote care, individuals in need of care and support may need less physical visits by care professionals in their homes (Mittelstadt et al., 2011). In that sense, their local privacy may be strengthened. If these aspects are considered at an early stage in the development process by those designing novel technology and by those responsible for integrating technology in health-care provision, mobile devices may serve to increase privacy by providing care recipients a larger set of care alternative to choose from and by offering them personal control over how their data are managed. At the same time, the value of access to different care solutions must be balanced against the stress the need to choose between alternatives may imply for some individuals. Once again, care recipients' level of digital literacy is important to consider. If the care recipient is ill trained to use self-care devices, there is a risk that he or she faces difficulties to protect their privacy in interaction with the technology.

24.4 Conclusion

24.4.1 Case Studies

In order to contextualize the rather limited assessment model presented earlier, a more encompassing ethical assessment conducted on home-help service device will be referenced together with a few other assessments focusing ethical aspects of novel home- and self-care technology. The first two assessments introduced in the following make use of scenarios in order to tease out ethical implications of novel technology in different contexts. The third assessment investigates a software device implemented in Swedish home-help care. All of them demonstrate the need to consider ethical aspects of remote care and ways in which this can be done.

24.4.1.1 Ethical Assessment in the Design of Ambient Assisted Living

In an article from 2008 "Ethical Assessment in the Design of Ambient Assisted Living," Veikko Ikonen and Eija Kaasinen propose a set of ethical guidelines that should be taken into account when designing applications and services that utilize micro–nano integrated platform for transverse ambient intelligence applications (MINAmI) platform for mobile-centric ambient intelligence. The guidelines also cover issues related to implementing the MINAmI platform itself. The ethical guidelines proposed in the article are built on six ethical principles that are selected based on the ethical assessment of MINAmI scenarios:

- Privacy: An individual should be able to control access to his or her personal information and to protect his or her own space.
- Autonomy: An individual has the right to decide how and to what purposes he or she is using technology.

- Integrity and dignity: Individuals should be respected and technical solutions shall not violate their dignity as human beings.

- Reliability: Technical solutions shall be sufficiently reliable for the purposes that they are being used for. Technology shall not threat user's physical or mental health.

- E-inclusion: Services should be accessible to all user groups despite their physical or mental deficiencies.

- Role of technology in the society: The society should make use of the technology so that it increases the quality of life and does not cause harm to anyone.

Key legal issues of ambient intelligence environments are privacy; data protection; intellectual property rights (IPR), particularly copyright (access to content; content adaptation); and contract law. The European legal system provides reasonable protection, for example, to the user privacy, but it should also be ensured that the legal system will not prevent useful services. It is stressed that privacy is well protected on the level of information privacy but that AmI solutions motivate protection of spatial and bodily privacy as well.

Ethical assessments, it is argued, should be conducted throughout the product life cycle and a holistic perspective on design is necessary. This means that a broader grip on stakeholders should be taken. Possible negative consequences should be identified and reported at an early stage. Failure to do so may be damaging from an economical perspective as well as from an ethical view. Furthermore, it is stressed that ethical issues must be addressed at a higher level as well (organizations, society) in order to assure that the best possible design decisions are chosen when conflicts appear between different stakeholders. A challenge identified in relation to ethical technology assessment is that of managing to involve system designers in the assessment and to increase their awareness of the meaning and value of ethical issues.

24.4.1.2 Ethics and Persuasive Technology: An Exploratory Study in the Context of Healthy Living (Page and Kray, 2010)

In the article "Ethics and Persuasive Technology: An Exploratory Study in the Context of Healthy Living" (2010), Rachel E. Page and Christian Kray investigate ethical implications of what they label persuasive technology. Persuasive technology is a combination of mobile and ubiquitous delivery mechanisms with the potential to reach and influence people everywhere at any time. More precisely put, persuasive technology is defined as "any interactive computing system designed to change people's attitudes or behaviour," which can be applied in a wide range of scenarios. Mobile (and ubiquitous) devices are very well suited for the delivery of persuasive content as they can sense contextual factors of relevance to a specific user (such as location and/or task) and tailor messages so that they are delivered in the most effective way. Considerable potential has been attributed to this technology in terms of helping people to change their behavior" (Page and Kray, 2010). An area within which this technology may get strong influence, it is argued, is that of healthy living. Persuasive technology could, for example, be applied to areas such as disease management and prevention (Page and Kray, 2010).

It is noted that while this technology may bring about important benefits, at the same time, it raises ethical considerations that, as of yet, are poorly understood. In their article, the authors aim to provide a better understanding for ethics in persuasive technology, and as such, it is the most important contribution. Focus group studies based on a set of scenarios are used by the authors to tease out possible ethical issues in persuasive technology employed within care provision.

The five scenarios used were as follows:

1. A food diary mobile phone application for teenage girls that provided incentives and motivated the users to follow a healthy diet. It was depicted as if created and influenced by a private company. The scenario was intended to instigate a discussion on the use of persuasive technology on a potentially vulnerable group—teenage girls (self-perception, body image, etc.) among the assessment participants.

2. A website for young adults designed to monitor their alcohol intake and be persuaded to drink less through social comparisons. The website was described as if provided by the government with the aim to support people to change their own behavior. Questions driving this scenario were whether the website would be perceived as an effective way of delivering persuasive content relating to healthy living together with the role of social comparisons.

3. An embedded device for morbidly obese adults to change their eating habits by delivering drastic messages such as "keep eating like that and you'll be dead soon" commissioned by their doctor. This scenario was designed to test extreme conditions and their impact on individuals' perception of what would be ethically acceptable. It was also meant to evoke an emotional response among the assessment participants and a discussion on any limits that possibly should be placed upon this kind of technology.

4. A text message system, similar to the warnings on cigarette packets, which were being sent to a mobile phone at the time when the user was about to have a cigarette. A commercial company was presumed to be the responsible agent behind the messages. Key features of this scenario were the proactive delivery of persuasion as well as the inclusion of contextual factors into the equation.

5. A game to persuade children to eat more fruit, encouraging them to ask their parents for fruit. The initiators in the scenario were the teachers of the children described in the scenario, who encouraged them to play the game. The main goal of this scenario was to investigate if manipulating children in this way was considered ethically acceptable by the assessment participants and whether or not *disguising* persuasion as a game would raise any concerns.

The usage of different scenarios with a number of different technology developers (senders) addressing several groups of stakeholders (users) aimed at teasing out the ethical acceptability of paternalistic coercion in the domain of health care. To what extent would potential system users from different stakeholder groups accept the coercive approaches? Based on the results of this study, the authors conclude that persuasion may backfire. The user must find motivation and reason on their own in order to accept a system that tries to change their behaviors and health patterns. Interestingly, three factors relevant for the perceived ethical acceptability of persuasive technology among the study subjects were identified: (1) the recipient, (2) the commissioner, and (3) the means of delivery. The authors stress that there is a link between the purpose of persuasion and the means used to persuade (Page and Kray, 2010). These findings should inform the design of future persuasive interventions.

24.4.1.3 An Interactive Ethical Assessment of Surveillance-Capable Software within the Home-Help Service Sector (Palm, 2013)

The third case study is an interactive ethical assessment of surveillance-capable software *iCare* used to control the care and support provided by home helpers in the Swedish home-help service sector (Palm, 2013). This assessment consists of a small-scale interview study on home helpers' experiences of and reactions to the implementation of the care software *iCare* in their workspace (covering their whole workday including all travels to and from the care recipients' homes). In the article, the interview serves as the starting point for an ethical analysis of the impact of the newly introduced care software *iCare* on key values within ethics: privacy, autonomy, and equality. The implementation and use case of surveillance-capable technology in a home-help service sector is assessed from the perspective of ethics and based on an ethical checklist extracted from ethical issues frequently occurring in relation to ICT in general and ICT in health care in particular. It is concluded that employees' level of awareness, access to sufficient and relevant information, and their chances of influencing surveillance conduct are significant for their acceptance of the surveillance regime. Surveillance in the home-help service setting has been investigated from the perspective of ethnology and organizational studies but not, as here, from the perspective of ethics. As such, this assessment contributes by teasing out conditions for the ethical acceptance of workspace surveillance. The assessment compares the introduction of *iCare* in two home-help units where the introduction process looked very different (communication strategies, reasons and conditions for implementations, etc.) and illustrates that the ways in which novel technology, and especially surveillance-capable such, is communicated when implemented in a workspace and in the homes of care recipients are of significant importance for values such as trust and privacy.

24.4.2 Future Direction of Research

Few thoroughgoing investigations of the ethical implications of mHealth technologies have been conducted. A model for eTA is proposed in this chapter. The aim of this model is to provide clear and concise ethical guidelines that could be used as checklists in the research, design, development, and implementation of mHealth solutions for future care provision. The proposed model suggests that nine ethical issues relevant to most any technologies should be used as a checklist to evaluate the ethical acceptability of novel mHealth technology and to guide a well-reflected introduction of such technology in the domain of health care. Unfortunately, the ethical assessment conducted within this chapter is of a limited kind. Only three of the nine ethical issues to be considered have been covered. Likewise, only a few of the stakeholders listed as worthy of consideration are addressed in this assessment. Hence, a more extensive eTA remains to be undertaken in order to properly encircle the many different aspects and implications of mHealth technology. Certainly, an encompassing ethical assessment that includes all issues and stakeholders may seem almost utopic. However, a utopic goal may result in that the assessments conducted become as inclusive as possible.

Importantly, eTAs should be conducted continuously to follow the life cycle of specific technologies. Rather than making one large-scale assessment at one point in time, it is preferable to undertake several smaller assessments repeatedly. By conducting several small-scale evaluations, an encompassing image can be had that covers the evolution of the technology at hand. Furthermore, the assessment model itself may need further improvement and revision and should be considered an iterative document rather than a complete assessment model. Future assessments may encircle ethical aspects that have not been covered by the present model.

As of yet, only a few more extensive ethical assessments of mHealth technology have been conducted. Three of them are described in the previous scenarios. Apart from these three assessments, ethical evaluations have typically focused the impact of mHealth devices on care recipients' (informational) privacy, risks related to data transfer, and the need for accurate data protection regimes to safeguard informational privacy. The model proposed here takes a much broader grip, inviting technology developers, care planners, care providers, care recipients, and other stakeholders to critically reflect on a broader range of ethically relevant aspects. Since the vast majority of the aspects listed in this inclusive assessment model have not been properly investigated yet, a most urgent challenge remains.

24.4.3 Conclusion

mHealth has been depicted as the care form of the future, enabling home-based care and a regime that to a greater extent can involve the care recipient in the health-care process, hence strengthening him or her as an active and well-informed care consumer (as opposed to a passive patient). In Sweden, the integration of mobile-based services and functions within the health-care sector has been seen as the most promising way to meet the increasing demands of an aging population. However, mHealth is still under development with prototypes piloted in test beds and living labs. A large number of devices have been developed, but a more integrated mHealth system has not been realized. This situation lends itself well to an eTA that preferably should be conducted at an early stage of development and implementation. Starting out from the eTA model suggested by Palm and Hansson (2006), three key features were selected for ethical analysis: *control, influence, and power, impact on social contact patterns*, and *privacy*. First, the ways in which mobile devices and an encompassing mHealth system may influence some of the stakeholders'—primarily care recipients—chances to control and influence the technology were considered. To some extent, the potential impact on informal care providers was discussed as well. Technology facilitating self-care and home-based care is often said to empower the care recipients, making them more active in and responsible for the care process. mHealth solutions enabling continuous health status updates may certainly render them better informed and more capable of making sound decisions regarding their health and provision of health services. However, the responsibility to, continuously, stay updated and make wise decisions may also generate a certain amount of stress. Encompassing and constant health monitoring may even alter the care recipients' self-understanding. In a worst-case scenario, they may come to think of themselves primarily as patients, that is, as someone in need of care and support, and their health status may become their overriding concern, preventing them from focusing on other aspects of themselves (thus impairing their personal autonomy). Also, if the intended users disagree with care planners and care providers' reasons for introducing certain care devices, they may perceive of technology-based support as unwelcome paternalism. Moreover, considering the impact of mHealth on social contact patterns, it was argued in the text that under an mHealth system, *the necessity* of physical meetings between care provider and care recipient would be reduced. At the same time, it was cautioned that a need for social encounters will remain. On the one hand, care recipients may benefit from a certain form of recognition and social contact that visits by the care provider imply. Care providers, on the other hand, may need the meetings with patients in order to develop within their profession—to become skilled care professionals. Importantly, for different reasons, both parties may depend on the traditional physical meetings for their well-being and development. If self-care is introduced as a substitute to direct encounters, this form of care may cause alienation and social isolation. Remote care may also increase the need for involvement of unqualified kin persons in the caring process. Without proper training and information regarding the function of self-care systems, kin persons may be uncomfortable, even stressed, if they must help the care recipient operating such device. Hence, the situation and interests of this category should be recognized by those introducing self-care systems in the home of care recipients. Mobile care solutions require a broader understanding for the car recipient and the context in which he or she will use this form of support. Spouses, partners, children, or other persons in the same household as the patient also deserve recognition in the planning of mobile-based care. Regarding privacy, it has been shown that (at least) three dimensions of privacy should be considered in relation to mHealth and that the technology may strengthen as well as circumscribe the many different aspects of privacy and that aspects other than informational privacy must be considered.

What can be learned from this, limited, eTA is that there are positive as well as negative consequences of a shift to mHealth. This brief inventory highlights aspects that should be avoided and promoted, respectively. Certainly, a more encompassing assessment covering all nine aspects of the eTA checklist would be ideal. Also, it would be desirable to systematically investigate the implications of a specific mobile device for each and all of the identified stakeholders. The main purpose of this analysis however was twofold: (1) to introduce eTA as a model for analyzing novel health-care technology and, (2) by means of this model, to highlight some of the ethical implications that mHealth may be clad with. Despite the limited range and scope of this particular assessment, some tentative conclusions can still be reached. It comes clear that a vision like the one expressed by SALAR—to replace traditional care with home care—motivated by values like personal security, independence, and well-being may carry undesirable aspects like uncertainty, social isolation, and stress. Hence, the development of mHealth requires a careful balancing of the many values and interests at stake. Importantly, mobile-based remote care should not be introduced as a means to replace traditional care. Social and ethical costs would most likely be too grave in terms of well-being and social isolation despite a potential economic gain. Rather, remote care may serve as a complement to traditional care and is likely to succeed if integrated in a framework of support from *flesh and blood* care providers. In order to provide ethically sound care, technology developers and care planners should be responsible for assessing novel technology in advance and care providers for continuously evaluating the technology once in use.

Questions

Some of the questions discussed are as follows:
1. How can the ethical implications of mHealth technologies be investigated?
2. How can negative aspects of novel technologies like mHealth systems be avoided?
3. What are the potential ethical implications of a shift from the traditional, institutionalized health care to care provision in the homes of care recipients?
4. What ethical aspects should be considered in relation to a shift?
5. What are the main assets of mHealth care from an ethical perspective?

References

A. Menarini Diagnostics. GlucoLog, B.T. device, Retrieved on August 9, 2014, http://www.en.menarinidiagnostics.se/Products/home_glucose_testing/GlucoLog-B.T.-device.

Andersson, G. Guided internet treatment for anxiety disorders. As effective as face-to-face therapies? *Studies in Health Technology and Informatics*, 181, (2012) 3–7.

Andersson, G., Carlbring, P., and Furmark, T. On behalf of the SOFIE Research Group. Therapist experience and knowledge acquisition in Internet-delivered CBT for social anxiety disorder: A randomized controlled trial. *PloS ONE*, 7(5) (2012b).

Andersson, G., Enander, E., Andrén, J., Hedman, P., Ljótsson, E., Hursti, B., Bergström, T. et al. Internet-based cognitive behaviour therapy for obsessive-compulsive disorder: A randomised controlled trial. *Psychological Medicine*, 42, (2012a) 2193–2203.

Boulos, Maged N. Kamel, Steve, W., Carloes, T., and Jones, R. How smartphones are changing the face of mobile and participatory healthcare: An overview, with example from eCAALYX, *Biomedical Engineering Online*, 10 (24) (2011). doi: 10.1186/1475-925X-10-24.

Brey, P. Disclosive computer Ethics: The exposure and evaluation of embedded normativity in computer technology, *Computers and Society*, 30(4) (2000) 10–16.

Burns, E. and Haslinger-Baumann, E. Evaluation of the nursing diagnosis "social isolation" and the use of evidence-based nursing. *Pflege* 21(1) (2008) 25–30.

Cacioppo, J., Hawkley, L., Crawford, E., Ernst J., Burleson, M., Kowalewski, R., Malarkey, W., Van Cauter, E., and Berntson, G. Loneliness and health: Potential mechanisms. *Psychosomatic Medicine*, 64 (2002) 407–417.

Coeckelbergh, M. Healthcare, capabilities, and AI assistive technologies, *Ethical Theory and Moral Practice* 13 (2) (2010).

Coeckelbergh, M. E-care as craftsmanship: Virtuous work, skilled engagement, and information technology in health care. *Medical Health Care and Philosophy* (e-publication ahead of print, accepted January 2013).

Collste, G. Under my skin: The Ethics of ambient computing for personal health monitoring. In S. Nagy Hesse-Biber (ed.) *The Handbook of Emergent Technologies in Social Research*. Oxford University Press: Oxford, U.K. (2011).

Comyn, G., Olsson, S., Guenzler, R., Özcivelek, R., Zinnbauer, D., and Cabrera, M. User Needs in ICT Research for Independent Living, with a Focus on Health Aspects, EUR Number: 22352 EN, October 2006, http://ipts.jrc.ec.europa.eu/publications/pub.cfm?id=1452.

Ericsson. Live Smart Mobile Health, Ericsson Mobile Health, Remote Patient Monitoring, EN/LZT 1380316/11 R6, Ericsson Nikola Tesla 2012, http://www.ericsson.com/hr/ict_solutions/e-health/emh/brochuraA5_EMH_R6_en.pdf.

Ericsson. Technology for Good Around the World. Retrieved on August 9, 2014, http://www.ericsson.com/thecompany/sustainability-corporateresponsibility/technology-for-good#!.

Essén, A. The two facets of electronic care surveillance: An exploration of the views of older people who live with monitoring devices. *Social Sciences & Medicine*, 67 (1) (2008) 128–136.

European Commission Information Society and Media Directorate-General (2011) Report on the public consultation on eHealth Action Plan 2012–2020, http://ec.europa.eu/information_society/activities/health/docs/policy/ehap2012public-consult-report.pdf (accessed August 28, 2013).

Fex, A. From novice towards self-care Expert—Studies of self-care among persons using advanced medical technology at home, Doctoral thesis, Division of Nursing Science, Department of Medical and Health Sciences, Faculty of Health Sciences. Linköping University: Linköping, Sweden (2010).

Giannetsos, T., Dimitriou T., Prasad N.R. People-centric sensing in assistive healthcare. *Security and Communication Networks* 4 (2011) 1295–1307.

Goulão, Maria de F. and Javier, F. Digital literacy and adults learners' perception: The case of a second chance to University. *Procedia—Social and Behavioural Science* 46 (2012) 350–355.

Istepanian, R. Guest editorial introduction to the special section on M-Health: Beyond seamless mobility and global wireless health-care connectivity. *IEEE Transactions on Information Technology in Biomedicine*, 8(4) (2004) 405–414.

Johnson, D. *Computer Ethics*. Englewood Cliffs, NJ: Prentice-Hall (1985); revised, 2nd edition., 1994; revised, 3rd edn., 2001; revised 4th edn., 2009.

Kubicek & Boye, iSelfhelp—Depression, The iSelfhelp depression test and depression app, 2011, https://itunes.apple.com/se/app/iselfhelp-depression/id481240621?mt=8.

Larsson, M. From hospital bed to e-Health, Developmental tendencies in Healthcare, UI code: 11071682, ISBN: 978-91-7164-558-6, July 5, 2010, http://webbutik.skl.se/sv/artiklar/fransjukhussang-till-e-halsa.html.

Mittelstadt, B., Fairweather, B., McBride, N., and Shaw, M. Ethical issues of personal health monitoring: A literature review. In: *ETHICOMP 2011 Conference Proceedings. Presented at Ethics and Information Technology the ETHICOMP 2011,* Sheffield, U.K. 13 (2011) 313–326.

Monahan, T. and Wall, T. Somatic surveillance: Corporeal control through information networks. *Surveillance & Society* 4(3) (2007) 154–173.

Ng, W. Can we teach digital natives digital literacy? *Computers and Education* 59(3) (2012) 1065–1078.

Page, R. and Kray, C. Ethics and Persuasive Technology: An Exploratory Study in the Context of Healthy Living. *1st International Workshop on Nudge & Influence through Mobile Devices Mobile HCI 2010* September 7–10, 2010, Lisboa, Portugal. http://ceur-ws.org/Vol-690/paper5.pdf.

Palm, E. När vården flyttar hem till dig—den mobila vårdens etik, *Etikk i Praksis/Nordic Journal of Applied Ethics* 4(2) (2010) 71–92.

Palm, E. A declaration of healthy dependence—The case of home care, *Health Care Analysis*, Online Publication (2012). http://www.springerlink.com/content/63l3771635v70v13/.

Palm, E. An interactive ethical assessment of surveillance-capable software within the home-help service sector. *Journal of Information, Communication and Ethics in Society*, 11(1) (2013a) 43–68.

Palm, E. Who cares? Moral obligations in formal and informal care provision in the light of ICT-based home care. *Health Care Analysis* 21(2) (2013b) 171–188.

Palm, E. and Hansson, S.O. The case for Ethical Technology Assessment (eTA), (written together with Sven Ove Hansson). *Technological Forecasting and Social Change*, 73 (2006) 543–558.

Palm, E., Nordgren, A., Verweij, M., and Collste, G. Ethically sound technology? Guidelines for interactive ethical assessment of personal health monitoring. In: S. Schmidt and O. Rienhoff (eds.) *Interdisciplinary Assessment of Personal Health Monitoring*. Studies in Health Technology and Informatics, (2013), ISBN 978-1-61499-255-4 (print), 978-1-61499-256-1 (online).

Psychology Lab CBT is a therapy tool support, Psykologilabbets, Retrieved on August 9, 2014, http://www.psykologilabbet.se/kbt-appverktyg-8941977.

Rössler, B. *Privacies: Philosophical Evaluations*. Stanford, CA: Stanford University Press (2004).

Rössler, B. *The Value of Privacy*. Cambridge, U.K.: Polity Press (2005).

OECD Health Policy Studies. Improving health sector efficiency. The role of information and communication technologies, June (2010). http://www.oecd.org/dataoecd/20/23/2431724.pdf (accessed August 28, 2013).

SALAR report, Sveriges Kommuner och Landsting "Från sjukhussäng till e-hälsa" (2010) SBN-nummer:978-91-7164-558-6.

Steg, H., Strese, H., Loroff, C., Hull, J., and Schmidt, S. *Europe is Facing a Demographic Challenge—Ambient Assisted Living Offers Solutions*. Berlin, Germany: VDI/VDE/IT (2006).

Szebehely, M. and Trydegård, G.-B. Home care for older people in Sweden: A Universal Model in Transition. *Health and Social Care in the Community*, (2011). doi/10.1111/j.1365- 2524.2011.01046.x/full.

UN. World population prospects: The revision 2008. United Nations, New York. http://www.economist.com/node/13888053 Commission of the European communities (2009).

Van Halteren, A., Bults, R., Wac, K., Konstantas, D., Widya, I., Dokovsky, N., Koprinov, G., Jones, V., and Herzog, R. Patient monitoring: The MobiHealth system. *The Journal on Information Technology in Healthcare* 2(5) (2004) 365–373.

World Health Organization. mHealth: New Horizons for Health Through Mobile Technologies, *Global Observatory for eHealth Series*, 3 (2011) http://www.who.int/goe/publications/goe_mhealth_web.pdf (accessed August 28, 2013).

Yakhlef, A. and Essén, A. Practice innovation as bodily skills: The example of elderly home care service delivery. *Organization*, 20(6) (2013) 881–903.

25

Rural Community Health in India: Problems and Solutions

Nandini Bondale, Sanjay Kimbahune, and Arun K. Pande

CONTENTS

25.1 Introduction

The definition of health adopted according to the Alma-Ata declaration is a state of complete physical, mental, and social well-being and not merely the absence of disease or infirmity [1]. This declaration has set up a vision for those providing primary health care to rural communities all over the world. Public health in developing countries is facing several challenges. Poverty, rapid population growth, poor public health delivery system to remote areas, and inadequate funds are some of those. The problems of public health are more severe in rural areas because of shortage of medical personnel serving in villages. In the shortage, the World Health Organization includes doctors, nurses, midwives, midlevel health workers, pharmacists, dentists, laboratory technicians, and community health workers, as well as managers and support workers [2]. The lack of basic infrastructure such as good roads, uninterrupted electricity supply, and quality education makes things difficult for those who are inclined to serve rural population. In addition, global warming, leading to sudden climate change, is creating additional problems for urban and rural public health systems. Although we are discussing about India in this chapter, the problems mentioned are representative of most of the developing countries [3].

In 2000, the United Nations (UN) announced the millennium development goals (MDGs) for a better world [4]. This brought world leaders together to meet concrete targets for advancing development and reducing poverty by 2015 or before. Out of the eight goals that are specified in the UN charter, three are related to health care. In response to the MDGs and part of an eleventh 5-year plan, the government of India set up the national rural health mission (NRHM) [5]. The mission was launched to address the problems related to primary health care in rural areas.

In this chapter, we describe how communities in rural areas can benefit from advances in information and communication technology (ICT) as well as from the advances in the robust and inexpensive medical devices, especially those based on mobile phones, for the primary health-care needs [6]. We describe in detail an ICT-based primary health-care delivery system called *mHealth primary health care (mHealth-PHC)*, developed by us. The design of *mHealth-PHC* is influenced by the NRHM infrastructure, problems faced by health-care delivery staff at NRHM, and needs of the community. This chapter is developed based on our work published earlier [7,8].

Our field trial shows that we could effectively meet the primary health-care requirements of the village community using our system. We describe how our system met the technology adoption challenges. Using our system, the doctors and medical students in city hospitals could participate in remote primary health-care delivery along with doctors stationed at primary health centers (PHCs) in villages. Such a collaborative approach between city and village public health systems, we believe, would go a long way in improving the quality of rural primary health care. Also, participation of city hospitals and medical students remotely would address the shortage of medical doctors and poor medical facilities in villages to some extent.

The field experience shows that health-care personnel involved at various levels in the PHC are comfortable in using the ICT system and are enthusiastic to employ such a system for efficient service to the community. Satisfaction of both the health personnel and the end beneficiaries is the first level of success for making a mark in the use of ICT in public health system.

25.2 Public Health and Millennium Development Goals

To address public health in rural areas, India has adopted the strategy of primary health-care approach as recommended by the Bhore committee since the inception of the 5-year plans [9]. Primary health care is defined as health care that is preventive, promotive, curative, and rehabilitative in nature. The concept of primary health care also implies the first contact with the health-care system, which is accessible, acceptable, affordable, and appropriate to all the people in the country. Primary health-care system has been established in all states of India, as health is an effective medium of achieving socioeconomic development.

Developing countries face an increasing incidence of noncommunicable chronic disease, along with the persistent threat of communicable disease. Diseases formerly concentrated in developed countries, such as hypertension, obesity, heart disease, and diabetes, are on the rise [10]. According to the UN Foundation report, globally, every minute, at least one woman dies from complications related to pregnancy or childbirth [11]. Maternal and child health care are the two focal points in MDGs.

25.2.1 Millennium Development Goals

In the UN millennium summit in 2000, 189 countries in the world agreed upon the eight goals to be achieved by 2015. Out of the eight goals, three goals relate to public health. They are as follows: (1) reduce child mortality, (2) improve maternal health, and (3) combat HIV/AIDS, malaria, and other diseases. A recent UN report mentions that the Asia-Pacific region, to which India belongs, has made big gains in reducing poverty and is moving fast toward other development goals, but still has high levels of hunger as well as child and maternal mortality [12].

25.3 Primary Health-Care System in India

Primary health care is a systems approach based on the principles of equitable distribution, intersectoral coordination, community participation, and appropriate technology. These principles are inculcated while implementing all national health programs. Primary health care as a system has a structure, a function, and protocols for documentation.

25.3.1 Structure

The structure of primary health care follows the revenue collection and distribution patterns existing in the country. Accordingly, one district is considered as a technical and administrative unit of primary health-care services. Each district is headed by two health professionals: a district health officer (DHO) and a civil surgeon. A DHO is in charge of the preventive health-care services in the rural areas, which include PHCs and subprimary health centers (SPHCs) (subcenters). A civil surgeon supervises the district hospital and rural hospital (RH) at taluka level. At the national level, the system is monitored through the Ministry of Health and Family Welfare. At the state level, there is the Directorate of Health Services, which plans, coordinates, and implements national health programs in all districts of the state.

For every 30,000 population in plain areas and 20,000 populations in tribal and hilly areas, one PHC has been recommended. Further, for every 3000 population in tribal areas and 5000 populations in plain areas, one subcenter is recommended. The objective is to reach the unreached with the preventive and promotive health-care services designed in the primary health-care system.

One PHC covers 25–30 villages and has about 6–8 subcenters providing basic health services to the villagers. The average distance between two PHCs is about 15–20 km. and that between subcenter and PHC is about 5–10 km. The interface between the community and PHC is entrusted to a female volunteer within the community, termed as accredited social health activist (ASHA). The four pillars of the public health system implemented through the PHC are an ASHA at village level, a female health worker called auxiliary nurse midwife (ANM), a male health worker called multipurpose worker (MPW) at the subcenters, and a medical officer (MO) at the PHC. The interaction between community and these key persons as well as the communication and coordination among these key persons has a great bearing on the effectiveness of the system.

25.3.2 Functions

Following are the major functions at PHC: (1) implementing national health programs, (2) implementing health surveys of the community to establish baseline health status, (3) ensuring safe water and environmental sanitation, (4) promoting health awareness, (5) monitoring disease surveillance and health-related data, and (6) doing epidemic investigations, containment, and research. All the activities at the PHC are documented on prescribed formats of reporting. The relevant health data are generated on a daily basis and is compiled and reported on a monthly basis. The PHC has both the center-based and outreach activities.

25.4 NRHM and Reproductive and Child Health Program

The NRHM is a program launched by the government of India in response to the MDGs, to improve the availability of and access to quality health care by the people, especially for those residing in rural areas, the poor, women, and children. The main goal of the NRHM is to reduce infant, child, and maternal mortality through promoting newborn care, immunization, antenatal care, institutional delivery, and postnatal care [5]. The NRHM is implemented through the primary health-care system in India. The NRHM has emphasized on the reproductive and child health (RCH) program as the top priority program implemented through the PHC and subcenter in all the states, leading to the achievement of the MDGs.

25.4.1 RCH Program

Among the various national health programs implemented through the PHC, RCH program focuses on preventive, promotive, and curative mother and child health services. This includes antenatal care, neonatal care, care about health of adolescent girls, prevention of reproductive tract infection, and population control. The ANM and ASHA conduct home visits to deliver antenatal and postnatal services to the mothers. They also record the health data of children below 5 years. The maternal mortality rate in the country is still high, which is 3 per 1000 live births, as against 1 per 1000, and the infant mortality rate ranges 50–70 per 1000 as against 30 per 1000, as prescribed by the MDGs. Early marriages and universality of marriages are known factors contributing to adverse maternal and child health. About 46% of women marry before the age of 18 years; 13% of married women have unmet needs for family planning. Less than 15% of women receive adequate ante-natal care. Three out of five women have home deliveries. About 16% of deliveries occur in the absence of a trained Dai and only 37% of women have access to postnatal checkups [13]. In the presence of these challenging circumstances, the RCH program becomes the key program of the primary health-care system.

25.5 Community Health Needs and Challenges

As mentioned earlier, each PHC covers 25–30 villages and has 6–8 subcenters providing basic health services to the villagers. The average distance between the subcenter and PHC is about 5–10 km. We are conducting *mHealth-PHC* field trials at a PHC in the Thane district of the Maharashtra state in India. There are 78 PHCs in Thane district, out of which 51 are tribal, catering to a majority of tribal population and 27 are nontribal. The PHC chosen for field trial is a tribal one. There are 8 subcenters attached to this PHC, covering 29 villages and 58 *Padas*. Maharashtra has 43,663 villages. Only 37.5% have the subcenter within the village; 77.7% of villages have a subcenter within 5 km and 8.9% have it beyond 10 km. The population of Thane district is 8,131,849 with 54% males and 45% females, the same ratio as that for Maharashtra state. Sex ratio in Maharashtra is 899 females to 1000 males in urban area and 948 in rural area [14].

Although the Government provides primary healthcare to rural community through NRHM, the community and the end user benefits get restricted due to the challenges faced by the community and PHC. We list some of them in general and for RCH in specific. Lack of transport and approach roads make it difficult for pregnant ladies to reach nearby PHC or RH. When complications arise, there is no easy way for ANM to seek advice of PHC doctors. Same is the case for PHC doctors; there is no easy and effective way to consult experts at the city hospitals. Secondary status of women in the community, preferences for male child, short birth intervals, inadequate immunization coverage, sociocultural practices, and poverty are other compounding factors adversely affecting maternal and child health.

The health information system at the subcenter is being operated manually. Hence, it causes inadvertent delay in transmitting the data recorded by ASHA and ANM to the PHC, located 5–10 km away from the subcenter. As a consequence, it delays the feedback from the PHC, and the mother is therefore deprived of prompt advice and intervention. The manual recording of data is subject to human errors, thus influencing health interventions unfavorably. Capacity building of health workers in terms of data management is lacking, and the health records are not interpreted epidemiologically. Thus, the interventions are differed till feedback is received from the PHC. For typical Indian rural conditions, there are other challenges such as good approach roads, safe drinking water, uninterrupted power supply, and availability of educational facilities for staff families to stay in villages. As a result, the posts of trained doctors remain vacant in many PHCs.

25.6 ICT Platform Innovation

Answering voice queries of the village health workers by call centers has been tested in few countries. In the absence of patient history and other observations, it is difficult for call centers to provide specific advice. With appropriate user interface and tested usability of the software, mobile phone can be a robust device in the hands of village health worker to provide quality health care to remote village community with the help of expert advice. Mobile phones are the most ubiquitous type of equipment in the world, with 3.3 billion people, that is, one of every two of earth's inhabitants, having at least one [10].

However, the use of mobile phone and Internet technology has its own challenges like database design, privacy of data, and human–computer interface along with scalability and computational problems [7]. Expert researchers in the field recommend that government support is critical for sustainability of mHealth projects [11]. *mHealth-PHC* design is based on the existing health-care infrastructure by the government.

25.6.1 *mHealth-PHC* Architecture

mHealth-PHC connects the rural patient to the doctor through an ANM or ASHA. The architecture of this platform reflects the processing of health-care services at different levels in the existing primary health-care infrastructure. Figure 25.1 shows the architecture of the *mHealth-PHC* platform.

The *mHealth-PHC* uses various technologies like mobile Internet, interactive voice response (IVR), Indian language font rendering, and usability frameworks to connect the remote patient to the doctor.

The design of *mHealth-PHC* tool is influenced by rural community requirement and challenges. The usability framework was considered while designing the interface. The following are few examples to illustrate our design approach. To ensure institutional delivery and knowledge of public health of the population in the village, an ASHA should be able to record her observation easily. A simple mobile phone and IVR technology were selected to record her visits. The transcription software was designed to convert her visit report from voice to text. The doctor is equipped with an electronic pen (E-pen) to write his/her prescription on regular paper. Blood and urine testing is needed for implementation of RCH as well as for diagnosis of other prevalent diseases in the local area. This required sophisticated yet inexpensive medical equipment. Also, the ICT system is required to work with 2G cellular network and power shortage.

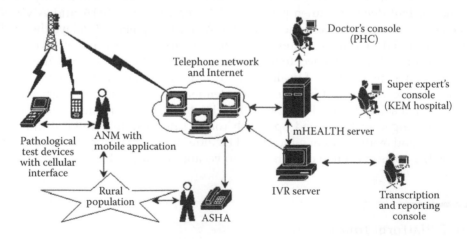

FIGURE 25.1
Components of the *mHealth-PHC* platform. (From Bondale, N. et al., mHEALTH-PHC: A community informatic tool for primary healthcare in India, in: *Proceedings of IEEE Conference on Technology and Society in Asia* (*T&SA*), Singapore, 2012, pp. 1–6.)

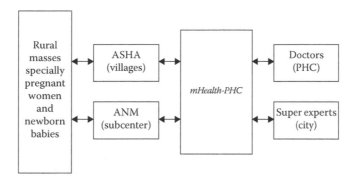

FIGURE 25.2
mHealth-PHC interaction among stakeholders.

The tool has five components: (1) software in local language on a mobile phone used at patient's end called client software; (2) servers in the secured data center; (3) doctor's console, which is viewed by a doctor/expert to suggest treatment/test and prescribe medicines to the patients; (4) an IVR application to record the information and observations; and (5) reporting console to view various reports related to rural health in a consolidated manner. The information flow while using *mHealth-PHC* is shown in Figure 25.2.

The ASHA's interface is IVR based. The ASHA is *eyes and ears* of the primary health-care setup as she goes house to house in a village. There is one ASHA for 1000 population. Conventionally, the ASHA maintains a register, in the paper format, which takes considerable time for processing and analysis. With *mHealth-PHC*, she can now record the details of her visits to families in villages on her interface and can be notified of any emergency or task using the same interface. A special template has been designed to cover the details of the ASHA's home visits, wherein she just selects items from the possible options to make the report. This saves lot of her time making her job easy and efficient. These voice records are then transcribed using transcription console and data are available in web-based form. Various reports can be generated from the data using reporting console. Figure 25.3 shows the quarterly report generated. These data can provide valuable inputs for policy makers to formulate and fine-tune the overall health policy.

The ANM at the subcenter provides preliminary health care to pregnant women and performs normal deliveries of babies. *mHealth-PHC* tool can aid the ANM, as it can be integrated with the portable, battery-operated medical test devices for blood and urine analysis. This facilitates inclusion of pathology test reports as part of *patient medical history*. The ANM can record a voice query through her interface. Wireless Internet makes it possible to upload patient's personal information and medical history to the server. A web console gives an integrated view of patient's medical history including the ANM's voice comments about the patient's illness and symptoms to the PHC doctor. Through the doctor's console, the doctor at the PHC can see the patient's details with medical history and listen to the ANM's recorded voice query. The doctor would give his/her advice in voice/text or may handwrite using E-pen as shown in Figure 25.4. The image of the prescription is captured and sent over wireless Internet to the ANM's mobile phone and the ANM can take action accordingly. If required, a specialist available at the RH or in city hospital could be contacted over the Internet and the entire case, including patient's medical history, could be referred through *mHealth-PHC* for expert advice.

Thus, *mHealth-PHC* connects ASHAs in villages, ANMs at subcenters, doctors at PHCs, and doctors at RHs or city hospitals through wireless, wireline Internet, and web technology.

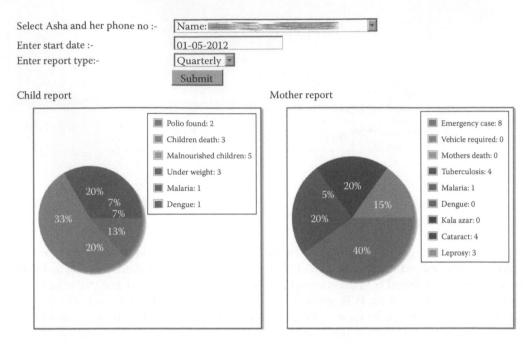

FIGURE 25.3
Sample quarterly report using *mHealth-PHC*.

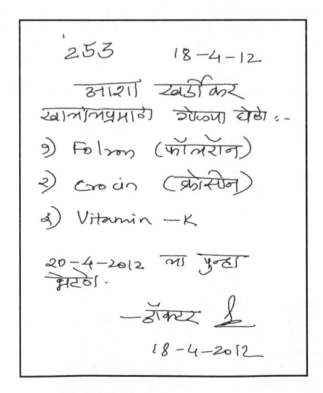

FIGURE 25.4
Doctor's prescription using E-pen in local language (Marathi).

25.7 ICT-Based Primary Health-Care Delivery

The first goal in implementing *mHealth-PHC* is to deliver high-quality primary health care. Hence, the mechanism to audit and take corrective actions if required is built in. Considering low literacy level of village health workers, we have provided the user interface in local language using multilanguage text rendering (MLTR) software. Entering lot of text is avoided by using voice input. Using mobile application, we have avoided power-related problem, which was one of the causes of failure of computer-based systems in villages. Medical devices are equipped with battery having reasonable capacity to work even in the absence of electrical power. Primary health care at affordable price is another goal. We have adopted software as service (SaaS) architecture. This would ensure that all hardware cost and maintenance issues are handled centrally at data center, and support client software will be eventually handled by MPW who would then be trained. The training would be useful to serve local community as well.

25.7.1 *mHealth-PHC* for RCH

The *mHealth-PHC* platform primarily digitizes *mother and child health card* as used conventionally in RCH program. The ANM registers a pregnant woman with the help of client software on mobile phone. A unique identity number (Id) is given to the mother as in the conventional registers called *R-15*. The newborn baby, when registered, gets unique Id as is given in the conventional *R-16* register. Id numbers of mother and child are linked in the system for better future reference. The platform connects two major users of the health system: beneficiary, that is, pregnant woman, via the ANM and the doctor. Hence, there are two major components of this tool: ANM interface and doctor's interface.

The ANM interface is on the mobile phone using local language. This application captures the entire pregnancy life cycle and postdelivery details. Based on the data, it automatically builds the history of mother beneficiary in the system. This software has the following functionalities: authentication, new patient registration, predelivery checkup module, predelivery trimester checkup module, postdelivery update, update of the history of registered patient, and making a health query. Figure 25.5 shows the mobile screenshots in local language.

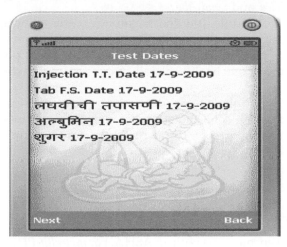

FIGURE 25.5
ANM interface: screen in English and local language (Marathi).

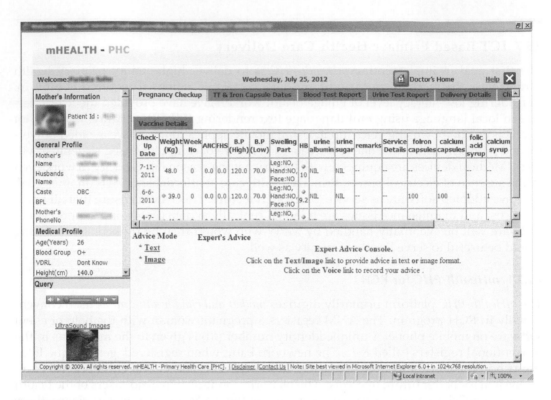

FIGURE 25.6
Screenshot of doctor's console.

The doctor's interface called as *doctor's console* is accessed on a computer, as the PHC infrastructure includes the facility of personal computer and Internet connection. Figure 25.6 shows the screenshot of the doctor's console. The beneficiary's personal profile as well as medical history is available to the doctor in the same format, as is currently available in the RCH program. The doctor's console has the following functionalities: authentication, *Check Queries, Give Advice,* and *Send Alerts.*

Using the *Give Advice* module, the doctor can advise by analyzing the beneficiary's profile, medical history, and test reports. The doctor can use text or voice for advising and use E-pen for prescribing. They can listen to various posts done by ASHAs to get rich information about health-related instances in a particular area and plan accordingly.

To communicate epidemic alerts or good practices to the larger community, the doctors can broadcast the message or send the alerts through the *Alerts* module. The doctor can select specific area or a group to send the alerts. The alerts can be in a local language.

25.8 Field Experience

For the patient, the first encounter is with an ASHA. For *mHealth-PHC* also, the first encounter is with an ASHA or ANM. We found both of them are willing to use the software and the feedback is encouraging.

25.8.1 Feedback on Adoption

A couple of weeks after training of ANMs to use *mHealth-PHC*, it was found that the ANMs' were using their application less frequently and the issues were related to the usability of the application. Major reasons were as follows:

a. The ANM would enter the data of the patient, but there was no feature to retrieve and check the data on their mobile phones. This would handicap them to check what they have entered.

b. To ask the question using voice interface, around six to seven clicks were needed in the menu to reach *Ask question*.

c. There was a mismatch between few terminologies and acronyms used in the software and that used by an ANM. For example, SPHC is termed as subcenter.

With extensive interactions with ANMs, we solved the aforementioned adoption challenges. A new call flow to their satisfaction was finalized, where the ANM can access desired functionality in less than three clicks. For data entry, a drop-down list was provided for easy and efficient functioning. With the growing field experience and feedback received, we are making suitable changes in the applications. It is an evolving process.

25.8.2 Significant Observations for Scalability

The following are the significant observations for scalability. (1) Key driver for application usage and adoption by health workers is using incentive-based remuneration in addition to the existing payment structure; (2) Charges for mobile usage, voice, or data are not seen as a burden in terms of expenses as compared to the travel expenses; (3) Health worker will be comfortable in using the new technology, only if it does not burden his/her time and is easy to handle; (4) Usability framework for both mobile- and web-based application is suitable for all key persons using *mHealth-PHC*; (5) Community and private participation is essential along with the government; and (6) Analysis using reporting application instantly gives perspective of the prevailing conditions in the field. This feature is well appreciated by planning and a decision-making authority, as it holds tremendous potential in preventive health care.

25.8.3 Data Analysis and the Case Studies

Data generated during the pilot study in the field have given a lot of insights. Analysis of the data showed appealing results. It was observed from hemoglobin (Hb) count of pregnant women that a large percentage of them had Hb count far below the accepted limit. Despite the fact that the mothers were given iron tablets, this was a cause of worry.

A root cause analysis including the parameters such as quantity, quality, and supply chain of medicines revealed interesting facts. One of the interesting observations was that pregnant women avoided taking iron tablets with a psychological fear of delivering a large, healthy baby. This data analysis helped to take the corrective actions that include improving intake of iron tablets, psychological counseling for the family, and conducting periodical workshops to address their problems.

Another case was that of *eclampsia*. With data analysis and the correlation among parameters like weight, blood pressure, and some pathological parameters, detection of eclampsia is now possible avoiding future problems.

25.9 Future Direction

mHealth-PHC will evolve as per the requirements of the community and stakeholders' needs. The convenience of use, benefits, and affordability would enable speedy adoption of the system. We see a role of advanced technologies to meet the aforementioned needs. The following are few examples.

25.9.1 Simple User Interface to Primary Health-Care System

For scaling up of the system and to capture the ASHA's reports into text, speech recognition technology can be incorporated for automatic conversion from speech into text. Similarly, for converting doctor's handwritten prescription into text, one may need to use handwritten script recognition. The earlier technologies would take away fear of operating ICT systems, improve the productivity of village health workers, and substantially reduce the investment required for training and capacity building of village health workers.

25.9.2 Small Ratio of Doctors to Patients

There is a shortage of qualified doctors and trained village health workers. The problem is more serious in rural areas. This problem can be handled to some extent by developing an expert system. An expert or intelligent system can be designed by capturing experts' diagnostic knowledge acquired through treatment of thousands of patients, analyzing millions of transactions and past decisions. It has been observed that an expert system could be useful for qualified but less experienced PHC doctor to take informed decision.

25.9.3 Assessment of Quality of Health Care and Effectiveness of Health-Care Policies

The government and NRHM would require functionality to assess the quality of health care and performance of PHCs and subcenters. The government has specified parameters to measure the quality of primary health care. The tracking of these parameters and analyzing them manually is often a tedious task. Data analytic techniques can analyze large data for observing specific health, behavior, cultural, and finance-related patterns.

25.9.4 Sustainable Operation on a Large Scale

In the pilot study, we demonstrated that ICT platform with wireless-enabled portable medical devices can deliver quality medical advice, promptly. By inducting micro health insurance and pharmaceutical companies, the model could be made sustainable by asking beneficiaries to pay according to the value perceived by them. We believe that public–private partnership will play an important role in expanding the sustainable model. The business model would require several functionalities in *mHealth-PHC* related to procurement, payment, audit, and so on, which can be provided by connecting *mHealth-PHC* system with existing hospital management system. Such an integrated system will also be useful to address legal and policy framework of the government for primary health care in the country, once it is adopted by the government.

Acknowledgments

Authors would like to acknowledge the management of Tata Consultancy Services, Tata Institute of Fundamental Research, Mumbai and G.S. Medical College and KEM hospital, Mumbai for their support. We thank the Project team at TCS Innovation Labs, Mumbai and Dr. R. Shinde, Dr. P. Sable, Dr. S. Shanbhag and Dr. M. Solanki (G.S. Medical College and KEM hospital, Mumbai). They also thank grass root healthcare workers, doctors and eco-partners for their continued support.

Questions

1. Describe the public health challenges in developing countries.
2. What are the MDGs?
3. Define primary health care.
4. Describe the primary health-care system in India.
5. Describe the various components of mHealth-PHC.
6. How is mHealth-PHC useful for the health-care needs of rural patients?
7. What are the benefits of mHealth-PHC system to different users?
8. How can mHealth-PHC be used by the governments to draft effective rural health policy?

References

1. World Health Organization, *International Conference on Primary Health Care*, Alma-Ata, Kazakhstan: World Health Organization, September 6–12, 1978.
2. World Health Organization, *Increasing Access to Health Workers in Remote and Rural Areas through Improved Retention: Global Policy Recommendations*, Geneva, Switzerland: World Health Organization, 2010.
3. Strasser, R., Rural health around the world: Challenges and solution. *Family Practice*, 20(4), 457–463, 2003.
4. United Nations, Millennium Developmental Goals 2015 and beyond, Last accessed date 25 June 2014, http://www.un.org/millenniumgoals
5. National Rural Health Mission, Mission document 2005 to 2015, Last accessed date 10 June 2013, http://mohfw.nic.in/NRHM/Documents/Mission_Document.pdf
6. Felix Jackson, Mobile medical devices an overview, Nov 2012 Last accessed at 25 June 2014. http://www.meddigital.com/what-is-a-mobile-medical-device
7. Pande, A., Kimbahune, S., Bondale, N., Shinde, R., and Shanbhag, S., Distributed processing and internet technology to solve challenges of primary healthcare in India, *Proceedings of International Conference on Distributed Computing and Internet Technology 2012 (ICDCIT 2012)*, Bhubaneswar, India: Springer, LNCS 7154, pp. 188–199, 2012.
8. Bondale, N., Kimbahune, S., and Pande, A., mHealth-PHC: A community informatic tool for primary healthcare in India, *Proceedings of IEEE Conference on Technology and Society in Asia (T&SA) 2012*, Singapore, pp. 1–6, 2012.

9. Park, J.E. *Textbook of Preventive and Social Medicine*, Jabalpur, Madhya Pradesh: Bhanodas Jalot Publications, 2009.
10. Kahn, J.G., Yang, J.S., and Kahn, J.S., "Mobile" health needs and opportunities in developing countries. *Health Affairs*, 29(2), 252–258, 2010.
11. Vital Wave Consulting, *mHealth for Development: The Opportunity of Mobile Technology for Healthcare in Developing World*, Washington, DC; Berkshire, U.K.: UN Foundation-Vodafone Foundation Partnership, 2009.
12. United Nations Development Program, Reports on Millennium development goals, Last accessed at 10 June 2013, http://www.undp.org/content/undp/en/home/librarypage/mdg/mdg-reports/asiapacific.html, 2013.
13. Balwar, R. (ed.), *Textbook of Public Health and Community Medicine*, Pune, Maharashtra: Armed Forces Medical College in collaboration with WHO India Office, 2009.
14. Census Of India, Office of Registrar and Census Commissioner of India, last accessed at 10 June 2013. http://www.censusindia.gov.in/2011-common/orgi_divisions.html.

26

Prospects of mHealth to Improve the Health of the Disadvantaged Population in Bangladesh

Fatema Khatun, Anita E. Heywood, Abbas Bhuiya,
Siaw-Teng Liaw, and Pradeep K. Ray

CONTENTS

26.1 Introduction

The introduction of information and communications technology (ICT) in health care, especially the application of mobile devices, commonly referred to as mHealth, has created the potential to provide efficient health services that are available anytime and anywhere (Kern and Jaron, 2003). Akter et al. (2010a) defined mHealth as "a personalized and interactive service whose main goal is to provide ubiquitous and universal access to medical advice and information to any user at any time over a mobile platform." mHealth, using mobile phone and multimedia technologies, has recently been successfully used to improve access to health care, timeliness of diagnosis, dissemination of health education, emergency medical response, and completeness of disease surveillance in resource-poor settings (Chang et al., 2011; Istepanian and Lacal, 2003; Rajatonirina et al., 2012; Seidenberg et al., 2012), and evidence in favor of its effectiveness is accumulating in a number of settings.

Bangladesh is the eighth most populous country in the world with 151 million people, of whom 30% are classified as extremely poor (Bangladesh Bureau of Statistics, 2011). Poverty is deep and widespread; the gross domestic product per capita in 2012 was US$1883 (purchasing power parity [PPP] current international dollar) (The World Bank, 2013). Despite continued poverty, Bangladesh has made significant progress in reaching the education and health indicators of the Millennium Development Goals (MDGs) (Chowdhury et al., 2013; National Institute of Population Research and Training [NIPORT] and MEASURE Evaluation, 2011). Bangladesh has a low ratio of formally trained, professional health workforce with recent estimates of only three doctors and three nurses/midwives per 10,000 population (Dawson et al., 2011), far short of the minimum standard of 22.8 per 10,000 recommended by the World Health Organization (WHO, 2010). As of 2007, the extent of the shortfall was estimated to be 50,000 physicians, 27,000 nurses, and 483,000 paraprofessionals to provide the minimum recommended level of health-care services to the population of Bangladesh (Bangladesh Health Watch, 2008). In addition, the health workforce ratio is lower in rural and remote regions, with the distribution of formally trained health workforce severely skewed toward urban and easily accessible areas. Given the reality of this acute shortfall and the growing demand for health-care services, the informal health-care sector has grown significantly and continues to grow. A study conducted in 2007 in Chakaria, a rural area of Bangladesh, revealed that only 4% of the health workforce in the area was formally trained (Rasheed et al., 2009). The quality of health care provided by the informal sector is unknown and mostly unregulated (Rasheed et al., 2009). In a community survey of health-care-seeking behavior, of those who sought care for illness, 64% had approached an informal health-care provider first for care (Mahmood et al., 2010). Beliefs regarding the quality of care, ease of access, low cost of treatment, and previous positive experience were some of the reasons cited for seeking care from informal health-care providers. Within this context, mHealth could emerge as a viable solution to serve the population's health-care needs through its easy access and low-cost mechanism by making quality health care more accessible, affordable, and effective across the country (United Nations Foundation and Vodafone Foundation, 2009).

The use of mobile phones has increased rapidly over the past decade, from 34% in 2005 to an estimated 91.2% penetration in 2012, with an estimated 6.8 billion mobile phone subscribers worldwide (International Telecommunication Union, 2013). In developing countries, the estimated penetration of mobile phones technologies reached 84% of the population in 2012 (International Telecommunication Union, 2013). In Bangladesh, uptake of mobile phone subscriptions has also increased rapidly, from 6% in 2005 to 64% in 2012, an estimated 97.1 million subscribers (International Telecommunication Union, 2013). In comparison, fixed telephone or landline subscriptions in Bangladesh remain below 1% of the population and have declined globally (International Telecommunication Union, 2013). The Government of Bangladesh (GoB) has endorsed ICT as a development tool in many sectors. For health care, ICT goals include improvements in health-care service delivery, increasing access to quality health care for hard-to-reach populations, and quality assurance of these services (Director General of Health Services, 2012). Consequently, the Ministry of Health and Family Welfare (MOHFW), nongovernment organizations (NGOs), research institutes, and private businesses have been implementing a small number of mHealth programs in Bangladesh.

To be effectively put into practice, mHealth must meet the community's needs and the community must also be ready to accept and use mHealth. Community readiness to accept mHealth programs depends on the recognition of value proposition and health literacy; the extent and patterns of use of the enabling technologies like mobile phones and personal digital assistants (PDAs); and adequate supporting infrastructure and concerns about ethical, legal, and socioeconomic issues associated with mHealth.

This chapter describes the current and future mHealth projects in Bangladesh with the aim to assess the potential of mHealth in providing services to the masses, especially the most disadvantaged, what is possible and what is not, and what infrastructure and regulatory frameworks are needed to optimize the potential effectiveness of mHealth in Bangladesh.

26.2 mHealth in Developing Countries

A number of academic papers and reports argue that ICTs can make a substantial contribution to improving health-care services in developing countries (Betjeman et al., 2013; Chetley, 2006; Lester et al., 2009, 2010; United Nations Foundation and Vodafone Foundation, 2009). Evidence from the mHealth projects around the world is emerging, demonstrating the potential of mobile communications to radically transform health-care services, even in some of the most remote and resource-poor environments (United Nations Foundation and Vodafone Foundation, 2009). Since landlines are not available in rural and remote areas, mHealth has the advantage of providing health-care services anytime anywhere to these areas but often has the disadvantage in terms of access to traditional health services.

Increasingly high rates of mobile phone access offer an opportunity to leverage the limited human resources in health currently available in developing countries to maximize health service reach. And it is also a low-cost option for health-care delivery for the countries where health budget is very inadequate (Lester et al., 2010). The short message service (SMS) network has improved health-care services for managing logistics, reporting events, and addressing emergencies that have reduced costs and saved time (Lemay et al., 2012). The effectiveness of mHealth behavior change communication interventions in sub-Saharan Africa, Ghana, and China shows that mHealth is a promising tool to foster behavior change in HIV/AIDS prevention, family planning, pregnancy care, tuberculosis, and improved reporting systems of HIV (Gurman et al., 2012; Seidenberg et al., 2012). In Kenya, a randomized control trail on text messages for adherence to antiretroviral therapy (ART) shows that patients who received SMS support had significantly improved ART adherence and rates of viral suppression compared with the control individuals (Lester et al., 2010). Mobile phones might be effective tools to improve patient outcome in resource-limited settings. Delivering text messages in this project had a marginal cost of about US$0.02/message and minimal fixed cost and suggests that this type of intervention is highly cost-effective (Lester et al., 2010).

Applications of mHealth for maternal, neonatal, and child health (MNCH) are currently being evaluated in a number of different countries (World Health Organization, 2011). For example, a systematic review of 34 studies using mHealth along the stages of continuum of care for maternal, neonatal, and child health found that the integration of mHealth for prenatal and newborn health services has shown positive outcomes such as facilitating urgent care during emergency obstetric referral and minimized time barrier. However, the sustainability and scalability of operations require further assessment and evaluation of programs that are ongoing (Tamrat and Kachnowski, 2011). mHealth is being used in Zanzibar, Tanzania, for pregnancy tracking, appointment reminders for prenatal care, SMS-based health education during pregnancy, point of care remote consultation, facilitating referral and access to health facilities, contacting health workers, mother and infant care during postpartum period, and immunization reminders. After introducing the project, the intervention group reached a skilled delivery attendance rate of 60%, compared with 47% in the control group (Lund et al., 2012; Tamrat and Kachnowski, 2011). A recent study in two districts of Bangladesh found that using mobile phones for tracking pregnant women for the delivery outcome, birth registration, and a reminder for childhood immunization has improved immunization coverage. With this intervention, Expanded Program on Immunization (EPI), vaccine coverage has increased from 60% and 67% in 2009 to 79% and 85% in 2010 in the two districts (Bill and Melinda Gates Foundation, 2012).

26.3 Health-Care Systems in Bangladesh

Health-care services in Bangladesh are delivered by public, private (for profit), and NGOs. The public health-care system is organized under the overall supervision of the MOHFW. At the central level, the MOHFW is responsible for applying, managing, coordinating, and regulating national health and family planning–related activities, programs, and policies through the Directorate General of Health Services (DGHS). The health-care delivery system in Bangladesh is based on the primary health-care (PHC) concept, a systems-based approach consisting equitable distribution of health care, community participation, health workforce development, multisectoral approach, and use of appropriate technology. The major functions of primary health care are to implement the national health program, promote health awareness, monitor disease surveillance, and collect data for reporting.

At the home and community level, each community has one community clinic, including one community health-care provider, one family welfare assistant, and one health assistant (HA), providing health care for 6000 people. At the union level, there is a union subcenter (USC) or health and family welfare center (HFWC). At the Upazila level, there is an Upazila health complex (UHC), which is the first referral level consisting of 31 hospital beds. The district hospital is the secondary referral level consisting of 100–250 hospital beds.

Along with the public sectors, proprofit private organization and NGOs are also major contributors to the health systems of Bangladesh and have contributed to the impressive outcome in terms of health indicators (Chowdhury et al., 2013). The multiplicity of different stakeholders and agents engaged in health production is referred to as *pluralism* in "Bangladesh: Innovation for Universal Health," a recent publication of *The Lancet* (Ahmed et al., 2013).

The use of mHealth in the health systems of Bangladesh starts from the community clinic level such as provision of laptops in community clinics, and every UHC provided one mobile phone for the emergency care, consultation, and referral. The MOHFW has expanded the network to all public health facilities including management information systems (MISs) to enable real-time data (Ministry of Health and Family Welfare and Government of the People's Republic of Bangladesh, 2011). A call to action toward innovation for universal health coverage in Bangladesh refers to the buildup of an interoperable eHealth information system by introducing electronic individual medical records and strengthening and ensuring interoperability of medical information systems in the public, private, and NGO sectors (Adams et al., 2013).

26.4 mHealth Initiatives in Bangladesh

This section takes a closer look at ongoing mHealth programs and research projects in Bangladesh and highlights the currently available data. In Bangladesh, a number of mHealth projects have been initiated, including in the provision of primary health care, in disease surveillance and the collection of routine data, for health promotion and disease prevention initiatives, and for health information system support tools for health workers and point of care services. The current projects are summarized in Table 26.1.

26.4.1 Primary Health Care

26.4.1.1 *Ministry of Health and Family Welfare*

In 2009, the MOHFW launched an mHealth project aimed at providing consultation services and enhancing health workforce training, by providing mobile phones and laptops to all district and subdistrict hospitals in Bangladesh. Under this project, one doctor remains available to receive calls from patients to provide medical advice and consultation, with the service available 24 h a day and 7 days a week (Director General of Health Services, 2012). The provision of laptops has enabled hospital staff to consult with each other via videoconferencing, which is helping health professionals who are delivering services in the rural and remote areas for consultation with specialist doctors. This will help to provide patients with better clinical decision making and management. This initiative will also assist in real-time data collection and update community health data for MIS reports on health service delivery, providing staff training such as subassistant community medical officer (SACMO) and community health workers (CHWs) (HNPSP 2012–2016).

TABLE 26.1

Ongoing mHealth Initiatives in Bangladesh

	Name of the Project	Nature	Target Population	Types of Services	Mode of Delivery	Outcome Indicator
Primary health care	Ministry of Health and Family Welfare	Public	All age groups	Health promotion	SMS, IVR	Increased utilization of health-care services
	Health line 789	Private/business	All age groups	Primary health care	Telemedicine services	Increased medical support and emergency care
	eClinic24	Private	All age groups	Primary health care, EMRs	Telemedicine services, medical hotline	Increased medical support and emergency care; improved medical data recording
Disease surveillance and data collection	A longitudinal study on the epidemiology of malaria in Bandarban district, Bangladesh	Research	Malaria suspects	Malaria case detection and management	SMS, phone call	Increased number of malaria case detection and treatment
Health promotion and disease prevention	Ministry of Health and Family Welfare	Public	All age groups	Health promotion messages, birth registration	SMS	Birth registration and increased immunization rate
	MAMA (Aponjon) project	Health program	Pregnant women and newborn	Health promotion messages	SMS, IVR	Decreased neonatal and maternal morbidity and mortality
	Expanded Program of Immunization (EPI)	Public	Infants	Birth registration, immunization services	SMS	Birth registration and increased immunization rate in underperformed areas
Health information systems, support toll for health workers, and point-of-care services	Evaluation of toll-free mobile phone for Maternal Newborn and Child Health (MNCH) care	Research	Pregnant women and newborn	MNCH services and referral through CHWs	Toll-free mobile phones uses	Emergency obstetric care, decreased maternal morbidity and mortality
Research projects	Future Health Systems	Research	Informal health-care providers	e-referral	Teleconsultation	Improving prescription pattern of informal health-care providers (decreased inappropriate drug prescription)
	Assessment of community readiness for mHealth intervention	Research	—	—	Survey and in-depth interviews on existing uses of mHealth services	Exploring communities' perception on the mobile phones' uses in health care

26.4.1.2 Telemedicine Reference Center Ltd.

Telemedicine Reference Center Ltd. (TRCL) is a privately held corporation from Bangladesh, which has been operating since 1999. TRCL is one of the longest serving companies in the telemedicine sector and initiated its medical call center project health line 789 with Grameenphone limited in 2006. The TRCL mHealth program is an extension of its eHealth platform designed for health system–based medical services. TRCL services include chronic disease management, medical hotlines, health communication tools, electronic medical record (EMR), and electronic prescription (e-prescription). In recent years, TRCL has launched a 24 h eClinic facility to make medical services from a qualified physician available to the callers (Telemedicine Reference Centre Limited).

26.4.1.3 Health Line 789

In 2006, Grameenphone initiated a unique telemedicine service, health line 789, a Global System for Mobile (GSM) communication infrastructure–based call center for Grameenphone's 10 million subscribers. Health line 789 provides a range of medical information facilities, emergency services (SMS-based laboratory reports and ambulance), and real-time medical consultations via mobile phone. The service provides remote access to personalized health services over the phone and delivers medical advice and support to all users for any medical situation, including medical emergencies. A panel of health professionals are available 24/7 at the physical front office (physician's interface), which is supported by a back office and a mobile network operator (for network management) to provide health information to the end users (see Figure 26.1) (Akter et al., 2010b). This initiative also provides medical information, consultation, diagnosis, treatment, and counseling from skilled health professionals and pharmacies' help, that is, drug information and laboratory test interpretation. It is a business model, in which users were charged from their mobile phone credit at a rate of US$0.38/call for 5 min. The Grameenphone health line 789 is the most successfully implemented mHealth initiative in Bangladesh to provide diagnostic, consultation, and treatment support (World Health Organization, 2011).

26.4.2 Disease Surveillance and Data Collection

Since June 2010, the use of mobile phones to report fever was added to the existing malaria surveillance platform conducted by the International Centre for Diarrheal Disease Research, Bangladesh (ICDDR, B) in collaboration with the Johns Hopkins Malaria Research Institute (Khan et al., 2011). This surveillance is ongoing to define the epidemiology of malaria in southeastern Bangladesh where malaria is hypoendemic. This surveillance is tracking febrile patient and diagnosed patients are provided with treatment under the national malarial control program. Participating households are provided with the phone numbers of project personnel to notify them via SMS or a phone call of any signs or symptoms of malaria, which include fever, frequent vomiting, altered mental status, convulsion or yellow coloration of the eye, as well as any serious medical concern. During the 2-year period from June 2010 to July 2012 in Bandarban district of Bangladesh, a total of 1046 people were identified for testing, of whom 265 (25%) tested positive for malaria. In the malaria cases reported during this period, mobile phones were used to initiate testing and treatment for 52% of diagnosed malaria cases (International Centre for Diarrhoeal Disease Research Bangladesh, 2012; Prue et al., 2013). The use of mobile phones assisted the study team in identifying

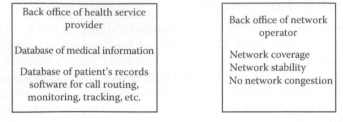

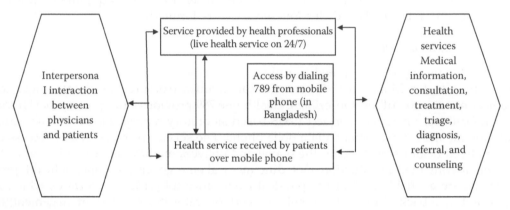

FIGURE 26.1
mHealth (hotline) service platform. (From Akter, S. et al., *Electronic Markets*, 20, 209–227, 2010b.)

cases and provided the community with increased access to health care and treatment for malaria through the national malaria program. Mobile phone–based case reporting has improved detection and treatment of malaria and could be used for detection and treatment of malaria and other febrile illnesses in hard-to-reach areas in similar contexts (International Centre for Diarrhoeal Disease Research Bangladesh, 2012; Prue et al., 2013).

26.4.3 Health Promotion and Disease Prevention

26.4.3.1 Ministry of Health and Family Welfare

The MOHFW has also launched other mHealth projects alerting people about upcoming health campaigns by sending SMS text messages to all mobile telephone numbers in the country. Alerts have been used to mobilize citizens for national immunization days, vitamin A week, national breastfeeding week, and national safe motherhood day (Director General of Health Services, 2012; World Health Organization, 2011). Although official evaluation of the impact of the SMS health messages has not yet been conducted, the ministry has reported wide acceptance of the immunization and vitamin A campaigns. Mass broadcasts of SMS health messages ensure that the estimated 97 million mobile telephone users in Bangladesh receive reminders and information regarding health campaigns. As a component of mHealth, the SMS-based pregnancy advice launched in March 2010 is expected to emerge as one of the pioneering programs of the Director General of Health Services (DGHS). The program includes registration of pregnant women through mobile phones. Upon registration into the program, pregnant mothers receive appropriate periodic antenatal, safe delivery, and postnatal care advice through SMS.

This program is aimed at contributing to the achievement of MDGs 4 and 5 of reducing maternal and child mortality. The international Text to Change organization conducting a project titled "A Largest Scale mHealth Project in Africa" provides information through text messages, fact sheets, and posters to disseminate maternal health information, increase service utilization, and obtain feedback from patients in Tanzania since 2012 (Text to Change). Another MOHFW initiative, the most well-publicized mHealth project in Bangladesh, delivers voice messages to mobile subscribers in rural areas with the aim of increasing utilization of community clinic health services for maternal and child health. Novel to this project is the use of the honorable prime minister's voice for this service. This is the first time in Bangladesh and across Asia that a prime minister has provided personal support for the use of an mHealth awareness campaign and the promotion of health services. This initiative sends voice calls through an automated system and targets 60 million subscribers (40% of total population) in Bangladesh. The primary target is the rural and periurban populations; subsequently, it will target urban population for supporting health services awareness and its utilization. This system using outbound dialer is an effective mobile-based voice campaign tool that provides fast, effective, and personalized communication. It is a cheap, effective, personalized, and automated service that can be utilized for community development and commercial purposes. This model can be used for other health promotion messages such as immunization reminders, smoking cessation, and promoting of healthy food habit and physical activity. The anticipated benefits of these mHealth activities are a positive impact on the health status of Bangladeshi's poor population with data related to this project pending (Ahmed, 2013).

26.4.3.2 Mobile Alliance for Maternal Action

Mobile Alliance for Maternal Action (MAMA) was launched in May 2011; this public–private partnership aims to improve maternal and child health by using mobile technology to deliver vital health information such as the importance of antenatal care, dispel myths, and misconceptions; highlight danger signs during pregnancy; reinforce breastfeeding practices; explain the benefits of family planning; and advise on proper nutritional intake to new and expectant mothers. The initiative is taking place in Bangladesh, India, and South Africa. In Bangladesh, the MAMA intervention was applied by D.Net, a social enterprise working at mainstreaming ICTs in critical development spheres including health (D.Net, Government of Bangladesh, and USAID, 2013). D.Net has developed a nonprofit social enterprise model called *Aponjon* with corporate and public sector supports, for the implementation of the MAMA intervention adapted to the Bangladesh context. The *Aponjon* information service has a business model but is not free for all. This low-cost service aims to reduce maternal and newborn morbidity and mortality. *Aponjon's* mission statement is to reach out to the poorest communities in remote areas of the country free of cost. Registration through SMS for *Aponjon* services is free. Charges for interactive voice recorder (IVR) or SMS charge is US$0.03 SMS or IVR. However, it is free for the subscribers from the poorest 20% (very low socioeconomic status) for whom *Aponjon* is supporting.

The service design of *Aponjon* includes the pregnant and new mothers and other key decision makers of a family, such as husbands, mothers, and mothers-in-law. The BCC messages encourage observance of recommended courses of action during pregnancy, delivery, postpartum, and beyond. The details of the behavior change communication (BCC) messages are accessible by registering the *Aponjon* intervention by dialing the number 162274142. The delivery of the BCC messages is being conducted using two different modes—IVR and SMS, as per the choice of the individual subscriber. In the case of IVR, the subscribers are able to listen to messages by dialing the short code or receive an automated call at a specified time, chosen by them. In the case of SMS, the intervention messages will be sent in text form (Bangali text in English letters). Since 42% of population in Bangladesh is illiterate (Bangladesh Bureau of Statistics, 2011), the IVR systems of audio messages will be able to leverage the disadvantages of the SMS. Along with these, *Aponjon* provides important weekly health messages to pregnant women, mothers of neonates, and their family members through voice- and SMS-based mHealth service after subscribing the Aponjon services. The service aims to provide mothers and their families with personalized, reliable, and specific messages based on the week of gestation or the age of the neonate. Aponjon services send valuable information including pregnancy symptoms, nutrition, and obstetric danger sign in twice weekly messages to subscribers through SMS or voice calls in line with the pregnancy calendar (D.Net.). At the birth of the baby, the subscriber must notify the Aponjon of the baby's date of birth to initiate baby care services including disease diagnosis, treatment and prevention, breastfeeding, imitating solid foods, and monitoring physical and developmental milestones. For the pregnancy information, there are two health messages (voice or SMS) per week, for 37 weeks of pregnancy; the messages start from the 6th week of pregnancy and continue till 42 weeks of gestation or the birth of the baby. Information for the new mother under this service, two health messages (voice or SMS) per week for 52 weeks, starts from the baby's birth till the baby is 1 year old. Moreover, one family member (husband or parent or parent-in-law) can get one health message (voice or SMS) per week along with the expecting mother/new mother, as the aforementioned family members are key decision makers for the any activities related to pregnancy, pregnancy outcome, and postnatal period in Bangladeshi culture.

26.4.3.3 Increasing Immunization Coverage in Bangladesh through Mobile Technology

The Bangladeshi EPI has achieved 96% reported coverage of the third dose of DPT (DPT3), the third dose of polio and one dose of measles vaccine (World Health Organization, 2012). Despite the high level of national coverage, the coverage among children living in rural hard-to-reach districts and urban street dwellers remains lower than national estimates at an estimated 42%–60% (NIPORT and MEASURE DHS, 2011). The main reasons for low immunization coverage in these groups are the absence of effective systems to track newborn children and remind parents/caregivers about immunization schedules and immunization sessions nearby. The Johns Hopkins University Global mHealth Initiative and ICDDR, B partnered with mPower Health to create an Android mobile–based application called *mTikka*, designed specifically to address these barriers to complete and timely immunization in Bangladesh. mTikka is used by HAs, the GoB cadre of workers responsible for EPI sessions in his or her catchment area. The application contains modules for registration of pregnant mother, birth notification, immunization scheduling, client reminders, assessing vaccine beliefs, session management, and reporting (Labrique, 2012). The study will assess the feasibility and effectiveness of the earlier stated birth registration and patient reminder systems through the use of mobile phones to improve child immunization coverage among hard-to-reach children (remote rural and urban street dwellers) with an aim to achieve equity in vaccination of Bangladeshi children and to reduce vaccine-preventable infectious diseases and child mortality. An evaluation of its outcome is currently being planned (International Centre for Diarrhoeal Disease Research Bangladesh, 2013).

26.4.4 Health Information System, Support Tools for Health Workers, and Point of Care Services

26.4.4.1 Evaluation of a Toll-Free Mobile Phone Service in Maternal and Neonatal Health Intervention in Rural Bangladesh

This service model has been tested in a subdistrict of Bangladesh, which aims to strengthen the health system, train community-based skilled birth attendants (CSBAs), and strengthen community support groups as part of demand-side financing project (i.e., to provide direct subsidies to the poor for access to specified health services) (Huq et al., 2012; Lira, 2012). This study follows a quasi-experimental design with a comparison group. In the intervention area, the CSBAs are provided with mobile phones to enhance their communication with the women's families and also with other health-care providers. Mothers and their family members are able to contact dedicated CSBAs free of charge to discuss maternal and neonatal concerns as well as emergencies using any mobile phone irrespective of their mobile carrier. CSBAs can also seek advice from skilled specialists using mobile phones if required (Huq et al., 2012).

26.4.5 Other Research Projects

26.4.5.1 Future Health Systems

Phase 2 of the Future Health Systems project of ICDDR, B in collaboration with the Telemedicine Reference Centre (TRCL) has been testing an intervention to augment ICDDR, B's efforts to franchise the village doctors as *SasthyaSena* (Iqbal et al., 2012), a cadre of dominant informal providers using pharmaceutical medication, to improve the quality of their prescribing services and establish accountability. The project components include registration of the village doctors with TRCL, which provides 24/7 access to the TRCL call center to get advice from qualified physicians using their mobile phone. The project has a built-in financial incentive for village doctors to use TRCL. Village doctors can charge an amount of Taka 30 (US$0.38) to the patients for whom the call is made. Of the Taka 30, Taka 12 (US$0.15) is kept by the village doctors and the remainder by TRCL. The village doctors receive a prescription from the TRCL physicians through text message, which is used in dispensing drugs to the patient (Bhuiya, 2013). This intervention was developed in response to earlier evidence from Phase 1 of the Future Health Systems project, which included the training of the village doctors on appropriate prescribing for major health problems (Mahmood et al., 2010). An assessment of the impact of this intervention revealed a statistically significant improvement of appropriate drug prescription; however, the size of the impact was less than expected (Hanifi et al., 2012). The main reason for the nonadherence of the village doctors to the prescription guidelines and the continued inappropriate dispensing of medications was the negative impact on their income from pharmaceutical sales, which are their main source of income (Wahed et al., 2012). Phase 2 study has commenced and followed a quasi-experimental design with a comparison group and baseline and end-line data collections to assess impact. Data available at this stage reveal that the demand for the use of the system is reasonable; however, uptake will depend on the efficient functioning of the TRCL call center.

26.4.5.2 Assessment Community Readiness for mHealth Intervention

Readiness is one of the most important factors for the success of eHealth initiatives (Information Technologies Group and Center for International Development at Harvard University, 2002; Jennett et al., 2003). Acceptance of new interventions in the target population is necessary for their uptake, continued use, and overall success. Multiple factors (social, political, organizational, and infrastructure) determine whether or not a new approach is successful; however, their interaction and relationships are complex and not well understood (The Alliance for Building Capacity, 2002). Although there are several mHealth projects ongoing in developing countries, findings related to the context and the mechanisms that encourage or discourage community acceptance of technological and organizational innovations are unavailable in developing countries' context. With this in mind, a study aiming to assess the current pattern of mobile phone ownership and usage by income level, education, and other socioeconomic indicators and the community awareness and perceived usefulness of mHealth has been conducted. This mixed methods study (quantitative survey and in-depth interviews), conducted in Chakaria subdistrict of Bangladesh in 2013, systematically examined community readiness and will add to the knowledge base of the determinants for the successful implementation of mHealth programs in countries like Bangladesh (Khatun and Akter, 2013). Data collection has recently been completed and reporting of the expected outcomes of how to best use mobile phones for health care including the perception of mHealth on health care, cost, trust, privacy, and intention of future uses of mobile phones for health care will be available in the near future.

26.5 Policy and Political Commitment

The Bangladesh government has shown interest in the use of ICT for national health and development with a mandate of the "Vision 2021 Bangladesh: A New Horizon" introducing the concept of "Digital Bangladesh." This commitment also aligns with the objective of Health, Population and Nutrition Sector Development Program (HPNSDP) (2011–2016) of the GoB—"Health Information Systems, eHealth (5.5.5, P 39)" (Ministry of Health and Family Welfare and Government of the People's Republic of Bangladesh, 2011). The objectives were to improve health information system through the development and the operation of population-based health information systems and to improve eHealth continuation and further development of mobile phone health service and other mHealth programs, expansion of telemedicine service, and expansion of videoconferencing.

Different health programs using accessible mobile phone technology and innovative community-based health promotion methods have already commenced through the MOHFW. However, the unavailability of evidence-based policy and management frameworks within national health strategies will be one of the most important challenges to the scaled adoption of mHealth services. Coordination between different government bureaus, research organizations, and commercial business is essential to establish and enforce guidelines on the content and technological design of services, exchange of data, and ICT infrastructure including network coverage. The policy frameworks will help to outline guidelines and protocols for health professionals in executing treatment based on support from remote consultations, to prescribe therapy despite their physical isolation, and to advise for timely referrals. In addition, further research and analytical evaluation are required to inform national health strategies and guide appropriate investment in and expansion of mHealth activities.

26.6 Opportunities and Challenges

The potential for and successful use of mHealth in providing health services to the masses, especially to the most disadvantaged, relate to multidimensional factors that can be broadly classified into the technical paradigm and the social paradigm.

26.6.1 Technical Paradigm

The technical paradigm consists of the infrastructure of IT systems in health-care services and the availability of the mHealth services in remote and rural settings, as well as data security and service quality.

26.6.1.1 Infrastructure, Network, and Application Design

Bangladesh is a developing country, and its infrastructure related to the mHealth services in rural and remote settings is still in the process of improvements for its application in all districts and subdistricts. Poor network coverage and the absence of a high-speed communication network may hinder the use of mobile telemedicine, which requires high-quality voice, video, and data sharing. Further, a major technological limitation is that mobile phones that are capable of sending and receiving text messages in Bangla—the official language spoken by more than 98% of the population as their first language—are not available to the entire population. Therefore, the SMS messages sent by the MOHFW and other health programs are mainly in English and therefore require translation (World Health Organization, 2011). The use of Bangla in text messages has begun but requires further software adaptation and availability of appropriate devices for trouble-free reading.

26.6.1.2 Data Security and Service Quality

There are concerns about the perceived quality of the mHealth services due to the lack of reliability and efficiency of the service delivery platform, knowledge, and competence of the provider; privacy and security of information; and, above all, their effects on satisfaction, future use intentions, and quality of life (Ahluwalia and Varshney, 2009; Angst and Agarwal, 2009; Ivatury et al., 2009; Kaplan and Litewka, 2008; Mechael, 2009; Norris et al., 2008; United Nations Foundation and Vodafone Foundation, 2009; Varshney, 2005). Akter et al. (2010c) evaluated the service quality of health hotline 789 in Bangladesh and found that patient perceived service quality has a strong significant impact on satisfaction, intention to continue using, and quality of life (Akter, 2012). To improve quality, safety, and efficiency of the health delivery system, a comprehensive management of health information across computerized systems and its secure exchange between consumers, providers, government, and quality entities are needed (Chaudhry et al., 2006).

26.6.2 Social Paradigm

Language barriers, illiteracy, SMS character restrictions, and lack of technical support in rural areas are some of the contextual factors described by the WHO that inhibit mHealth usage (World Health Organization, 2011). Health literacy and cognitive disabilities such as vision and hearing loss among elderly are also crucial for mHealth's effectiveness (Kwan, 2013).

26.6.2.1 Ownership and Inequity in Access

Mobile phone ownership is one important factor for the success of mHealth. Despite the high number of subscribers, mobile phone ownership is greatest among the upper socioeconomic status population, especially mobile phones with enhanced technologies, Internet access, and wide screens (*smartphones*). Associated cost is the primary reason for not owning a mobile phone (Noordam et al., 2011). Given the key use of mHealth to deliver maternal and child health, the lower mobile phone ownership rates by women are also a challenge to the success of these programs. Globally, women are 21% less likely to own a mobile phone than men, and this gender disadvantage is highest in South Asia, followed by sub-Saharan Africa (GSMA Development Fund and Cherie Blair Foundation for Women and Vital Wave Consulting, 2010). In Kenya, disparities exist in mobile phone ownership with respect to gender, age, education, literacy, urbanization, and poverty (Zurovac et al., 2013). Data related to the ownership of mobile phones among patients/caregivers in respect of socioeconomic status and gender are crucial for Bangladesh. However, no data are available on characteristics of mobile phone ownership in Bangladesh. The community readiness study will reveal this information.

26.6.2.2 Education and Health Literacy

Illiteracy, lack of technical support in rural areas, and lack of technical knowledge in how to use mobile phones and associated applications add complexity to the contextual factors such as the language barriers and restrictions to 160 characters per SMS that inhibit mHealth usage in developing countries' context (World Health Organization, 2011). Since 42% population of Bangladesh is not able to read or write, an IVR system of voice message and recording in the MOHFW and Aponjon project has allowed to reach them. An important concept is health literacy, the ability to access, understand, appraise, and apply health information. Health literacy is crucial for the implementation of mHealth program and the best use of mHealth in developing countries including Bangladesh.

26.6.2.3 Trust and Privacy

A recent study on WHO's mHealth initiatives reported that service quality perceptions are mainly influenced by privacy issues in developing countries (World Health Organization, 2011). Privacy and security refer to the confidentiality of a patient's personal health information, which is a large concern for eHealth records (EHRs). In this regard, the WHO expressed its concerns that "security and privacy issues are especially critical in low and lower-middle income countries, where mobile phones are often shared among family and community members, leading to potential challenges with protecting confidential health information, particularly in the case of conditions like HIV/AIDS, tuberculosis which remain highly stigmatized" (World Health Organization, 2011).

26.6.2.4 *Application Design and Effect of Disability on mHealth*

Application design and uses of mobile phones could limit the mHealth program usage by the elderly. The selection of user communication types such as text, audio, or video should take into account the attitude, cognitive status, and disability of the user. Handsets with larger screens and larger buttons that are easy to see and touch for those with impaired vision may be required to better meet their needs (Kwan, 2013). In Bangladesh, the number of elderly population is increasing and started epidemiological transition from communicable diseases to the noncommunicable diseases or chronic diseases. mHealth has potentials to serve the patients with chronic disease by reminding medication intake through SMS and thus increasing the medication compliance (Fischer et al., 2012; Katz et al., 2012). Till now, no study has focused on the use of mHealth in older adults and those with disabilities in Bangladesh.

26.6.2.5 *Affordability and Acceptability*

The scaling up of the mobile telemedicine services depends on its ability to provide a cost-effective service. The cost of mobile telemedicine is too high (US\$0.38 for each call for 5 min consultation) for the rural areas in comparison to the per capita income in Bangladesh (Ivatury et al., 2009). There is a need to reduce the cost of mHealth services accessible for poor and disadvantaged populations. Toll-free mHealth services or free of cost services for the population who is in low socioeconomic condition would be one option for better uptake (Huq et al., 2012; MAMA). mHealth service can ensure broader reach by ensuring and delivering easy, cost-effective, and useful solutions for health services.

GoB through MOHFW is offering mHealth services for free to the entire population. The advantages of mHealth services are more applicable for those who are living in rural and remote areas where health-care systems are unavailable or poorly accessible. The *Aponjon* project is offering the maternal and child health care free for the population living below the poverty level (bottom 20%). Toll-free private/business mHealth services addressing the financial benefit and equity for the poor may play a key role to serve the disadvantaged population in Bangladesh. A sustainable mHealth model is likely to be based on the Aponjon model where those who can afford will pay and subsidize the costs for the poorer section of the population in order to achieve equity in health-care services.

There are a number of financial challenges to the use of mHealth that need to be addressed by the local and international providers of innovative mHealth solutions. These include how to find sustainable financing solution to allow the project scaling up in a business model and how to introduce innovative simple solutions through mHealth in a complex health system. The contexts and mechanisms that encourage or discourage community acceptance of the use of technological and organizational innovations are challenges that may hinder the success of mHealth in developing countries including Bangladesh.

26.7 Future Direction

mHealth programs in Bangladesh are growing rapidly, with the potential to benefit primary health-care services, increase care seeking from public facilities, improve surveillance systems, and increase immunization coverage. At this stage, more comprehensive and rigorous empirical evidence on mHealth in terms of health and socioeconomic benefit is not available but is essential to gather. mHealth research in public health domain ideally should draw from medical and clinical expertise, behavioral and social science, user interface design, and computer science to improve health outcome. This field is growing fast and in an unprecedented rate with the changing technology. In the future, the use of smartphones by the mass people will help imply new app design in health-care and public health program. Since compared to other technology the ownership and use of mobile phones are more prevalent among people of low socioeconomic status, the use of mobile phones may reduce the impact of digital divide inherent in any web-based health interventions. This is the first technology where the industry has documented a reverse trend toward a *digital divide*—the gap between people with effective access to digital and information technology and those with very limited or no access at all (Krishna et al., 2009). This increases the likelihood of successfully delivering health improvement interventions to traditionally hard-to-reach populations. Innovative partnerships are crucial to scale mobile initiative in a promising stage. Health-care regulation needs to be conducted to encourage innovation in the marketplace and at the same time needs to ensure service quality and protection of privacy. Future mHealth program will depend on the evidence-based critical knowledge that will enable health administrators and policy makers to make better informed decisions about the investment of the limited health resources in technology. These aforementioned critical issues will determine the success of achieving the agenda of "Vision 2021 Bangladesh: A New Horizon" that introduced the concept of "Digital Bangladesh."

26.8 Conclusion

In Bangladesh, mHealth programs are growing rapidly. Numerous mHealth programs have already been implemented in Bangladesh with the aim of facilitating major benefits to PHC services, increasing care seeking from public facilities, improving surveillance systems, and increasing immunization coverage. Evaluations of the impact of these mHealth projects are pending and are vital to the provision of an evidence base for the use of mHealth in Bangladesh. Comprehensive and rigorous empirical evidence on the health and socioeconomic benefit of mHealth is essential to justify the costs of current programs and implement future mHealth programs in Bangladesh.

Questions

1. Why mHealth is needed in a developing country?
2. What types of mHealth program is ongoing in a developing country—Bangladesh?
3. How are the poor getting benefits?
4. What are the barriers to implement the mHealth in developing countries?

References

Adams, A. M., Ahmed, T., El Arifeen, S., Evans, T. G., Huda, T., Reichenbach, L., and Lancet Bangladesh, Team. (2013). Innovation for universal health coverage in Bangladesh: A call to action. *Lancet, 382*(9910), 2104–2111. doi: 10.1016/S0140-6736(13)62150-9.

Ahluwalia, P. and Varshney, U. (2009). Composite quality of service and decision making perspectives in wireless networks. *Decision Support Systems, 46*, 541–551.

Ahmed, N. U. (2013). Personal Communication. Use of mobile voice call for improving knowledge, awareness, disease prevention and utilization of health services for reducing mortality of newborn, child and mother though Community Clinic Project, MOHFW in Bangladesh. Ethics Advanced Technology Limited. Dhaka, Bangladesh.

Ahmed, S. M., Evans, T. G., Standing, H., and Mahmud, S. (2013). Harnessing pluralism for better health in Bangladesh. *Lancet, 382*(9906), 1746–1755. doi: 10.1016/S0140-6736(13)62147-9.

Akter, S. (2012). Service quality dynamics of mHealth in a developing country. (PhD), The University of New South Wales, Kensington, New South Wales, Australia.

Akter, S., D'Ambara, J., and Ray, P. (June 6–9, 2010c). User perceived service quality of mHealth services in developing countries. Paper presented at the *18th European Conference on Information Systems*, Pretoria, South Africa.

Akter, S., D'Ambra, J., and Ray, P. (2010b). Service quality of mHealth platforms: Development and validation of a hierarchical model using PLS. *Electronic Markets, 20*, 209–227.

Akter, S. and Ray, P. (2010a). mHealth—An ultimate platform to serve the unserved. *Yearbook of Medical Informatics*, 94–100. http://www.ncbi.nlm.nih.gov/pubmed/20938579.

The Alliance for Building Capacity. (2002). Framework for rural and remote readiness in telehealth. Project report for CANARIE, June 2002. See http://www.fp.ucalgary.ca/telehealth/Projects-Canarie-Final%20Report,%20June%202002.htm. Last accessed April 22, 2012.

Angst, M. C. and Agarwal, R. (2009). Adoption of electronic health records in the presence of privacy concerns: The elaborate likelihood model and individual persuasion. *MIS Quarterly, 33*, 339–370.

Bangladesh Bureau of Statistics. (2011). Report of the household income and expenditure survey 2010. Dhaka, Bangladesh: BBS.

Bangladesh Health Watch. (2008). The state of health in Bangladesh 2007: Health workforce in Bangladesh; Who constitutes the healthcare system? Dhaka, Bangladesh.

Betjeman, T. J., Soghoian, S. E., and Foran, M. P. (2013). mHealth in Sub-Saharan Africa. *International Journal of Telemedicine and Applications, 2013*, 1–7. doi: 10.1155/2013/482324.

Bhuiya, A. (2013). Use of mHealth to improve the quality of services of the village doctors research brief. Bangladesh International Centre for Diarrhoeal Disease Research, Dhaka, Bangladesh.

Bill and Melinda Gates Foundation. (2012). Gates foundation announces winner of inaugural gates vaccine innovation award. From http://www.gatesfoundation.org/press-releases/Pages/gates-vaccine-innovation-award-winner-120124.aspx. Last accessed July 18, 2012.

Chang, L. W., Kagaayi, J., Arem, H., Nakigozi, G., Ssempijja, V., Serwadda, D., Quinn, T. C., Gray, R. H., Bollinger, R. C., and Reynolds, S. J. (2011). Impact of a mHealth intervention for peer health workers on AIDS care in rural Uganda: A mixed methods evaluation of a cluster-randomized trial. *Aids and Behavior, 15*(8), 1776–1784. doi: 10.1007/s10461-011-9995-x.

Chaudhry, B., Wang, J., Wu, S., Maglione, M., Mojica, W., Roth, E., Morton, S. C., and Shekelle, P. G. (2006). Systematic review: Impact of health information technology on quality, efficiency, and costs of medical care. *Annals of Internal Medicine, 144*(10), 742–752.

Chetley, A., Davies, J., Trude, B., McConnell, H., Ramirez, R., Shields, T., Drury, P., Kumekawa, J., Louw, J., Fereday, G. and Nyamai-Kisia, C. (2006). Improving health, connecting people: The role of ICTs in the health sector of developing countries: InfoDev. Working paper no 1, 2007.

Chowdhury, A. M., Bhuiya, A., Chowdhury, M. E., Rasheed, S., Hussain, Z., and Chen, L. C. (2013). The Bangladesh paradox: Exceptional health achievement despite economic poverty. *Lancet, 382*(9906), 1734–1745. doi: 10.1016/S0140-6736(13)62148-0.

Dawson, A., Howes, T., Gray, N., and Kennedy, E. (2011). *Human Resources for Health in Maternal, Neonatal and Reproductive Health at Community Level: A Profile of Bangladesh.* Sydney, New South Wales, Australia: The University of New South Wales.

Director General of Health Services. (2012). Mobile phone in health service. From http://www.dghs.gov.bd/index.php?option=com_content&view=article&id=79&Itemid=98&lang=en. Last accessed May 16, 2012.

DNet, Government of Bangladesh, and USAID. (2013). Power of Health in every MAMA's hand. From http://www.aponjon.com.bd/index1.php. Last accessed May 10, 2012.

Fischer, H. H., Moore, S. L., Ginosar, D., Davidson, A. J., Rice-Peterson, C. M., Durfee, M. J., MacKenzie, T. D., Estacio, R. O., and Steele, A. W. (2012). Care by cell phone: Text messaging for chronic disease management. *The American Journal of Managed Care, 1*(18), 2–7.

GSMA Development Fund and Cherie Blair Foundation for Women and Vital Wave Consulting. (2010). Women and mobile: A global opportunity. A study on the mobile phone gender gap in low and middle-income countries. From http://www.cherieblairfoundation.org/uploads/pdf/women_and_mobile_a_global_opportunity.pdf. Last accessed June 1, 2012.

Gurman, T. A., Rubin, S. E., and Roess, A. A. (2012). Effectiveness of mHealth behavior change communication interventions in developing countries: A systematic review of the literature. *Journal of Health Communication: International Perspectives, 17*(suppl 1), 82–104. doi: http://dx.doi.org/10.1080/10810730.2011.649160.

Hanifi, S. M. A., Urni, F., Mamun, A.A., and Iqbal, M. (2012). Impact of SasthyaSena intervention. In T. Wahed, S. Rasheed, and A. Bhuiya (Eds.), *Doctoring the Village Doctors.* Dhaka, Bangladesh: International Centre for Diarrhoeal Disease Research (ICDDRB).

Huq, N. L., Koehlmoos, T. P., Azmi, A. J., Quaiyum, M. A., Mahmud, A., and Hossain, S. (2012). Use of mobile phone: Communication barriers in maternal and neonatal emergencies in rural Bangladesh. *International Journal of Sociology and Anthropology, 4*(8), 226–237.

Information Technologies Group and Center for International Development at Harvard University. (2002). *Readiness for the Networked World: A Guide for Developing Countries.* Cambridge, MA: Center for International Development, Harvard University. From http://www.readinessguide.org. Last accessed June 1, 2012.

International Centre for Diarrhoeal Disease Research Bangladesh. (2012). Cell phone use for detection and treatment of malaria cases in an endemic, remote district of Bangladesh. *Health Science and Bulletin 10*(3), 15–19.

International Centre for Diarrhoeal Disease Research Bangladesh. (2013). Increasing immunisation coverage in Bangladesh through mobile technology. from http://www.icddrb.org/media-centre/news/4129-increasing-immunisation-coverage-in-bangladesh-through-mobile-technology. Accessed 12 January, 2014.

International Centre for Diarrhoeal Disease Research Bangladesh. (2013). Increasing immunisation coverage in Bangladesh through mobile technology. From http://www.icddrb.org/media-centre/news/4129-increasing-immunisation-coverage-in-bangladesh-through-mobile-technology. Last accessed January 12, 2014.

Iqbal, M., Hoque, S., Moula, A., Rahman, M., Choudhury, S., Rasheed, R., and Bhuiya, A. (2012). The SasthyaSena intervention: An experiment in social franchising. In T. Wahed, S. Rasheed, and A. Bhuiya (Eds.), *Doctoring the Village Doctors.* Dhaka, Bangladesh: International Centre for Diarrhoeal Disease Research (ICDDRB).

Istepanian, R. and Lacal, J. (2003). Emerging mobile communication technologies for health: Some imperative notes on m-health. Paper presented at the *25th Silver 59 Anniversary International Conference of the IEEE Engineering in Medicine and Biology Society,* Cancun, Mexico.

Ivatury, G., Moore, J., and Bloch, A. (2009). A doctor in your pocket: Health hotlines in developing countries, innovations: Technology, governance & globalization. *MIT Press Journal (Online), 4*(1), 119–153.

Jennett, P., Jackson, A., Healy, T., Ho, K., Kazanjian, A., Woollard, R., Haydt, S., and Bates, J. (2003). A study of a rural community's readiness for telehealth. *Journal of Telemedicine and Telecare, 9*(5), 259–263.

Kaplan, B. and Litewka, S. (2008). Ethical challenges of telemedicine and tele health. *Cambridge Quarterly of Healthcare Ethics, 17*, 401–416.

Katz, R., Mesfin, T., and Barr, K. (2012). Lessons from a community-based mHealth diabetes self-management program: "It's not just about the cell phone." *Journal of Health Communication, 17*(Suppl 1), 67–72. doi: 10.1080/10810730.2012.650613.

Kern, S. E. and Jaron, D. (2003). Healthcare technology, economics and policy: An Evolving balance. *IEEE Engineering in Medicine and Biology, 22*(1), 16–19.

Khan, W. A., Sack, D. A., Ahmed, S., Prue, C. S., Alam, M. S., Haque, R., Khyang, J. et al. (2011). Mapping hypoendemic, seasonal malaria in rural Bandarban, Bangladesh: A prospective surveillance. *Malaria Journal, 10*, 124. doi: 10.1186/1475-2875-10-124.

Khatun, F. and Akter, S. (2013). mHealth research across Australia and Bangladesh. Paper presented at the *mHealth Workshop 2013*, The University of New South Wales, Sydney, New South Wales, Australia. http://www.asb.unsw.edu.au/research/asia-pacificubiquitoushealthcareresearchcentre/resources/Pages/default.aspx. Last accessed June 30, 2014.

Krishna, S., Boren, S. A., and Balas, E. A. (2009). Healthcare via cell phones: A systematic review. *Telemedicine Journal and E-Health, 15*(3), 231–240.

Kwan, A. (2013). Using mobile technologies for healthier aging: mHealth alliance. United Nations Foundation, Washington, DC.

Labrique, A. (2012). mTikka: A virtual, interactive, mobile-phone based immunization record to improve vaccination rates in rural Bangladesh. From http://www.jhumhealth.org/resources/mtikka-summary.

Lemay, N. V., Sullivan, T., Jumbe, B., and Perry, C. P. (2012). Reaching remote health workers in malawi: Baseline assessment of a pilot mhealth intervention. *Journal of Health Communication: International Perspectives, 17*(Suppl 1), 105–117. doi: http://dx.doi.org/10.1080/10810730.2011.649106.

Lester, R. T., Mills, E. J., Kariri, A., Ritvo, P., Chung, M., Jack, W., Habyarimana, J. et al. (2009). The HAART cell phone adherence trial (WelTel Kenya1): A randomized controlled trial protocol. *Trials, 10*, 87. doi: 10.1186/1745-6215-10-87.

Lester, R. T., Ritvo, P., Mills, E. J., Kariri, A., Karanja, S., Chung, M. H., Jack, W. et al. (2010). Effects of a mobile phone short message service on antiretroviral treatment adherence in Kenya (WelTel Kenya1): A randomised trial. *Lancet, 376*(9755), 1838–1845. doi: 10.1016/S0140-6736(10)61997-6.

Lund, S., Hemed, M., Nielsen, B. B., Said, A., Said, K., Makungu, M. H., and Rasch, V. (2012). Mobile phones as a health communication tool to improve skilled attendance at delivery in Zanzibar: A cluster-randomised controlled trial. *BJOG, 119*(10), 1256–1264. doi: 10.1111/j.1471-0528.2012.03413.x.

Mahmood, S. S., Iqbal, M., Hanifi, S. M., Wahed, T., and Bhuiya, A. (2010). Are "Village doctors" in Bangladesh a curse or a blessing? *BMC International Health Human Rights, 3*, 10–18.

MAMA: Mobile Alliance for Maternal Action. From http://www.mobilemamaalliance.org/mama-bangladesh. Last accessed June 30, 2014.

Mechael, P. (2009). The case for mHealth in developing countries. *In Technology, Governance, Globalization*. 4 (1), 103–118.

Ministry of Health and Family Welfare and Government of the People's Republic of Bangladesh. (2011). Health, Population and Nutrition Sector Development Program (2011–2016), Program implementation plan. Dhaka, Bangladesh. Retrieved from http://www.dghs.gov.bd/index.php/en/mis-docs/important-documents/item/hpnsdp-2011-16?category_id=115 Last accessed June 30, 2014.

National Institute of Population Research and Training (NIPORT) Mitra and Associates, MEASURE DHS, and ICF International. Dhaka, (2013). Bangladesh demographic and health survey 2011. Dhaka, Bangladesh. URL: http://dhsprogram.com/pubs/pdf/FR265/FR265.pdf.

Noordam, A. C., Kuepper, B. M., Stekelenburg, J., and Milen, A. (2011). Improvement of maternal health services through the use of mobile phones. *Tropical Medicine & International Health, 16*(5), 622–626. doi: 10.1111/j.1365-3156.2011.02747.x.

Norris, T., Stockdale, R., and Sharma, S. (2008). Mobile health: Strategy and sustainability. *The Journal of Information Technology in Healthcare, 6*(5), 326–333.

Prue, C. S., Shannon, K. L., Khyang, J., Edwards, L. J., Ahmed, S., Ram, M., Shields, T. et al. (2013). Mobile phones improve case detection and management of malaria in rural Bangladesh. *Malaria Journal, 12*, 48. doi: 10.1186/1475-2875-12-48.

Rajatonirina, S., Heraud, J. M., Randrianasolo, L., Orelle, A., Razanajatovo, N. H., Raoelina, Y. N., Ravolomanana, L. et al. (2012). Short message service sentinel surveillance of influenza-like illness in Madagascar, 2008–2012. *Bulletin of the World Health Organization, 90*(5), 385–389. doi: 10.2471/BLT.11.097816.

Rasheed, S., Iqbal, M., and Urni, F. (2009). Inventory of facilities. In A. Bhuiya (ed.), *Health for the Rural Masses: Insights from Chakaria*. Vol. ICDDR, B Monograph, pp. 25–38. Dhaka, Bangladesh: International Centre for Diarrhoeal Disease Research.

Seidenberg, P., Nicholson, S., Schaefer, M., Semrau, K., Bweupe, M., Masese, N., Bonawitz, R., Chitembo, L., Goggin, C., and Thea, D. M. (2012). Early infant diagnosis of HIV infection in Zambia through mobile phone texting of blood test results. *Bulletin of the World Health Organization, 90*(5), 348–356. doi: 10.2471/BLT.11.100032.

Tamrat, T. and Kachnowski, S. (2011). Special delivery: An analysis of mHealth in maternal and newborn health programs and their outcomes around the world. *Maternal and Child Health Journal, 16*(5), 1092–1101. doi: DOI 10.1007/s10995-011-0836-3.

Telemedicine Reference Centre Limited. From http://trclcare.com/. Accessed June 30, 2014.

Text to Change. (2014). Largest scale mHealth project in Africa. Retrieved January 15, 2014. From http://projects.texttochange.org/en/project/779/.

United Nations Foundation and Vodafone Foundation. (2009). mHealth for Development: The opportunity of mobile technology for healthcare in developing world. From http://www.vitalwaveconsulting.com/insights/mHealth.htm. Last accessed May 10, 2012.

Varshney, U. (2005). Pervasive healthcare: Applications, challenges and wireless solutions. *Communications of the Association for Information Systems, 16*(3), 57–72.

Wahed, T., Alamgir, F., and Sharmin, T. (2012). Perceptions of SasthyaSena : Talking to VDs, villagers and community leaders. In T. Wahed, S. Rasheed, and A. Bhuiya (Eds.), *Doctoring the Village Doctors*. Dhaka, Bangladesh: International Centre for Diarrhoeal Disease Research (ICDDRB).

The World Bank. (2013). From http://data.worldbank.org/indicator/NY.GDP.PCAP.CD. Accessed September 15, 2013.

World Health Organization. (2010). *World Health Statistics 2010*. Geneva, Switzerland: World Health Organization. Accessed 12 September, 2013.

World Health Organization. (2011). mHealth: New horizons for health through mobile technologies. *WHO Global Observatory for eHealth Series*. Geneva, Switzerland: World Health Organization.

World Health Organization. (2012). EPI Fact Sheet, Bangladesh 2011. From http://www.searo.who.int/entity/immunization/data/bangladesh_epi_factsheet_2011.pdf. Last accessed September 12, 2013.

Zurovac, D., Otieno, G., Kigen, S., Mbithi, A. M., Muturi, A., Snow, R. W., and Nyandigisi, A. (2013). Ownership and use of mobile phones among health workers, caregivers of sick children and adult patients in Kenya: Cross-sectional national survey. *Global Health, 9*, 20. doi: 10.1186/1744-8603-9-20.

27

Mobile Health in the Developing World: Needs, Facts, and Challenges

Arwa Fahad Ababtain, Deana Ahmad AlMulhim,
Faisel Yunus, and Mowafa Said Househ

CONTENTS

27.1 Introduction

The developing countries are struggling to provide adequate health care to the needy people, especially in the rural areas. Availability of accessible and quality health-care services has been the key challenges in the developing world. mHealth technology is considered to be the fastest-growing phenomena in health care and represents a new growing area within the Arab world (Pillai, 2012). Eighty percent of rural areas in the developing world have access to the mobile networks (Kaplan, 2006; Donner, 2008; Jeannine, 2011). Although mHealth is still in its early stages of development, it has already started to transform health-care delivery due mainly to the success of mHealth applications and programs that have been implemented in the developing world. The United Nations Foundation (UNF) reported on more than 50 mHealth applications and projects on health education, awareness, data collection, remote monitoring, disease tracking, and diagnostic support in the developing countries. The goals of implementing these mHealth applications are to enhance the efficiency and the accessibility of health-care systems and to reduce the mortality in the developing countries (United Nations Foundation, 2009). Much work has been done in the developed world (e.g., United States, Britain, Canada), but there has not been as many initiatives carried out in the Arab world, especially in Saudi Arabia. With a high number of mobile phone users in the Arab world, it is important to consider the role of mHealth in improving health-care services in this part of the world (United Nations Foundation, 2009; Pillai, 2012).

mHealth can help in increasing the access to health-care information and services, enhancing efficiency with low cost, and improving communication between patients and health-care providers in a timely matter. These features have the potential of greatly improving the health-care systems in the developing countries including the Arab world, which will improve the quality of life for millions of people (Patrick et al., 2008; Kahn et al., 2010). Rajput et al. (2012) developed and implemented a mobile device–based system for home-based care in the developing countries and concluded that using such system was viable and cost-effective for data collection and facilitating the work that then leads to improvement in health-care outcomes. Currently, the health-care system in the developing world is witnessing a cultural shift from a traditional approach that relied on the patients following the doctors' orders without their involvement in decision making to a more patient-centered approach where the patient is equipped with information and knowledge gained from easy access to modern technology, as reported by The Boston Consulting Group (2012). Given the increasing use and reliance on the mobile technology, we believe that using mHealth in the developing countries, and the Arab world in particular, will positively impact the health-care outcomes and will significantly increase the access to the health-care systems.

In the light of the aforementioned reasons, we focus on mHealth and the need for such an innovation in improving health-care delivery in the developing world with a specific focus on Saudi Arabia in this chapter. We explore various successful mHealth applications in the Arab world, mHealth challenges, and the future and emerging trends for mHealth technology and discuss the solutions for better implementation in such technology.

27.2 Need for mHealth in the Developing World

The primary purpose of a health-care system is to address people health needs and provide high quality of services. The technologies that are currently being used in health-care sector have provided much hope for a significant improvement in population health on account of their wide-ranging accessibility. mHealth is one of the most effective communication technologies that support health-care services in many countries all over the world. In the developing world, the main factors that are conducive to the implementation and use for such tools are poor access to health care, higher health-care costs, emerging diseases, and substandard health-care quality (Jeannine, 2011; Piette et al., 2012; Shin, 2012). According to Jeannine (2011), the factors that have led to the current rapid development of mHealth in the developing world are the growth of mobile phone users, the need to treat many people in the rural areas, the lack of timely disease control, decreasing the cost of mobile phones, the cultural factors, the shortage of health-care workers, and the lack of resources and infrastructure.

The mobile market in the developing world is considered to be the most rapidly growing sector—from the world's 5.3 billion people that used mobile phones in 2010, 70% were from the developing world (Qiang et al., 2012). In India, a country with 1.21 billion people, the number of mobile subscribers increased from 652 million in 2010 to 858 million in 2011 with better coverage in the rural areas (Garai and Candidate, 2011). The number of mobile users in Africa has jumped from 280 million in 2007 to around 600 million in 2012. This has made the mobile phones to be the most widely used computing devices for Africans (Waruingi and Underdahl, 2009; Danis et al., 2010). Furthermore, 49% of Nigeria's population had used mobile phones in 2009 (Pyramid Research, 2010). The Arab world, and Saudi Arabia in particular, has witnessed a significant expansion in the field of technology and communication. Saudi Arabia is considered as the second biggest market for mobile phones in the Middle East. The usability of mobile phones with their attractive capabilities is on the rise especially among the teenagers. Saudi Arabia recorded a high mobile penetration rate (186%) compared to the developing countries average of 73% and the developed countries average of 116%. Recently, the advanced wireless fourth-generation (4G) technology has been made available in Saudi Arabia with its large bandwidth and a faster transfer rate, and it is expected that Saudi Arabia will be the leading 4G market in the Middle East by 2016. Additionally, the amount of Internet usage in the Arab world in general and Saudi Arabia in particular has risen over the past few years to about 41% at the end of 2010, which is one of the highest in the developing world (Abanumy and Mayhew, n.d.; Alsenaidy and Tauseef, 2010; CITC, 2010; Rehalia and Kumar, 2012).

In view of this rapid expansion in the usage of mobile phones in Asia, the Middle East, and Africa, mHealth can be the most effective and accessible way to solve many issues in the health-care sector of the developing world. According to Qiang et al. (2012), the world attention to health has been dedicated on controlling communicable and chronic diseases such as HIV/AIDS, malaria, tuberculosis, heart diseases, and diabetes. mHealth can play a major part in halting the spread of these diseases and improving the quality of life for millions, for instance, by expanding the health-care services, increasing the awareness, and encouraging behaviors that limit infections. In addition, maternal and child health is another major concern in the developing world. For instance, in Bangladesh, the possibility of surviving to the age of 10 is 24% for children who lose their mothers compared to 89% for children with living mothers. mHealth can deliver valuable information to the people who struggle with this problem (United Nations Foundation, 2009; Danis et al., 2010; Kahn et al., 2010; Qiang et al., 2012). Furthermore, mHealth can be used to monitor epidemic diseases. As stated by the Pyramid Research (2010), nongovernment organizations (NGOs) are monitoring the spread of pandemics and viruses, in Africa, Asia, and Brazil, through the use of mobile networks and satellite communications. By undertaking this kind of health surveillance via mHealth tools, the progress of these diseases can be delayed or stopped, the financial load of the diseases can be reduced, and the quality of patient life can be improved (Waruingi and Underdahl, 2009).

Although several countries in the developing world have limited resources, still they are able to employ mHealth tools to ensure the high quality of health-care services. mHealth can support the monitoring of health workers and can help in controlling the distribution and use of fake drugs. In addition, mHealth can provide timely information for medical, educational, and emergency situations, thereby helping in the allocation of the right resources for the right individuals. For example, a call center in India is using mobile phones to collect information from patients trying to find emergency care and then sends ambulances with the needed equipment for each case. Another example is the Nokia pilot project in Brazil that supports NGOs and health organizations in gathering, communicating, and saving data using smartphones with Internet connectivity (Pyramid Research, 2010; Bolton, 2012; Qiang et al., 2012). Some countries in the Muslim developing world have specific religious values concerning interaction of males and females. mHealth can be an effective tool to service people without infringing on their religious beliefs. For example, women can use mobile devices to contact their health providers without presenting for the face-to-face contact. Additionally, some matters with cultural sensitivity such as family planning, can be handled secretly via mHealth (Qiang et al., 2012). mHealth applications can also allow elderly as well as disabled people to communicate with health-care workers without leaving their place, which improves the delivery of care for such cases (Qiang et al., 2012). For instance, Philips Lifeline medical alert product has permitted aging individuals to access the emergency services with one click of a button (Pyramid Research, 2010). The shortage of the human resources has increased the burden on developing world's health-care systems. According to the UNF (2009), the countries in the developing world have an acute shortage of health-care workers. mHealth can be the solution in this case too as the health care can be accessible and effective even with a reduced number of the health-care workers (Akter et al., n.d.).

Short message service (SMS)—an important feature of mobile communication—can be used for a quick transfer of health information such as delivering reminders. As stated in Piette et al. (2012), 60% of the high-income countries and 30% of the low- and middle-income countries are using SMS to improve the patient treatment compliance. In a trial related to smoking cessation, the percentage of people quitting smoking doubled after 6 months of receiving *txt2stop* messages on their mobiles (Jeannine, 2011; mplushealth, n.d.; Piette et al., 2012). In the Arab World, growing cost, shortage of trained staff, and increasing prevalence of lifestyle-related diseases such as diabetics all have negatively affected the quality of health-care systems (Byrne, 2011; mplushealth, n.d.; Pillai, 2012). According to the World Health Organization (WHO) data, the rate of chronic illnesses, such as diabetes and heart diseases, in the Arab countries is higher than African and Asian countries (Alastair, 2010). mHealth can provide a reasonable solutions and real efforts are underway to improve the health-care services in the Arab world. The mplushealth conference is conducted annually, which is the only platform where health-care specialists, insurance providers, telecommunication decision makers, and government managers are meeting to discuss the potential, issues of concern, and rollout of mHealth in the Middle East. Qtel group is working to help patients manage their health easily and efficiently in the Middle East. It is delivering a few mHealth innovative solutions and also undertaking various studies on the need for developing new mHealth applications in the region. In 2011, Qtel group announced partnership with Mobile Health Company in Saudi Arabia with the intention of providing guidelines and health-care instructions through mobile phones. Also, patients in Iraq and Kuwait are supported with different SMS/multimedia messaging service (MMS) information, educational services, and health tips from Qtel group (Byrne, 2011; mplushealth, n.d.; Pillai, 2012). In Saudi Arabia, the readiness of the citizens to adopt new technology, the accessible and increasing use of mobile phones, and the availability of good technical mobile infrastructure have presented a worthy opportunity for mHealth applications that are able to provide high quality of services for all citizens and residents (Abanumy and Mayhew, n.d.).

The developing world has a clear need for innovative and effective solutions for its health-care issues. Based on the preceding narrative, mHealth is essential in sorting out the health-care issues of the developing world and can strengthen the health-care systems (Akter, 2012; United Nations Foundation, 2009). mHealth can provide both clinicians and patients with a better and flexible way to manage and improve the health and has supported the delivery of quality health care to many people in the remote areas (mplushealth, n.d.; United Nations Foundation, 2009; Waruingi and Underdahl, 2009; Garai and Candidate, 2011). We will witness a rapid development in mHealth technologies over the coming few years. This expansion will help in bridging the gap in health care based on several factors: (1) mHealth technology will be widely available at a reasonable price and high performance, (2) mHealth is public centered and can be accessed from anywhere at any time, (3) it provides timely and high-quality health care, (4) mHealth and its applications are empowering people as well as enhancing preventive care, and (5) it fills the gap in health funding and the shortage of health-care workers. The potential of mHealth, however, is wide ranging and is not limited to the aforementioned five factors (International Telecommunication Union, 2009).

27.3 mHealth in Saudi Arabia

Saudi Arabia is the second biggest market for mobile phones among the Middle Eastern countries. According to the United Nations Conference on Trade and Development (UNCTAD), Saudi Arabia has the largest number of mobile phone users worldwide. Between 1990 and 2011, there has been a massive increase in the use of the Internet, and the number of fixed phone line subscribers has decreased (Alsenaidy and Ahmad, 2010). Worldwide, the users of mobile phones are increasing every year especially for smartphones. The smartphone penetration has increased from 2% to between 5% and 10% within the last 2 years in the Middle East. In Saudi Arabia, smartphone market share is around 12% of the total markets in the Arab world (Alastair, 2010). The smartphone penetration rate is expected to increase from 25% to around 49% between 2011 and 2016. Additionally, Saudi Arabia and a few other countries in the Middle East are actively implementing 4G networks. It is estimated that Saudi Arabia will be the leading 4G market with just above 11 million 4G subscriptions by 2016 (Alsenaidy and Ahmad, 2010). The telecommunication infrastructure is quickly growing in Saudi Arabia. The Saudi Arabian government has been privatizing the telecommunication sector, which has led to increased availability of the Internet and high mobile penetration. The government has also implemented an mGovernment portal to facilitate communication between the government organizations and also to improve information sharing with the citizens and businesses through mobile technologies (Abanumy and Mayhew, n.d.; Alsenaidy and Ahmad, 2010).

 The Saudi Arabian government is heavily investing in the health-care sector in order to improve the quality of health-care services. For example, in 2010, $16.7 billion was allocated to health care with more than $6 billion for eHealth initiatives (Alastair, 2010). The availability of wireless technology in Saudi Arabia has played a major role in the adoption of mHealth services (Dolan, 2011). As stated by the CEO of Mobily, a leading mobile telecommunication company in Saudi Arabia, the implementation of mobile technology offers a straight and widespread platform for various applications that support and improve health all over the kingdom (Mobily, n.d.). Currently, various mGovernment services and mHealth applications are offered in Saudi Arabia, which are considered suitable and useful. One such example is the Health Mobile service, which is an interactive mobile phone service that sends daily text messages (SMS) with medicine, health, and disease prevention information from the Ministry of Health (MOH) to all citizens (Abanumy and Mayhew, n.d.;

Alsenaidy and Ahmad, 2010). Another example is *Mobile Baby* service that was launched by Mobily and sends ultrasound stills, 3D images, and video clips to the parents' mobile device. In addition, Saudi Arabia is one of the largest markets for the Bluetooth-enabled blood glucose meters. There are plans to offer another innovative mHealth service, by the Saudi Arabian operators Mobily and Ericsson, to help in remotely monitoring the body vital signs and providing relevant data and alerts to the health-care providers (Dolan, 2011; Mobily, n.d.). In addition, Qualcomm is currently testing various mHealth solutions in Saudi Arabia with the aim of improving access to health care and monitoring and diagnosing diseases over the wireless networks (Alastair, 2010; Mobily, n.d.).

27.4 Examples of mHealth Applications in the Developing World

A study by UNF and Vodafone Foundation has listed 51 mHealth programs that are operating in 26 developing countries all over the world. These programs and projects focus on six main areas: treatment and support services, health education and awareness services, data collection and remote monitoring services, disease surveillance and drug adherence services, health information systems and point of care services, and emergency medical services (Vital Wave Consulting, n.d.; United Nations Foundation, 2009; Akter, 2012). A brief overview of these programs is given in the following.

27.4.1 Diagnostic Treatment and Support Services

These mHealth services provide support, guidance, and consultation on patient diagnosis and treatment activities to health-care workers located in remote areas. These are available round the clock and are accessible through the Internet over the mobile phones (Blynn and Aubuchon, 2009; Akter, 2012). *Health Hotline* service is an example of this type. It is a medical call center that offers health-related information, guidance, referrals, and prescriptions to callers through phone lines. Health professionals, such as nurses, paramedics, or physicians, receive the calls and follow certain standard protocols in order to evaluate the medical situation and based on that provide the accurate information and advice. *Health Hotline* service has already been implemented in four developing countries and has served around 10 million people. *HealthLine* in Bangladesh, *HMRI* in India, *TeleDoctor* in Pakistan, and *MedicalHome* in Mexico are using such services. As a result of this service, the health-care situation in these countries has improved and has provided support to people who were previously unable to access health care at a reasonable cost (Akter, 2012). Another example is *OpenMRS*, which is collaborative open-source software. It has been designed to support the health-care delivery in the developing countries after the serious need for advance HIV treatment in Africa. *OpenMRS* was established with the purpose of providing electronic medical record system that supports all kinds of medical treatment for health-care providers as well as helping in sharing ideas among software users. This service can be available in any poor resources environment and can be modified based on the organization's needs without any advance programming. Currently, over 124 research and clinical centers are engaged in *OpenMRS* projects all over the world (Bolton, 2012).

27.4.2 Health Education and Awareness Services

These services usually use one-/two-way communication programs that send text messages to mobile subscribers to support public health and behavior change movements (Vital Wave Consulting, n.d.). At present, various developing countries, such as South Africa, India, Tanzania, Uganda, and the Philippines, are implementing this service to prevent HIV/AIDS, sexually transmitted diseases, TB, diabetes, and heart diseases (Akter, 2012). *Text to Change* (*TTC*) is an awareness and education program that was piloted in Uganda in 2008. It offers HIV/AIDS awareness through an SMS-based interactive quiz with the aim of increasing the HIV/AIDS awareness, educating participants about HIV/AIDS safe behaviors, encouraging people to search for HIV tests and treatment, and improving attendance at the HIV consulting and services centers. The participants in this service are sent various questions on their mobile phones to assess their knowledge and awareness. The participants with high accuracy and participation rate are given free airtime, prepaid mobile phone minutes, and other awards. Within 3 months of its start, the TTC has provided HIV/AIDS awareness to 15,000 mobile phone prescribers in Uganda. This pilot project has improved HIV testing and consultation services by 40% in comparison to previous years. Due to its success, Uganda MOH has used TTC platform to launch similar awareness projects like *Texting4Change* in 2009 (United Nations Foundation, 2009; Danis et al., 2010; Pyramid Research, 2010). The *Project Masiluleke* has been designed to fight HIV/AIDS in South Africa. The *Project Masiluleke*, which means "lending a helping hand," started in 2008. This project has two components: the advertisement for the National AIDS Helpline using *please call me* messages and the TxtAlert. The mobile operator provides *please call me* service to the subscribers who like to request a call from another person. In this service, the message contains the word *please call me* in addition to the sender phone number as well as space for additional advertisements' text. Once the patient calls the helpline center, he or she will receive information about testing services and their locations from the hotline representatives. The second component, TxtAlert, aims to improve the attendance of antiretroviral (ARV) therapy patients. This service sends an SMS reminder to ARV patients to attend for their appointments at the clinic and also to take their medicines as prescribed. If the patients are unable to attend the scheduled appointment, another SMS can be sent in order to reschedule the appointment in a simple and easy way. Currently, over one million educational text messages are sent per day through this service in South Africa. The *Project Masiluleke* has improved the attendance level at the participating hospitals where it has reached 95%–100% within 2 years as result from TxtAlert messages (United Nations Foundation, 2009; Pyramid Research, 2010).

27.4.3 Data Collection and Remote Monitoring Services

These services provide access to specific disease-related patient data to the health-care providers (Vital Wave Consulting, n.d.). An example of this service is the *Nokia Data Gathering system*, which offers fast and efficient data collection to control the dengue virus spread in Brazil. The first full implementation of this system was undertaken in 2008 by the Amazonas Health Department with a goal of saving time to save lives. Customized questionnaires can be created in this system based on specific needs and are then distributed to health agents' mobile phones. Once the surveys are completed by the field workers, they are sent back to the server through wireless connection. This information is integrated with the organization's existing systems for direct analysis. In addition, each record can be located using GPS services in the Nokia Data Gathering system. Encouraging results have been realized from this system. It has reduced the time spent for data gathering and has increased end-user acceptance. Over 400 completed surveys were gathered by 20 professionals in only 2 days compared to up to 2 months before this system was implemented (United Nations Foundation, 2009; Bolton, 2012). *EpiHandy* project in Bergen, Norway, is another example of this service where it provides mobile data collection and record access tool through open-source and free use of the application. The main purpose of this project is to support the data collection and monitoring by health workers and researchers in rural areas who provide child nutrition, breastfeeding, and HIV prevention services. Currently, *EpiHandy* is deployed in Uganda, Zambia, Nepal, and South Africa. The 5-year evaluation of this system shows improvement in data entry errors and overall user acceptance. The system is also cost-effective in large surveys compared to traditional paper-based method (Shin, 2012).

27.4.4 Disease Surveillance and Drug Adherence Services

These services provide one-/two-way communication to observe patient health condition, maintain health-care worker's appointments, and guarantee accurate medication treatment adherence (Vital Wave Consulting, n.d.). *SIMpill* is an innovative technology that is used for drug adherence and disease surveillance. The medicine is given to the patient in a SIMpill bottle with mechanically programmed medication schedules. Once the bottle is opened by the patient, an SMS message is automatically sent from the bottle's SIM card to the patient's clinic in order to document that the medication has been taken. If the patient disregards taking the medication on the scheduled time, an SMS message is sent to the patient's mobile phone. If he or she still ignores the SMS, another SMS is sent to the health-care worker, family member, or a friend notifying them that the patient has not taken their medication. Since the device tracks the patient's compliance, *SIMpill* results can be analyzed very easily. The system has been successful throughout the trial period in South Africa for the tuberculosis patients because of its simplicity: it has recorded 94% compliance rate and 92% treatment rate (Blynn and Aubuchon, 2009). Another example of this service is the *Real-Time Biosurveillance* Program. The main purpose of this system is to analyze mHealth systems in order to develop early detection and notification system specific for disease and epidemic occurrence. This program has started in India and Sri Lanka where frontline health-care workers were selected to digitalize the patients' data at health-care centers. Later, the patients' data are transferred to central servers using mobile phones that contain customized application called *mHealth Survey*. After the statistical analysis, the results are sent to regional and local health officials via mobile phones for notification purposes (United Nations Foundation, 2009; Rehalia and Kumar, 2012).

27.4.5 Health Information Systems and Point of Care Services

These services provide central health-care IT systems that can be accessed and integrated with mHealth applications (Vital Wave Consulting, n.d.). One example of this service is the Uganda *Health Information Network (UHIN)* project. It was launched in 2003 to collect and transmit data into the health information system (HIS) using 350 handheld computers personal digital assistants (PDAs) connected through the local GSM cellular network. To exchange data with the MOH system, health workers in remote villages gather and input data in their PDAs using digitalized MOH forms. The PDAs are frequently and regularly updated with new data, weekly health recommendations, and the latest articles. In addition, each PDA is tracked to check if it gets any new information that is necessary for the MOH. The UHIN program functions in 5 regions with 174 health-care centers, 600 health-care workers, and the local populations. The program has resulted in improving the compliance rate, the data quality has improved, and it has been found cost-effective (Vital Wave Consulting, n.d.). Furthermore, *ChildCount+* project is another example. This project started in 2009 in Sauri, Kenya, with a purpose of empowering communities to increase child survival and maternal health. The community health workers can use any standard phone to send text messages and register patients in the central web dashboard to monitor their health by specialized health-care team and reduce gaps in treatment. Using *ChildCount+* program, the health-care workers can easily track children who need nutritional or medical involvement. More than 9500 children under 5 years were observed for community-based management of serious malnutrition by 100 community health-care workers using mobile applications. The UNICEF is in the process of developing *ChildCount+* in Senegal (Jeannine, 2011; Rehalia and Kumar, 2012).

27.4.6 Emergency Medical Services

The emergency medical services use mobile device applications with GPS capabilities to send and receive data about disease occurrences and geographical range of public health emergencies (Vital Wave Consulting, n.d.). *FrontlineSMS* is one such free service. It is based on an open-source software that is web independent and compatible with simple mobile devices that are commonly used in the developing world. Various organizations can use this software to distribute and collect information through SMS. In addition, users can connect a range of mobile devices to a computer so they can send and receive SMS text messages. Once it is installed on the computer, the mobile phones can connect to it via USB cables, and then users can send many text messages as well as health-care workers can send SMS back to the main database. The *FrontlineSMS* scalability and sustainability have made it very effective software to record various disease occurrences and provide the needed emergency medical services. *FrontlineSMS* has developed *Medic Mobile*, which is first launched in Malawi to enable health workers to document patient data in real time and more precisely. The patients' data can be accessed and remotely updated. This leads to low costs and promptly provides the needed health-care services (Blynn and Aubuchon, 2009; Bolton, 2012).

27.5 Examples of mHealth Applications in the Arab World

The developing countries including those in the Arab world have major problem with high maternal mortality. In 2008, around 358 women died in the developing countries because of pregnancy or childbirth complications. Etisalat Company has provided a solution for this issue. Etisalat is considered the largest operator in the Middle East and operates in 17 countries across the Middle East, Africa, and Asia (Held, n.d.). Etisalat *Mobile Baby* program is deployed in UAE, Saudi Arabia, and Nigeria. A complete suite of services are offered for birth attendees and midwives with the aim of improving rates of safer pregnancies. The program provides remote monitoring of the pregnancy development through ultrasound images, offers protocols to recognize warning signs during labor and delivery, affords money through the phone in order to pay for emergency transportation, and provides communication and referral facility. The Etisalat *Mobile Baby* process involves three steps.

First, the pregnant mother information is registered in the Mobile Baby application by the traditional birth attendants (TBAs). Then, the pregnancy status and danger signs are documented through her pregnancy in the Mobile Baby application and reported to the medical facility by the TBAs. Lastly, TBAs offer support to mothers in labor to ensure that both the mother and the baby are healthy. The application also offers portal for communication with the on-call doctor in case of complications during the delivery. The Etisalat Mobile Baby program has trained more than 500 birth attendants and midwives and has provided support to over 20,000 pregnant ladies. There are plans to implement Mobile Baby program in other developing countries (Held, n.d.).

Various mHealth applications are currently in use in Saudi Arabia. As part of eGovernment implementation, mGovernment applications are implemented in the MOH hospitals and clinics. *Physician Appointment Reminder* is a mobile application that reminds patients about their appointments with the physicians. The application sends SMS messages with date, time, and clinic location to the patient's mobile phone. As a result of this system, the number of *no-show* patients has reduced (Abanumy and Mayhew, n.d.). In addition, MOH offers *Health Mobile* interactive service with the purpose of keeping the mobile subscribers up to date with the new medicine and health and disease prevention information as well as providing specialized courses that focus on pregnancy, diabetes, etc. A daily message is prepared by MOH specialists and consultants, which is then sent to the subscribers. Presently, service officials are in the process of providing new particular courses via SMS to educate the subscribers about the most widespread diseases in Saudi Arabia (Alsenaidy and Ahmad, 2010; Ministry of Health, 2011).

A few mobile companies are also providing various mHealth services and applications in Saudi Arabia. The Saudi Telecom Company (STC) organized a free participation for a week in *Your Health* services corresponding to the World Health Day. A monthly *Health Magazine* is sent to the participants' mobile with various topics and news, the latest medical research and health studies, and special monthly health information related to women. There are 15 specialized channels in *Your Health* services including diabetes services and multimedia channel. Any customer, who wants to receive information regarding any of these channels, can send a blank text message to a specified number (5010) and get information on how to participate in the preferred services. By enrolling in *Your Health* services, a free electronic medical file is created for the participants (https://www.facebook.com/pages/Sehhatak-STC/405513362829022). The participants can then calculate their body mass index, save their health information and retrieve it any time they want, and help the doctor to take the right medical decision (Okaz news, 2011). In addition, another Saudi Arabian telecom operator, Mobily, has collaborated with French Health Company to launch an mHealth application that supports diabetes patients. Through this smart application, patients can share their data and receive instant reports from their doctors. In this way, diabetes patients can avoid long-term complications, have more control over their disease, and take the right diabetes-related decision (Global Telecoms Business, 2012). Lastly, Mobily has also partnered with Qualcomm and Dr. Sulaiman Al-Habib Medical Group to launch *Baby Ultrasound MMS* Service. The OB/GYN Clinic in Dr. Sulaiman Al-Habib Hospital offers the *Baby Ultrasound MMS* application as a free service to any parent who likes to share their baby images with their family and friends. The Baby Ultrasound MMS application is well matched with any ultrasound machine and provides high-quality 2D and 3D images and videos. It also enables doctors to make instant analysis of the sonograms at anytime and at anywhere. After visiting Dr. Sulaiman Al-Habib Hospital, parents will be able to receive their baby's ultrasound images and videos through MMS. This service saves a lot of time where instead of printing out the images or saving the video in a DVD, it just transmits the video straight from the ultrasound machine to the parents' smartphone or e-mail (Mobily, n.d.; Dolan, 2011).

27.6 mHealth Challenges in the Developing World

As noted in Section 27.4, mHealth can provide a great opportunity to solve several issues in the current health-care systems and can help in improving the quality of health care in the developing world. On the other hand, various challenges and barriers can also affect the success of this new health-technical innovation.

First, although the mHealth mobility feature has made it easy to access tools in the developing world, yet many network challenges remain that can reduce its effectiveness. The Internet is unreliable and/or expensive in many of the developing countries, and that presents a real barrier to the mHealth implementation. As a result of the limited Internet connectivity, various countries are not able to access health information (Mishra and Singh, 2008; Blynn and Aubuchon, 2009; Bolton, 2012). Furthermore, the voice recognition mobile phone application can be a very supportive tool for providing health care in countries with the high levels of illiteracy, but this kind of application requires a third-generation (3G) network that is not currently available in many of the poor-resource countries in the developing world.

Second, mHealth can be a good solution for communities with unique culture and customs such as the developing world, but at the same time, this cultural diversity may restrict the implementation of mHealth applications. People in these countries have different cultural values, beliefs, and customs that affect their health behaviors and reduce their ability to take control of their health. One of the solutions for that diversity is to customize mHealth applications so they can meet the community needs (Waruingi and Underdahl, 2009; Qiang et al., 2012). Besides, illiteracy, as well as a range of spoken languages in the developing world, is another barrier that could reduce the benefit of mHealth tools such as SMS text massaging (Kaplan, 2006).

Third, the security and privacy of the transferred information are another important challenge (Akter et al., n.d.; Rehalia and Kumar, 2012). A serious concern about violations when dealing with electronic communication was revealed by the 1996 Health Information Portability and Accountability Act (HIPAA), which is responsible for protecting the personal health information (Waruingi and Underdahl, 2009). Some personal devices, such as the cellular phones, smartphones, and PDA, have limited security features that cannot restrict unauthorized access to confidential information, or they have restricted and difficult security mechanisms that reduce the accessibility of the device itself (Shin, 2012). In addition, many people are worried about losing their devices that contain personal materials and medical information and predisposing them to exploitation by illegal individuals (Patricia et al., 2010; Pillai, 2012; West, 2012).

Fourth, another barrier is the cost of mHealth devices. While mobile devices are becoming affordable for many people in the developing world, they are still expensive for a large majority especially in the rural areas (Kaplan, 2006). In addition, customers in the developing world are mainly using mobile phones with chargeable minutes: additional minutes are required for health consultation or personal health monitoring that might be unaffordable for numerous people in the developing world (Waruingi and Underdahl, 2009). Also, the cost of mobile access is more expensive than fixed-line access in the developing world, and as a result, the users of this kind of connection are decreased (Kaplan, 2006). In South Asia and Africa, women are less likely to own a mobile phone compared to men due to the cost, illiteracy, and poor electricity. People usually own a mobile phone for business needs, while others borrow a phone for a short time (Noordam et al., 2011). The developing countries also have low adoption rate for mHealth applications as it is difficult to measure the return on investment (ROI) on such applications but something that is of much value to the health vendors (Yu et al., 2006; Patricia et al., 2010).

Fifth, the lack of required education, the needed knowledge, and the availability of relevant training are also important hurdles in the effective use of mHealth applications. The low education and a few health literacy skills in some countries of the developing world lead to low utilization of mHealth applications. The health professionals are also required to have a high level of knowledge and skills to use the mHealth technology safely and effectively. Moreover, health-care workers must receive the required training in the mHealth technical capabilities and boundaries (Mishra and Singh, 2008). In general, old people are slower to accept any new technology compared to the young ones. For that reason, good education and strong training can increase the old users' acceptance of the mHealth technology in the developing world (Yu et al., 2006). Besides, the availability of many mobile devices with lack of service support can add extra challenge for both patients and clinicians. All these issues result in users' resistance or medical treatment errors (Mishra and Singh, 2008; Pillai, 2012).

Sixth, the current policy environment and the absence of mobile phone usage guidelines and standards also impact the expansion of the mHealth projects in the developing world (Jeannine, 2011; Rehalia and Kumar, 2012). Usually, there is almost never a single *owner* of the health systems elements that affect the integration and the interoperability. A need for an agreed-upon mHealth technical architecture as well as data exchange standards is essential to overcome this barrier. Moreover, the governments in the developing world prefer to wait for broadband networking before implementing health-related applications. This leads to unnecessary investment barrier to the mHealth deployment. To solve this issue, policies are needed to indicate the mHealth goals and objectives and take benefit of the current commercial wireless networks (Patricia et al., 2010). Specific policies must be added in order to use the advance features of the mHealth technologies. Additionally, all parties must be involved in the designing and implementation phases in order to develop the most effective mHealth applications for their countries (Jeannine, 2011).

Seventh, another important barrier in the developing world is the development of mHealth applications based on the community needs and resources. Commonly, in the developing world, rural and urban areas have dissimilar resources that may affect the mHealth structure and content. Rural areas suffer from poor infrastructure and lack of health-care workers, thus requiring high-quality medical products and services to overcome this shortage. On the other hand, urban areas have their own specific problems. Higher population number, unequal distribution of resources, and unhealthy diets all lead to different needs and expectations from the health system (Lewis et al., 2012; Qiang et al., 2012). In addition, some countries in the developing world have shortage of capacity and resources for the health research institutions that limit their ability to design and develop mHealth applications based on population needs (Garai and Candidate, 2011).

Lastly, the quality of mobile services in the developing world is another major challenge. The mobile service quality is influenced by a range of things such as the mobile device; the mobile network; the information system; the information itself; the reliability, flexibility, and usability of mHealth applications; cost; security; and cultural factors (Akter et al., n.d.; Patricia et al., 2010; Akter, 2012). In addition, the service quality includes the battery life and memory storage. By controlling these factors, the efficiency of mHealth services will be greatly enhanced (Yu et al., 2006; Patricia et al., 2010).

To summarize, although mHealth presents with many opportunities in the developing world, various significant challenges and barriers remain. These issues require serious consideration and research as well as efficient involvement from the governments, other organizations, various stakeholders, and health providers to benefit from this much-needed technology.

27.7 mHealth Solutions and Recommendations for the Developing World

In order to have a successful implementation of mHealth applications and services in the developing world, various steps, procedures, and strategies must be followed right from the beginning to overcome the current challenges and barriers.

Before developing any new mHealth application, an in-depth and comprehensive research and assessment must be undertaken. Identifying the local conditions, such as current health-care infrastructure, coverage of mobile network, policy environment, literacy levels, language requirements, and cultural practices, is vital to counter the main threats to a successful implementation. In addition, consideration of population's health-care needs and sustainability and scalability of the mHealth application is very helpful for the most suitable mHealth solution (Jeannine, 2011; Noordam et al., 2011; Déglise et al., 2012). The initial assessment should also include selecting the right hardware and software that can operate on a large-scale, emerging long-term funding plan, developing clear guidelines and workflows for the implementation, and considering the long-term national health goals. Covering all requirements from the start will ensure a successful pilot project that is capable of expansion at the national level and likely to receive official recognition and acceptance (United Nations Foundation, 2009; Jeannine, 2011). In addition, clear objectives of what the mHealth program is trying to achieve as well as recording of the desired health outcomes in a particular community are critical for the success of a project. For example, *Mwana* project in Zambia and *SMS for Life* in Tanzania have effectively delivered the key elements of their programs that have resulted in solid justification for the real need of such projects at the population level, thus assuring the long-term sustainability of such projects (Jeannine, 2011; Pillai, 2012).

Developing precise policies and specific standards for planning, designing, implementing, and integrating any mHealth system will guarantee its positive impact on the health-care outcome. Every country has specific policies and standards for their existing health-care systems, and therefore, involving them in the preparation process for the mHealth program will ensure achieving their overall health outcome goals. Some resource-poor countries in the developing world may not have any policies or standards as yet. So, there is a need for up-to-date, elaborate, and specific policies and standards to help the stakeholders in the planning, implementation, and preservation phases. Without effective policies and the standards, the community can continue to have several issues related to the privacy and confidentiality of patient information, medical treatment workflow and procedures, data exchange, and systems integration (Jeannine, 2011; Noordam et al., 2011; Pillai, 2012). The governments must enforce precise policies in any mHealth application that ensure system encryption, confidentiality protocols, and hardware passwords in order to secure the patients' data (Patricia et al., 2010).

Many countries in the developing world are considered low-income countries, so instead of creating new applications that require a lot of new investment, time, and resources, it is better to take the best practice applications from the high-income countries, use their successful experiences, and customize the existing systems according to the community needs. The duplication of efforts can weaken the efficiency of mHealth and prevent the program from gaining the needed funds. For that reason, the goal of any organization must be collaboration with similar developed organizations to strengthen the existing mHealth efforts. It is more efficient and cost-effective to see what other players are offering and if there are other solutions available in the market (Jeannine, 2011). Thankfully, various mHealth projects are currently operating accordingly and sharing the best practice technologies and applications with other organizations in the developing countries, which is good news for the success of mHealth (United Nations Foundation, 2009). A proper implementation of a project also needs collaboration with the local implementation partners such as local health-care agencies, community health workers, social marketing organizations, and community leaders. In this way, the project will be able to overcome the local barriers and reservations and would be able to better integrate the local languages and themes (Jeannine, 2011). Moreover, in order to overcome any network and technical issues, the project initiatives need to have industry-related partners, for instance, mobile network operators and technology companies. These partners can offer their technical knowledge, expertise, and resources that can increase the probability of having successful and long-term project. Also, this partnership can be done with the local stakeholders and the MOH in order to integrate the project with the existing national health-care systems, thereby encouraging the policy makers to support the mHealth project and approving its expansion across broader areas (Jeannine, 2011).

Human resources are one of the core requirements in any mHealth project. Therefore, involving the technical IT specialist in the deployment, consuming, and maintaining phases of the mHealth application is critical to the long-term success of the project (Jeannine, 2011). All relevant public and private health care and IT stakeholders must be engaged in and informed about the policy and business environment in order to have an effective mHealth program (Patricia et al., 2010). Additionally, to ensure the application acceptance and usage from the end users, they have to be involved in the development of the mHealth intervention. Local stakeholders as well as local populations should have the required education and training that help them to know how to use the mHealth system to their benefit (Jeannine, 2011). Moreover, understanding the end users' work environment is essential when designing applications and devices. For example, an easy-to-use application is necessary in the mHealth environment (United Nations Foundation, 2009). Likewise, identifying what elements of the program inspire end users and making sure that they are taken care of are important for a consistent and effective use of mHealth (Jeannine, 2011). Hence, communication and engagement with all the relevant parties of mHealth project during all the phases are required to understand their needs, overcome any difficulties, and ensure the successful result (United Nations Foundation, 2009).

For an effective and sustainable implementation of an mHealth program, it needs to start as a pilot project on a small scale with preferably simple cellular technology and open-source software that covers specific areas. Later on, after rigorous evaluation of the project, the project can be deployed on a wider area. In this way, devices and infrastructure costs can be controlled, cultural diversity can be recognized, security issues can be overcome, and various issues and difficulties can be known and avoided before the full-scale implementation (Blynn and Aubuchon, 2009). Furthermore, within any community, usually there are people with different health needs; therefore, integrating multiple solutions via mHealth to help the project team solve more than one issue and covering more than a single health objective are recommended (Jeannine, 2011).

Any mHealth application, even the successful ones, needs continuous monitoring and evaluation using predefined, specific, and measurable metrics. This will help in ensuring the long-term success of the project and avoiding any likely failure in reasonable time. This can be done by monitoring and collecting data periodically to evaluate the project efficiency in achieving the target outcomes and meeting the organizational health-care needs. According to a recent WHO report, only 7% of the mHealth projects have been evaluated in the developing world. For that reason, more data are required to update the current policies, create an effective environment, and enhance the mHealth application features (United Nations Foundation, 2009; Jeannine, 2011). Furthermore, since most countries in the developing world are facing the same challenges and barriers, establishing a global network involving all the main official players is highly recommended to introduce a complete global approach that is able to address the cost cutting, interlink common issues, and support the expansion of mHealth projects in the developing world (Jeannine, 2011).

In Saudi Arabia, although the launch of mGovernment initiative has provided clear benefits to the citizens and the government agencies, certain challenges still remain that affect successful implementation of this initiative. For instance, public acceptance of such initiatives and high infrastructure costs are the main issues. The mGovernment applications are also at an early stage of their development. Only SMS technologies are currently used in the public sector because of its low cost. In view of that, an in-depth and wide-ranging research is needed to indentify the factors that can negatively influence the implementation of the mGovernment initiative, thereby reducing the use of mHealth services in Saudi Arabia. In addition, a comprehensive collaboration with all related parties, including government, MOH, telecommunication companies, and customers, is necessary to ensure the development of a suitable application that can be adopted and used by various customers on regular basis. As Saudi Arabia is geographically a large country with various hospitals, patients, and health-care providers, starting with a pilot project is essential to identify and rectify any technical and nontechnical issues in order to ensure a successful long-term mHealth system that adequately serves different users' needs (Abanumy and Mayhew, n.d.; Alsenaidy and Ahmad, 2010).

To conclude, mHealth has a great potential to revolutionize the delivery of quality health care at an affordable cost and in a timely manner in the developing countries. By ensuring that all the required characteristics, such as usability, accessibility, mobility, affordability, and flexibility, are considered at the development stages of any mHealth application, the challenges and the barriers can easily be overcome resulting in acceptable, consistent, and effective mHealth services that achieve the required health-care outcomes in the developing countries.

27.8 Case Studies

27.8.1 The Working Mother

In 2006, Maha received a gift on her graduation from the King Saud University in Riyadh, Saudi Arabia. It was a wonderful smartphone. She was able to access the Internet, check her e-mails, and download many useful applications. Maha graduated as an application programmer and was interested in exploring why the hospitals in Saudi Arabia were not taking the full advantage of the new mobile technology and applying such technology in the health-care field.

Shortly afterwards, in 2007, Maha got married and moved with her husband Khalid to the city of Khobar, where Khalid was working for Saudi Arabian American Oil Company (ARAMCO) Company. The company employees and their family members received free medical treatment at the local hospital. When Maha started using medical services at the ARAMCO hospital, she discovered a new world: almost everything could be done through her small smartphone. The hospital provided Maha with the mobile application SAMCO APP that was connected to the hospital system database and could be downloaded to any smartphone across various platforms. With the help of this mHealth application, Maha was able to book new appointments, select her family doctor, change or delete appointments, request for prescription refill, and read her basic laboratory test results. SMS was also a tool provided by the app that allowed the patients to manage their health. Through this tool, different SMS reminders were sent to Maha's smartphone to remind her with any upcoming and new appointments, medication refill, new vaccination, and a laboratory test. There was also a provision within the SMS where Maha was able to request any needed health-care services.

In 2010, Maha was pregnant with her first baby. During her follow-up appointments with Dr. Sahar—her OB/GYN doctor—Maha started receiving educational SMS and videos to her mobile about the pregnancy, delivery, and breastfeeding. Dr. Sahar also recommended her to download pregnancy application "You and Pregnancy" (انت والحمل) in order to read and be aware of her bodily changes each week, monitor her food, and count the baby kicking in the latter months of pregnancy instead of writing each information down on the paper. This helped Maha present her progress to the doctor at each appointment. After Maha delivered her baby boy Abdullah, her dietitian recommended her to download the mobile calorie counter application "Directory Calories" (دليل السعرات الحرارية) to organize and monitor her food and weight in a healthier and balanced way. Also, Maha's family doctor suggested using medication reminder application "Medication Alarm" (منبه الدواء) to ensure that all her family members were taking their medication on time with the correct dose. The electronic mail (e-mail) service was also activated to organize Maha's schedule. For any new appointment, medication refill, or new baby vaccination, an e-mail was sent to Maha with the date and time, which could be directly and seamlessly saved on the smartphone calendar.

These mHealth applications and services helped Maha to manage her family health more effectively, accurately, and in an easy and accessible way. She was absolutely delighted that her life was progressing in an organized manner, especially with an infant, thanks to all the mHealth services.

27.8.2 mHealth at Work

Ali worked as an administrator at a busy outpatient surgical clinic at the National Guard Health Affairs (NGHA) in Riyadh, Saudi Arabia. Ali was on regular prescription medicines for his long-standing hypertension and seasonal allergy. Due to his workload, he used to manage all his health-related matters, like booking follow-up appointments with the family physician, changing or deleting any appointments due to his shift work or holidays, requesting for repeat prescriptions, or keeping a track of his laboratory test results, especially kidney functions, through his mHealth hospital application. However, in early September 2012, NGHA's web and mobile services were shut down due to a virus attack. Hospital's website and mHealth applications were not accessible. This incident had an obvious and noticeable effect on Ali's life: he needed to return to what he was doing 7 years ago at his previous work, that is, making all the appointments and medication requests in person. When he discussed this issue with other staff members who were being treated at the same hospital, they were not concerned about the shutdown of electronic services as some of them were not using those web and mobile technology at all due to acceptance issue, education and awareness level, or privacy concerns.

The web problem at NGHA was finally fixed, and all the advanced tools were back online within a couple of weeks. Ali was very pleased on the restoration of the much-needed services and noticed that he could not live without mHealth tools. He hopes that many hospitals in Saudi Arabia will start to implement and use mHealth services in the very near future in order to improve their quality of care and increase the awareness level in addition to simplifying the health-care management for patients as well as the health providers.

27.9 Future Directions

There is a clear and distinct difference among countries in the developing world at the economic level, availability of the technical infrastructure, the network generations, the cultural practices, and the literacy stages. However, the future of mHealth in all of these countries is likely to be bright. As stated by UNF (2009), the main technology trends in mHealth will be the same as what has been categorized in the technology improvement in the past four decades, that is, reducing diseases, enhancing speed of networks, and decreasing costs. The mHealth intervention has clearly shown a great enhancement of health and lives for numerous people in the developing world. In the future, more pilot projects will be implemented based on specific needs of the countries. If a project considers the community needs, uses available technology, improves the control of individual's health, and collaborates with local organizations, there will be a great chance for it to be officially accepted and used (Blynn and Aubuchon, 2009).

Vital Wave Consulting (n.d.) has revealed that, in the next 15 years, the health-care policy makers and providers in the developing world will focus more on prevention instead of treatment of noncommunication diseases in addition to the aging population issues. Although it is expected that these countries will have shortage or unequal distribution of health-care workers, mHealth will be used to overcome these issues due to its benefits in the areas of relatively low-cost, broader-coverage, and timely health solutions. The advance technologies, such as new network generations, WiMAX, intelligent mobile devices, and dedicated wireless devices, will support the mHealth applications to reduce issues and provide more valuable solutions (Vital Wave Consulting, n.d.). Some countries will soon be able to use 4G technology that will provide additional benefits to ensure cost-effective, high data transmission rates and larger bandwidth that support reducing the gap in medical care (Mishra and Singh, 2008; Rehalia and Kumar, 2012). Text messaging applications will remain with large-scale deployment in other new areas of health care. Various reports have shown that the number of smartphone users will grow rapidly and as a result, additional smartphone applications will be developed to be used in many health-care services such as home management of chronic diseases or supporting people to adopt a healthy lifestyle (Bäck and Mäkelä, 2012).

According to West (2012), in the future, doctors and patients will use mobile devices to manage numerous health issues, and for that reason, the remote monitoring will heavily depend on the mHealth market all over the world including the developing world. Besides, mHealth applications will use images and video tools in addition to text messages for specific health fields such as diagnosis, consultation, and follow-up. Different sensors in the mobile phones will be used in the health-care field, for instance, the data will be provided by camera, GPS, and acceleration sensors and processed by custom mobile applications that will then be sent using SMS or MMS tools. Moreover, some telemedicine applications will be developed to send visual information via mobile phone cameras (Bäck and Mäkelä, 2012). Also, speech recognition technologies will be integrated with the mHealth application to increase the use of interactive speech by both doctors and patients (Yu et al., 2006).

With regard to the mHealth economic impact, mobile health will be expanded in the number and type of initiatives. It is expected to reach a multibillion dollar industry by 2017. According to West (2012), the PwC report has estimated that annual revenues will reach $23 billion worldwide including $6.8 billion in Asia, $1.6 billion in Latin America, and $1.2 billion in Africa. As soon as communication and delivery services are improved and transmission errors are reduced, the mobile device will play an active part in the economic growth and health-care improvement throughout the world including the developing world. In addition, the use of mHealth applications will help consumers and health-care workers in the developing world to become more educated and proactive in looking for the best health-care services that have the potential to improve their health and support their economics (Qiang et al., 2012).

In the Middle East, mHealth has a glowing future. According to Pillai (2012), it is predicted that the size of health-care market in the Middle East will increase to $100 billion in the next 15 years. The governments are heavily investing in the health-care sector to enhance the services and improve the quality of health care. For that reason, new and advanced technologies, such as mHealth, will certainly be a part of daily living (Pillai, 2012). In Saudi Arabia, the availability of advanced technical and economical infrastructures in addition to the usability of various resources will greatly support the implementation of mHealth. It is estimated that by 2016, Saudi Arabia will be the leading 4G market with just above 11 million 4G subscribers. Moreover, it is expected that smartphone penetration rate will increase from 25% in 2011 to 49% in 2016 (Alsenaidy and Ahmad, 2010). As a result of these services, the data mobility will be enhanced, the delivered information will be up to date, and the citizen's services will be more effective and efficient. All citizens will be able to access the Internet over wireless networks using smartphone applications and other wireless devices to manage their personal health information, monitor and diagnose their health conditions, and improve the quality of their life (Abanumy and Mayhew, n.d.; Alsenaidy and Ahmad, 2010). The strong and stable implementation of wireless technologies supported by the availability of educated generation and efficient financial planning will increase the adoption of various mHealth applications in Saudi Arabia (Mobily, n.d.; Househ, 2011; Almutairi et al., 2012; Househ, 2012a; Househ, 2012b; Househ et al., 2012a; Househ et al., 2012b; Ababtain, 2013; Alajmi, 2013; Aldabbagh, 2013; Alghamdi, 2013; Househ, 2013; Househ et al., 2013; Paton, 2013; Househ, 2014).

27.10 Conclusions

The published literature and the technical papers have clearly documented the status and challenges of mHealth in the developing world. There is an essential need for mHealth to overcome various health-care issues. Facts revealed that there is current existence of various challenges and barriers; however, good planning and effective solutions will enable the developing countries to reap the maximum benefit of mHealth and improve their health-care services as well as enhance the quality of people's life. We believe that mHealth offers an effective solution for various health-care issues in the developing countries. In spite of the challenges, the developing countries have the capability to control these issues. With time, the mHealth applications will be used by all people regardless of their education level or social classes. They will remotely monitor their health information, consult their doctors, see their high-quality images and videos whenever they want, and use valuable applications to control their health at home, which will result in healthier communities in the developing world. Since Middle Eastern countries, especially the Gulf Cooperation Council (GCC) countries, are considered middle-income countries, they will have the highest opportunity to implement successful mHealth applications that are capable to meet their health-care needs.

Questions

1. Explain the current situation of healthcare systems in the developing world.
2. Explore the revolution of mobile technology in the developing world, especially in Saudi Arabia.
3. Define mHealth and highlight the potentials and the importance of this technology in the future of health care in the developing countries.
4. Identify the main factors that are responsible for expansion of mHealth in the developing world and the Middle East in particular.
5. Define the main types of mHealth services and present the most successful projects and applications in various developing countries.
6. Explore mHealth challenges and barriers in the developing world.
7. Discuss the recommended solutions to overcome current mHealth challenges.
8. Outline the future and emerging trends of mHealth in the developing world.

Key Terms and Definitions

- Arab world: Consists of 22 countries that speak Arabic language and extends from the Atlantic Ocean in the west to the Arabian Sea in the east and from the Mediterranean Sea in the north to the Indian Ocean in the southeast.
- Communicable diseases: Also known as transmissible diseases or infectious diseases. These are clinically evident illnesses (i.e., with characteristic medical signs and/or symptoms of disease) resulting from the infection, presence, and growth of pathogenic biological agents in an individual host organism.

- Developed country: Also known as more-developed country (MDC); is a sovereign state that has a highly developed economy and advanced technological infrastructure relative to other less-developed nations.

- Developing country: Also known as a less-developed country; is a state with a low living standard, underdeveloped industrial base, and low Human Development Index (HDI) relative to other countries.

- eHealth: Electronic health, which is a combination of medical informatics, public health, and business administration and aims to improve health through technology all around the world.

- mHealth: Mobile health, which is a service delivered through a mobile device and is meant to support provision of health care.

- Middle East: Regions around the southern and eastern shores of the Mediterranean Sea and west Asia. It extends from Morocco to the Arabian Peninsula, Iran, and beyond.

- Ministry of Health (MOH): A government advisor on health, their main function being to take care of public health issues.

- Noncommunicable diseases (NCDs): Medical conditions or diseases that by definition are noninfectious and nontransmissible among people. NCDs may be chronic diseases of long duration and slow progression, or they may result in more rapid death such as some types of sudden stroke.

- SMS/MMS: Short message service (SMS) is a text messaging service component of phone, web, or mobile communication systems, using standardized communications protocols that allow the exchange of short text messages between fixed-line and mobile phone devices. Multimedia messaging service (MMS) is a standard way to send messages that include multimedia content to and from mobile phones.

- United Nations Foundation (UNF): Founded in 1945 after the World War II in San Francisco when 51 countries signed the UN Charter to create an organization facilitating cooperation in international law, international security, economic development, social progress, human rights, and the achievement of world peace.

- Vital Wave Consulting: A respected partner of some of the world's global companies, foundations, and multilateral organizations. The company has a proven track record of anticipating ICT market movements and testing ideas through solution and program pilots, modeling, and research.

- World Health Organization (WHO): Special organization within the United Nations system with the main role of directing and coordinating public health issues.

References

Abanumy, A. and Mayhew, P. (n.d.). M-government implications for E-government in developing countries: The case of Saudi Arabia. In *EURO mGOV*, vol. 2005, pp. 1–6. 2005.

Ababtain, A. F., Almulhim, D. A., and Househ, M. S. (2013). The state of mobile health in the developing world and the Middle East. *Studies in Health Technology and Informatics, 190*, 300–302.

Akter, S. (2012). Service quality dynamics of mHealth in a developing country. *Journal of Information Technology and Software Engineering*, 394. Australia: The University of New South Wales; 2012. [*PhD Thesis*] http://www.unsworks.unsw.edu.au/primo_library/libweb/action/dlDisplay.do?vid=UNSWORKS&docId=unsworks_10545&fromSitemap=1&afterPDS=true webcite

Akter, S., D'Ambra, J., and Ray, P. (n.d.). User perceived service quality of mHealth services in developing countries. *18th European Conference on Information Systems*, Pretoria, South Africa, p. 13.

Alajmi, D., Almansour, S., and Househ, M. (2013). Recommendations for implementing telemedicine in the developing world. *Studies in Health Technology and Informatics, 190*, 118–120.

Alastair, S. (2010). Qualcomm plans Mideast mobile health rollout. Retrieved October 30, 2012 from http://in.reuters.com/article/2010/08/01/idINLDE67001H20100801.

Aldabbagh, D., Alsharif, K., and Househ, M. (2013). Health information in the Arab World. *Studies in Health Technology and Informatics, 190*, 297–299.

Alghamdi, E., Yunus, F., and Househ, M. (2013). The impact of mobile phone screen size on user comprehension of health information. *Studies in Health Technology and Informatics, 190*, 154–156.

Almutairi, M. S., Alseghayyir, R. M., Al-Alshikh, A. A., Arafah, H. M., and Househ, M. S. (2012). Implementation of computerized physician order entry (CPOE) with clinical decision support (CDS) features in Riyadh hospitals to improve quality of information. *Studies in Health Technology and Informatics, 180*, 776.

Alsenaidy, A. M. and Ahmad, T. (2010). A review of current state m government in Saudi. *Global Engineers and Technologists Review, 2*(5), 5–8.

Bäck, I. and Mäkelä, K. (2012). Mobile phone messaging in health care—Where are we now? *Journal of Information Technology and Software Engineering, 01*(02), 1–8. doi:10.4172/2165-7866.1000106.

Blynn, E., & Aubuchon, J. (2009). Piloting mHealth: A research scan. Cambridge, MA: Management Sciences for Health. Retrieved from https://wiki.brown.edu/confluence/download/attachments/9994241/mHealth+Final.pdf

Bolton, L. (2012). Helpdesk research report: Mobile telephony for improved. *Health Service and Data Management, 74*, 1–11.

Byrne, M. (2011). Mobile healthcare solutions. Retrieved October 30, 2012 from http://www.smeadvisor.com/2011/06/mobile-healthcare-solutions/.

Communications and Information Technology Commission (CITC). (2010). Annual Report 2010.

Danis, C. M., Ellis, J. B., Kellogg, W. A., Beijma, H. V., Hoefman, B., Daniels, S. D., and Loggers, J.-W. (2010). *Mobile Phones for Health Education in the Developing World: SMS as a User Interface*, London, *ACM DEV*, U.K., p. 9.

Déglise, C., Suggs, L. S., and Odermatt, P. (2012). Short message service (SMS) applications for disease prevention in developing countries. *Journal of Medical Internet Research, 14*(1), e3. doi:10.2196/jmir.1823.

Dolan, B. (2011). Mobile operator launches Baby Ultrasound MMS. Retrieved October 30, 2012 from http://mobihealthnews.com/10066/mobile-operator-launches-baby-ultrasound-mms/.

Donner, J. (2008). Research approaches to mobile use in the developing world: A Review of the literature. *The Information Society, 24*(3), 140–159. doi:10.1080/01972240802019970.

Garai, A. and Candidate, M. (2011). Role of mHealth in rural health in India and opportunities for collaboration. *Technology, 4*(October), 1–5.

Global Telecoms Business. (2012). Mobily launches diabetes app with Sanofi. Retrieved October 30, 2012 from http://www.globaltelecomsbusiness.com/Article/3099495/Sectors/25196/Mobily-launches-diabetes-app-with-Sanofi.html.

Held, G. (n.d.). Etisalat mobile baby. *Presentation.* http://www.ictet.org/downloads/Mob_ejtJpe_jfnJ.pdf. Accessed November, 2012 .

Househ, M. (2011). Sharing sensitive personal health information through Facebook: The unintended consequences. *Studies in Health Technology and Informatics, 196*, 616–620.

Househ, M. (2012a). Mobile social networking health (MSNet-Health): Beyond the mHealth frontier. *Studies in Health Technology and Informatics, 180*, 808–812.

Househ, M. (February 1–4, 2012b). Re-examining perceptions on healthcare privacy—Moving from a punitive model to an awareness model. HEALTHINF 2012—*Proceedings of the International Conference on Health Informatics*, SciTePress, Vilamoura, Portugal, pp. 287–291.

Househ, M., Ahmad, A., Alshaikh, A., and Alsuweed, F. (2012). Patient safety perspectives: The impact of CPOE on nursing workflow. *Studies in Health Technology and Informatics, 183*, 367–371.

Househ, M., Borycki, E. M., Kushniruk, A. W., and Alofaysan, S. (2012). mHealth: A passing fad or here to stay? *Telemedicine and E-Health Services, Policies and Applications: Advancements and Developments,* 151.

Househ, M. (2013). The use of social media in healthcare: Organizational, clinical, and patient perspectives. *Studies in Health Technology and Informatics, 183,* 244–248.

Househ, M., Alsughayar, A., and Al-Mutairi, M. (2013). Empowering Saudi patients: How do Saudi health websites compare to international health websites? *Studies in Health Technology and Informatics, 183,* 296.

Househ, M., Borycki, E., and Kushniruk, A. (2014). Empowering patients through social media: The benefits and challenges. *Health Informatics Journal, 20*(1), 50–58.

International Telecommunication Union (2009). *Mobile eHealth Solutions for Developing Countries.* Geneva, Switzerland: Telecommunication Development Bureau.

Jeannine, L. (2011). Scaling up mobile health elements necessary for the mhealth in developing countries. *Actavis Consulting Group,* (December), 23.

Kahn, J. G., Yang, J. S., and Kahn, J. S. (2010). "Mobile" health needs and opportunities in developing countries. *Health Affairs, 29*(2), 252–258. doi:10.1377/hlthaff.

Kaplan, W. A. (2006). Can the ubiquitous power of mobile phones be used to improve health outcomes in developing countries? *Globalization and Health, 2,* 9. doi:10.1186/1744-8603-2-9.

Lewis, T., Synowiec, C., Lagomarsino, G., and Schweitzer, J. (2012). E-health in low- and middle-income countries: Findings from the Center for Health Market Innovations. *Bulletin of the World Health Organization, 90*(5), 332–340. doi:10.2471/BLT.11.099820.

Ministry of Health. (2011). Health awareness: Mobile health. Retrieved October 30, 2012 from http://www.moh.gov.sa/en/HealthAwareness/Pages/Mobile.aspx.

Mishra, S. and Singh, I. P. (2008). mHealth: A developing country perspective. *Making the e-health connection. Bellagio, Italy* (2008): 1–9.

Mobily. (n.d.). Mobily—Saudi Arabia's leading mobile operator—GSM, 3.5G services, roaming, business solutions. Retrieved October 30, 2012 from http://www.mobily.com.sa/portalu/wps/wcm/connect/mobilycontent/ee/company+news/en/news132.

mplushealth. (n.d.). mHealth in the developing world. Retrieved October 30, 2012 from http://mplushealth.com/SiteRoot/MHAH/Overview/mHealth-in-the-Developing-World/.

Noordam, A. C., Kuepper, B. M., Stekelenburg, J., and Milen, A. (2011). Improvement of maternal health services through the use of mobile phones. *Tropical Medicine and International Health, 16*(5), 622–626. doi:10.1111/j.1365-3156.2011.02747.x.

Okaz news. (2011). STC offers free subscription for a week in the "health" SMS. Retrieved October 30, 2012 from http://www.okaz.com.sa/new/Issues/20110417/Con20110417412606.htm.

Paton, C., Househ, M., and Malik, M. (2013). The challenges of publishing on health informatics in developing countries. *Applied Clinical Informatics, 4*(3), 428–433.

Patrick, K., Griswold, W. G., Raab, F., and Intille, S. S. (2008). Health and the mobile phone. *PMP, 35*(2):177–181. doi: 10.1016/j.amepre.2008.05.001.

Patricia, M., Hima, B., Nadi, K., Sarah, S., Ada, K., Adina, G., Lin, F., and Ossman, J. (2010). *Barriers and Gaps Affecting mHealth in Low and Middle Income Countries: Policy White Paper.* Washington, DC: mHealth Alliance, pp. 1–79.

Piette, J. D., Lun, K. C., Moura, L. A, Fraser, H. S. F., Mechael, P. N., Powell, J., and Khoja, S. R. (2012). Impacts of e-health on the outcomes of care in low- and middle-income countries: Where do we go from here? *Bulletin of the World Health Organization, 90*(5), 365–372. doi:10.2471/BLT.11.099069.

Pillai, P. (2012). mHealth: The future of health is mobile. Retrieved October 30, 2012 from http://middleeasthospital.com/Jan2012lo.pdf.

Pyramid Research. (2010). The impact of mobile services in Nigeria, Abuja, Nigeria.

Qiang, C. Z., Yamamichi, M., Hausman, V., Miller, R., and Altman, D. (2012). *Mobile Applications for the Health Sector.* Washington, DC: World Bank.

Rajput, Z. A., Mbugua, S., Amadi, D., Chepng, V., Saleem, J. J., Anokwa, Y., Hartung, C. et al. (2012). Evaluation of an Android-based mHealth system for population surveillance in developing countries. *Journal of the American Medical Informatics Association, 19*(4), 655–659. doi:10.1136/amiajnl-2011–000476.

Rehalia, A. and Kumar, R. (2012). A review on mHealth system and technologies. *International Journal of Current Research and Review, 04*(03), 53–58.

Schweitzer, J. and Christina, S. (2012). The economics of eHealth and mHealth. *Journal of Health Communication, 17*(suppl. 1), 73–81.

Shin, M. (2012). Secure remote health monitoring with unreliable mobile devices. *Journal of Biomedicine and Biotechnology*, 2012, 546021. doi:10.1155/2012/546021.

The Boston Consulting Group. (2012). The socio-economic impact of mobile health. February 28, 2012.

United Nations Foundation. (2009). *mHealth for Development: The Opportunity of Mobile Technology for Healthcare in the Developing World*. Washington, DC; Berkshire, U.K.: Vital Wave Consulting.

Vital Wave Consulting. (n.d.). *mHealth in the Global South*. Washington, DC: United Nations Foundation and Vodafone Group Foundation.

Waruingi, M. and Underdahl, L. (2009). Opportunity in delivery of health care over mobile devices in developing countries. *AJFAND, 9*(5), 1–11.

West, D. (2012). How mobile devices are transforming healthcare. *World Health*, (18), 1–14. https://vacloud.us/sandbox/groups/5069/wiki/a69cb/attachments/1ddb8/Brookings%20-%20How%20Mobile%20Devices%20are%20Transforming%20Healthcare.pdf. Accessed December, 2012

Yu, P., Wu, M. X., Yu, H., and Xiao, G. C. (2006). The challenges for the adoption of m-health. *SOLI*, Shanghai, China, pp. 181–186.

Section VIII

mHealth Technology Implications

28

Enhancing Decision Support in Health-Care Systems through mHealth

Michael Mutingi

CONTENTS

28.1 Introduction

Mobile health (mHealth) is a recent technological evolution in electronic health (eHealth) that incorporates smartphone technology in a health-care industry environment (Maeder, 2012; Varnfield et al., 2012). The innovation, mHealth, is an eHealth practice that is widely supported by mobile devices and smartphones meant to capture and provide health-related information from various sources. This implies that health-care information can be stored, retrieved, and analyzed in a more sophisticated way. As such, the mHealth setting enables or enhances real-time as well as off-line decision analysis. In this connection, decision makers and policy makers can make use of this development to make informed decisions, in the short, medium, and long terms. By and large, mHealth is a system that goes beyond health-care coordination and information storage, to enhance decision support (WHO, 2011). Therefore, mHealth should be viewed as more than just a collection of devices and applications; it extends the utility and efficiency of health-care infrastructures through integration with enterprise systems (Dolan, 2010; Krohn and Metcalf, 2012).

To aid decision makers in developing informed real-time and off-line decisions, decision support systems can be developed, while relying on the readily available health-care information from mHealth technologies. An appreciable number of decision support systems have been developed, including multiagent systems and other hybrid systems in a home health-care setting, such as home health-care network (H2N) (Islam et al., 2011), Home Care Hybrid Multiagent Architecture (HoCaMa) (Fraile et al., 2009a), and Geriatric Ambient Intelligence (GerAmi) (Corchado et al., 2008). However, these applications have been focused on real-time or very short-term decisions. We argue that short-term, medium-term, as well as long-term decision makers can be enhanced by the availability of accurate and specific information obtained from mHealth applications. In other words, decision support systems can be developed to aid managers at strategic, tactical, and operational levels, based on the mHealth information.

The mobile communication technology is the most widely used communication infrastructure in the world. Most of the world's population has access to some form of mobile communication (Ascari et al., 2010). To date, mHealth is an area of innovation with the potential to make a huge difference in the health-care status of the society. For instance,

- SMS alerts are mHealth innovations that can be utilized to remind individual patients to take their prescription drugs at the appropriate time
- Remote diagnosis as well as treatment can be utilized for patients without the need for visiting a physician in a hospital
- Remote health monitoring devices can be used to track and report patients' conditions upon which further decisions can be made
- Data collected through mHealth technologies can be aggregated and analyzed to assist health-care managers when making medium- to long-term decisions

Therefore, this technology can offer societies an opportunity to transform the health-care services significantly, specifically in decision making, at various levels. Table 28.1 shows some of the basic enabling innovations that can enhance decision support in a health-care setting.

TABLE 28.1

Enabling Innovations for Enhancing Decision Support in Health-Care Delivery

Traditional Health Care	mHealth Delivery Innovations
Patient visits hospital when feeling serious symptoms.	Patient interacts with caregiver any time as soon as symptoms appear.
Physician conducts hands-on examination.	Physician communicates with patient and obtains information from biometric sensors.
Physician prescribes medicine, trusting that the patient will take it as prescribed.	Timely SMS reminders ensure that patient takes medicine as prescribed; reports are fed back on real time.
Patient health improves but goes back to potentially unhealthy lifestyle.	Patient can be monitored remotely and advised if conditions worsen.
Potential unavailability of medical talent.	Patient can access global medical expertise in a connected world.

Researchers and practitioners have focused their attention on real-time dynamic decision making. For instance, Islam et al. (2011) developed a home health-care network (H2N), a health-care aide system integrating sensor devices with wireless networks. The aim is to improve, besides basic care-giving functionalities, the social aspects of elderly care. In the same vein, Fraile et al. (2009a) presented HoCaMA, a hybrid multiagent architecture that facilitates remote monitoring and care services for disabled patients in a home care environment. The system combines multiagent systems and web services to facilitate the communication and integration with multiple health-care systems. It incorporates a system of alerts through SMS and MMS mobile technologies.

Multiagent systems have been found to be very effective in dynamic real-time decision making. Corchado et al. (2008) developed a multiagent system coupled with radio-frequency identification (RFID), Wi-Fi technologies, and handheld devices, aimed at providing new possibilities into building integrated and distributed intelligence software applications. The intelligent software system's key agent, geriatric agent (GerAg), is able to dynamically schedule nurses' tasks, report on their activities, and monitor patient care. Furthermore, Tapia et al. (2006) developed a dynamic multiagent system for monitoring Alzheimer patients' health care in execution time in geriatric residences. The multiagent system architecture implements autonomous deliberative case-based agents with reasoning and planning capabilities. Using RFID technologies, the agents are equipped with the ability to operate in wireless devices and capable of obtaining information about the environment. The agents can respond to events, take the initiative according to their goals, communicate with other agents, and interact with users, making use of past experiences to find the best decision in order to achieve the set goals.

Fraile et al. (2009b) presented a case study of the HoCa hybrid multiagent architecture aimed at improving the health-care status of dependent people in home care environment. The HoCa architecture utilizes a set of distributed components to provide solutions to the needs of the assisted people. Its main components are software agents that interact with the environment through a distributed communications system.

Though there is much awareness on the application of mHealth in real-time decisions, there is need to extend this technology to operational, tactical, and long-term decision making, relying on the availability of accurate and specific information from mHealth applications. The aim is to provide enhanced decision systems, so that decision makers can make quick and intelligent decisions even in complex and dynamic environments, based on available information. The focus of this chapter is to explore the opportunities of mHealth in enhancing decision support in the health-care industry. In this regard, this chapter seeks to achieve the following:

1. Identify the common problems and complexities at different levels of health-care organizations.
2. Highlight the various decisions that have to be generated at different levels of the health-care organization.
3. Analyze the potentials of mHealth in providing decision support in the health-care sector.

The rest of this chapter is structured as follows: the next section presents a general discussion on decision support and the use of mHealth technologies. Section 28.3 identifies, at strategic, tactical, and operational levels, the decision problems found in health-care organizations. Section 28.4 highlights how mHealth technologies assist decision support at the patient level and the health-care worker level. This chapter concludes with Section 28.5.

28.2 Decision Support and mHealth Technologies

Decision support systems are software algorithms that can advise health providers concerned with patient diagnoses through the interaction of patient data and medical information, such as prescribed drugs (Ascari et al., 2010; WHO, 2011). This can be facilitated by using mobile devices and other mHealth devices to input relevant patient data and receive targeted health information. A few empirical studies have demonstrated the usefulness of mHealth technologies in health-care decision support.

28.2.1 Research Methodology

In this chapter, an exploratory investigation is conducted based on a wide selection of sources of existing empirical studies compiled from the information on mHealth-related applications. Sources of literature include a wide search in online databases, such as EBSCO Inspec, ISI Web of Science, Ei Compendex, ScienceDirect, online journals, and conference proceedings. The keywords carefully selected for the search process included *mHealth, telemedicine, mHealth decision support, mHealth, eHealth,* and personal health care. In this respect, the careful selection of keywords helped reduce the number of irrelevant studies and eliminated those studies in which the mHealth decision support was not the major research focus.

28.2.2 Major Research Findings

Research findings in this chapter reveal six major recent empirical applications in the literature, involving mHealth technologies (Tapi et al., 2006; Corchado et al., 2008; Fraile et al., 2009a,b; Bajo et al., 2010; Islam et al., 2011). Table 28.2 provides a listing of these empirical findings. It can be seen from these results that most of the applications are

TABLE 28.2

Recent Empirical Studies Concerned with Decision Support in Health Care

Research Focus	Purpose of Study	Technologies	References
Health-care aide for the elderly	A home health-care network (H2N) system aimed at improving social aspects of elderly care, besides basic care-giving functionalities	Integrated sensor devices and wireless networks	Islam et al. (2011)
GerAmi: a multiagent system	To provide an intelligent environment that dynamically schedules nurses' tasks, reports on their activities, and monitors patient care, on real time	Wi-Fi, RFID, multiagent system technologies, and handheld devices	Corchado et al. (2008)
HoCaMA, a hybrid multiagent architecture	(i) To facilitate remote monitoring and care services for disabled patients in a home care environment (ii) To facilitate integrated communication with multiple health-care systems	Integrated multiagent systems, web services, SMS and MMS mobile technologies	Fraile et al. (2009a)
A dynamic multiagent system	(i) To and dynamically monitor Alzheimer patients' health care in execution time in geriatric residences (ii) To provide real-time intelligent decisions	Autonomous case-based agents, Wi-Fi, RFID, and wireless technologies	Tapia et al. (2006)
HoCa hybrid multiagent architecture for home care	To improve the health-care status of dependent people in home care environments so as to provide solutions to the needs of the assisted people	Hybrid multiagent architecture, software agents, and distributed communication network technologies	Fraile et al. (2009b)
THOMAS architecture in home care	(i) To supervise and monitor dependent patients at home (ii) To provide functionalities, such as automatic reasoning and planning mechanisms for scheduling medical staff, location, tracking, and identification	Multiagent system, detection sensors, access control mechanisms, door opening devices, and video cameras	Bajo et al. (2010)

focused on short-term real-time decisions and off-line operational decisions, facilitated by intelligent systems that incorporate multiagent systems, hybrid multiagent systems, RFID technologies, Wi-Fi technologies, and other mobile devices (Islam et al., 2011; Rossi et al., 2012).

A closer look at the identified studies and their technologies reveal that a wider approach to decision making, over short-term to long-term horizons, is possible and crucial for managers and policy makers in health-care organizations and related stakeholders. Decision making can be enhanced across the whole health-care value chain consisting of a network of health-care drug medical device or equipment manufacturers, mobile networks, medical stuff in health-care institutions, patients, and health-care operations analysts.

28.3 mHealth Value Chain

An mHealth value chain is a network of health-care-related processes or activities that add value to the health-care status of the society. This includes medical drug supplies, medical device supplies, mobile network development, data collection and storage, health-care analytics, and medical intervention. In the mHealth value chain, decision support involves the use of computational techniques to assist decision makers in developing appropriate decisions (Nealon and Moreno, 2003). As such, mHealth technologies can play a central role in making sound health-care decisions, all enhanced by the readily available information through mHealth technologies. It is vital to create secure, unobtrusive, and adaptable environments, equipped with intelligent systems capable of interacting with humans, for monitoring and optimizing health-care decisions (Nealon and Moreno, 2003; Bajo et al., 2010). In fact, the technologies can be utilized across the health-care supply chain, consisting of the patient, the health-care organization, the manufacturers who supply medical devices and drugs, as well as the sponsors and related stakeholders. Figure 28.1 shows the mHealth value chain, illustrating the major players in the chain, and how their decision making can be enhanced with the mHealth technologies.

Patient data collection is significantly enhanced through the help of mobile network service providers (Ascari et al., 2010; Dolan, 2010). The availability of data storage and retrieval capabilities enables further analysis of health-care problem situations on a short-tem, medium-term, as well as long-term bases. Personal health care is greatly enhanced through the availability of timely feedback from medical staff who is well informed by the available health-care information (Varnfield et al., 2012; Krohn and Metcalf, 2012). With the abundance of accurate information, decisions on medical interventions and decisions can be improved significantly. Health-care analytics assist clinicians to make appropriate health-care decisions and interventions on individual patients, following the trend analysis of the health-care status of the individual patients (Rossi et al., 2012). To a large extent,

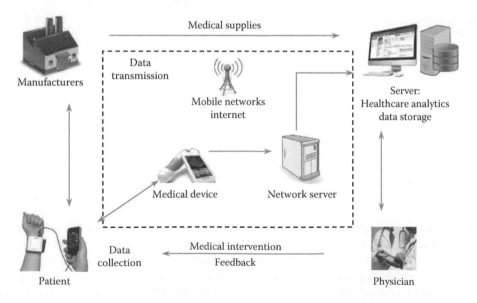

FIGURE 28.1
An mHealth value chain system.

the availability of adequate data may help to accelerate medical research and innovation activities. Better decisions can be made by the use of health-care analytics so as to manage diseases in the most effective way. For instance, one may need to investigate the variance in drug dose–response within patient demographics so as to come up with better decisions in research and development activities (WHO, 2011). In so doing, manufacturers and suppliers of medical drugs may also utilize the medical research information and the health-care analytics for further decisions on their production and manufacturing decisions. Aggregated data obtained through the utilization of the mHealth technologies can be used by the health-care organization, the manufacturers, and the sponsors for medium- to long-term decisions. Therefore, by and large, the availability of accurate information gathered through the application of mHealth technologies is quite handy across various players in the mHealth value chain and at all decision levels.

The next section discusses the decision problems that are of common occurrence in health-care systems, at different levels of decision making, from an operations management viewpoint.

28.4 Decision Problems in Health-Care Systems

Deriving from the explorative study of empirical cases concerned with decisions in health care, operations management decision problems in health care can be broadly classified into three interrelated categories: long-term, medium-term, and short-term decisions (Mutingi and Mbohwa, 2013a). Accordingly, operations management–type decisions can be grouped into integrated plans, involving strategic level plans, tactical level plans, and operational level plans (Chahed et al., 2009). This section presents the derived classification and framework for decision making in health-care organizations.

Figure 28.2 provides a framework for integrated decision making in health-care organizations. The forward flow of operations activities involves the flow of long-term plans and policies from the strategic level to the tactical level (Mutingi and Mbohwa, 2013a). Subsequently, medium-term plans obtained from the tactical level are passed on to the operational level for implementation. In the feedback loop, realized operational variances are fed back to the tactical level. From the tactical level, any realized tactical variances are fed back to the strategic level. The feedback mechanism is crucial for continuous improvement purposes.

28.4.1 Decisions at Strategic Level

Strategic decisions are concerned with the establishment of long-term plans and policies that support the long-term objectives of the health-care organization, typically within a 1–5-year horizon (Matta et al., 2012). In this case, decisions should include formulation of strategic objectives, market strategy, capacity planning, districting, and development of staffing policies.

28.4.1.1 Formulating Strategic Objectives

The formulation of strategic objectives is an important stage that affects the rest of the decisions at lower levels. As a result, the objectives have to be defined meaningfully and clearly. Aggregated trend analysis of health-care service demand and other service patterns have to be carefully analyzed so as to come up with workable long-term objectives (Rossi et al., 2012).

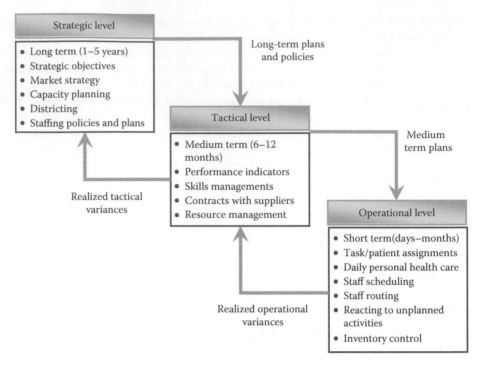

FIGURE 28.2
Integrated decision levels for health-care organizations.

28.4.1.2 Formulating the Market Strategy

The health-care sector is inundated with uncertainty, especially in the absence of detailed demand and service information (Matta et al., 2012). In such environments, it is difficult to define a long-term market strategy. Therefore, health-care organizations have no option except to rely on mHealth technologies for the capturing of up-to-date information for planning purposes. Oftentimes, market strategy considers the effectiveness of the delivered services, the profiles of patients, and the geographical area covered. The mHealth technologies are a key application in providing adequate information pertaining to the services rendered, the demand for services, the patient profiles, and the location (WHO, 2011; Varnfield et al., 2012). Thus, mHealth is essential for enhancing decision support in health-care systems.

28.4.1.3 Capacity Planning

Service capacity defines the number of available professionals for specific types of health-care services. To meet the service demand correctly, the amount of resources available for the provision of that particular service should be determined accurately based on adequate knowledge of service demand patterns. The presence of adequate data enables health-care systems with mHealth technologies to store sufficient data for making informed decisions in regard to capacity building (WHO, 2011). Uncertainties in service demand are reduced significantly, thereby enhancing decision support when making aggregate service capacity decisions and other related resource planning decisions.

28.4.1.4 Staffing Policies

Recruitment and staffing policies over the long term are largely influenced by the predominant service demand patterns. To understand the demand patterns, service demand trends for specific health-care needs are aggregated over time in order to formulate sustainable staffing policies. In the absence of sufficient demand information, it is difficult to formulate staffing strategies, and developing staffing policies is uncertain. A careful analysis service demand patterns will enhance decision making in regard to staffing and recruitment policies.

28.4.1.5 Developing Districting Plans

Based on the aggregated demand information obtained through the use of mHealth technologies, policy makers in the health-care sector can group small geographical areas into larger clusters, called districts, so as to enhance resource assignment decisions. Service demand data are used to assign a patient to a particular district and then to a specific health-care operator available in that district. The decision process can be enhanced periodically, especially in an mHealth environment. The availability of detailed data, in this instance, minimizes uncertainties, which enhances planning.

28.4.2 Decisions at Tactical Level

Tactical level decisions are concerned with the direct implementation of the top-level or strategic decisions, usually within a 6–12-month horizon (Matta et al., 2012). Major decision problems include development of performance indicators, skills management, staff scheduling, development of supplier contracts, and resource management and planning (Mutingi and Mbohwa, 2013a).

28.4.2.1 Performance Indicators

Defining performance indicators involves selection of an appropriate set of performance measures and the choice of measurement methods. Performance metrics should measure daily health care, social care, and logistical activities, in an aggregated form (Islam et al., 2011). The adequacy of performance evaluation is so much dependent on the availability, quality, and accuracy of the information captured for the analysis.

Capturing daily information through mHealth enables health-care decision makers to manage the individual care worker as well as organizational performance (Ascari et al., 2010; Dolan, 2010). Unlike in traditional health-care systems, all the data relevant for performance measurement can be obtained with the most desired frequency and accuracy (Maeder, 2012). Therefore, using appropriate performance measurement tools, aggregate performance information can be captured easily and at all times. This greatly enhances decision making at tactical levels.

28.4.2.2 Skills Management

The improvement of the skills of health-care professionals is imperative. Howbeit, successful specializations or diversification of skills can only be realized when the relevant data are made sufficiently available. Information regarding the training needs can be based on the behavioral trend patterns of specific pathologies and service quality both at individual worker and organizational levels (Matta et al., 2012). The use of sophisticated expert systems and decision support systems with learning capabilities can be implemented to assist decision makers in regard to skills management and development at aggregate levels.

28.4.2.3 Contracts with Suppliers and Partners

Establishing contracts with suppliers and partners involves the definition of agreements for the supply of medications, medical equipment, and other consumables for the smooth operation of the health-care organization (Matta et al., 2012). Notably, to guarantee sufficient care coverage and care continuity, the supply requirements should be adequate. However, the determination of the requirements depends on the accuracy of demand analysis (Mutingi and Mbohwa, 2013a). With accurate trend analysis of the demand, and close tracking and monitoring of demand behavior, health-care operations managers can make informed decisions on adequate supply contracts for care continuity and coverage. This approach is potentially viable and robust in an mHealth environment.

28.4.2.4 Resource Management

Decision problems in this category pertain to medium-term plans in connection with material as well as human resources needed in each planning horizon (WHO, 2011; Mutingi and Mbohwa, 2013a). Material resources need to be distributed and allocated to each section, department, or location, in an attempt to satisfy uncertain demand from each point of need. The problem becomes complex in the case of multiple resource allocation, where optimizing the coordination process becomes a serious challenge (Trautsamwieser et al., 2011; Mutingi and Mbohwa, 2013a). However, human resource management, the management of the number of staff personnel to be allocated to each section, is the major challenge, yet the most important task (Akjiratikarl et al., 2007). This entails the management of contract work professionals and permanent or regular workers. Oftentimes, coordination of these scarce human resources among various demand points is a challenge, especially in the absence of sufficient information (Ascari, 2010).

The main challenge in resource management involves the estimation of the demand for service. The complexity of the problem increases with the number of demand points. With scarcity of the required demand information and lack of decision tools, generating tactical decisions in regard to resources is a complex task. As such, mHealth provides an opportunity for capturing up-to-date information storage for further analysis of demand trends, based on decision tools. The data for trend analysis can be obtained directly from the short-term decision analysis at the operational level.

28.4.3 Decisions at Operational Level

Operational decisions are concerned with establishing coordinated flows and activities of the health-care organization (Matta et al., 2012). These decisions are directly impacted by the available information and decision tools on a daily basis. Some of the decisions include staff scheduling and rescheduling, task assignment, patient assignments, daily personal health care, reacting to unplanned activities, and inventory control (Akjiratikarl et al., 2007; Corchado et al., 2008; Bajo et al., 2010; Mutingi and Mbohwa, 2013a).

28.4.3.1 Staff Scheduling

Staff scheduling for health-care workers involves the allocation of shifts to care workers so as to cover the service requirements over the planning period (Mutingi and Mbohwa, 2013a). Thus, service requirements are determined by the demand analysis using past data that are obtained from the mHealth demand statistics. Information regarding patient treatment at home, admission, and demission, and staff capacity affect staff schedule decisions (Akjiratikarl et al., 2007; Corchado et al., 2008; Bajo et al., 2010). This suggests that mHealth technologies are useful in enhancing the accurate capture of daily datasets from care workers and individual patients. As such, daily datasets provide a record of the patient health-care needs deriving from the personal health-care records and how the care workers cope with the health-care demands, of which the data can be recorded and kept up to date using mHealth applications.

28.4.3.2 Task/Patient Assignment

This is a daily operational activity that involves the finding a suitably qualified health-care worker to assign tasks or patients (Tapia et al., 2006; Akjiratikarl et al., 2007; Fraile et al., 2009b). Patient assignment is more prevalent in a home health-care setting where care continuity is most desired (Mutingi and Mbohwa, 2014; Corchado et al., 2008). Task assignment in home care considers the evolving patient preferences on time windows during which the patient care must commence preferences of the care workers, the travel times, and the relative workload of the workers. Now, individual preferences and requests are always evolving over time and, therefore, should be updated dynamically to enhance informed assignment decisions based on current information. To avoid burnout, care workers should be assigned tasks considering individual stress levels as well as the prevailing workload. The task assignment problem is especially challenging in the home health-care setting where routing is involved.

28.4.3.3 Staff Routing

Health-care staff routing involves sequencing of activities to be executed by each worker, where each worker has to travel in order to visit a set of patients (Akjiratikarl et al., 2007). This is a complex problem that necessitates the application of dynamic real-time assignment of tasks using sophisticate decision support systems. In a dynamic environment where the patient assignment is done on real time, time window constraints and resource restrictions make the routing problem even more complex. Multiagent systems and expert systems are a suitable application in such complex environments (Mutingi and Mbohwa, 2013b). Information on the status of workers, workloads, patients' requests and their health status, travel times, and resource availabilities can be recorded and shared for effective decision making on real time.

28.4.3.4 Daily Personal Health Care

Personal health care entails self-management for elderly individuals or patients with chronic diseases (Maeder, 2012). Monitoring devices and associated software are needed to aggregate real-time data to provide feedback to the user via telecare station (Varnfield et al., 2012). A decision support system can be implemented to inspect the recorded datasets and apply the necessary trend analysis so as to provide warnings in regard to any anomalous situations. With machine learning capabilities, the detection of repeated patterns, and trends long-term behavioral status can be enhanced (Rossi et al., 2012). Therefore, with mHealth technologies, the storage of information and self-management of information for early stages of chronic diseases such as diabetes, cardiovascular, or respiratory diseases are made easy.

28.4.3.5 Reacting to Unplanned Decisions

Operationally, planned assignments may not work out due to changes on the ground, both in the hospital and home care settings (Mutingi and Mbohwa, 2013a). Unplanned activities often arise due to emergent behavior or status of patients and care workers alike; patient requests and worker requests have to be taken into account on a continual basis (Tapia et al., 2006). Decision support systems that enhance planning and replanning have to be put in place for effective reactions to emergent situations, control of disruptions, redefining routes, reassignments, and rerostering or rescheduling (Mutingi and Mbohwa, 2013b). Thus, accurate real-time information storage and update via mHealth technologies are crucial for improving the decision-making process.

28.5 mHealth and Enhanced Decision Support

Learning from the previous sections, mHealth technology applications can enhance decision support at the organizational level, the care worker level, and the patient level.

28.5.1 mHealth at Organization Level

At the top level, that is, at the organizational level, decisions are taken at aggregate level based on available data. For instance, staff forecasting planning decisions strongly rely on accurate trend analysis of health-care demand variation across geographical regions and over different time horizons and for different service needs (Mutingi and Mbohwa, 2013b). With accurate, reliable, and readily available data through adoption of mHealth technologies, decision support is enhanced significantly (Ascari et al., 2010; WHO, 2011).

28.5.2 mHealth at Care Worker Level

At employee level, care workers can utilize mHealth technologies to make decisions regarding individual patients. Care workers can easily access and update the patient's eHealth data necessary for health-care delivery (Rossi et al., 2012). Additionally, care workers can communicate effectively with specific patients using electronic data. For instance, care workers can use the tablet computers to gain access to tools such as decision flow charts that may be needed for delivering their particular functions (Maeder, 2012). Moreover, such mHealth applications can be used at care worker team levels.

28.5.3 mHealth at Patient Level

Personal health-care decision support, at patient level, can be enhanced through the application of mHealth technologies (Maeder, 2012). At patient level, mHealth can support self-management of personal health care for elderly people or patients with chronic diseases such as diabetes, cardiac, or respiratory diseases (Rossi et al., 2012; Varnfield et al., 2012). Using this development, the patient can be actively involved in addressing potentially adverse situations much earlier.

Personal health-care decision support can be improved through mHealth. For example, a patient can use tablet computers to effectively interact with care workers using electronic media in form of chats, blogs, or e-mails (Maeder, 2012). This allows for advanced implementation of decision support tools that can analyze data and provide useful feedback. By applying a suitable decision support system, data inspection and trend analysis can help to highlight warnings for any uncharacteristic tendencies (Corchado, et al., 2008; Maeder, 2012). In addition, long-term behavioral variations can be detected on time through intelligent or machine learning software.

28.5.4 Health-Care Worker Integration

Patients with advanced stage chronic diseases are often managed with multiple health-care workers who collaborate as a team. The use of tablet computers in an mHealth environment can be handy, so much that the team of care workers can share and access the patient's information on real time (Corchado et al., 2008; Maeder, 2012). The patient's care plan and eHealth records can be downloaded conveniently in a web-based setup. Features such as videoconferencing and other multiparty conferencing capabilities can be utilized. Given the appropriate decision and analysis software, further analysis can be done and displayed, while maintaining the integrity of the information records (Maeder, 2012; Varnfield et al., 2012). The team of workers can dynamically update the patient's eHealth records in a team environment. Hence, mHealth technologies can enable the care team to effectively and efficiently perform their work in a more integrated way.

28.6 Limitations in mHealth Application

Opportunities of mHealth in enhancing decision support in complex health-care systems are explored in this chapter. However, in practice, there are some limitations associated with the utilization of these technologies. Some of these limitations are as follows:

- Health-care systems that incorporate mHealth technologies may become too complex for patients, clinicians, and decision makers. Therefore, this may call for provision of extra training. In addition, the technologies may be too expensive to implement.

- Since most mHealth technologies would involve the use of intelligent software for decision making, clinicians and patients are concerned about the possible further weakening of the patient–physician relationship.

- Though there is much enthusiasm on the use of mHealth technologies, there is a high risk of challenges to the appropriate use and validation of mHealth technologies. Such technologies are often subject to perversion, which ultimately raise ethical and security concerns.

28.7 Conclusions

To enhance decision support at strategic, tactical, and operational levels in health-care systems, the addition of mHealth technologies is imperative. The main reason for this assertion is that informed decisions rely on the dynamic capture, storage, analysis, and retrieval of information on a long-term, medium-term, as well as short-term bases. This chapter presents an analogy of different decision levels that encompass specific activities in different time horizons. Examples of decision problems were identified, highlighting the need for mHealth technologies for decision support in personal health care and how this may be utilized for higher-level decision support.

Strategic activities in the health-care environment require accurate information trends derived from lower levels, that is, the tactical and operational levels. Realized tactical and operational variances are fed back to the strategic level, so that strategic objectives, marketing strategies, capacity planning, districting, and staff planning policies can be adjusted effectively. Aggregated long-term trend analyses of the patient needs can be derived from the realized variances, using suitable intelligent/expert methods. The ensuing long-term plans have a tactical bearing in the medium-term plans.

Tactical activities can be enhanced through the use of accurate trend analysis of the behavioral information derived from the application of mHealth technologies at operational levels. Realized operational variances in the implementation of the operational plans, such as routing and task assignment, are utilized in adjusting performance indicators, staff schedules, establishing contracts with suppliers, and resource management. In turn, the medium-term plans generated at this level will influence the operational activities in the next planning horizon.

Operational activities in health-care systems are centered on the care worker and the patient. Such activities involve short-term decisions in line with task assignment, patient assignment, routing and scheduling (rostering), and reacting to unplanned plans, among others. These activities actively require dynamic storage, update, processing, and sharing of information for decision-making purposes. Thus, mHealth comes in handy for such dynamic information exchange between care workers and patients, on real time. As such, mHealth enables decision makers at operational level to make informed decisions efficiently, even on real time. With advanced machine learning features, decision makers are able to devise behavioral trends of the personal health status of individual patients with improved accuracy. Overall, mHealth innovations can enhance decision support, especially in a dynamic health-care environment.

Acknowledgments

I appreciate the reviewers for their invaluable comments on the earlier version of this chapter. I am grateful to my dear friend, Mercy Idi Mutingi, who greatly supported me in the preparation of this chapter.

Questions

1. Suggest any limitations associated with the application of mHealth technologies for decision support in health-care systems.

2. How may data collected through the use of mHealth technology devices be useful for long-term decisions?

3. Explain the linkage between the strategic, tactical, and operational levels in regard to integrated decision making in a health-care system.

References

Akjiratikarl, C., Yenradee, P., and Drake, P. R. 2007. PSO-based algorithm for home care worker scheduling in the UK. *Computers & Industrial Engineering,* 53, 559–583.

Ascari, A., Bakshi, A., and Grijpink, F. 2010. *mHealth: A New Vision for Healthcare,* McKinsey & Company, Inc., and GSMA, New York.

Bajo, J., Fraile, J. A., Pérez-Lancho, B., Corchado, J. M. 2010. The THOMAS architecture in home care scenarios: A case study. *Expert Systems with Applications,* 37, 3986–3999.

Chahed, S., Marcon, E., Sahin, E., Feillet, D., and Dallery, Y. 2009. Exploring new operational research opportunities within the home care context: The chemotherapy at home. *Health Care Management Science,* 12, 179–191.

Corchado, J. M., Bajo, J., Abraham, A. 2008. GerAmi: Improving healthcare delivery in geriatric residences. *IEEE Intelligent Systems,* 23(2), 19–25.

Dolan, B. 2010. Philips suggests dedicated mHealth spectrum. http://mobihealthnews.com/7178/philips-suggests-dedicated-mhealth-spectrum/. Accessed March 23, 2013.

Fraile, J. A., Bajo, J., Abraham, J., and Corchado, J. M. 2009a. HoCaMA: Home care hybrid multiagent architecture. In A. E. Hassanien et al. (eds.), *Pervasive Computing: Innovations in Intelligent Multimedia and Applications, Computer Communications and Networks,* Springer-Verlag London Limited 2009, Dordrecht, the Netherlands, pp. 259–285.

Fraile, J. A., Tapia, I. D., Rodríguez, S, and Corchado, J. M. 2009b. Agents in home care: A case study. In E. Corchado et al. (eds.), *Hybrid Artificial Intelligence Systems 2009,* LNAI 5572, pp. 1–8, Springer-Verlag, Berlin, Heidelberg, Germany.

Islam, R., Ahamed, S. I., Hasan, C. S., and O'Brien, C. 2011. Home-Healthcare-Network (H2N): An autonomous care-giving system for elderly people. In A. Holzinger and K. M. Simonic (eds.), *USAB,* LNCS 7058, pp. 245–262, Springer-Verlag, Berlin, Heidelberg, Germany.

Krohn, R. and Metcalf, D. 2012. mHealth: From smartphones to smart systems. http://www.mhimss.org/sites/default/files/resource-edia/pdf/CHAPTER%202_mHealth.pdf. Accessed on August 19, 2013.

Maeder, A. J. 2012. Tablet computers for mhealth: opportunities for personal healthcare, *Proceedings of the IASTED International Conference, Health Informatics (AfricaHI 2012),* Gaborone, Botswana, pp. 355–359.

Matta, A., Chahed, S., Sahin, E., and Dallery, Y. 2012. Modelling home care organisations from an operations management perspective. *Flexible Services and Manufacturing Journal.* doi: 10.1007/s10696-012-9157-0.

Mutingi, M. and Mbohwa, C. 2013a. Home care staff planning and scheduling: An integrated operations management perspective. *Proceedings of the 13th Botswana Institution of Engineers Biennial Conference October 15–18, 2013,* Gaborone, Botswana, pp. 124–130.

Mutingi, M. and Mbohwa, C. 2013b. A Home healthcare multi-agent system in a multi-objective environment. *The 25th Annual Conference of the Southern African Institute of Industrial Engineering July 9–11 2013*, Stellenbosch, South Africa, pp. 1–10.

Mutingi, M. and Mbohwa, C. 2014. Home healthcare staff scheduling: A clustering particle swarm optimization approach. *Proceedings of the 2014 International Conference on Industrial Engineering and Operations Management January 7–9, 2014*, Bali, Indonesia, pp. 303–312.

Nealon, J. L. and Moreno, A. 2003. Applications of software agent technology in the health care domain. In A. Moreno and J. L. Nealon (eds.), *Whitestein series in Software Agent Technologies*, Vol. 212, Birkhäuser Verlag AG, Basel, Germany.

Rossi, D. D., Lorussi, F., and Tognetti, A. 2012. Wearable systems for e-health: Telemonitoring and telerehabilitation. *Proceedings of the IASTED International Conference, Health Informatics (AfricaHI 2012)*, Gaborone, Botswana, pp. 335–338.

Tapia, D. I., Bajo, J., De Paz, F., and Corchado, J. M. 2006. Hybrid multi-agent system for alzheimer health care. In S. O. Rezende and A. C. R. da Silva Filho (eds.), *Proceedings of HAIS 2006*, Ribeirao Preto, Brazil.

Trautsamwieser, A., Gronalt, M., and Hirsch, P. 2011. Securing home health care in times of natural disasters. *OR Spectrum*, 33, 787–813.

Varnfield, M., Ding, H., and Karunanithi, M. 2012. Use of mobile phone based health applications in home care delivery of cardiac rehabilitation. *Proceedings of the IASTED International Conference, Health Informatics (AfricaHI 2012)*, Gaborone, Botswana, pp. 360–365.

WHO. 2011. *mHealth: New Horizons for Health through Mobile Technologies: Second Global Survey on eHealth*, World Health Organization, Geneva, Switzerland.

29

mHealth Technology Implication:
Shifting the Role of Patients from Recipients
to Partners of Care

Muhammad Anshari and Mohammad Nabil Almunawar

CONTENTS

29.1 Introduction

A technological *paradigm shift* where customers are equipped with powerful mobile information and communication devices, such as tablets or smartphones, to access information easily is reshaping many industries, including health care. Consequently, they are ready to be empowered to actively participate in health-care businesses and decision-making processes. Some health-care services are being transformed from physical-based services where a complete physical presence is necessary to mobile-based services in which health-care activities can be performed at anytime, anywhere. Mobile-based services greatly help in preventing diseases, promoting health, and offering services that can be accessed according to customers' comfort times and places (Haux et al., 2002). As the penetration of mobile technology is high and growing, the adoption of this technology in health-care information management is highly expected as it can help improve the quality of services provided.

A successful health-care provider provides and maintains good-quality services to meet customers' (patients) demand (Almunawar et al., 2012). In a competitive environment, the effort to attract a customer (patient) can take a month or more, but it is easy to lose one. Therefore, there must be values added to pleasing a customer, satisfying the customers' need, and building long-lasting relationship between customers and the organization that serves them (Low, 2002). The advancement of mobile technology has brought a possibility to extend health-care services by enabling clients to participate more actively in health-care processes. For instance, health-care providers may offer mobile health (mHealth) promotion and education through social networks embedded in apps. Patients with their smartphones can easily access the mobile services provided and participate actively in relevant social media where they can share and discuss their concerns regarding health and health care.

Accessing health-care services through smart mobile devices (mHealth) encourages patients to take more responsibility for their own health, such as actively participate in decision making about their health and help each other through information and knowledge sharing. Furthermore, mHealth provides patients with the ability to access and control information that fits with their personalized needs. mHealth is extending health-care services through smart mobile devices with multiple types of online interactions: interactions between health-care staff and patients, interaction among patients, and patients' self-interaction through mobile applications (apps).

Presently, there is a lack of literary discussion on the roles of patients on the issue of mHealth. Hence, there is a knowledge gap in addressing the development of a mechanism(s) in encouraging patients' responsibility to take greater roles in health care and delivery arrangements. Having a larger role in health care will ultimately empower patients, especially through the accessibility and control of information and knowledge using smart mobiles devices that they currently have. This chapter discusses patients' expectations of mHealth services and the way these services can empower them through their active participations in various health-care processes. The main goal of this chapter is to introduce a promising future research direction, which may shape the future of mHealth. In this chapter, we examine patients' participation concerning the process of empowerment through mHealth to make patients more proficient in dealing with their own health-care issues. We start from analyzing various related issues through literature study. We then propose a model based on the analysis. In Section 29.2, we present a literature review of related work. Section 29.3 contains the methodology, while Section 29.4 discusses implications of the proposed model, followed by case studies in Section 29.5. The future direction is discussed in Section 29.6. Section 29.7 presents the conclusion.

29.2 Literature Review

Recent discussion of patients' participations is supported in many health literatures, and it has been used for patients and health-care services over the past decade. However, examining and extending the idea into a more specific technological adoption like mobile technology are pivotal. Recent developments of mobile technology are very interesting to study in the context of health care as it offers highly attractive features for mHealth. The main advantage of mHealth is the ability to connect to the Internet anywhere, anytime, allowing patients to perform activities outside the bounds of physical offices and time constraints and consequently enables them to remain continuously in touch with their health-related social networks. mHealth provides health-care organizations with the ability to broaden services beyond its usual practices and thus provides an ample opportunity to achieve complex health-care goals especially in dealing with patients' education and delivering high-quality services. mHealth will help patients have greater control in the process of interaction between them and their health-care organizations, and among patients themselves, hence empowering them in various health-care processes and decision making. The idea of empowerment in health care emerged as a response to the rising concern that patients should be able to play a critical role in improving their health. In traditional health-care practices, a patient is mostly perceived as a recipient of care as well as medical decisions. However, a paradigm shift has been taking place in health-care industry from patients who receive care to those who actively participate in their health care. This change has spread throughout the entire health-care industry as a social movement characterized by the right to act based on informed choice, active participation, self-help perspective, and full engagement in critical processes (Kieffer, 1984).

29.2.1 Empowerment

Many researchers have discussed the issue of empowerment in health-care organizations. Empowerment can be analyzed from three distinct perspectives: the perspective of patient–health-care provider interactions (Skelton, 1997; Paterson, 2001; Dijkstra et al., 2002; van Dam et al., 2003), the patient self-interactions (Anderson et al., 1995; McCann et al., 1996; Davison et al., 1997; Desbiens et al., 1998), and patient-to-patient interactions (McWilliam et al., 1997; Golant et al., 2003; Maliski et al., 2004). However, research that specifically discusses the issue of empowerment in the domain of mHealth is still very limited. Australia, a pioneer in this matter, has adopted a personally controlled electronic health record (PCEHR) system, which stands as an example for empowerment through e-health services (NEHTA, 2013). A significant element of patient empowerment has been achieved by allowing them to view their medical information electronically. However, PCEHR has only enabled patients to view their electronic medical record (EMR); it does not allow collaboration and conversation between patients or between patients and their health-care providers.

The idea of empowerment of patients in health care emerged as a response to the rising concern that patients should be able to play a critical role in improving their health. In traditional health-care practices, a patient is the recipient of care as well as medical decisions. However, as mentioned earlier, a paradigm shift has taken place in the health-care industry, which allows active participation of patients in dealing with their own health.

Empowerment is well supported in the health-care literature and related to customers and health-care services over the past decade (Skelton, 1997; Paterson, 2001; Dijkstra et al., 2002; van Dam et al., 2003). In an organizational context, empowerment implies the provision of necessary tools to staff in order to be able to resolve, on the spot, most problems or questions faced by customers. Besides, the staff can deal with customers directly hence may reduce the number of dissatisfied customers who would otherwise have complained but now, simply switch brands (Low, 2002).

In the conventional system, health-care providers (through their health-care professionals) decide on the treatments given to their patients. Hence, patients are treated as objects with very little or no roles in the existing health-care processes. There will always be circumstances in which patients choose to hand over the responsibility of decision making to health-care providers due to their difficulty in selecting available options or lack of time to understand their health problem and the options. However, this does not undermine the proposition that patients' empowerment benefit both patients and health-care providers as it will promote efficiency and more importantly, decisions should be made from the perspective of patients (Segal, 1998).

According to McWilliam et al. (1997), empowerment is a result of both interactive and personal processes, where the emergence of *power* (or potential) is facilitated by caring relationships. Empowerment as an interactive process suggests that power is *transferred* by one person to another, whereas empowerment as a personal process suggests that power is *created* by and within the person. Although the expected outcome is the same, which is the gain of more power over one's life, the nature of the two processes is very different (Aujoulat et al., 2006). The first course entails that power can emerge through active cocreation and collaboration in an empowering relationship. In the second case, the process of empowerment is perceived from the point of view of the customers, which is considered as a process of personal transformation.

Gibson (1991) defined empowerment as a process of helping people to assert control over the factors that affect their lives. It encompasses both the individual responsibility in health care and the broader institutional or societal responsibilities in enabling people to assume responsibility for their own health. McWilliam et al. (1997) view empowerment as the result of both an interactive and a personal process, where the emergence of *power* (or potential) is facilitated by a caring relationship, not merely given by someone, nor created within someone. In other words, the emergence of a person's potential because of an empowerment process may be viewed as a cocreation, within a true partnership (Low, 2002).

Unfortunately, there are not many literatures that discuss the issue of empowerment that integrates the individual, social, and medical aspects in mHealth's model. Therefore, there is a knowledge gap in addressing the issue of how health-care providers can develop a mechanism(s) in encouraging patient's responsibility to take a greater role in deciding about their own care and delivery arrangements that will empower them in health-care service delivery to meet the increasing demand and expectation of customers equipped with mobile technology.

The proposed model is developed to enhance existing theory of empowerment in e-health business processes with the help of recent web technology. The model integrates the wider scope of empowerment in health care (personal, social, and medical), value configuration of e-health's business process, EMR, and adoption of Web 2.0. The model is expected to contribute in determining dimensions of e-health business process with the possible perspective of empowerment.

29.2.2 Value Configuration in Mobile Services

Understanding a business process helps health-care organizations to appropriately deliver the strategic plan in providing mHealth service. It is important to examine each business process as a layer of value to the customers (patients/patients' family). Customers (patients) place a value in these services according to the quality of outcome, quality of service, and price. The value of each layer depends on how well they are performing. When a health-care provider cannot achieve its strategic objectives, it needs to engineer its activities to fit business processes with the strategy (Michael, 1994). If the business processes do not fit the strategy, it will diminish the value. For example, the value of a health promotion is reduced by a delayed response to a patient's query or poor communication skills. The value of e-consultation in the mHealth service will reduce by a late response by the person in charge.

FIGURE 29.1
Health-care value chain. (From Porter, M., *Competitive Advantage: Creating and Sustaining Superior Performance*, Free Press, New York, 1985.)

Figure 29.1 originated from Porter's value chain analysis. Each activity adds value to patients. These are registration, patient care, discharge, marketing, and service. Porter (1985) proposed value chain framework for the analysis of organizational level competitive strengths and weaknesses. It is a method for decomposing the firm into strategically important activities and understanding their impact on cost and value. According to Porter, the overall value, creating logic of the value chain with its generic categorical activities, is valid in all industries. Later, Stabell and Fjeldstad (1998) proposed three distinct value configuration models and argued that the value chain is not valid in all industries. There are groups of industries that differ in business nature and resource allocation. The three distinct generic value configuration models are value chain, value shop, and value network. Those are required to understand and analyze firm-level value creation logic across a broad range of industries. With the identification of alternative value creation technologies, value chain analysis sharpens the value configuration analysis to gain competitive advantage. The study revealed that the implementation of the model experienced serious problems and is unsuitable in a number of service industries. Stabell and Fjeldstad (1998) suggested that the value chain models activities in long-linked technology, while the value shop is created by mobilizing resources and activities to resolve a particular customer problem. The value network creates value by facilitating a network relationship between their customers using a mediating technology. Health care, schools and universities, and consulting firms are examples of value shop that rely on an intensive technology. The generic value shop diagram appears to be a diagnosis-focused shop.

29.2.3 Social Networks

Mobile technology with social network enables multiple connections, which allow patients to gain additional knowledge. It offers more diverse value generation resulting from informal conversations with customers within social networks. A health-care provider must consider the essential role of the conversation among patients as recent information and communications technology (ICT) enables them to share experiences without barriers (Anshari et al., 2012). For example, the conversation that takes place among patients on a social network of health-care services may influence the image of the health-care provider (Almunawar et al., 2012).

The social networks in mHealth allow a patient to communicate and share their experience with other patients as well as communicate with health care professionals for consultations. The connection of patients and health-care professionals in social networks is greatly significant. This factor will be inherited in the mHealth that we propose. Social networking can generate a way to strengthen the relationship between organizations and their customers. It can be used as enablers in creating close and long-term relationships between an organization and its customers (Askool and Nakata, 2010). In addition, social networks can play a significant role in managing customer relationship and may stimulate fundamental changes in consumer behavior (Greenberg, 2009). The communication revolution generated by the web, especially Web 2.0, has broad and deep impact on inter-personal relationship in all areas, including health care. The booming number of social networking groups and support groups for patients on the Internet and their influence on health behavior is only beginning to be explored (Rimer et al., 2005) and remains an important area for future research. The concept of a social network defines organization as a system that contains objects such as people, groups, and other organizations linked together by a range of relationships (Askool and Nakata, 2010). Some organizations are building online social networks to engage customers and exchange ideas, innovations of new services or products, quick feedback, and technologies from people outside the orga-nization (Lafley and Charan, 2008).

29.3 Methodology

The proposed model is developed to examine the implications of mHealth to enhance the theory of empowerment and participation in health-care services with the help of recent mobile technology. The model integrates the wider scope of empowerment into three dis-tinct roles as personal, social, and medical aspects of patients. The model is expected to contribute in drawing dimensions of mHealth and its implication towards extending the role of patients from recipients of care to partners of care with the possible perspective of empowerment and participation. We developed a model of mHealth to give a direction for any health-care providers in providing mHealth systems.

29.4 Implications

Patients' empowerment and participation in health-care service can be grouped into per-sonal, social, and medical. Figure 29.2 covers the dimensions of mHealth in mobile personal (mPersonal), mobile medical (mMedical), and mobile social (mSocial). mPersonal can be composed of personal information (account administration) and personal health activities. Activities in administration include setting account, privacy setting, changing password, update contact, and any personal matters that patients are given full control over informa-tion. Personal health activities are any health-related activities that can be done by any patient. mMedical entails the ability of patients to access medical records or information that is authorized by the health-care provider. mMedical facilitates the interaction between health-care providers and patients. mSocial presents a facility for patient-to-patient interac-tions like WhatsApp and BlackBerry Messenger (BBM). Details of each component will be highlighted when we discuss the model.

mHealth services are expected to improve medical services and experience, personaliza-tion in services, and social support and sharing through social networks. These are at least three implications of introducing mHealth in health-care organizations.

29.4.1 Redefining Business Process

Based on the literature review, we proposed a business model for mHealth, as shown in Figure 29.3. The figure provides our perspective of mHealth in dealing with three layers of

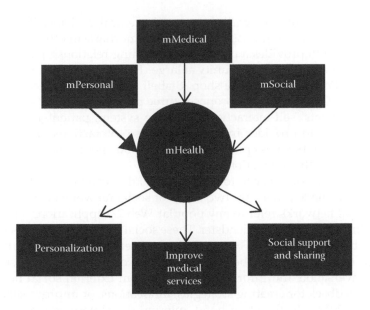

FIGURE 29.2
mHealth features.

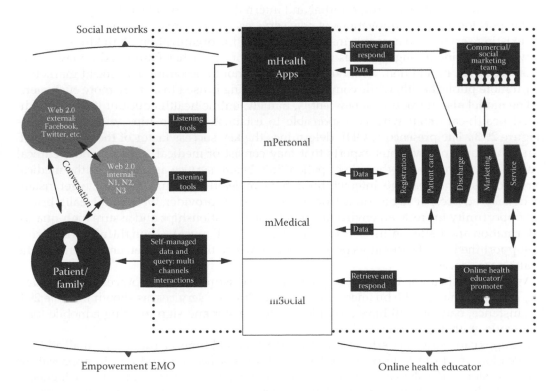

FIGURE 29.3
The business model of mHealth.

patient interactions as mentioned earlier (personal, medical, and social). The model offers a starting point for identifying possible theoretical mechanisms that might account for ways in which mHealth provides a platform for building relationships between a health-care provider, patients, and the community at large. The model accommodates the concept of value chain by Porter (1985), value shop (Stabell and Fjeldstad, 1998), and social networks (Greenberg, 2009). The model proposes three points of patients' interactions. Those interactions are patients' self-interactions with the system, patient–patient interactions, and patients–health-care provider interactions. These interactions obviously will affect the existing health-care business processes, and new business processes may need to be engineered to embrace the interactions.

The term social networks refer to tools in Web 2.0 that can be used by patients to converse online. In our model, there are two kinds of social networks: external and internal. The external social networks refer to any popular Web 2.0 applications such as Facebook, Twitter, LinkedIn, MySpace, and Friendster. These social networks can be used as platforms for interactions between patients. The external social networks connect with mHealth systems, although the networks are beyond the control of the mHealth system. However, constructive conversation and information shared from external social networks should be captured as feedback for creating strategies, innovations, or improve services (Anshari and Almunawar, 2012). On the other hand, internal social networks are operated, managed, and maintained within the health-care organization's infrastructure. The pivotal targets of these networks are conversations among patients or their families on similar health problems and illnesses. For example, patients with diabetes will be motivated to share their experiences, information, and knowledge on treatments with other diabetic patients.

In general, the aim of having external and internal social networks is to engage patients and export ideas, foster innovations of new services, and ensure quick response/feedback for existing services from people inside and outside the organization. Both kinds of social networks provide a range of roles for the patient or his or her family (Almunawar and Anshari, 2011). In addition, a marketing strategy should accommodate social marketing to promote public health, while commercial marketing is used to acquire more customers.

The model also introduces a new entity, namely, online health educator (OHE), a dedicated health-care staff who is responsible to establish relationships with the patients (Figure 29.3). The presence of OHE determines the key success factor of the system. OHE can be a team from various experts that may consist of medical, clinical, and technical teams. The team ensures the role of patient in e-health service as partner in the medical care process. When patients interact through the mHealth system, they want to get instant feedbacks, advises, or suggestions from the health-care provider. OHE in mHealth brings the opportunity to create an environment of trusted relationships and assure high-quality information and better online services for patients. OHE must be thoughtfully considered to support the idea of patients as partners of care where they have been empowered in the health-care process.

As mentioned earlier, mHealth may change the existing business processes. For example, mHealth changes the business process in health-care services, as shown in Table 29.1. For instance, patients will have a simple way to register and sign in using a mobile form that works with any mobile device.

The apps can use jQuery effects as well as AJAX for real-time username or e-mail address verification. It can include field validation, admin notification, and admin approval for available slots. mHealth can empower patients in determining the time they will choose on available slots. More scenarios that are complex can be derived from the proposed model mentioned earlier (Figure 29.3).

TABLE 29.1

Comparison of Health-Care Business Process in Registration's Module

	Conventional Health-Care Service	mHealth Service
Registration process	Physical presence.	Mobile registration by patient.
Data verification	Health-care admin.	System.
Consultation time	Determined by staff or by appointment.	Patient may choose available slot in system.
Accessibility	Information belongs to the health-care provider that makes difficult to be accessed by patient.	Information are available to both parties (health-care provider and patients).

29.4.2 Empowering Patients

Empowerment is an important feature that can be used as a strategy in mHealth services to improve health literacy and customers' satisfaction. As discussed in the previous section, empowerment refers to the process of gaining influence or control over events and outcomes of importance (Fawcett et al., 1994). Gibson (1991) defines empowerment as a process of helping people to assert control over the factors that affect their lives. It encompasses both the individual responsibility in health care and the broader institutional or societal responsibilities in enabling people to assume responsibility for their own health.

Empowering patients can be seen as an effort to build loyalty and trust between a health-care organization and its patients. Empowerment encourages patients to take responsibility for their own health and make decisions about their health-care affairs. Patients are empowered in the sense of controlling the process of interaction to their health-care providers and among patients themselves. However, providing empowerment in any state of interaction levels to patients in mHealth condition is a challenging task.

Three dimensions of interaction proposed in the previous section are further explained in a circular model, as shown in Figure 29.4. The model draws the business scope and process of mHealth. Based on the model, health-care organizations may vary in implementing empowerment in their services. For instance, patients are able to generate contents for health records. The empowered patients can then make records themselves that can help health-care staff get better views of the patients' condition.

Figure 29.4 depicts three domain areas of object-oriented design for the mHealth service that could be considered in the process of decomposing mHealth scenarios. The model extends the role of patients into three distinct functions as individual, social, and medical. Each role comprises a set of objects that detail the function and arrange activities within mHealth's context between customers of a health-care provider. We adopt the object-oriented approach as it supports extendibility, so that new objects or entities can be easily derived from the existing ones if they are needed in the future.

There are three main objects in the model, namely, mPersonal, mSocial, and mMedical. mPersonal is an object for personal activities that records health status and services. mPersonal consists of object identity/profile (ID), object personal habits (HB), object exercise (EX), object emotional and spiritual (SE), object personal health plan (HP), object personal account (AC), and more objects that can be extended depending on the need and urgency. mSocial consists of all objects that relate to social networks and media. These are object conversation (CS), object knowledge management (KM), and object resolution (RS). Finally, mMedical consists of objects that encompass activities of checkups, I/P treatment, and O/P treatment. These activities range from e-appointment (EA), examination (XM), treatment (TM), and e-prescription (EP). Each object also consists of sub-submodules. For instance,

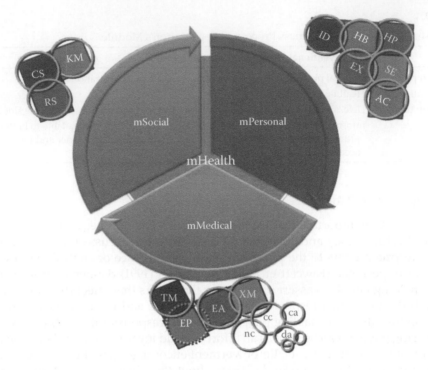

FIGURE 29.4
mHealth business scope and process.

object XM may be made up of chronic disease (cc) and nonchronic disease (NC). Chronicle diseases comprise diabetes (DA), cancer (CA), obesity (ob), etc.

29.4.2.1 mPersonal

At the individual level, empowerment can be achieved through a process of recognizing, promoting, and enhancing one's personal abilities to meet one's own needs, solve one's own problems, and mobilize the necessary resources in order to feel in control of one's lives. mPersonal is an individual habit and lifestyle that affects one's overall health status. Empowering in this level comprises basic habits, physical exercises, and emotional factors. Patients may have full control over any mHealth activities in this module. For instance, customers can update, edit, and delete their own exercise history through their mHealth system.

29.4.2.2 mSocial

Patient-to-patient interactions can affect patients' participations, interactions, and empowerment in mHealth services. In this business process (Figure 29.4), patient-to-patient interaction is represented as mSocial. mSocial enables interactions, conversations, and networking among patients and between health-care provider and patients at the same time. The networking creates the value of network such as multiway conversations and sharing of experience and knowledge. One central feature of network empowerment is making use of individual competence to collectively initiate changes. Interaction between patients is part of social life; it engages them to share experiences, symptom histories, treatment strategies, type of medicines consumed, long-run planning, etc. Getting feedback from others may be beneficial because it can be seen as a means of support and encouragement and demonstrates collective knowledge sharing.

29.4.2.3 mMedical

mMedical defines interaction between patients and their health-care provider in dealing with medical care activities. The activities comprise checkups, outpatient treatment, and inpatient treatment with the purpose to give comprehensive personal medical views and histories. Additionally, checkups, inpatient treatment, and outpatient are the activities that customers mostly have direct interaction with the provider for consultation or physical treatment. In many cases, the health-care provider provides limited access to patients in accessing their EMRs. mMedical facilitates patients to have certain rights and authorization in accessing medical records. For instance, the health-care provider may grant access to customers to view their examination (XM)'s records.

29.4.3 Redefinition of Data Ownership

The third implication from the proposed model is a redefinition of data ownership. Data (records) ownership and accessibility in electronic health records (EHRs) or EMRs have drawn attention to provide an in-depth analysis of the work of EMRs/EHRs showing its relevance to contemporary health-care services in this study. Determining data ownership is based on the function delineated from the model.

The first role derives the function of mPersonal where an individual poses as a personal health actor. The personal health actor means that patients are in control of their own personal health data and information. Data ownership falls in this category as it belongs to the person himself or herself. Those data are personal identity (ID), personal habits of patient (HB), exercise activities (EX), spiritual and emotional activities (SE), personal health plan (HP), personal account information (AC), and so on. For instance, ID contains personal information such as name, address, phone number, e-mail address, log-in ID, and password in the system. HB is the daily habit of the individual that can include eating, sleeping, and any other habits that may have an effect on personal health. The EX is routine exercise activities of individual that may be beneficial when recorded in the systems. The customers own all subclasses in this category where they should manage by themselves. The process will replace a conventional approach where health-care organization owns patient's data and health-care staffs are used to input patients' information into their systems. mPersonal enables patients to take care of their own data. Figure 29.5 proposes a data flow diagram of mHealth's scope and coverage. The diagram can explain data ownership of mHealth. Data in mPersonal consist of personal health administration and personal health. Personal health administration refers to personal records of the patient (date of birth, address, account identification, phone, etc.), and personal health refers to the personal health activities as discussed in the previous section. Ownership of both data (personal administration and health) belongs to the patients, and health care provider may have partial or full access over the data.

mSocial considers the individual (patient) as a social health agent for others. It is considered as the lively feature of mHealth where customers are involved in social networks where they can share and discuss with others concerning their experiences in dealing with health-care matters. Nowadays, people use social networks daily. If a patient updates his or her status in social networks, it may trigger conversation among their circle of friends. Bringing this scenario into mHealth services is not only interesting but also challenging.

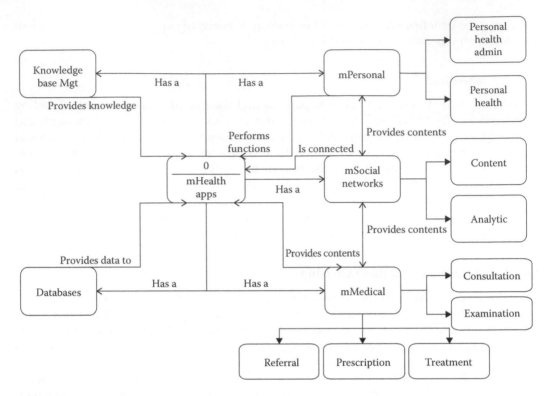

FIGURE 29.5
Flow diagram of mHealth.

For instance, patients with the same illness like diabetes may share their experiences with other patients in social networks. Conversations in social networks may lead to virtual supporting groups that can enrich and strengthen patients' motivations to fight for better health. Patients as social health agents provide a broad range of empowerment to utilize social networks facilities in mHealth services. Components of mSocial are conversation (CS), chat, update status, forum, wikis, blog, knowledge management (KM), personal knowledge, group knowledge, and asking for a specific service (RS). CS is standard social network activities such as sharing and conversation in social media. We propose that data ownership in this category is open to all parties. It means either patients or their health-care provider have full control over the data and contents in the module.

mMedical implements the individual (patient) as a medical care partner in the process of health care. It is composed of objects examined (XM), e-appointment (EA), e-prescription (EP), and e-treatment (TM). XM is an online consultation between patients and medical staffs that can lead to the generation of EMRs for the patients. Many health-care providers prevent patients from accessing their EMRs. This makes patients unable to access their medical history. Consequently, whenever a patient visits a different doctor for opinions, a redundant diagnosis may need to be performed. One of the aims of e-health as stated by WHO (2005) is to educate patients regarding their health status, condition, and history. Therefore, mMedical proposes to redefine data ownership to educate patients and to improve awareness of their health condition by having the ability to access their medical histories and statuses. We propose that data ownership in mMedical can be partially given to patients. For instance, access to e-prescription (EP) a subset of object medical

TABLE 29.2

Summary of Data Ownership

Module	Submodule	Data Ownership	
		Patient	Provider
mPersonal	Administration	Full	Partial
	Personal health	Full	Partial
mMedical	Consultation	Partial	Full
	Examination	Partial	Full
	Treatment	Partial	Full
	Prescription	Partial	Full
	Referral	Partial	Full
mSocial	Content	Full	Full
	Analytic	No access	Full

can be given to patients. The availability of EP to patients will speed the management of prescription since they are able to access and learn anything related to their medication. In summary, object class medical may shift the role of a patient from recipient of care to partner of medical care as they are empowered to have access to relevant information and knowledge. Table 29.2 summarizes data ownership of mHealth.

We distinguish between full and partial ownership. Full ownership means that the individual or groups have full control over the data to add, delete, and update, while partial ownership is specific authorization granted to the users (patients or health-care provider).

29.5 Case Studies

The following real cases are disputes and complaints to health-care providers, which are not handled properly as the health-care providers have no mechanism and proper system to address them. Our mHealth model can help handle the cases and in general, it contributes in resolving many problems between customers and their health-care providers.

29.5.1 Social Networks in Healthcare Setting

Prita Mulyasari, a housewife and a mother of two, was a patient at Omni International Hospital, Tangerang, Jakarta, for an illness that was eventually misdiagnosed as mumps. Her complaints and dissatisfaction about her treatment, which started as a private e-mail to her friends in September 2008, were made public. It rapidly distributed across forums via online mailing lists. Once the e-mail became a public knowledge, Omni responded by filing a criminal complaint and a civil lawsuit against Prita. Then, a verdict against Prita, at Banten District Court on May 13, 2009, she was sentenced to 6 years in jail and fined 204 million rupiah (US$ 20,500). A support group on Facebook as well as Indonesian blog has attracted a considerable amount of support. A mailing list and Facebook group called *koin untuk Prita* (coins for Prita) started raising money for her throughout Indonesia. People began collecting coins to help Prita to pay the fine. Seeing the huge support for Prita, Omni International Hospital dropped the civil lawsuit. Significant pressure eventually led Prita to being released from detention on June 3, 2009.

The aforementioned case illustrates how the provider is unable to handle a simple case of complaint from its services. The case blown up at the national scale, forcing the provider to back down, unable to handle a huge scale of public pressure. The result is bad reputation of the provider and losing the trust from its customer at the national level (Indonesia).

This can be easily addressed by mSocial part of the model. The customer (Prita) will be given access to the internal social network to pose her complaints and to share her concern either with the provider or with the member of the network. Having listened from the customer (Prita), the provider should act accordingly to solve the problem. Prita will not share her problem to the community at large if the provider can address her problem nicely. In fact, she may spread the good news about how the provider treats her professionally starting from her conversations with the provider (through the provider's health educator) in the internal social network within *mSocial* subsystem of mHealth.

29.5.2 Customer Service in Demand

I recently encountered something upsetting with the RIPAS hospital. A few weeks ago, I phoned the RIPAS hospital and asked to speak with a doctor at one of the specialist divisions. When the receptionist picked up the phone, I told her that I wanted to speak with a doctor. To my surprise, she said that I could not even leave a message as their office was about to close (that was at 3:30 pm). When I requested again that I wanted to leave a message, she continued to decline, repeating that they were about to close. I was puzzled at the staff's attitude. Was she uncooperative because she was lazy or was it because she did not care enough to accommodate my request? Hospital employees should be more caring and not disregard people who call in. What if it had been an emergency? This could really ruin the reputation of the hospital (HDJ @ Bandar Seri Begawan, Brunei Darussalam, 2009).

The second case is about the bad service of health-care provider (in this case, a public hospital). Due to time limit (office hour), the hospital was unable to extend their service to its patients. This classical case can be easily handled by extending service via online channel, possibly using mobile devices. In our model, the subsystem *mMedical* can handle various queries from patients online. A dedicated medical staff will be assigned to handle and answer the queries immediately or consult with a specialist first then answer the queries. Patients can send a message online and get response professionally.

29.6 Future Direction

Based on the proposed model, we have developed a prototype of mHealth, namely, Clinic 2.0. Though the system is not complete modules as proposed in the business architecture, nevertheless, it extends multiways of patients' interactions. Three possible relationships involved in the mHealth system are patient self-interactions (patient with the systems interaction), patient-to-patient interactions, and patient–health-care provider interactions. It draws the interface graphical user interface (GUI) of the Clinic 2.0. To assess how the model works in practice, we will conduct the testing for mHealth in many possible scenarios and settings. Figure 29.6 is a screenshot of Clinic 2.0. The main menu is designed to be user friendly with a clear navigation design to allow users to find and

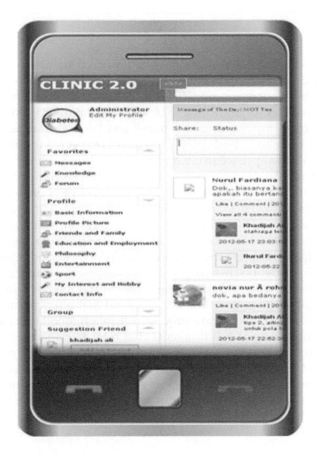

FIGURE 29.6
Screenshot mHealth of Clinic 2.0.

access information effectively. The top menu consists of a search box, My Health, Medical Record, Logout, and Message of The Day (MOTD), status, conversation, profile, and group. The search allows visitors to quickly find the information they need. My Health is a representation of object personal in the business process design. It represents personal health activities. mMedical is a representation of object medical or medical activities. The quick logout component adds a new menu type to Clinic 2.0 that allows a single-click *logout* menu item (without requiring user confirmation to logout).

One of the special features in the home menu is the MOTD, a reminder message to each user (patients or staff) managed by the health educator and is customized based on the need of each user. This means that MOTD is uniquely different for each patient. For instance, the patient with diabetes will likely receive a note or reminder from the healthcare educator based on his or her health condition. The OHE updates the MOTD regularly. The notification for the user appears when there is a friend request or new message in the Inbox from other users. The next section is status updates also known as a *status*, which allows users to post personal messages for their friends to read or share with others. In turn, friends can respond with their own comments, as well as clicking the "Like" button. A user's most recent updates appear at the top of their Timeline/Wall and is also noted in the "Recently Updated" section of a user's friend list. The purpose of this feature is to allow users to inform their friends of their recent *status*.

29.7 Conclusion

The model accommodates empowerment into integrative participation and interaction that is beneficial for both customers and health-care organization. Furthermore, the integrated approach can help health-care organizations in defining which scope of empowerment they will implement in the organization. The modular approach will assist health-care organizations to initiate empowerment by stages and later on to measure the empowerment process and performance modularly. The study is expected to contribute in drawing dimensions of mHealth comprehensively and its implication toward extending the role of patients from recipients of care to partners of care with the possible perspective of patients' participation. mHealth provides a comprehensive perspective of patients as an individual health actor, social health agents, and medical care partners. The model identified possible theoretical mechanisms that might account for ways in which mHealth provides a platform for building relationships between a health-care provider, patients, and the community at large. mHealth is composed of mPersonal, mMedical, and mSocial. The model is developed from social networks, listening tool, and health-care value configuration. There are three types of descriptors used to highlight the central themes found throughout the model. First, personal health actor refers to mPersonal that refers to individual or patient who takes an active role on health-care processes through mobile services. Secondly, the system views patients as social health agents that refer to patients' activities related to social support through embedded social networking apps. Third, the core concepts of the system are grounded in sets of principles where patients are seen as medical care partners. It consists of modules that encompass activities of checkups, I/P treatment, and O/P treatment that accommodate mobile technology in relation to medical care process.

Questions

1. What are the implications of mHealth in health-care service to customers?
2. What are types of empowerment that customers can expect from mHealth?
3. What are the challenges of introducing social networks in mHealth?
4. In your opinion, what will be the implications of empowerment in mHealth initiative?

References

Almunawar, M.N. and Anshari, M. Improving customer service in healthcare with CRM 2.0. *GTSF Business Review*, 1(2) (2011):228–234.

Almunawar, M.N., Wint, Z., Low, P.K.C., and Anshari, M. Customer's expectation of e-health systems in brunei. *Journal of Health Care Finance*, 38(4) (2012):36–49.

Anderson, R.M., Funnel, M.M., Butler, P.M., Arnold, M.S., Fitzgerald, J.T., and Feste, C. Patient empowerment: Results of a randomised control trial. *Diabetes Care*, 18(1995):943–949.

Anshari, M. and Almunawar, M.N. Framework of social customer relationship management in e-health services. *Journal of e-Health Management*, 2012(2012):1–14.

Anshari, M., Almunawar, M.N., Low, P.K.C., and Wint, Z. Customer empowerment in healthcare organisations through CRM 2.0: Survey results from brunei tracking a future path in e-health research. *ASEAS—Austrian Journal of South-East Asian Studies*, 5(1) (2012):139–151.

Askool, S.S. and Nakata, K. Scoping study to identify factors influencing the acceptance of social CRM. *Proceedings of the 2010 IEEE International Conference on the Management of Innovation and Technology (ICMIT)*, Singapore, pp. 1055–1060, 2010.

Aujoulat, I., d' Hoore, W., and Deccache, A. Patient empowerment in theory and practice: Polysemy or cacophony? *Patient Education and Counseling*, 66(1) (2007):13–20.

Davison, B.J. and Degner, L.F. Empowerment of men newly diagnosed with prostate cancer. *Cancer Nursing*, 20(1997):187–196.

Desbiens, S.A., Wu, A.W., Yasui, Y., Lynn, J., Alzola, C., Wenger, N.S., Connors, A.F., Phillips, R.S., and Fulkerson, W. Patient empowerment and feedback did not decrease pain in seriously ill hospitalised patients. *Pain*, 75(1998):237–248.

Dijkstra, R., Braspenning, J., and Grol, R. Empowering patients: How to implement a diabetes passport in hospital care. *Patient Education and Counseling*, 47(2002):173–177.

Fawcett, S.B., White, G.W., Balcazar, F., Suarez-Balcazar, Y., Mathews, R., Paine, A., Seekins, T., and Smith, J. A contextual-behavioral model of empowerment: Case studies involving people with disabilities. *American Journal of Community Psychology*, 22(1994):471–496.

Haux, R., Ammenwerth, E., Herzog, W., and Knaup, P. Health care in the information society. A prognosis for the year 2013. *International Journal of Medical Informatics*, 66(3) (2002):3–21.

Gibson, C.H. A concept analysis of empowerment. *Journal of Advanced Nursing*, 16(1991):354–361.

Golant, M., Altman, T., and Martin, C. Managing cancer side effects to improve quality of life: A cancer psycho education program. *Cancer Nursing*, 26(2003):37–44.

Greenberg, P. CRM at the speed of light. *Social CRM 2.0 Strategies, Tools, and Techniques for Engaging Your Customers*, 4th ed. New York: McGraw-Hill Osborne Media, 2009.

Kieffer, C. H. *Citizen empowerment: A developmental perspective*. Prevention in human services, 3(2–3), 9–36, 1984.

Lafley, A.G. and Charan, R. *The Game-Changer: How You Can Drive Revenue and Profit Growth with Innovation*. New York: Crown Business, 2008.

Low, P.K.C. *Strategic Customer Management: Enhancing Customer Retention and Service Recovery*. Singapore: BusinesscrA F T, 2002.

Maliski, S.L., Clerkin, B., and Letwin, M.S. Describing a nurse case manager intervention to empower low-income men with prostate cancer. *Oncology Nursing Forum*, 31(2004):57–64.

McCann, S. and Weinman, J. Empowering the patient in the consultation: A pilot study. *Patient Education and Counseling*, 27(1996):227–234.

McWilliam, C.L., Stewart, M., Brown, J.B., McNair, S., Desai, K., Patterson, M.L., Del Maestro, N., and Pittman, B.J. Creating empowering meaning: An interactive process of promoting health with chronically ill older Canadians. *Health Promotion International*, 12(1997):111–123.

Michael, K.B. *Strategy and Architecture of Health Care Information Systems*. New York: Springer-Verlag, pp. 43–50, 1994.

NEHTA, PCEHR Dispense Record v1.0. 2013. Accessed on July 24, 2014. http://www.nehta.gov.au/implementation-resources/clinical-documents/EP-1323.

Paterson, B. Myth of empowerment in chronic illness. *Journal of Advanced Nursing*, 34(2001):574–581.

Porter, M. *Competitive Advantage: Creating and Sustaining Superior Performance*. New York: Free Press, 1985.

Rimer, B. K., Lyons, E. J., Ribisl, K. M., Bowling, J. M., Golin, C. E., Forlenza, M. J., & Meier, A. How new subscribers use cancer-related online mailing lists. *Journal of medical internet research*, 7(3), 2005.

Segal, L. The importance of patient empowerment in health system reform. *Health Policy*, 44(1998):31–44.

Skelton, A.M. Patient education for the millennium: Beyond control and emancipation? *Patient Education and Counseling*, 31(1997):151–158.

Stabell, C.B. and Fjeldstad, Ø.D. Configuring value for competitive advantage: On chains, shops, and networks. *Strategic Management Journal*, 19(1998):413–437.

van Dam, H.A., van der Horst, F., van den Borne, B., Ryckman. R., and Crebolder, H.P. Provider–patient interaction in diabetes care: Effects on patient self-care and outcomes. A systematic review. *Patient Education and Counseling*, 51(2003):17–28.

World Health Organization. Global observatory for eHealth series. Retrieved March 4, 2011 from http://www.who.int/goe/publications/en/, 2005.

30

Understanding User Privacy Preferences for mHealth Data Sharing

Aarathi Prasad, Jacob Sorber, Timothy Stablein, Denise L. Anthony, and David Kotz

CONTENTS

30.1 Introduction

Mobile health (mHealth) technologies, including health text messaging, mobile phone apps, remote monitoring, wearable devices, and portable sensor devices, have grown rapidly in the past 5 years and are expected to play an important role in improving access to health information, resources, and clinical care. mHealth devices can be used to monitor activities (Fitbit [9]), sleep (Wakemate [35]), emotions (Affectiva [1]), vital signs like blood pressure (Withings blood pressure cuff [38]), or fetal conditions (HeartSense [14]). Users can collect their personal health and physical and social activity information and upload it to a vendor website, social networking website, a personal health record (Microsoft HealthVault [21] or, formerly, Google Health [11]), or a health-provider-operated electronic health record (EHR). Once the data is uploaded, users can share the information with health providers who help diagnose their illness or monitor their treatment. Family and friends can motivate them as they work toward a healthier lifestyle. People can also share their experiences with their peers (e.g., others suffering from similar medical conditions) and provide support while in recovery [10]. New mHealth technologies might also enable users to share health information with pharmacists, insurance companies, drug companies, employers, or others involved in their health care.

Studies of mobile location tracking applications show that at least some users vary in their willingness to share depending on the benefits [25], place and context [3], or by recipient [7]. We set out to examine whether similar variations in patterns of use existed with mHealth technologies. We conducted exploratory focus-group discussions to gain preliminary understanding about users' privacy preferences. We discovered that if users are not comfortable with the way their information is being collected or shared, they may not use mHealth technologies at all, or use them in limited ways, thereby reducing the potential for mHealth technologies to improve health and health care. In addition, users' preferences for collecting and sharing information are likely to vary depending on the types of information, the types of recipients, and how the information was to be used.

Prior work, including the preliminary study we mentioned earlier, studied users' sharing behavior through focus groups, surveys, and interviews; in these cases, study participants either were given hypothetical scenarios about health data management, had a brief opportunity to use a health device, or were assumed to have experience with collecting and sharing health information. People's stated privacy preferences and concerns, however, may differ from their actual sharing behavior [6,16]; thus, it is important to examine actual sharing behavior with real mHealth devices that can share real data with real people.

To study how new mHealth users share different types of personal information with different recipients over time, we conducted a user study with $n = 41$ participants. To the best of our knowledge, ours is the first study that explores users' privacy concerns by requiring them to *actually* share the information collected about them using mHealth devices; our subjects could decide whether to share the information and if so, how much information to share with others. The device we used for our study is one of the most popular devices, a fitness device called Fitbit. None of the participants in the user study had ever used a Fitbit prior to the study. At least 10 participants had previously used a pedometer but had never uploaded its data to a website or online application.

In this chapter, we first describe users' privacy preferences based on the results of the focus group. Then we use the findings from our user study to answer the following questions:

- Did the participants share different types of personal or sensed information more or less frequently?
- Do participants' decisions about sharing health information differ across types of sharing partners (family members, friends, third parties, and the public)?
- Does sharing behavior change over time; are participants' privacy preferences dynamic?

We confirmed that people's sharing behavior depends on the type of information being shared and the sharing recipient. Our results showed that the participants were generally less willing to share personal demographic information or context information collected by the mHealth device than about sharing the health information that the device is meant to collect. Our results also showed something surprising—study participants were more willing to share some information with *strangers* than with their own family and friends; among strangers, they were more willing to share some information with specific third parties than with *the public* at large. We also confirmed that people's privacy behavior is dynamic; participants' sharing behavior changed during the course of our study. It is important to understand people's willingness to share, so that mHealth devices can provide patients with the controls to share their information in a manner such that they can enjoy the benefits provided by the device without disclosing more information than is necessary.

In this chapter, we use the term *user* to denote the mHealth device user and *sharing partner* to denote the person(s) with whom the user shares her fitness information.

30.2 Focus Groups

We conducted exploratory focus-group discussions to gain a preliminary understanding about users' privacy preferences. The focus groups were approved by Dartmouth's Institutional Review Board. We conducted eight focus-group sessions with 3–7 participants each, who were college students (aged 19–30), hospital outpatients (aged 80–85), or residents of a retirement community (aged 65–100). Each focus group lasted for not more than 90 min and all the participants were paid for their time. We chose these groups since we wanted to talk to users who have some health experiences—some who have been recently hospitalized and others who are monitored continuously outside the hospital—and users who have limited health-care-related experiences.

Since mHealth devices are not yet common, the focus-group participants were presented with hypothetical scenarios where mHealth devices were used. There were four scenarios in which an mHealth device was used to collect a user's personal information (measuring medication intake, diet and exercise, location or social interactions); the collected data was uploaded to a private website and then shared with health providers, family, or friends. The scenarios for the young and the old differed in the age of the protagonists and their medical condition but were similar in every other aspect like the information collected and the manner in which it was collected, stored, and shared; the scenarios are available in the technical report [26].

We presented each scenario to the participants, after which they were asked about the advantages and disadvantages of using mHealth sensors in that scenario. They discussed their concerns regarding the collection of the particular health information in each scenario and whether there were certain times and places when they did not want to collect that information. The participants talked about why they would want to share certain health information types with health providers, caretakers, family members, and friends. They also raised some concerns regarding storage and transmission of the collected information.

We recorded the discussions. We coded the discussions manually and grouped the statements into categories—privacy concerns were broadly classified into three categories and challenges for reducing privacy concerns were classified into seven categories.

30.2.1 Privacy Concerns

Unintended disclosure. One student brought up the issue of how someone could make sensitive inferences from seemingly trivial information, "You can draw some kind of analogy or trend [from the collected health information] that could be misused." Some participants were worried about the information being sent to a website, via the Internet. A student was concerned about "the level of encryption and transmission [of information from] the device and [to] the website. Also, how is [the website] categorizing [the user]? His name, date of birth, social security number?" Another student was more open to using the device if it did not have any Internet connectivity; he said, "If you connect to the Internet, *I* start to become skeptical in terms of privacy, the information has the ability to leave the device." Another student was worried about losing the device; she asked, "What if you misplace the device? Is there security on the device."

Misuse of information. A majority of the participants were worried that their personal information might be used by people they had not intended to share it with. A few students were worried that potential employers might not hire them, if they wear the device to a job interview. Some participants were worried about discrimination by insurance companies. After hearing the scenario about Jack who uses an mHealth device to track his medication intake, an elderly participant said, "Insurance companies might not want to insure Jack if he is lax about taking his medication." A student was worried about the information being misused by the government; he said, "I'm not too into the government knowing where I am going and what I am doing." Some participants were worried about their information being used for marketing purposes. A student commented, "Wouldn't people want [our personal information] for other things to sell products and to target [a specific] audience?" Another student was concerned about stalkers, she said, "If someone can hack into the website, then someone can track you like a stalker."

Change in trust relationships. A student was concerned about using a device to monitor patients' adherence to their treatment because it meant that "the doctor didn't trust [the patients] to be honest." An elderly participant said that he would not use the mHealth device unless he trusted his doctor; he said, "The more the information you collect, the more trust you need to have that information is secure." A few students felt that constant patient monitoring would improve the doctor–patient relationship; according to one student, "They are working together, it's like a partnership." Another student pointed out that "[The device] *holds* [the patient] accountable a lot more, compared to when she could lie to her doctor and say that [the treatment] is not working." Some elderly participants were concerned about sharing information with their family. One said, "[Suppose you share sensitive information] with one family member, then there is a family gathering and they discuss [the medical situation]." Another elderly participant said that if a patient's wife was to constantly monitor his location and his activities, "it would destroy the trust [he had] in [his] wife." One elderly participant was open to sharing his information only with his daughter, since she took care of him. One student, on the other hand, pointed out that sharing health information with family would lead to "more arguments in the family." Most students agreed that they would trust their doctor with their health information more than their family. One student said, "If I like had a medical condition, I would feel obligated to talk to a doctor but less obligated to talk to a sibling about it on a daily basis." Most students were not open to sharing their health information with their friends and some felt that if they had to share their health information with their friends, they would trust only their closest friends.

30.2.2 Benefits versus Privacy Trade-Off

Some participants wanted to use the devices, because they understood the benefits of the device, since they or their family or friends had suffered from a similar condition and they agreed that they would wear the device at all times. One student said, "If I'm being tracked and for my own benefit, I'll keep it on whenever I can and as long as it is with my doctor and utmost with my family." Another student said, "If you want [the device], it makes you a bit more willing to put more information up there. If it is something that is forced upon you, you might not respond well to it." A few participants felt that they would not be concerned about privacy if they were using the device to get better; one student pointed out, "If I was really concerned about the disorder, I think I would definitely not be concerned about the privacy."

30.2.3 Challenges to Reduce Privacy Concerns

Need for clarity of information collection and usage. Most participants expressed the need to be aware of what information was being collected by the device. A student pointed out that his privacy concerns depended on what information was being collected; he said, "[If the device takes] into account details of someone's life, that is going to affect the way they act and get into privacy issues."

Ignorant users. One student could not understand why anyone would steal or misuse her health information. A user who cannot comprehend the consequences of a privacy breach might end up disclosing more information than necessary.

Providing adequate control to the user. All the participants wanted the control to decide what information to share and with whom and under what circumstances. Some of them felt that having the control to turn the device on or off or to take the device off would defeat the purpose of using the device. After hearing the scenario about a patient using an mHealth device to track his medication intake, one student said, "If you have to remember to turn it on or off, it becomes optional. It's like taking your medication in the first place." Another student wanted the control to delete some information collected by devices before it was shared with others. An elderly participant said she wanted complete control over all the decisions she made; she said, "People [at the retirement home] like to have control over their lives as much as possible. Unless I became incapable, I will consider everything intrusive unless I can choose what to do with the information."

Flexibility of privacy controls, based on recipients. We discovered that privacy concerns varied with the data recipient. One student wanted to share the data with someone who could help him understand the data that was collected. A few students said they would share their information only if it could not be traced back to them by strangers; one student said, "I wouldn't mind if [my information] wasn't associated to my name in any way, if I was purely a number." Some participants wanted to share information only with sharing partners who they felt could offer some medical help. One student said sharing decisions "depend on kind of help [family and friends] can give based on the position I'm in." Most participants were more open to sharing their health information with doctors than with their family. One student said, "Your doctor has your health in mind. Your parents have, like, so many other interests in mind," while another student said, "What the parents view as social norm, whereas the doctor views it from medical point of view." On the other hand, one elderly participant said, "We want to be independent [from our family] as long as we can. We just want to be dependent on people [at the retirement home]. But I will be okay sharing it with caregivers." One elderly participant said, "I didn't want my wife to know [about my stroke] since I didn't want her to worry, since she was [out of town]," while another elderly participant said, "I would tell her, [she] would worry less if [she] knew early." According to a few students, health information should not be shared with family unless the patient could not make decisions on their own, for example, if the patient was a minor or an elderly person.

Flexibility of privacy controls, based on information type. We discovered that privacy concerns varied with the type of information that was being collected and shared. After hearing about the different scenarios, most participants felt that they would be more open to sharing their diet and exercise information with others than medication, location, and social interactions, though one student said that even though she would base her sharing decisions on who the information was being shared with, she would be most concerned about sharing her location and her diet with others. Some students said that collecting information about location and social interactions would be desirable if the patients had a criminal record, if they were suspected to be terrorists or if they were in prison.

Laws. A few students were concerned about privacy laws. One student was worried about their complexity, "Some things are difficult to be explained to people, especially the huge privacy laws," while another student was worried whether his information will still be protected when laws change in the future, "Laws change. [Suppose] right now, no third party can [access the information collected using the mHealth devices]. What if ten years down the road, the supreme court says [the third parties] have the right?"

Device form factor. A majority of the students said that they would not wear the device if it was conspicuous because they were worried about being judged by others. A student said, "If it's like conspicuous, you know, people would always be like asking, what does that device do?" Elderly participants were concerned about physical comfort (one participant gave an example of his watch: "I used to wear it 24 hours a day, now it keeps me awake, so I take it off but forget to put it on") and they did not want the device to disrupt their normal routine, with notifications.

30.3 Fitbit User Study

During the focus groups, the participants voiced concerns that they thought they might have about the collection and sharing of their health information, based on the hypothetical scenarios that we presented to them. To better understand what concerns people might have when they actually share their health information, we conducted a social experiment to examine users' decisions to share particular types of information with various types of information recipients (and requesters) over a 5-day period. The participants carried a device that collected their personal health information and shared the collected information with family, friends, and third parties. From the focus groups, we found that exercise was considered to be the least sensitive type of personal health information when compared to medications, location, and social interaction. So we decided to conduct a user study where users would use a device that collects exercise information, to understand whether users would have privacy concerns when sharing seemingly insensitive information like steps, calories, and sleep.

30.3.1 Study Design

During the study, users were asked to carry a Fitbit [9], a popular mHealth device that uses an accelerometer to estimate a user's calories burned, steps taken, distance traveled, and sleep quality. During the 5 days of the study, each subject was asked to wear the Fitbit at all times, except when swimming, bathing, or any time they felt uncomfortable wearing it. They were asked to upload the collected data at least daily to fitbit.com. Unfortunately,

TABLE 30.1

Participants

	Male	Female	Total
Students	8	13	21
Working	5	7	12
Retired	0	8	8
Total	13	28	41

fitbit.com only provided users with limited coarse-grained data sharing options, and no mechanism for monitoring sharing behaviors. So, we developed a custom web interface that displayed both uploaded Fitbit data and personal traits and allowed users to share data with others. The participants used this interface (instead of fitbit.com) to view their data and make sharing decisions, throughout the study.

The goal of the user study was to understand people's willingness to share their personal health/fitness information with family, friends, third parties, and the public. Previous work has shown that young and old people have different views about sharing health information [8,15,34]. So, we recruited a sample of college students, working adults, and retirees for the user study. We recruited 21 undergraduate students, 12 adult workers from the local area including Dartmouth employees, and 8 female elderly residents of a local retirement home, as shown in Table 30.1. It was not an aim of the study to understand the influence of gender and occupation on privacy concerns, so we did not focus on the distribution of participants among these categories. The recruitment flyer presented the study as a study of a new device to help individuals trying to lose weight and/or improve fitness and health. To avoid self-selection bias, the participants were not told that the study was about privacy. The subjects were required to own a computer, to be injury-free, and to be able to walk and carry the device with them during the 5 days. Study participants were paid for their time.

We did not retain any sensitive information, like the participants' fitness information collected using the Fitbits, after the study; we stored only the sharing settings that they chose. Study participants were debriefed after the study to make them aware of the deception used in the study and to inform them that the goal of the study was to understand their privacy concerns and not just to collect their activity data and that we shared their data only with the people they chose as their sharing partners. This study, including our use of deception and subsequent debriefing procedure, was approved by Dartmouth's Institutional Review Board.

To study participants' willingness to share secondary information, apart from the primary sensed information, we also collected other related personal information about each participant, including his or her age, gender, height, weight, health goals, overall activity level, and academic major. Henceforth, we refer to these seven characteristics as *traits*.

To understand how the participants share information with real people, we asked them to select family members and friends to receive their shared information. Throughout the study, the participants also received requests to share data with specific third parties. These specific third parties represent academic researchers, medical labs, private companies, and the government. The third parties were real, but the requests were fake (e.g., one of the e-mail requests was from a fictitious group of students at Harvard University, requesting activity data for use in a machine-learning class). Each participant also had a *public profile* available on the website with their Fitbit data; we told the participants that this profile was visible to anyone who had the URL (unless they changed the setting, as described in the following).

The website provided opt-out sharing settings; by default, all the collected information was shared in the finest detail unless the participants changed the settings to *opt-out* of sharing. We used an opt-out policy (instead of opt-in) to be consistent with the majority of online applications now available in which the default privacy setting is to share, with the option to *opt-out* of sharing. We understand, though, that people's sharing decisions are influenced by default settings [31]. Although we agree that opt-in settings give more control to the user, for our study, we used opt-out to provoke action (visit website and change settings) that we could observe among those with privacy preferences.

We wanted to study people's willingness to share their information, and not how they adapted to the device and controls on the website. So, the study was divided into two phases: a learning phase, in which participants were given 2 days to get used to the device and website, and the study phase, in which we observed participants' sharing settings during days 3–5 of the study.

The learning phase—day 1 and 2. The researchers met with the participants, individually, on both days to answer any questions they had about using the device and the website. On the first day of the study, the participants were asked to select at least one family member and two friends with whom to share their information. An e-mail was sent to these sharing partners, informing them about the study and asking them to be a part of the study.

On the second day, we told the participants that their information would be shared with their family and friends from the next day onward and that they could decide, by using the controls on the website, whether and what they wanted to share with their family members and friends. We also informed the participants that over the next few days, they might get requests from third parties to share their information but that they could use the controls on the website to limit sharing of their information; they were required to visit the website for each third-party request so that we could observe their sharing choices. Similarly, we informed them that their data on the website would be open to the public but they could use controls to opt-out of sharing. We did not tell the participants that the third-party requests were fake or that their information was not actually exposed to the public.

The sharing phase—days 3–5. On the third day of the study, an e-mail was sent to the family members and friends with a link to a webpage where they could see the participant's shared Fitbit information and traits. Throughout the study, the participants could change the sharing settings for each type of information and for each sharing partner. The Fitbit-collected activity data (i.e., steps, calories, and sleep) could be shared in 5 min, hourly, 6 hourly, or daily summaries, or it could be not shared at all. By default, activity data was shared at the maximum setting, that is, 5 min granularity. The participants were also able to share or hide their personal traits (age, height, weight, gender, activity level, health goals, academic major), independently by type and for each sharing partner. The participants also received e-mails everyday during the 5-day period of the study, containing status information. From the third day of the study onward, these mails explained who was receiving their information and what sharing settings they had chosen for each information type for each sharing partner. The message also contained a link to the site where they could change these settings at any time.

The web interface showed different views of the same information. The default view, shown in Figure 30.1a, was the participant's view. The interface also had views corresponding to each sharing partner (see Figure 30.1b). On these views, the participant could decide what information she wanted to share with her sharing partner, make changes with the click of a button, and observe exactly what her sharing partner will be able to

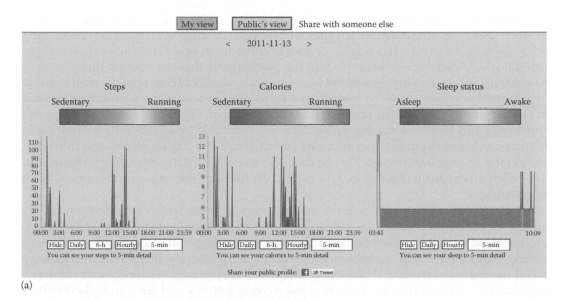

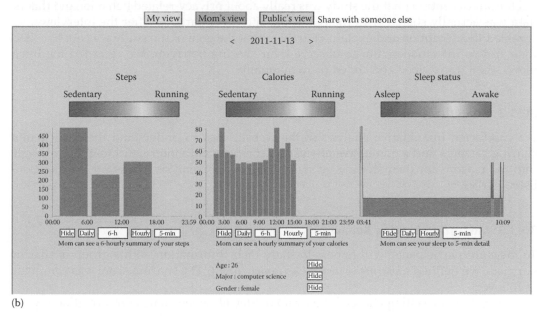

FIGURE 30.1
Screenshots of views on the study website. (a) Own view. (b) Sharing partner's view (Mom's view).

see; the goal of our website design was to reduce the disconnect between sharing controls and the information being shared.

The e-mails from third parties, requesting to access participant information from the website, including all information and their e-mail addresses, were sent by us as if from six different organizations/groups. Each third-party e-mail explained who the group was and why they wanted the data. The groups identified were college students, a research lab, a government agency, an engineering company, a wellness institute, and a pharmaceutical company. Here, we do not examine differences in willingness to share between these different types of third parties. Instead, we analyze sharing behavior with all third parties as a category to compare to other sharing partner categories (family, friends, the public). We told the participants that the data on the website to be shared with third parties was not anonymous (e-mail address was shared), but that their e-mail address would be used by the third parties only if the researchers needed to contact them. The e-mail address was, however, not visible on their public profile and hence not shared with the *public*. The order of third-party e-mail requests was randomized, and the number of requests varied across the days (three on the third day, one on the fourth, and two on the last day). We did not actually share the participants' data with any third-party organizations, but crafted the messages to be as believable as possible.

We monitored participants' website activity through logs, which recorded when they logged in, when they looked at their own data view or the views of their various sharing partners, and when they changed the sharing settings. To understand the reasons behind participants' sharing behavior, we conducted poststudy interviews in which we asked the participants several questions: whether they ever took off the device to hide any information, whether they ever changed the sharing settings and why, and whether they fell for our deception (that the study was really about privacy-related behavior and that no data was actually shared with the public or with third parties). After the interviews, we revealed the deception and explained to them that the goal of the study was not to collect their fitness information, but to observe their sharing behavior. We recorded the interviews, then transcribed and coded them manually.

30.3.2 Analytic Methods

We conducted quantitative analysis of the website logs to understand the participants' sharing settings and a qualitative analysis of the poststudy interviews to give us insight into the reasons for their sharing behavior. We use the analysis to answer the questions listed earlier.

30.3.3 Measuring Sharing Behavior

To measure participants' willingness to share their information with a group of sharing partners, we defined a *sharing score* for each participant. The sharing score was computed as follows:

$s(u, p, t, d)$ is the setting chosen by user (u) on day (d) when sharing information type (t) with sharing partner (p). For steps, calories, and sleep, the setting can be 0, 1, 2, 3, or 4, which corresponds to the five possible settings: hide, share daily summary, share 6 hourly summary, share hourly summary, and share 5 min detail, respectively. For the other information types, the setting can be either 0 (hide information) or 1 (share information). By default, $s(u, p, t, d) = 4$ for t = {steps, calories, sleep} and $s(u, p, t, d) = 1$ for t = {age, gender, health goals, height, activity level, academic major, weight}. That is, the default setting is the maximum sharing setting. We normalize each score by dividing by max(t):

$$\max(t) = \begin{cases} 4 & \forall\, t \in (\text{steps}, \text{calories}, \text{sleep}) \\ 1 & \text{otherwise} \end{cases} \qquad (30.1)$$

To represent the amount of information shared with a category of sharing partners (i.e., family, friends, third parties, the public), we computed an overall sharing score for that category, labeled as *group sharing score*. Equation 30.2 defines the group sharing score for participant (u) on day (d) as the mean of all sharing scores for type (t) for each member of the sharing partner group (g). For example, if a participant identified two friends then the friend group sharing score for weight would be the mean of the sharing scores for weight for the two friends:

$$\text{grpscore}(u,g,t,d) = \frac{1}{|g|} \sum_{\forall p \in g} s(u,p,t,d) \qquad (30.2)$$

To measure how personal traits were shared, we also computed a combined sharing score for traits, which is the average of the sharing scores of all the personal traits, labeled *traits*.

We focus most of our analysis on two snapshots: the initial sharing score, $\text{grpscore}(u, g, t, 2)$ for the last setting on day 2 for each type t, and the final sharing score, $\text{grpscore}(u, g, t, 5)$. We normalize each score by dividing by max(t), on day 5. Unless mentioned otherwise, we used t-tests to compute the difference between normalized sharing scores, to understand how different information types were shared with the different groups; paired sample t-tests were used to compare different behaviors of a single subject, whereas independent sample t-tests were used to compare one subject to another. We used *analysis of variance* (ANOVA) post hoc testing using Bonferroni's method to compare the sharing behavior of students, employees, and retirees; whereas the t-test is used to compare the means between two groups, ANOVA is a statistical procedure used to compare means between three or more groups.

30.4 Results

We present in the succeeding text the results from the quantitative and qualitative analysis. We highlight only some of the important results in this paper; full results are available in the technical report [26].

30.4.1 Information Sharing Analysis

Twenty-seven of the 41 participants identified both a family member and friends with whom to share information. Among the 27 participants, the mean number (and standard deviation) of family members and friends that the participants chose were 1.26 (0.44) and 2.11 (0.5), respectively. Two participants did not identify a family member, but did identify friends, while 11 participants did not select any friends or family members (the reasons given by these respondents included privacy concerns, not wanting to bother them, and expectation of lack of interest, as discussed in Section 30.4.2). These 11 participants, however, were not considered when we computed sharing scores for friends and family. Similarly, when comparing sharing scores between family and the public, we used the scores of all participants who selected at least one family member.

TABLE 30.2

Mean and (Standard Deviation) of Final Normalized
Sharing Scores for Family

Information	Family
Steps	0.96 (0.19)[a]
Calories	0.96 (0.19)
Sleep	0.96 (0.19)
Activity	0.93 (0.26)
Age	0.96 (0.19)
Gender	0.93 (0.26)
Goals	0.86 (0.36)[a]
Height	0.93 (0.26)
Major	0.95 (0.21)
Weight	0.84 (0.36)[a]

[a] Sharing score for sensed information (steps, calories, and sleep) is significantly higher than for weight and goals, sharing score for weight is significantly less than the score for most personal traits (age, gender, academic major, height, and activity level), and sharing score for goals is significantly less than the score for age, $p \leq 0.1$.

Did the participants share different types of personal or sensed information more or less frequently? Table 30.2 shows the final normalized sharing scores for family used to evaluate within-subject differences in sharing different types of information.

Table 30.2 shows that with family members, the study participants shared weight and health goals less than age, academic major, activity level, and the sensed information (steps). It is not surprising that the participants were willing to share obvious information known to their family members such as age and gender, but at least some participants seemed to consider information like weight and health goals as more private. In the post-study interviews, some participants were reluctant to share this information because they worried that family members might judge them and even reprimand them. A web interface might influence sharing behavior, but in the web interface we built, however, sensed information was more prominent than personal traits, so the sensitivity of weight and health goals had nothing to do with the interface layout.

Do participants' decisions about sharing health information differ across types of sharing partners (family members, friends, third parties, and the public)? Table 30.3 shows the normalized final group sharing scores for sensed information (steps, calories, sleep), traits (the combined score), weight, goals, and academic major across three comparison categories: family versus friends, family versus public, and public versus third parties. Recall that a score of 1 implies that the information has been shared to its maximum with all the sharing recipients in that group.

More information shared with family than friends. Subjects shared weight with family significantly more often than with friends. The sharing scores for friends were marginally less than that for family members for all other information types as well, but the differences are not statistically significant. From the interviews, we learned that some participants were more concerned about sharing their information with their friends because they were worried of being judged by their friends more than by their family, especially in the case of students, since they see their friends every day.

More information shared with family than with the public. Table 30.3 shows that the participants shared more information about their steps, calories, and sleep with family than with the

TABLE 30.3

Final Normalized Sharing Scores for Family versus Friends, Family versus Public, and Public versus TP

	Family	Friends	Family	Public	Public	TP
Steps	0.96	0.94	0.96	0.91[a]	0.89	0.89
	(0.19)	(0.14)	(0.19)	(0.22)	(0.27)	(0.25)
Calories	0.96	0.94	0.96	0.89[a]	0.89	0.89
	(0.19)	(0.17)	(0.19)	(0.25)	(0.27)	(0.25)
Sleep	0.96	0.95	0.96	0.87[a]	0.87	0.87
	(0.19)	(0.14)	(0.19)	(0.30)	(0.30)	(0.29)
Traits	0.91	0.83	0.91	0.78[a]	0.80	0.83
	(0.23)	(0.26)	(0.23)	(0.29)	(0.29)	(0.32)
Goals	0.85	0.73	0.86	0.61[a]	0.68	0.82[a]
	(0.36)	(0.42)	(0.36)	(0.50)	(0.48)	(0.38)
Major	0.94	0.91	0.95	1.00	0.98	0.82[a]
	(0.21)	(0.25)	(0.21)	(0.00)	(0.16)	(0.38)
Weight	0.83	0.64[a]	0.84	0.54[b]	0.59	0.74[a]
	(0.37)	(0.46)	(0.36)	(0.51)	(0.50)	(0.43)
	$n = 27$		$n = 28$		$n = 37$	

TP, Third parties.

[a] Final scores for the two groups are different, $p \leq 0.05$.

[b] Final scores for the two groups are different, $p \leq 0.01$.

public. The participants said that they felt uncomfortable sharing sensed information with strangers because it made them feel like they were *being watched*. Not surprisingly, the participants also shared personal traits significantly more with family than with the public.

Less information shared with the public than with third parties. Table 30.3 shows that the participants were generally more open to sharing weight and health goals with specific third parties than with the public. In the poststudy interviews, some participants said this was because they perceived some benefit in sharing information with specific third parties. Third-party request e-mails contained a reason for wanting the participants' data, and at least some participants apparently expected the third parties to use their data for the purposes mentioned in the e-mail. In contrast, some participants expressed concern about who among the public would be accessing their information or how they might use it.

Surprisingly, the participants were less willing to share academic major with the specific third parties than with the public. Some participants said during the poststudy interview that they shared information with specific third parties because they thought that the information would be useful, based on the purpose stated in the third-party request. Some of them felt that academic major was not relevant to the request.

Given the comparison on personal traits, we were surprised to see no difference in sharing of sensed information between the public and third parties.

Does sharing behavior change over time; are participants' privacy preferences dynamic? In Table 30.4, we show select traits and sensed information by the initial (end of day 2) and final (end of day 5) sharing scores for three sets of sharing partners: family, friends, and the public.

Sharing sensed information. For family, the participants did not change sharing behavior of sensed information over the course of the study, while for friends, there was a slight (nonsignificant) change. For the public, however, we found that there was a statistically

TABLE 30.4

Initial and Final Normalized Sharing Scores for Family, Friends, and the Public

	Family		Friends		Public	
	Initial	Final	Initial	Final	Initial	Final
Steps	0.96	0.96	0.97	0.95	0.94	0.88[b]
	(0.19)	(0.19)	(0.10)	(0.14)	(0.22)	(0.26)
Calories	0.96	0.96	1.00	0.95	0.93	0.88
	(0.19)	(0.19)	(0.02)	(0.17)	(0.23)	(0.27)
Sleep	0.96	0.96	0.98	0.95[b]	0.93	0.87[a]
	(0.19)	(0.19)	(0.09)	(0.16)	(0.22)	(0.30)
Traits	0.95	0.91	0.93	0.84[b]	0.92	0.80[c]
	(0.19)	(0.23)	(0.13)	(0.26)	(0.17)	(0.29)
Activity	0.96	0.93	0.95	0.84[a]	0.98	0.85[b]
	(0.19)	(0.26)	(0.20)	(0.36)	(0.16)	(0.36)
Goals	0.93	0.86	0.95	0.75[b]	0.83	0.68[b]
	(0.26)	(0.36)	(0.29)	(0.41)	(0.38)	(0.47)
Weight	0.89	0.84[a]	0.80	0.63[c]	0.83	0.61[c]
	(0.31)	(0.36)	(0.35)	(0.46)	(0.38)	(0.49)
	$n = 28$		$n = 29$		$n = 41$	

[a] Initial and final scores are different, $p \leq 0.1$.
[b] Initial and final scores are different, $p \leq 0.05$.
[c] Initial and final scores are different, $p \leq 0.01$.

significant reduction in sharing scores for steps and sleep. Some participants felt uncomfortable sharing their steps and sleep, as the study progressed; they said that they felt like they were being watched.

Sharing traits. Similarly, in the case of the trait information, there was a slight (nonsignificant) reduction in sharing scores for family. However, there was a statistically significant difference between the initial and final sharing scores for friends and for the public. We learned from the poststudy interviews that some participants were embarrassed to share certain personal traits with friends and concerned about sharing their personal traits with strangers; they might have realized it only seeing their data over time. More details of the poststudy interviews are given in Section 30.4.2.

Demographic differences. Though the study was not designed to examine differences in sharing by characteristics like occupational status or gender, we did find some differences across these characteristics and so present them as preliminary findings that are suggestive of future study. As shown in Table 30.5, independent sample *t*-tests revealed that females shared traits (the combined score), weight, and goals with friends, significantly less than did male subjects. Table 30.5 shows only the difference in sharing with friends, but there was a statistically significant difference in the extent that personal traits, weight, and activity level were shared with the public and third parties as well [26].

As shown in Table 30.6, students shared their weight with family more than employees shared with family. In contrast, they shared their health goals less with the public than employed adults did with the public. Some of the employees said they did not want to share weight information with their family to avoid discussion about weight management. Students considered health goals to be sensitive and did not want to share this information with the public, but surprisingly, employees were more willing to share this information with the public.

TABLE 30.5

Final Normalized Sharing Scores Based on Gender

	Female	Male
grpscore(u, Friends, Traits, 5)	0.78	0.95[a]
grpscore(u, Friends, Weight, 5)	0.47	0.95[b]
grpscore(u, Friends, Goals, 5)	0.64	0.95[c]

[a] Sharing scores of females and males are different, $p \leq 0.1$.
[b] Sharing scores of females and males are different, $p \leq 0.01$.
[c] Sharing scores of females and males are different, $p \leq 0.05$.

TABLE 30.6

Final Normalized Sharing Scores—Students versus Employees

	Students	Employees
grpscore(u, Family, Weight, 5)	0.90	0.50[a]
grpscore(u, Public, Goals, 5)	0.48	0.83[a]

[a] Sharing scores of students and employees are different, $p \leq 0.1$.

Students were much more concerned than retirees about sharing their personal traits, weight, and goals with the public, as shown in Table 30.7. (*Caveat:* Recall that we had only female retirees; given that females shared less than males with friends in Table 30.5, we are not sure why female retirees shared more with the public than students.) There are several possible reasons for this difference in behavior: we speculate that the retirees are not used to the technology or do not want to bother their family and friends by sharing with them information that the retirees feel will not be of interest; when they do share this information with others, they are less concerned about the information than students. We expect students, on hand, to be used to the technology and used to sharing information electronically with others. We speculate that they might have changed the default settings either because they were curious about the different settings or because they were really concerned about what they were sharing with others. We expect students and employees to have more reasons to be worried about their activities and to hide it from their family and friends than retirees, either because they were embarrassed about some information, maybe their weight, or they wanted to hide some information, like partying or sexual activity. Students and employees were more engaged in the study than retirees. In Section 30.4.2, we present anecdotal evidence of such behavior and concerns.

To summarize, we found that weight and health goals appeared to be most sensitive among the information collected during the study. The participants exhibited disparate and dynamic sharing behavior of this information. We found some evidence that sharing behavior might vary with occupational status and gender. After observing the

TABLE 30.7

Final Normalized Sharing Scores—Students versus Retirees

	Students	Retirees
grpscore(u, Public, Traits, 5)	0.74	1.00[a]
grpscore(u, Public, Weight, 5)	0.48	1.00[b]
grpscore(u, Public, Goals, 5)	0.48	1.00[b]

[a] Sharing scores of students and retirees are different, $p \leq 0.1$.
[b] Sharing scores of students and retirees are different, $p \leq 0.05$.

participants' sharing behavior, we wanted to understand the reasons for their sharing behavior, as discussed in the next section.

30.4.2 Poststudy Interviews

To give us insight into the reasons for participants' sharing behavior, we conducted post-study interviews. We discuss the answers to the following questions: "Did you change the sharing controls on the interface at any point? If so, what influenced that decision? Did you change the sharing controls for X? If so, why?" For answers to other questions, please refer to our technical report [26]. We recorded the interviews. We coded the interviews manually and grouped the statements into categories. We discuss in the succeeding text the reasons for the participants' sharing behavior.

Amount of information collected. Three participants mentioned that they felt the information was not sensitive because the device collected information only for 5 days. One female student said that the reason she shared the information with third parties was because "it was a study and it wasn't very long."

Context of data collection. A few students were concerned about sharing information that was collected by devices while they were at parties or staying up late. A male employee asked, "Would [the device] be something you would keep on during sexual activity or when you go to the bathroom?"

Sensitivity of the data. One male student shared all his information with others, but said that he might have had concerns about sharing "If maybe I was someone who [was] trying to exercise more and I exercised less." A few female students did not want to share their activity information because they felt like they were being watched. One of them said, "They can see every step I take, that was just a little weird."

Information utility. Some participants decided to share their information with others depending on how they would use the information. One female student said, "I think I hid my weight from almost everybody, except for people who actually needed it for medical purposes." One employee, being a researcher, was open to sharing her health information with other researchers. Some participants considered the third parties differently, based on their reason for requesting the data. One female student said, "I was fine with sharing things [with universities]; for some reason, they felt a lot more legitimate, you know what they would be doing, studying. It was random people that I didn't know what they were doing that I [did not want to share my information with]." One male student said he would share information with everyone, as long as it would not affect him in the future when he was applying for insurance or jobs. He said, "I would be fine with all of those, with the exception if that has an impact on the ability to apply for insurance or something of that nature, in which case I would start to worry."

Anonymity. A few students felt their information would not be linked to them (even though they were aware that their e-mail was being shared), so they were comfortable sharing it with third parties. According to a female student, "They don't know who I am, they are just doing research." One male employee felt that his identity is linked more to his name than his e-mail. He said, "It's my name, but I control [my e-mail]. I control what I get, I can change my e-mail address."

Sharing partners. Most participants said they would share information depending on who it was being shared with. One female student said "If there was someone who was a lot heavier than me, I probably would have given them the 5 minute calories, because they might feel bad that I used so many calories throughout the day. With friends who were less active than me, I would have shared less."

Partner involvement. One male student was happy to share his activity information with others; he said when you share activity information, "You feel like other people are in this with you, it makes it easier to keep going," and he said encouraging feedback from his friends made him feel good about sharing his information. A female student shared her Fitbit information with third parties, because according to her, "When I was wearing [the device] and getting data requests all the time, it felt like what I was doing was important," and she was disappointed that the requests were fake, and to her, it meant that "no one actually cares about your data and no one's going to use it, it was all for nothing kind of thing." Some students did not want to share certain information due to fear of being judged by others. A female student said, "With my friends, I wasn't sure whether to share my height and weight, because sometimes especially if I am sharing with my girlfriends, oh they are like you are heavier than me, lighter than me." Another female student said, "I don't mind articulating [my health information] in person, but on a website, I feel it is more easily judged in the wrong way that I can't fully explain what is going on."

Negative experiences. A female student was concerned about sharing information with the government, because she grew up in a country where "everything was monitored by the government." One female employee was sharing all her information with the third parties during the study, until she noticed that she started getting spam about weight loss, which must have been coincidental.

Relationships. One elderly participant was not comfortable with sharing her activity information with her children. She said, "I didn't want them to have to encourage me to walk more. They don't need to know. We are very, very close but they don't need to know how much I walk." A male student said, "I told [my mom] I would tell her of any results of any significance, but I told her that I was hiding the data and I wasn't going to let her see it. Honestly, my friends didn't care about the data."

Some students were more comfortable sharing their information with family than their friends. One said, "If it was someone I didn't know I would share everything. Friends they know you, but you are not close enough to share everything with them. I shared everything with my mom."

Some female students were more comfortable sharing personal information with their family and third parties than with their friends. One student said "I might have left height and weight with family, but friends don't need to know that. I shared it with companies and researchers, because I think it is pertinent."

Some students were more comfortable sharing their information with family and friends than third parties. Two of them said, "I didn't share any information with the extra researchers. I don't know who they are and I have no affiliation with them," and "I don't know [the requesters]. I don't think it is weird that they were asking for [the information], but it was weird sharing with them. From teammates, hid my weight and my health goals. From my mom, I didn't hide anything."

Some students wanted to share their information with third parties more than with people they knew, like their family and friends. They said they wanted to share less information with family and friends: one participant said "because I know them personally, whereas the third parties they seem, not that personal … so I felt like more of a pressure to hide more specific activity levels from [my family and friends]." Another said "Because my parents are people who are big on exercise. If I don't do much exercise, they wouldn't like that," while a third participant said "A bunch of researchers looking at the data, I don't care. But I might think twice about some people I know, depending on who they are." Another participant said, "People who don't know me, it would be fine. My age doesn't bother me, it would be mostly my weight. It all depends on who gets it, what is the purpose. If it is somebody studying what is the better way to do things."

Some participants did not want to share information with private companies. A male student said, "I'm against corporations. I probably wouldn't want any of them [to have access to my information], except students." A male employee said, "Oh yeah, I would share that info [with students]. With individuals, with family members, or friends who are interested and people doing research, I have no problem. It is just third-party companies [that I wouldn't want to have the data]."

One elderly retiree was not tech savvy; her husband was helping her manage her Fitbit account and he might have had an influence on her sharing decisions.

Information types. We asked the participants whether they would use a mobile device that collects personal health information like their heart rate, breathing rate, pulse rate, medication, diet and exercise, location, and social interactions, if it gave them similar sharing controls as the Fitbit in our study. Most students considered medications to be most sensitive. A male student said, "Just like bodily functions, you can't really use that against you, whereas medication you are taking, that's something like, there are some medications people don't want other people finding out that they are taking." Some of them were worried about sharing location and social interactions. Students who were athletes were concerned about sharing their vital signs and exercise information. One female employee was open to sharing any information "as long as [she] could control who saw what."

30.5 Discussion

In addition to the focus groups, the user study helped us identify the key factors that influence privacy decisions regarding health and fitness information collected using mHealth devices. The findings from this study can help guide mHealth device and application developers and privacy advocates to build flexible privacy controls for mHealth devices, with sensible defaults and expressive controls for users to change the settings thereafter.

We discovered that people were concerned about sharing more information than necessary with their sharing partners. They based their decisions to share information on the type of information being shared and people the information was being shared with, that is, the sharing partners. While making the sharing decision, people take into account the volume of information that is shared, why the sharing partner needs the information, the context in which the information was collected, and how sensitive the information is to them. People considered a certain type of information to be sensitive if sharing partners could misuse the information to cause harm to them or if sharing the information could cause a change in the relationship that the people shared with the sharing partner. So their decision to share information with sharing partners also depended on how they expected the sharing partners to use the information and whether their relationship with the sharing partners would be affected if the sharing partner was aware of this information. On the other hand, they shared information if it was already known to the sharing partners; for example they shared age and gender with family and friends.

People were more willing to share their health information with sharing partners who had a positive influence on their health decisions, while they were hesitant to share it with sharing partners with whom they have had prior negative interactions. People seemed to be willing to share their personal information when they felt that the sharing partner would use the data for the betterment of society, for example, by using the data for research; in such cases, the sharing did not directly benefit them, but it benefited the society. Even in these instances, some people ensured that they would not be harmed; they were particular about their anonymity and wanted to ensure that their health information could not be linked to their identity. On the other hand, people also took into account their prior experiences with sharing partners when making the decision to share their information; some participants had prior negative experience with the government and refused to share any information with government agencies.

We also observed that the participants changed their sharing settings during the course of the study after receiving (negative and positive) feedback from sharing partners. For example, one female employee was sharing all her information with the third parties during the study, until she noticed that she started getting spam about weight loss; she assumed it was from one of the third parties involved in the study and stopped sharing her data with all third parties.

When designing sharing controls for a system, especially a system that handles sensitive information like personal health information, it is important to ensure that the controls are easy to use, flexible, and convenient and account for ignorant users and changing privacy laws. Care should be taken that only adequate control is given to the patient since they should not restrict the sharing partner from accessing information that could be crucial to their health care; determining how much control is adequate remains a challenge.

Our study revealed three interesting findings about people's privacy concerns regarding their sensed health information:

1. *Demographic information shared less than sensed information.* The study revealed that the participants were less willing to share the demographic information we collected than the activity information that was sensed by the device. For example, initial sharing scores for weight and health goals were less than the initial sharing scores for other information types, including the sensed information. We expect users to be concerned about sharing certain context information, depending on how it might affect the value they perceive in the information being shared. For example, a user might share her location information when it is being collected by her asthma sensor and shared with her mother, but she might not want to share her location, when it is collected as part of her activity information by her fitness device.

2. *Information shared more with strangers than their own family and friends.* We discovered that sharing scores for friends were lower than scores for family, while sharing scores for friends were lower than for third parties. However, sharing scores for the public were mostly the same or less than for friends. The focus groups and the poststudy interviews revealed different reasons for people's sharing behavior, including their relationship with the sharing recipients. The participants were more willing to share if they perceived benefits in sharing, especially when it came to sharing with specific third parties, as opposed to the public.

3. *Dynamic sharing behavior.* We confirmed that privacy concerns are not static; mHealth device users may change their sharing decisions over time

Given these findings, we elaborate on two recommendations that will help guide the development of flexible privacy controls that enable users to express their sharing preferences easily:

Flexible controls need to support both fine- and coarse-grained approaches to sharing. Throughout the study, we observed a wide diversity in sharing behavior, which was in accordance with the varied privacy preferences of the focus-group participants. Some study participants used a very coarse-grained approach, while others took the time to fine-tune their privacy settings. Sensible default settings are required to support those users who never change their sharing settings, either because they are busy, lazy, not tech savvy, or want immediate benefits from the system. The availability of granular controls encouraged the participants, who were averse to sharing everything, to share some information, instead of hiding all information. The participants expressed disparate sharing preferences and exhibited dynamic sharing behavior in our study, which implies that default *one-size-fits-all* settings are not enough.

We observed contradicting behavior among participants; some participants shared more with their friends than with family, while others shared more with their family than with their friends. We also observed dynamic sharing behavior; some participants changed the amount of information they shared during the course of the study. Granular levels of sharing and expressive controls, which we discuss next, can help such users change their sharing setting easily to map their preferences.

Reducing disconnect between information and granular controls. Users make their sharing decisions based on what information they are sharing and who they are sharing it with. Since sharing decisions are dynamic, the information should be clearly presented and the controls flexible and easy to use, to allow the participant to map their privacy preferences easily. Narrowing the gap between settings and what is actually shared can help users change their behavior easily to suit their sharing preferences. For example, in our study, the website home page for every participant was divided into different views, one view corresponding to one sharing recipient, where the participant could decide what information she wanted to share with that recipient. View for sharing recipient "Mom" on the participant's home page displayed exactly what Mom would see as the participant's health information. By combining the information and the granular controls, the interface made it possible for the participant to observe what Mom would see for different choices of sharing settings and finally choose the setting that best mapped her privacy preferences. We did not test the usability of the system, so we cannot claim that our design is the best way to provide granular controls for sharing health information. Designing an interface for an mHealth device and application that collects a large amount of sensed, personal demographic, and context information and whose user has the option to choose a large number of sharing recipients is an interesting and challenging problem.

User studies, like ours, could benefit from a bigger sample size, better population sampling, and longer duration. Nevertheless, the study helped us understand people's willingness to share and their dynamic sharing behavior. We expect these findings to hold broadly for other mHealth devices and applications as well. A general privacy setting for all mHealth devices is not possible, given the disparate sharing behavior among users for even a single mHealth device. We recommend seeking a general approach to health information visualization: a flexible design that supports all mHealth devices and allows users to also visualize how they are sharing their health information with others.

30.6 Related Work

Previous work has looked at people's willingness to share information with others. Previous work suggests that users will change behavior when the context of information sharing varies [23]. For example, studies of location tracking show that at least some users will vary their willingness to share depending on place and social context [3], time since the start of information sharing [28], recipients [7], and their closeness to the recipients [36].

Other studies of context show that users are more likely to reveal information when the reward from the exchange increases [25] but less likely to do so when risk of identity theft increases [4]. Our findings confirm that these results hold even for people's willingness to share their health information.

Willingness to share. Previous work has shown that people make privacy decisions based on the information being shared and the person they are sharing it with. It has been shown, through surveys and interviews, that users share location information based on the sharing recipient, why the recipients want the information, what would be useful to them, and whether the users want to disclose that information with them; during the study, users received hypothetical sharing requests from family and friends [7]. (Our findings confirmed that people use the same sort of logic to make privacy decisions with health information.) An online survey showed that the participants do not understand the value of sharing location information and their privacy decisions depend on the sharing recipient [33]. The aforementioned two studies, however, never gave the participants an opportunity to actually share the information with real people but just learned about their sharing preferences through interviews and surveys. People may not be aware of real privacy risks until they actually share the information with others and receive feedback about the sharing [6]; our study gave the participants the opportunity to actually share the information with real people. Our study showed that people did have privacy concerns about sharing certain information types, but they changed their sharing settings during the course of the study. Our findings confirmed results from previous work, which showed that the participants change their privacy policy decisions with time, but the participants in that study knew the information was not being shared with real people [20]. The manner in which people think they might share their information changes once their information is actually shared with real people; the findings from our study are more valid than previous work, because our study participants actually shared information with real people. Also, the other studies were focused on understanding privacy concerns while sharing other types of information; we wanted to understand how sharing behavior changes when it comes to health and fitness information. Certain types of health information might be more sensitive than other types of personal information, like location, so it is important to study people's privacy preferences regarding health information.

Health information. Maitland and Chalmlers conducted interviews to understand the role of peers in weight management and what information people are willing to disclose to their peers [17]. Caine and Henania studied people's desires for sharing their health information, using questionnaires, card tasks, and interviews [5]. We, too, wanted to understand users' willingness to share their fitness and health information, but we gave users an opportunity to actually share their own information with family, friends, and third parties. Olson et al. conducted surveys with employees (median age of 35) to study people's willingness to share their personal information, including pregnancy and health status, with others and they identified similarities in what people wanted to share and who they wanted to share it with [24]. Our work is different from theirs in that we conducted a study with young students, employees, and retirees, where subjects collect information about themselves and actually share that information with real people (or in some cases, believe that their data is being shared with actual people).

Information collected using sensors and mobile devices. Klasnja et al. study the privacy concerns of patients using a fitness device by conducting interviews [18]. We also study privacy concerns of patients using a fitness device, but we focus on their willingness to share the collected information. Raij et al. showed that people are more aware of privacy risks once they receive feedback about their shared health information, collected using mobile sensors, and have a stake in the data, that is, if the shared health data is their own [27]. The study participants (in this case, students) filled out a survey after seeing feedback about their information for 10–15 min. In our study, we allow the participants to share the collected information with real individuals chosen by them and study how willing they are to share their activity and sleep information with friends, family, and third parties. To the best of our knowledge, ours is the first study that explores users' privacy concerns by giving them the opportunity to actually share the information collected about them using mHealth devices.

Our work focuses on the privacy concerns that people will have when they share their health information; we expect people to have these concerns irrespective of the system used to share the health information. In our study setup, health information was collected using a mobile device called Fitbit and uploaded to a private web server and shared with sharing partners through the web. However, health information can also be shared via EHRs and social networks. In the following, we highlight research that focuses on the privacy concerns that users have when sharing health information using these systems. The following research complements our work since it addresses privacy concerns that people have when using systems, other than private websites, to share their health information.

Electronic health records. In the context of EHRs, several studies have showed similar results as ours. Shaw et al. conducted a literature review of 21 articles (with publishing dates between 1994 and 2008), which focused on the privacy concerns that patients and health-care providers have about EHRs in various countries [30]. The study revealed that patients' willingness to share health information depended on how sensitive they considered the information to be; sexual and mental health were considered to be most sensitive. Patients were willing to share their health information with health-care providers involved in their care but were against sharing it with other medical and nonmedical individuals. In a recent study, Caine and Henania showed that sharing preferences varied by type of information and recipient, and overall sharing preferences varied by participant [5]. Haas et al. presents a set of design requirements for privacy-preserving EHRs; their requirements, however, are for controls to manage privacy policies and, unlike ours, will not be able to support the actual sharing behavior of patients [12]. Researchers have also highlighted the need for a balance to protect individuals from potential harm that may be caused by exposing personal information and the quality and safety expected of the health-care system [29]. Finally, Alemán et al. presents a survey of all the security and privacy issues and proposed solutions for EHRs [2].

Social networks. Prior research describes the privacy risks introduced by social networking in health-care domain [13,19,37]. Newman et al. conducted interviews to understand health interactions on online health communities and social networks and propose design recommendations for future systems that will support health-oriented social interactions [22]. Thompson et al. discovered how medical professionals violated patient privacy by posting protected health information on their publicly available social networking sites [32].

30.7 Case Studies

This chapter described a study conducted to understand privacy preferences and sharing behavior of mHealth device users.

Raij et al. and Caine and Henania are two other studies with similar goals [5,27]; the first one was conducted to understand privacy risks that might arise with the use of mHealth devices and other wearable sensors, while the second one was conducted to understand the privacy preferences of patients who use electronic medical records.

30.8 Summary

To provide flexible and expressive privacy controls, it is important to understand users' willingness to share their personal health information collected using the device. Other researchers used interviews and surveys to understand people's willingness to share; their results might not reflect real privacy concerns, since people remain unaware of real privacy risks until they actually share the information with others. We conducted a user study to understand how willing users were to share their personal health information that they collected using an mHealth device that they carried with them at all times for 5 days. We recommend a flexible design for sharing controls that narrows the gap between controls and the information being shared, allowing patients to visualize how they are sharing their health information with others.

Acknowledgments

This research results from a research program at the ISTS, supported by the NSF under Grant Award Number 0910842 (TISH) and Award Number 1143548 (PC3) and by HHS (SHARP program) under Award Number 90TR0003-01. We also thank undergraduate research interns Alexandra Della Pia and Tina Ma and our colleagues in the Dartmouth TISH group, for their valuable feedback.

Questions

1. Briefly describe an incident where a privacy breach occurred with an mHealth device. How was it resolved?

2. What can you learn from the controversy, and how would you ensure a similar situation will not happen with an mHealth app that you develop?

3. State five instances in which information collected using mHealth devices can end up disclosed more than necessary?

4. Do you think the recommendations from the study can be applied to mHealth apps collecting nonfitness information? Design an interface, highlighting the available sharing settings (feel free to exclude values) for an app that monitors onset of asthma. The app asks the patient to blow into a spirometer twice a day and suggests them to use an inhaler. It monitors when the patient uses the inhaler and shares with the patient's mom. The app also monitors where the patient goes and the dust and pollen count in the atmosphere in the locations at those times when the readings were taken.

5. How would you extend the study, if the goal were to understand sharing behavior based on gender?

References

1. Affectiva, Inc., 2014. http://www.affectiva.com/, January 2011.
2. J. L. F. Alemán, I. C. Señor, P. Á. O. Lozoya, and A. Toval. Security and privacy in electronic health records: A systematic literature review. *Journal of Biomedical Informatics*, 46:541–562, 2013.
3. D. Anthony, T. Henderson, and D. Kotz. Privacy in location-aware computing environments. *IEEE Pervasive Computing*, 6:64–72, 2007.
4. D. Baumer, J. Earp, and J. Poindexter. Quantifying privacy choices with experimental economics. In *Proceedings of Workshop on Economics of Information Security (WEIS)*, June 2–3, 2005, Cambridge, MA, pp. 1–16, 2005.
5. K. Caine and R. Hanania. Patients want granular privacy control over health information in electronic medical records. *Journal of the American Medical Informatics Association*, 20(1):7–15, January 2013.
6. K. Connelly, A. Khalil, and Y. Liu. Do I do what I say?: Observed versus stated privacy preferences. In *Proceedings of Human-Computer Interaction (INTERACT)*, Vol. 4662, Lecture Notes in Computer Science, Chapter 61, pp. 620–623, Springer, New York, 2007.
7. S. Consolvo, I. E. Smith, T. Matthews, A. LaMarca, J. Tabert, and P. Powledge. Location disclosure to social relations: Why, when, & what people want to share. In *Proceedings of the SIGCHI Conference on Human Factors in Computing Systems (CHI)*, pp. 81–90, ACM, New York, 2005.
8. G. Demiris, B. K. Hensel, M. Skubic, and M. Rantz. Senior residents' perceived need of and preferences for smart home sensor technologies. *International Journal of Technology Assessment in Health Care*, 24(1):120–124, 2008.
9. Fitbit, Inc., 2014. http://www.fitbit.com, January 2011.
10. J. H. Frost and M. P. Massagli. Social uses of personal health information within PatientsLikeMe, an online patient community: What can happen when patients have access to one another's data. *Journal of Medical Internet Research*, 10(3):e15+, May 2008.
11. Google Health, Inc., 2013. http://www.google.com/intl/en-US/health/about/, January 2011.
12. S. Haas, S. Wohlgemuth, I. Echizen, N. Sonehara, and G. Müller. Aspects of privacy for electronic health records. *International Journal of Medical Informatics*, 80(2):26–31, 2010.
13. C. Hawn. Take two aspirin and tweet me in the morning: How Twitter, Facebook, and other social media are reshaping health care. *Health Affairs*, 28(2):361–368, 2009.
14. iBaby Labs, Inc., Heartsense. 2013. http://www.ibabylabs.com/product/heartsense, August 2013.
15. C. J. Hoofnagle, J. King, S. Li, and J. Turow. How different are young adults from older adults when it comes to information privacy attitudes and policies? Social Science Research Network Working Paper Series, April 2010. doi: http://dx.doi.org/10.2139/ssrn.1589864.
16. C. Jensen, C. Potts, and C. Jensen. Privacy practices of internet users: Self-reports versus observed behavior. *International Journal of Human-Computer Studies*, 63(1–2):203–227, July 2005.
17. J. Maitland and M. Chalmers. Designing for peer involvement in weight management. In *Proceedings of the SIGCHI Conference on Human Factors in Computing Systems (CHI)*, ACM, New York, 315–324, 2011.
18. P. Klasnja, S. Consolvo, T. Choudhury, and R. Beckwith. Exploring privacy concerns about personal sensing. H. Tokuda, M. Beigl, A. Friday, A.J. Brush, Y. Tobe (Eds.) In *Proceedings of the International Conference on Pervasive Computing (Pervasive)*, Springer-Verlag, New York, May 2009.
19. J. Li. Privacy policies for health social networking sites. *Journal of the American Medical Informatics Association*, 20:704–707, 2013.
20. M. L. Mazurek, P. F. Klemperer, R. Shay, H. Takabi, L. Bauer, and L. F. Cranor. Exploring reactive access control. In *Proceedings of the SIGCHI Conference on Human Factors in Computing Systems (CHI)*, pp. 2085–2094, ACM, New York, 2011.
21. Microsoft, Microsoft HealthVault. https://www.healthvault.com/us/en, January 2011.

22. M. W. Newman, D. Lauterbach, S. A. Munson, P. Resnick, and M. E. Morris. "It's not that I don't have problems, I'm just not putting them on Facebook": Challenges and opportunities in using online social networks for health. In *Conference on Computer Supported Cooperative Work and Social Computing*, ACM, New York, 341–350, 2011.

23. H. Nissenbaum. *Privacy in Context*, Stanford University Press, Palo Alto, CA, 2010.

24. J. S. Olson, J. Grudin, and E. Horvitz. A study of preferences for sharing and privacy. In *Extended Abstracts on Human Factors in Computing Systems (CHI EA)*, pp. 1985–1988, ACM, New York, 2005.

25. S. Patil, G. Norcie, A. Kapadia, and A. J. Lee. Reasons, rewards, regrets: Privacy considerations in location sharing as an interactive practice. In *Proceedings of the Eighth Symposium on Usable Privacy and Security, SOUPS '12*, ACM, New York, 5:1–5:15, 2012.

26. A. Prasad. Exposing privacy concerns in mHealth sensing. Technical Report TR2012-711, Dartmouth College, Hanover, NH, February 2012. http://www.cs.dartmouth.edu/reports/TR2012-711.pdf.

27. A. Raij, A. Ghosh, S. Kumar, and M. Srivastava. Privacy risks emerging from the adoption of innocuous wearable sensors in the mobile environment. In *Proceedings of the SIGCHI Conference on Human Factors in Computing Systems (CHI)*, ACM, New York, 11–20, 2011.

28. N. Sadeh, J. Hong, L. Cranor, I. Fette, P. Kelley, M. Prabaker, and J. Rao. Understanding and capturing people's privacy policies in a mobile social networking application. *Personal Ubiquitous Computing*, 13(6):401–412, August 2009.

29. A. Shachak and A. R. Jadad. Electronic health records in the age of social networks and global telecommunications. *Journal of the American Medical Association*, 303(5):452–453, 2010.

30. N. T. Shaw, A. Kulkarni, and R. L. Mador. Patients and health care providers' concerns about the privacy of electronic health records: A review of the literature. *Electronic Journal of Health Informatics*, 6(1):e3, 2010.

31. R. H. Thaler and C. R. Sunstein. *Nudge: Improving Decisions about Health, Wealth, and Happiness*, Yale University Press, London, U.K., 2008.

32. L. A. Thompson, E. Black, W. P. Duff, N. P. Black, H. Saliba, and K. Dawson. Protected health information on social networking sites: Ethical and legal considerations. *Journal of Medical Internet Research*, 13(1):e8, 2011.

33. J. Tsai, P. Kelley, L. Cranor, and N. Sadeh. Location-sharing technologies: Privacy risks and controls. In *Proceedings of the Research Conference on Communication, Information and Internet Policy (TPRC)*, Arlington, VA, 2009. Accessed on September 25–27, 2009.

34. M. van der Velden and K. E. Emam. "Not all my friends need to know": A qualitative study of teenage patients, privacy, and social media. *Journal of the American Medical Informatics Association*, 20(1):16–24, January 2013.

35. Perfect Third Inc. WakeMate, 2011. http://wakemate.com/, January 2011.

36. J. Wiese, P. G. Kelley, L. F. Cranor, L. Dabbish, J. I. Hong, and J. Zimmerman. Are you close with me? Are you nearby? Investigating social groups, closeness, and willingness to share. In *Proceedings of the International Conference on Ubiquitous Computing (UBICOMP)*, September 17–21, 2011, Beijing, China, pp. 197–206, 2011.

37. J. Williams. Social networking applications in health care: Threats to the privacy and security of health information. In *Proceedings of the 2010 ICSE Workshop on Software Engineering in Health Care, SEHC '10*, pp. 39–49, ACM, New York, 2010.

38. Withings, Inc., Withings blood pressure cuff. 2014. http://www.withings.com/en/bloodpressuremonitor, November 2011.

31

House of Quality and Comparative Assessment of Health-Care Services

Saradhi Motamarri, Pradeep K. Ray, and Chung-Li Tseng

CONTENTS

31.1 Introduction

In manufacturing, finance, and other service industries, continual innovation has brought new forms of delivery, giving rise to new forms of services (Hwang and Christensen 2008). Competition in health-care delivery has led to newer forms of services like outpatient surgery centers, executive wellness programs, independent nursing group practices, hospitals, nursing homes, intermediate care facilities, and home health-care programs (Lim and Zallocco 1988). mHealth is one of the emerging alternatives for the delivery of health-care services.

In the extant literature, there is very little focus on health-care services design. The hierarchy of *quality function deployment* (QFD) matrices provides a robust service design framework in translating/transforming customer needs progressively into operational functional targets. This chapter focuses on identifying the differentiating service characteristics of mHealth vis-à-vis alternative health-care services, a precursor step toward mHealth service design. The prominent multivariate technique, *discriminant analysis* (DA), is employed to assist in the quantitative comparative evaluation of services. Lastly, this chapter, as a case study, presents the application of these principles in unearthing differentiating characteristics of a commercial mHealth service.

31.2 mHealth: A Viable Element in Health-Care Provision

Globalization, ever-increasing competitive climate, and continual technological sophistication have been placing tremendous pressure on organizations to deliver more with fewer employees. Support business functions like accounting, personnel, and information management have become integral processes and no longer ancillary functions anymore (Mazur 1993). The situation is no different with respect to service industries like health care. Globally all nations, including the developed economies, are facing escalating costs of health care at a time when it is increasingly difficult to match budgets to keep up with the service demands (PC 2011). For example, the total expenditure on health in Australia as per the latest estimate for 2010–2011 stands at A\$130.3 billion, while a decade ago, the same figure was A\$77.5 billion (AIHW 2012). A recent study by the *Institute of Medicine* (IOM) of the United States on the grave concerns of escalating health-care costs observed that the current expenditure on health care at \$2.5 trillion, which is 17% of America's gross domestic product (GDP), is unsustainable (IOM 2010). These reports further extrapolate and predict that health care might consume as much as 25% of the US GDP by 2037 (Gale 2012). Ironically, IOM estimates that of all spending on health care, as much as \$750 billion per annum is wasteful expenditure.

In sharp contrast to these issues of health-care status in the developed world, Table 31.1 provides a substantial gap of health-care delivery in the developing world through a sample of health system indicators for selected countries. These indicators highlight the grave

TABLE 31.1

Sample Health System Indictors for Selected Countries

Country	Births Attended by Skilled Health Personnel (%) 2005–2011	Hospital Beds per 10,000 Population 2005–2011	Physicians per 10,000 Population 2005–2010
Bangladesh	27	3	3
India	50	9	6.5
Mexico	95	16	19.6
Pakistan	45	6	8.1
Russia	100	97	43.1
United Kingdom	99	33	27.4
United States	99	30	24.2

Source: Modified from Dieter, G. and Schmidt, L., *Engineering Design*, 4th ed., McGraw-Hill, New York, 2008.

shortage of health-care professionals per 10,000 populations among developing countries. The demand for health care is rising faster than the supply of health-care professionals. The problem is acute in poor countries (*The Economist* 2012; World Health Organization 2012). *The Economist* (2012) observes that while labor productivity in the United States has increased by 1.8% annually, over the last two decades, productivity in the health sector has declined by 0.6%. *The Economist* also notes that the shortage of health workers is a universal issue. In spite of these sectorial challenges, it is in developing countries that innovations for newer solutions are emerging. And it is not surprising that mHealth is transforming health care in some of the developing countries. Though mHealth is in its infancy, it is becoming a distinct player in developing countries due to its affordability and right time and right place availability (Lewis et al. 2012).

31.3 Need for Health-Care Services Design Framework

Considering the broad agenda of better health care for all, the dire situation of health-care status in developing countries (Table 31.1), and the vast potential and promise of mHealth (Motamarri et al. 2012), there is a significant opportunity for the research community to direct attention toward health-care services design and operation. The big question is on how to do more with fewer people and lesser resources without sacrificing the quality of health care (Mazur 1993). To address similar competitive challenges of the manufacturing sector, Mizuno and Akao have developed tools and techniques that later came to be known as *QFD* (Mazur 1993; Mizuno and Akao 1978). QFD has enabled the *voice of the customer* (VoC) to resonate across the levels of the organization end to end, that is, from planning to production. While IT Infrastructure Library (ITIL) (OGC 2007) comes in handy with its systematic framework to address the services operation phase, there is a need to bring the patients' perspective to service providers and guide them in devising health-care services.

Chan and Wu (2002b) reviewed over 650 publications on QFD and categorized QFD's diverse applications such as product development, quality management, customer needs analysis, product design, planning, engineering, decision making, management, teamwork, transportation and communication, electronics and electrical utilities, software systems, manufacturing, services, education and research, and other industries. *House of quality* (HoQ) is the first phase of the QFD approach and is fundamental and strategically important (Chan and Wu 2005). HoQ is basically a design tool. Hauser and Clausing's (1988) classic paper on HoQ has brought its significance to the western community. With its widespread success in bringing together various functional divisions of manufacturing, HoQ has been applied in various forms and to various degrees of sophistication in product development, manufacturing, engineering, and, subsequently, the service industry to design and develop quality *services* (Chan and Wu 2002b; Mazur 1993; Ramaswamy 1996; Ray 2003; Shahin 2008). HoQ interlinks customer requirements (CRs), their rankings, engineering characteristics (ECs), performance measures, and competitive products/services and thereby elicits in a single diagram the areas of improvements required to win in the market (Chan and Wu 2002a, 2005; Hauser 1993; Hauser and Clausing 1988).

By propagating VoC across the organization and across the technical specialties, QFD became the sole quality system to echo CRs in the process of products/services design

(Mazur 1993). Laudon and Laudon (2013, p. 521) summarize that: "Rationalization of procedures is often found in programs for making a series of continuous quality improvements in products, services, and operations, such as Total Quality Management (TQM) makes achieving quality an end in itself and the responsibility of all people and functions within an organization." Charteris (1993, p. 1) states that: "QFD provides a systematic means of identifying customer requirements and translating them into achievable product characteristics." QFD and HoQ are the tools of TQM that have facilitated cascading the *VoC* down the layers of an organization (Ramaswamy 1996; Ray 2003). A similar mechanism needs to be identified to echo VoC in the design of health-care services, which is essentially the goal of this chapter.

31.4 House of Quality

The core structure of QFD is HoQ and, in its most comprehensive form, consists of eight sections, as shown in Figure 31.1 (Dieter and Schmidt 2008; Ramaswamy 1996; Ray 2003; Wulan 2011). Product/service planners express their understanding of the product/service, through these eight sections, commonly referred to as *rooms*. Starting

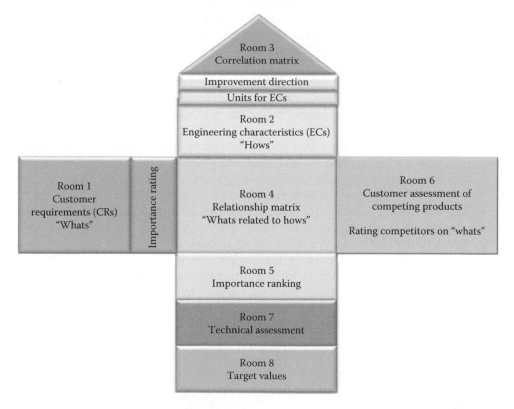

FIGURE 31.1
HoQ. (From Dieter, G. and Schmidt, L., *Engineering Design*, 4th ed., McGraw-Hill, New York, 2008.)

at Room-1, expressing the customer needs in customer's own language, the HoQ is progressively built and refined by successively developing the rest of the rooms. The construction of the full house demands elaborate information and is progressively constructed in a step-by-step process. This eight-step process is detailed below, taking hints from Wulan (2011).

Step 1: Identify CRs in Room-1. The customer or end user needs are compiled as CRs in Room-1 in the form of CRs and their importance ratings. These are the initial input of an HoQ. Room-1 is also called the *Whats* section. Ideally these are as stated by the customers in their own language, for example, low price, tastes good, and appetizing in appearance. CRs are categorized by customer's groups. Hauser and Clausing (1988) refer to the CRs as attributes and groups as attribute bundles. Based on a customer's assigned priority, the *Weight/ Importance* rating for each CR is computed and the *Weight/Importance* column is populated.

Step 2: Identify ECs in Room-2. ECs describe the product's performance as a whole and its functional features to meet CRs. The ECs are also known as functional requirements. At this stage, the CRs of Room-1 are expressed in parameters, design variables, and constraints, such as weight, size, and thickness. These inputs of Room-2 are also referred to as *Hows*.

Step 3: Build a correlation matrix of the ECs in Room-3. The triangular roof of HoQ or Room-3 is utilized to establish the correlation matrix on how the ECs support or impede one another.

Step 4: Build a relationship matrix between CRs and ECs in Room-4. At this step, the main body of HoQ—a 2D relationship matrix—is constructed with individual CRs to the ECs. Each cell is marked with a symbol that indicates the strength of the combination between the CR, or its row, and the EC of the column. It implies how significant the EC is in satisfying the CR.

Step 5: Rank the importance of the identified ECs (CRs) in Room-5. At this step, the focus is to find which ECs are of critical importance to satisfy the CRs stated in Room-1. Naturally, the ECs with highest rating are given special consideration because these ECs have the greatest effect upon customer satisfaction. The ranking can be either absolute or of relative importance.

Step 6: Analyze the product with competitive products in the market in Room-6. This step involves competitive analysis or customer assessment of competing products. A table is designed to record how the top competitive products rank with respect to the CRs listed in Room-1. This information may come from customer surveys, industry consultants, and marketing departments.

Step 7: Estimate the technical advantage and difficulty of each EC in Room-7. Organizations need to ensure that their design is competitive, with respect to the competing products in the market, prior to investing in the development of a new product or service (Cohen 1995). So Step-7 of Room-7 deals with the technical assessment, or benchmark, of important ECs with that of competitive products/ services.

Step 8: Assign the target value to each EC in Room-8. The final step in Room-8 involves setting target values for ECs and that is the ultimate output of HoQ for the design of the product/service.

31.5 Hierarchy of QFD Matrices

HoQ interlinks CRs, their rankings, ECs, performance measures, and competitive products/services and thereby elicits in a single diagram the areas of improvement required to win in the market. In practice, HoQ is developed over an iterative process, commonly referred to as the *hierarchy of HoQ matrices*. Starting with customer needs and customer assessment of competitive products/services, a series of HoQ matrices are built where the output of the preceding matrix becomes an input to the succeeding matrix as shown in Figure 31.2. While different conventions exist for the representation of the *hierarchy of QFD matrices*, this is the most common convention and referred to as the *American Supplier Institute* (ASI) model or the *Clausing model* (Clausing 1994; Ramaswamy 1996).

The matrix labeled 1 in this diagram is the HoQ matrix described in the previous section. The service requirements obtained from customers are translated into design characteristics and service performance targets in this stage. The design characteristics derived in Matrix-1 are the inputs to Matrix-2, which is interpreted as the *service/process matrix* (Ramaswamy 1996) as it is applied to services design. The service design attributes are partitioned into process design attributes and associated process performance targets. In Matrix-3, called the *process/subprocess matrix*, analysis is drilled down to establish design characteristics for the subprocess that makes up the process. These requirements are also standards to which the service operations should be managed once the design is implemented (Ramaswamy 1996). Finally in Matrix-4, the *subprocess/function matrix*, design requirements for individual subprocesses and functions are computed. By linking the output of each matrix to the input of the following matrix, the hierarchy of matrices cascades VoC to drive the design of service down to the most detailed level. Thus, HoQ interlinks CRs, their rankings, ECs, performance measures, and competitive products/services and thereby elicits in a single diagram the areas of improvements required to win in the market. For mHealth to be successful, it is essential to address its competitive strengths vis-à-vis with other conventional health-care delivery systems.

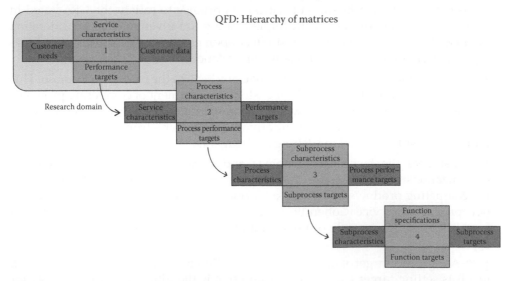

FIGURE 31.2
Hierarchy of QFD matrices for service design. (From Ramaswamy, R., *Design and Management of Service Processes: Keeping Customers for Life*, Addison-Wesley Publishing Company, Boston, MA, 1996.)

31.6 Comparative Assessment of mHealth

A robust service design is an elaborate and painstaking process of translating/transforming customer needs progressively into operational functional targets as per the HoQ process described in Section 31.4. In view of the breadth and depth involved in the application of QFD for health-care services, this chapter restricts itself to the Matrix-1 where customer evaluation of competing services is used as the basis to build the service characteristics. In summary, the scope of this chapter is limited to a subset of the theme of health-care services design, that is, evaluation of competing services. As such, the rest of the discussion focuses on *Room-6: evaluation of competing services* of the HoQ matrix (Figure 31.1).

In the broadest sense, patients' perceptions and comparison of competing services can be performed at both qualitative and quantitative levels. Motamarri (2013b) provided a qualitative comparison of health-care delivery systems, public hospital (PH), general practitioner (GP), traditional medicine (TM), and mHealth. The common service offered across the range of these delivery systems is consultation and advice.

The qualitative comparison of the alternatives was performed on a qualitative scale of low–medium–high and along the dimensions of nine attributes, namely, *accessibility, availability, ease of use, privacy, empathy, promptness, capacity, range of services,* and *end-to-end medical needs.* The comparative model is shown in Table 31.2.

Motamarri (2013b) has proposed a *six-cell services comparison model (SCSCM) for healthcare* that portrays the range of comparative segments on a 2D frame, consisting of the dimensions *system* (single system, intrasystem, and intersystem) versus *comparison type* (qualitative and quantitative). The resultant model consists of six quadrants, as shown in Figure 31.3. While Table 31.2 provides the comparative advantages of mHealth over the conventional services, as per the SCSCM, there are opportunities to carry on a quantitative comparative assessment of mHealth with conventional services. Figure 31.3

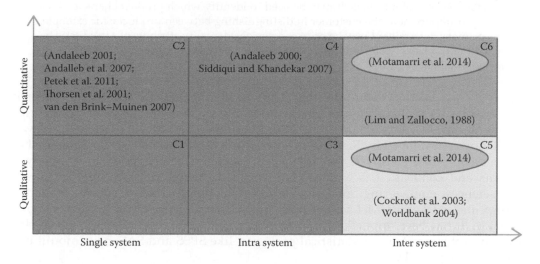

FIGURE 31.3
Quadrant model for comparison of services. (From Motamarri, S., *Distinguishing mHealth from Other Health Care Alternatives in Developing Countries: A Study on Service Characteristics*, The University of New South Wales, Sydney, NSW, Australia, 2013; Motamarri, S., A six cell services comparison model for healthcare, in *Third Australasian Symposium on Service Research and Innovation (ASSRI'13)*, University of New South Wales, Sydney, NSW, Australia, 2013.)

TABLE 31.2

Qualitative Comparison of Health-Care Service Delivery Alternatives

Attribute	PH	GP	TM	mHealth
Accessibility	Low	Medium	Medium	High
Availability	Medium	Medium	Medium	High
Ease of use	Low	Medium	Medium	High
Privacy	Low	Medium	Medium	High
Empathy	Low	Medium	Medium	High
Promptness	Low	Medium	Medium	High
Capacity	Static	Static	Static	Dynamic
Range of services	High	Medium to low	Low	Limited
End-to-end medical needs	High	Low	Low	Low

Source: Modified from Hair, J.F.J. et al., *Multivariate Data Analysis, A Global Perspective*, 7th ed., Prentice Hall, Upper Saddle River, NJ, 2010.

also depicts some selected studies pertaining to these comparative quadrants. Motamarri (2013) provides comprehensive quantitative analysis of mHealth. For a detailed discussion of this quantitative study, readers can refer to (Motamarri et al. 2014). The next sections detail the methods employed to achieve quantitative comparison of mHealth with other services.

31.6.1 Services Comparison and Discriminant Analysis

Charteris (1993, p. 15) summarizes the role of DA in QFD as:

> DA is used to classify products into two or more categories using a set of predictor variables spaced at intervals. It may be used to identify which product characteristics are most important to the customer in distinguishing between products; for example, the primary determinant may be price or a perceived quality attribute of a product. The mathematical basis of DA is similar to regression analysis but differs in that DA caters for nominal data.

DA is a classification technique that helps in identifying the factors (or independent variables [IVs]) that differentiate the cases into various categories of a categorical dependent variable (DV) (Hair et al. 2010; Malhotra 2004; McLachlan 1992; Schwab 2006). DA was developed by R.A. Fisher in 1936 as a multivariate classification method (Fisher 1936). Subsequently it was enhanced, extended, and applied to a variety of problems and contexts. It is widely used in diverse fields: physical, biological, and social sciences, engineering, medicine, image detection (pattern recognition, remote sensing), and marketing (McLachlan 1992). In DA, the existence of the groups is known as a priori. It has been applied by information systems (IS) researchers to aid in classification problems. Popular statistical packages, like SPSS and SAS, have inbuilt procedures to carry out DA.

DA constructs the model based on the variation of the observational units. On the basis of this model, new observational units or cases are classified into groups or categories (Motamarri et al. 2014). DA may also be used to characterize group separation based upon a reduced set of variables and analyze the original variable's contribution to separation and the degree of separation (Savic 2008). Savic et al. (2008, p. 29) observe that:

> DA gets its name from the way the model is constructed. Each of the groups in the dependent variable must have a set of measurements. For example, if there are five different brands of mobile phones, each brand must has its own set of ratings by a sample of respondents. Each respondent could evaluate only the brands he or she has used or separate samples of respondents could rate each brand. On the basis of collected data discriminant model calculates set of coefficients for each brand separately. The set of coefficients for each brand distinguishes or discriminates the brand among the others. With five brands of mobile phones, computer develops five sets of coefficients. Multiple regression method will develop single set of coefficients for the same problem.

The comparative assessment intends to find the characteristics that distinguish various health-care services based on the ratings of patients. This essentially requires a computational scheme that classifies individual cases into various categories. Mathematically, this is a situation where the IVs are metric and the DV is categorical. DA is naturally suited for such classification problems, for example, to differentiate the categories of health-care services (DV) with a set of metrical rating of questions (IVs) about the health-care services. *DA estimates a linear relationship between categorical health-care services (DV) and linear combinations of one or more metrical ratings (IVs).* DA is capable of handling more than two groups for DV. When DV has more than two categories, the technique is commonly referred to as *multiple discriminant analysis (MDA)* (Hair et al. 2010; Malhotra 2004). Both DA and logistic regression (LG) are appropriate statistical techniques when DV is categorical. LG, or logit analysis, is a specialized form of regression that is formulated to predict and explain a binary (two-group) categorical variable rather than a metric DV.

31.6.2 Discriminant Analysis

DA attempts to find a linear combination of predictor variables that best separate individual cases into prior specified groups (Mika et al. 1999). These combinations, or the variates, are usually referred to as *discriminant functions* (DFs). A typical DF looks like the following:

$$Z_{jk} = a + W_1 X_{1k} + W_2 X_{2k} + \cdots + W_k X_{nk} \tag{31.1}$$

where
   Z_{jk} is the discriminant Z score of DF j for object k
   a is the intercept
   W_i is the discriminant weight for IV i
   X_{ik} is the IV i for object k

DA is widely used in marketing and is relatively new to information services, especially scaling to four groups. This study is an attempt to formulate DF(s) to characterize health-care services, especially mHealth. Hair et al. (2010) provided a detailed step-by-step six-stage *DA decision process* to conduct analysis. A simplified version of this process is presented in Figure 31.4. Schwab (2006) provided a systematic computational procedure to analyze data through DA. For a detailed treatment on DA, readers can refer to these excellent sources (Hair et al. 2010; Malhotra 2004; Schwab 2006).

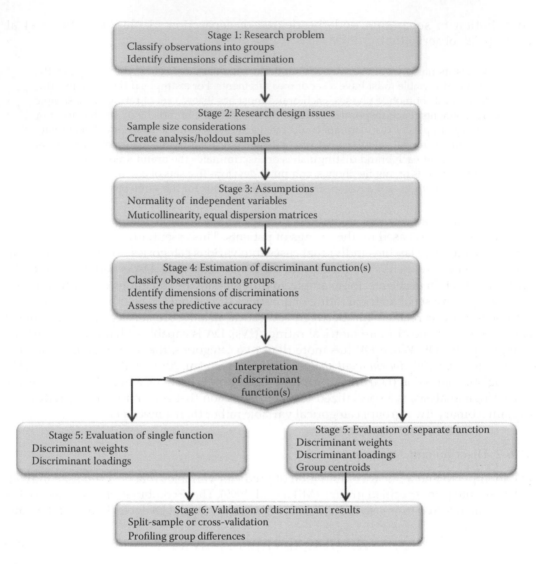

FIGURE 31.4
DA process framework. (Adapted from Hair, J.F.J. et al., *Multivariate Data Analysis, A Global Perspective*, 7th ed., Prentice Hall, Upper Saddle River, NJ, 2010.)

31.6.3 Discriminant Functions

DA combines (weight) the variable scores into a single new composite variable (Equation 31.1), the *discriminant score*. Different weight combinations associated with the variables may produce different scores, meaning different functions. At the end of the DA process, each group will have a normal distribution of discriminant scores. The degree of overlap between the discriminant score distributions (Figure 31.4) can be a measure in determining the model's success in classifying the cases into groups (Hair et al. 2010). As depicted in Figure 31.5, the top two distributions overlap too much, while the bottom one has less overlap. As such, the top model misclassifies too many cases, while the bottom model has minimal misclassification of cases (Hair et al. 2010). Consequently, the bottom model is much better than the top model. Standardizing the variables eliminates scale differences

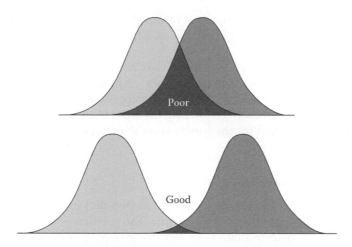

FIGURE 31.5
Discriminant distributions. (Adapted from Hair, J.F.J. et al., *Multivariate Data Analysis, A Global Perspective*, 7th ed., Prentice Hall, Upper Saddle River, NJ, 2010.)

between the variables. And absolute weights can be used to rank variables in terms of their discriminating power; the larger the weight, the most powerful in differentiating the groups.

31.6.4 Model Estimation Procedures Using DA

This section explains, in more detail, the DA method, following the guidelines noted in Schwab (2006). As noted earlier, DA is utilized to find relationships between a categorical (nonmetric) DV and metric or dichotomous IVs. DA attempts to use the IVs to distinguish among the categories or groups of the DV (Hair et al. 2010; Malhotra 2004; McLachlan 1992). The DA model's usefulness is assessed based on its predictive accuracy rate, or the ability to predict the known group memberships. DA computes a new variable called the *DF score*, using eigenvalues, which is used to predict group membership. As shown in Equation 31.1, a DF looks similar to a regression equation where IVs are multiplied with coefficients and summed to produce a score. Discriminant scores are standardized; as such, the sign of the score determines to which group a case belongs to. A DF can be imagined as a boundary between groups. Thus, to distinguish two groups, one statistically significant function is required. Inductively, it means if the DV has *n groups*, $(n - 1)$ significant function(s) are required to distinguish *n* categories. If a DF is able to distinguish among groups, it means that at least one IV must have a strong relationship with it (Hair et al. 2010; Malhotra 2004; Schwab 2006).

Once significant DFs are identified, the next step involves the interpretation of each of these functions. The relationship between IV and DV is interpreted through the way a DF distinguishes groups and the role of an IV in each function. SPSS provides *functions at group centroids* that indicate which groups are separated by which functions.

In order to understand the relationships, DA produces a table called *structure matrix* similar to its counterpart in factor analysis. It identifies the loading or correlation between each IV and each DF. This lets the researcher interpret which variables to interpret for each function based on the loading, and the role of each IV is interpreted on the function it loads most. DA provides two alternatives to analyze the role of IVs on DV groups. The researcher either can simultaneously load all the variables or enter the variables in a stepwise manner. The former approach is referred to as *simultaneous* DA and the latter approach is referred to as *stepwise DA* (Hair et al. 2010; Schwab 2006). The next subsection briefly describes this procedure.

For the case of simultaneous DA, as all the variables are entered together, only those variables whose loadings are 0.30 or higher on one or more of the DFs are interpreted. In the case of stepwise DA, only those variables that meet the statistical test for inclusion are interpreted. A variable can have a high loading on more than one function. In such situations, the variable is interpreted for the function on which it has the highest loading (Hair et al. 2010; Schwab 2006).

31.6.5 Stepwise Discriminant Analysis

This section uses the guidelines for stepwise DA, as detailed in Malhotra (2010). Stepwise DA is analogous to stepwise multiple regression as the predictors entered are sequentially based on their ability to discriminate between the groups. A univariate analysis is conducted for each predictor, treating the groups as a categorical variable, and an *F* ratio is calculated. The predictor with the highest *F* ratio is the first to be included in the DF. A second predictor is added, based on the highest adjusted or partial *F* ratio, taking into account the predictor already selected. Each predictor selected is tested for retention based on its association with other predictors selected. This process is repeated until all predictors meeting the criteria for inclusion and retention are satisfied. Standard computer packages, like SPSS, provide a summary of the predictors entered or removed. The stepwise selection procedure is based on optimizing criteria called the *Mahalanobis procedure*, named after its inventor. The Mahalanobis distance is a generalized measure of the distance between the two closest groups (Malhotra 2010).

31.6.6 DA, Analysis Sample, and Holdout Sample

Another significant difference of DA with other multivariate techniques is that the researcher can subdivide the entire sample into two groups, namely, an analysis sample and a holdout sample. Thus, we can say that the DA gives rise to two possible scenarios:

1. DA model based on the entire sample
2. DA model based on the analysis sample

In the case of *b*, the DA model built with an analysis sample is reapplied to classify the holdout sample, thereby making it feasible to test the predictive power of the model on a dataset that is not used to build the original model. In this case, the *classification matrix* contains an additional section where it shows the performance of the DA model on the holdout sample. In this case, for the sake of simplicity, the author has presented the analysis based on the entire sample. However, an important point to note in either case, DA actually evaluates the performance of the model by recursively testing the model on holding out a case, building the DA model on the remaining cases, and then tests the predictive power of the model. This output is enumerated as a *cross-validated model* in the *classification matrix*.

31.6.7 Selective Applications of DA in Health Care

Lim and Zallocco (1988) studied, for the first time, intersystem competition by analyzing consumer attitudes toward divergent health-care systems, namely, hospitals, home health care, nursing homes, and outpatient clinics. They have applied MDA to identify the distinguishing factors and the most favorite health-care delivery system, that is, home health care. Andaleeb (2000) used DA to study the quality of services provided by public and private hospitals in Bangladesh. He was able to identify the predictors for the patients' choice of hospitals. Andaleeb et al. (2007) compared services of public and private hospitals in Bangladesh and then compared private hospitals with their foreign counterparts, from the perspective of Bangladesh patients. Again, they applied DA to identify the intrasystem characteristics between government and private hospitals and private hospitals and foreign hospitals. Based on their analysis, they concluded that the overall quality of service was better in foreign hospitals than the private hospitals in Bangladesh in all factors, including the *perceived cost* factor.

Kwak et al. (2002) have applied DA for classifying and predicting the symptomatic status of HIV/AIDS patients. They have identified that the total patient admission (AMDTOT) variable is the most significant factor classifying between HIV and AIDS status. They observed that classifying and predicting patients into groups are important in the provision of health care. They also noted that DA may produce better results than the traditional parametric and nonparametric methods. Verdessi et al. (2000) have studied DA in refining customer satisfaction assessment with the objective of turning routine customer information into a more accurate decision-making tool. DA helped in identifying physician care as a significant variable in differentiating the satisfaction of patients. Thus, DA is a generic and powerful technique that enables identification of group differences, as well as finding the critical factors that contribute to group separations.

31.6.8 DA Model Accuracy

DA consists of two stages: (1) DFs are derived in the first stage, and (2) DFs are used in the second stage to classify the cases. DA does compute correlation measures; however, these correlations measure the relationship between IVs and their discriminant scores. DA provides a mechanism to assess the utility of the discriminant model, called *classification accuracy*. Classification accuracy is computed as a ratio of the predicted group membership to the known group membership of respective cases (Hair et al. 2010; Schwab 2006). A typical benchmark for classification accuracy is that it should be at least 25% better than the rate of accuracy achievable by chance alone. Even if IV has no relationship to DV groups, one would still expect a certain percentage of predictive accuracy, and this is referred to as *by chance accuracy*. Thus, to ascertain model usefulness, the researcher needs to assess whether the cross-validated accuracy rate is 25% more than the proportional by chance accuracy rate (Hair et al. 2010; Malhotra 2004; Schwab 2006).

The cross-validated accuracy rate is a one-at-a-time holdout method that classifies each case, based on a discriminant solution for all the other cases in the analysis. It is a more realistic estimate of the accuracy rate because DA may inflate accuracy rates when the cases classified are the same cases used to derive the DFs (Hair et al. 2010; Schwab 2006). DA also has predictive abilities, whereby the model constructed to explain the phenomenon can also be used to classify new cases or predict to which group they belong to. Thus, DA suits the current research objectives of finding factors that differentiate a group of services and develop a predictive model, so as to serve as a policy guideline and also assist in mHealth service design.

31.7 Case Study: Outcome of mHealth Comparison

Motamarri (2013) has conducted an assessment of mHealth over conventional services in Bangladesh, based on the data collected by Asia Pacific Ubiquitous Healthcare Centre (APuHC) at the University of New South Wales, Sydney. The study has been limited to comparative assessment and extraction of service characteristics with an objective to improve health-care services design, with specific reference to mHealth. This research has positively established significant factors that differentiate mHealth vis-à-vis conventional services, PH, GP, and TM, within a developing country context. Figure 31.6 presents the HoQ model for mHealth service design. A detailed technical analysis of the application, extraction of the model, and validation is discussed in Motamarri et al. (2014). The section provides a brief summary of salient information from DA, and for a detailed discussion, the reader is referred to Motamarri et al. (2014).

31.7.1 Dataset

The survey instrument consisted of demographic details and 20 items grouped into four service quality dimensions as shown below (Table 31.3). The prominent health-care delivery platforms in Bangladesh are PH, GP, and TM in addition to the commercial mHealth service offered by *Grameenphone* (Motamarri et al. 2012; Motamarri et al. 2014). The patients are asked to rate the four services with respect to the 20 service elements on a 7-point Likert scale. The sample consisted of four groups pertaining to each health-care service category. Each group contained 50 cases each (25% of total sample); thus, the total sample size is 200.

31.7.2 DA Classification Results

The dataset was systematically analyzed following the DA process depicted in Figure 31.4. The analysis identified three significant DFs. The DF scores applied in classifying the

Attribute bundles	Customer attributes	Importance	Service measures	Customer perceptions			
				PH	GP	TM	mHealth
Ubiquity	Ease	1		2.88	3.34	5.12	6.66
	Accessibility	2		2.70	3.00	3.08	6.28
	Promptness	4		3.30	4.48	5.20	6.42
	Confidence	6		3.12	4.94	3.20	6.26
	Orderliness	9		3.12	4.68	3.22	6.26
	Completeness	11		3.20	5.02	3.76	6.06
Information quality	Up-to-date	5		3.56	5.04	3.12	6.28
	Safety	8		4.08	5.20	3.34	6.20
Value	Cost	3		5.26	2.80	5.58	6.34
	Helpful	7		3.48	5.14	5.24	6.36
	Empathy	10		3.26	4.90	5.02	6.28

(Design characteristics. Service measures: Out of scope of current research.)

FIGURE 31.6
HoQ model for mHealth service design.

TABLE 31.3

Service Quality Dimensions

Perceived Systems Quality	Perceived Interaction Quality	Perceived Information Quality	Perceived Outcome Quality
1. Reliability	1. Helpful	1. Completeness	1. Ease
2. Accessibility	2. Promptness	2. Accurate	2. Convenience
3. Availability	3. Courtesy	3. Up to date	3. Cost
4. Safety	4. Empathy	4. Orderliness	4. Confidence
5. Efficiency			5. Enjoyable
6. Privacy			
7. Usefulness			

TABLE 31.4

Classification Results

	Health Service	Total Cases Cases	%	PH Cases	%	GP Cases	%	TM Cases	%	mHealth Cases	%
					Predicted Group Membership						
Original	PH	50	25.0	35	70.0	8	16.0	4	8.0	3	6.0
	GP	50	25.0	11	22.0	36	72.0	1	2.0	2	4.0
	TM	50	25.0	5	10.0	3	6.0	35	70.0	7	14.0
	mHealth	50	25.0	0	0.0	0	0.0	1	2.0	49	98.0
		200	100								
Cross-validated	PH	50	25.0	33	66.0	9	18.0	5	10.0	3	6.0
	GP	50	25.0	11	22.0	36	72.0	1	2.0	2	4.0
	TM	50	25.0	5	10.0	4	8.0	34	68.0	7	14.0
	mHealth	50	25.0	0	0.0	0	0.0	2	4.0	48	96.0
		200	100								

Source: Modified from Ramaswamy, R., *Design and Management of Service Processes: Keeping Customers for Life,* Addison-Wesley Publishing Company, Boston, MA, 1996.

original dataset and output of DA are summarized as classification results shown in Table 31.4. The table shows that the DFs had successfully classified 77.5% of the original cases accurately. Table 31.4 also shows that the model achieved a cross-validation accuracy of 75.5%. These accuracies shall be better than the baseline accuracy needed. As the dataset has four equal groups, the by chance classification accuracy is 25%. In order to ascertain the validity of the model, the classification accuracy shall be 25% more than by chance accuracy, meaning it is 31.25%. Thus, the DFs' performance was far better than the required baseline accuracy.

31.7.3 DA Interpretation of Discriminant Functions

Having established the validity of the DFs and their ability in classifying cases, the next stage of DA was to interpret the DFs and thereby assign meaningful names to them. To achieve this, DA provided standardized canonical DF coefficients and the structure matrix as shown in Table 31.5. These statistics summarized the final set of factors entered into the

TABLE 31.5

Standardized Canonical DF Coefficients and Structure Matrix

	Function					
	DF Coefficients			Structure Matrix		
Variables	1 Ubiquity	2 Information Quality	3 Value	1 Ubiquity	2 Information Quality	3 Value
Ease	0.394	−0.399	−0.047	0.728*	−0.290	−0.160
Accessibility	0.525	0.145	0.187	0.703*	0.166	0.036
Promptness	0.306	−0.084	−0.349	0.584*	−0.143	−0.393
Confidence	0.270	0.631	−0.125	0.511*	0.458	−0.371
Orderliness	−0.162	0.512	0.220	0.492*	0.382	−0.278
Completeness	−0.374	−0.360	−0.296	0.409*	0.289	−0.393
Up to date	0.497	0.329	0.081	0.470	0.529*	−0.256
Safety	−0.119	0.376	0.378	0.362	0.495*	−0.148
Cost	0.246	−0.224	0.795	0.396	−0.359	0.629*
Helpful	0.033	−0.424	−0.342	0.455	−0.034	−0.504*
Empathy	−0.093	−0.428	−0.322	0.465	−0.027	−0.469*

Source: Modified from Hair, J.F.J. et al., *Multivariate Data Analysis, A Global Perspective*, 7th ed., Prentice Hall, Upper Saddle River, NJ, 2010.

* Largest correlation between each variable and any discriminant function.

DF model and their relationships. The canonical correlation coefficients measured the association between the DFs and the selected factors. The structure matrix provided the important information about the factors and their loadings on each DF. The factor loadings assisted in identifying the most influential factors associated with each function. This in turn helped in assigning a meaningful name to each DF (Hair et al. 2010; Malhotra 2004; Schwab 2006). The DA identified 11 variables (IVs). To improve the readability, the variables were grouped in order of their association with the DF and then were sorted by their influential loading.

The DFs one to three are named as *ubiquity, information quality, and value*, respectively. These three dimensions consisted of 11 variables out of the 20 variables as shown in Table 31.3. As shown in Table 31.5, ubiquity consists of the variables *ease, accessibility, promptness, confidence, orderliness*, and *completeness*. The variables *update-to-date and safety* form as information-quality dimension. Finally, the value dimension consists of the variables *cost, helpful, and empathy*. Thus, DA assisted in identifying the service characteristics that distinguish one health-care service from the other.

31.7.4 DA Computation of House of Quality

DFs and their corresponding dimensions constitute as the *customer assessment of competing healthcare services* as shown in Figure 31.1. However, we moved a step forward to propagating these important outputs into various cells of the HoQ matrix. And the resultant matrix is shown in Figure 31.6. DFs serve as the attribute bundles and the respective dimensions as customer attributes. Mean ratings of all these 11 variables are computed, and they form as the customer perceptions or ratings of each characteristic versus the respective health-care service. Figure 31.7 is the graphical extension usually pinned to the right of the HoQ matrix of Figure 31.6. Now it is apparent from Figure 31.7 that mHealth in all means surpassed in performance against its competition (i.e., PH, GP, and TM).

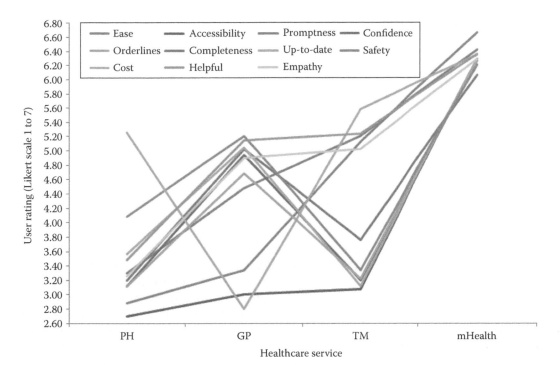

FIGURE 31.7

HoQ model for mHealth service design–service characteristics. Note: PH, public hospital ; GP, general practitioner; TM, traditional medicine; mHealth, mobile health.

31.8 Future Directions

Future research will focus on enhancing the current HoQ model with the computation of performance targets (area earmarked as out of scope in Figure 31.6), and then cascade these outcomes toward the development of process characteristics, subprocess characteristics, and function specifications and function targets. The lowest level of the QFD matrix in essence is the operational characteristics and operational targets.

The outcome of the DA model for health-care services can also be viewed as the patients' perception of service operation. The three dimensions and the associated 11 factors are to be ingrained in *services design* in order to improve *services operation*. ITIL is developed by the United Kingdom's Office of Government Commerce (OGC) as a response to systematically execute services management in a five-phase model. ITIL is the de facto industry standard for IT services management (OGC 2007). It is possible to integrate higher phases of QFD matrices to the ITIL operational framework, in order to aid in the *health-care services design* as well as *operation*.

With the burgeoning penetration of mobile phones, the offering of cost-effective medical services through mobile phones has naturally been viewed positively by patients as the conventional services lack capacity and attitude to improvise the services environment (Andaleeb 2001; Ivatury et al. 2009; Mechael 2009). A well-designed and well-delivered service will naturally generate better satisfaction for a consumer. As noted earlier, consumers generally opt for the services that are easy to use, accessible, promptly served, organized, and fulfill satisfactorily their original need.

We thus believe that service design is an important antecedent to achieve well-performing health-care services. We hope that service providers and planning agencies will progressively work toward addressing the existing shortcomings with their services portfolio and also recognize the complementary nature of mHealth in achieving better health care for all.

31.9 Summary

mHealth, in comparison to other existing health-care services, is much easier to use. The patient or his or her care provider has to simply dial a prescribed number from his or her mobile phone or from a designated community phone. From the IS perspective, the user acceptance (UAT) models empirically ascertain that usefulness and accessibility will influence the acceptance of IT (Davis 1989; Venkatesh et al. 2003). In terms of *ubiquity*, to avail the service of PH, GP, or TM, a patient has to make a physical trip to these places. Furthermore, the trip is only meaningful during the operating hours of the provider, while the mHealth helpline can be reached from a location convenient to the patient, and that too at the very moment, he or she requires the service, that is, it could be midnight, a weekend, or on a holiday (Akter and Ray 2010). *QFD* and *HoQ* have been employed in manufacturing, not only to echo the *voice of customer* across the organization but also to address the mounting cost pressures. This discussion made a preliminary attempt to compute the *comparative analysis of healthcare services* or *Room-6 of HoQ* (Figure 31.1) and extracted *service characteristics*, a key element to move forward with health-care services design.

Questions

1. Who popularized QFD and HoQ?
2. How many rooms are there in the standard HoQ matrix and name each of them?
3. What is the core philosophy of QFD/HoQ in terms of services design?
4. Were there any comparative assessments of mHealth vis-à-vis conventional health-care delivery alternatives?
5. How are DA and HoQ related?

References

AIHW. 2012. *Health Expenditure Australia 2010–11*, AIHW, Canberra, ACT, Australia, pp. 183.

Akter, S. and Ray, P. 2010. *mHealth—An Ultimate Platform to Serve the Unserved*, Schattauer, Stuttgart, Germany.

Andaleeb, S. S. March 1, 2000. Public and private hospitals in Bangladesh: Service quality and predictors of hospital choice, *Health Policy and Planning* (15:1), 95–102.

Andaleeb, S. S. 2001. Service quality perceptions and patient satisfaction: A study of hospitals in a developing country, *Social Science & Medicine* (52:9), 1359–1370.

Andaleeb, S. S., Siddiqui, N., and Khandakar, S. July 1, 2007. "Patient satisfaction with health services in Bangladesh," *Health Policy and Planning* (22:4), 263–273.

Chan, L.-K. and Wu, M.-L. 2002a. Quality function deployment: A comprehensive review of its concepts and methods, *Quality Engineering* (15:1), 23–35.

Chan, L.-K. and Wu, M.-L. 2002b. Quality function deployment: A literature review, *European Journal of Operational Research* (143:3), 463–497.

Chan, L.-K. and Wu, M.-L. 2005. A systematic approach to quality function deployment with a full illustrative example, *Omega* (33:2), 119–139.

Charteris, W. 1993. Quality function deployment: A quality engineering technology for the food industry, *International Journal of Dairy Technology* (46:1), 12–21.

Clausing, D. P. 1994. *Total Quality Deployment: A Step-by-Step Guide to World-Class Concurrent Engineering*, ASME Press, New York.

Cohen, L. 1995. *Quality Function Deployment: How to Make QFD Work for You*, Addison-Wesley, Boston, MA.

Davis, F. D. 1989. Perceived usefulness, perceived ease of use, and user acceptance of information technology, *MIS Quarterly: Management Information Systems* (13:3), 319–339.

Dieter, G. and Schmidt, L. 2008. *Engineering Design*, 4th ed., McGraw-Hill, New York.

The Economist. June 2, 2012. Squeezing out the doctor, http://www.economist.com/node/21556227

Fisher, R. A. 1936. The use of multiple measurements in taxonomic problems, *Annals of Human Genetics* (7:2), 179–188.

Gale, S. F. 2012. The human element, *PM Network*, (26:12), 26–32.

Hair, J. F. J., Black, W. C., Babin, B. J., and Andersoon, R. E. 2010. *Multivariate Data Analysis, A Global Perspective*, 7th ed., Prentice Hall, Upper Saddle River, NJ.

Hauser, J. R. 1993. How Puritan-Bennett used the house of quality, *Sloan Management Review* (34:3), 61–70.

Hauser, J. R. and Clausing, D. 1988. The house of quality, *Harvard Business Review* (May–June), 63–73.

Hwang, J., and Christensen, C. M. 2008. "Disruptive Innovation In Health Care Delivery: A Framework For Business-Model Innovation," *Health Affairs* (27:5), 1329–1335.

IOM. 2010. *The Healthcare Imperative: Lowering Costs and Improving Outcomes: Workshop Series Summary*, The National Academic Press, Washington, DC.

Ivatury, G., Moore, J., and Bloch, A. 2009. A doctor in your pocket: Health hotlines in developing countries, *Innovations: Technology, Governance, Globalization* (4:1), 119–153.

Kwak, N. K., Kim, S. H., Lee, C. W., and Choi, T. S. 2002. An application of linear programming discriminant analysis to classifying and predicting the symptomatic status of HIV/AIDS patients, *Journal of Medical Systems* (26:5), 427–438.

Laudon, K. C. and Laudon, J. P. 2013. *Management Information Systems Managing the Digital Firm*, 13th ed., Pearson Education Limited, Essex, England.

Lewis, T., Synowiec, C., Lagomarsino, G., and Schweitzer, J. May 1, 2012. E-health in low- and middle-income countries: Findings from the Center for Health Market Innovations, *Bulletin of the World Health Organization* (90:5), 332–340.

Lim, J. S. and Zallocco, R. June, 1988. Determinant attributes in formulation of attitudes toward four health care systems, *Journal of Health Care Marketing* (8:2), 25–30.

Malhotra, N. 2004. *Marketing Research: An Applied Orientation*, 4th ed., Pearson Education, Upper Saddle River, NJ.

Malhotra, N. 2010. *Marketing Research: An Applied Orientation, 6/E*, 6th ed., Pearson Education, Upper Saddle River, NJ.

Mazur, G. H. 1993. QFD for service industries—From voice of customer to task deployment, in *The Fifth Symposium on Quality Function Deployment*, June 20–22, 1993, Novi, MI.

McLachlan, G. J. 1992. *Discriminant Analysis and Statistical Pattern Recognition*, Wiley, New York.

Mechael, P. N. 2009. The case for mHealth in developing countries, *Innovations: Technology, Governance, Globalization* (4:1), 153–168.

Mika, S., Ratsch, G., Weston, J., Scholkopf, B., and Mullers, K. R. 1999. Fisher discriminant analysis with kernels, *Neural Networks for Signal Processing IX, 1999. Proceedings of the 1999 IEEE Signal Processing Society Workshop*, August 23–25, 1999, Madison, WI, pp. 41–48.

Mizuno, S. and Akao, Y. 1978. *Quality Function Deployment: A Company Wide Quality Approach*, JUSE Press, Japan.

Motamarri, S. 2013a. *Distinguishing mHealth from Other Health Care Alternatives in Developing Countries: A Study on Service Characteristics*, The University of New South Wales, Sydney, NSW, Australia.

Motamarri, S. 2013b. A six cell services comparison model for healthcare, in *Third Australasian Symposium on Service Research and Innovation (ASSRI'13)*, University of New South Wales, Sydney, NSW, Australia, pp. 573.

Motamarri, S., Akter, S., Ray, P., and Tseng, C.-L. 2012. mHealth: A better alternative for healthcare in developing countries, in *PACIS-2012 Proceedings*, July 11–15, 2012, Ho Chi Minh City, Vietnam, Paper 29.

Motamarri, S., Akter, S., Ray, P., and Tseng, C.-L. 2014. Distinguishing 'mHealth' from other healthcare services in a developing country: A study from the service quality perspective, *Communications of the Association for Information Systems* (34:1), 669–690.

OGC. 2007. ITIL, Version 3, Office of Government Commerce, The Stationery Office, London, U.K.

PC. 2011. *PC Update: Caring for Older Australians*, Productivity Commission, Canberra, Australia, pp. 32.

Ramaswamy, R. 1996. *Design and Management of Service Processes: Keeping Customers for Life*, Addison-Wesley Publishing Company, Boston, MA.

Ray, P. K. 2003. *Integrated Management from E-Business Perspective: Concepts, Architectures and Methodologies*, Kluwer, New York.

Savic, M., Brcanov, D., and Dakic, S. 2008. "Discriminant analysis-Applications and Software Support," *Management Information Systems* (3:1), 029–033.

Schwab, J. A. 2006. "Discriminant Analysis–Basic Relationships," University of Texas, Austin. http://www.utexas.edu/courses/schwab/sw388r7_spring_2006/SolvingProblems/0_SolvingHomeworkProblems_spring2006.htm

Shahin, A. 2008. Quality function deployment: A comprehensive review, in *Total Quality Management— Contemporary Perspectives and Cases*, T. P. Rajmanohar, ed., ICFAI University Press, Isfahan, Iran, pp. 47–79.

Siddiqui, N. and Khandaker, S. A. 2007. Comparison of services of public, private and foreign hospitals from the perspective of Bangladeshi patients, *Journal of Health Population and Nutrition* (25:2) June, 221–230.

Venkatesh, V., Morris, M. G., Davis, G. B., and Davis, F. D. 2003. User acceptance of information technology: Toward a unified view, *MIS Quarterly: Management Information Systems* (27:3), 425–478.

Verdessi, B., Jara, G., Fuentes, R., Gonzalez, J., Espejo, F., and de Azevedo, A. C. 2000. The role of discriminant analysis in the refinement of customer satisfaction assessment, *Revista de Saúde Pública* (34), 623–630.

World Health Organization. 2012. *World Health Statistics 2012*, World Health Organization, Geneva, Switzerland, pp. 180.

Wulan, M. 2011. MAE Design Model: QFD. http://www.qfdonline.com/.

32

Economic Approaches to Assessing Benefits of mHealth Technology: A Guide for Decision and Policy Makers

Thomas Martin and Paul Solano

CONTENTS

32.1 Introduction to Assessing Cost in mHealth

mHealth is often presented as a low-cost option for communication between patients and medical care providers, including the sharing of data, also referred to as patient health information (PHI). This communication is conducted with mobile technologies (mobile phones, personal digital assistants [PDAs], laptops, radio-frequency identification devices, etc.). The integration of these technological devices into medical care practice to achieve broad population health benefits and the potential to decrease or mitigate rising health-care costs is considered by some health-care specialists to be increasing over the near future (Dobkin and Dorsch 2011). The economics of mobility and mHealth is increasingly an item of interest, since specifically health-care organizations are seeing material productivity gains and/or tangible savings for their mobile enterprise investments (Galvin 2011).

The perceived advantages of mobile technologies rest on their design features of portability (anywhere), immediacy (any time), and convenience (easy access); these features, together with their comparatively low unit cost to the user, account for their pervasiveness in society (Norris et al. 2009). Besides these features, the utility and efficacy of mHealth devices related to health care lie in their informational capabilities that are provided to medical practitioners as other users. mHealth devices are vehicles for the dissemination of information of different types and content that not only yield medical care knowledge but also information for the diagnoses and prescription of medical treatment. Devices can be differentiated by their different purposes that result in variance in content, type, and level of information among devices. More pertinently, devices designed for the same decision-making purpose(s) can and do also vary by types of medical health information and informational capability embedded in the separate devices.

In addition to their design features, medical care users are expected to choose an mHealth device because of its purpose and the informational capabilities provided (Ryan 2004). Given that devices for the same purpose(s) vary by features and informational capabilities, users must trade off these dimensions when they select a device. However, users' valuation of any mHealth device can be understood within the context that a device is an intermediary instrument for health-care delivery. More specifically, it is very likely that users' preference and valuation of a device are based on how the device facilitates patient care and contributes to improved patient health outcomes that are the resulting benefits of an mHealth technology. Potential users for consideration in the assessments include the perspectives of providers or patients.

One of the advantages of the methods discussed in this chapter includes the ability for the analyst to broadly evaluate various technologies in the mHealth space. These valuations can be conducted across a number of diverse products within the mHealth space that include health and wellness offerings and the use of mobile technology to augment other critical health-care infrastructure(s). Of course, a user's valuation and choice of devices would be motivated not only by the potential benefits that can be derived but also of the costs that would be incurred for purchasing devices.

The latter issue points to the role of economic drivers that could foster the acquisition of mHealth technologies. Understanding the underlying economic drivers of mHealth represents a fundamental basis of transitioning mHealth technology from emerging disruptive technology into common practice. The goal is to create a broad knowledge base that fosters acceptance. This understanding involves consideration of several issues that would determine the adoption of mHealth technology. A need is present in the mHealth environment to appreciate incentive structures and business models that propagate the use of mHealth. Decision makers in organizational and governmental roles must also be concerned with the expected costs as well as the potential gain from mHealth technology, a result that can be appraised by quantifying the benefits of such technology. While mHealth has the potential to overcome traditional obstacles to the delivery of health services to the poor in lower- and middle-income countries—issues related to access, quality, time, and resources—there is little evidence as to whether the expected benefits and savings can be actualized on a large scale (Schweitzer and Synowiec 2012). One method to assess and provide evidence on benefits is by conducting cost-benefit analysis (CBA). By using CBA, an ancillary advantage of employing existing economic methods in mHealth, is that we work towards establishing a body of knowledge that can serve as a persuasive instrument for mHealth analysts and decision makers.

Several major economic methods are available as evaluation tools for appraising decision-making alternatives, but only a few can be employed to assess and provide evidence on expected benefits of mHealth devices. The purpose of this chapter is to review various available economic evaluation methods for their applicability as assessment tools of the economic impact of mHealth technologies. The central focus is upon two different economic "stated preferences" approaches for conducting CBA, that is, contingent valuation analysis (CVA) and conjoint analysis (CA), and a brief consideration of comparative effectiveness research (CER), cost-effectiveness analysis (CEA), cost utility analysis (CUA), and cost-minimization analysis (CMA). The two CBA methods can facilitate the assessment of the valuation that organizational managerial decision makers and actual technology users assign to mHealth devices as potential consumers of such technology. In particular, CVA and CA can ascertain consumers' preferences for mHealth technology as well as aid in the determination of costs and appraisal of the role of economic incentives for the purchase of mHealth-focused services. By gaining an understanding of consumers' valuations and thus their preferences associated with particular mHealth attributes, decision makers can formulate strategies to address areas of opportunity to expand the overall adoption of mHealth.

32.2 Cost-Benefit Analysis: An Overview

As viewed in the economics literature, CBA is a comprehensive research method *intended* for the assessment of whether a proposed public policy, program, or project would enhance societal welfare. This objective is achieved by comparing (1) the value of all consequences for societal members impacted by the initiative that are quantified in monetary terms to measure benefits with (2) all the financial cost incurred as well as the social (nonmarket) costs associated with the initiative (Boardman 2010). If benefits are greater than costs, then society realizes a gain in economic efficiency, or within the context of mHealth utilization, a net improvement in health care. While designed for public policy, CBA has applicability and can be expanded to include approaches to decision making in various for-profit arenas. In a very general manner, studies in the for-profit sector would focus on increased productivity or aid in determining consumer preference to maximize firm profitability.

Despite their differences in goal maximization, a fundamental CBA concept common to both private sector and public sector decision makers and analysts is an understanding of market demand for mHealth devices in addition to the need to generate welfare gain or profits. Market demand—or drivers associated with the establishment of cost/pricing—for an mHealth device is derived from four factors: (1) the financial cost of the device to the buyer/use that would be equal to the price paid offset by any financial subsidies received by the buyer, (2) the actual and desirable attributes of a device that is consistent with potential users' preferences, (3) potential purchasers' income that indicates the ability to pay for a device, and (4) the value that the user could yield in health-care improvements with the device. Together all four factors stimulate a consumer's willingness to pay (WTP) for the technology. An individual's WTP is the monetary amount (or actual net price paid) that a potential user would give up to purchase a device. When sales and purchases of mHealth are realized through the understanding of consumers' demand, social value in the form of societal welfare enhancement and private sector profit could be produced. In the next fol-lowing subsections, we outline approaches employed in CBA to measure WTP from which consumer demand can be derived.

32.2.1 Contingent Valuation Analysis

One approach to assess WTP in CBA research is CVA. The number of health-care contin-gent valuation studies has grown rapidly with the majority undertaken in the context of CBA (1994). In many contingent valuation studies, WTP methods have been applied in health-care settings to investigate the demand of various initiatives or services so as to further the understanding of their pricing.

With CVA, surveys have become the major tool in assigning monetary values (through price determination) on nonmarket goods and amenities that are traditionally difficult to quantify (Howe et al. 1994). Historical examples include applications in the environmental economics area where assessment of improvements in welfare are challenging to quantify. For contingent valuation surveys, a random sample is drawn from among the population of potential consumers, and respondents are asked to consider a hypothetical scenario where a demand exists for the outcomes and concomitantly the benefits of the public pro-gram are evaluated (Diener et al. 1998).

The scenario is comprised of a description of two very pivotal dimensions that are used to elicit respondents' WTP to acquire the good to be evaluated. First, the respondents are constrained to appraising a particular good—for example, a specific form factor of mHealth, be it devices or software services—with a set of particular characteristics and informational capabilities (attributes). For example, an organization is considering a move-ment away from pagers or other simple message service (SMS) into smartphone-based communication. The developing business unit or organization is concerned with offset-ting the cost of development of such a tool or platform. The organization could undertake a study to determine both WTP for the platform and the willingness to accept for trade-offs in functionality when moving from SMS to native smartphone application–based technology.

Second, the payment vehicle for purchasing/obtaining the good is described to the respondent. The vehicle specifies how the respondents' cost for the particular mHealth device is to be paid. In CBA studies, differences in payment vehicles used for the same have resulted in differences in individuals' WTP (Boardman 2010). For instance, payment vehicles could vary from 100% out-of-pocket costs by respondents to various ranges in subsidies from their employer. There can be multiple elicitations of respondents' WTP for the described good by varying the payment vehicle and therefore changing the financial burden of the respondent. The resulting respondents' WTP monetary estimates can be used by (1) a for-profit firm to compare the respondents' price valuation with the firm's cost and potential/actual market price and (2) public entities to compare with the trans-formed into benefits of a device (transformed/calculated for the WTP estimates) that can be compared with either the financial or social cost or both produced by program implementation.

There are several types of contingent valuation models. All seek to elicit an individu-al's price preference for the determination of WTP for a good or service through differ-ent types of questions that can be deployed in the CVA survey. Price can be elicited by directly seeking an individual respondent's maximum WTP; indirect models extrap-olate the WTP by constructing a demand curve from separate (but not maximum) valuations of all the individual respondents. For a more in-depth discussion of methods for elicitation, readers are directed to texts by Drummond (2010) and Boardman (2010).

One direct method is an open-ended question where the respondents are asked their maximum WTP (price) for the mHealth device or service. In the example survey below, the price for smartphone applications is presented and the respondent is directly asked to provide a stated maximum price (see insert below) for a good or service. For some generally large-scale and costly goods, there is concern with this type of question that respondents would provide very large WTP estimates because of the absence of a reference point for valuation; consequently, some studies have provided a sequential list (or ladder) of ascending (or descending) order of incremental price values to anchor the respondents' estimates. A second direct method is a closed-ended iterative bidding. Individual respondents are asked whether they would pay a specified amount for the good (a device). If yes (no) is the response, then increasing (decreasing) amounts are given in follow-up questions until a negative (positive) response is given to reveal the maximum WTP. A third form of direct elicitation is the contingent ranking method. The respondents are asked to rank combinations of the quantity or quality of the good (e.g., different attributes of a device) and the monetary payment for each mixture; afterward, the analyst must assign WTP values for the ordinal rankings of the combinations.

The referendum method is an indirect approach to the elicitation of the WTP of respondents. With the referendum method, respondents are subjected to a yes/no (of dichotomous choice) question focused on whether they are willing to pay a particular price to obtain a good or policy (Boardman 2010), and no follow-up questions are applied. The answers are viewed by the analyst as bid prices from which a resulting demand curve of WTP prices among all respondents is derived. These types of questions are appropriate for assessing an individual's willingness to accept complements or substitutes. For example, the referendum model could focus on revealing a preference for incentives that cover the cost of developing apps by clinicians' providers or managers within an organization. Organizations could also assess a preference for incentives to cover the cost of tablets versus smartphones by clinicians, providers, or managers within an organization. Some large hospital organizations could assess certain business units for the purchase or use of mobile technologies.

For the open-ended question and the referendum approaches, mail and internet surveys can be employed; the other two cited methods require the interaction of the researcher with the respondent, a process that increases the financial cost of the investigation. Overall, there is wide variation among health-care contingent valuation studies in terms of the types of questions being posed and the elicitation formats being used. Therefore, the classification and appraisal of the literature are difficult because reporting of methods and researcher relationship with the conceptual framework of CBA is quite broad (Diener et al. 1998).

As a method, CVA is useful to frame a valuation study (Boardman 2010). This is true for applications in mHealth and as potential methods to assess apps for purchase and innovative business models in the digital health space where consumer preference is of high importance. Finally, there is an emerging practice of utilizing advanced statistical models with combined mixed methods for the analysis of contingent valuation surveys. Analysts interested in statistical methods are directed to Howe et al. (1994) and Pankratz et al. (2002). For an example of open-ended WTP question, see the following box.

Example of direct elicitation of preference: What is the maximum price you are willing to pay to purchase and download a health-care-related app? (Please answer in $USD amounts. What is the maximum price you are willing to pay for an app? Please answer in numerical format only 0.00.)

32.2.2 Conjoint Analysis in mHealth

CA is another survey approach that employs hypothetical market situations to determine individuals' WTP for mHealth technology. CA is based on the premise that any good or service can be described by its characteristics (or attributes) and that the extent to which an individual values a good or service depends on the levels of these characteristics (Ryan and Farrar 2000). More specifically, CA allows the comparison of the two different alternatives that have the same objective with similar attributes, each of which vary in degrees across the alternatives. For example, two different mHealth devices that have the same purpose and the same attributes/characteristics would differ in the scope of their capability and properties, for example, access to information in a clinical context, complexity and ease of download, more and less extensive informational content on a topic, and range of cost (Ryan and Farrar 2000).

Most CAs in health-care studies have utilized the discrete choice method to determine survey respondents' WTP for a good/service between two alternatives (e.g., devices) with the same purposes based on the differences in attribute levels. Respondents are presented with two choices and asked to choose their preferred good or service (Ryan and Farrar 2000). The choices are evaluated by advanced statistical methods that yield WTP estimates of each attribute from which total WTP can be calculated for each choice. The power of CA lies in its ability to find out (1) the absolute monetary value respondents place on an attribute and thus allow determination of the relative importance of each attribute as well as (2) trade-offs potential consumers are accepting every time they make decisions (Steblea et al. 2009). As with the WTP of contingent valuation(s), the WTP generated from CA can be compared to both the financial costs and social costs associated with the devices.

Some methodological variations prevail for conjoint evaluations. A number of elicitation methods exist that include the direct methods and the indirect methods. It can be argued that given the recent emergence of mHealth applications and the limited research in the area, it is preferential to use an open-ended format that allows respondents to indicate directly their maximum WTP to receive or purchase a commodity (Ratcliffe 2000). However, the discrete choice method may also be advantageous. Discrete choice experiments (DCEs) are regularly used in health economics to elicit preferences for health-care products and programs (Lancsar and Louviere 2008). These methods can, therefore, inform the service design process (Cunningham et al. 2010), which is a key barrier to adoption of mobile technologies in health care or a challenge in articulating business models to investors. Nevertheless, in designing conjoint studies, analysts should consider multiple methods of elicitation of preference, especially with the broad and complex business models associated with mHealth. For example, respondents could be asked to evaluate trade-offs between "onetime download fees" and other business models such as the licensing of software. See the following box for trade-offs based on inclusion of mHealth as a component of the Meaningful Use program underway in the United States.

Series of questions employing CA assessing impacts change in policy:

Q1. Are you aware of the Meaningful Use program for electronic health records?

- Yes
- No

Q2. Would you support the inclusion of mobile apps as a component of Meaningful Use as an integral approach to the provision of care?

- Yes
- No
- No opinion
- No knowledge of the impact

Q3. If providers were reimbursed for the use of apps in care delivery, as a component of Meaningful Use, would you pay more for an app?

- Yes
- No

Q4. How much more, based on your original payment statement, would you be willing to pay for an app, if it was a Meaningful Use objective? (In $USD; for example, if you responded $5 in your original statement and would now be willing to pay $7, answer $7. Please answer in numerical format only 0.00.)

32.2.3 Study Design and Limitations

There are trade-offs to the use of CVA and CA in the estimation of the value revealed preferences expressed in the form of WTP. There is a well-documented limitation to the use of CVA and CA methods that impinges on the reliability of the study through the interjection of bias (Goldar and Misra 2001; Boardman 2010; Drummond 2010). Research suggests that, when asked to place a monetary value on something people currently do not have to pay for, they tend to overestimate the amount they would be willing to pay (Griffin 2011). The potential to overestimate the true WTP represents the opportunity to interject bias into a traditional WTP study. WTP studies are routinely criticized as biased as respondents may overestimate WTP (Goldar and Misra 2001). An advantage of CBA in comparison with other evaluation techniques such as CEA is that costs and benefits for a program can be compared directly because they are both in money units; the obvious disadvantage is the difficulty inherent in obtaining valid and reliable estimates of WTP for CVA and CA (Diener et al. 1998). From an assessment of conducting CBA associated with mHealth, obtaining estimates of economic benefits is challenging at this stage.

To increase the reliability of a study, a mixed methods approach could be utilized, combining structured interviews with a large-scale collection of data to improve the accuracy of results. CA especially poses some problems in determining appropriate attributes to assess. A possible reason for respondents' reluctance to choose is that some of the trade-offs do not have a logical pathway between their components (Wainwright 2003). This ability to fully understand a survey is called "cognitive effort." The cognitive effort of a survey model for mHealth is assessed by conducting a pilot survey to assess understanding of questions presented to the respondent. Directly asking for WTPs especially for complex and unfamiliar goods is a cognitively challenging task for respondents (Brown et al. 1996), making the piloting of large surveys with complex conjoint or contingent questions associated with mHealth a critical step in the design of the survey tool.

32.2.4 Summary of Contingent Valuation and Conjoint Analysis in mHealth

Both CVA and CA studies encompass CBA. Both methods are useful for decision makers and policy makers when assessing mHealth. CVA and CA studies provide information related to consumer preference that is important when undertaking program or project implementation; as such, they can assist decision makers in the consideration to adopt particular mHealth technology(s). An example of potential applications of CV and CA studies includes assessing trade-offs between mHealth technologies of similar or differing qualities.

32.3 Comparative Effectiveness Research: Role and Challenges of CER in the Field of mHealth

The principal goal of CER is to allow decision makers (patients, clinicians, purchasers, and policy makers) to make informed decisions on specific health practices (Sox et al. 2010). CER requires the development, expansion, and use of a variety of data sources and methods to conduct timely and relevant research and disseminate the results in a form that is quickly usable by clinicians, patients, policy makers, and health plans and other payers (AHRQ). A major gap persists between the best available information on therapeutic effectiveness and safety (Avorn and Fischer 2010) especially in the emerging field of mHealth. Technology and associated applications in health care are becoming major sources to fill the gap in moving research findings from paper to the bedside. At the same time, the clinical safety of these interventions is also coming under scrutiny.

In the United States, CER was given a boost by the Patient Protection and Affordable Care Act (PPACA) in 2010 and the American Recovery and Reinvestment Act in 2009. The U.S. Agency for Healthcare Research and Quality (AHRQ) plans to involve hospitals in conducting studies on patients of diverse population segments, and comparative effectiveness is one method available to researchers to assess differences in outcomes related to direct cost (Aston 2010). Overseas, CER studies and findings are widely utilized in single-payer systems or domestically in the United States within integrated delivery networks (IDNs). In general, the current thrust of CER articulated by government policy is to improve decision making (Levy et al. 2009). Some of the challenges associated with CER research includes the likelihood that providers or regulating bodies will adopt practices shown to be efficacious while saving costs (Levy et al. 2009; Aston 2010; Sox et al. 2010). CER includes building a body of evidence consisting of prospective randomized trials including pragmatic trials and observational research using data obtained in the course of regular practice (Sox et al. 2010). However, there is a challenge with disseminating CER findings. What will be the role of CER studies in assessing and amassing a body of knowledge around the field of mHealth?

Technology is often overlooked in the delivery of CER study results to the point of care or bedside. The increased use of "accessory technology" such as electronic health records (EHRs) and mobile technologies is opening the gates for more "on-demand" information at the bedside. Multiple industry surveys point to providers increasingly accessing medical reference apps or other decision-making platforms from mobile devices. Arguably mHealth could play an important role in delivering information to decision makers on courses of action. Furthermore, mHealth will be the subject of CER studies in the delivery of care or information to patients covered elsewhere in this book.

Examples of CER-focused research will evolve over time. Medication Adherence, "App Cohorts," and other studies will aid in the adoption of cost-effective and high-quality treatments. However, a number of fundamental or foundational approaches to conducting CER research in the field of mHealth are needed to aid the formation of a robust evidence base.

32.3.1 Foundations for CER Studies with mHealth Applications

Since CER studies vary widely across established methods, understanding and improving CER design at a fundamental level can advance the adoption of mHealth. As mHealth evolves, the assessment of outcome studies will increase in profile for decision makers. A number of potential considerations are provided for analysts or researchers designing CER studies for mHealth-based interventions:

1. First steps should be to evaluate acceptance. The technology acceptance model was initially developed by Davis et al. (1989) and warrants additional use to assess mHealth. Enhanced acceptance and positive attitudes toward mHealth technologies will aid in creating a backbone of research for additional and more rigorous studies.

2. mHealth provides a platform to collect both social and quantitative observations. The mobile device is unique in its abilities to interact with the user by sending prompts and notifications to collect information. Designing an intervention that collects both quantitative and qualitative data sets will yield improved findings. This also aids in establishing a basis for the assessment of cost ex post. Furthermore, mobile number portability in many countries helps to minimize loss to follow-up. In addition, devices are becoming increasingly secure. However, privacy and security laws and regulations differ from country to country, which could inhibit collection on a national scale.

3. Assess engagement and participation from provider and patient perspectives. Adherence studies—compliance with medical regimen(s)—are improved with engagement by both providers and patients (Vollmer et al. 2011). Applications of mHealth CER studies include adherence strategies. Research methods should seek to include perspectives of both providers and patients as the transition from "fee for service" to "payment for quality" occurs.

4. Consider mobile devices for the dissemination of findings including considerations for mHealth-enabled repositories serving query retrieve functions. One likely avenue for the dissemination of future CER will be continuing medical education programs for physicians and other health professionals (Avorn and Fischer 2010). However, consideration should be given to new applications and models that provide mobile-based access to CER research and findings in a simple format. Numerous studies have shown that quick reference apps and drug interaction apps are highly utilized by providers (Franko Oi 2012; Payne et al. 2012; Williams 2012). A number of localized apps exist for providers that could expand offerings to include summary findings of CER studies.

32.3.2 CER Summary

Similar to CBA, CER studies encompass numerous methods that could aid in assessing the impacts of mHealth. The first step in mHealth CER studies is to establish adoption preferences and models by consumers. This work is already underway with a number of studies available providing information on consumer or provider preference for devices (Johnson and Mansfield 2008; Franko Oi 2012; Payne et al. 2012). As mHealth-focused CER studies evolve and become more rigorous, a simplified approach to sharing information is needed. Currently, CER studies are widely evidenced in mHealth by the use of drug interaction apps for both consumers and providers. However, a much broader approach is needed to provide quick reference–style apps leveraging CER studies. Quick reference apps remain a primary source of information for providers. Thus, consideration should be given to dissemination of CER information and research via mobile applications. Furthermore, programs like QUEST Comparative Effectiveness & Innovation Program (QCEIP) or the AHRQ should be used as a model for localized CER research. Organizations should continue to provide resources and searchable information optimized for mobile technologies when presenting CER findings. Considerations for mHealth "optimization" for national repositories of CER studies should also be considered to aid in the dissemination of findings.

32.4 Three Other Alternative Methods

Three other economic evaluation methods—CEA, CUA, and CMA—may seem to be potential candidates to assess the preferences for various mHealth technologies. All three methods have many methodological commonalities and share some of the same weaknesses. Consequently, the methods have similar limitations in their applicability for assessing the benefits of mHealth technology. Each of the methods is reviewed briefly for their basic requirements, and then they are considered collectively for their applicability and appropriateness for assessing mHealth technology choices. Readers can find more extensive discussion of the three methods in the works of Brent (2003), Ryan (2004), Muennig (2008), Boardman (2010), Drummond (2010), and Brazier (2007).

32.4.1 Cost-Effectiveness Analysis

CEA is very widely used and is perhaps the most prevalent, economic evaluation method for assessing not only health-care and health policy issues but also research topics in most other (nonhealth) fields of study. It is also viewed by many evaluation researchers as the major substitute for CBA, inclusive of CVA and CA. In essence, CEA is employed to compare an outcome of competing mutually exclusive policy/program alternatives, with the outcome representing potential enhancement in economic efficiency, that is, gains in value consistent with individuals' preferences. CEA can be and is employed by researchers for one of two reasons (Boardman 2010). First, the most central outcome or any major outcomes that capture the value of benefits produced by the various alternatives either cannot be monetized, or substantial difficulty in monetization is encountered. Second, because of moral or social reasons, decision makers and researchers may be resistant to assigning monetary values to the impacts to program/policy alternatives. In these situations, the goods/services for which policy/program alternatives are to be evaluated do generate outcomes that are measurable and are considered surrogates for the "immeasurable" benefits that such alternatives would produce.

Central to conducting a CEA is the calculation of cost-effectiveness ratios that provide the bases of choosing among alternatives. Two different metrics are used to compile a ratio. First, costs are determined for each selected alternative. Second, a single (and thus the same) effectiveness measure common to all the alternatives is selected; the effectiveness measure is comprised of an interval scale with a range of values that allows significant comparisons of differences among program/policy outcomes. The determination of a cost-effectiveness ratio provides an estimate of the dollar cost per unit of effectiveness. The separate cost and effectiveness values for each alternative are estimated on the basis of available empirical evidence about how much cost would be incurred to undertake the alternative and the numerical score on the effectiveness scale that would be produced by the cost of the selected alternative. The smaller the cost-effectiveness ratio, the lower is the cost per unit of effectiveness, and thus the more favorable the outcome is.

Since competing alternatives are to be compared, the cost differences of the alternatives are expected to produce differential gains measured by the effectiveness metric. Thus, CEA entails consideration of marginal or incremental values that require choosing the appropriate base from which to compute effects and costs of alternatives. If only one (or single) policy change is evaluated, then the alternative is compared with the status quo; consequently, a positive ratio value would indicate that increased effectiveness results from increase in dollar cost. If more than one alternative is assessed, the differentials in the cost-effectiveness ratios among the alternatives are calculated. Because the outcome of the competing alternatives is measured along a single interval scale for effectiveness, then the estimated cost-effectiveness ratios can be ranked in terms of their differential values of cost per unit of effectiveness. Thereafter, choices among the alternatives can be made by decision makers consistent with the value of the cost-effectiveness ratios and the budget (resource) constraint that is imposed on the selection of proposed program/policy changes.

32.4.2 Cost-Minimization Analysis

CMA has been used in a multitude of health-care studies, especially pharmacoeconomics analyses. With respect to the latter, CMA has been primarily applied to determine and thus compare costs of different drugs for the same therapeutic purpose with equal efficacy and equal tolerability (Rascati 2011). The basic requirement of CMA is to determine the lowest cost among competing products/designs/goods that have the same purpose but also the same outcome. The methodological purpose of CMA is to obtain a predetermined outcome at the least cost of the alternatives assessed; this purpose is in contrast with CBA and CEA by which benefits are to be maximized with the limited amount of available resources. The outcome of the each alternative considered is to be equally effective, that is, the consequences of competing alternatives are the same and only the costs of inputs are the important bases of decision.

32.4.3 Cost Utility Analysis

CUA is viewed by some health economists as either a form (Boardman) or a special case (Drummond, Brazier) of CEA. This method has been confined essentially to the evaluation of medical care interventions. Like CEA, alternatives are defined, their costs are calculated, outcomes are measured, ratios are compiled, differences in marginal values are computed for the ratios, and the ranking of ratios are determined. However, two or more treatment alternatives are assessed for their impacts of more complex effectiveness measures than used in CEA. The composite measure is to assess the improvement in the health state of individuals who are subjects in the intervention; it is comprised of two dimensions. In addition to a quantity dimension that is employed with CEA, the outcome includes a quality dimension.

Most commonly in applications of CUA, both dimensions have incorporated into the "utility" measure of either quality-adjusted life years (QALYs) or the disability-adjusted life years (DALYs). For both QALYs and DALYs, the quantity dimension is the total number of additional years of life and the quality dimension is the quality life during the additional years of life, each of which would be expected to be induced by the intervention. A most straightforward usage of a QALY is when the consequences of treatment alternatives encompass a trade-off between mortality (the quantity of the length of life) and morbidity (the quality of life). The quantity and quality dimensions are combined to make them commensurate as a utility measure in a two-step procedure for deriving values on a numerical interval scale for each alternative. First, for every alternative, the numbers of expected years of life for subjects due to the intervention are estimated and then their health states for each year are defined. Second, the different health states of the described subjects are valued through the use of a rating scale. In past studies, four methods have been popular: the health rating scale, the time trade-off approach, the standard gamble method, and the health index scale.

The resulting numerically scaled values for each individual subject in each alternative (intervention) are transformed into cost/utility ratios whose scores measure the (dollar-valued) cost per utility unit gained. These estimated QALY or DALY scores permit the ranking of the utility values of alternatives so that the differences in costs among alternatives are compared for their change in utility, and consequently the most "beneficial" alternatives can be chosen. The most beneficial QALYs or DALYs are indicated by the lowest (dollar-valued) cost per utility unit. Once the rankings have occurred, a threshold value of a QALY or DALY value—a cost per unit of utility—can be established for budgetary purposes; any ratio for a particular intervention above the threshold value that intervention would not be supported for implementation as a policy/program/practice is defined by the analyst.

Because of the valuation component, for example, assigning value to health states/status over the expected life years—for example, by the health rating method—QALYs and DALYs are often called preference-based measures of health. Utility scores are to provide direct proxies for individuals' utility for a positive outcome of an alternative, and thus, it is expected to reflect individuals' preferences and, consequently by inference, indicate their valuation of benefits for an alternative. Thus, the ranking process facilitates a comparison of the marginal utility of the individual choices for policy/program alternatives.

For the cost/utility ratios to be representative of utilities, the analysts/researchers must consider two important measurement issues that affect the validity of the ratios. There must be consideration of the quality of the empirical evidence upon which the health state and life expectancy by the interventions estimates are derived. Also, a decision must be rendered about the individuals who are to participate in the determination of the life years and health states and the valuations of the health states. In particular to the last one, the validity of valuations is a result of preferences regarding the choice made by selecting the intervention. For example, tradeoffs associated with telehealth or remote monitoring vs. in person care. In the research of CUA, studies have relied upon singularly as well as a mix of medical experts, patients, and the general population. From an mHealth perspective, all three areas are of concern and provide the potential for further research.

32.4.4 Impact on mHealth Assessments

With respect to mHealth, any two or more devices could be evaluated as competing items using CMA and CEA. The major concern with these two methods is what is to be assessed or the outcome to be measured. The analysts/researchers should clarify the purpose of the investigation within this context. For instance, several devices could have the same purpose or/and similar attributes or characteristics: Does the evaluation objective assess the capabilities of the devices for only purpose or attributes important to gain adoption? The first step would be to define or select the alternatives. The alternatives would be the particular mHealth devices that have the same objective and are expected a priori to attain a positive outcome. A second step would involve the determination of the costs of the devices. Here the analysts must affirm whether only financial cost or both social and financial costs are included in the alternatives and ensure that the costs are valued consistently across the alternatives. Finally, the central outcome that is expected from the devices should be selected; thereafter, the effectiveness measure should be chosen in which the outcome is measured on an interval scale that indicates lesser to greater impacts of the alternatives. For example, devices or applications could be appraised for their differences in the extent of their informational content or ease of interpretation by providers against productivity gains.

If the selected mHealth devices or interventions did have the same informational capabilities, it could be argued that equivalency of effectiveness is applicable to the outcomes of health care for which the mHealth technology is employed. That is, it is assumed by the consumer choosing among competing alternatives that the same amount of benefits would be realized by each alternative. In effect, CMA shifts the main focus of evaluation to cost and away from outcomes. The stipulation of CMA that mHealth devices to be compared should be equivalent in effectiveness or equally efficacious for the alternatives could be an obstacle to evaluation by sellers and buyers of mHealth technology. Given that mHealth and health care at large operate in a dynamic market of innovation, it may be difficult to obtain competing mHealth devices with ostensibly the same purpose that have identical attributes and capabilities.

Since CMA is limited in the scope of analysis, some researchers have argued that it is not a full economic evaluation (Drummond 2010). In addition, some health economists consider that CMA is no longer useful (Briggs and O'Brien 2001) because improvements in research methods, data collection, and statistical techniques permit the application of more comprehensive analytical approaches such as CUA, CEA, and CBA.

Even if the outcomes of the alternatives do vary in quantity of effectiveness, CEA as an evaluation method would be problematic. mHealth technologies have multiple attributes and capabilities that have or serve the same purposes. The consideration of only one effectiveness measure representing the central outcome does not allow the trade-off among the various attributes and capabilities and their levels, and thus, it hinders the determination of the relative importance of different attributes and capabilities of mHealth devices by potential consumers.

In principle, this problem could be avoided or minimized in one of three ways. Potential outcomes could be incorporated or integrated into the central effectiveness measure (and thus a composite measure would be formed). Because of the diversity of attributes or capabilities of mHealth devices, however, it is difficult to combine the different attributes and capabilities in one composite measure since the effectiveness measures of different attributes and capabilities are unlikely to be commensurate metrics. It could also be argued that, instead of focusing on a central outcome, each important outcome of a policy/program alternative should be assessed separately. However, with one cost estimate for an alternative, it is difficult to ascertain how much of that cost is attributable to each outcome; thus, a determination of a straightforward cost per unit of any outcome would be very unlikely. Finally, to capture the contribution of other noncentral outcomes, one could estimate the monetary value of at least some of the major ones and include these as benefits in the numerator of the cost-effectiveness ratio. These benefit estimates would offset and thereby reduce the costs of the alternatives and perhaps make them more attractive for purchase. Such a procedure would enhance the accuracy of the assessment of the value of mHealth devices or interventions. It would also shift the analysis from one of technical efficiency, in which CEA measures the unit cost of a desirable dimension of mHealth device, to allocative efficiency, in which the choice among alternatives is based on the amount of benefits generated by a device. However, if such an effort to identify the benefits of outcomes is to be undertaken, then a major question arises: Why not directly undertake a CBA instead of a CEA?

Several other issues prevail regarding the efficacy of CEA as an evaluation method. First, in the economic evaluation literature, it is unclear what is an appropriate value of a cost-effectiveness ratio that warrants the adoption of an alternative. Put differently, is there a threshold value that makes an alternative an acceptable policy/program choice? This criticism is applicable to CUA. Second, who should determine the central outcome for a CEA? Should it be potential users of mHealth, potential consumers of mHealth, or analysts acting independently or in cooperation with potential users and consumers? How do these decisions facilitate or hinder the selection of mHealth devices consistent with users' preferences and thus one that is most likely to be most beneficial among the choice of alternatives? In a related point, cost-effectiveness can be criticized for excluding the appraisal of the utility that individuals would derive from using mHealth by focusing only on the change in health outcomes and not other gains identified within CBA.

As described earlier, CUA is an economic evaluation method that seeks to include individual utility in policy/program choices. It does so by a composite measure of the quantity and quality in which individuals' health states over time are valued through the use of rating scales. It would seem that the utility score would be a constructive and practical basis for choosing among mHealth alternatives given that they would seemingly provide information on preferences of individuals for the mHealth alternatives and thus indicate the value of the contribution of the alternatives to health care. (This ignores the issue of any appropriate threshold of cost utility ratios.) The major question for conducting mHealth evaluations is: How can CUA be applied to or adapted for the assessment of mHealth technology or other technologies in the delivery of health care? The answer is not readily apparent. Moreover, the task of applying CUA would appear to be very difficult and cumbersome, if the two-step procedure to derive utility scores was followed to evaluate mHealth alternatives.

One major difficulty lies in identifying the population that would be impacted by the alternatives. Once determined, the analyst would have to establish the life years and health states of the target population that are produced by the alternatives. Then the health status of the target population would have to be valued. Putting aside which rating scale should be used and the methodology to be employed, a central issue is who should do the ratings of the health states. In other words, whose preference and thus whose valuations count? Should the ratings be conducted by analysts/researchers, medical experts, the target population, or the potential users of the mHealth device? Each group may give different valuations.

A more fundamental obstacle is encountered by evaluating mHealth technology with CUA. The critical issue is that the evidence of positive impacts on health and health care by mHealth is limited at the present time (Schweitzer and Synowiec 2012). Conducting a CUA means that gathering empirical evidence on how mHealth affects medical and health-care delivery and improves health outcome of individuals. In other words, the knowledge of the impacts of mHealth technology is encompassed by a great deal of uncertainty. This knowledge and data acquisition to clarify and reduce the uncertainty requires a major and long-term research effort. To a certain extent, this approach is somewhat reflected in the creation of the Patient Centered Outcomes Research Institute (PCORI) that seeks to create clinical data research networks. The use or focus of mobile technologies represents a major opportunity to create such a network for ad hoc assessment(s). In the meantime, the mHealth market is burgeoning and undergoing seemingly rapid changes and development. Moreover, there is a need to assess mHealth technology simply because it is readily available and it is utilized quite prevalently by members of society.

It is questionable whether CEA and CMA are better methods for assessing the impacts of mHealth devices. The same impediment of limited knowledge about mHealth impacts also pertains to the efficacy and appropriateness of evaluating mHealth devices by both CEA and CMA. Because of the limited evidence on mHealth impacts, the outcomes for which policy/program alternatives are evaluated by CMA and CEA have ambiguous connection to benefits that are considered or are seemingly immeasurable. Thus, the outcomes and the concomitant effectiveness measures of chosen alternatives can be viewed as surrogates for a final but desirable undefined outcome(s). In addition, CMA and cost-effectiveness do not permit the revelation of consumer preferences for different features and attributes of mHealth devices that the consumers/users would view as important to the purpose for which the technology is to be used. Consequently, as evaluation tools, the two economic methods are limited in scope and capability for assessing mHealth technology. They can be characterized as being suboptimal approaches to evaluate the benefits of program/policy mHealth technological alternatives intended to improve health care and health.

Although there are some weaknesses, CBA provides an alternative to the limitations of CMA, CEA, and CUA. One advantage of CBA is that the costs and the benefits for a program can be compared directly because they are both in money units. Another advantage is that CBA encompasses several methods to estimate benefits of policy/program alternatives. In particular, as covered in earlier sections, the two evaluation methods of CVA and CA can be utilized to estimate the monetary value of impact under uncertainty (Boardman 2010).

CBA facilitates the measurement of individuals' WTP for mHealth devices so that benefits can be derived indirectly when there are potentially desirable outcome(s). Both CVA and CA extract valuations by asking mHealth users who are experts in medical and health care and obtain their assessment of the expected contribution of technologies through their putting a value on the purchase of the devices, that is, they indicate what is worth based on what the users believe the devices can do in health-care delivery. In doing so, the two methods elicit users' (as consumers) preferences and their valuation for different attributes and capabilities of devices based on their expert knowledge of health care and health states that they judge likely to occur if the devices were adopted. Such valuation(s) allows a more comprehensive appraisal of a device and permit a more informed basis for trade-offs and different attributes and capabilities, a procedure for which CA is better suited than CVA. Moreover, both methods also facilitate easy incorporation of other dimensions that shape preferences and valuation such as the role of income and payment vehicles. As with any methods, the obvious disadvantage of both approaches is the difficulty inherent in obtaining valid and reliable WTP estimates, given the state of methodological limitations (Diener et al. 1998). Finally, it is difficult to state what the role of CVA and CA will be in the long term when research reveals more empirical findings about the direct impacts of mHealth technology on health and medical delivery and improvement in individuals' health states.

32.5 Future Direction

Understanding the economic drivers and potential changes to policy will aid mHealth adoption. This is facilitated by the creation of a larger research base in the field of CBA and other methods that tie cost to the utility of use of mHealth technologies. Increasingly, the role of technology in health care is coming into question from a value proposition perspective. According to the U.S. Bureau of Labor and Statistics, the overall use of technology in health-care settings remains low when compared to other sectors like finance, banking, and other professional services (1994). mHealth technologies may eventually leapfrog other technologies such as the EHR. However, this has not widely occurred due to ill-defined or spurious improvements in "health status" or improvements in clinical outcomes. In economic parlance, improved outcomes or health status can be quantified as "costs" and analyzed. Analysis of cost is only beginning to occur in the mHealth field on a small scale with limited variables making extrapolation and generalization of results difficult.

A continued focus on assessing benefits and costs associated with mobile apps and technology is needed. Understanding the WTP for goods at a rudimentary level, inclusive of potential challenges associated with the suggested methods, is important for the assessment and creation of policy change. CBA has proven useful in analyzing "soft" benefits to society in other fields like environmental economics. In a similar vein, there is a limited or unknown return on investment associated with the use of mHealth technologies in clinical settings. This remains a significant barrier to adoption revealed in other applications of health-care technology. In the United States, widespread adoption of technology among health-care providers is generated largely through the Meaningful Use program that provides funding for the use of EHR technology. Similar programs and projects exist in other countries and the application of CBA will prove useful to decision makers in both public and private roles.

Overall, the future of economic analyses associated with mHealth will encompass multiple research methods using multiple perspectives. The ability to collect social, behavioral, and physical information on a single device presents a large opportunity for researchers. Many challenges addressed and outlined elsewhere in the book present significant barriers to widespread adoption. Approaching mHealth from a cost-benefit perspective reflects a further refinement of the ability to properly discern value realized by users of mHealth technology to aid in adoption.

32.6 Summary

The aim of this chapter was to provide background on various methods for assessing the economic impact of mHealth technologies. Advancing the adoption and use of mHealth will require an understanding of consumer preferences associated with the adoption of technology. Decision makers can leverage existing methods described earlier to begin to quantify benefits associated with the use of mHealth. As noted earlier, the benefits associated with mHealth may not always present themselves as a direct improvement in outcomes with a reduction in cost. In certain cases, modified approaches to assessing cost, utility, and benefits are required. Modified applications of the methods in this chapter, optimized for the mHealth ecosystem, will further aid decision makers.

A number of researchers have advocated for the creation of new methods to explore mHealth (Nilsen et al. 2012). When appropriate, modifications of existing methods such as CER, CMA, and CBA will suffice, leading to a better understanding of the impact of mHealth on the delivery of health care.

The poor economic rationale for pursing large scale projects associated with mHealth is often invoked as a major barrier to widespread adoption of the technology. Establishing an understanding of costs and benefits can assist decision makers with the formulation and creation of policy at both the governmental and firm level. Furthermore, the exploration of social and quantitative data sets easily obtained from mobile devices represents a major opportunity to assess both cost and outcomes associated with use of advanced technology.

It is important to note that individuals have different preferences associated with the decision to adopt mHealth technologies. The amount of disagreement across individuals will vary by attribute (Johnson and Mansfield 2008) and can impact the design and deployment of the methods described in this chapter. The methods outlined in this chapter explore the advantages and present the disadvantages of deploying survey research and economic modeling to assess consumer or provider preference for mHealth apps and technologies. Understanding the WTP for goods at a rudimentary level, inclusive of potential challenges associated with the suggested methods, is important for the assessment of policy change. Many of the methods presented in this chapter are directly applicable to the field of mHealth and provide the ability to gain an understanding of the characteristics important to consumer demand for products.

For more robust discussions on the formal role of CBA, utility theory, or other fundamentals associated with conducting such assessments, readers are directed to *Cost-Benefit Analysis* by Boardman (2010). Readers are also directed to *Applied Methods of Cost–Benefit in Healthcare* by McIntosh et al. (2010).

32.7 Case Studies

32.7.1 Assessing Trade-offs Associated With the Cost of Technology

A public hospital is considering a transition from pagers to provider-owned smartphone devices with messaging service supported by a smartphone "app." The hospital employs a robust bring your own device (BYOD) policy and supports multiple devices and platforms. The hospital employs a mobile device management (MDM) platform that aids in keeping information protected, private, and secure. The information technology support staff is concerned about how providers will react to the transition from pagers to smartphones. Furthermore, the executive management team is concerned about covering the cost of acquiring or developing the app. An outside vendor suggests the organization undertake an exercise in CBA. The analysis will determine attributes important to providers with current systems (paging system) in place. The analysis will evaluate potential needs or attributes as the organizations transition to mHealth-focused apps. Finally, to satisfy the needs of the executive team, the analysis will explore provider and business units' WTP for an "app-enabled" service.

32.7.2 Assessing Willingness to Pay

A public health organization is planning a public health service enabled by mobile devices. The end user will be provided with medical support via a mobile app. The public health organization would like to understand the tendencies of an established user base that utilizes similar apps but charges a fee. The public health organization needs to understand both the attributes important to use and user WTP for advanced support.

Questions

1. CBA is useful for which of the following?
   - (a)   Governmental programs
   - (b)   Private sector projects or programs
   - (c)   Assessing consumer preference
   - (d)   All of the above
   - (e)   None of the above
2. Willingness to pay is?
3. Which type of study is useful for assessing consumer or user trade-offs as a result of a hypothetical market?
   - (a)   Conjoint analysis
   - (b)   Comparative effectiveness research
   - (c)   Cost minimization
   - (d)   None of the above
4. What four elements should an analyst consider including in mHealth comparative effectiveness research?
5. True or false: Cost minimization is useful for assessing or comparing mHealth apps.

References

Aston, G. (2010). Comparative effectiveness. *Trustee: The Journal for Hospital Governing Boards* **63**(1): 13–14.

Avorn J. and M. Fischer. (2010). 'Bench to behavior': Translating comparative effectiveness research into improved clinical practice. *Health Affairs (Project Hope)* **29**(10): 1891–1900.

Boardman, A. E. (2010). *Cost-Benefit Analysis: Concepts and Practice*. Upper Saddle River, NJ: Prentice Hall.

Brazier, J. (2007). Measuring and valuing health benefits for economic evaluation. From http://site.ebrary.com/id/10167512. Accessed on June 28, 2014.

Brent, R. J. (2003). *Cost-Benefit Analysis and Health Care Evaluations*. Cheltenham, U.K.: Edward Elgar.

Briggs, A. H. and B. J. O'Brien (2001). The death of cost-minimization analysis? *Health Economics* **10**(2): 179–184.

Brown, T. C., P. A. Champ et al. (1996). Which response format reveals the truth about donations to a public good? *Land Economics* **72**(2): 152–166.

Cunningham, C., K. Deal et al. (2010). Using conjoint analysis to model the preferences of different patient segments for attributes of patient-centered care. *The Patient: Patient-Centered Outcomes Research* **1**(4): 317–330.

Davis, F. D., R. P. Bagozzi, and P. R. Warshaw (1989). User acceptance of computer technology: A comparison of two theoretical models. *Management Science* **35**(8): 982–1003.

Diener, A., B. O'Brien et al. (1998). Health care contingent valuation studies: A review and classification of the literature. *Health Economics* **7**(4): 313–326.

Dobkin, B. and A. Dorsch (2011). The promise of mHealth: Daily activity monitoring and outcome assessments by wearable sensors. *Neurorehabilitation and Neural Repair* **25**(9): 788–798.

Drummond, M. F. (2010). *Methods for the Economic Evaluation of Health Care Programmes.* Oxford, U.K.: Oxford University Press.

Franko Oi, T. T. F. (2012). Smartphone app use among medical providers in ACGME training programs. *Journal of Medical Systems* **36**(5): 3135–3139.

Galvin, C. (2011). Transforming healthcare with mHealth solutions: The opportunities, efficiencies, and ROI of mobile technology. From http://www.marketinfogroup.com/transforming-health-care-with-mhealth-solutions-the-opportunities-efficiencies-and-roi-of-mobile-technology/. Accessed on June 28, 2014.

Goldar, B. and S. Misra (2001). Valuation of environmental goods: Correcting for bias in contingent valuation studies based on willingness-to-accept. *American Journal of Agricultural Economics* **83**(1): 150–156.

Griffin, R. (2011). Willingness to pay. *Training Journal-Ely-*(May): 27–30.

Howe, C. W., B.-J. Lee et al. (1994). Design and analysis of contingent valuation surveys using the nested Tobit model. *The Review of Economics and Statistics* **76**(2): 385–389.

Johnson, F. R., and C. Mansfield (2008). Survey-design and analytical strategies for better healthcare stated-choice studies. *The Patient: Patient-Centered Outcomes Research.* **1**: 299+.

Lancsar, E., and J. Louviere (2008). Conducting discrete choice experiments to inform healthcare decision making: A user's guide. *Pharmacoeconomics* **26**(8): 661–677.

Levy, A., B. Harrigan, K. Johnston, and A. Briggs (2009). Comparative effectiveness research through the looking glass. *Medical Decision Making: An International Journal of the Society for Medical Decision Making* **29**(6): 6–8.

McIntosh, E., P. Clarke et al. (2010). *Applied Methods of Cost-Benefit Analysis in Health Care.* Oxford, U.K.: Oxford University Press.

Muennig, P. (2008). *Cost-Effectiveness Analyses in Health: A Practical Approach.* San Francisco, CA: Jossey-Bass.

Nilsen, W., S. Kumar, A. Shar, C. Varoquiers, T. Wiley, W. T. Riley, M. Pavel, and A. A. Atienza (2012). Advancing the science of mHealth. *Journal of Health Communication* **17**(suppl): 5–10.

Norris, A. C., R. S. Stockdale et al. (2009). A strategic approach to m-health. *Health Informatics Journal* **15**(3): 244–253.

Pankratz, M., D. Hallfors et al. (2002). Measuring perceptions of innovation adoption: The diffusion of a federal drug prevention policy. *Health Education Research* **17**(3): 315–326.

Payne, K. B., H. Wharrad, and K. Watts (2012). Smartphone and medical related App use among medical students and junior doctors in the United Kingdom (UK): A regional survey. *BMC Medical Informatics and Decision Making* **12**(1): 121.

Rascati, K. (2011). Essentials of pharmacoeconomics. *Value in Health* **14**(4): 616–617. http://www.amazon.com/Essentials-Pharmacoeconomics-Lippincott-Williams-Wilkins/dp/1451175930

Ratcliffe, J. (2000). The use of conjoint analysis to elicit willingness-to-pay values proceed with caution? *International Journal of Technology Assessment in Health Care* **16**(1): 270–290.

Ryan, M. (2004). Discrete choice experiments in health care: NICE should consider using them for patient centred evaluations of technologies. *British Medical Journal* **328**(7436): 360–361.

Ryan, M. and S. Farrar (2000). Using conjoint analysis to elicit preferences for health care. *British Medical Journal* **320**(7248): 1530–1533.

Schweitzer, J. and C. Synowiec (2012). The economics of eHealth and mHealth. *Journal of Health Communication* **17**: 73–81.

Sox, H. C., M. Helfand et al. (2010). Comparative effectiveness research: Challenges for medical journals. *Journal of Clinical Epidemiology* **63**(8): 862–864.

Steblea, I., Steblea, J., & Pokela, J. (2009). Healthcare's best-kept secret: use of conjoint analysis in healthcare marketing research. *Marketing Health Services* **29**(4), 1–10.

Vollmer, W. M., A. Feldstein et al. (2011). Use of health information technology to improve medication adherence. *American Journal of Managed Care* **17**: SP79–SP87.

Wainwright, D. M. (2003). More 'con' than 'joint': Problems with the application of conjoint analysis to participatory healthcare decision making. *Critical Public Health* **13**(4): 373–380.

Williams, J. (2012). The value of mobile apps in health care. *Healthcare Financial Management: Journal of the Healthcare Financial Management Association* **66**(6): 96–101.

Section IX

mHealth Cloud Applications

33

mHealth and Pharmacy in Health Care

Kevin A. Clauson, Zaher Hajar, Joshua Caballero, and Raymond L. Ownby

CONTENTS

33.1 Introduction

Irrational use of medications is a global challenge, with the World Health Organization (WHO) estimating that "more than half of all medicines are prescribed, dispensed, or sold inappropriately." This situation is exacerbated by half of all patients not taking their medications as intended, with nonadherence rates in disorders like schizophrenia reaching over 70%. Medication nonadherence and polypharmacy (i.e., use of multiple medications) are massive barriers to optimal medication management and are the source of substantial health and financial costs. Mobile health (mHealth) offers myriad methods to address these challenges, ranging from simple text message reminder systems to ecological momentary interventions (EMIs) to leveraging interface versatility for eHealth literacy solutions. In this chapter, current scientific evidence will be used to explore the role of mHealth in enhancing pharmacy-related outcomes in varying health-care settings and diverse populations. Current and future developments in mHealth will be explored through the lens of select disease states and patient populations.

33.2 Challenges with Medication Adherence and the Role of mHealth

One of the most daunting and costly challenges in health care today is medication adherence, as medication nonadherence can reach a prevalence rate as high as 60%–80% in chronic disorders (AlGhurair et al., 2012; Garner, 2010). Problems with adherence also transcend population and disease state. The effects of nonadherence can lead to a deleterious cascade of events that renders medical efforts suboptimal, compromises patient health outcomes, and ultimately places an economic burden on society.

According to the WHO, there are five dimensions that affect medication adherence and form the WHO multidimensional adherence model (Figure 33.1) (World Health Organization, 2003). These include health-care team/health system (e.g., provider quality of services, insurance), therapy (e.g., adverse effect, number of medications), condition (e.g., comorbid disorders, severity), patient (e.g., cognition, perceptions), and socioeconomic (e.g., health literacy, social support) related barriers.

These barriers to adherence have been explored in the literature from multiple perspectives including behavioral (e.g., persuasive technology), cultural, environmental, economic, and pharmaceutical. In order to improve adherence and overall health outcomes, it is imperative that interventions employ a multifactorial approach. While some interventions can be applied to patients with differing disease states or those from diverse backgrounds, not every intervention can be indiscriminately applied to all populations. For example, a computer-based intervention to improve medication adherence in human immunodeficiency virus (HIV) had to be culturally adapted for Spanish-dominant Hispanics (Jacobs et al., 2013). Therefore, finding novel and creative opportunities to increase adherence and improve health outcomes is crucial. Recently, pharmacists have been directly providing reimbursable services to optimize therapeutic health outcomes for patients. This is commonly referred to as medication therapy management (MTM). The core elements of MTM services focus on comprehensive medication reviews, development of a personal medication list, medication action plan, intervention and referral, and documentation

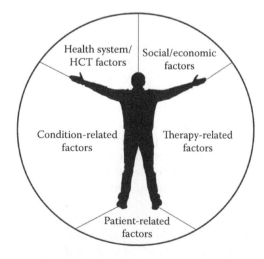

FIGURE 33.1

Five dimensions of adherence. *Note*: HCT=health care team. (World Health Organization, *Adherence to Long-Term Therapies: Evidence for Action, World Health*, World Health Organization, 2003, retrieved from http://www.who.int/chp/knowledge/publications/adherence_report/en/.)

TABLE 33.1

mHealth Tools to Address Factors in Medication Nonadherence

Dimension	Factor	mHealth Tool(s)
Social and economic	Limited English language proficiency	Medical encounter app (MediBabble), mobile medical translator (Jibbigo)
Health-care system	Lack of positive reinforcement	Mango Health, MEMOTEXT
Condition related	Depression	Mobilyze
Therapy related	Complexity of medication regimen	eMedonline, Helius, Pillboxie, Rite Aid NowClinic Online Care
Patient related	Alcohol or substance abuse	Alcohol–Comprehensive Health Enhancement Support System (A-CHESS mobile platform)

Source: American Society on Aging, and American Society of Consultant Pharmacists Foundation, *Adult Meducation: Improving Medication Adherence in Older Adults*, 2006, pp. 1–96, retrieved from http://www. adultmeducation.com/downloads/Adult_Meducation.pdf. Accessed on January 15, 2014.

and follow-up. These services are tailored to different disease states and are focused in communicating medication-related issues such as duplication of therapy, lack of efficacy, adverse effects, and nonadherence. Working with the patient to improve nonadherence can improve their outcomes and serve as an extension of the primary care provider who prescribed specific pharmacotherapy. One subset of MTM includes a focus on the dimensional factors of medication adherence via use of mHealth tools (Table 33.1).

Additionally, a burgeoning interest in leveraging technology to improve medication adherence has been increasingly observed in research and entrepreneurial arenas. Multiple methods of implementation have been tested, including mHealth tools enhanced by interactive voice response (IVR) as well as capitalizing on the availability of using the mobile phone as a trigger for EMIs (Heron and Smyth, 2010; Kaplan and Stone, 2013). mHealth is empowering by the fact that it uses a ubiquitous technology, mobile phones, which are widely utilized across all socioeconomic strata. In the United States, according to a recent survey of 2252 adults by Pew Research Center, more than 90% of the adult population uses a mobile phone (Rainie, 2013). Additionally, the mobile phone has ascended to become the most quickly adopted consumer technology in the history of the world (Rainie and Wellman, 2012).

mHealth also holds multiple advantages over other connected health-care technologies. Mobile phones have a significant outreach across geographic boundaries and at a low cost. Likewise, interventions can be conveniently delivered directly to people and can be easy to use by the patient or consumer. This overall health-care model is practical and generalizable and can be cost effective. One example of an mHealth intervention is text messaging or short message service (SMS). Text messages can be used as convenient reminders multiple times a day or as signals to drive behavioral change in contextual environments as with smoking cessation or delivering deteriorating air quality alerts in the case of asthma (Fogg, 2002). These interventions can also be tailored to the patient's characteristics, such as demographic information, environmental contexts, concomitant medications taken, and other variables (Cole-Lewis and Kershaw, 2010). For example, if during the MTM process a pharmacist determines that a patient has a lack of efficacy due to nonadherence, implementing SMS could be an effective way to use mHealth to improve adherence and outcomes.

Particularly challenging medication adherence scenarios can be observed in disease states where the patient is required to take more than one medication on multiple instances daily, such as diabetes and HIV/acquired immunodeficiency syndrome (AIDS)

(World Health Organization, 2003). The use of text messaging to improve medication adherence has been explored in diabetes and HIV. Quinn et al. (2011) published the first study in the United States that evaluated the effect of mobile text messaging and patient–provider web portals on the overall blood glucose control in type 2 diabetes measured by glycosylated hemoglobin levels (HbA1c). The intervention consisted of automated, real-time educational and behavioral messages in tailored format based on blood glucose measurements, current diabetes therapy, and lifestyle behaviors. The investigators reported a significant reduction of 1.2% in the average plasma glucose concentrations (95% confidence interval [CI], 0.6%–1.8%; p < 0.001) in the intervention group compared to the usual care group over a 12-month period.

In lower socioeconomic countries such as those in sub-Saharan Africa, SMS interventions have also been tested to address the therapy-related dimension and remind HIV-infected patients to take their medications (Horvath et al., 2012). The results of multiple studies on this type of mobile-mediated intervention have been grouped and analyzed. Compiled results confirmed that the intervention group receiving weekly text reminders demonstrated a lower risk of nonadherence at 50 weeks (relative risk, 0.78; 95% CI, 0.68–0.89).

While there have been examinations of mHealth applications (apps) for medication adherence (Dayer et al., 2013) and for related purposes such as prescribing and pharmacology education (Haffey et al., 2013), it is surprising that more studies have not been published exploring the direct impact of mHealth on medication adherence given the importance of the problem and its impact on health expenditure. While increasing investment in the development of these technologies is regularly observed in mainstream and mHealth news venues, most of this work is not rigorously explored nor disseminated in peer-reviewed clinical journals.

33.3 Participatory Medicine Model Driving Change in mHealth

Health-care reform, redesign, and revolution are underway globally and driven by factors including cost considerations, advances in health information technology, and, in some cases, mandated patient-centered care (Brown, 2009; Kitson et al., 2013; Locock, 2003). However, an additional driver for change is generated by the patients themselves, who are sometimes referred to as ePatients (Ferguson and e-Patient Scholars Working Group, 2007). Health care largely is currently conducted via its legacy system of a top-down, paternalistic medicine model. ePatients are advocating the adoption of a participatory medicine model, in which they are active, engaged partners in their health care; this is perhaps best illustrated with the process of shared decision making in clinical care (Elwyn et al., 2012; Lundberg, 2009). However, these patients and patient advocates are not alone in their assertion for change. Luminaries such as the former director of the U.S. National Institutes of Health (NIH), Elias Zerhouni, M.D., has stated, "As opposed to the doctor-centric, curative model of the past, the future is going to be patient-centric and proactive" (Zerhouni, 2007). Support for the tenets of participatory medicine has also come from leaders in clinical informatics like cardiologist Eric Topol, M.D., who has spent many years advocating the advantages of mHealth solutions in conjunction with patients (Topol, 2012).

As mentioned previously, health literacy is also a factor under the socioeconomic dimension that influences medication adherence. Poor health literacy has been associated with nonadherence in several disease states. Lower levels of health literacy have been linked to minority populations, lower socioeconomic status, and advanced age. A study in patients with diabetes found that poor health literacy was associated with nonadherence and that this factor contributed to significantly decreasing the effect race (i.e., black) had on adherence (Osborn et al., 2011). A study in HIV patients suggested poor health literacy was associated with lower adherence to antiretroviral therapy (Kalichman et al., 1999). Finally, a study in over 1500 elderly patients with cardiovascular-related diseases suggested that health literacy (and number of medications) might be associated with nonadherence (Gazmararian et al., 2006).

When the variable of the ePatient is added to an mHealth solution system, this concern regarding the impact of health literacy shifts to include similar concerns about eHealth literacy (Norman and Skinner, 2006). Pharmacy personnel are well suited, based on their training and accessibility in many communities, to serve some of the roles of the eNavigator in addressing these issues for patients considering mHealth-mediated interventions or programs (López and Grant, 2012). This proposed role may be particularly enhanced in specialty considerations, such as with certified diabetes educators (CDEs). Clinicians such as pharmacists or nurses who are CDEs can also build upon their existing framework to act as eNavigators. In this role, those clinicians can seek to increase adoption of mHealth tools and consequently reduce health disparities, as one pilot underway for Latinos with type 2 diabetes (T2DM) has attempted to demonstrate (López and Grant, 2012).

33.4 Applications and Devices for Engaging Patients and Improving Medication Adherence

In the United States, over 120 million people own smartphones, which represents an estimated 55% mobile market penetration (ComScore, 2013). While capitalizing on smartphones' high penetration rate and superior processing power, a growing interest has been demonstrated in using these minicomputers as a cost-effective solution to the challenges of medication nonadherence. Based on a recent review published in the *Journal of the American Pharmacists Association*, there are approximately 160 apps available on Apple iTunes store, Android Marketplace, and BlackBerry App World that varyingly address medication adherence (Dayer et al., 2013). Of those, 12 apps are available on both iOS and Android platforms. Perhaps surprisingly to some, the Android Marketplace contains the highest number of medication adherence apps, while iOS stores have more multilingual medication adherence apps than any other platforms. Table 33.2 provides a list of popular medication adherence apps and platform availability of each (Dayer et al., 2013).

The majority of adherence apps generate reminders that look like an SMS text message or a *push notification* alert that informs the patient of when to take their medications. Some apps require a box to be checked verifying that the patient took the dose or skipped it (e.g., Med Agenda, RxmindMe, Dosecast, MedsIQ, PillManager, and MediMemory). While the features of medication adherence apps are generally aimed to be easy to use, their main disadvantage may be the complexity and time needed to enter medication regimens by patients, especially if the regimens are complex. As a solution to this issue, providers can use app-companion portals (MedActionPlan; http://www.medactionplan.com/) to build patient regimens and push them to the patient's device. This approach can help alleviate the burden of data entry and minimize errors. For example, if a patient has difficulty with

TABLE 33.2

Popular Medication Adherence Applications by Platform

Application Name	iPhone	Android
MyMedSchedule	×	×
MyMeds	×	×
MedSimple	×	×
Med Agenda	×	
RxmindMe Prescription	×	
Dosecast	×	×
MediMemory	×	
PillManager	×	×
MedsIQ		×

Source: Dayer, L. et al., *J. Am. Pharm. Assoc.*, 53(2), 172, 2013, doi:10.1331/JAPhA.2013.12202.

nonadherence or eHealth literacy, a clinical pharmacist providing MTM services could serve as an eNavigator with the patient and partner with configuring the app, tailoring it to the patient's needs (e.g., inserting all medications and schedules of when to take), and educating the patient on its use of any relevant incentives or disincentives.

Some in pharmacy have taken a more direct role regarding app development as a pilot at St. John's University between computer science and pharmacy demonstrated (MacKellar and Leibfried, 2013). In one health information technology (HIT) course at the university, pharmacy students were brought into the class to serve as domain experts for a task analysis session in collaboration alongside HIT students. The ultimate goal of the group was to design and create user-friendly apps for the community pharmacy and hospital pharmacy settings. Surveyed students reported that the collaboration was beneficial and that they found mobile development to be a useful skill.

Alternately, several entrepreneurial efforts have taken the medication adherence gap to the stage of product development for mobile devices in order to create solutions and make medication regimen management more patient friendly as detailed in the succeeding text.

GlowCap, a device produced by Vitality, is designed as a prescription medication vial cap. Its internal sensor is able to detect when the patient opens the vial to take a pill. If the patient misses a dose, GlowCap illuminates and plays music to remind the patient to take his medication. If the patient is not prompted to action by this method, GlowCap will trigger a system to call the patient's phone in order to remind them to take their medication. GlowCap also hosts a prescription refill button. When pressed, the device then initiates an order to the patient's registered pharmacy to process the needed refill. GlowCap can be provided by insurers or purchased directly by patients at community pharmacies (e.g., CVS) (Comstock, 2013).

Proteus or *Helius* is a silicon wafer embedded or affixed to an ingestible pill. The chip generates electricity when it comes in contact with stomach fluids, which then sends signals to a disposable patch on the patient's skin. The disposable patch can communicate this information to the patient's smartphone or health-care provider. Information gathered includes the time the *smart pill* is taken and physiologic disposition like heart rate and temperature.

AdhereTech is a company that developed a pill bottle that is able to detect the number of medications or how much liquid medication is inside the bottle and relay this information to the cloud via data synchronization over mobile networks. The patient is also reminded of their medication regimen via configurable phone call or text message. Going forward, these devices could become part of the standard of care by pharmacists as part of MTM services in order to improve therapy and monitor efficacy.

33.5 mHealth and Medication Management: Focus on Chronic Disease in the Outpatient Setting

In order to provide optimal medication management (e.g., MTM services) via mHealth, it is necessary to identify and understand the factors among the dimensions discussed earlier that may influence nonadherence (e.g., race and number of disease states and/or medications). For example, a study in elderly patients with diabetes, coronary heart disease, and hypertension found nonadherence rates as high as 41% (Marcum et al., 2013). It was also noted that despite using different measures of adherence, blacks were at a higher risk for nonadherence; however, other minorities were not studied. Other factors that appeared to be associated with increasing nonadherence included history of falls, sleep complications, hospitalization within 6 months, being unmarried, poor self-rated health, and lower income. Race and income appear to be associated with lower levels of adherence in other disease states as well. Another study linked unemployment and race as factors associated with poor adherence in HIV patients (Kyser et al., 2011). Additionally, a meta-analysis demonstrated that minorities had substantially greater odds of developing nonadherence to statin therapy compared to Caucasians (Lewey et al., 2013). Medication adherence has also been evaluated in attention deficit hyperactivity disorder (ADHD) in which researchers observed a nonadherence rate of over 45% when measured using medication event monitoring systems (MEMSs); notably, these were higher than rates reported by patients (18%) or clinicians (32%) (Yang et al., 2012). More disturbing are reports that suggest over 75% of discontinuation rates of ADHD pharmacotherapy occurs within the first 3 months (Toomey et al., 2012). Another trial in uncontrolled hypertensive patients with diabetes demonstrated nonadherence rates of almost 50% (Ledur et al., 2013). Unfortunately, nonadherence rates in nondeveloped countries can be even higher. Data show nonadherence rates of 75% in antihypertensive treatments in countries such as Gambia and Seychelles (World Health Organization, 2003). Additionally, psychiatric comorbidities can contribute to an increase in nonadherence. A 9-year observational trial found that a comorbidity of depression in hypertensive males led to a higher rate of 1.52 of *days-not-treated* after matching for other factors (e.g., socioeconomic status). Within the realm of psychiatry itself, the rates of nonadherence are dire. For example, rates of antipsychotic nonadherence have been reported as almost 75% within the first 1.5 years of therapy (Lieberman et al., 2005). Additionally, another study suggests that over half of patients diagnosed with schizophrenia take less than 70% of their medications (Goff et al., 2010). Overall, the impact of nonadherence to antipsychotic medications is estimated to cost the United States between $1.4 and $1.8 billion just in rehospitalizations (Sun et al., 2007).

mHealth apps have an important role to play in the management of chronic conditions. The management of chronic diseases typically requires ongoing monitoring of the condition as well as sustained adherence to treatment recommendations. Conditions such as hypertension and diabetes, for example, can be monitored via mobile apps that automatically send status information to providers by way of electronic health records (Källander et al., 2013; Kaufman, 2010; Logan, 2013). Up-to-date information about patients' blood pressure or pattern of glucose measurements can also give healthcare providers essential information about treatment effectiveness that is far superior to that obtained in single visits or through sporadic patient self-report.

Another key role for mHealth apps in managing chronic disease is their capacity to provide patients with ongoing coaching on the management of their conditions. At its simplest, coaching can take the form of reminders to take medications or record blood pressure or glucose. More complex apps have the capacity not only to remind but also to create more complex interactions with patients. In a chronic disease such as HIV infection, apps have been developed not only to remind but also to provide ongoing feedback about treatment response and even encouragement about patient behavior (Muessig et al., 2013). Monitoring can thus be two way not only in HIV but in a number of other conditions, with apps providing both providers and patients with feedback about long-term trends in treatment effectiveness or disease progression.

mHealth apps can also be useful in health promotion activities and the prevention of chronic diseases. Applications have been developed to reduce the occurrence of high-risk sexual behaviors that increase the possibility of becoming infected with sexually transmitted infections (Muessig et al., 2013) or to allow for more informed decision making by sexually active adults (e.g., Hula; https://www.hulahq.com/). Mobile apps are also commercially available to monitor activity level (e.g., Withings; Nike Fuel Band); apps to monitor dietary intake are among the most popular for mobile phones (e.g., myFitnessPal).

Finally, another area in which mHealth apps may have an impact on chronic diseases is their potential to link patients with each other as well as their providers via social media (Merolli et al., 2013). Given the improving browsing capabilities of mobile devices, sites such as PatientsLikeMe (http://www.patientslikeme.com/) are available on mobile devices even without site-specific apps.

33.6 Leveraging the Connectivity of mHealth to Reduce Medication-Related Hospital Readmissions

Every year in the United States, $2 trillion is spent on health care, with one-third consumed by costs associated with hospitalizations (Wilding, 2013). About one hospitalization out of nine is actually a readmission, defined as a patient being readmitted within 30 days of prior admission. According to the Centers for Medicare and Medicaid Services (CMS), Medicare spends $17.5 billion on readmissions every year with an average expenditure of $9600 per readmission (HIMSS Media, 2013). These distressing numbers are usually tied to poor continuity of care after patient discharge and lack of coordination and education of the patient (Stauffer et al., 2011).

As a recent measure to reduce readmissions, the Hospital Readmissions Reduction Program (HRRP), under the Affordable Care Act (ACA) delineated measures to penalize hospitals for excess readmissions. Moreover, CMS will be posting hospital readmission rates on Medicare's Hospital Compare list (HIMSS Media, 2013). Since the inception of the HRRP, hospitals and health systems have been increasingly involved in developing solutions to minimize readmissions.

One of the solutions that can address elements of hospital readmissions is mHealth. Support from the provider side is underlined by the findings of a recent survey that revealed that 89% of clinicians use laptop computers for their daily activities, 53% use smartphones, 49% use cellular phones, and 47% use tablet computers (HIMSS Analytics, 2012). As a result, the concept of *care anywhere* was derived from the trending ubiquity of mobile devices in the clinical field. mHealth may be incorporated in the solution in multiple ways ranging from infrastructure redesign at the hospital to patient-tailored solutions at home.

Cloud-based mobile app is one of the solutions whereby patient information including time of admission, time of discharge, laboratory data, follow-up appointments, and other related information can be shared with the health-care team before and after admission in any health-care setting. One example is the *Kaiser Permanente* app, which allows its nine million patients to access lab test results and personal health records (PHRs). The app also features messaging doctors, scheduling appointments, refilling pharmacy prescriptions, and locating medical facilities (Dolan, 2012). Again, these apps can be leveraged by clinicians such as pharmacists to communicate with providers and streamline medication regimens and treatment. For example, elderly patients on average see multiple providers for their care (e.g., cardiologist for heart condition, endocrinologist for diabetes, pulmonologist for asthma) and take between 8 and 15 medications, creating an increased danger of negative consequences from polypharmacy (Farrell et al., 2011). Sometimes, this can lead to a duplication of therapy or medications that can interact causing a lack of efficacy for one or more diseases. Given that prescribers sometimes do not communicate effectively with each other, an intermediary can elect to use mobile-mediated communications to bridge the communication gap between prescribers, effectively manage patient medications, and coordinate care.

Another issue contributing to hospital readmission is poor medication adherence after patient discharge. Reasons include a change in therapy or medication regimen from the regular administration by the nurse in the hospital to discharge on multiple medications that the patient has to self-administer; these can be further problematic in conjunction with sudden changes in medication regimens. Several mHealth solutions that may tackle medication adherence have been discussed in Section 33.4, including apps and devices such as GlowCaps, Helius, and AdhereTech.

Finally, conferencing between expert providers and patients using mobile devices in addition to remote monitoring may benefit patients at high risk of readmission (e.g., patients with an acute myocardial infarction, heart failure, or pneumonia) (de Waure et al. 2012; Smith, 2013). In one study, the Geisinger Monitoring Program (GMP) was used as an adjunct to case management in a Medicare population. The GMP used a range of technologies including interactive voice recording (IVR) and data pairing and storage within the electronic medical record (EMR). Patients in the program received scheduled phone calls issued by the automated system to ask patients questions related to their health condition. The patients' responses were stored in the EMR with notifications based on priority to health-care providers to call back and take action as necessary. Those patients in the GMP group experienced a 44% reduction in 30-day hospital readmissions postimplementation (95% CI, 23%–60%; p = 0.0004) (Graham et al., 2012) compared to the control group that received standard of care.

33.7 Legal and Policy Barriers within Pharmacy-Related mHealth

The ubiquity of mobile devices and networks and the falling costs for select technologies have created enormous opportunities in pharmacy-related mHealth, aspects of which include mPharmacy (Doukas et al., 2011). However, there is an accompanying opportunity cost in terms of resources that have to be allocated to navigate legal and policy challenges of many mHealth initiatives. Similarly, since significant components of mHealth can occur outside the conventional health-care system (much like health itself does), the need to maintain the integrity of patients' protected health information (PHI) has led some to suggest that any HIT will lead to encroachment on privacy (Norman et al., 2011). The debate between protecting patient privacy and avoiding stifled innovation also continues to rage in mHealth, with some corners asserting that *fear of the unknown* should not be a guiding principle (Farr, 2013; Miliard, 2012; Shapiro, 2013).

In the United States, the speed of change of HIT, eHealth, and mHealth has necessitated modifications to privacy and security protections of patient health information via the Health Insurance Portability and Accountability Act (HIPAA) (Department of Health and Human Services, 2013). Changes, including those to the privacy, security, and enforcement rules, were amended to allow for modifications about Breach Notification for Unsecured Protected Health Information and other exigencies. Breach, in which the electronic patient health information (ePHI) is compromised, remains a threat through loss or theft of mobile devices (e.g., mobile computers, phones, flash drives). Aggregated data from the U.S. Department of Health and Human Services (HHS) on breaches affecting over 500 or more individuals reveal that laptop computers and other portable electronic devices account for up to 44% of all occurrences of breach since HHS began tracking (Figure 33.2) (Department of Health and Human Services, 2012).

The ePHI lost through these breaches includes patient identifying information along with their specific medication regimens, medication costs, lab results, etc. This type of specific medication information can be particularly sensitive as the name of the medication itself can reveal socially stigmatized medical conditions (e.g., bipolar disorder, HIV/AIDS) about the patient that can have far-ranging consequences.

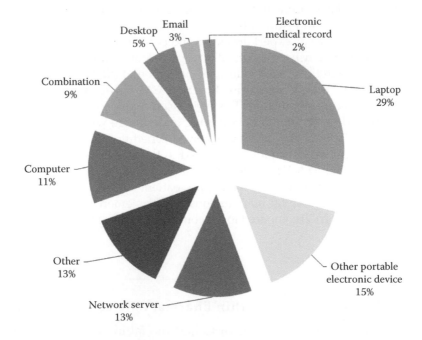

FIGURE 33.2
Breach of ePHI by media type.

These developments have not gone unnoticed and may serve as another threat to adoption and development of mHealth efforts.

Improvements in security and implementation efforts can be facilitated with meticulous planning, training of health workers and support staff, and creating an institutional culture of mindfulness that touches all stakeholders. Alternately, in clinical practice, pharmacy personnel providing MTM services can counsel patients and introduce this technology in an auspicious environment with reduced concern regarding time constraints. These issues must be closely attended, especially in vulnerable and marginalized patient populations and/or when eHealth literacy issues could negate informed choice by a patient (Clauson et al., 2013). With personalized or targeted health services enabled by mobile devices including federal health promotion, corporate wellness programs, and extension services advanced by nongovernmental entities, privacy must be paramount (World Health Organization Global Observatory for eHealth, 2011).

33.8 Case Studies

33.8.1 The Smart Pill Bottle to Improve Medication Adherence

One emerging technology that addresses multiple WHO-outlined dimensions of adherence (i.e., system, patient) is the smart pill bottle. A patented smart pill bottle by AdhereTech (Figure 33.3) can be configured to alert a patient to take their medication via use of on-bottle lights and chimes as well as integrating automated phone calls, text messages, or e-mails that can notify the patient or caregiver about scheduled and missed doses (Stanford Medicine X, 2013). Additionally, the smart pill bottle can serve as a data hub for bidirectional communication tied to medication adherence in order to evaluate patient motivations or behaviors when scheduled doses are not taken (e.g., a cancer patient choosing not to take a dose due to fears of debilitating nausea). Notably, pharmacists and pharmacy

FIGURE 33.3
Image of smart pill bottle. (Courtesy of Adhere Tech., Inc., New York.)

technicians are extremely well positioned at the point of dispensing or discharge to adopt the aforementioned roles of eNavigators to guide and enhance these potential benefits with a smart pill bottle. The AdhereTech smart pill bottle also uses a wireless code division multiple access (CDMA) module to automatically send two types of encrypted, HIPAA-compliant information to the cloud each time the bottle is accessed by a patient to take their medication. The first datum logs the time when the bottle cap is removed and when it is replaced. Second, sensors in the walls of the bottle measure capacitance, with the level of reduced capacitance corresponding to the number of pills (or amount of liquid medication) removed by the patient (Pharmacy Times, 2013). These data generated by the bottle could be used by clinicians, data or behavioral scientists, or the patient for optimizing MTM using established approaches or predictive analytics.

The impact of AdhereTech's mHealth technology on medication adherence is being examined in clinical studies including in HIV-positive patients at Weill Cornell Medical College (2013) and in patients with type 2 diabetes at Walter Reed National Military Medical Center (University of Alabama in Huntsville, 2013). mHealth technologies like this smart pill bottle also offer promise for increasing medication adherence and managing cost containment in the burgeoning area of specialty pharmacy. Leveraging AdhereTech's open application programming interface (API) would also allow for integration with other data streams and solutions to advance this type of persuasive technology.

33.8.2 Automated Dosing Reminders and Personal Health Records for Glaucoma

Microsoft HealthVault is a web-based system designed to store and maintain users' health and fitness information in the form of a PHR. MEMOTEXT is a patient adherence solution that uses condition-specific voice, text, and other mobile interventions to improve patient outcomes of therapy. In a challenge to ensure patient adherence to prescribed medication regimens, Johns Hopkins University in collaboration with Microsoft HealthVault conducted a clinical study evaluating the impact of automated dosing reminders on medication adherence in patients with glaucoma (Chang et al., 2010). The intervention consisted of a Microsoft HealthVault app that sends patient preferences to MEMOTEXT, which alerts patients to take their medication via a short message or a scheduled IVR call. Interim analysis was conducted on 428 patients in March 2011. Medication adherence was significantly improved with daily medications. The intervention group demonstrated a statistically significant increase in medication adherence from 51% to 67%. The control group with no intervention showed no change in medication adherence (MEMOTEXT, 2011).

33.9 Future Directions of Technology in mHealth and Pharmacy

Remote monitoring as well as prescribed and informational apps tied to pharmacogenomic and behavioral data are all on the cusp of helping optimize the delivery of pharmaceutical care. While medication nonadherence is one of the primary challenges of health care in the present, it does not show signs of abating in the near future. However, in the coming years, advances in pharmacy-related mHealth technologies will result in an improved arsenal of tools that clinicians can use to attack the medical and financial costs that medication nonadherence is responsible for. Already, we are seeing emerging technologies such as ingestible event markers, also termed digital medicines (or more commonly *smart pills*). One such digital medicine (i.e., Helius), tied to a remote monitoring functionality for the provider and an app for the patient, has already entered the market demonstrating positive early returns in mental disorders, kidney transplants, as well as overall cost reductions in treating infections (Au-Yeung and DiCarlo, 2012; Eisenberger et al., 2013; Kane et al., 2013). Similarly, the wider availability of pharmacogenomics data and the ability to incorporate it in a contextual manner through a streamlined mechanism like Illumina's MyGenome app could allow for prescribing and MTM to be augmented with real-time personalized medicine information. Initial medication selection and modifications could be determined, in part by responses to the medications that the patients are genetically predisposed to having. Also, since genetic factors have been attributed as controlling factors in medication disposition (in some cases to up to 95% of the variability), putting that type of information at a clinician's fingertips via a mobile device could prove invaluable (Hoppe et al., 2011). Finally, as the field of mobile pharmacy-focused persuasive technology advances, an individual's aggregated behavioral data from an increasing number of streams (often in conjunction with social media channels) combined with approaches including gamification, incentivization, and disincentivization represent the potential of EMIs and patient–provider to develop health action plans (Fogg and Eckles, 2007; Heron and Smyth, 2010). Remote monitoring devices, mobile apps, and mobile generated and aggregated pharmacogenomics and behavioral data offer a very real promise for the future of mHealth and pharmacy.

Questions

1. Breach of ePHI has been associated most commonly with which of the following?
   (a) Desktop
   (b) E-mail
   (c) Laptop
   (d) Network server
2. An example of an mHealth tool targeting the health-care system dimension of medication adherence is
   (a) Jibbigo
   (b) Mango Health
   (c) MediBabble
   (d) Pillboxie
3. Forgetfulness is a factor from which domain of medication adherence that has been successfully targeted by SMS?
   (a) Condition related
   (b) Health-care system
   (c) Patient related
   (d) Therapy related

4. Elements including comprehensive medication reviews, development of a personal medication list, medication action plan, intervention and referral, and documentation and follow-up refer to the process of

(a) Medication therapy management (MTM)

(b) Mobile medication management (MMM)

(c) Optimal medication management (OMM)

(d) Optimal medication analysis (OMA)

5. Which digital distribution platform offers the most multilingual medication adherence apps?

(a) BlackBerry World

(b) Google Play

(c) iOS App Store

(d) Windows Phone Store

6. Which mobile solution is the only one that provides actual medication adherence data rather than data that serve as a proxy for adherence data?

(a) Dosecast

(b) GlowCap

(c) Helius

(d) MEMS

References

AlGhurair, S. A., Hughes, C. A., Simpson, S. H., and Guirguis, L. M. (2012). A systematic review of patient self-reported barriers of adherence to antihypertensive medications using the world health organization multidimensional adherence model. *Journal of Clinical Hypertension*, 14(12), 877–886. doi:10.1111/j.1751-7176.2012.00699.x.

American Society on Aging, and American Society of Consultant Pharmacists Foundation. (2006). Adult meducation: Improving medication adherence in older adults, pp. 1–96. Retrieved from http://www.adultmeducation.com/downloads/Adult_Meducation.pdf. Accessed on January 15, 2014.

Au-Yeung, K. Y. and DiCarlo, L. (2012). Cost comparison of wirelessly vs. directly observed therapy for adherence confirmation in anti-tuberculosis treatment. *International Journal of Tuberculosis and Lung Disease*, 16(11), 1498–1504. doi:10.5588/ijtld.11.0868.

Brown, C. V. (2009). Healthcare information technology: A necessary step for healthcare reform. *MD Advisor*, 2(4), 33–35.

Chang, D. S., Friedman, D. S., Frazier, T. C., Miller, R., Plyler, R. J., and Boland, M. V. (2010). *Predictors for Poor Adherence with Glaucoma Therapy in Electronically Monitored Patients The Automated Dosing Reminder Study (ADRS)*. In American Academy of Ophthalmology, Chicago, IL.

Clauson, K. A., Elrod, S., Fox, B. I., Hajar, Z., and Dzenowagis, J. H. (2013). Opportunities for pharmacists in mobile health. *American Journal of Health-System Pharmacy*, 70(15), 1348–1352. doi:10.2146/ajhp120657.

Cole-Lewis, H. and Kershaw, T. (2010). Text messaging as a tool for behavior change in disease prevention and management. *Epidemiologic Reviews*, 32(1), 56–69. doi:10.1093/epirev/mxq004.

ComScore. (2013). comScore Reports January 2013 U.S. Smartphone Subscriber Market Share. Retrieved October 09, 2013.

Comstock, J. (2013). GlowCaps now sold through CVS, new randomized control trial launches. *mobihealthnews*. Retrieved from http://mobihealthnews.com/20750/glowcaps-now-sold-through-cvs-new-randomized-control-trial-launches/. Accessed on January 15, 2014.

Dayer, L., Heldenbrand, S., Anderson, P., Gubbins, P. O., and Martin, B. C. (2013). Smartphone medication adherence apps: Potential benefits to patients and providers. *Journal of the American Pharmacists Association*, 53(2), 172–181. doi:10.1331/JAPhA.2013.12202.

De Waure, C., Cadeddu, C., Gualano, M. R., and Ricciardi, W. (2012). Telemedicine for the reduction of myocardial infarction mortality: A systematic review and a meta-analysis of published studies. *Telemedicine Journal and E-Health*, 18(5), 323–328. doi:10.1089/tmj.2011.0158.

Department of Health and Human Services. (2012). Health information privacy. Breaches affecting 500 or more individuals. Retrieved August 23, 2013, from http://www.hhs.gov/ocr/privacy/hipaa/administrative/breachnotificationrule/breachtool.html.

Department of Health and Human Services. (2013). Modifications to the HIPAA Privacy, Security, Enforcement, and Breach Notification rules under the Health Information Technology for Economic and Clinical Health Act and the Genetic Information Nondiscrimination Act; other modifications to the HIPAA rules. *Federal Register*. Retrieved September 15, 2013, from http://www.ncbi.nlm.nih.gov/pubmed/23476971.

Dolan, B. (2012). Kaiser Permanente offers patients Android app for EMR access. *mobihealthnews*. Retrieved October 09, 2013, from http://mobihealthnews.com/16047/kaiser-permanente-offers-patients-android-app-for-emr-access/.

Doukas, C., Maglogiannis, I., Tsanakas, P., Malamateniou, F., and Vassilacopoulos, G. (2011). mPharmacy: A system enabling prescription and personal assistive medication management on mobile devices. In J. C. Lin and K. S. Nikita (eds.), *Wireless Mobile Communication and Healthcare*. Springer, Berlin, Germany, pp. 153–159. doi:10.1007/978-3-642-20865-2_20.

Eisenberger, U., Wüthrich, R. P., Bock, A., Ambühl, P., Steiger, J., Intondi, A., Kuranoff, S. et al. (2013). Medication adherence assessment: High accuracy of the new ingestible sensor system in kidney transplants. *Transplantation*, 96(3), 245–250. doi:10.1097/TP.0b013e31829b7571.

Elwyn, G., Frosch, D., Thomson, R., Joseph-Williams, N., Lloyd, A., Kinnersley, P., Cording, E. et al. (2012). Shared decision making: A model for clinical practice. *Journal of General Internal Medicine*, 27(10), 1361–1367. doi:10.1007/s11606-012-2077-6.

Farr, C. (2013). Health app makers to feds: Dithering on regulation is stifling innovation. *VentureBeat*. Retrieved July 12, 2013, from http://venturebeat.com/2013/03/19/health-app-makers-to-feds-dithering-on-regulation-is-stifling-innovation/.

Farrell, B., Szeto, W., and Shamji, S. (2011). Drug-related problems in the frail elderly [letter]. *Canadian Family Physician*, 57(2), 168–169.

Ferguson, T. and e-Patient Scholars Working Group. (2007). e-patients: How they can help us heal healthcare. Patient advocacy for health care quality, pp. 1–126. http://e-patients.net/e-Patients_White_Paper.pdf.

Fogg, B. (2002). *Persuasive Technology: Using Computers to Change What We Think and Do*. 1st edn., p. 312. Morgan Kaufmann, San Francisco, CA.

Fogg, B. and Eckles, D. (eds.). (2007). *Mobile Persuasion: 20 Perspectives on the Future of Behavior Change*. 1st edn., pp. 1–166. Stanford Captology Media, Palo Alto, CA.

Garner, J. B. (2010). Problems of nonadherence in cardiology and proposals to improve outcomes. *American Journal of Cardiology*, 105(10), 1495–1501. doi:10.1016/j.amjcard.2009.12.077.

Gazmararian, J. A., Kripalani, S., Miller, M. J., Echt, K. V, Ren, J., and Rask, K. (2006). Factors associated with medication refill adherence in cardiovascular-related diseases: A focus on health literacy. *Journal of General Internal Medicine*, 21(12), 1215–1221. doi:10.1111/j.1525-1497.2006.00591.x.

Goff, D. C., Hill, M., and Freudenreich, O. (2010). Strategies for improving treatment adherence in schizophrenia and schizoaffective disorder. *Journal of Clinical Psychiatry*, 71(Suppl 2), 20–26. doi:10.4088/JCP.9096su1cc.04.

Graham, J., Tomcavage, J., Salek, D., Sciandra, J., Davis, D. E., and Stewart, W. F. (2012). Postdischarge monitoring using interactive voice response system reduces 30-day readmission rates in a case-managed Medicare population. *Medical Care*, 50(1), 50–57. doi:10.1097/MLR.0b013e318229433e.

Haffey, F., Brady, R. R. W., and Maxwell, S. (2013). Smartphone apps to support hospital prescribing and pharmacology education: A review of current provision. *British Journal of Clinical Pharmacology*, 77(1), 31–38. doi:10.1111/bcp.12112.

Heron, K. E. and Smyth, J. M. (2010). Ecological momentary interventions: Incorporating mobile technology into psychosocial and health behaviour treatments. *British Journal of Health Psychology*, 15(Pt 1), 1–39. doi:10.1348/135910709X466063.

HIMSS Analytics. (2012). HIMSS analytics survey demonstrates widespread use of mobile devices to support patient care activities. Retrieved October 09, 2013, from http://www.himssanalytics. org/about/NewsDetail.aspx?nid = 81558.

HIMSS Media. (2013). *Leveraging Mobile Technology to Reduce Hospital Readmissions*, p. 4. New Gloucester. Retrieved from http://www.mhimss.org/sites/default/files/resource-media/ pdf/Qualcomm White Paper.FINAL_pdf. Accessed on January 15, 2014.

Hoppe, R., Brauch, H., Kroetz, D. L., and Esteller, M. (2011). Exploiting the complexity of the genome and transcriptome using pharmacogenomics towards personalized medicine. *Genome Biology*, 12(1), 301. doi:10.1186/gb-2011-12-1-301.

Horvath, T., Azman, H., Kennedy, G. E., and Rutherford, G. W. (2012). Mobile phone text messaging for promoting adherence to antiretroviral therapy in patients with HIV infection. *Cochrane Database of Systematic Reviews*, 3, CD009756. doi:10.1002/14651858.CD009756.

Jacobs, R., Ownby, R., Caballero, J., and Kane, M. (2013). Development of a linguistically appropriate and culturally relevant intervention to increase health literacy in HIV-positive Spanish-speaking Hispanics in the United States. In *6th International Conference of Education, Research and Innovation (ICERI)*, pp. 3174–3184. Seville (Spain).

Kalichman, S. C., Catz, S., and Ramachandran, B. (1999). Barriers to HIV/AIDS treatment and treatment adherence among African-American adults with disadvantaged education. *Journal of the National Medical Association*, 91(8), 439–446.

Källander, K., Tibenderana, J. K., Akpogheneta, O. J., Strachan, D. L., Hill, Z., ten Asbroek, A. H. A., Conteh, L., Kirkwood, B. R., and Meek, S. R. (2013). Mobile health (mHealth) approaches and lessons for increased performance and retention of community health workers in low- and middle-income countries: A review. *Journal of Medical Internet Research*, 15(1), e17. Retrieved from http://www.jmir.org/2013/1/e17/. Accessed on January 15, 2014.

Kane, J. M., Perlis, R. H., DiCarlo, L. A., Au-Yeung, K., Duong, J., and Petrides, G. (2013). First experience with a wireless system incorporating physiologic assessments and direct confirmation of digital tablet ingestions in ambulatory patients with schizophrenia or bipolar disorder. *Journal of Clinical Psychiatry*, 74(6), e533–e540. doi:10.4088/JCP.12m08222.

Kaplan, R. M. and Stone, A. A. (2013). Bringing the laboratory and clinic to the community: Mobile technologies for health promotion and disease prevention. *Annual Review of Psychology*, 64, 471–498. doi:10.1146/annurev-psych-113011-143736.

Kaufman, N. (2010). Internet and information technology use in treatment of diabetes. *International Journal of Clinical Practice. Supplement*, 166, 41–46.

Kitson, A., Marshall, A., Bassett, K., and Zeitz, K. (2013). What are the core elements of patient-centred care? A narrative review and synthesis of the literature from health policy, medicine and nursing. *Journal of Advanced Nursing*, 69(1), 4–15. doi:10.1111/j.1365-2648.2012.06064.x.

Kyser, M., Buchacz, K., Bush, T. J., Conley, L. J., Hammer, J., Henry, K., Kojic, E. M., Milam, J., Overton, E. T., Wood, K. C., and Brooks, J. T. (2011). Factors associated with non-adherence to antiretroviral therapy in the SUN study. *AIDS Care*, 23(5), 601–611. doi:10.1080/09540121.2010.525603.

Ledur, P. S., Leiria, L. F., Severo, M. D., Silveira, D. T., Massierer, D., Becker, A. D., Aguiar, F. M., Gus, M., and Schaan, B. D. (2013). Perception of uncontrolled blood pressure and non-adherence to anti-hypertensive agents in diabetic hypertensive patients. *Journal of the American Society of Hypertension*, 7(6), 477–483. doi:10.1016/j.jash.2013.07.006.

Lewey, J., Shrank, W. H., Bowry, A. D. K., Kilabuk, E., Brennan, T. A., and Choudhry, N. K. (2013). Gender and racial disparities in adherence to statin therapy: A meta-analysis. *American Heart Journal*, 165(5), 665–678, 678.e1. doi:10.1016/j.ahj.2013.02.011.

Lieberman, J. A., Stroup, T. S., McEvoy, J. P., Swartz, M. S., Rosenheck, R. A., Perkins, D. O., Keefe, R. S. et al. (2005). Effectiveness of antipsychotic drugs in patients with chronic schizophrenia. *New England Journal of Medicine*, 353(12), 1209–1223.

Locock, L. (2003). Healthcare redesign: Meaning, origins and application. *Quality and Safety in Health Care*, 12(1), 53–57.

Logan, A. G. (2013). Transforming hypertension management using mobile health technology for telemonitoring and self-care support. *Canadian Journal of Cardiology*, 29(5), 579–585.

López, L. and Grant, R. W. (2012). Closing the gap: Eliminating health care disparities among Latinos with diabetes using health information technology tools and patient navigators. *Journal of Diabetes Science and Technology*, 6(1), 169–176.

Lundberg, G. D. (2009). Why health care professionals should practice participatory medicine: Perspective of a long-time medical editor. *Journal of Participatory Medicine*, 1(1), e3. Retrieved from http://www.jopm.org/opinion/commentary/2009/10/21/why-health-care-profes-sionals-should-practice-participatory-medicine-perspective-of-a-long-time-medical-editor-2/. Accessed on January 15, 2014.

MacKellar, B. and Leibfried, M. (2013). Designing and building mobile pharmacy apps in a healthcare IT course. In *Proceedings of the 13th Annual ACM SIGITE Conference on Information Technology Education—SIGITE'13*. pp. 153–154. ACM Press, New York. doi:10.1145/2512276.2512304.

Marcum, Z. A., Zheng, Y., Perera, S., Strotmeyer, E., Newman, A. B., Simonsick, E. M., Shorr, R. I., et al (2013). Prevalence and correlates of self-reported medication non-adherence among older adults with coronary heart disease, diabetes mellitus, and/or hypertension. *Research in Social and Administrative Pharmacy*, 9(6), 817–827. doi:10.1016/j.sapharm.2012.12.002.

MEMOTEXT. (2011). Case studies. Retrieved from http://www.memotext.com/case-studies/. Accessed on January 15, 2014.

Merolli, M., Gray, K., and Martin-Sanchez, F. (2013). Health outcomes and related effects of using social media in chronic disease management: A literature review and analysis of affordances. *Journal of Biomedical Informatics*, 46(6), 957–969. doi:10.1016/j.jbi.2013.04.010.

Miliard, M. (2012). mHealth industry "in learning mode" for privacy and security. *Government Health IT*. Retrieved July 12, 2013, from http://www.govhealthit.com/news/mhealth-industry-learning-mode-privacy-and-security.

Muessig, K. E., Pike, E. C., Legrand, S., and Hightow-Weidman, L. B. (2013). Mobile phone applications for the care and prevention of HIV and other sexually transmitted diseases: A review. *Journal of Medical Internet Research*, 15(1), e1. Retrieved from http://www.jmir.org/2013/1/e1/. Accessed on January 15, 2014.

Norman, C. D. and Skinner, H. A. (2006). eHealth literacy: Essential skills for consumer health in a networked world. *Journal of Medical Internet Research*, 8(2), e9. doi:10.2196/jmir.8.2.e9.

Norman, I. D., Aikins, M. K., and Binka, F. N. (2011). Ethics and electronic health information technology: Challenges for evidence-based medicine and the physician-patient relationship. *Ghana Medical Journal*, 45(3), 115–124.

Osborn, C. Y., Cavanaugh, K., Wallston, K. A., Kripalani, S., Elasy, T. A., Rothman, R. L., and White, R. O. (2011). Health literacy explains racial disparities in diabetes medication adherence. *Journal of Health Communication*, 16(Suppl 3), 268–278. doi:10.1080/10810730.2011.604388.

Pharmacy Times. (2013). Startup incorporates smartphone tech into Rx bottles. *Pharmacy Times*. Retrieved December 21, 2013, from http://www.pharmacytimes.com/publications/issue/2013/May2013/Startup-Incorporates-Smartphone-Tech-Into-RX-Bottles.

Quinn, C. C., Shardell, M. D., Terrin, M. L., Barr, E. A., Ballew, S. H., and Gruber-Baldini, A. L. (2011). Cluster-randomized trial of a mobile phone personalized behavioral intervention for blood glucose control. *Diabetes Care*, 34(9), 1934–1942. doi:10.2337/dc11-0366.

Rainie, L. (2013). Cell phone ownership hits 91% of adults. *Pew Research Center*. Retrieved August 09, 2013, from http://www.pewresearch.org/fact-tank/2013/06/06/cell-phone-ownership-hits-91-of-adults/.

Rainie, L. and Wellman, B. (2012). *Networked: The New Social Operating System*. 1st edn. The MIT Press., Cambridge, MA.

Shapiro, G. (2013). OP-ED: Don't let privacy concerns stifle innovation. *Nextgov*. Retrieved July 12, 2013, from http://www.nextgov.com/emerging-tech/2013/06/op-ed-dont-let-privacy-concerns-stifle-innovation/65195/.

Smith, A. C. (2013). Effect of telemonitoring on re-admission in patients with congestive heart failure. *Medsurg Nursing*, 22(1), 39–44.

Stanford Medicine X. (2013). Stanford medicine X 2013—Startup presentation—Josh Stein. Retrieved December 13, 2013, from http://www.youtube.com/watch?v = ULAZdWS_5Qs.

Stauffer, B. D., Fullerton, C., Fleming, N., Ogola, G., Herrin, J., Stafford, P. M., and Ballard, D. J. (2011). Effectiveness and cost of a transitional care program for heart failure: A prospective study with concurrent controls. *Archives of Internal Medicine*, 171(14), 1238–1243. doi:10.1001/archinternmed.2011.2744444.

Sun, S. X., Liu, G. G., Christensen, D. B., and Fu, A. Z. (2007). Review and analysis of hospitalization costs associated with antipsychotic nonadherence in the treatment of schizophrenia in the United States. *Current Medical Research and Opinion*, 23(10), 2305–2312. doi:10.1185/030079907X226050.

Toomey, S. L., Sox, C. M., Rusinak, D., and Finkelstein, J. A. (2012). Why do children with ADHD discontinue their medication? *Clinical Pediatrics*, 51(8), 763–769. doi:10.1177/0009922812446744.

Topol, E. (2012). *The Creative Destruction of Medicine: How the Digital Revolution Will Create Better Health Care*. 1st edn., p. 320. Basic Books, New York.

University of Alabama in Huntsville. (2013). Smart pill bottle invented at UAH heads to clinical trials. Retrieved December 13, 2013, from http://www.uah.edu/news/research/6404-smart-pill-bottle-invented-at-uah-heads-to-clinical-trials#.Us7JxPRDtXw.

Weill Cornell Medical College. (2013). Weill cornell to test AdhereTech's wireless pill bottle, thanks to NYC grant. Retrieved December 13, 2013, from http://weill.cornell.edu/news/deans/2013/07_25_13.shtml.

Wilding, M. (2013). 5 ways healthcare providers can reduce costly hospital readmissions. *HealthWorks Collective*. Retrieved October 09, 2013, from http://healthworkscollective.com/ecaring/93606/5-ways-healthcare-providers-reduce-costly-hospital-readmissions.

World Health Organization. (2003). Adherence to long-term therapies: Evidence for action. *World Health*. World Health Organization, Geneva. Retrieved from http://www.who.int/chp/knowledge/publications/adherence_report/en/. Accessed on January 15, 2014.

World Health Organization Global Observatory for eHealth. (2011). *mHealth: New Horizons for Health through Mobile Technologies* (Vol. 3). World Health Organization, Geneva, Switzerland. Retrieved from http://www.who.int/goe/publications/goe_mhealth_web.pdf. Accessed on January 15, 2014.

Yang, J., Yoon, B.-M., Lee, M.-S., Joe, S.-H., Jung, I.-K., and Kim, S.-H. (2012). Adherence with electronic monitoring and symptoms in children with attention deficit hyperactivity disorder. *Psychiatry Investigation*, 9(3), 263–268. doi:10.4306/pi.2012.9.3.263.

Zerhouni, E. A. (2007). The promise of personalized medicine. *NIH Medline Plus*, Winter, 2–3. doi:10.1037/e450962008-002.

34

Health Web Science: Facilitating Health Care and Well-Being Using Examples in Urinary Incontinence, Medical Education, and Diabetes

Grant P. Cumming, Joanne S. Luciano, Sandra MacRury,
Douglas McKendrick, Kate Stephen, and Andrew Chitty

CONTENTS

34.1 Need for Change in Health Service Delivery

Health service delivery is undergoing astonishing change. There is a consensus that current modes of health-care delivery are unsustainable, in both the developed and developing world (Crisp 2010, Sánchez-Serrano 2011). From a global perspective, humanity faces profound questions about how our planet can sustain nine billion people by 2050. Current dietary choices and lifestyles are contributing to an unprecedented burden of chronic noncommunicable diseases. Life expectancy has increased, and environmental and climate change pose additional new challenges. These factors, taken together with a shortage of health professionals, contribute to the challenge of health-care delivery on a global scale (Crisp 2010, WHO 2011). Infectious diseases, the main challenge of the nineteenth and twentieth centuries, have given way to the prevalence of chronic disease (Wanless 2002, Crisp 2010), but nevertheless remains a challenge.

For more than a century now, the model for health care has been a biomedical model (i.e., derived from the germ theory of disease) underpinned by the controlled clinical trial (Wooton 2007) aimed at addressing infectious diseases, the main cause of death in western medicine. However, chronic conditions are closely related to lifestyle choices that need a wholly different approach. In response, medicine is changing to embrace a biopsychosocial model that emphasizes patient-centered care delivered by interdisciplinary provider teams (Johnson 2012). This biopsychosocial model is a call to change (include) the role of the patient and expand the domain of medical knowledge to address directly the needs of each patient (Borrell-Carrió et al. 2004) rather than the needs of a population. The future of health care in this era of chronic disease requires increasing effort directed toward improving personal choices regarding life risks (Keeney 2008) and requires the full engagement of people in their own health care and lifestyle decisions (Wanless 2002, Crowley and Hunter 2005, Bell 2006).

34.2 Role for Information and Communication Technologies

In the 1990s, health information and communication technologies (ICTs) first offered promise to help mitigate against the problems facing the delivery of health care (Ranck 2012). However, it required the cultural shifts that social media and mobile devices have catalyzed since, together with the recognition that many health-care systems are now at tipping point (Crisp 2010) to galvanize communities working in ICTs and health to integrate the Internet and related technologies into the delivery of health care (Ranck 2012). However, despite the widespread belief in the usefulness of mobile technology in health, there is still a paucity of study-based evidence (Chigona et al. 2013). In addition, because of the problems in implementing eHealth projects, in the United Kingdom (Fawdry et al. 2011) and other parts of the world (Kaplan and Harris-Salamone 2009), the health-care ICT industry has been perceived as overpromising and underdelivering, destroying trust and causing skepticism (Smith 2011). We assert that new methodologies and approaches are required to design and evaluate mHealth within the context of both Health Web Science (HWS) and Medicine 2.0, from their experience gained as a digital health-care cluster comprising companies, nongovernment organizations, health boards, and academia locally and internationally.

34.3 Digital Health-Care Ecosystem: P4 + C^n = eIMT

Figure 34.1 illustrates the digital health ecosystem that is being developed in the north of Scotland with its ethos on codesign and collaboration underpinned by the disciplines of HWS and Medicine 2.0 to mitigate against the fixed milestones and rigid methodologies of previous eHealth innovation (Greenhalgh et al. 2010a,b).

34.3.1 Health Web Science

HWS, a subdiscipline of web science (Berners-Lee et al. 2006), studies the web in relation to all aspects of human health. HWS emerged from the observation that there was a profound increase and a predominant use of the web for health-related activity that spanned a spectrum from basic science research to clinical research to lay people seeking social support communities and answers to health-related questions and concerns. A need therefore arose to understand through the academic lens the interplay between health and well-being and the web/Internet, in this case, the lens of web science. HWS discussions began during a web science curriculum meeting in the summer of 2010 at the University of Southampton, England, and continued in the summer of 2011 with a foundational workshop in Koblenz, Germany, under the aegis of the Association of Computing Machinery (Brooks et al. 2011).

HWS aims to interrogate the mutual interplay among the World Wide Web, the health data it contains, and the patients, practitioners, and researchers who utilize it. It is envisaged that HWS will help engineer the future web and web-related technologies to facilitate

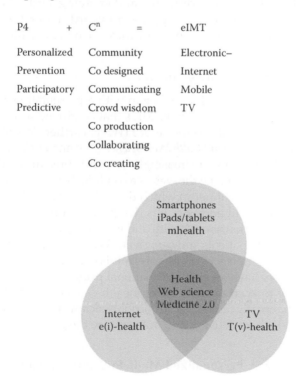

P4	+	C^n	=	eIMT
Personalized		Community		Electronic–
Prevention		Co designed		Internet
Participatory		Communicating		Mobile
Predictive		Crowd wisdom		TV
		Co production		
		Collaborating		
		Co creating		

FIGURE 34.1
P4 + C^n = eIMT digital health ecosystem.

health-related endeavors and empower health professionals, patients, health researchers, and lay communities. HWS is therefore multidisciplinary (Luciano et al. 2013). HWS complements and overlaps with other disciplines, most notably Medicine 2.0 and Health 2.0 (Eysenbach 2008, Hughes et al. 2008).

34.3.2 Medicine 2.0

Since its inception, the web has undergone significant transformations. Initially, the web enabled documents (web pages) to be found and read through a web browser. This web, "Web 1.0," was "read only." Next, individuals began to use the web to author their own content and to interact with one another, "Web 2.0." Increasing use and further development of web technology and infrastructure enabled metadata to be used to give meaning to data and to link data along with web pages and people, "Web 3.0." "Web 3.0" is often referred to as the "Semantic Web." Terms such as "Health 2.0" and "Medicine 2.0" are now in the literature (Van De Belt et al. 2010). These terms reflect the use of computer software and web technologies to support and enable interaction with and creation of user-generated content relevant to health care. Medicine 2.0 has been used to denote the social web in health, medicine, and biomedical research including mHealth (Medicine 2.0 Congress 2014). There is no general consensus as to these definitions; definitions are influenced by the different stakeholders' agendas, of which there are many (Van De Belt et al. 2010). Nevertheless, there seems to be agreement that using the suffix 2.0 denotes the ability of the technology to provide two-way interaction that supports user-generated content, the hallmark of Medicine 2.0. The lack of a specific definition reflects in part that the technology, and how we use it, is dynamic and evolving, and this in turn highlights the requirement for a multidisciplinary approach toward understanding the impact of the current technology on health care, communities, individuals, and society as a whole.

34.3.3 Health Web Science and Medicine 2.0

The dialog between HWS and Medicine 2.0 began with a presentation at the 2012 Medicine 2.0 Conference in Boston, MA (Luciano et al. 2012), that continued formally in the literature (Luciano et al. 2013). This nascent discipline of HWS is further described and developed in the book *Foundations and Trends in Health Web Science* (Luciano et al. in press). The essential motivations behind HWS are many: increasingly, scientific breakthroughs are powered by advanced computing capabilities using the web as a conduit; therefore, it is important to understand and describe the distinct manner in which the web is used for health research, clinical research, and clinical practice. In addition, it is desirable to support consumers who utilize the web for gathering information about health and well-being and to elucidate approaches to provide social support to both patients and caregivers. With worldwide motivation to improve both the effectiveness and efficiency of health care, this is particularly timely. Quality improvement and cost containment have become international priorities as governments, employers, and consumers struggle to keep up with rising health-care costs. The HWS remit includes governance and provenance (Luciano et al. 2013), not typically discussed within the context of Medicine and Health 2.0 where delivery and outcome are the primary foci.

34.3.4 Digital Health Care, Personalized Medicine, and Digital P4™ Medicine

Digital health care involves the use of ICTs to help address the health problems and challenges faced by patients. These technologies include both hardware and software solutions and services. Personalized medicine customizes medical treatment for patients, based on their classifications into subpopulations that share characteristics indicative of susceptibility to a particular disease or response to a specific treatment. Digital P4 medicine more broadly defines personalized medicine as a health-care paradigm, with four key attributes, predictive, personalized, participatory, and preventative (P4), and uses a range of technologies from the fields of ICT, medical equipment, and pharmaceutical devices to deliver P4 medicine (HIE 2012). In 2003, Leroy Hood introduced the term P4; his vision was that it transforms the practice of medicine, moving it from a largely reactive discipline (with an emphasis on sickness and treatment) to a proactive (prevention, self-caring) one (P4 Medicine Institute 2012). The ambition is that patients will benefit from better diagnoses leading to individually targeted and thus more effective treatments as a consequence of new forms of active participation by patients and consumers in the collection of personal health data, such that a virtual data cloud of billions of health-relevant data points will surround each individual patient, thereby accelerating discovery science and simplifying treatment selection (P4 Medicine Institute 2012).

34.3.5 Internet of Things

The phrase the "Internet of Things" was proposed by Kevin Ashton in 1999 (Ashton 2009). The Internet of Things refers to the objects that can be uniquely identified via the Internet. These objects, when connected to the Internet, enable people and things to be connected with each other, anytime, anyplace, with anything and anyone, ideally using any path/ network and service (European Commission 2009). The Internet of Things is becoming realized in the health domain. Already, movements such as the quantified self (the incorporation of technology to collect personal daily life data using wearable sensors), smart cities (a movement that utilizes Internet-enabled, broadband, wireless, and digital equipment in the management by a city of their infrastructure systems for electricity, water supply, waste), and smart homes (a movement that uses automation technologies to provide home owners feedback information by monitoring many aspects of a home) are gaining momentum and further pushing adoption and development.

Handling the quantity of "big data" generated from the Internet of Things through curation, visualization, and interpretation of data is a fertile area for innovation. Analysis of "big data" will be instrumental in bringing about the "health-care singularity" (Gillam et al. 2009) when the speed of medical innovation enables research findings to be put into clinical practice instantaneously (in theory). This finding-to-practice lag time has continued to decrease, linearly, with the current lag approximately 17 years from research finding to accepted clinical practice (Gillam et al. 2009).

34.3.6 Collaboration, Codesign, and the Need for an Agile Iterative Approach

Underpinning the paradigm shift from a treatment model to self-caring medicine are collaborative, cocreative, codesign principles that encompass an agile iterative methodology across multiple platforms that aim to maximize engagement with the user. Too many eHealth projects have assumed that adoption would be automatic once the system was deployed. However, when the systems were deployed, adoption did not automatically follow and indeed did not happen (Greenhalgh et al. 2010a,b). We need to understand individual behaviors and motivations in the new biopsychosocial model of lifestyle choices. New approaches are therefore required to understand the complexities and enable design and codesign of innovative approaches using digital, web-enabled mobile technologies in health care.

Participatory action research (PAR) is a methodology based on reflection, data collection, and action. It aims to improve health and reduce health inequities by involving the people who, in turn, will be motivated to take actions to improve their own health (Baum et al. 2006). Cooperative inquiry comes under the umbrella of PAR (Heron and Reason 2006). The aim of cooperative inquiry is to "research 'with' rather than 'on' people." It emphasizes that all active participants are fully involved in research decisions as coresearchers. These approaches lend themselves well to the agile, iterative methodology, whereby each iteration or cycle of development is evaluated and the lessons learned then fed into the next cycle. It is increasingly being utilized as a methodology from a patient-centered perspective and is being used in the north of Scotland in building digital solutions.

34.4 Need for Evidence-Informed Medicine and Mixed Method Evaluation

Internet-delivered interventions have the potential to combine the tailored approach of individual face-to-face interventions while maintaining the scalability of public health interventions with very low marginal costs per additional user. The main benefits are realized when these applications are used by a large number of people. Using Internet-based interventions can substantially reduce the cost and increase the benefit when the potential "wisdom of the crowd" is enabled to emerge from the health ecosystems (Luciano et al. 2013). In addition, because people interact with each other, what emerges are at this moment "the unknown unknowns." Undoubtedly, they will take many forms and in themselves are properties that need to be studied. Internet-delivered health care, development, engineering, and how it is used need therefore to be understood both at the micro level, that is, building and testing applications, and the macro level, that is, studying the use of the microsystem by many users interacting with one another (Hendler et al. 2008).

Modern medical care has been influenced by two paradigms, evidence-based medicine (EBM) and patient-centered medicine (PCM). They focus on different aspects of medical care and have little in common (Bensing 2000). EBM tends to be disease orientated and doctor centered rather than patient centered (Sweeney et al. 1998). EBM trials do not necessarily reflect what happens in practice, and the trials may only involve a small percentage of people who meet the inclusion criteria. PCM recognizes the patient as the expert living with the condition who has a contribution to bring to the table (Mezzich et al. 2010, 2011). A patient-centered approach therefore recognizes that the patient is a source of data and knowledge and an integral part of the development team (codevelopment) when designing an intervention. In medicine, there is often no correct treatment; the right treatment for a patient involves factors not often articulated or considered within the health-care systems. For example, some patients prefer natural approaches, while others prefer the latest technology; some patients are risk takers and others are conservative (Groopman and Hartzband 2011). In each of these dimensions, the degree to which the preference is held is a factor in settling on the correct treatment choice. The EBM paradigm is therefore shifting to evidence-informed medicine and the formulation of personalized models of care. Notably, the personalized model of care is informed by, but not based on, the E of EBM (Miles and Loughlin 2011). There is a shift to evidence-informed medicine and an increasing recognition that mixed method approaches are needed, that is, those that involve both qualitative and quantitative methodologies in the evaluation of the digital delivery of health care (Lilford et al. 2009, Cresswell et al. 2011). Indeed, increasingly, mixed method design is being used to address the challenges of effective patient care and to understand the role of consumerism in promoting high-quality health-care services.

With these new approaches, there is a growing awareness that Internet interventions work, but as yet, it is unclear for whom, for what behaviors, and for which medical conditions (Murray 2012).

34.5 First Do No Harm, Do No Evil: The Need for Health Web Science and Medicine 2.0

The promise of mHealth is as yet unrealized. An ambition of mHealth is that both patients and health-care providers will benefit from more targeted and effective treatments, with patients being more informed and therefore more responsible for their health. An optimistic view envisions that individuals will have increased choice and control over their health, with reassurance of health and early detection of disease. A skeptical view is that proactive health care will create ethical dilemmas; increase anxiety and needless confusion, not least in the already "worried well"; and thereby lead to "overdiagnosis" and unnecessary invasive procedures with their inherent risks.

While Internet access is freely available in most "first-world" countries, rural areas in these countries can be disadvantaged; the United Kingdom provides one example (Ofcom 2013). The advent of fourth-generation long-term evolution (4G LTE) mobile phone technology* is hoped to address many of these inequalities and allow universal coverage and availability of mHealth (Vodafone 2014). This inequality in Internet access is even more pronounced in "third-world" countries that have poor Internet coverage and speeds. In many of these countries, traditional Internet subscriptions have been bypassed by mobile phone subscriptions as the dominant method of Internet access. For example, in Botswana, Internet coverage is poor with only 269,000 users or 12.8% of the population (Internet World Stats 2012) and extremely slow with an average user bandwidth of 0.93 Mbps (ranked 156th fastest in the world) (oAfrica 2011).

The increasing mobile network coverage has the potential to facilitate health-care delivery and allow health-care projects that use mobile networks and devices researched and developed in the first world to be fully applicable to the third world and vice versa. mHealth's ability to reach populations who face barriers to accessing health services such as lack of time, lack of money, or embarrassment at their health condition is attractive for companies and individuals to exploit without an evidence base for their efficacy.

To enable mobile delivery without proper study may violate the Hippocratic Oath, historically taken by physicians and other health-care professionals, to practice medicine honestly and not cause harm in the name (or practice) of doing good. In a similar vein, a central pillar of Internet giant Google's corporate identity and core values is the mantra "Do the right thing; don't be evil." The Google "oath" is to do good things for the world even if it must forgo short-term gains. It is thus up to health web scientists together with colleagues in Medicine 2.0 to demand that mHealth applications undergo evaluation and the public is aware of the findings. In doing so, these communities remain true to Hippocrates' (the father of medicine) mantra of *first do no harm* and Google's maxim, *do no evil*.

---

* 4G LTE is a standard for Internet-based wireless communication of high-speed data for mobile phones. 4G LTE is an acronym for fourth-generation long-term evolution.

34.6 Social Networking

Peer-to-peer health care acknowledges that patients and caregivers are knowledgeable about themselves, about each other, and about treatments. Many people want to share what they know (whether correct or incorrect) to help other people. This community-powered health care through social networks taps into two disparities (Fox and Jones 2009):

1. The mobile difference: People with smartphones are more likely to share, create, and participate, not just consume information.
2. The diagnosis difference: Having a chronic disease significantly increases the likelihood that Internet users both contribute to and consume peer-generated health content, thereby learning from each other, not just from institutions.

Networks are not inevitably good for health: via social networks, obesity can arguably be "spread" by person-to-person interaction and the normalization of obesogenic habits (Christakis and Fowler 2007). Social norms affect other health areas: a student's risk of smoking is doubled if more than half of his or her social network smokes (Christakis and Fowler 2007). Facebook and other online social networking platforms may potentially be used to incentivize good health behaviors and reverse these adverse health behaviors. Health-related "apps" that can be downloaded to smartphones and tablets may provide another alternative to promote positive health behaviors.

Gaming in health care aims to bridge the gap between medicine, entertainment, technology, and education and should entertain while achieving a given health benefit. Social games are web-based games that can be played with others and include interactive elements or content that can be shared online. There is an increasing interest in online gaming because games can be integrated with social networks; the popularity of downloadable apps on smartphones is growing. A key driver is the competitive element associated with sharing achievements publicly (Zaidi 2011). Read and Shortell (2011) highlighted the potential of Internet games to promote behavior change in prevention and treatment, and a more recent report suggests that "gamification" may improve the effectiveness of traditional health interventions to motivate behavior change and can thereby lead to better health outcomes (Gotsis et al. 2013). The Games for Health Journal: Research, Development, and Clinical Applications, launched in 2011 (Mary Ann Liebert, Inc.), is a bimonthly peer-reviewed journal dedicated to the development, use, and applications of game technology for improving physical and mental health and well-being and the first to address the application of gaming in health care.

34.7 mHealth in a World of Bring Your Own Device

The travel, hospitality, financial services, retail, and telecommunications sectors have quickly utilized mobile technology to provide users with platforms that enable them to manage their affairs and social interactions. The health-care sector has been slow to adopt the Internet protocol (IP) technologies and is only now beginning to provide users with enabling platforms. As the other sectors have found, at each stage of their digital transformation, a key driver is the rapid adoption by citizen consumers of both the telecom devices and the telecom services required to connect them to the network. Thus, many citizens now live in a bring your own device (BYOD) world. However, devices and connectivity are not evenly distributed across populations, and the development of mHealth services must be conducted in the context of contemporary data on consumer technology adoption within target populations. Most see the smartphone as the pivotal access device for mHealth services. Thirty-one percent of smartphone users in the United States used their phones to look for health or medical information online (Duggan and Rainie 2012). When this is considered alongside the trend of uptake of smartphones, which approached 50% in regions of the United Kingdom in 2012 (Ofcom 2012), and is expected to continue to increase across the world (Schmidt and Cohen 2013), it is clear that mHealth is a tool that offers opportunities to provide health information to vast populations across the world. However, although 80% of 18–29-year-olds in America have a smartphone, only 18% of people over 65 years own one (Duggan and Rainie 2012), and while the 60–70 age group has adopted smartphones, there is a significant drop-off in adoption beyond 75 years as there is with fixed line broadband and home Wi-Fi access (Ofcom 2012). The focus group findings described earlier may help to explain the reticence in older people to adopt smartphone technology, and this may result in increased future demand for tablets that may be more acceptable and easier to use for older people. However, it is possible to overstress this barrier to usage among older groups and for it to be used as a reason not to explore mHealth opportunities. For instance, texting, an activity often talked of as the preserve of the young, is widely used by over-65s. A survey by the U.K. telecom regulator Office of Communications (OFCOM) in 2012 found that 21% of people over 65 sent a text message every day (Ofcom 2012), that is, their mobiles were used more for texting than for mobile calls. Evidence is emerging that tablet devices with their larger screens and touch-based interfaces may be particularly important for mHealth services that target older groups. The majority of the 12.1 million tablet devices in the United Kingdom are now owned by people over 45 with those over 55 representing the largest group (Newsworks 2013). This profile is almost unique in the adoption of new technology devices. It remains true that generally older (and poorer) populations in developed economies are less avid adopters and consumers of mobile technologies but this is a matter of degree, not kind, and is changing relatively rapidly. Even those researchers who are concerned that multiple and mobile connectivity may create a second "digital divide" recognize that though there are fewer "next-generation Internet users" over 65 than under 65, there is an increasing proportion of these multiply connected, mobile, always-on users in every demographic group (Dutton and Blank 2011).

The brave new world of technology raises important ethical issues. Health-care providers must be sensitive to use technology to be an enabler, rather than a controller of older adults, who may not desire externally imposed health and safety monitoring systems (Rogers and Fisk 2010).

34.7.1 Examples from the North of Scotland

This section presents three examples of mHealth in the north of Scotland within its digital health framework program illustrating points made in this chapter. Specifically, the role of evidence-informed medicine, mixed method evaluation, gamification, increasing uptake of Twitter in a medical community, and the potential benefits of mHealth in diabetes with the start of an evaluation in a remote and rural location involving codesign will be highlighted.

34.7.1.1 mHealth and Urinary Incontinence

The populations who suffer from urinary incontinence have been characterized as individuals who may delay or neglect to seek help from a health professional (Koch 2006). One way to address this is to provide support via a smartphone app. For patients with a smartphone, the process of accessing an app requires no face-to-face interaction and can be performed anywhere and at any time. This may suit individuals who neglect to make an appointment with a health professional or who are uncomfortable/embarrassed speaking to anyone about their incontinence. Funding was obtained to support a PhD for the development of an eHealth tool for the management of urinary incontinence: the role of apps for pelvic floor muscle exercise (PFME). The research undertaken for this thesis comprises three main methodologies: focus groups, a randomized controlled trial, and interviews by telephone, that is, mixed methodology evaluation with a view to evidence-informed medicine. The evidence shows that PFME helps mitigate against urinary incontinence, PFME being the first line of treatment for urinary incontinence (NICE 2006).

As part of the PhD research study (dissertation in preparation, Stephen K. 2014), focus groups were held in the north of Scotland to gauge the views of women regarding apps for PFME. Individual participants were given an iPod touch for the duration of the focus group to allow them to test a small number of different apps. Separate focus groups were held for young mothers and older women. The younger women were more accepting of the apps and could see the benefits of having this type of app on their smartphone. The majority of older women found the iPod touch too small for comfortable use; they struggled to read from the small screen and found the touch screen difficult to navigate, especially the on-screen keypad. Many of the older women suggested that the content of the apps would be useful on a platform they were familiar with, namely, TV, DVD, or desktop computer.

The iPod touch devices were subsequently used in a PFME randomized controlled trial in Moray, in the north of Scotland. The use of iPod touches in this trial, as opposed to smartphones, may have impacted the results because individuals in this study were not in the habit of carrying the device and the devices were not enabled to allow Internet access (which some features of the apps required). The results revealed that most of the older participants in the intervention group could not "get to grips" with the technology and others quickly got bored of or frustrated with the apps. For the participants who found the apps helpful in exercising their pelvic floor muscles, the app helped them to concentrate on the exercise (duration of muscle contractions and number of repetitions). Others found the apps fun or relaxing to use and this had a positive impact on their motivation to exercise.

These findings suggest that where individuals engage with the technology, it can be effective in supporting motivation and adherence. It cannot be assumed, however, that all smartphone users will find all apps useful, and so the ability for individuals to choose apps that suit them or to personalize their app can be important in the effectiveness of the technology. Subtle differences such as the font and the color scheme and even the accent of the voice giving instructions can act as barriers and promoters for the use of an app. However, if it is difficult or takes too long to adjust the settings of the app, the benefits that could come from personalization may never be realized. For some, the gamification of PFME added incentivization.

34.7.1.2 mHealth, Twitter, and Medical Professionals

Twitter is emerging as a potentially valuable means of real-time mass communication of health-care information, for example, disaster alerts, drug safety warnings, and tracking disease outbreaks and alerts (Terry 2009). Social networking is increasingly being recognized by health-care organizations as a valuable method of sharing text, pictures, video, and audio with patients. However, social networking's raison d'être is a two-way interaction creating an ideal arena for patient-centered care, whereby patients communicate with and receive information and advice relevant to their specific health-care needs from a variety of sources including health-care services. Social networking provides a new opportunity for patients using mobile devices to provide real-time feedback to health-care providers on the care they have received by using forums such as the Patient Opinion on Twitter (https://twitter.com/patientopinion) and Facebook.

Twitter, a microblogging tool, is easily accessed via a number of platforms, namely, the Twitter website, applications developed for smartphones and tablets, and via short message service (SMS) from mobile phones (in certain countries). Less than half of the tweets posted are through the website, with the majority of users preferring to use mobile apps on their smartphones or tablets (Sysomos 2009) making it an ideal medium for mHealth.

Twitter is not purely a tool for social networking with limited use to medical professionals. Twitter is finding traction in business and is beginning to be used in academia for many purposes including rapid sharing and dissemination of information and for citing articles. Organizers, delegates, and speakers each have found tweeting to be beneficial in their own domains as well as a tool for online discussion at meetings and conferences. This function of Twitter is achieved by the use of a digital "back channel," which is a nonverbal, real-time projection of the tweet. During digital back channel use, the speaker presents in the usual traditional manner in the "front" area, and simultaneously, the audience and people distant from the meeting can communicate with each other using the "back" area. This use of Twitter and other social media has the potential to change the health communications space associated with conferences.

A report describing the uptake and use of a Twitter stream as an integral part of the communication structure of an anesthetic conference (McKendrick et al. 2012) demonstrated a 530% increase in the number of tweets posted and a 300% increase in the number of people tweeting, compared with those that "erupted" spontaneously at a reference anesthetic conference (McKendrick 2012). Most anesthetists still have to adopt this new educational tool.

There are many ways to use Twitter at a medical conference to enhance the experience for both those at the conference and those unable to attend. Attendees at this conference used Twitter to take notes from presentations. Because of the small length of tweets (140 character maximum), these usually take the form of learning points or salient messages. This can be a useful way to remember the little gems of information learned at a congress, and by looking at what other people at the congress have tweeted or retweeted reinforces the "real" learning points from each talk or session. Although not used at the conference described, Twitter can be used to pose questions to speakers. This could lead to a Twitter debate to argue, discuss, seek clarification, and post questions and answers, during a presentation. This is usually done on mobile devices during the talk and is often far less disruptive than whispering to one's neighbor. By viewing posts, asking questions, and joining discussions, Twitter can even allow delegates to participate in parallel sessions at the same time. The speakers at this conference used Twitter to promote their subject or "take-home" messages, but could also have received feedback from the delegates via tweets on their presentation, where comments about the composition and legibility of slides, for example, could influence future presentations. An archive of the tweets posted can be retained as a method of record keeping for the purposes of recollection and continuous professional development and demonstrate reflection during the meeting for revalidation. Twitter can be further used as a method of "amplifying" the congress to a wider audience. Tweets posted by delegates can be read by people who were not present at the meeting, who then in turn retweet the message to their followers; this creates a second tier of information spreading. This retweeting can continue for many tiers. Some Twitter users have hundreds of thousands of followers and information can be disseminated very quickly. The social element of Twitter can be used to promote the official conference dinner, lost-and-found items, unofficial social events and dinners, tips for accommodation, or places to go for food and entertainment. Physical meetings can be arranged with other tweeters at a conference (a "tweetup").

Although there are many advantages to Twitter, there are some potential pitfalls. By default, all tweets made are public unless an individual changes the settings or sends the message as a direct message (DM). As an individual posting tweets, it is critically important not to broadcast any information or views on Twitter that might conflict with or defame employers, colleagues, students, academics, researchers, and other university stakeholders. Health-care professionals have concerns about the use of Twitter that need to be addressed, including patient, personal, and other health-care professional's privacy. Such concerns are increasingly being recognized and discussed (McCartney 2012), and guidance is now being produced by organizations such as the British Medical Association to provide practical and ethical advice to assist doctors (British Medical Association 2011).

34.7.1.3 mHealth and Diabetes

Almost every month, it seems, someone is coming up with a new mHealth device or system to help people living with diabetes manage their health. As a chronic illness, the management of both type 1 and type 2 diabetes is an area where mobile devices may help given the opportunity for data exchange on a routine basis. Blood glucose measurements, insulin dose adjustment, and other physiological data including weight and diet history are most commonly collected through devices such as blood glucose meters, tablets, personal digital assistants, and smartphones via wired and wireless connections to websites to permit sharing. These data albeit real time or historical, recorded, or tracked could contribute to short- and long-term benefits, through empowerment to active self-management.

While increased glucose testing is a frequent output, it is not necessarily translated into increased knowledge or improvements in blood sugar control (Cafazzo et al. 2012). mHealth and telemedicine are most likely to be effective in diabetes when linked to education and behavior advice (Farmer et al. 2005). Mobile phone use can include texting education and motivational input, with treatment reminders and links to healthy living management tools. Nevertheless, personalized education or decision support is often missing in diabetes apps (Chomutare et al. 2011), although overall, evidence is limited that phone messaging in long-term conditions aids self-management (de Jong et al. 2012).

However, it is significant that there is growing evidence that mHealth solutions that require patient/user input and thus provide feedback to the user are significantly more effective than those where patient data are sent automatically from sensors (Agboola et al. 2013). This supports the benefits of implementing an active learning model that provides user feedback and contrasts with mHealth solutions with traditional teleHealth approaches to home monitoring. Indeed, one of the authors, Joseph Kvedar of Boston-based Partners HealthCare, has subsequently suggested that an advantage of mHealth will be to exploit users "addiction" to their mobile devices to capture and represent data (Kvedar 2013).

Diabetes apps available on mobile platforms are a huge area for growth, albeit with the caveat that most apps are platform specific and connectivity may be an issue limiting access, thus highlighting the need for common platforms. While products often appear easy to use and score well in patient satisfaction and usability evaluations, it is becoming clear that codesign at an early stage in the prototype design process is important to obtain specific end-user input and to meet user expectations (McCurdie et al. 2012).

Data collection, to be of best use to patients and health-care professionals, needs to have a secure audit trail with the capability to be incorporated, preferably by autopopulation, into the electronic patient record. Despite this, few studies have addressed privacy and sensitivity and the need for regulatory approval around patient data and confidentiality and treatment recommendations need to be explored. Studies to date have for the most part had low sample sizes and short intervention times that preclude effective integration of mHealth options into normal health-care practice in diabetes (Holtz and Lauckner 2012). More rigorous testing is thus required with larger clinical trials to provide clear evidence that clinical benefits are derived (Liang et al. 2011, Baron et al. 2012) in addition to cost effectiveness and sustainability.

In due course, mHealth could be a major contributor to smart health (sHealth), informing predictive and preventative health care through, for instance, smart sensing, with multidimensional analysis of the cumulative big data providing improved health outcomes.

From the earlier discussion, we believe that mHealth in diabetes is integral to the digital approach to diabetes care and as part of the virtual clinic in the north of Scotland complementing conventional contact practice. Virtual clinics can take several forms, with or without the patient present, utilizing digital technology for viewing results and images and communication between health professionals and patients. Ultimately, the goal is to triage and prioritize attendance at conventional clinics and to reduce unnecessary travel and appointments. This has particular relevance to the remote and rural agenda, offering patient-centered options for support that addresses a number of the issues raised by current service delivery models. A study is therefore underway to determine the usability and acceptability of a variety of digital communication modes (WebEx home video conferencing (VC), text, e-mail, and phone) between the specialist diabetes team and patients with type 1 diabetes. Based on an elicitation questionnaire derived from a model of technology acceptance, the initial codesign stage has generated a diabetes technology acceptability questionnaire. This will now be used in the main clinical study alongside an evaluation of methods for assimilation of digitally collected or transferred clinical communications and glucose results into patient records. These results will then inform integration of the digital/virtual diabetes clinic at scale.

34.8 Case Study

The consequences of not adapting an agile iterative approach and coproduction methodology in implementing HealthSpace in the National Health Service (NHS).

HealthSpace was a free, secure online personal health organizer provided by the U.K. NHS that would help people manage their health, store important health information securely, and find out about NHS services near them. However, the project failed.

If engagement with the public, professionals, and patients is not part of the design process, "the risk that the service will be abandoned or not adopted at all is substantial." These were the conclusions drawn from the English HealthSpace experience (Greenhalgh et al. 2010). The problems with implementing HealthSpace and similar projects not only in the United Kingdom (Fawdry et al. 2011) but also in other parts of the world (Littlejohns et al. 2003, Kaplan and Harris-Salamone 2009) have been to see the health-care IT industry as overpromising and underdelivering, destroying trust, and causing skepticism (Smith 2011). Robertson et al. (2012) concluded that "if a single lesson is drawn to be drawn from the NHS IT project then, it is that high-level political leadership should create the framework to kick-start IT-based reform, but that thereafter it should be professionally led with a realistic timescale, recognizing that experience with new systems, conflicting interests, and changing circumstances will require careful negotiations along the way." It is increasingly being recognized that the new world of health care is more "patient centric" and about partnership with a flattening of traditional top-down hierarchical approaches as the Salzburg statement to shared decision making attests whereby patients and clinicians are called on to work together to be coproducers of health (Elwyn 2011).

The agile iterative approach (derived from principles of software development) recognizes that "unpredictable emergent properties" can result as patients, public, and professionals interact with the technology and with each other. Therefore, the development process/research program must be flexible, adaptive, and evolutionary in order to accommodate these unexpected consequences. This approach contrasts with a rigid research program that can be likened to "waterfall" software development where complex systems are built as originally commissioned, without going back and revisiting requirements in the light of development and possible changing conditions. As a consequence, by the end of the process, the systems are not fit for purpose, a criticism that has been labeled at the "fixed milestones" of the HealthSpace project (Greenhalgh 2010).

34.9 Future Directions

This chapter makes the case that if you want to understand the World Wide Web structure, searchability, and the Internet of Things, you must first understand the web's networks. If you want to understand eHealth of which mHealth is a part, you will need the emerging disciplines of HWS and Medicine 2.0.

Because current models of health care are unsustainable, new digital mHealth frameworks such as $P4 + C^n = eIMT$ are needed and new tools of incentivization and evaluation are required. These tools may include network analysis using graph theory, game theory, and behavioral theory. Big data generated from the Internet of Things will need to be curated, visualized, and disseminated with new algorithms being designed and developed. To address the problem of information overload, improved text mining and semantic technologies will be required.

34.10 Conclusion

The future of medicine is progressively becoming P4, that is, preventative, participatory, personalized, and predictive, with cocreation/collaboration across multiple platforms. The use of mHealth is increasing in health-care delivery. Understanding its integration, use, and emergent properties is integral to understanding the benefits of mHealth to health care. It is incumbent on those developing health technologies that are being delivered via the Internet to embrace new methodologies of evaluation and recognize the limitations of those already utilized. The disciplines of, and related to, HWS and Medicine 2.0 have potential to provide frameworks and leadership in integrating mHealth into mainstream health-care delivery.

Questions

Determine whether the statements under the following topics are true or false and discuss:

1. Health professionals are facing the following challenges for the delivery of health care:
   (a) Increased fertility rate and decrease in chronic disease (noncommunicable diseases).
   (b) Paradigm shift from a proactive model of health care to a reactive model.
2. ICT and personalized medicine
   Are ICT and personalized medicine currently embraced universally?
3. HWS
   (a) Overall, there is robust evidence for the efficacy of health-care interventions.
   (b) Agile, iterative approaches offer flexibility in the design of software systems.
   (c) Emergent properties are unpredictable.
   (d) Delivery of health care utilizing the web may be a form of social engineering.
4. Social networking
   (a) People with smartphones are less likely to share, create, and participate and therefore be consumers of information.
   (b) For women in developing countries, use of ICT has direct positive impacts on life satisfaction.
   (c) Social networks are always good for health.
   (d) Social gaming is a potential avenue for promoting behavior change and treatment.
   (e) Guidance for medical practitioners is being offered by professional bodies to assist in use of social media, for example, Twitter.
5. When considering mobile devices for health
   (a) iPads and tablets should be considered.
   (b) The ability to test for glucose results in improvement of control and better understanding of the disease.

References

Agboola, S., Havasy, R., Myint, K. et al. 2013. The impact of using Mobile-enabled devices in remote monitoring programs. *J Diabetes Sci Technol.* 7(3): 623–629.

Ashton, K. 2009. That "Internet of Things" thing. http://www.rfidjournal.com/articles/view?4986 (accessed January 14, 2014).

Baum, F., MacDougall, C., and Smith, D. 2006. Participatory action Research. *J Epidemiol Community Health*. 60: 854–857.

Baron, J., McBain, H., and Newman, S. 2012. The impact of mobile monitoring technologies on glycosylated haemoglobin in diabetes: A systematic review. *J Diabetes Sci Technol*. 6(5): 1185–1196.

Bell, A. 2006. Wanless III—Engagement 0? The public's health. *Br J Healthcare Manage*. 12(11): 347.

Bensing, J. 2000. Bridging the gap. The separate worlds of evidence-based medicine and patient-centered medicine. *Patient Educ Counsel*. 39: 17–25.

Berners-Lee, T., Hall, W., and Hendler, J.A. 2006. A framework for web science. Foundations and Trends(r) 1, 11–130.

Berners-Lee, T., Hall, W., Hendler, J., Shadbolt, N., and Weitzner, D. 2006. Creating a science of the web. *Science*. 313(5788): 769–771.

Borrell-Carrió, F., Suchman, A.L., and Epstein, R.M. 2004. The biopsychosocial model 25 years later: Principles, practice, and scientific inquiry. *Ann Fam Med*. 2: 576–582.

British Medical Association. 2011. Using social media: Practical and ethical guidance for doctors and medical students. http://www.medschools.ac.uk/SiteCollectionDo,cuments/social_media_guidance_may2011.pdf (accessed August 28, 2013).

Brooks, E.H., Cumming, G.P., and Luciano, J.S. 2011. Health web science: Application of web science to the area of health education and health care. In *Proceedings of the Second International Workshop on Web Science and Information Exchange in the Medical Web*, Glasgow, U.K. (pp. 11–14). Association Computing Machinery, New York.

Cafazzo, J.A., Casselman, M., Hamming, N. et al. 2012. Design of an mHealth app for the self-management of adolescent type 1 diabetes: A pilot study. *J Med Int Res*. 14(3): e70.

Chigona, W., Nyemba-Mudenda, M., and Metfula, A.S. 2013. A review on mHealth research in developing countries. *J Community Inform*. 9(2). http://ci-journal.net/index.php/ciej/article/view/941/1011 (accessed January 14, 2014).

Chomutare, T., Fernandez-Luque, L., Arsand, E. et al. 2011. Features of mobile diabetes applications: Review of the literature and analysis of current applications compared against evidence-based guidelines. *J Med Int Res*. 13(3): e65.

Christakis, N.A. and Fowler, J.H. 2007. The spread of obesity in a large social network over 32 years. *N Engl J Med*. 357: 370–379.

Creswell, J.W., Klassen, A., Plano Clark, V.L., and Smith, K. 2011. *Best Practices for Mixed Methods Research in the Health Sciences*. Bethesda, MA: Office of Behavioral and Social Sciences Research, National Institutes of Health.

Crisp, N. 2010. *Turning the World Upside Down. The Search for Global Health in the 21st Century*. 1st edn. London, U.K.: Royal Society of Medicine Press Ltd.

Crowley, P. and Hunter, D.J. 2005. Putting the public back into public health. *J Epidemiol Community Health*. 59: 265–267.

de Jongh, T., Gurol-Urganci, I., Vodopivec-Jamsek, V. et al. 2012. Mobile phone messaging for facilitating self-management of long-term illnesses. *Cochrane Database Syst Rev*. 12: CD007459.

Duggan, M. and Rainie, L. 2012. Cell Phone activities. Pew Internet American Life Project http://pewinternet.org/~/media//Files/Reports/2012/PIP_CellActivities_11.25.pdf (accessed August 28, 2013).

Dutton, W.H. and Blank, G. 2011. Next generation users: The Internet in Britain. Oxford Internet Survey 2011 Report, OII, University of Oxford, Oxford, U.K.

Elwyn, G. 2011. Salzburg statement on shared decision making. *BMJ*. 342: d1745.

European Commission. 2009. Internet of Things strategic research roadmap. CERP-IoT. http://www.grifs-project.eu/data/File/CERP-IoT%20SRA_IoT_v11.pdf (accessed January 14, 2014).

Eysenbach, G. 2008. Medicine 2.0: Social networking, collaboration, participation, apomediation, and openness. *J Med Internet Res*. 10(3): e22. http://www.jmir.org/2008/3/e22/ (accessed January 14, 2014).

Farmer, A., Gibson, O.J., Tarassenko, L. et al. 2005. A systematic review of telemedicine interventions to support blood glucose self-monitoring in diabetes. *Diabet Med*. 22(10): 1372–1378.

Fawdry, R., Bewley, S., Cumming, G., and Perry, H. 2011. Data re-entry overload: Time for a paradigm shift in maternity IT. *J R Soc Med*. 104: 405–412.

Fox, S. and Jones, S. 2009. The social life of health information. Pew Internet and American Life Project'. http://www.pewInternet.org/~/media//Files/Reports/2009/PIP_Health_2009.pdf (accessed August 28, 2013).

Gillam, M., Feied, C., Handler, J. et al. 2009. The healthcare singularity and the age of semantic medicine. Hey, T., Tansley, S. and Tolle, K. (Eds.) In: *The Fourth Paradigm: Data-Intensive Scientific Discovery*. Redmond, WA: Online: Microsoft Research. pp. 57–64. http://research.microsoft.com/en-us/collaboration/fourthparadigm/ (accessed January 14, 2014).

Gotsis, M., Wang, H., Spruijt-Metz, D. et al. 2013. Wellness partners: Design and evaluation of a web-based physical activity diary with social gaming features for adults. *JMIR Res Protoc*. 2(1): e10. http://www.researchprotocols.org/2013/1/e10/ (accessed January 14, 2014).

Greenhalgh, T., Hinder, S., Stramer, K. et al. 2010a. Adoption, non-adoption, and abandonment of a personal electronic health record: Case study of HealthSpace. *BMJ*. 341: c5814.

Greenhalgh, T., Stramer, K., Bratan, T. et al. 2010b. *The Devil's in the Detail: Final Report of the Independent Evaluation of the Summary Care Record and HealthSpace Programmes*. London, U.K.: University College London.

Groopmand, J. and Hartzband, P. 2011. *Your Medical Mind: How to Decide What Is Right for You*. New York: Penguin Books.

Hendler, J., Shadbolt, N., Hall, W., Berners-Lee, T. et al. 2008. Web Science: An interdisciplinary approach to understanding the web. *Commun. ACM*. 51: 60–69.

Heron, J. and Reason, P. 2006. The practice of co-operative inquiry: Research 'with' rather than 'on' people. In P. Reason and H. Bradbury (eds.), *Handbook of Action Research*. London, U.K.: SAGE Publications Ltd. pp. 179–188.

HIE (Highlands and Islands Enterprises). 2012. P4 Digital Healthcare Scoping Study. http://www.hie.co.uk/common/handlers/download-document.ashx?id=fe60f4e4-b9eb-4745-933b-bff13b6eab3c (accessed January 14, 2014).

Holtz, B. and Lauckner, C. 2012. Diabetes management via mobile phones: A systematic review. *Telmed J E Health*. 18(3): 175–184.

Hughes, B., Joshi, I., and Wareham, J. 2008. Health 2.0 and medicine 2.0: Tensions and controversies in the field. *J Med Internet Res*. 10(3): e23. http://www.jmir.org/2008/3/e23/ (accessed January 14, 2014).

Internet World Stats. Internet Usage Statistics for Africa. 2012. http://www.internetworldstats.com/stats1.htm#top (Archived by WebCite® at http://www.webcitation.org/6Iz8yNQRs) (accessed August 28, 2013).

Johnson, S.B. 2012. Medicine's paradigm shift: An opportunity for psychology. http://www.apa.org/monitor/2012/09/pc.aspx (accessed January 14, 2014).

Kaplan, B. and Harris-Salamone, K.D. 2009. Health IT success and failure: Recommendations from literature and an AMIA workshop. *J Am Med Inform Assoc*. 16(3): 291–299.

Keeney, R.L. 2008. Personal decisions are the leading cause of death. *Operat Res*. 56: 1335–1347.

Koch, L.H. 2006. Help-seeking behaviors of women with urinary incontinence: An integrative literature review. *J Midwifery Women's Health*. 51(6): e39–e44.

Kvedar, J. 2013. Making health addictive. *Healthcare IT News* http://www.healthcareitnews.com/blog/making-health-addictive (accessed January 14, 2014).

Liang, X., Wang, Q., Yang, X. et al. 2011. Effect of mobile phone intervention for diabetes on glycaemic control: A meta-analysis. *Diabet Med*. 28: 455–463.

Lilford, R.J., Foster, J., and Pringle, M. 2009. Evaluating eHealth: How to make evaluation more methodologically robust. *PLoS Med*. 6(11): e1000186.

Littlejohns, P., Wyatt, J., and Garvica, L. 2003. Evaluating computerized health information systems: Hard lessons still to be learnt. *BMJ*. 326: 860–863.

Luciano, J.S., Cumming, G.P., Kahana, E. et al. 2012. Health Web Science—Definition, scope, and how it complements Medicine 2.0. Med 2.0'12, In *Fifth World Congress on Social Media, Mobile Apps and Internet/Web 2.0 Boston. iProceedings Medicine 2.0*, September 15–16. Boston, MA. pp. 210–211.

Luciano, J.S., Cumming, G.P., Kahana, E., Wilkinson, M.D. et al (in press). Foundations and Trends in Health Web Science. Foundations and Trends® in Web Science.

Luciano, J.S., Cumming, G.P., Wilkinson, M.D. et al. 2013. The emergent discipline of Health Web Science. *J Med Internet Res.* 15(8): e166. http://www.jmir.org/2013/8/e166/ (accessed January 14, 2014).

Medicine Congress. 2014. Medicine 2.0: Social Media, Mobile Apps, and Internet/Web 2.0 in Health, Medicine and Biomedical Research. http://www.medicine20congress.com/ocs/index.php/med/med2014 (accessed January 14, 2014).

Mezzich, J.E., Snaedal, J., van Wheel, C. et al. 2010. From disease to patient to person: Towards a patient-centered medicine. *Mt Sinai J Med.* 77: 304–306.

Mezzich, J.E., Snaedal, J., van Wheel, C. et al. 2011. Introduction to patient-centered medicine: From concept to practice. *J Eval Clin Pract.* 17: 330–332.

Miles, A. and Loughlin, M. 2011. Models in the balance: Evidence-based medicine versus evidence-informed individualized care. *J Eval Clin Pract.* 17: 531–536.

McCartney, M. 2012. How much of a social media profile can doctors have? *BMJ.* 344: e440.

McCurdie, T., Taneva, S., Casselman, M. et al. 2012. The case for user-centered design. *Biomed Instrum Technol.* Fall(Suppl.): 49–56.

McKendrick, D. 2012. Smartphones, Twitter and new learning opportunities at anesthetic conferences. *Anaesthesia.* 67: 438–439.

McKendrick, D.R., Cumming, G.P., Lee, A.J. 2012. Increased use of Twitter at a medical conference: A report and a review of the educational opportunities. *J Med Internet Res.* 14(6): e176. http://www.jmir.org/2012/6/e176/ (accessed January 14, 2014).

Murray, E. (2012). Web-Based Interventions for Behavior Change and Self-Management: Potential, Pitfalls, and Progress Med 2.0 2012; 1(2):e3. http://www.medicine20.com/2012/2/e3/.

NICE. 2006. National Institute for Health and Clinical Excellence. *The Management of Urinary Incontinence in Women.* London, U.K.: The Royal College of Obstetricians and Gynaecologists.

Newsworks. 2013. Tablet ownership and behaviour. Results from YouGov survey 28th–31st December 2012. http://www.slideshare.net/newsworks/tablet-ownership-and-behaviour-16016360 (accessed January 14, 2014).

oAfrica. Household Internet speeds for 25 African countries, cities. 2011. http://www.oafrica.com/broadband/household-internet-speeds/ (Archived by WebCite® at http://www.webcitation.org/6IzE78oxq) (accessed August 28, 2013).

Ofcom. 2012. Communications market report 2012. Ofcom, London, U.K.

Ofcom. 2013. Average UK broadband speed continues to rise. http://media.ofcom.org.uk/2013/08/07/average-uk-broadband-speed-continues-to-rise/ (accessed January 14, 2014).

P4 Medicine Institute. 2012. P4 medicine. http://p4mi.org/p4medicine (accessed January 14, 2014).

Ranck, J. 2012. *Connected Health: How Mobile Phones, Cloud and Big Data Will Reinvent Healthcare.* San Francisco, CA: GigaOM.

Read, J.L. and Shortell, S.M. 2011. Interactive games to promote behavior change in prevention and treatment. *JAMA.* 305: 1704–1705.

Robertson, A., Cornford, T., Barber, N. et al. 2012. The NHS IT project: More than just a bad dream. *Lancet.* 379: 29–30.

Rogers, W.A. and Fisk, A.D. 2010. Toward a psychological science of advanced technology design for older adults. *J Gerontol B: Psychol Sci Soc Sci.* 65B: 645–653.

Sánchez-Serrano, I. 2011. *The World's Health Care Crisis: From the Laboratory Bench to the Patient's Bedside.* Amsterdam, the Netherlands: Elsevier.

Schmidt, E. and Cohen, J. 2013. *The New Digital Age, Reshaping the Future of People Nations and Business.* Great Britain, U.K.: John Murray.

Smith, R. 2011. Can Information technology improve healthcare? http://blogs.bmj.com/bmj/2011/11/22/richard-smith-can-information-technology-improve-healthcare/ (accessed January 14, 2014).

Stephen, K. Motivation and Adherence to Pelvic Floor Muscle Exercise and the Role of Smart Phone Apps (in press doctoral thesis, University of The Highlands and Islands, 2014).

Sweeney, K.G., MacAuley, D., and Pereira Gray, D. 1998. Personal significance: The third dimension. *Lancet*. 351: 134–136.

Sysomos. 2009. Inside Twitter Clients, November 2009. http://www.sysomos.com/insidetwitter/clients/ (Archived by WebCite® at http://www.webcitation.org/67Bpdq9TT) (accessed August 28, 2013).

Terry, M. 2009. Twittering healthcare: Social media and medicine. *Telemed J E Health*. 15(6): 507–510.

Van De Belt, T.H., Engelen, L.J., Berben, S.A.A. et al. 2010. Definition of health 2.0 and medicine 2.0: A systematic review. *J Med Internet Res*. 12(2): e18. http://www.jmir.org/2010/2/e18/ (accessed January 14, 2014).

Vodafone. 2014. Helping Rural New Zealand do their thing better. http://www.vodafone.co.nz/network/rural/ (Archived by WebCite® at http://www.webcitation.org/6MhTCq6bq) (accessed January 14, 2014).

Wanless, D. 2002. Securing our future health: Taking a long-term view. http://si.easp.es/derechosciudadania/wp-content/uploads/2009/10/4.Informe-Wanless.pdf (accessed January 14, 2014).

Wooton, D. 2007. *Bad Medicine: Doctors Doing Harm since Hippocrates*. New York: Oxford University Press.

World Health Organization. 2011. Health statistics. http://www.who.int/whosis/whostat/2011/en/index.html (accessed January 14, 2014).

Zaidi, A. 2011. Social gaming trends for 2011. http://econsultancy.com/uk/blog/7022-social-gaming-trends-for-2011 (Archived by WebCite® at http://www.webcitation.org/6JAD3CWIP) (accessed August 28, 2013).

35

mHealth and Medical Imaging

Chandrashan Perera and Rahul Chakrabarti

CONTENTS

35.1 Introduction

The ubiquity of the camera-equipped smartphone has led to a growing body of evidence assessing the capacity for smartphones to perform as tools to facilitate diagnosis and aid clinical management. The increasing resolution of smartphone cameras accompanied with the portability, accessibility, and built-in Internet connectivity has resulted in a device that is well positioned to capture and transmit medical images. Additional benefits include a relatively low cost of image acquisition and widespread patient access to these devices.

Current-generation smartphones have excellent cameras that meet the standards of image-quality requirements for medical photography. Even previous generations of smartphone cameras that are commonly found in developing nations are adequate for certain diagnostic purposes in specialized disciplines [1]. Also, the high resolution of mobile displays has afforded the ability to assess clinically significant pathology on images transferred and reviewed on a mobile device. The value of the smartphone camera can also be extended greatly with the use of adapters that mount the smartphone onto other devices such as microscopes [2] and ophthalmic slit lamps [3].

There is an emerging evidence base growing in an increasing number of specialties, including radiology, ophthalmology, dermatology, and public health. Furthermore, there is a strong interest in mobile health (mHealth) medical imaging in rural and remote areas where access to specialist services may be limited and accurate diagnosis and timely referral of emergent pathology are required [4,5].

It is also important to recognize and address potential pitfalls to this subfield of medical imaging. Clinicians must be aware of ethical and legal protocols applying to images of patients taken on personal digital devices. It is imperative that regulatory bodies provide guidance in these matters [6]. Also, a rigorous evidence base needs to be established to ascertain diagnostic equivalence compared to existing technologies.

This chapter presents evidence from a review of the published literature surrounding the use of mobile devices for medical imaging.

35.2 Benefits of Utilizing mHealth for Medical Imaging

35.2.1 Health Economics

Where mHealth is poised to make the greatest benefit is through improving affordability of medical services to low-resource settings. The 2009 Global Observatory for eHealth study by the World Health Organization (WHO) showed that 49% of all member states, including low-resourced countries, were actively involved in mobile telemedicine practices either as pilot projects or informally [7]. By using affordable smartphone cameras to triage medical care, there is a potential to improve the allocation of medical services.

35.2.2 Apps

The emergence of mobile technologies equipped with applications has improved the versatility of such devices to partake in medical imaging. These applications, which are commonly referred to as *apps*, are specialized software for a particular task. A wide variety of applications exist already, and these are rapidly growing. In the area of mHealth imaging, applications are usually specialized toward image analysis or function to facilitate the transfer of images.

Dermatology is an area where automated image processing is popular, though further work is required to enhance the quality of the algorithms used to formulate diagnoses [8]. It is foreseeable in the future that such applications would increase in quality over time, possibly to become useful in the clinical setting.

Another benefit of using an app-based imaging solution is that the process can be tailored to ensure health insurance and portability act (HIPAA) compliance, following principles of privacy, and security, especially when images are being transferred between devices.

Valid concerns regarding apps include the safety of patient data storage, regulation of app development (particularly in the context of content accuracy and validity), and the dependency on some for wireless network access. Nevertheless, with the growing coverage of the Internet, and their simplicity of use by multidisciplinary health care professionals and patients, apps have undoubtedly improved access to medical imaging at all levels of the health system.

35.2.3 Connectivity

A key benefit of imaging on an mHealth platform is the connectivity that is inherent to the technology. Most mobile phones have ready access to high-speed Internet access, and many other devices used in mHealth either have their own connection to the Internet or are able to link to other devices to help connect them online. It is important to note that connectivity does not only include high-speed Internet access but also local connectivity options, such as Wi-Fi, Bluetooth, global positioning system (GPS), and near field communication (NFC).

The most important aspect to mHealth connectivity in the context of imaging is high-speed Internet access. Standard 3G (also known as universal mobile telecommunications system [UMTS]) is capable of speeds up to 2 Mbps, which can be improved to 10 Mbps through high speed downlink packet access (HSDPA) technology. Long term evolution (LTE) further builds on this with speeds up to 100 Mbps. This rapid rise in speed has led to the ability to transfer images rapidly, in a useful time frame. To put this into perspective, the amount of time required to transfer a 1 MB picture over general packet radio service (GPRS) networks requires approximately 70 s. In clinical practice, especially when dealing with emergency situations, this represents a delay, which would be unacceptable to most clinicians, especially if multiple images are to be transmitted. However, with even the use of basic 3G (UMTS) data transfer, the same image would take approximately 4 s, which is a much more acceptable delay for any user.

Local connectivity options such as Wi-Fi, Bluetooth, and NFC also offer a number of benefits in certain situations. Wi-Fi is an excellent method to transfer large amounts of data. For example, if a device has been used to collect a large amount of clinical photos in an environment with poor Internet connectivity, when the device then is returned to an area with Wi-Fi availability, the images can be transmitted rapidly to a central repository online, assuming a suitable Internet connection is available. Similarly, both Wi-Fi and Bluetooth can be used to transfer images between devices in a rapid fashion. Unfortunately, one of the difficulties with peer-to-peer image transfer lies in software compatibilities. In order to send images directly between devices using such local connectivity options, the devices need a way of linking to each other, through either a specially designed application or software built into the operating system.

Mobile phone connectivity is ubiquitous in a number of different populations and offers an opportunity to connect with populations that were once inaccessible. In Africa, there are approximately 1 billion people and 750 million phones. In South Africa, there are 52 million subscriber identity module (SIM) cards shared among 47 million people. Even though the majority of these are basic handsets, the ability to take photos is becoming an increasingly standard feature. This represents an amazing source of potentially valuable self-reporting clinical photos.

35.2.4 Self-Reporting

By its nature, mobile technology tends to be closely associated with the user, unlike traditional computing devices. Smartphones are usually carried within convenient reach, and a number of new devices are being developed that are designed to be worn by the user.

In the realm of imaging in mHealth, this opens up a number of possibilities to use integrated imaging devices, which allow patients to self-document and self-report their medical conditions and health.

This is especially helpful in dermatology, where patients often need to self-monitor lesions. In psoriasis, patient compliance is a serious issue, with compliance rates as low as 60%. In order to improve this, a study in 2009 created a system for patients to photograph their lesions and report symptoms that were then forwarded to their treating physician at regular intervals [9]. They found that the pictures taken by patients were of adequate quality for the monitoring lesions, was well tolerated by patients, and only required about 10 min for the patient to complete a "report." It is interesting to note that even in 2009, the authors chose to use a specially developed application to transmit the images between the mobile phone and the physician interface to maintain security and user simplicity rather than using multimedia messaging system (MMS).

In order to address the obesity epidemic, smartphone applications such as the National Health Service's (NHS's) "My Meal Mate" are helping users to monitor their dietary intake [10]. As early as 2006, researchers were putting together solutions to make use of cameras with the ability to transfer data through wireless connections [11,12], which was found to be useful when assessing the nutritional information of a meal. With many behavior change strategies, an important element is the ability to monitor existing behavior and then to reexamine the changed behavior. One automated way to monitor meals is to use a camera phone that is constantly taking photos at a set interval, with a third party able to review these images for relevance. One study showed that at an interval of 10 s, they were able to adequately monitor patient's food intake [13].

Expanding on this, another concept is *lifelogging*, whereby data are collected in an automated fashion, and then these data can be processed later for various purposes, such as monitoring behaviors, implementation of lifestyle changes, and safety monitoring. With recent improvements in software design and battery technology, it has now become possible to use a smartphone as a *wearable camera*, which takes photos at regular intervals. These images are then processed by an automated program and are able to identify activities performed. Gurrin et al. showed that the use of a smartphone for such purposes is possible and is as reliable as dedicated hardware for this purpose [14]. Currently, there are few programs that have been refined for clinical use. While gathering a large amount of data is potentially useful, this also creates difficulty in analyzing the data to produce clinically useful information.

35.3 Current Evidence for mHealth Imaging in Medical Specialties

The benefits afforded by mHealth technologies as outlined in the previous section have resulted in a growing body of published evidence. mHealth has grown to encompass a broad range of functions including diagnostic purposes, facilitating triage of emergent conditions, improving communication between healthcare providers to expedite patient care, and providing a feasible system for patient follow-up and self-monitoring. The ever-increasing capacity of mobile technologies has also enabled their purpose to expand beyond simply an image-capturing tool. This has been reflected in a spectrum of basic sciences, medical and surgical specialties, and public health that are using the imaging capabilities of mHealth to improve access and delivery of patient care.

35.3.1 Imaging in the Laboratory and Basic Sciences

A novel application of mHealth is its role as a diagnostic aid in the laboratory. Traditionally, laboratory tasks such as microscopy, fluorescent imaging, and enzyme-linked immunosorbent assays (ELISA) require sophisticated and expensive equipment. In this context, the use of mobile technology has been explored as a viable adjunct for laboratory imaging. This is important as it offers the potential to overcome the lack of cost-effective and simple diagnostic measures that have previously delayed important diagnoses. In the context of the detection of an ovarian cancer biomarker, Wang et al. developed a cell phone–based charge-coupled device (CCD) coupled with an inexpensive microchip to perform ELISA that allowed colorimetric analysis of urine samples [15]. This system had a sensitivity of 89.5% and specificity of 90% to detect the HE4 biomarker. Importantly, this eliminated the need for conventional bulky, expensive spectrophotometers that are currently used.

The use of mHealth in the laboratory has also paved the way for improving the proverbial bench-to-bedside care of patients. Importantly, novel applications of mHealth imaging have the capacity to improve care for some of the most prevalent and life-threatening health conditions. A poignant example of this is the management of tuberculosis, which remains one of the most common infectious diseases worldwide. While low-cost nonproprietary testing for tuberculous specimens has been developed, these still require skilled personnel to observe and grade the microscopic slides. Zimic et al. demonstrated that smartphone cameras were adequate to capture images of pathologic plates taken through a standard microscope that were then sent to experienced, trained personnel who were off-site to comment on the image [16]. This way, rapid diagnosis of pathology specimens could be performed even in the most remote areas, thereby overcoming limitations of having trained personnel experienced in plate reading.

The development and application of adaptors for mobile phones has also significantly enhanced their potential as laboratory imaging devices. Tseng et al. developed a 38 g, compact, holographic microscope adaptor that transformed the conventional cell phone light-emitting diode (LED) into a digital microscope [17]. In this case, blood samples of patients with *Giardia lamblia* (which contributes to significant morbidity worldwide from giardiasis) were loaded from the side of the adaptor, and incoherent LED light was scattered from each microscopic object to coherently interfere with background light. This created a lens-free hologram of each microscopic object that was reconstructed through digital processing on the phone itself. This has set the platform for the emergence of fluorescent microscopic and flow cytometric techniques to be performed using mobile phones. Zhu et al. showed that a compact, lightweight, disposable microfluidic channel positioned above the existing camera of a cell phone was of comparable sensitivity to a conventional hematology analyzer in rapidly imaging body fluids for conducting cell counts [18]. The microfluidic device acted as a optofluidic waveguide to guide the excitation of light as it passes through a fluid specimen. The cell phone camera was able to record the fluorescent movies of the specimen as it passed through the external microfluidic channel. This has significant applications in monitoring conditions such as viral loads and CD4+ lymphocyte counts in HIV and measuring the presence of waterborne parasites that affect millions worldwide.

Clearly, with advances in technology, the capacity for mobile technology to provide important laboratory services at the point of care will be enhanced. While further refinements and studies will only enhance the diagnostic accuracy of these interventions, the greatest benefit will be to bring feasible and reliable advanced laboratory measures to the remote areas to improve timely diagnosis of life-threatening conditions.

35.3.2 Radiology

In the field of radiology, the applications of mHealth and imaging largely relate to the ability to transfer and interpret clinical investigations on mobile devices. The ability to accurately display images on mobile devices for interpretation on mobile devices has improved the communication between healthcare providers and allowed timely management of patients.

Commonly used scenarios relate to trauma (head and neck), abdominal pathology, and investigation of acute coronary syndromes. Abdominal computerized tomography (CT) is commonly used for the surgical evaluation of patients with suspected appendicitis. Choudhri et al. showed a 98% concordance in the diagnosis of acute appendicitis when abdominal CT scans of 25 patients with right-sided abdominal pain were interpreted on a smartphone (iPhone) using a digital imaging and communications in medicine (DICOM) viewer application compared to a conventional computer workstation [19]. Similarly, interpretation of CT scans of patients with possible cervical spine (neck) injuries on a smartphone has shown high sensitivity for detection of clinically significant fractures, within similar time frame for interpretation on a larger computer-station monitor [20]. In patients with chest pain being explored for cardiac pathology, LaBounty et al. also demonstrated that interpretation of coronary CT angiography images on a portable smartphone possessed high diagnostic accuracy for detection and exclusion of clinically significant narrowing of the coronary arteries. Furthermore, Yaghmai et al. showed that in patients with head trauma, transmission of CT images could be rapidly and reliably performed using commercially available software on a personal digital assistant (PDA) or smartphone [21]. In their particular study, the authors demonstrated that the entire process from image capture to review by a specialist took on average 11.5 min, which facilitated rapid referral to trauma radiologists and neurosurgeons. A novel expansion of this technique is the popular use of using smartphone video capacity (video multimedia service, video MMS) to capture and send multiple CT scan slices in an efficient manner. This application is particularly useful when transferring a large volume of images that are all required to form a full clinical judgment [22], for example, in neurosurgery where a whole series of neuroimaging was sent to off-site consultants after a video was taken from a series of CT slices being shown on a computer monitor [23]. These examples highlight a growing utility of smartphones to display images of sufficient quality for remote diagnosis of clinically significant pathology to be performed.

There is a growing interest in the application of smartphones as a diagnostic instrument for radiological purposes. Huang demonstrated that a smartphone connected to a 10 MHz ultrasonic surface transducer was able to perform spectrographic analysis of Doppler ultrasound [24]. The study demonstrated the potential of using the smartphone as a portable medical ultrasound device. Further developments of this nature would have significant benefit in bringing important clinical investigations to the bedside.

35.3.3 Dermatology and Plastic Surgery

One of the most prolific applications of mHealth in medical imaging is in the field of dermatology. Dermatology is a perfect candidate specialty for the application of mHealth as it necessitates an intervention that can encompass diagnostic ability, monitoring, and aftercare of potentially malignant and chronic skin lesions that relies on using an objective measure.

The greatest application of smartphone cameras has been in the field of mobile teledermoscopy. Fundamentally, teledermoscopy is defined as images of the skin taken using a mobile device, which may be enhanced with a modified magnifying lens attachment. A growing body of research has demonstrated that teledermoscopy using mobile phones is a reliable approach for the assessment and monitoring of suspicious, potentially malignant skin lesions. Kroemer et al. in a randomized study of 278 patients demonstrated high sensitivity and specificity and strong correlation of mobile teledermoscopy compared to conventional clinical examination in the assessment of benign and malignant pigmented and unpigmented skin lesions [25]. Similar studies have consistently demonstrated adequate diagnostic accuracy of clinically significant skin lesions such as skin cancers [26,27]. The utility of this method is that it overcomes the barriers to accessing specialist dermatologists as images can be sent remotely for expert opinion.

Beyond diagnostic ability, teledermatology can also provide an important avenue for patient follow-up in the setting of chronic skin diseases. Berndt et al. trialed a system of teledermoscopy comprised of two main components: (1) dermoscopy performed using an application on a mobile phone that enabled the skin image and patient data to be transferred to a clinician, and (2) an online portal for the care providers by which to interact with the patients themselves [28]. The study showed a feasible and safe system of teledermoscopy that could be performed for the management of chronic skin conditions such as eczema, psoriasis, and monitoring of postoperative wounds. Similarly, Pirris et al. demonstrated that the status of surgical wounds in a cohort of neurosurgical patients could be reliably monitored remotely through patients sending photographs taken using cell phones to the treating medical team [29]. Particularly for people residing in remote areas, this platform of teledermatology offers an important clinical adjunct without the inconvenience of travel and costs.

The popularity of mHealth imaging has extended to the field of plastic surgery in which wound care is paramount. A growing number of observational studies have shown that mHealth in plastic surgery has improved management of wound care, assessment of burns, and real-time monitoring of free-flap grafts. In the context of trauma, Hsieh et al. showed that images of a broad range of hand and limb injuries presenting to a trauma hospital captured on a mobile phone were of sufficient quality to facilitate teleconsultation with the consultant surgeon in order to establish initial surgical management plans [30]. Importantly, the management decision was modified in only 15% of cases when the patient was reviewed by the surgeon. Further studies have demonstrated the utility of mHealth imaging in time-critical plastic surgical cases such as postoperative monitoring of free-flap grafts. Hwang et al. demonstrated that monitoring and rapid notification of free flaps using smartphone camera and communicating the images with clinical history using a messenger application increased the threatened flap salvage rate from 50% to 100% [31]. This was partly facilitated by shortening the interval to operating theater for patients requiring surgical reexploration of potentially compromised grafts. Further studies have shown that this concept of *real-time* monitoring of free flaps using smartphone photography and transferring images over a wireless network not only achieves comparable diagnostic accuracy to having a specialist on-site but improves patient care by shortening the response time if wound-threatening changes develop [32].

The sustainability of mHealth imaging in dermatology and plastics is also dependent on the ease of use of the techniques described. In most cases, studies have shown that with minimal additional training, nonspecialist medical workers (doctors in training, nurses, allied health workers) can be trained to perform simple tasks of image capture. Florczak showed that nurses were able to satisfactorily monitor pressure wounds, recognize changes in wound status, and determine the risk of severe skin damage in elderly residents of a nursing home using a smartphone camera [33]. Similarly, in the setting of teledermoscopy, Chung showed that the technique could be easily taught to nonspecialists such as medical students and nurses and yet reproduce gradable images [34].

Despite the practicalities and benefits of mHealth in dermatology and plastic surgery, there are potential pitfalls that need to be considered particularly when discussing the feasibility of using such methods for population screening. While there is a growing number of automated applications that can claim to accurately interpret images to categorize as benign or malignant [35,36], Wolf highlighted in a study of 60 melanoma and 128 control (pigmented, nonmalignant lesions) comparing 4 popular applications that there existed large variability in the diagnostic accuracy. Three of the four smartphone applications failed to identify 30% of true melanomas [8]. Additionally, variability in color accuracy between existing applications is another technical barrier to accurate diagnosis. In this context, current evidence suggests that the most appropriate use of mHealth in dermatology is for image capture rather than automated diagnosis. In this way, the benefits of teledermoscopy as an accessible and affordable tool for patients can be combined with safe clinical practice.

35.3.4 Ophthalmology

In the field of ophthalmology, the application of mHealth has similarly increased the potential to facilitate timely diagnosis and expedite management for sight-threatening conditions. A growing body of literature has highlighted the versatility of smartphones to aid clinical assessment. This is particularly relevant for accurate diagnosis of the most common causes of vision impairment and blindness.

Globally, refractive error remains the leading cause of all vision impairment. To address this, a novel adaptor to a conventional smartphone was developed. The pilot study by Bastawrous et al. demonstrated the ability for refraction to be assessed using a pinhole adaptor to an iPhone [37]. The authors showed there was no statistically significant difference in the Near Eye Tool for Refractive Assessment (NETRA) adaptor compared to standard subjective assessment of a patient's refraction (spherical equivalence). Additionally, the low cost of these adaptors (estimated market price of US$30) demonstrated the potential for such innovations to be cost effective particularly in low-resource settings where optometrists or trained refractionists are not readily available.

Smartphones have also been explored as adjuncts to clinical examination for ophthalmic pathology. Commercially available slit-lamp adaptors have been demonstrated to be able to capture adequate quality imaging of the anterior chamber of the eye [38]. Lord et al. have previously described retinal imaging using an iPhone and indirect lens [3]. However, the sensitivity and specificity for imaging posterior segment (retina and optic nerve) disease using this technique is yet to be validated. Nevertheless, the capacity to display high-resolution images from a retinal camera for reading by a specialist on a smartphone screen has been successfully demonstrated. Lamirel et al. demonstrated that retinal images taken using a nonmydriatic camera in an emergency department setting were able to be transferred and graded by an off-site ophthalmologist with high accuracy on an iPhone display to identify emergent retinal and optic disk pathology [1]. Similarly, Kumar et al. showed that retinal images of patients with diabetic retinopathy when read by a specialist on a smartphone achieved high concordance with images read on a standard office computer workstation [39]. These examples suggest that smartphones have the potential to be a valuable tool for remote photographic assessments. Furthermore, given that smartphone cameras have an ever-increasing image resolution, it is conceivable that in the near future, smartphones may provide fundus images to facilitate early referral of patients in low-resource settings.

From these examples in the field of ophthalmology, it is clear that advances in telemedicine and validation of smartphone technology will improve access to eye care worldwide. The greatest benefit will be in low- and lower-middle-income countries where a major challenge is the provision and accessibility of trained eye-care professionals. Further validation studies of these novel devices will be an important next step for enhancing the role of mHealth to provide timely diagnosis and treatment for underserved populations.

35.3.5 Expanding Role of mHealth Imaging in Other Medical Specialties

The utility of mHealth in medical imaging has expanded to almost all disciplines of health care. While a large body of evidence currently exists in the aforementioned specialties, there are a growing number of examples in other specialties that serve a diverse range of purposes.

In cardiology, there are increasing applications of mobile technologies to monitor clinically important parameters. Bolkhovsky et al. demonstrated the video recordings of finger tip pulsation using a smartphone camera could be used to monitor heart-rate variability using similar principles to conventional photoplethysmography (PPG) [40]. Specifically, the PPG signal components around 0.1 Hz corresponded to the sympathetic component of the heart rate [41,42]. Tahat extended this concept to demonstrate that mobile phones could receive electrocardiogram (ECG) data from skin leads and record and transmit the information dynamically via MMS to clinicians [43]. The clinical implication of this would be of particular benefit for remote monitoring of people with cardiac arrhythmias and those with angina or coronary atherosclerotic disease who would benefit from ambulatory monitoring of their pulse rhythm and rate without wearing cumbersome equipment as occurs at present.

There is an increasing shift to applying mHealth imaging technologies to facilitate procedures. In neurosurgery, freehand placement of ventricular catheters into the brain has varying accuracy. As most patients will have had MRI or CT of their brain, Thomale et al. showed that entering basic dimensions using a smartphone-based algorithm assisted the surgeon in precise entry into the ventricles based on the individual's specific anatomy [44]. Similarly, in the context of approaching the patient with significant laryngeal edema requiring advanced airway management, smartphone recording of video laryngoscopy was demonstrated as a useful adjunct to plan conventional intubation methods [45]. The portability of smartphone cameras has also been shown to be of advantage in the management of dental emergencies. In a comparison between a special-purpose oral camera, digital single-lens reflex (SLR), and smartphone camera to capture images of the oral cavity, the smartphone camera provided the most convenient method while preserving quality necessary for initial diagnosis even at high compression ratios [46].

The growing interest of integrating mHealth into medical imaging is supported by the potential cost savings and affordability compared to current technologies while preserving highest standards of clinical care to the patient. For example, Sohn et al. demonstrated that the resolution of images obtained from gastroscopy onto a smartphone (iPhone) was considered of comparable resolution to a traditional high-definition camera in order to make a clinical judgment. Yet, the smartphone was reported to be 3000 times cheaper than the HD camera [47]. Countless examples are provided whereby ambulatory patients can all be monitored for the risk of common postoperative complications such as bruising, blood-stained bandaging, allergic reaction, bandage being too tight, and gait disturbances by transmitting images taken on smartphones from home by the patient [48–50]. These examples have significant benefit to the patient in receiving closer attention from medical professionals, improves patient satisfaction, and avoids the costs associated with travel and unnecessary hospital visits.

35.4 Technology Considerations

35.4.1 Image Sensor

The imaging sensor is the part of mobile devices that allow them to convert light (photons) into an electrical signal that can be processed and used to recreate an image. There are two main types of sensors that are commonly used, CCD and complementary metal oxide semi conductor (CMOS). A CCD chip converts an entire image into one sequence of voltage, whereas with CMOS, each pixel has its own circuitry to create its own voltage sequence. Due to the much lower power requirements of CMOS sensors, these are far more commonly found in most smartphones.

There are two basic aspects of the sensor that affect image quality. It is a combination between the number of individual light collecting pixels on the sensor and the physical size of each pixel. By having more light collecting pixels, the *megapixel* (i.e., total number of pixels) count increases. However, it is essential to understand that in practice, image quality is not determined by megapixel count. Mobile devices have limited space available for imaging sensors, and as such by packing excessive numbers of pixels into a fixed-sized sensor, image quality is degraded as each individual pixel is smaller and of a lower quality. Other factors such as lens quality, dynamic range, backside illumination, and image stabilizers are all features that are equally important toward producing an excellent image.

35.4.2 Image Processing

Cameras on smartphones and mobile devices have a key limitation when compared to other camera designs—physical space limitation. However, they also have a key advantage over other cameras, the amount of computational power available. By having a more powerful image-processing chip coupled with a powerful processor, one is not only able to enhance the image taken but also able to customize it for the task at hand. For example, cell phone–based rapid diagnostic tests (RDT) to detect various diseases such as malaria, tuberculosis, and HIV have been developed by using a cell phone camera, a special adapter, and specially designed RDT strips [51]. After capturing the image of the test strip, the software is then able to ready the image for processing by converting to grayscale and enhancing the image. An algorithmic calculation is then performed on this image to produce a diagnosis.

A particularly exciting area of software development has been to reduce blur from camera motion. Regular cameras counteract this with *image stabilizers*, which move the camera sensor in the operate direction of motion. However, many smartphones are equipped with accelerometers and gyroscopes. By using data from these, and the significant processing power available, new algorithms have been developed that allow for significantly sharper images when longer exposure times are required [52]. Currently, these algorithms require approximately 6 s on a Samsung Galaxy S2 to process, but with refinement in both the hardware and software, this could be a promising method of improving smartphone photos.

35.4.3 Extensibility

Other adapters have also been developed to augment and enhance the imaging capabilities of the mobile devices. A number of adapters have been developed to utilize smartphones as field microscopes [2,17,53,54] and other lab equipment [15,16,18,55–59]. Among the ophthalmic community, the placement of a smartphone at the eyepiece of a slit-lamp microscope allows digital capture of images [38,60,61], and even dedicated ophthalmo-scopes are being designed for use with smartphones [62]. With the increasing availability of 3D printers, we foresee a number of new adapters being developed for specialized-use scenarios. While these adapters are designed to make use of the existing imaging sensor on the camera, some adapters allow imaging beyond the capabilities of built-in devices. For example, mobile phone ultrasound probes are being developed that allow ultrasound imaging to be performed on a smartphone, using the excellent processing power and high-resolution displays that are featured in many new smartphones [24,63,64].

35.4.4 Compression

Images from most devices are stored in a compressed format, JPEG (Joint Photographic Experts Group). By default, most smartphones take images at their highest-quality settings and minimize compression to maintain maximum quality. However, for medical use, one needs to be mindful that images need only be of sufficient quality for clinical purposes. While the JPEG compression standard is excellent at low compression ratios, significant artifact is seen at higher compression ratios. JPEG2000 is a wavelet-based compression algorithm that was designed to maintain better image quality than standard JPEG under higher image compression ratios [65,66]. Depending on the intended clinical use of images, appropriate compression ratios and algorithms should be chosen. For example, with standard fundus photography for diabetic retinopathy, JPEG compression ratios of 1:20–1:12 provide images that are of sufficient quality for their intended purpose of screening diabetic retinopathy [67]. While data connectivity is often excellent in developed nations, in developing nations, and areas with poor cellular reception, data transfer rates are not as quick, and it is especially in these environments that maximum compression of images is essential to allow mHealth-based imaging to operate. In other situations such as high-resolution radiology applications, it may be beneficial to use images of higher quality, and further studies are needed to ascertain what the *acceptable* compression is for each particular use.

35.4.5 Display Technology

Beyond their imaging capabilities, new smartphones are also bundled with very-high-quality displays, which are often of higher quality than typical desktop displays. This is of special significance to radiologists and clinicians viewing radiological tests on their mobile device [19–21,68–75]. The two most commonly used technologies in screens are in-plane switching liquid crystal displays (IPS LCDs) and organic light-emitting diode (OLED). Due to the wide variety of commercially available smartphones with varying display qualities, it is outside the scope of this chapter to assess each. However, there are a number of important factors to consider when assessing the quality of a display. An important determinant of perceived sharpness is the number of pixels on the screen per unit area, which is usually measured in pixels per inch (PPI). While a higher PPI generally leads to a sharper image, there is a point where increasing the PPI does not have any value, which is due to the maximum resolving power of the eye, approximately 30 arc s [75]. For a device viewed at 10″, this approximates to 310 PPI, beyond which point extra resolution would be unresolved by the human eye. Color and contrast calibration is also important, with some displays having excellent calibration from the factory and others with poor calibration. It is important to assess various parameters such as contrast ratios and reflectance in both high and low ambient environments to get a true reflection of practical utility.

Finally, the ability to reproduce the standard color gamut (sRGB being the most commonly used gamut) is also important when viewing medical photos.

As technology evolves in mobile devices, there are a number of features that are likely to improve. First, photographic quality will improve, not only through better image sensors but also improved image signal processors and algorithms. In 2014, the latest-generation smartphone displays are often better than many standard computer monitors; however, they also will continue to increase in capability. Moving forward, the biggest advantage of evolving technology is the lowering of cost of various components, which will result in high-quality mHealth imaging more accessible.

35.5 Ethico-Legal Issues

Given that mHealth imaging has only become popularized in the recent few years, it is not surprising that a 2012 review did not identify any clear guidelines specifically addressing imaging on smartphones [76]. Should images be transmitted from one clinician to another via mobile devices, it is important that both the sending and receiving clinicians be aware of the relevant issues.

35.5.1 Consent

Informed consent needs to be a guiding principle for clinical photography in general. Consent can be obtained in several ways and may be expressed or implied. Some argue that photography for documenting conditions is part of the medical record, and so through the patient engaging in a consult, and allowing a picture to be taken, consent is implied [77]. However, it is important to note that images taken on mobile devices are often taken with the aim of communicating the information with other medical professionals. As such, it is prudent to record verbal consent in the medical record and the intended use of images as discussed with the patient.

35.5.2 Privacy

Depending on the pathology being imaged, privacy is a principle that should be maintained, and where possible, patient-identifying features should be omitted from images. As a general principle, features to avoid are the eyes in full-face photography, unique tattoos, scars, and body piercings. This is understandably difficult in certain circumstances, such as documenting lesions on the face [78].

35.5.3 Secure Storage

An advantage of smartphones over regular digital cameras is the encryption that is available. Most devices now have the option of protecting their contents with a password, and newer devices are integrating fingerprint readers to increase protection [79]. Furthermore, data stored on devices can be stored in an encrypted manner to further secure the data. If a clinician's smartphone containing medical images is misplaced, and no PIN/passcode protection was enabled on the device, the clinician would find themselves open to disciplinary action. While details of local privacy principles vary geographically, most governing bodies require *reasonable measures* for protection of sensitive information, which is clearly breached in the previous example [78].

35.5.4 Practical Guidelines

There are number of *good practice* points that clinicians should follow [6]. First, the principles of informed consent are required at all times, with clear documentation of the consent process when images are taken. It is important to ensure patients are aware of the purpose and intended audience of the photograph. Where possible, avoid identifiable features such as the face if they are not clinically necessary. Images taken should be securely sent to the intended audience, and a copy should also be saved for the patient's permanent medical record. Once the image is securely stored in the patient's permanent medical record, it is advisable to delete photographs from personal devices at the earliest opportunity.

Clinicians must be aware that the same ethical and legal protocols apply to images of patients taken on personal digital devices as those applying to any medical record. We believe that the further guidance is needed from governing bodies to advise doctors how to best utilize the imaging capabilities of mHealth devices yet maintain important ethical and legal principles.

35.6 Future Direction

Modern medical practice is undergoing a transformation in the way it communicates and provides health care. Contributing to this evolution have been the advances made in telecommunication devices that have facilitated the integration of the telephone, Internet, text messaging, photo, and video. A growing body of literature has demonstrated a spectrum of purposes where mHealth has been applied across a myriad of medical subspecialties, particularly in the context of medical imaging [80]. Additionally, there is a steadily increasing market for smartphone-based applications encompassing a variety of purposes such as diagnostic aids, continuity of patient care, and facilitating communication between healthcare providers and patients. While mHealth technologies have raised much interest in the medical community, there is a necessity for scientific rigor and sound principles of evidence-based medicine must also apply to mHealth.

Despite the growing body of literature regarding mHealth, recent meta-analyses have shown that there are currently few studies based on high methodological quality [80,81]. Limitations included the lack of objective clinical outcomes and the heterogeneity between reported outcomes among studies with similar interventions. Additionally, it was noted that the majority of studies were conducted in high-income countries. This limited the extent to which the results can be contextualized to low-resource settings, where mHealth is positioned to make the biggest impact.

While the use of mHealth in developing nations has many potential benefits, there are significant challenges to implementation that need to be addressed. Due to the novel nature of mHealth technologies and interventions, there are few regulations or policies in place to ensure adequate quality in the delivery of mHealth, especially in developing nations. This will necessitate collaboration between government, nongovernment organizations, and telecommunication companies to deliver equitable access and service. In addition, it is imperative that the integrity of legal principles of confidentiality and data security, as with any health record, is preserved. Issues surrounding grading of mHealth research, validation of interventions, standardization of technologies, and establishing guidelines for mHealth use must also be addressed.

Despite this, mHealth has shown several excellent examples by which it can facilitate health care. Particularly, in the low-resource setting, the use of mHealth for medical imaging is well suited to address many of the challenges faced by healthcare providers and consumers. The existing concerns regarding limited high-quality research in mHealth should guide the planning of studies that address methodological deficiencies in previous studies. This will ultimately enable mHealth to be accepted as an important adjunct to the delivery of health care globally.

Questions

1. What are the three important ethico-legal elements to consider with smartphone imaging?
2. What are the ways that images can be transferred from mobile devices wirelessly?
3. What are the current medical specialties with the strongest evidence for imaging in mHealth?
4. What are the ranges of potential applications of mHealth imaging?
5. Is mHealth imaging feasible in low-resource settings?

References

1. Lamirel C, Bruce BB, Wright DW et al. Nonmydriatic digital ocular fundus photography on the iPhone 3G: The FOTO-ED study. *Archives of Ophthalmology* 2012;**130**(7):939–940. doi: 10.1001/archophthalmol.2011.2488 [published Online First: Epub Date].
2. Bogoch II, Andrews JR, Speich B et al. Mobile phone microscopy for the diagnosis of soil-transmitted helminth infections: A proof-of-concept study. *The American Journal of Tropical Medicine and Hygiene* 2013;**88**(4):626–629. doi: 10.4269/ajtmh.12-0742 [published Online First: Epub Date].
3. Lord RK, Shah VA, San Filippo AN et al. Novel uses of smartphones in ophthalmology. *Ophthalmology* 2010;**117**(6):1274–1274 e3. doi: 10.1016/j.ophtha.2010.01.001 [published Online First: Epub Date].
4. Chakrabarti R, Perera C. Smartphones in stroke. *Neurosurgery* 2012;**71**(4):E910; author reply E10-1. doi: 10.1227/NEU.0b013e3182662136 [published Online First: Epub Date].
5. Demaerschalk BM, Vargas JE, Channer DD et al. Smartphone teleradiology application is successfully incorporated into a telestroke network environment. *Stroke* 2012;**43**(11):3098–3101. doi: 10.1161/strokeaha.112.669325[published Online First: Epub Date].
6. Chakrabarti R, Perera CM. Click first, care second photography. *The Medical Journal of Australia* 2013;**198**(1):21.
7. World Health Organisation. *mHealth—New Horizons for Health through Mobile Technologies.* Global Observatory for eHealth, Geneva, Swizterland: WHO, 2011.

8. Wolf JA, Moreau JF, Akilov O et al. Diagnostic inaccuracy of smartphone applications for melanoma detection. *JAMA Dermatology* 2013;**149**(4):422–426.
9. Hayn D, Koller S, Hofmann-Wellenhof R et al. Mobile phone-based teledermatologic compliance management—Preliminary results of the TELECOMP study. *Studies in Health Technology and Informatics* 2009;**150**:468–472.
10. Carter MC, Burley VJ, Nykjaer C et al. "My Meal Mate" (MMM): Validation of the diet measures captured on a smartphone application to facilitate weight loss. *The British Journal of Nutrition* 2012;**109**:539–546. doi: 10.1017/S0007114512001353 [published Online First: Epub Date].
11. Kikunaga S, Tin T, Ishibashi G et al. The application of a handheld personal digital assistant with camera and mobile phone card (Wellnavi) to the general population in a dietary survey. *Journal of Nutritional Science and Vitaminology* 2007;**53**(2):109–116.
12. Wang DH, Kogashiwa M, Ohta S et al. Validity and reliability of a dietary assessment method: The application of a digital camera with a mobile phone card attachment. *Journal of Nutritional Science and Vitaminology* 2002;**48**(6):498–504.
13. Arab L, Winter A. Automated camera-phone experience with the frequency of imaging necessary to capture diet. *Journal of the American Dietetic Association* 2010;**110**(8):1238–1241.
14. Gurrin C, Qiu Z, Hughes M et al. The smartphone as a platform for wearable cameras in health research. *American Journal of Preventive Medicine* 2013;**44**(3):308–313.
15. Wang S, Zhao X, Khimji I et al. Integration of cell phone imaging with microchip ELISA to detect ovarian cancer HE4 biomarker in urine at the point-of-care. *Lab on a Chip* 2011;**11**(20):3411–3418.
16. Zimic M, Coronel J, Gilman RH et al. Can the power of mobile phones be used to improve tuberculosis diagnosis in developing countries? *Transactions of the Royal Society of Tropical Medicine and Hygiene* 2009;**103**(6):638–640.
17. Tseng D, Mudanyali O, Oztoprak C et al. Lensfree microscopy on a cellphone. *Lab on a Chip* 2010;**10**(14):1787–1792.
18. Zhu HY, Mavandadi S, Coskun AF et al. Optofluidic fluorescent imaging cytometry on a cell phone. *Analytical Chemistry* 2011;**83**(17):6641–6647.
19. Choudhri AF, Carr TM, Ho CP et al. Handheld device review of abdominal CT for the evaluation of acute appendicitis. *Journal of Digital Imaging* 2012;**25**(4):492–496.
20. Modi J, Sharma P, Earl A et al. iPhone-based teleradiology for the diagnosis of acute cervicodorsal spine trauma. *The Canadian Journal of Neurological Sciences* 2010;**37**(6):849–854.
21. Yaghmai V, Salehi SA, Kuppuswami S et al. Rapid wireless transmission of head CT images to a personal digital assistant for remote consultation. *Academic Radiology* 2004;**11**(11):1291–1293.
22. Scandrett K. Video multimedia messaging system (MMS) supporting referral for an acute subdural haemorrhage: A case report. *Journal of Mobile Technology in Medicine* 2012;**1**(2):28–30.
23. Shivapathasundram G, Heckelmann M, Sheridan M. Using smart phone video to supplement communication of radiology imaging in a neurosurgical unit: Technical note. *Neurological Research* 2012;**34**(3):318–320.
24. Huang CC, Lee PY, Chen PY. Implementation of a smart-phone based portable doppler flowmeter. *IEEE International Ultrasonics Symposium* 2012, pp. 1056–1059.
25. Kroemer S, Fruhauf J, Campbell TM et al. Mobile teledermatology for skin tumour screening: Diagnostic accuracy of clinical and dermoscopic image tele-evaluation using cellular phones. *British Journal of Dermatology* 2011;**164**(5):973–979.
26. Massone C, Hofmann-Wellenhof R, Ahlgrimm-Siess V et al. Melanoma screening with cellular phones. *PloS One* 2007;**2**(5):e483.
27. Tran K, Ayad M, Weinberg J et al. Mobile teledermatology in the developing world: Implications of a feasibility study on 30 Egyptian patients with common skin diseases. *Journal of the American Academy of Dermatology* 2011;**64**(2):302–309.
28. Berndt RD, Takenga MC, Kuehn S et al. Development of a mobile teledermatology system. *Telemedicine Journal and e-Health* 2012;**18**(9):668–673.
29. Pirris SM, Monaco EA, Tyler-Kabara EC. Telemedicine through the use of digital cell phone technology in pediatric neurosurgery: A case series. *Neurosurgery* 2010;**66**(5):999–1004.

30. Hsieh CH, Tsai HH, Yin JW et al. Teleconsultation with the mobile camera-phone in digital soft-tissue injury: A feasibility study. *Plastic and Reconstructive Surgery* 2004;**114**(7):1776–1782.

31. Hwang JH, Mun GH. An evolution of communication in postoperative free flap monitoring: Using a smartphone and mobile messenger application. *Plastic and Reconstructive Surgery* 2012;**130**(1):125–129.

32. Engel H, Huang JJ, Tsao CK et al. Remote real-time monitoring of free flaps via smartphone photography and 3G wireless Internet: A prospective study evidencing diagnostic accuracy. *Microsurgery* 2011;**31**(8):589–595.

33. Florczak B, Scheurich A, Croghan J et al. An observational study to assess an electronic point-of-care wound documentation and reporting system regarding user satisfaction and potential for improved care. *Ostomy/Wound Management* 2012;**58**(3):46–51.

34. Chung P, Yu T, Scheinfeld N. Using cellphones for teledermatology, a preliminary study. *Dermatology Online Journal* 2007;**13**(3):1087–2108.

35. Doukas C, Stagkopoulos P, Kiranoudis CT et al. Automated skin lesion assessment using mobile technologies and cloud platforms. *Proceedings of the Annual International Conference of the IEEE Engineering in Medicine and Biology Society*, San Diego, USA, 2012, pp. 2444–2447.

36. Ramlakhan K, Shang Y. A mobile automated skin lesion classification system. *Proceedings—International Conference on Tools with Artificial Intelligence*, Boca Raton, Florida, USA, 2011, pp. 138–141.

37. Bastawrous A, Leak C, Howard F et al. Validation of near eye tool for refractive assessment (NETRA)—Pilot study. *Journal of Mobile Technology in Medicine* 2012;**1**(3):6–16.

38. Barsam A, Bhogal M, Morris S et al. Anterior segment slitlamp photography using the iPhone. *Journal of Cataract and Refractive Surgery* 2010;**36**(7):1240–1241.

39. Kumar S, Wang EH, Pokabla MJ et al. Teleophthalmology assessment of diabetic retinopathy fundus images: Smartphone versus standard office computer workstation. *Telemedicine Journal and e-Health* 2012;**18**(2):158–162.

40. Bolkhovsky JB, Scully CG, Chon KH. Statistical analysis of heart rate and heart rate variability monitoring through the use of smart phone cameras. *Proceedings of the Annual International Conference of the IEEE Engineering in Medicine and Biology Society*, San Diego, USA, 2012, pp. 1610–1613.

41. McManus DD, Lee J, Maitas O et al. A novel application for the detection of an irregular pulse using an iPhone 4S in patients with atrial fibrillation. *Heart Rhythm* 2013;**10**(3):315–319.

42. Jonathan E, Leahy MJ. Cellular phone-based photoplethysmographic imaging. *Journal of Biophotonics* 2011;**4**(5):293–296.

43. Tahat AA. Mobile personal Electrocardiogram monitoring system and transmission using MMS. *International Journal of Telemedicine and Applications* 2008:118–122.

44. Thomale UW, Knitter T, Schaumann A et al. Smartphone-assisted guide for the placement of ventricular catheters. *Child's Nervous System* 2013;**29**(1):131–139.

45. Newmark JL, Ahn YK, Adams MC et al. Use of video laryngoscopy and camera phones to communicate progression of laryngeal edema in assessing for extubation: A case series. *Journal of Intensive Care Medicine* 2013;**28**(1):67–71.

46. Park W, Kim DK, Kim JC et al. A portable dental image viewer using a mobile network to provide a tele-dental service. *Journal of Telemedicine and Telecare* 2009;**15**(3):145–149.

47. Sohn W, Shreim S, Yoon R et al. Endockscope™: Using mobile technology to create global point of service endoscopy. *Journal of Endourology* 2013;**27**:1154–1160.

48. Martinez-Ramos C, Cerdan MT, Lopez RS. Mobile phone-based telemedicine system for the home follow-up of patients undergoing ambulatory surgery. *Telemedicine Journal and e-Health* 2009;**15**(6):531–537.

49. Salibian AA, Scholz T. Smartphones in surgery. *Journal of Healthcare Engineering* 2011;**2**(4):473–486.

50. Wang Q, Yu Y, Lv YQ et al. Mobile phone as pervasive electronic media to record and evaluate human gait behavior. *Journal of Innovative Optical Health Sciences* 2012;**5**(1).

51. Mudanyali O, Dimitrov S, Sikora U et al. Integrated rapid-diagnostic-test reader platform on a cellphone. *Lab on a Chip* 2012;**12**(15):2678–2686.

52. Sindelar O, Sroubek F. Image deblurring in smartphone devices using built-in inertial measurement sensors. *Journal of Electronic Imaging* 2013;**22**(1):011003.

53. Breslauer DN, Maamari RN, Switz NA et al. Mobile phone based clinical microscopy for global health applications. *PloS One* 2009;**4**(7):e6320.
54. Frean J. Microscopic images transmitted by mobile cameraphone. *Transactions of the Royal Society of Tropical Medicine and Hygien* 2007;**101**(10):1053.
55. Martinez AW, Phillips ST, Carrilho E et al. Simple telemedicine for developing regions: Camera phones and paper-based microfluidic devices for real-time, off-site diagnosis. *Analytical Chemistry* 2008;**80**(10):3699–3707.
56. Tapley A, Switz N, Reber C et al. Mobile digital fluorescence microscopy for diagnosis of tuberculosis. *Journal of Clinical Microbiology* 2013;**51**(6):1774–1778.
57. Preechaburana P, Macken S, Suska A et al. HDR imaging evaluation of a NT-proBNP test with a mobile phone. *Biosensors and Bioelectronics* 2011;**26**(5):2107–2113.
58. Lee D, Chou WP, Yeh SH et al. DNA detection using commercial mobile phones. *Biosensors and Bioelectronics* 2011;**26**(11):4349–4354.
59. Zhu H, Yaglidere O, Su TW et al. Cost-effective and compact wide-field fluorescent imaging on a cell-phone. *Lab on a Chip* 2011;**11**(2):315–322.
60. Chhablani J, Kaja S, Shah VA. Smartphones in ophthalmology. *Indian Journal of Ophthalmology* 2012;**60**(2):127–131.
61. Ye YF, Jiang H, Zhang HC et al. Resolution of slit-lamp microscopy photography using various cameras. *Eye and Contact Lens* 2013;**39**(3):205–213.
62. Blanckenberg M, Worst C, Scheffer C. Development of a mobile phone based ophthalmoscope for telemedicine. *Proceedings of the Annual International Conference of the IEEE Engineering in Medicine and Biology Society*, 2011, pp. 5236–5239.
63. Huang CC, Lee PY, Chen PY et al. Design and implementation of a smartphone-based portable ultrasound pulsed-wave Doppler device for blood flow measurement. *IEEE Transactions on Ultrasonics, Ferroelectrics, and Frequency Control* 2012;**59**(1):182–188.
64. Wojtczak J, Bonadonna P. Pocket mobile smartphone system for the point-of-care submandibular ultrasonography. *The American Journal of Emergency Medicine* 2013;**31**(3):573–577.
65. Wang HY, You XD. Study of Jpeg2000 on android. Wavelet Active Media Technology and Information International Conference. 2012:57–61.
66. Krishnan K, Marcellin MW, Bilgin A et al. Prioritization of compressed data by tissue type using JPEG2000. *Proceedings—SPIE The International Society for Optical Engineering* 2005;**5748**:181–189.
67. Basu A, Kamal AD, Ilahi W, Khan M, Stavrou P, Ryder RE (2003). Is digital image compression acceptable within diabetic retinopathy screening?. *Diab Med.* 20(9):766–71.
68. John S, Poh AC, Lim TC et al. The iPad tablet computer for mobile on-call radiology diagnosis? Auditing discrepancy in CT and MRI reporting. *Journal of Digital Imaging* 2012;**25**(5):628–634.
69. Laughlin PM, Neill SO, Fanning N et al. Emergency CT brain: Preliminary interpretation with a tablet device: Image quality and diagnostic performance of the Apple iPad. *Emergency Radiology* 2012;**19**(2):127–133.
70. Goost H, Witten J, Heck A et al. Image and diagnosis quality of x-ray image transmission via cell phone camera: A project study evaluating quality and reliability. *PloS One* 2012;**7**(10):e43402.
71. McNulty JP, Ryan JT, Evanoff MG et al. Flexible image evaluation: iPad versus secondary-class monitors for review of mr spinal emergency cases, a comparative study. *Academic Radiology* 2012;**19**(8):1023–1028.
72. Choudhri AF, Radvany MG. Initial experience with a handheld device digital imaging and communications in medicine viewer: OsiriX mobile on the iPhone. *Journal of Digital Imaging* 2011;**24**(2):184–189.
73. Reponen J, Niinimaki J, Kumpulainen T et al. Mobile teleradiology with smartphone terminals as a part of a multimedia electronic patient record. *International Congress Series* 2005;**1281**:916–921.
74. LaBounty TM, Kim RJ, Lin FY et al. Diagnostic accuracy of coronary computed tomography angiography as interpreted on a mobile handheld phone device. *JACC: Cardiovascular Imaging* 2010;**3**(5):482–490.

75. Edirisinghe Y, Crossette M. Accuracy of using a tablet device for the use of digital radiology manipulation and measurements. *Journal of Mobile Technology in Medicine* 2012;**1**(2):23–27.

76. Payne KF, Tahim A, Goodson AM et al. A review of current clinical photography guidelines in relation to smartphone publishing of medical images. *Journal of Visual Communication in Medicine* 2012;**35**(4):188–192.

77. Scheinfeld N. Photographic images, digital imaging, dermatology, and the law. *Archives of Dermatology* 2004;**140**(4):473–476. doi: 10.1001/archderm.140.4.473[published Online First: Epub Date].

78. Kunde L, McMeniman E, Parker M. Clinical photography in dermatology: Ethical and medico-legal considerations in the age of digital and smartphone technology. *Australasian Journal of Dermatology* 2013;**54**(3):192–197.

79. Perera CM. Principles of security for the use of mobile technology in medicine. *Journal of Mobile Technology in Medicine* 2012;**1**(2):5–7. doi: doi:10.7309/jmtm.10 [published Online First: Epub Date].

80. WHO. *mHealth: New Horizons for Health through Mobile Technologies. Based on the Findings of the Second Global Survey on eHealth.* Global Observatory for eHealth. Geneva, Switzerland: World Health Organization, 2011.

81. Free C, Phillips G, Watson L et al. The effectiveness of mobile-health technologies to improve health care service delivery processes: A systematic review and meta-analysis. *PLoS Medicine* 2013;**10**(1):e1001363. doi:10.1371/journal.pmed.63.

82. Free C, Phillips G, Galli L et al. The effectiveness of mobile-health technology-based health behaviour change or disease management interventions for health care consumers: A systematic review. *PLoS Medicine* 2013;**10**(1):e1001362. doi:10.1371/journal.pmed.62.

36

3R (Retiming, Regeneration, Reshaping) Dataflow Engine to Enable Online Professional Health Care

Anpeng Huang

CONTENTS

36.1 Introduction

Noncommunicable diseases (NCDs) (or chronic diseases) are of long-duration and slow-progression symptoms [1,2]. As classified by the World Health Organization (WHO), there are some typical NCDs, such as cardiovascular diseases, cancer, chronic respiratory diseases, diabetes, and strokes [3,4]. In the WHO official file Noncommunicable Diseases Country Profiles 2011 [5], it is reported that NCDs are the leading mortality causes, comprising 63% of the annual deaths in the world. In China, there are 260 million chronic disease patients according to the Chinese government statistic report [6]. In other words, one in every five Chinese is affected with at least one kind of chronic diseases. Most importantly, major organs, for example, brain, heart, and kidneys, are subjected to malfunctions and even impairments from NCDs. Moreover, NCDs are regarded as the major risk source of physical disabilities. For example, as the most common cause of disability in the United States, arthritis is causing one in every three working-age adults (aged 18–65 years) to report work limitations [7]. In addition, NCDs also lead to many kinds of mental illness, for example, stress, anxiety, and terror. Besides physical and mental health problems, NCDs create huge socioeconomic losses, as shown in the report "Preventing Chronic Diseases: a Vital Investment" from WHO [8]. If only considering the three major kinds of NCDs, that is, heart disease, stroke, and diabetes in China, the national gross domestic product (GDP) loses US$550 billion during the recent decade.

To solve the huge socioeconomic problem, the emerging trend is to change the care delivery pattern of NCDs, for example, home care [9]. Actually, if patient recovery can be self-cared at home with a proper control of disease progression, the chronic care and peripheral expense may be largely saved in contrast to the hospitalized care. Thereto, home care patients feel less stressed in the familiar environment with family support. In terms of home care, daily routines for NCDs mainly include medication instruction, diet control, and behavior adjustment. For example, it is helpful for cardiovascular disease patients if they are monitored for any possible adverse drug reaction and take light diet with appropriate exercises. To prevent an acute attack from NCD-affected home care persons, health conditions should be intensively monitored and tracked for professional examine if necessary.

To reach the aforementioned goal, health conditions of NCD-affected people must be collected and delivered to a professional health-care center online without unexpected disruption and distortion. As affirmed in the WHO official file [9], information and communication technology (ICT) is a promising tool to change health-care delivery today and will be at the core of responsive health systems in the future. Specifically, ICT is to fulfill information processing and communications functions by primarily using computer science, telecommunication, and peripheral technologies. When ICT is applied to health-care services, the greatest challenge is how to perform clinical perception (CP) online.

36.2 Virtual Clinical Perception for Online Professional Services

In medical science, clinical diagnosis is still a semiempirical science due to the native 3I (incomplete, indeterminate, and individual) features of biosignals. Furthermore, the 3I features are time varying even for the same person. In reality, doctors make clinical decisions by combining their medical professional knowledge with skillful experiences. In other words, to see a doctor means the doctor has to sense a disease in his/ her patient in person. As a result, CP plays a decisional role in medical diagnosis and treatment. CP is defined as the ability to observe, recognize, discriminate, and interpret clinical evidences from physical examination and clinical observation [10,11]. Medical professionals obtain cumulating sensory perceptual experience when they investigate the clinical phenomena themselves. Because creative activity is involved, CP is an integration of clinical features and underlying medical symptoms. Please note that CP is an active process, not a passive reception of observational data. Thus, the semiempirical CP is of *making sense and judgment* of *a disease in a patient* rather than *diseases in data*. In an ICT-based health system, digital data turned from electrophysiological signals are delivered for health-care purposes, in which any kinds of collected biosignals must also be converted into digital data for digital transmission [1]. In digital transmission, a patient is detached from the *direct sense and judgment* of a professional, although they are connected online. As a consequence, the real and essential CP process gets lost. Therefore, the left choice is to provide the virtual reality of CP for an ICT-based health system, in which the delivered digital information can be restored into the conventional medical patterns for professional investigation. Furthermore, the challenge is greatly amplified due to the natural 3I-featured legacy in CP that is thoroughly transferred to the health-care network (CARE-NET). Thus, in this study, the fundamental goal is about how to realize effective *virtual clinical perception* (*VCP*) for online health-care users who are tied to an ICT-based health system. To achieve this goal, we propose a 3R (retiming, regeneration, and reshaping) dataflow engine, which is the key to enable an ICT health system to offer *VCP* for online professional health-care services, including home care.

36.3 CARE-NET (Health-Care Networks): Overall ICT Architecture for Health-Care Services

As mentioned earlier, it is promising to extend ICT into online health-care services [12]. Driven by this promising future, a number of ICT-based health-care systems are already demonstrated, which can be classified into three types.

The first type focuses on signal sensing and acquisition [13–16]. In [13], the authors proposed an enhancement of the common mode rejection ratio (CMRR) for higher amplifier gain and lower noises when extracting physiological signals. In [14], an acquisition function of multiphysiological parameters was embedded into an ICT system. In the literatures of [15,16], the authors discussed the sensor power consumption issue with complementary metal oxide semiconductor (CMOS) and radio-frequency (RF) circuit techniques. In [17], the developed wearable efficient telecardiology system (WE-CARE) offers a 7-lead electrocardiogram (ECG) real-time monitoring service over mobile networks, which is objective to collect adequate clinical ECG information while considering user-mobility needs.

The second type aims at signal transmission [18–20]. In [18], wireless communication technology is applied for a distributed monitor system. In [19], the authors tested ZigBee and Bluetooth technologies for local robust wireless connection. In [20], the authors proposed a Best-fit Carrier Dial-up (BCD) algorithm to guarantee medical image and video transmission in mobile networks.

The third type is to interpret the acquired medical signals in ICT health-care applications [21,22]. In [21], Bayesian theorem and decision trees are used to construct a web-based decision support system. In [22], a decision support system is conceived for a specific disease of diabetes.

In all kinds of the aforementioned systems, there are a number of major common enabling technologies. This motivates us to devise a basic health-care network framework based on the combination of ICT technologies and health-care applications, called CARE-NET, which can be evolved to any existing and future ICT-based health systems. The conceived CARE-NET architecture consists of three layers: sensing layer, network layer, and application layer, as shown in Figure 36.1.

In CARE-NET, the sensing layer is served as the front end, in which biological electrode sensors and biosensors are used to detect, monitor, and control physiological signals, such as temperature, blood pressure, blood oxygen saturation, pulse rate, respiration, ECG, and electroencephalogram (EEG). Besides these biosensors, there are some new approaches for information acquisition. For example, radio-frequency identification (RFID) is broadly used in hospitals for patient tracking, ownership identification, drug control, etc. Typically, the sensing layer collects and displays physiological signal/data and performs signal/data processing if necessary. For the state-of-the-art constraints of chip performance and

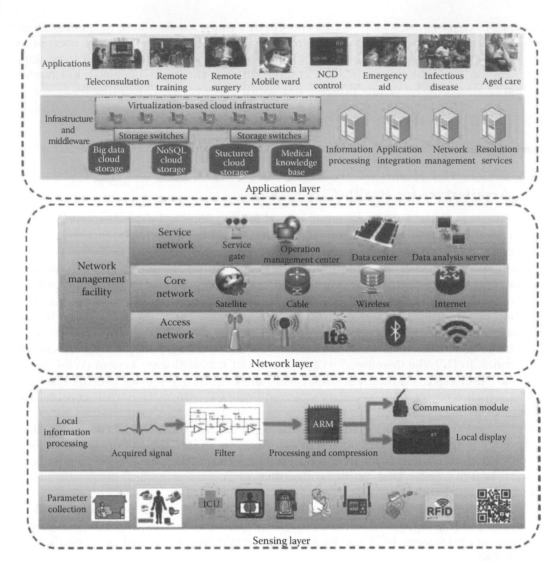

FIGURE 36.1
The architecture of CARE-NET.

computing complexities at sensing terminals, local processing should be simplified as much as possible. This is also the reason that main data processing functions are carried out in the application layer.

To make the best use of collected digital data, they should be available upon an authorized request and shared by granted agents. Therefore, the sensing data need to be carried and transmitted over networks at a required reliability. In the network layer, there are wired and wireless transmission technologies in terms of access media. Naturally, a wireless technology is preferred for CARE-NET because it can offer services without time and location limitations. In other aspects, a wireless radio channel suffers from severe interference, fading, path loss, shadow, and other negative effects, so transmission reliability is more critical in wireless mobile networks. In terms of networking protocols, there are Internet protocol (IP) computer networks, telecommunication protocol mobile networks, etc. While digital data are transmitted in all-IP pattern over mobile Internet, delay is a considerable problem since the "better-than-best effort" is desired to provide quality benchmarked health-care services. Additionally, other related topics in this layer include authentication, authorization, and accounting. In CARE-NET, the network layer transmits information at required reliability and survivability.

Naturally, the collected data should be used for given health-care purposes, for example, home care, disease control and prevention, emergency rescue, allergy test, genetic disease tracking, and privacy-sensitive treatment of sexually transmitted disease. In the application layer, enabling technologies essentially include application infrastructure and middleware, information processing, application integration, and service management. With health-care applications emerging and the collected data cumulating, to effectively analyze and structurally archive the digital data, cloud computing may be a feasible enabling technology in terms of the available advanced options. It can provide dynamic deployment and smoothen the expansion of computing, storage, communications, and other resource requirements at a low-cost way.

As explained earlier, the devised CARE-NET architecture can cover existing ICT health-care system technologies and is also open to incorporate any existing or future advanced ICT technologies, for example, decision support system, life-cycle health management, nanochip sensor, and quantum communications. Taking advantages from CARE-NET, people can access online professional health-care services, which helps NCD-affected persons to live in a normal lifestyle. In fact, the digital data in CARE-NET are absolutely different from a conventional professional recognition approaches. As highlighted before, even in reality, clinical diagnosis is still a semiempirical science because the acquired health information is always 3I. In addition, the 3I features themselves are also of the time-varying nature. For instance, the 3I-featured data of a patient in the evening may be quite different from those captured in the morning. If the collected digital data could not be translated into correct professional knowledge, the purpose of ICT-based health systems is totally lost. In response, a 3R dataflow engine is proposed to turn the captured digital data into the conventional patterns under the constraints of medical criteria. Considering real requirements in health applications, the purpose-designed engine needs to be enhanced with privacy-protection access control and anomaly detection functions as well.

36.4 3R Dataflow Engine: A Key to Online Professional Health Care

Now, we will explain the design motivation and principles of 3R dataflow engine. As highlighted in Table 36.1, the 3R dataflow engine is designed delicately to handle challenges from the special 3I features of collected health data, in order to revive CP from CARE-NET, which was vanished due to the digital transmission mechanism. As shown in Figure 36.2, the proposed 3R dataflow engine consists of the core 3R functions and peripheral components, including access control and application programming interfaces (APIs). For the expected *VCP*, the 3R core module aims at converting the captured digital data into the medical conventional patterns. In the 3R core module, a necessary plug-in function is

TABLE 36.1

Motivations and Design Principles of 3R Dataflow Engine in CARE-NET

3I Features of Real CP in Health-Care Applications		3R Purposes for Online Health Care	
Requirements	Natures	Functions	Objectives
Health monitor, digital data sampling	Incomplete data, unpredicted anomalies	Clock extraction, data synchronization	Retiming: to provide an exactitude timing basis
Semiempirical clinical diagnosis, conventional patterns	Individual character biosignals, time varying	Recovery from digital data, anomaly discrimination	Regeneration: to locate anomalies and augment reality of VCP
High fidelity, fine accuracy, to match with different resolutions	Indeterminate information, uncertainties in applications	Noise clearing, distortion correction, resolution adjustment	Reshaping: to tune the recovered biosignals for VCP requirements

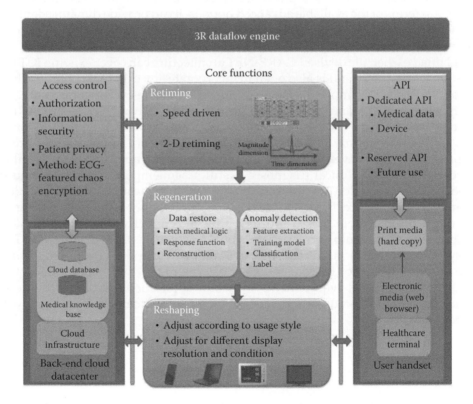

FIGURE 36.2
The advised structure of 3R dataflow engine.

to locate abnormal signals from the collected medical data at a required accuracy level. Otherwise, the professionals are lost in the large cumulating data. In terms of peripheral functions of the dataflow engine, the major concern is privacy protection. For privacy health information protection, it is worthwhile to furnish the engine with a feasible admission control. From the point of software design view, the designed 3R dataflow engine is a network middleware between the back-end cloud data center and a user handset, which can fulfill the purpose of *VCP* in CARE-NET.

In the following, we will illuminate the working schemes of 3R dataflow engine from the aspects of access control, 3R core function, and API, respectively.

36.4.1 Access Control

To enable privacy-sensitive applications in CARE-NET, the collected data must be encrypted for information security. Obviously, an encryption algorithm should be effective and simple enough for time-sensitive health-care applications. To devise an encryption algorithm, the key generation method is the major part. Motivated by this, a novel solution, an ECG-featured chaos encryption mechanism, is designed for CARE-NET applications, as shown in Algorithm 36.1. Currently, biological natures are broadly used for biometric identification [23–25]. Usually, biodata captured from a person is individual, unique, and time varying, which is qualified to be an encryption key in CARE-NET. In our study, the R wave to R wave (RR) interval nature* abstracted from ECG signals is chosen for the initial value of key generation. Compared with fingerprint or image recognition, the RR interval can be straightly coencoded with other digital biosignal data. In contrast with pulse rate, the RR-interval feature is of high-level confidential at lower complex. The raw RR interval must be mapped to a key form for encryption application. In this study, considering the low complexity requirement and the 1-D feature of digital ECG data, logistic chaos mapping function† is used to generate an encryption key from RR intervals.

In nature, logistic mapping is a quadratic polynomial mapping, which is a classic example of chaotic phenomena by a simple nonlinear equation as follows:

$$x_{n+1} = rx_n(1-x_n) \quad r \in [0,4], x_n \in [0,1] \tag{36.1}$$

where
  r is the logistic parameter, which decides the distribution pattern of the equation
  $x_n \in [0, 1]$ is the condition of a chaotic state that is sensitive to the initial condition x_0

When $3.5699456 < r < 4$, especially, when r is closing to 4, the output sequence is of the pseudorandom character after a limited number of iterations. The obtained pseudorandom sequence is preferred for the key generation since it is nonperiodic nonconvergent and has an avalanche effect. Please note, at the special case $r = 4$, chaos phenomenon may be still at a disorder state, but it will contain some implicit information (e.g., cyclical character) if a large enough number of initial values were tried. This is why $r = 4$ is not applied for the mapping in practice.

To apply the generated key into the collected data in CARE-NET, an encoding and decoding method is needed. The popular-cited standardized advanced encryption standard (AES) is robust enough for this purpose [27,28]. To realize this purpose, the key is embedded into the collected data with exclusive-OR (XOR) operation in a round mode, in which the raw collected data are replaced by the encrypted data with a nonlinear function,

---

* RR interval: R wave to R wave interval or R wave length connecting two QRS complex. It represents the time required from a ventricular depolarization to the next ventricular depolarization.
† For the chaos approach, there are several variations: logistic mapping for 1-D data, He'non mapping for 2-D data, Lorenz mapping for 3-D data, and other options [26].

and then the encrypted data are cyclically shifted and mixed in rows and columns. The designed encryption process is shown in Algorithm 36.1.

Algorithm 36.1 The conceived algorithm of ECG-featured chaos encryption.

// **Given preset key array** $k\bar{e}y$.
// **Given preset encrypted data array** *encrypt_data*.
// *raw_ data* **represents digital date before en encryption.**
// *ECG_data* **is a period of ECG waveform used for RR-interval extraction.**
// r **is defined as the parameter in Logistic mapping.**
// n **is defined as the iteration time in Logistic mapping.**
// *RR_ len* **is defined as the R-R interval length used for key generation.**

1: **procedure** ENCRYPTION (*raw_data,ECG_data*)
2: //key generation
3: *RR_len extract_ RR(ECG_data)*
4: //set intial vales as RR-interval length
5: *key*(1) ← RR_len
6: **for** i ← *1,n* **do**
7:      *key(n+1)* ← $\gamma * key(n) * (1-key(n))$
8: **end for**
9: //AES Step 1: XOR raw data with key
10: *encrypt_data* ← *Add Key(raw_data, key(n))*
11: //AES Step 2: look up corresponding byte and replace
12: *encrypt_data* ← *SubShift (encrypt_data)*
13: //AES Step 3:cyclic shift
14: *encrypt_data* ← *Shift Rows(encrypt_data)*
15: //AES Step 4:linear conversion to mix each line
16: *encrypt_data* ← *MixColumns(encrypt_data)*
17: **return** *encrypt_data*
18: **end procedure**

With the embedded encryption, the access control process monitors and manages two connection modes in different deployment scenarios. One mode is a socket connection for real-time monitoring applications. For example, in a point-of-care field, all captured data should be restored into the conventional medical patterns in real time and archived in the cloud data center concurrently. The other connection is the hypertext transport protocol (HTTP) mode, which is typically for general purpose applications, for example, offline backtracking historical anomalies and periodical data collection of 24/7 home care services*. In these general-purpose applications, when an urgent request or health-risk alert happens, the HTTP mode can be automatically switched to the socket communication mode for real-time monitoring.

In both socket and HTTP connections, users employ digital signatures to enhance access authentication, so as to prevent any illegal access or hacker invasion. Additionally, all

---

* The listed two examples are very useful in reality. For example, historical medical data are very valuable to support clinical diagnoses. In 24/7 home care applications, the captured data may be uploaded to cloud data centers in a preset period through the HTTP mode, in which the period size can be customized by users.

requests are recorded in an audit log in the cloud database. If an authorized connection is set up, dataflow transmission between the user and its data center is activated by the 3R dataflow engine. In the following, we will explain how the 3R dataflow engine restores digital data into its medical conventional patterns for the *VCP*, in which health-risk alert function is embedded for discriminating anomalies from normal data.

36.4.2 3R Dataflow Engine

Let us take ECG as an example to explain the 3R functions in the following (in the rest of this chapter, the ECG is studied if no specific notice is mentioned). In reality, ECG information is the typical electrophysiological signal that is popular in most clinical diagnoses, not only limited to cardiovascular diseases. In medical science, ECG signals are recognized in time and magnitude dimensions. But in CARE-NET, the ECG data are turned into 1-D bit streams for digital transmission. Is it possible to reproduce 2-D medical signals from digital bit streams? If it is feasible, the restored signal outcome must be matched with medical diagnostic criteria without any loss of necessary clinical information.

In the proposed 3R dataflow engine, the core function consists of three threads, that is, retiming thread, regeneration thread, and reshaping thread. Because of limited space, please refer to [1] for more details about 3R dataflow engine design.

36.4.3 API for Supporting Terminal Applications

While physiological signals are restored to their medical conventional forms with health-risk alert labels, the data presentation needs to match with different application requirements. In the 3R dataflow engine, there are two kinds of APIs, that is, dedicated API and reserved API, which facilitates real-time record and historical backtracking functions. In terms of dedicated APIs, they are further classified into two types, device-specific APIs and data-specific APIs. The device-specific APIs send data to dedicated interfaces of smartphone, tablet PC, TV, printer, and so on. The data-specific APIs are sensitive to data types in applications, and different data may be carried by different data protocols, for example, FTP file or text message. Furthermore, the other major function of APIs can adjust data delivery modes according to specific medical requirements. Let us return to the ECG example. The standard ECG waveform has a default 10 mm/mV sensitivity and 25 mm/s chart driving speed, but under different conditions, sensitivity of 5 or 2.5 mm/mV and chart driving speed of 12.5 and 6.25 mm/s may be requested through APIs.

36.5 Case Study: Clinical Trials

To evaluate our proposal, it is necessary to define a reasonable performance metric. To obtain optimal results of the proposal, the difference between VCP and CP should be minimized. Naturally, the difference is a desired evaluation metric, defined as augmented reality margin (AR margin). The question is which parameter is suitable to measure the difference. To answer the question, let us look at characters of input and output in the CP process first. With respect to the output of the CP, they are full of uncertainties because creative activities of professionals are involved. As a result, it is not suitable to take the output into account of the evaluation, due to its unquantifiable feature and professional intervention. Thus, we will only investigate the determinate input, namely, physical examination information, which is used as the parameter of the defined performance metric earlier.*

In CARE-NET, the captured digital data have to be restored into the medical conventional pattern, which is the input source of the VCP. As mentioned earlier, to guarantee the quality, the input of VCP should be the same or close to the input of the CP. Thus, we need to check the recovery performance of digital data in CARE-NET. Specifically, we incorporated the designed 3R dataflow engine into our WE-CARE [17], which is a typical implementation of CARE-NET (please see [17] for more details). In nature, ECG is a transthoracic (across the thorax or chest) interpretation of the electrical activity of the heart over a period of time, as detected by electrodes attached to the outer surface of the skin and recorded by a device external to the body [30]. It is the objective indicator of excitement, propagation, and recovery process of the heart. It is proved that certain features of ECG can indicate the happening or early signs of many NCDs [31,32]. Since cardiovascular disease is ranked as one of the leading mortality cause by WHO [5], it is significant to take ECG signals in our clinical trials. Besides the collected ECG signals, the WE-CARE in [17] is extended to contain some typical biosignals with numeric features, including blood pressure, heart rate, body temperature, blood oxygen saturation, and pulse rate. In our performance evaluation, the collected digital signals from WE-CARE are compared with the medical data directly from hospitals in terms of their recognition quality. The clinical trial samples are listed in Table 36.2.

To verify the accuracy in physiological signal recovery in WE-CARE, we compared the original signals and the signals recovered through the 3R dataflow engine. The original signals are obtained by medical measurement instruments in hospitals, for example, ECG machine and sphygmomanometer.

Accordingly to the results shown in Figure 36.3, the recovered ECG waveform is compared with the original waveform. We can observe that the recovered waveform has all

TABLE 36.2

Sample Statistics in Clinical Trials

Group No.	Age		Average Height (cm)		Average Weight (kg)	
	Range	Average	Male	Female	Male	Female
Group 1	20–30	26.74	167.33	175.32	66.4	58.7
Group 2	30.40	35.68	165.25	175.24	66.2	58.9
Group 3	40.50	48.25	164.88	174.58	67.7	59.3
Group 4	50.60	57.32	164.50	173.98	67.4	59.6
Group 5	>60	69.49	162.84	168.96	66.3	58.7

---

* Actually, there are two types of input information to the CP comprehensive process, physical examination and clinical observations. In nature, physical examination is the typical hard-knowledge input to the CP, which are gathered from biochemical tests (e.g., blood test, urinalysis), medical measurement (e.g., temperature, weight), electrophysiological detection and sensing (e.g., ECG and EEG signals), medical scanning (e.g., ultrasonic scopes, magnetic resonance imaging, computed tomography scanners), biodata captured from various sensors, etc. As for the soft-knowledge input of the CP, clinical observation, its information quality is depending on a professional, which is affected by many uncertain factors, for example, mood, personalities, skills, environment, and interaction with patients [29]. In this study, only the determinate input, namely, physical examination information, is considered for quantifiable comparison while leaving the clinical observation input for the future study.

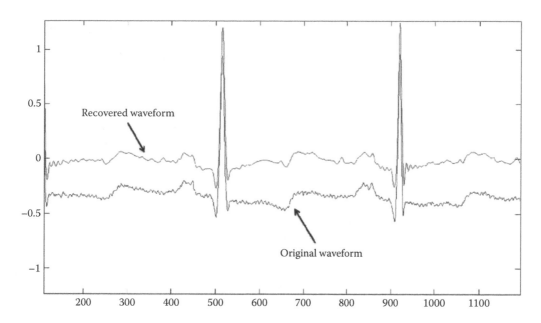

FIGURE 36.3
Clinical trial results in terms of AR margin between original and recovered ECG waveforms.

necessary medical details with an allowed shift in the ECG baseline. These results are obtained by analyzing the position of R wave in ECG signals, in which the confidence interval is [−0.020, +0.020] s considering the length of the standard display format (the smallest grid of the background is 0.040 s) and the recognition ability of visual acuity (the recognition error of visual acuity is ±0.5 small grid). In these trials, the key for encryption algorithm can be fast generated after 300 iterations, in which the RR-interval length is used as the initial value with $r = 3.98$.

In these trials, except for ECG signals, other related signals are directly recovered to numerical values, as shown in Table 36.3. From results in the table, we can see that AR

TABLE 36.3

AR Margin Results for Biosignals with Numeric Characters

Sample No.	Heart Rate	Blood Pressure	Pulse	Temp.	Respiration	Oxygen
1	0.38	1.04	2.67	0.42	0.53	1.18
17	1.44	3.46	3.12	0.97	1.18	2.24
25	0.25	1.98	0.47	0.3	1.57	0.19
29	0.15	0.37	0.43	0.38	0.58	0.31
34	1.56	0.38	0.41	1.33	0.44	0.66
46	0.68	1.75	0.84	1.16	0.25	1.41
52	1.16	1.23	0.63	1.27	0.32	0.33
53	0.33	0.64	1.06	0.31	1.55	1.13
64	0.77	0.51	0.28	0.19	0.86	0.77
77	0.03	0.11	2.77	1.35	1.29	1.72
89	0.46	1.16	0.32	0.6	0.32	0.31
93	1.04	0.49	1.11	1.88	1.64	1.19
Average	0.69	1.09	1.18	0.85	0.88	0.95

margins are comparatively lower, in which the average AR margin of heart rate is at 0.69%, blood pressure at 1.09%, pulse at 1.18%, body temperature at 0.85%, respiration at 0.88%, and blood oxygen saturation at 0.95%. They are very helpful in clinical applications. On average, the AR margin of these numeric biosignals can be minimized to 0.94%, which is slightly higher than the results from complex ECG data. This is because that different biosignal capture manners are deployed here. In our tests, the WE-CARE is the dedicated system for ECG capture at a required high fidelity, while the captured numeric biosignals are collected by wearable sensors and transmitted together with ECG data in a chosen mobile network. Consequently, there are more distortions in the collected numeric biosignals due to the wearable sensor manners, whose measurement accuracy is lower than that of the professional devices in hospitals.

To the end, the proposed 3R dataflow engine can properly address challenges rooted from 3I-featured health-care information and allow CARE-NET to perform online health care. The clinical trials prove that the technology matches with the medical requirements in health-care applications.

36.6 Conclusions and Future Topics

In this chapter, we focused on the greatest challenge in CARE-NET, 3R dataflow engine design for offering online professional health-care services. Because of digital transmission, the natural *CP* is lost in CARE-NET. Unfortunately, the challenge is greatly magnified due to the native and time-varying 3I features that are totally inherited by the captured digital data in CARE-NET. To deal with this challenge, we proposed a novel 3R dataflow engine to enable effective *VCP* in the CARE-NET, where 3I-featured digital data are properly addressed. To perform clinical trials, the purpose-designed 3R engine is incorporated into our developed WE-CARE, which is a typical implementation of CARE-NET. The results show that the proposal can restore the digital data into the original medical pattern with limited distortions, in which the AR margin can be minimized to 0.68% in terms of the complex ECG data. There are more open problems in CARE-NET for future study, for example, lower-power-consumption sensor chip in the sensing layer, self-adaptive health data transmission in the network layer, data mining for clinical decision–support system, and life-cycle health management in the application layer.

Questions

1. What are CP and VCP?
2. What are the 3I characteristics for clinical information?
3. What are the basic components of the CARE-NET? Explain the role of ICT in the CARE-NET.
4. How to design a clinical trial for the CARE-NET test?
5. List out some of the typical applications for the CARE-NET. What are the major functions in the WE-CARE system?

References

1. Y. Zhang, A. Huang, D. Wang, X. Duan et al., A 3R dataflow engine for restoring electrophysiological signals in telemedicine cloud platforms, *Proceedings of 2012 IEEE 14th International Conference on e-Health Networking, Applications and Services (Healthcom 2012)*, Beijing, China, pp. 486–489, October 2012.
2. Council of the European Union, 3053rd Employment, Council conclusions Innovative approaches for chronic diseases in public health and healthcare systems, *the Social Policy Health and Consumer Affairs Council Meeting*, Brussels, U.K., December 2010.
3. J. Bousquet, J. M. Anto, P. J. Sterk, I. M. Adcock et al., Systems medicine and integrated care to combat chronic noncommunicable diseases, *Genome Medicine*, 3(7), 1–12, July 2011.
4. World Health Organization, 2008–2013 Action plan for the global strategy for the prevention and control of non communicable diseases, WHO report, Geneva, Switzerland, March 2009.
5. World Health Organization, Noncommunicable diseases country profiles 2011, WHO report, Geneva, Switzerland, September 2011.
6. Ministry of Health of China, *2011 China Health Statistics Yearbook*, Ministry of health in China, Beijing, China, January. 2012.
7. Centers for Disease Control and Prevention, Arthritis: The Nation's most common cause of disability, http://www.cdc.gov/chronicdisease/resources/publications/aag/arthritis.htm, Accessed January 26, 2014.
8. World Health Organization, Preventing chronic diseases: A vital investment, WHO report, Geneva, Switzerland, October 2005.
9. World Health Organization, mHealth: New horizons for health through mobile technologies, 2011, http://www.who.int/goe/publications/goemhealth web.pdf, Accessed January 26, 2014.
10. K. Cox, Teaching and learning clinical perception, *Medical Education*, 30(2), 90–96, Accessed March 13, 2009.
11. A. Shaw, J. Latimer, P. Atkinson et al., Clinical perception and clinical judgment in the construction of a genetic diagnosis, *New Genetics and Society*, 22(1), 3–19, 2003.
12. F. Collins, The real promise of mobile health apps, Scientific American, http://www.scientificamerican.com/article.cfm?id=real-promise-mobile-health-apps, July 1, 2012.
13. L. Woojae, C. Min-Chang, and C. S. Hwan, CMRR enhancement technique for IA using three IAs for bio-medical sensor applications, *2010 IEEE Asia Pacific Conference on Circuits and Systems (APCCAS)*, Kuala Lumpur, Malaysia, pp. 248–251, December 2010.
14. A. Huang, W. Xu, Z. Li., M. Sarrafzadeh et al., System light-loading technology for mHealth: Manifold-learning-based medical data cleansing and clinical trials in WE-CARE project, *IEEE Journal of Biomedical and Health Informatics*, to be published.
15. A. C. Paglinawan, Y. H. Wang, S. C. Cheng et al., CMOS temperature sensor with constant power consumption multi-level comparator for implantable bio-medical devices, *Electronics Letters*, 45(25), 1291–1292, December 2009.
16. C. Chiung-An, L. Ho-Yin, C. Shih-Lun et al., Low-power 2.4-GHz transceiver in wireless sensor network for bio-medical applications, *2007 IEEE Conference in Biomedical Circuits and Systems (BIOCAS 2007)*, Montreal, Quebec, Canada, pp. 239–242, November 2007.
17. A. Huang, C. Chen, K. Bian, X. Duan et al., WE-CARE: An intelligent mobile telecardiology system to enable mHealth applications, *IEEE Journal of Biomedical and Health Informatics*, 18(2), 693–702, March 2014.
18. C. Yu and W. Xi, Design of distributed intelligent physiological parameter monitor system based on wireless communication technology, *International Seminar on Future BioMedical Information Engineering, 2008 (FBIE'08)*, Hubei, China, pp. 403–406, December 2008.

19. I. H. Mulyadi, E. Supriyanto, N. M. Safri, and M. H. Satria, Wireless medical interface using ZigBee and bluetooth technology, *Third Asia International Conference on Modelling & Simulation (AMS'09)*, Bali, Indonesia, pp. 276–281, June 2009.

20. Y. Zhang, A. Huang, D. Wang, X. Duan et al., To enable stable medical image and video transmission in mobile healthcare services: A Best-fit Carrier Dial-up (BCD) algorithm for GBR-oriented applications in LTE-A networks, *Proceedings of IEEE International Conference on Communications 2013 (ICC 2013)*, Budapest, Hungary, pp. 4368–4372, June 2013.

21. C. C. Chang and L. Hsueh-Ming, Integration of heterogeneous medical decision support systems based on web services, *Ninth IEEE International Conference on Bioinformatics and BioEngineering (BIBE'09)*, Taichung, Taiwan, pp. 415–422, June 2009.

22. L. Chang-Shing and W. Mei-Hui, A fuzzy expert system for diabetes decision support application, *IEEE Transactions on Systems, Man, and Cybernetics, Part B: Cybernetics*, 41(1), 139–153, February 2011.

23. A. K. Jain, R. Bolle, and S. Pankanti, *Biometrics: Personal Identification in Networked Society*, Kluwer Academic Publishers, New York, pp. 1–41, 2002.

24. A. Jain, L. Hong, and S. Pankanti, Biometric identification, *Communications of the ACM*, 43(2), 90–98, February 2000.

25. J. Bringer, H. Chabanne, and B. Kindarji, Identification with encrypted biometric data, *Security and Communication Networks*, 4(5), 548–562, June 2010.

26. J. C. Sprott, *Chaos and Time-Series Analysis*, Oxford University Press, New York, pp. 2003.

27. Federal Information Processing Standards Publication 197, Announcing the ADVANCED ENCRYPTION STANDARD (AES), United States National Institute of Standards and Technology (NIST), November 2001.

28. D. Selent, Advanced encryption standard, *Rivier Academic Journal*, 6(2), 1–14, Fall 2010.

29. P. M. Hildreth and C. Kimble, The duality of knowledge, *Information Research*, 8(1), 16–24, October 2002.

30. P. W. Macfarlane, A. V. Oosterom, O. Pahlm, P. Kligfield et al., *Comprehensive Electrocardiology*, Springer-Verlag London Limited, London, U.K., pp. 3–48, 2010.

31. F. Azizi, A. Ghanbarian, A. A. Momenan, F. Hadaegh et al., Prevention of non-communicable disease in a population in nutrition transition: Tehran Lipid and Glucose Study phase II, *Trials*, 10(1), 5, January 2009.

32. S. Lindeberg and B. Lundh, Apparent absence of stroke and ischaemic heart disease in a traditional Melanesian island: A clinical study in Kitava, *Journal of Internal Medicine*, 233(3), 269–275, 2009.

Answers to the End-of-Chapter Questions

Chapter 2

1. *Primary prevention*

   Depression prevention intervention in adolescents

   Weight management intervention for overweight and obese adults using SMS

   Secondary prevention

   SMS reminders for postpartum women to test for type 2 diabetes after they had been diagnosed with gestational diabetes mellitus

   Sending SMS to men who have sex with men to motivate them undergo HIV testing

   Tertiary prevention

   Sending SMS to increase patient retention in care after HIV diagnosis

2. Norris et al. highlighted phases for a sustainable strategy toward mHealth from the survey of senior executives, strategists, planners, and managers in New Zealand. The three phases were identifying suitable applications, channel development activity, and confirming activity for sustainability. The objective was to produce high-quality mHealth modalities.

3. The purpose of the formative research after the conceptualization of the intervention has been to inform the development of the intervention itself. It has been done using focused-group discussions and surveys.

   The aim of the qualitative follow-up after community-based randomized controlled trials has been to inform the improvement of the intervention. It has been done using semistructured interviews.

4. There are some possible reasons for the interest in quitting smoking by mHealth researchers and developers:

   1. The act of quitting is a tangible primary outcome that can be assessed with confidence and low risk of bias.

   2. A date can be set for quitting, and the behavior change (which can be supported by mHealth interventions) can be focused on that specific date.

   3. Smoking is a global risk factor for many noncommunicable diseases, so the positive effects of cessation can be assessed not only immediately for at-risk individuals but also later for secondary health outcomes.

   4. Behavioral change theories have been historically a fundamental part of research on smoking cessation. So the inclusion of mHealth interventions has been assessed with these theories in mind

5. Pellowski et al. systematically reviewed the literature to assess advances in technology-based interventions for HIV-positive people. They found that most studies have been focused on medication adherence and have used different technologies such as SMS, cell phones, or computers for providing specific interventions.

   They also identified several gaps in the literature, specifically lack of enough evidence on the engagement of HIV-positive people and sexual risk reduction.

6. de Jongh et al. conducted a systematic review to assess the effects of SMS and MMS in helping people self-manage their long-term illnesses. They showed no significant effects for SMS or MMS in controlling glycosylated hemoglobin as an index of diabetes control over a prolonged period of time. They concluded that long-term effects of interventions should be assessed considering the enthusiasm with the expansion of mHealth modalities.

7. Free et al. conducted a systematic review to assess the effectiveness of mHealth interventions on health behaviors of consumers in the United Kingdom. One of their aims in reviewing 26 trials was to assess if SMS was effective in increasing adherence to ART. They showed that communicating with SMS decreased the viral load with a relative risk of 0.85; however, the intervention should be tested in other contexts along with its cost-effectiveness. Later reviews on this piece of evidence showed that technology-based approaches will be helpful and effective for changing health behavior and managing diseases.

8. Guerriero et al. assessed the cost-effectiveness of SMS smoking cessation support using cost per quitter, cost per life year gained, and cost per quality-adjusted life years (QALY) gained as the main outcomes of interest from the TXT2Stop trial. By assessing treatment costs and using health state values of lung cancer, stroke, myocardial infarction, chronic obstructive pulmonary disease, and coronary heart disease, they showed that text-based smoking cessation support would be cost saving; that is, with a cost of 16,120 pound sterling per 1,000 smokers, 18 life years and 29 QALYs would be gained.

    Junod Perron showed that SMS reminders were cost-effective in reducing missed appointments in academic clinics, which could be considered saving in terms of the costs of missed appointments on clinician time and administration time and cost from the health-service perspective.

Chapter 3

1. Users of mobile health and fitness applications differ from patients by not having chronic clinical conditions yet but being at risk for that condition. They differ from athletes in many respects, most importantly by having a sedative lifestyle and not performing physical exercise on an expert level.

ANSWER TO CASE SCENARIO 1

This software integrates several sensor readings that are relevant for action. Overheating is associated with injury risk because users might faint. Therefore, appropriate warnings to stop jogging immediately can prevent adverse effects.

ANSWER TO CASE SCENARIO 2

The *theory* should cover general parameters of health behaviors and users' physiology. The *model* should cover task-specific parameters for jogging and users similar to Mr. Smith. The *tutor* should produce tailored and specific guidance for Mr. Smith based on theory, modeling, and simulation.

2. There is an increase in chronic conditions predicted that increase the number of potential users. At the same time, the number of available sensor increases based on technological progress. In contrast, the number of physicians per population is constant for years and not expected to increase.

3. Most people find it difficult to understand diagrams and complex systems. Many people also display an insufficient level of health literacy. This implies that they will not be able to understand those data and take action based on provided information. Most users do not want to learn facts but want support for changing their health behaviors.

4. Without theory, there are no guiding concepts when designing mobile applications. Current applications display very low levels of system intelligence and mostly little guidance for users. Theories can help fix that problem based on concepts and evidence.

5. Health behavior theories spell out global parameters that hold for all kinds of health behaviors and users. In contrast, models define specific parameters and simulate a user for a specific task of interest.

6. Simulation allows exploring complex systems, like health, in a safe environment, without time constraints, transparency, and without real consequences. In simulations, users can try out actions and observe consequences, often much faster than in real life.

Chapter 4

1. Diabetes is an illness in which there is an abnormally high level of glucose in the blood. It is caused by our genes and lifestyle (*environmental risk*) factors.

2. The most important parameter to be monitored in diabetic patients is their blood glucose level.

3. If the glucose level is not controlled, it can lead to some major complications like cardiovascular diseases, blindness, nerve disorders, and kidney failures. Thus, it should be monitored effectively to avoid these complications.

4. Current diabetes monitoring methods do not meet patients' needs and are not effective in helping people control the disease.

5. What is needed is an effective self-management wireless solution that collects, stores, and transmits important parameters (e.g., glucose level) in real time to the health professionals and receives feedback correspondingly.

6. WBAN is a special wireless network that consists of small intelligent devices attached to clothing, on the body, or even implanted under the skin that are capable of establishing a wireless communication link.

7. The wireless nature of these networks, and the increasing sophistication of implantable and wearable medical devices and their integration with wireless sensors, provides promising applications in medical monitoring systems and real-time feedback to the user or medical personnel to improve health care. They allow patients to experience greater physical mobility, where they are able to conduct their daily activities in a normal living environment. Consequently, patients are no longer compelled to stay in hospital, and their quality of life improves considerably. The sensors of a WBAN are lightweight, low-powered, and able to measure important parameters such as body temperature, heartbeat, and glucose level.

Location-based information can also be provided if needed. For example, WBAN can be an essential part of an efficient self-management tool for diabetic patients.

8. GPS is a satellite-based radio navigation system that provides reliable positioning, navigation, and timing (PNT) information for devices equipped with appropriate GPS receivers.

9. Andriod is a user-friendly and versatile operating system. It was developed by the Open Handset Alliance, an open source initiative piloted by Google. The open system architecture with customizable third-party development and debugging environment features helps the users to manipulate the features and create their own customizable applications.

10. REST is a distributed system framework that uses web protocols and technologies. The REST architecture involves client and server interactions built around the transfer of resources. It is often used in mobile applications and social networking websites.

Chapter 5

1. Key points are detected using the scale-invariant feature transformation (SFIT) key point detector.

2. Key points on the food image are first detected, and then local features extracted at key points are matched with a codebook of code words. The histogram of frequency of code words matched in the image is used to encode the input image.

3. The LBP describes the texture of an image.

4. The 1970s.

5. Any two from: 24 h recall, diet history interview, food frequency questionnaire, and food record (food diary).

6. The use of mobile technologies is very well suited to the capture of food portion size information with the opportunity for automatic identification of both the food in the photograph and the volume of the portion being consumed. It is also useful for real-time data capture of eating occasions as being investigated with the use of senor technology.

Chapter 6

1. Wellness apps are important to the society as they help consumers achieve their health behavior goal that will improve their well-being and lifestyle. In an attempt to prevent the increased incidences of diseases and illness in developed world societies, wellness apps are used as a good health prevention tool that is low cost and has a wide reach, to disseminate health information, services and also provide an avenue for health service providers to form relationships with their consumers. In sum, wellness apps perform a preventative health function that is socially beneficial.

2. The challenges are as follows:

    • Currently, the classification of wellness apps is poorly curated and as such, wellness apps may get *lost* within each classification, as consumers have to spend time searching for a wellness app that suits them.

    • To shorten the search time for wellness apps, it is likely that most people based their search on app rankings (e.g., Top 20 apps in any category). While it is understandable to shorten search time, it is also limiting, as consumers will automatically default to this list rather than have the capability to search for a wellness app based on their own search criteria and terms (which is currently lacking in any app stores).

- There are currently no regulations that exist to legitimize the information that are offered through wellness apps. Because of the lack of governance, some wellness apps may offer suggestions and tips that may not be in accordance with preventative health practices. As such, consumer confidence level of using wellness apps may be compromised and as a result, consumers default to selected wellness apps that are only endorsed by credible health organizations.

3. The four steps toward designing an effective wellness app are as follows:

    1. (Re) Define the behavior goal orientation offering of the wellness app.
    2. Market scan for similar type wellness apps and use it as a template.
    3. Design effective wellness app through co-creating the app development process using a user-centered approach. The five core app elements that developers should focus are usability, personalization, aesthetics, sociability, and trustworthiness.
    4. Finally, evaluate the wellness app and decide if there is a need to expand the app functionalities to fit other market segments.

Detailed explanation of these steps is stated from Section 1.1.2.

4. Fogg behavioral grid put forth that there are 15 ways in which a behavior can change and the purpose of this grid is to assist developers to think carefully about their market segments they will be targeting with their wellness apps. This is because inevitably each segment (or subsegments) will consist of people with different behavioral orientations. As such, the design of the app needs to match users' behavioral goal orientation in order to successfully trigger the healthy behavior that they want to achieve. To this end, it is critical that for the design of any wellness apps, behavioral goal segmentation must be considered and overlay on top of our demographic segmentation variables (e.g., age, gender, and so on) in order to design effective wellness apps.

5. A user-centered approach, an approach endorsed by the World Health Organization, is required to co-create the design process with consumers to ensure effective outcomes. Specifically, a user-centered approach can lead to vast improvements in app capability and design. From a methodological perspective, Agile method echoes the user-centered approach to design wellness apps. For a detailed explanation of this methodology, please see www.agilemethodology.org.

Chapter 7

1. The usage of the mobile application, social network, and its recommendation algorithm is the main component and advantage of the presented system model that differentiates it from other health-care systems. This component provides a new perspective in the use of information technologies in pervasive health care and makes the presented system model more accessible to users. Broadband mobile technology provides movements of electronic care environment easily between locations and Internet-based storage of data allows moving location of support. The use of a social network, on the other side, allows communication between users with the same or similar condition and exchange of their experiences.

2. COHESY has three basic layers. The first layer consists of the bionetwork and mobile application, the second layer is presented by the social network, and the third layer enables interoperability with the primary/secondary health-care information systems.

3. The mobile application in COHESY performs readings by using various sensors regarding user's health parameters and the parameters of the physical activities he or she performs. By using their mobile phone and the installed application, the users have access to the medical personnel at any time. The medical personnel can remotely monitor the patient's medical condition, reviewing the medical data (blood pressure, blood-sugar level, heart rate) arriving from the mobile application of the patient. The installed mobile application has access to the social network where it can store users' data and read average data readings on health and physical activities of all users. In this way, the mobile application within COHESY provides a tool for a complete personal health care.

4. The specific security requirements are data storage and transmission, data confidentiality, authentication, access control, and privacy concerns.

5. The proposed algorithm generates recommendation for a specific activity that the users should perform in order to improve their health. The basic idea is to find out which physical activities affect change (improvement) of the value of health parameters. This dependence continues to be used by the algorithm to recognize the same or similar health situations found in another user with similar characteristics. If there is information in the users' history that after performing some physical activity their health condition has improved, the algorithm accepts this knowledge and proposes the activity to other users with similar health problems.

6. The algorithm is based on the dependence between the values of the health parameters (e.g., heart rate, blood pressure, arrhythmias) and the user's physical activities (e.g., walking, running, biking). So the recommendation algorithm uses a variety of data, such as health parameters, physical activity parameters, health records and personal information of the user, as well as prior knowledge gained on the basis of previous experience of other users who have similar characteristics with the considered user.

Chapter 8

1. Behavioral scientist who can advise on techniques of behavior change, communication specialist to tailor messages, doctors, nurses, or public health professionals that work in obesity intervention research.

2. Timing might be inconvenient for participants; messages are not tied to any real events/behaviors/locations, etc.; messages have not been redeveloped from the original intensive face-to-face intervention to fit an SMS protocol; target group was not involved in the development of the messages.

3. Adjust timing: ask participants when they would like to receive messages, invite a social scientist to collaborate to work with participants using qualitative methods to develop new messages that they find acceptable in an SMS environment, invite a communications expert to help develop the messages, and use sensors to detect behavior and develop timing schemata based on when people are active or inactive.

4. Dr. Zemi could add sensors to detect when people are active, and base timing of messages on information from the sensors on when people are active or inactive. He could ask the participants when they want to receive messages. He could bring in a social scientist to help participants to generate messages that they find acceptable.

5. Adaptive interventions are personalized interventions that respond to participant data. They might be personalized or adapted in one (or several) way, including personalized feedback, messaging, timing and dose, in reaction to data collected from participants at chosen intervals.

6. No.

Chapter 9

1. Recovery state. Mrs. Smith progresses during the episode from intention state all the way to recovery state.

2. *Fit!* seems to assess basic health values and behavior change status.

3. The description does not mention risk and resource communication. The option to explore features might indicate action planning support. The strict guidance indicates action initiative support. Given conditions of her relapse, it is unlikely that *Fit!* has provided support for relapse prevention. It is also unlikely that *Fit!* provides appropriate recovery support given the feedback after returning to the application.

4. During the first days, Mrs. Smith demonstrates action self-efficacy by evaluating the program as easy that seems to continuously decrease during the following period of time. She displays very little coping self-efficacy on the rainy day. She does not display much recovery self-efficacy after relapse, but at least she does not give up completely.

5. Taking pictures of meals and tracking activity automatically indicate passive mode of operation. Strict guidance indicates passive instruction.

6. Updating the system after entering demographics and goals can indicate global intelligence if the system adjusts to those values. Not integrating weather information in exercise recommendations and interpreting no entries as fasting days indicate little local intelligence.

Chapter 10

1. Disease management (DM) has often failed to provide the promised return on investment (ROI). We need to enter a new paradigm for population management through psychological profiling and automated outreach. We must abandon the costly call-center-based outreach and the boilerplate communication to large groups of individuals.

   DM companies could enhance their offerings by deploying personalized communication providing fully scalable and more effective intervention programs. They should identify individual patient's compliance risks based on the individual's psychological and current emotional profile. All this would enable them to overcome the retrospective character of current sophisticated but ineffective targeting methods.

2. Health insurance companies ultimately rely on optimizing the member behavior in order to achieve ever-increasing cost reduction goals. They are challenged by a

complex array of requirements, with responsibilities both within and outside their companies. They lack analysis of current interventions and communication methods in terms of their effectiveness among subpopulations. They desperately require compliance improvement for selected member segments, such as the chronically ill, through type-specific interactions that stimulate their self-motivation.

3. Customer loyalty provides the foundation for a successful business in the highly competitive diagnostics and pharma industry. However, most loyalty programs are indirect and costly and fail to provide customer interaction, resulting in limited product differentiation.

    Companies need to better understand their customer down to the individual or at least major segment level. They need to perform compliance analysis for entire customer segments based on individual psychographic characteristics, emotional state, and motivation. They require new product positioning plans targeting user groups through improved effectiveness and sustained long-term use.

    Pharmaceutical companies can deploy new types of customer monitoring services and self-management support to enhance their attractiveness with health plans and ultimately influence formulary decisions.

4. The key interest of physicians and their associations is their patients' engagement and therapy adherence. This requires an effective and cost-efficient patient management system for therapy regimens, which they do not offer today.

    ACOs and also individual physicians should start to implement patient classification via analysis of individual psychological characteristics and emotional disposition of their entire populations. They could identify successful candidates for certain programs, studies, or procedures. This can lead to a personalized communication process alongside long-term treatment procedures. Ultimately, one could develop design guides for image campaigns leading to enhanced organization and practice positioning.

5. Employers seek improved ways to reduce the cost of the health benefits provided to their employees. Most of currently deployed wellness programs are ineffective and their outcome benefits unproven.

    In order to improve their employees' health and productivity, they need to individualize their offerings and perform motivational and psychographic profiling of the workforce. This will allow a better fit of their health-care and wellness program offerings.

6. In recent decades, the fields of health care and patient treatment have seen enormous technological progress and success. Conversely, advances in physician/patient communication and support of patient self-management are minor and rare. Health plans and disease management organizations are still contacting their large member populations with boilerplate information or via costly call centers. All in all today's direct member communication is limited and too inefficient.

Chapter 11

1. Behavioral change in relation to a chronic illness or condition is when an individual adopts new practices and changes existing behaviors to those best suited to living with a normally debilitating chronic condition.

2. An individual's behavior and self-management requirements can be highly personalized to cultural context; thus, it is difficult to define one single approach.

Both behavioral change and self-management are also highly complex fields, and research is needed to better understand the specific behavioral change and self-management techniques in relation to cultural and disease-specific contexts. Due to this complexity, it has been difficult to combine and deliver both self-management and behavioral change into a personalized platform that will have a widespread effect across various individual users. To reach a better framework for common adoption will require improved translation of information between all stakeholders in the primary care ecosystem, to understand and overcome not only the barriers to mHealth adoption but also how best to deliver psychoeducation learning programs to patients.

3. Three from any of the following: providing information and feedback, goal setting, tailoring the user experience, prompting, problem solving, plans for social change, contract for change, relapse planning, preventative coping strategies, giving data an emotional hook, and developing social interactions with peer support.

   Examples—Any of the following: lead boards among peer users on an application to increase motivation; better tracking of progress and exercise programs as part of a feedback system; setting exercise and diet goals in obesity; providing breathing technique videos as coping strategies to help prevent exacerbations in COPD; personalizing an application to the sociodemographic profile of the user; and providing information to peer support connections via GPS mapping.

4. A large proportion of devices and systems are often designed by those who may not have an appreciation for the difficulties that certain users face when using technology, such as physical impairment and a lack of experience with technology. Furthermore, many people are excluded from using products as a result of designers not taking account of the end user's functional capabilities. As such, it is important to gain a good understanding of the problems faced by users when designing for them.

   There are a variety of methods that have been employed to design for older people or those with disabilities, such as universal design, inclusive design, and accessible design. These can come under the umbrella of UCD, which mandate that users are significantly involved in the design process from as early as possible. This is to ensure that the product designed as a result of this process is usable by that group, as well as incorporating the needs of the group into the design of the end product.

   However, in addition to the needs of chronically ill people being catered for in the technology itself, the methods used to assess these needs must also be tailored for this user group. Issues such as the length of interviews, attendance, question design, and terminology used during interviews and requirements gathering need to be well planned when chronically ill and particularly older adults are involved.

   In an effort to encourage more understanding of the needs of these different user groups, many of whom live with one or more chronic conditions, there are studies and techniques that have tried to mimic the impairments and constraints of people with a variety of impairments. These include simulating visual impairments in software so designers can experience the impairments for themselves and *Third Age Suit*, developed by the Ford Motor Company to simulate reduced joint mobility so designers understand what it feels like to be older.

5. One of the issues that may arise with combining patient and clinician data is clinically validating the data received by the clinician from the patient. Furthermore,

can it be incorporated in the electronic health record used by the clinician and the data displayed to the clinician in a usable and meaningful format?

Other issues include ethical and privacy issues regarding these data—who owns it, who is allowed access to it (and who grants that access), where is it stored, can (should?) it be anonymized and used in broader research, or should this even occur?

Chapter 12

1. Neuropsychiatric disorders can be directly treated through mobile technologies by teaching and providing feedback related to emotion regulation strategies.

2. That emotion dysregulation is a primary factor in the development and maintenance of psychopathology.

3. As a set of functional and adaptive coordinated responses to environmental demands.

4. When the adaptive functions of emotions fail to fit with the demands in the environment or context, the result is dysregulated behavior and decreased well-being.

5. Flexibility is a *difference* measure between how effectively an individual can upregulate emotion and how effectively they can downregulate emotion. The ability to move between these two extremes is what defines flexibility. Flexibility is related to well-being by way of the functional view of emotion. That is, given that emotions have adaptive functions, the ability to flexibly deploy them in the right context will lead to more successful coping in a variety of contexts.

Chapter 13

1. The World Health Organization defines health as "a state of complete physical, mental and social well-being and not merely the absence of disease or infirmity." This definition has not been amended since 1948 but has been enhanced by a number of declarations and charters.

2. Cultural lag theory suggests that a period of maladjustment occurs when the nonmaterial culture is struggling to adapt to new material conditions. Ogburn in explaining his theory distinguishes four stages: innovation, accumulation, diffusion, and adjustment. By applying cultural lag theory to mHealth, it is possible to state that mobile phone technology has been partly successful in the process of accumulation and diffusion in the area of health. Yet it is progressing through full adjustment process in order to be used safely and efficiently in many areas of health now and in the future.

3. mHealth can enhance the communication between doctors and patients through the flow of information. Increasingly, literature in this area indicates the growing direct use of mobile phones as an alternative to face-to-face patient/doctor visits. Thus, smartphone can add a distinctive opportunity for the exchange of further information leading to better communication between doctors and patients with much improved outcomes.

4. Within the general division of labor in society, certain occupations are significant sources of injury and disease. For example, the sexual division of labor has tended to concentrate men in the occupations in which such hazards are greatest. These include construction, mining, waterside work, and farming, which are mainly done by men. However, young men and adolescents are more likely to engage in a range

of dangerous activities such as risky driving, contact sports, and physical aggression. Consequently, males suffer higher rates of nonintentional and intentional injury.

Whereas the health hazards which apply to environments where majority of workers are women, such as office job, nursing, or household work are often harder to detect and may be underestimated.

5. The smartphone such as *Aurora*, which was developed by Komosion, and supported by NSW Department of Family Services, can assist women who are in dangerous situations. By pushing one button, women alert police immediately. Minister Goward in launching this phone said: "This app is the first of its kind in the world to combine information with access to help services, the ability to create a trusted network of friends who can be easily contacted with an agreed message and a GPS system to 'call-for-help' alert."

6. Factors influencing health equity in Australian society include the following:

   • Every aspect of government policies affects health and health equity directly or indirectly. These include finance, education, economy, housing, transport, and health, to name a few. For example, while health may not be the main aim of finance policies of the government, it has a strong bearing on health and health equity.

   • Another major factor causing health inequity is the inconsistency of government policies for health. For example, government's trade policy encourages production and consumption of food high in fat and sugar. This is in contradiction with the government official health policy.

   • The third major factor leading to inequity in health relates to the fact that there is a visible lack of an individuals' empowerment to challenge and change the unfair health distribution resources in society. Therefore, there is a need to empower vulnerable people to challenge the existing health inequity and demand organized and systematic improvement of health status of population.

7.

   • Approach to mHealth should be interdisciplinary by bringing the required experts together including specialists with backgrounds in interpersonal and mass communications, human factors and social sciences, clinical sciences, health informatics and IT, computer science and engineering, public health/population sciences, and health management and policy.

   • The second focus should be on the efficient use of mHealth within the patient-centered needs of each individual, as well as the needs of the community.

   • Digital divide or access to the Internet through smartphones between social groups and within the countries needs to be bridged.

Chapter 14

1. (1) Technology, (?) system, (3) service, (4) information, (5) individual factors, (6) team, organization and environment, and (7) rules, regulations, and policies.

2. (1) Usability; (2) clean ability, upgrades, maintenance, and repair processes and effectiveness; (3) flexibility, adjustability, and adaptability; (4) technology interactions; and (5) accessibility.

3. (1) Complexity/ease of use, (2) time response, (3) ease of learning, (4) usefulness, (5) security, (6) data processing, (7) technical support, and (8) communicability.

4. (1) Reliability, (2) system support, (3) assurance, (4) quick responsiveness, (5) empathy, (6) follow-up service, and (7) technical support and training.

5. (1) Urgency, (2) currency, (3) accuracy, (4) completeness, (5) relevancy, and (6) consistency.

6. (1) Demographic factors (age, gender, ethnicity, socioeconomic, and cultural), (2) cognitive factors (memory, attention, vigilance, reaction time, information processing, problem solving, error propensity, recognition, and recovery), (3) psychological factors, (4) self-efficacy, (5) experience, and (6) preferences and expectations.

7. As MHSs are playing a critical role in human life, assessment is required for this kind of systems in order to prove their success and failure. Traditional design analysis typically focuses on the system assessment from the technology point of view, whereas sociotechnical analysis looks at how the technology will incorporate into the user's activities. From the perspective of sociotechnical aspect, social and technical issues are complementary and they cannot be separated. The key sociotechnical factors influence the performance and success of MHS. The factors must be studied in relation to each other and they cannot be viewed independently. In our model (Figure 14.4), some components are more tightly coupled than others. For instance, technology and system, service, and information are dependent on one another and other social components mostly depend on these components. Most of the components in the proposed model have a positive effect on the use of MHS as well as user satisfaction from MHS. Based on DeLone and McLean (2003), the use of the system and user satisfaction will affect the success of the system positively. Therefore, we can conclude that the sociotechnical components in the proposed model influence the performance and success of MHS. Based on the previous researches, we clarify the relationship between all the sociotechnical components in the proposed model and success of MHS. We find a positive relationship between those components and success of MHS. The positive relationship between sociotechnical components and success of MHS is illustrated in Table 14.6.

Chapter 15

1.

(1) Improve timely data collection and transfer in contexts where poor road and network infrastructure is common.

(2) Improve the ability to identify and manage disease outbreaks and epidemics.

(3) Facilitate better adherence to medicines (i.e., antiretrovirals, DOTS programs, ACT) and medical protocols (i.e., IMCI).

(4) Provide better support and supervision for remote and community health workers.

(5) Facilitate the work of lay-level cadres, who often differ in terms of experience, training, and literacy level.

2.

(1) Is mHealth the most appropriate tool for this particular intervention?

(2) What level of access should each of the stakeholders be given to individual patient data?

(3) Who has access to the phone? Is there any sensitive information transmitted that could cause issues if accessed by someone else?

3.

    (1) Ministry of Health:

        (a) Strengths: Provides guidance on national policies and ensures that the design of the project is aligned to their needs as a Ministry.

        (b) Weakness: May lack technical expertise. Must balance a number of different priorities and health demands under restricted financial resources.

    (2) Nongovernmental organizations/civil society:

        (a) Strengths: Can assist with the coordination of national mHealth policies. Have a strong on the ground presence, often in very remote parts of LMICs. Have strong project implementation track records.

        (b) Weaknesses: NGOs can sometimes exacerbate the problem of uncoordinated and paralleled programming in LMICs.

    (3) Academics:

        (a) Strengths: Strong research skills, which can help to develop an evidence base for the use of mHealth in LMICs.

        (b) Weaknesses: Less experienced in terms of program implementation.

    (4) Communication regulatory bodies:

        (a) Strengths: As they award licenses to mobile network operators, they can be helpful to negotiate with MNOs.

        (b) Weaknesses: The priority of a regulatory body may not be to improve health outcomes.

    (5) Mobile network operators:

        (a) Strengths: Provide a service that is imperative for the implementation of mHealth (data storage, transfer, encryption).

        (b) Weaknesses: As for-profit companies, a strong business model must be presented to secure buy-in.

    (6) End-users:

        (a) Strengths: Can provide important insight and feedback on the mHealth project—what works, what doesn't, and where improvements are needed.

        (b) Weakness: Likely to lack technical expertise.

4.

    (1) mHealth is an extension of human capacity and must cater to the existing capacities of the end-user. Because of the great variability in end-user literacy and previous experience with mobile phones, mHealth application has to be perceived to provide tangible benefits to the end-user, both in terms of time and cost.

    (2) mHealth can only be employed to the extent one is motivated to use it.

5.

    (1) Technological challenges:

        (a) Choice of contextually appropriate mHealth hardware and software. Taking into account factors such as cost/affordability, susceptibility to theft, local capacity to repair and replace parts, familiarity, and phone literacy.

(b) Keeping hardware and software operating effectively, and providing the support to resolve any issues and problems that may occur in a timely manner.

(c) Providing flexibility in hardware and software to allow for user flexibility and localization, and managing any localization that may occur of the hardware or software.

(2) Environmental challenges:

(a) Poor and variable power supply can make charging mobile phones problematic.

(b) Poor road and transport infrastructure can make travel very difficult, expensive, and time consuming.

(c) Ethical challenges, including who controls the data collected by the mHealth system and how such data is used. There are also ethical challenges arising from whether or not patient data should be encrypted, and how individual patient data can be securely and safely stored.

(d) Funding the mHealth initiative, including hardware, software, training, support, and associated infrastructure needed.

(e) Development of business models that allow for the funding of mHealth initiatives on an ongoing basis.

(3) Social and human factors:

(a) Building local capacity to use and support mHealth systems can be difficult where there are low levels of education and literacy.

(b) Lack of capacity to generate the evidence that mHealth improves health outcomes in a cost-effective way.

(c) Motivating and training users can be difficult, costly, and time consuming.

6.

(1) A socio technical approach recognizes the interrelatedness of technology, people, and the society in which people live and work. Such an approach views any information system as a social system that cannot be transferred physically in the same way as a software application or a piece of hardware. Thus, mHealth implementers have to consider a wide variety of technological, environmental, social, political, and cultural factors when adopting a socio technical approach.

(2) A socio technical approach does not assume that mHealth hardware and software will simply fit into any LMICs social, cultural, and political environment. Any mHealth implementation should not be viewed as simply the transfer of just the technology, such as a mobile phone. It is the transfer of an entire set of social, political, cultural, and technical values, and many of these values are likely to have originated in more developed countries.

(3) Thus, the mHealth implementer will have to carefully consider how the technology that is being introduced will affect the social systems present in the LMIC and how these can be made to fit together in a way that facilitates the adoption of the mHealth initiative.

Chapter 16

1. (b)
2. (a)
3. (e)
4. (c)
5. (d)
6. (b)
7. (c)
8. (c)
9. (a)
10. (e)

Answers for Case Studies

1. (a) and (d)
2. (a)
3. (a)

Chapter 17

1. We believe that current mHealth applications and devices are not adapted. There is a wide heterogeneity of data regarding the eHealth situation of the person and a clear lack of collaboration, coordination, and interoperability between services and data sources. This is particularly true when we focus on specific contextual data such as the information about the persons' dependency and their real need of services.

2. The dependency evaluation tools are similar. They are all based on some of the ADLs and classification by grouping needs into some groups (iso-resource groups). The AGGIR is an example of such methods that we studied to help the reader to understand the objective of our work and as a proof of concept of our approach.

3. Of course, the person's activities and dependency level can be easily considered since the proposed framework is ready to integrate all these contextual information and that can come from different heterogeneous sources.

4. We believe that in order to make relevant and personalized mHealth services, we have to improve our knowledge about the context of person. All of the different profiles include important dimensions about the context of the person and should not be ignored.

5. According to our experience in this field, mHealth devices and applications (including different sensors) provide usually some basic and specific services. Having seen the wide heterogeneity of needs of assistance and health care, we believe that the best approach is to create new services based on the existing elementary services. This provides more flexibility and avoids developing new service for each new profile. The gained flexibility helps also to personalize the provided services and assistance and to be as closer as possible to the person's context.

Chapter 18

1. The implementation of mobile device applications as information and communications technology (ICT) tools for the educational process of imparting knowledge through the field concerned with teaching and learning and for training, which involves instruction or practice. As such, these tools are used for assessment, health-care awareness, health-care employee training, self-study, and health-care knowledge collaboration.

2.

    1. To retrieve the latest medical knowledge through regular LMSs

    2. As simulations for the patient case studies

    3. To increase the health-care awareness; for example, with diabetes information that can be effectively sent using an SMS

3. By adopting a PBL instructional stance, it affords the trainees to practice in a real-life scenario that is based on a constructivist (andragogy) cognition model. The process starts by identifying a problem as a case study model, which may have different scenarios, aligned with the instructional objectives. Next, there is an examination of the learner's performance using an assessment instrument such as questioning, which is based on the expected training outcomes.

    It is proposed here that when PBL strategies are implemented using mHealth education/training, they will assist to center the learner, focus on specific tasks, and enhance the collaborative learning research knowledge base. They will also present a clear example of the efficiency and effectiveness of combining the ICT tools and the PBL delivery modes and also enable the health-care students to be independent. In addition, they become more autonomous learners using the research tools to search for their medical knowledge and to select, analyze, evaluate, and present the information.

4. The first aspect regarding difficulties for mHealth deployment is about knowing how to apply the correct theoretical background to the health-care practices during the training sessions.

    The second aspect relates to the time flexibility of using mHealth educational tools and the cost to the individual trainee.

    The third aspect relates to the likelihood of the following personal barriers to adoption of mHealth education/training programs, such as individual learning preferences, whether they can afford attending the training programs; the individual's (cognitive) approach to organize and process the information; external constraints, which may affect them, such as family commitment and full-time work; the ability to resolve and manage their participation needs for the mHealth education class as required; and negotiation with the management to resolve any outcome issues.

    The fourth aspect stems from the trainee/student's reflection and interaction with the course content, which includes their willingness to interact with their actual health-care practice and experience, with their new instructional/training topics. If this is the case, the mHealth tools should assist them to achieve that end.

Chapter 19

1. The nature of culture is pervasive and broad that requires a detailed study of language, knowledge, laws, religions, food, customs, music, art, technology, work patterns, products, and other artifacts that give the society its distinctive flavor. A great number of definitions are provided regarding this. Ferraro (1994) and Merchant and Merchant (2011) affirm that, "as early as 1952, researchers identified more than 160 definitions of culture, and in 1994, it was estimated that culture has been defined in approximately 400 ways." By having diverse definitions, culture affects people's deeds in the society due to their ideas, values, attitudes, and normative or expected patterns of behavior. "Culture is not genetically inherited, and cannot exist on its own, but is always shared by members of a society" (Hall, 1976). Moreover, Hofstede (1991) gives another definition of culture as "the collective programming of the mind which distinguishes the members of one group from another" (p. 1), which is passed from generation to generation. Culture is the society personality and it helps to distinguish one society from another (Schiffman and Kanuk, 2000). De Mooij and Hofstede (2010) refer to culture as the glue that binds groups together. He mentioned that culture includes whatever has worked in the past such as shared beliefs, attitudes, norms, roles, and values that are usually transferred from generation to generation within the groups. According to the objective of consumer behavior studies (the influence of culture on consumer behavior), Schiffman and Kanuk (2000) defined culture as "the sum total of learned beliefs, values, and customs that serve to direct the consumer behavior of member of a particular society" (p. 322).

2. Culture has five dimensions including power distance, individualism/collectivism, masculinity/femininity, UA, and long-term orientation.

3. The range of avoidance of uncertainty and ambiguity among people is related to this part of national culture value. People with high UA culture care about rules and they do not like to take risk, while people with low UA culture more willingly accept risk. They feel that there should be as few rules as possible and they are open to have different behaviors when needed. "Uncertainty avoidance reflects the degree of comfort members of a culture feel in unfamiliar or unstructured situations and the extent to which a society tries to control the uncontrollable" (Hofstede, 2001).

4. Trust has been defined as "a set of beliefs that other people would fulfill their expected commitments under conditions of vulnerability and interdependence" (Rousseau et al., 1998). Alali and Salim (2011) used trust as one of the factors in their *virtual communities of practice* success model, noting that trust influences user satisfaction as well as intention to use (Alali and Salim). Trust measurement turns out to be an important indicator of support for a health-care system (Gilson, 2005; Schee et al., 2006; Abelson et al., 2009). Because of the importance of trust in the health-care area, Lord et al. (2010) evaluated "the effect of patient physician trust on how British South Asian (BSA) and British White (BW) patients cope when diagnosed with cancer" (Lord et al., 2010). Wright et al. (2004) identified trust as the most significant factor in doctor/patient communication (Wright et al., 2004).

5. Mobile health-care anxiety is "the main reason for prevention of computer technology adoption is defined as trepidation, fear, concern and hesitation of damaging the device, being embarrassed" (Heinssen et al., 1987). Jen and Chao (2008) defined mobile health-care anxiety as "a high anxious response towards interaction with the mobile patient safety information system" that has a negative effect on the quality of care that physicians provide for the patient (Jen and Chao, 2008). As the results show, information quality, system use, user satisfaction, mobile health-care anxiety, impact on the individual, and impact on the organization are all factors that are significantly connected with mobile patient safety service system success. This study concluded that mobile patient safety ISs play a role in improving patient services and reducing patient risk, as well as showing that the use of these communication systems may unintentionally be a factor in physician anxiety (Jen and Chao, 2008). Abelson et al. (2009) defined anxiety as the antonym of trust.

6. Monochronic/polychronic time can be evaluated as one of the culture dimensions in the acceptance of new technologies in health care (Zakour, 2004). Hall and Hall (1969) divided societies into two categories based on the time use: monochronic or polychronic. According to them, individuals from monochronic time-use societies tend not to focus on more than one activity at the same time. They prefer to have control on their time and stick to a schedule. In contrast, individuals from polychronic time-use societies have more tendencies to focus on more activities in a block of time and are willing to be free in terms of switching between activities. Thereupon, the first group will accept IT more in order to control their time and schedule, while the second group will accept it less as it might confine their freedom of using time comfortably (Lindquist and Kaufman-Scarborough, 2007).

Chapter 20

1. Population aging is defined as the process through which older individuals become a proportionally larger share of the total population, as a result of declining fertility rates and increasing life expectancy. The prevalence of chronic diseases common in old age is expected to increase significantly due to population aging, having a significant impact on health-care expenditure and provision. The current curative, procedure-driven, provider-centered approach to health care has proven economically unsustainable and ineffective to cater for the demands of an aging population. Instead, our health-care system should focus on preventative strategies, where users are at the center and are empowered to take active control over their health.

    mHealth technologies have the potential to support health-care providers provide coordinated, timely, cost-effective, and accessible services to the wider population, which is particularly relevant to the management of chronic conditions common in old age. Moreover, mHealth technologies have the potential to provide older adults—and their caregivers—tools that can assist them in taking more active control over their health and self-manage their conditions more effectively. mHealth solutions also can support older adults' safety and independence, allowing them to remain at home for longer. Finally, mHealth solutions have the potential to help older people stay connected with their communities, which is particularly significant given the recently established link between loneliness, functional decline, and death among older populations.

2. mHealth solutions can directly support different areas of older adults' lives including their health and wellness, safety, and ability to remain socially connected. Health and wellness solutions can support older adults by monitoring physiological and mental health status, assisting the prevention and management of chronic conditions. Examples of solutions in this category include vital signs monitors and body-worn sensors that can monitor activity levels and other physiological measures.

    Safety solutions are designed to support safety and independence of older adults who are beginning to develop functional and cognitive limitations but wish to remain in their homes. This category includes motion and environmental sensors that can raise an alert when unusual activities or events are detected. PERS and GPS tracking devices also fall in this category and allow the older person to call for help and be located at any time, in case of emergency.

    Social connectedness solutions aim to support adults to remain connected with their families and communities. They may include mobile applications that offer more user-friendly interfaces to mobile devices, as well as social networking platforms for people suffering from specific chronic conditions, which allow users to connect, share experiences, and learn from other people in similar circumstances.

3. mHealth technologies can help family caregivers make more informed decisions by providing access to relevant educational content and information about specific services. Different mobile tools can also help caregivers become more efficient in their role, by allowing them to track and manage daily tasks, medication schedules, doctor appointments, and billable work as well as to keep all medical information centralized and easy to access.

    mHealth solutions can also assist caregivers to improve communication and care coordination with other family members and professionals. One example is shared calendar applications, where caregivers can allocate tasks to other family members, friends, or professional caregivers, who can then input information about how they got on with their tasks.

4. mHealth solutions can support professional caregivers by making their workflow more efficient and facilitating better care coordination. These solutions are designed to facilitate patient data collection, record keeping, and communication with other health-care professionals and family members. Many mHealth tools for professional caregivers have the ability to analyze the data collected and trigger automatic alarms when abnormalities are detected, enabling them to provide timely assistance to their clients.

5. Technology design issues remain a challenge to wider mHealth adoption. Existing solutions are often difficult to use, nonintuitive, disruptive to user's lifestyle, stigmatizing, or perceived as not addressing a real need and for this reason tend to be abandoned by users over time. These issues have been associated to the fact that many technology developers don't take into consideration the huge variations in terms of needs, preferences, and abilities among older adults, as well as the lack of sufficient end user involvement during the solution's development stages.

In order to design technology that is relevant, usable, and desirable for older adults, it is recommended that mHealth developers work closely with end users throughout the design process to fully understand not only their needs, skill level, and physical/cognitive abilities but also their preferences, lifestyles, and aspirations. Designing for "us as we age" is another recommended approach to encourage designers to pay more attention to the look, feel, and desirability of the technologies produced.

Chapter 21

1. Some of the factors are: system/app reliability, usability, user-friendliness, functionality, lack of face to face interaction, and information privacy.

2. Preventing or monitoring chronic illnesses, dementia, and fall detection are the treatments that mobile app solutions have shown to be the most successful.

3. Some mobile applications have been developed to address the aging population's isolation challenges. For example, some applications have enabled just-in-time access to family members and care providers. As for the future applications, games, social media tools, and other communication tools could help elderly find comfort and support when they need them.

Chapter 22

1. Effective IS design needs to allow for flexible HCI heuristics that afford individual delivery and/or input modes. This means design for a decrease of sensory/perceptual processing skills, response speed, motor skills, and cognitive abilities, for instance, to increase the usability of mHealth programs when older users require changed textual/graphical image sizing.

2. Appropriate mHealth service development needs to include older people in web design focus groups to tease out their requirements. Persona and scenario development can assist in tailoring mHealth to the needs of older people.

3. The best solution for installing an mHealth standards commission would be to include all stakeholders, including older people, government agencies, and the industry sector to ensure more accountability for the IS design process.

4. Maintaining mHealth quality control will be possible through a government-led commission that has power to regulate accountability for IS design for older people.

5. Older Australian's access and equity to mHealth programs can be addressed at all levels of government, at a local level through local council (financial) participation and support.

6. Ethical issues that arise with mHealth adoption surround the need to protect an older person's individual right to privacy, especially regarding their general health and welfare issues.

Chapter 23

1. Although different individual aspects related to legislation, technology, and acceptance of the users are a real issue for the full integration of mHealth within the HCSs, the biggest challenge is to address each one of these aspects from a joint point of view, where all the key players are involved.

2. mHealth represents a better management of resources, reduction of unnecessary appointments, need of less infrastructure, continuous monitoring, and quick interventions. Therefore, a reduction in the patient expenditure can be achieved because of different aspects such as smaller waiting times and less inactive days.

3. Among the advantages of implementing raising awareness campaigns through SMSs are

- Higher penetration potential of the campaign, as penetration rate of mobile cellular phones in Latin America region is over 100%
- Low cost compared with systems requiring wider bandwidths
- Easy implementation, as it is not necessary to deploy any additional infrastructure

The disadvantages are as follows:

- Size and content of the amount of data are limited to 160 characters.
- Reliability. It is not possible to verify that information was received.
- Privacy. The developer cannot ensure privacy of the information.

4. Different considerations should be made in order to address a specific health-care scenario through mHealth; among the most important are

- Target population
- Traffic profile of the needed mHealth services
- Scenario constraints (e.g., rural, urban, mobile)
- Available infrastructure (technological, material, and human)

5. Latin America's EP presents a high incidence of noncommunicable diseases like type 2 diabetes. Thus, several of the mHealth projects are focused on this kind of diseases.

Chapter 24

1. A model for eTA serves to tease out implications of mHealth technologies and the effects on different stakeholders.

2. If conducted at an early stage before the technology is entrenched in society, an eTA can highlight positive aspects to be promoted and drawbacks to be avoided. Ideally, an ethical assessment should be conducted in close cooperation between technology developers and ethicists.

3. Such a shift may affect the relationships between professional care providers— care recipients—and informal care providers. The effects of these agents should be considered carefully when implementing home-based care by means of mHealth systems. Effects on (1) the many dimensions of privacy, (2) control and self-government, and (3) social contact patterns should be of concern.

4. Ways in which negative effects on (1) the many dimensions of privacy, (2) control and self-government, and (3) social contact patterns can be avoided should be investigated.

5. Wisely designed and implemented, such technology can empower individuals and allow them to remain at home, maintaining an as *ordinary* as possible life despite a need for care provision and support. In many ways, mHealth systems can strengthen local and decisional privacy.

Chapter 25

1. Public Health in developing countries is facing several challenges. Poverty, rapid population growth, poor public health delivery system to remote areas, and inadequate funds are some of those. The problems of public health are more severe in rural areas because of shortage of medical personnel serving in villages. In the shortage, the World Health Organization includes doctors, nurses, midwives, mid-level health workers, pharmacists, dentists, laboratory technicians, and community health workers, as well as managers and support workers. The lack of basic infrastructure such as good roads, uninterrupted electricity supply, and quality education makes things difficult for those who are inclined to serve rural population. In addition, global warming, leading to sudden climate change, is creating additional problems for urban and rural public health systems.

2. In the UN millennium summit in 2000, 189 countries in the world agreed upon the eight goals to be achieved by 2015, which are known as the MDGs. They are as follows:

    1. End poverty and hunger.

    2. Universal education.

    3. Gender equality.

    4. Child health.

    5. Maternal health.

    6. Combat HIV/AIDS.

    7. Environmental sustainability.

    8. Global partnership.

    Out of the eight goals, three goals relate to public health. They are as follows: (1) reduce child mortality, (2) improve maternal health, and (3) combat HIV/AIDS, malaria, and other diseases.

3. Primary health care is defined as health care that is preventive, promotive, curative, and rehabilitative in nature. The concept of primary health care also implies the first contact with the health-care system, which is accessible, acceptable, affordable, and appropriate to all the people concerned.

4. To address public health in rural areas, India has adopted the strategy of primary health-care approach. Primary health-care system has been established in all states of India, as health is an effective medium of achieving socioeconomic development. Primary health care as a system has a structure, a function, and protocols for documentation.

    The structure of primary health care follows the revenue collection and distribution patterns existing in India. Accordingly, one district is considered as a technical and administrative unit of primary health-care services. Each district is headed by two health professionals: DHO and civil surgeon. The DHO is in charge of the preventive health-care services in the rural areas, which include PHCs and SPHCs (subcenters). The civil surgeon supervises the district hospital and RH at the taluka level. At the national level, the system is monitored through the Ministry of Health and Family Welfare. At the state level, there is Directorate of Health Services that plans, coordinates, and implements national health programs in all districts of the state.

    For every 30,000 population in plain areas and 20,000 populations in tribal and hilly areas, one PHC has been recommended. Further, for every 3000 population in tribal areas and 5000 populations in plain areas, one subcenter is recommended. One PHC covers 25–30 villages and has about 6–8 subcenters providing basic health services to the villagers.

The interface between the community and PHC is entrusted to a female volunteer within community, termed as ASHA. The four pillars of the public health system implemented through the PHC are an ASHA at village level; a female health worker called ANM; a male health worker called MPW, at the subcenters; and an MO at the PHC.

The functions of the PHC are as follows: (1) implementing national health programs, (2) implementing health surveys of the community to establish baseline health status, (3) ensuring safe water and environmental sanitation, (4) promoting health awareness, (5) monitoring disease surveillance and health-related data, and (6) doing epidemic investigations, containment, and research. The PHC has both the center-based and outreach activities.

5. mHealth-PHC connects the rural patient to the doctor through an ANM or female ASHA. The mHealth-PHC uses various technologies like mobile Internet, IVR, Indian language font rendering, and usability frameworks to connect the remote patient to the doctor.

The tool has five components: (1) software in local language on a mobile phone used at patient's end called client software; (2) servers in the secured data center; (3) doctor's console, which is viewed by a doctor/expert to suggest treatment/test and prescribe medicines to the patients; (4) an IVR application to record the information and observations; and (5) reporting console to view various reports related to rural health in a consolidated manner.

The following figure shows the architecture of the *mHealth-PHC* platform.

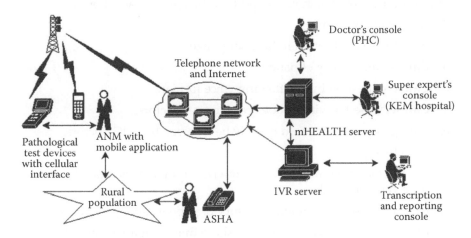

The ASHA's interface is IVR based. With *mHealth-PHC*, she can record the details of her visits to families in villages on her interface and can be notified of any emergency or task using the same interface. With the help of a specially designed template, the ASHA can easily make the home visit reports. The voice records can be transcribed using transcription console and data are available in web-based form. Various reports can be generated from the data using reporting console. The transcription software converts her visit report from voice to text.

The ANM at the subcenter provides preliminary health care to pregnant women and performs normal deliveries of babies. *mHealth-PHC* tool can aid the ANM, as it can be integrated with the portable, battery-operated medical test devices for blood and urine analysis. This facilitates inclusion of pathology test reports as part of the *patient medical history*. The ANM can record a voice query through her interface. Wireless Internet makes it possible to upload patient's personal information and medical history to the server. A web console gives an integrated view of patient's medical history including the ANM's voice comments about the patient's illness and symptoms to the PHC doctor. Through the doctor's console, the doctor at the PHC can see the patient's details with medical history and listen to the ANM's recorded voice query. The doctor would give his/her advice in voice/text or may handwrite using an E-pen. The image of the prescription is captured and sent over wireless Internet to the ANM's mobile phone and the ANM can take action accordingly. If required, a specialist available at the RH or in city hospital could be contacted over Internet, and the entire case, including patient's medical history, could be referred through *mHealth-PHC* for expert advice.

Thus, *mHealth-PHC* connects ASHAs in villages, ANMs at subcenters, doctors at PHCs, and doctors at RHs or city hospitals through wireless, wireline Internet, and web technology.

6. *mHealth-PHC* solution effectively bridges the last mile connection between rural patients and PHC doctors. It enables meaningful data-driven asynchronous interactions between health workers and doctors. The integrated pathological devices provide data for better diagnostics. The virtual remote prescription helps in providing clear and storable evidence of advice.

7. The different users or stakeholders and their benefits are as follows:

   1. Remote health workers

      a. Easy to use application in local language.

      b. Can see the history of patient.

      c. Can ask question in free flowing voice format.

      d. Can serve better with remote guidance of doctors.

      e. Emergency queries could quickly be resolved by PHC doctors.

   2. PHC doctors

      a. Asynchronous communication (queries in inbox), facilitate to use doctor's time very efficiently.

      b. He/she can see the complete history of patients and give advice.

      c. For critical issues, can do broadcasts so that the masses can take corrective actions.

   3. Patients in remote areas

      a. They get effective advice and help at their location rather than traveling. Traveling may be difficult since road conditions and transportation could be bad or unavailable. Patients save on traveling cost and time.

      b. Remote and quick advice is more valuable in case of emergencies. It is a potential life saver.

4. Health administrators and policy makers
   a. Get the on-field and current data on how the interventions are performing in the field.
   b. Analytics helps in predicting the trends and taking corrective action.

8. The *mHealth-PHC* platform enables collection of data from ASHA workers, ANM, MPW, PHC doctors, and medical experts. Since different actors of the primary health-care system have different functions to perform, data generated by them are rich in variety. One can say that the status of rural health is *hidden* in these data. The ASHA goes house to house to note down pregnancy and illness cases. Her visit report is captured in the form of voice records in the database. The ANM registers the patients and captures detail information about patients and their family details. These data are captured on mobile and transferred to the database on the server. Similarly, PHC doctor's prescriptions written using E-pen are also stored in the database. *mHealth-PHC* provides facility for interaction between city doctors and PHC doctors. This interaction captures the quality of medical advice and is stored in the database. Thus, database of the server provides rich information for analysis.

   Each data point in the database has time and location stamp. With simple analysis, drawing graphs of health conditions of patient clusters against time and locations can indicate status of pregnancies, new births, and spreading of particular disease during specific season, say monsoon. The government health department can use this analysis for immediate actions and for drafting rural health policy as a long-term measure.

Chapter 26

1. mHealth is an ultimate platform to serve the underserved (Akter and Ray, 2010). The shortfall of health human resources and wide spread of mobile phones have created a ubiquitous opportunity for the mHealth in the developing countries with its full potentials. It's easy access, lower cost, and service to the hard-to-reach population have opened a new horizon in the public health.

2. There are different mHealth services in Bangladesh that are ongoing. A number of mHealth projects have been initiated including in the provision of primary health care, in disease surveillance and the collection of routine data, for health promotion and disease prevention initiatives, and for health information system support tools for health workers and point-of-care services. Among them, some are health programs and some are related with the research and testing of different models of mHealth and its usefulness. In Bangladesh, the largest mHealth program is the health hotline 789, which is running under a business model and serving millions of people in Bangladesh since 2006.

3. Since compared to other technology the ownership and use of mobile phones is more prevalent among persons of low socioeconomic status, the use of mobiles phones may reduce the impact of digital divide inherent in any web-based health interventions. This is the first technology where industry has documented a reverse trend toward a *digital divide*—the gap between people with effective access to digital and information technology and those with very limited or no access at all (Krishna et al., 2009). This increases the likelihood of successfully delivering

health improvement interventions to traditionally hard-to-reach populations. mHealth services with toll-free or subsidies will help to reach its full potentials and serve the poorest of the poor.

4. The barriers of implementing mHealth in developing countries are cost related to afford, inequity in access, ownership of mobile phones, lower level of health literacy.

Chapter 27

1. Generally, healthcare systems in developing countries have weaknesses in term of accessibility, availability, affordability, and continuity of services. A wide range of chronic and communicable diseases are straining the systems, especially in Africa, followed by Asia and Latin America. Arab countries are undergoing transition in their healthcare systems, which has impacted the health of society in different ways.

2. In the developing world, mobile phone technology reached farther than any other communications technology. A recent 80% growth in mobile phone subscriptions has been driven by developing countries. Saudi Arabia is considered as the second biggest market for mobile phones in the Middle East and has the largest number of mobile users worldwide.

3. mHealth is the way of providing healthcare information and services through mobile technologies such as mobile phones. This field came into existence as a result of the intersection of health, technology, and finance with the support of governments. mHealth systems can be classified depending on the target group into three types: mHealth for hospital patients, mHealth for healthy people, and mHealth for chronically ill patients. Although mHealth development is still in its early stages, it represents a cost-effective solution to increase the access to healthcare information in order to improve the healthcare delivery and quality of care as well as increasing the ability to diagnose, treat, educate, train, and track diseases in a timely manner, especially for people in remote rural areas such as in developing countries.

4. The main factors behind the current rapid development of mHealth in the developing world are the following: increase in the number of mobile phone users, the need to treat many people in rural areas, lack of timely disease control, decreased cost of mobile phones, cultural factors, shortage of healthcare workers, and lack of resources and infrastructure. In the Middle East, growing cost, shortage of capacity, and increase in lifestyle-related diseases such as obesity and diabetes have all affected the quality of healthcare systems negatively and increased the need for mHealth development in this region as well as in other parts of the developing world.

5. The mHealth programs and projects in the developing world are on six main services: diagnostic and treatment support services, health education and awareness services, data collection and remote monitoring services, disease surveillance and drug adherence services, health information systems and point-of-care services, and emergency medical services.

   The most successful projects and applications in the developing countries to support these six services are the following: Health Hotline, Text to Change (TTC), Nokia Data Gathering system, SIMpill, Uganda Health Information Network (UHIN), and FrontlineSMS.

6. Various challenges and barriers affect the success of mHealth development in the developing world, these include network challenges, cultural values, beliefs and customs, security and privacy of transferring information, and the cost of mHealth devices. Generally in the developing world, the cost of mobile access is more expensive than fixed line access. Additionally, lack of education needed knowledge, and strong training; the current policy environment; and the absence of mobile phone guidelines and standards are contributing toward these challenges. Moreover, rural and urban areas have dissimilar resources, which may affect the structure and content of mHealth. Finally, the current quality of health services in the developing world is influenced by mobile device; mobile network; information systems; availability of appropriate information; reliability, flexibility, and usability of mHealth applications; cost; security; and cultural factors.

7. To overcome current challenges, various procedures and strategies should be developed at the early stages of implementation. A comprehensive research must be conducted to identify local conditions, such as healthcare infrastructure, mobile network, and literacy levels. In addition, clear objectives of what mHealth programs are trying to achieve and the desired health outcomes should be identified. Structuring precise policies and specific standards for planning, designing, implementing, and integrating mHealth systems must be done. Making use of the best practice applications from high-income countries and customizing existing systems according to community needs instead of creating new applications in order to save money, time, and resources should be encouraged. Moreover, involving the IT specialist in the deployment, consumption, and maintenance phases and continuous monitoring and evaluation by using specific measurable metrics are necessary. By ensuring that all the required characteristics, such as usability, accessibility, mobility, affordability, and flexibility, are considered in the application, most of the challenges and the barriers can be overcome.

8. It is expected that in the next 15 years, healthcare policy makers and providers in the developing world will focus more on preventive healthcare instead of just disease treatment. Advance technologies and tools will be deployed to allow doctors and patients to use mobile devices for remote monitoring and to manage numerous health issues. Moreover, mHealth applications will use images and video tools in addition to text messages for specific health fields such as diagnostics, consultation, and follow-up. Different sensors in mobile phones will be used in the health field: the data will be provided by camera, GPS, acceleration sensors, processed by custom mobile applications and then sent using SMS or MMS application. Also, speech recognition technologies will be integrated with the mHealth application in order to increase the use of interactive speech. Additionally, advance technologies, such as new network generations, WiMAX, intelligent mobile devices, dedicated wireless devices, will support mHealth applications to reduce issues and provide more valuable solutions. mHealth has a great future in the Middle East; it is predicted that the size of the healthcare market will rise to $100 billion in the next 15 years. Governments invest huge amounts in the healthcare sector in order to enhance the services and improve the quality of care. For that reason, new and advance technologies such as mHealth will have a considerable role in supporting the vision of adequate and affordable healthcare for all people.

Chapter 28

1. Some of the limitations are as follows: (1) increased complexity of health-care system, which may demand provision of extra training (the technologies may be too expensive to implement); (2) clinicians and patients may be concerned about the weakening of patient–physician relationship; and (3) possibilities of technology perversion, which may raise ethical and security concerns.

2. Data collected can be analyzed further using appropriate statistical methods such as trend analysis, or intelligent methods such as neural networks, or dynamic methods such as system dynamics, in order to make forecasts that are useful for making long-term plans.

3. A feed forward mechanism should be implemented in the decision structure of a health-care system such that strategic information is used to perform tactical decisions, which in turn are used to determine operational decisions. In addition, a feedback mechanism is used to make informed adjustments. Thus, any realized discrepancies from lower levels are used to modify or improve upper-level decisions.

Chapter 29

1. At least implications of mHealth are empowering patients to have personalized services, changing of business process in health-care service, and redefining data ownership.

2. Customers can expect personal empowerment where patients can personalize services, social empowerment where they are able to make conversation with other patients or medical staffs in social networks, and finally medical empowerment where patients have the ability to access or update their medical records by their own.

3. The challenges are at least reliability and quality of health or medical information in social networks to be practically applicable for customers.

4. The answers can vary from customer satisfaction, improve health literacy, participations, decision making, etc.

Chapter 30

1. The sexual activity of Fitbit users could be found using Google search since their user profiles that listed their activities were public and searchable. After this situation was publicized, Fitbit changed their privacy policy to hide all activity records of users from other Fitbit users and search engines, as well as requested Google, Microsoft, and Yahoo to remove all indexed Fitbit user profile pages from their search engines.

2. The chapter "Understanding Users' Privacy Preferences for mHealth Data Sharing" gives recommendations on how to design sharing controls in order to protect the privacy of mHealth device users.

3. In the following are five instances when unintended disclosure could happen:
   1. When communication between devices is unencrypted
   2. When an unauthorized person can access the data on devices
   3. When the devices provide only coarse-grained sharing controls
   4. When the user is confused by complex fine-grained sharing controls
   5. When a user is lazy and needs immediate benefits and hence keeps the default settings

4. See the following figure

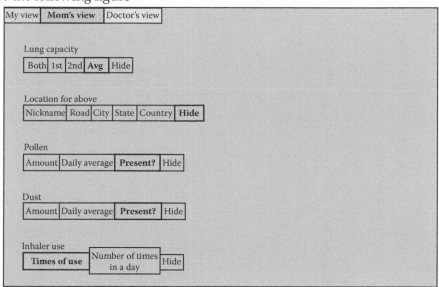

5. The number of participants should be more, and focus should be on having an almost equal number of male and female participants for a particular age group. That is, for example, there should be at least 20 male students and 20 female students.

Chapter 31

1. John R. Hauser, Don Clausing.
2. There are eight rooms in the HoQ matrix. The rooms are named as (1) CRs, (2) ECs, (3) correlation matrix, (4) relationship matrix, (5) importance ranking, (6) customer assessment of competing products, (7) technical assessment, and (8) target values.
3. The central idea of QFD/HoQ is to resonate the VoC across the organization and bring together different technical specialties of an organization onto a single diagram.
4. Motamarri (2013) made a maiden attempt to compare mHealth vis-à-vis conventional services like PH, GP, and TM from the perspective of patients.
5. DA is a classification technique that attempts to classify cases into defined groups. The core element of the HoQ is Room-6, that is, comparative assessment of competing products. DA can be used to extract the distinguishing characteristics of a service in relation to its competition.

Chapter 32

1. (d)
2. Answer: The monetary amount (or actual net price paid) that a potential user would give up to purchase a device, good, or service.
3. (a)
4. (1) _____ (Evaluate acceptance)
   (2) _____ (Collect social and quantitative findings)

(3) _____ (Assess from multiple perspectives)

(4) _____ (Consider mobile devices as a means of dissemination of findings)

5. True

Chapter 33

1. (c)

2. (b)

3. (c)

4. (a)

5. (c)

6. (c)

Chapter 34

1. (a) False

(b) False

Discussion

The demographic profile is transitioning from a pyramid shape, where the largest part of the population is on the bottom, representing the young, to a kite shape, where the largest part of the population is now on the top representing the ageing population seen particularly in developed countries. This is a consequence of a declining fertility rate. In addition to the challenges presented by more people living into old age, there is an increasing burden of chronic disease. To mitigate against this burden of chronic disease, there are growing numbers of initiatives that aim to move the health-care model from a reactive (treatment) model to a proactive (preventive) model. These efforts are aimed at increasing the focus on citizens to take more responsibility and maximize their own health on all parts of the continuum between life and death. Simultaneously, there are growing initiatives where patients utilize the Internet to obtain the information and support they need to address health concerns.

2. False

Discussion

There is a growing interest in the concept of personalized, predictive, participatory, and preventive health care to counteract the increasing burden of chronic disease to health-care systems. The use of the Internet to deliver health care is regarded by proponents of the technology as a means of complementing current models of health care by enabling the individualization/tailoring of health care. However, to build systems capable of delivering this kind of care and expect people to engage is flawed. A bottom-up approach using principles of cocreation and collaboration is required.

ICT has been successfully integrated into commerce, airlines, and manufacturing. However, health-care ICT has not been so straightforward. This is a global experience. Utilizing coproduction and cocreation approaches in the development of businesses has proved beneficial, and these approaches are finding traction within the NHS. Engagement with all stakeholders in design and delivery of health-care projects is important.

3. (a) False

(b) True

(c) True

(d) True

Discussion

Evidence for the efficacy of eHealth interventions from a variety of studies, including randomized controlled trials, are generally lacking despite grand claims and implementation. There is an increasing literature arguing for different if not new methodologies to evaluate these novel interventions. eHealth interventions need to be studied both at the micro level (the actual intervention itself) and the macro level (the way people interact with each other). Because properties arise that are unexpected, flexibility is needed in the design. As data are collected from online health-care delivery, it is theoretically possible to personalize health care by using different behavioral techniques to motivate people to adopt certain health behaviors. As people are unaware of this, it can be regarded as a form of social engineering.

4. (a) False

(b) True

(c) False

(d) True

(e) True

Discussion

ICT has an indirect, enabling, and empowering role that leads to increased life satisfaction; this has been particularly seen in the population of women in developing countries. A reason postulated is the technology counteracts cultures where women have more socially controlled roles, a lower sense of freedom, and autonomy and therefore well-being. There is increasing evidence that people who are happier have more positive health outcomes. Smartphone users are more likely to share, create, and participate online and thus are contributors to the "wisdom of the crowd." There is evidence that networks cannot only promote health but can do the opposite by normalizing obesogenic and smoking habits.

5. (a) True

(b) False

Discussion

One size does not fit all—the use of smartphone apps in the older user may be hindered by the size of the device, for example, the screen and keyboard are too small. This may be rectified by the use of tablets/iPads.

Home testing for glucose does not necessarily equate to an increased understanding of the disease (diabetes). mHealth is more likely to be effective when linked to education and behavior advice.

Chapter 35

1. Key principles for the ethical use of mobile devices for medical images:
   - Consent—document consent discussions with patient and the intended use of images.
   - Privacy—avoid identifying features.
   - Security—keep data secure through the use of a PIN/passcode.
2. Many mobile devices are connected to cellular networks, and this often represents the most convenient method of image transfer. Other options include Wi-Fi, Bluetooth, and NFC.
3. As of 2013, the strongest evidence is in laboratory settings, dermatology, plastics, radiology, and ophthalmology. Overall, there are very few high-quality studies published in this field, with most studies being pilot or case studies.
4. Mobile devices can be used in numerous health situations:
   - Diagnostic—either through direct imaging of pathology or indirectly through adapters such as microscope adapters, rapid diagnostic kits, or ultrasound probes.
   - Communication—images taken on mobile devices can be easily communicated between healthcare providers for advice and between patients and clinicians.
   - Viewing—the high-quality displays on newer smartphones are an excellent way to view both clinical photos and radiological studies.
5. Important barriers need to be overcome before medical imaging on mobile devices grows in utility:
   - Access—even in developing nations, there is a high uptake of mobile phones. However, the infrastructure is not designed for high-speed data transfer, and can be a limitation for mHealth imaging.
   - Affordability—by utilizing commercially mass produced products that are a fraction of the cost of current imaging modalities, mHealth offers a more cost-effective method of implementing certain programs.
   - Evidence—health policy makers need to evaluate the evidence for mHealth-based technologies, given the relative paucity of data in 2013.

Chapter 36

1. Please refer to Section 36.2.
2. Please refer to Sections 36.2 and 36.3.
3. Please refer to Section 36.3.
4. Please refer to Section 36.4.
5. Please see the Refs. [1,17].

Index

A

Access control, 3R dataflow engine
  advanced encryption standard (AES), 675
  cloud database, 677
  dataflow transmission, 677
  ECG-featured chaos encryption, 675–676
  exclusive-OR (XOR) operation, 675
  hypertext transport protocol (HTTP)
    mode, 676
  logistic mapping, 675
  privacy-sensitive applications in
    CARE-NET, 675
  pseudorandom sequence, 675
  socket connection, 676
Accredited social health activist (ASHA), 454,
    456–457, 460
Activities of daily living (ADLs), 298, 405
Adoption challenges, in nursing
  Bar Code Medication Administration
    (BCMA), 288
  device value, 289
  FDA regulation, 288–289
  price of apps, 289
  security, 289–290
  smartphones and tablets, 287–288
Advanced encryption standard (AES), 290, 675
African Medical and Research Foundation
    (AMREF), 328–329
Agency for Healthcare Research and Quality
    (AHRQ), 598
Agile iterative approach, 635–636, 644
Aging populations
  collaborative care, 397
  culturally and linguistically diverse
    (CALD), 395
  dementia, 395
  education and professional development,
    394–395
  family, primary carers, 394
  health insurance, 396
  Heath Workforce Australia (HWA), 397
  ICT (*see* Information and communications
    technology (ICT))
  mobile applications, 403–406
  nature of culture, 394
  nutrition and living standards, 392
  personality development, 394

private and public settings, 395
residential care, 393–394
technologies, 393
WHO slogan, 396
AHRQ, *see* Agency for Healthcare Research and
    Quality (AHRQ)
American Recovery and Reinvestment Act, 598
AMREF, *see* African Medical and Research
    Foundation (AMREF)
AMREF Health Africa program, 329
Aponjon information service, 473–474
Application programming interfaces (APIs)
  and access control, 673
  for supporting terminal applications, 677
Arab world
  Baby Ultrasound MMS service, 494–495
  Etisalat *Mobile Baby* program, 493–494
  health-care systems, 486
  Internet usage, 487
  maternal mortality, 493
  MOH hospitals and clinics, 494
  Your Health services, 494
ASHA, *see* Accredited social health activist
    (ASHA)
Augmented reality margin (AR margin)
  biosignals with numeric characters,
    679–680
  clinical trial results, 678–679
  definition, 677
Automated dosing reminders, 624
Autonomy Gerontology Iso-Resources Group
    (AGGIR) model, 298, 303–304, 311
Auxiliary nurse midwife (ANM)
  home visits, 454
  interface, mobile phone, 459
  mHealth-PHC, 456
  pregnant women and babies, 457, 459
  training, 461

B

Baby Ultrasound MMS service, 494–495
Bangladesh health-care systems
  affordability and acceptability, 479
  application design and disability, 479
  data security and service quality, 477
  digital and information technology, 480